THEORY OF
OSCILLATORS

THEORY OF OSCILLATORS

by

A. A. ANDRONOV, A. A. VITT
AND S. E. KHAIKIN

TRANSLATED FROM THE RUSSIAN BY
F. IMMIRZI

THE TRANSLATION EDITED AND ABRIDGED BY
W. FISHWICK
PROFESSOR OF ELECTRICAL ENGINEERING
University College of Swansea

DOVER PUBLICATIONS, INC.
NEW YORK

Copyright © 1966 by Pergamon Press Ltd.
All rights reserved under Pan American and International Copyright Conventions.

Published in Canada by General Publishing Company, Ltd., 30 Lesmill Road, Don Mills, Toronto, Ontario.

This Dover edition, first published in 1987, is an unabridged and unaltered republication of the edition published by Pergamon Press, Oxford, in 1966 as Volume Four of *The International Series of Monographs in Physics*. It is reprinted by special arrangement with Pergamon Books, Ltd., Headington Hill Hall, Oxford OX3 OBW, England.

This book is a translation of *Теория Колебаний* (Teoriya Kolebanii) published by Gosudarstvennoye Izdatel'stvo Fiziko-Matematicheskoi Literatury, Moscow.

Manufactured in the United States of America
Dover Publications, Inc., 31 East 2nd Street, Mineola, N.Y. 11501

Library of Congress Cataloging-in-Publication Data

Andronov, A. A. (Aleksandr Aleksandrovich), 1901–1952.
Theory of oscillators.

Translation of: Teoriia kolebaniĭ.
Reprint. Originally published: Oxford ; New York : Pergamon Press, 1966. (International series of monographs in physics ; v. 4)
Bibliography: p.
Includes index.
1. Oscillations. I. Vitt, A. A. (Aleksandr Adol'fovich), d. 1937. II. Khaĭkin, S. È. (Semen Émmanuilovich) III. Fishwick, Wilfred. IV. Title.
QA871.A52   1987   531'.322   87-24354
ISBN 0-486-65508-3 (pbk.)

CONTENTS

PREFACE TO THE SECOND RUSSIAN EDITION   xiii

NOTE FROM THE ENGLISH EDITOR   xiv

INTRODUCTION   xv

I. LINEAR SYSTEMS

§ 1. A linear system without friction (harmonic oscillator)   1

§ 2. The concept of the phase plane. Representation on the phase plane of the totality of motions of a harmonic oscillator   4
    1. The phase plane — 2. Equation not involving time — 3. Singular points. Centre — 4. Isoclines — 5. State of equilibrium and periodic motion

§ 3. Stability of a state of equilibrium   11

§ 4. Linear oscillator in the presence of friction   15
    1. Damped oscillatory process — 2. Representation of a damped oscillatory process on the phase plane — 3. Direct investigation of the differential equation — 4. Damped aperiodic process — 5. Representation of an aperiodic process on the phase plane

§ 5. Oscillator with small mass   36
    1. Linear systems with half degree of freedom — 2. Initial conditions and their relations to the idealization — 3. Conditions for a jump — 4. Other examples

§ 6. Linear systems with "negative friction"   50
    1. Mechanical example — 2. Electrical example — 3. Portrait on the phase plane — 4. Behaviour of the system for a variation of the feedback

§ 7. Linear system with repulsive force   64
    1. Portrait on the phase plane — 2. An electrical system — 3. Singular point of the saddle type.

II. NON-LINEAR CONSERVATIVE SYSTEMS

§ 1. Introduction   74

§ 2. The simplest conservative system   75

CONTENTS

§ 3. Investigation of the phase plane near states of equilibrium ... 79

§ 4. Investigation of the character of the motions on the whole phase plane ... 88

§ 5. Dependence of the behaviour of the simplest conservative system upon a parameter ... 98
 1. Motion of a point mass along a circle which rotates about a vertical axis — 2. Motion of a material point along a parabola rotating about its vertical axis — 3. Motion of a conductor carrying a current

§ 6. The equations of motion ... 118
 1. Oscillating circuit with iron core — 2. Oscillating circuit having a Rochelle salt capacitor

§ 7. General properties of conservative systems ... 125
 1. Periodic motions and their stability — 2. Single-valued analytic integral and conservativeness — 3. Conservative systems and variational principle — 4. Integral invariant — 5. Basic properties of conservative systems — 6. Example. Simultaneous existence of two species

III. NON-CONSERVATIVE SYSTEMS

§ 1. Dissipative systems ... 146

§ 2. Oscillator with Coulomb friction ... 151

§ 3. Valve oscillator with a ∫ characteristic ... 157

§ 4. Theory of the clock. Model with impulses ... 168
 1. The clock with linear friction — 2. Valve generator with a discontinuous ∫ characteristic — 3. Model of the clock with Coulomb friction

§ 5. Theory of the Clock. Model of a "recoil escapement" without impulses ... 182
 1. Model of clock with a balance-wheel without natural period — 2. Model of clock with a balance-wheel having a natural period

§ 6. Properties of the simplest self-oscillating systems ... 199

§ 7. Preliminary discussion of nearly sinusoidal self-oscillations ... 200

IV. DYNAMIC SYSTEMS WITH A FIRST ORDER DIFFERENTIAL EQUATION

§ 1. Theorems of existence and uniqueness ... 209

§ 2. Qualitative character of the curves on the t, x plane depending on the form of the function $f(x)$ ... 212

§ 3. Motion on the phase line ... 213

§ 4. Stability of the states of equilibrium ... 215

§ 5. Dependence of the character of the motions on a parameter ... 218
 1. Voltaic arc in a circuit with resistance and self-induction — 2. Dynatron circuit with resistance and capacitance — 3. Valve relay (bi-stable

trigger circuit) — 4. Motion of a hydroplane — 5. Single-phase induction motor — 6. Frictional speed regulator

§ 6. Periodic motions   234
     1. Two-position temperature regulator — 2. Oscillations in a circuit with a neon tube

§ 7. Multivibrator with one RC circuit   246

V. DYNAMIC SYSTEMS OF THE SECOND ORDER

§ 1. Phase paths and integral curves on the phase plane   254
§ 2. Linear systems of the general type   257
§ 3. Examples of linear systems   267
     1. Small oscillations of a dynatron generator — 2. The "universal" circuit
§ 4. States of equilibrium and their stability   272
     1. The case of real roots of the characteristic equation — 2. The characteristic equation with complex roots
§ 5. Example: states of equilibrium in the circuit of a Voltaic arc   281
§ 6. Limit cycles and self-oscillations   287
§ 7. Point transformations and limit cycles   291
     1. Sequence function and point transformation — 2. Stability of the fixed point. Koenigs's theorem — 3. A condition of stability of the limit cycle
§ 8. Poincaré's indices   300
§ 9. Systems without closed paths   305
     1. Symmetrical valve relay (trigger) — 2. Dynamos working in parallel on a common load — 3. Oscillator with quadratic terms — 4. One more example of non-self-oscillating system.
§ 10. The behaviour of the phase paths near infinity   324
§ 11. Estimating the position of limit cycles   332
§ 12. Approximate methods of integration   340

VI. FUNDAMENTALS OF THE QUALITATIVE THEORY OF DIFFERENTIAL EQUATIONS OF THE SECOND ORDER

§ 1. Introduction   351
§ 2. General theory of the behaviour of paths on the phase plane. Limit paths and their classification   353
     1. Limit points of half-paths and paths — 2. The first basic theorem on the set of limit points of a half-path — 3. Auxiliary propositions —

4. Second basic theorem on the set of the limit points of a half-path — 5. Possible types of half-paths and their limit sets

§ 3. Qualitative features of the phase portrait on the phase plane. Singular paths 363

1. Topologically invariant properties and topological structure of the phase portrait — 2. Orbitally stable and orbitally unstable (singular) paths — 3. The possible types of singular and non-singular paths — 4. Elementary cell regions filled with non-singular paths having the same behaviour — 5. Simply connected and doubly connected cells

§ 4. Coarse systems 374

1. Coarse dynamic systems — 2. Coarse equilibrium states — 3. Simple and multiple limit cycles. Coarse limit cycles — 4. Behaviour of a separatrix of saddle points in coarse systems — 5. Necessary and sufficient conditions of coarseness — 6. Classification of the paths possible in coarse systems — 7. Types of cells possible in coarse systems

§ 5. Effect of a parameter variation on the phase portrait 405

1. Branch value of a parameter — 2. The simplest branchings at equilibrium states — 3. Limit cycles emerging from multiple limit cycles — 4. Limit cycles emerging from a multiple focus — 5. Physical example — 6. Limit cycles emerging from a separatrix joining two saddle-points, and from a separatrix of a saddle-node type when this disappears

VII. Systems with a Cylindrical Phase Surface

§ 1. Cylindrical phase surface 419
§ 2. Pendulum with constant torque 422
§ 3. Pendulum with constant torque. The non-conservative case. ... 427
§ 4. Zhukovskii's problem of gliding flight 436

VIII. The Method of the Point Transformations in Piece-wise Linear Systems

§ 1. Introduction 443
§ 2. A valve generator 446

1. Equation of the oscillations — 2. Point transformation — 3. The fixed point and its stability — 4. Limit cycle

§ 3. Valve generator (the symmetrical case) 461

1. The equations of the oscillations and phase plane — 2. Point transformation — 3. Fixed point and limit cycle.

§ 4. Valve generator with a Biassed $\int$ characteristic 468

1. The equation of the oscillations — 2. Point transformation — 3. Fixed points and limit cycles — 4. The case of small values of a and γ.

CONTENTS

§ 5. Valve generator with a two-mesh RC circuit 480
  1. The phase plane — 2. The correspondence functions — 3. Lamerey's diagram — 4. Discontinuous oscillations — 5. Period of self-oscillations for small values of μ.

§ 6. Two-position automatic pilot for ships' controller 501
  1. Formulation of the problem — 2. The phase plane — 3. The point transformation — 4. Automatic pilot with parallel feedback — 5. Other automatic controlling systems

§ 7. Two-position automatic pilot with delay 517
  1. Ship's automatic pilot with "spatial" delay — 2. Automatic ship's pilot with pure time delay

§ 8. Relay Operated Control Systems (with dead zone backlash and delay) 536
  1. The equations of motion of certain relay systems — 2. The phase surface — 3. The point transformation for $\beta < 1$ — 4. Lamerey's diagram 5. Structure of the phase portrait — 6. The dynamics of the system with large velocity correction

§ 9. Oscillator with square-law friction 555

§ 10. Steam-engine 559
  1. Engine working with a "constant" load and without a regulator — 2. Steam-engine working on a "constant" load but with a speed regulator — 3. Engine with a speed dependent load torque

IX. NON-LINEAR SYSTEMS WITH APPROXIMATELY SINUSOIDAL OSCILLATIONS

§ 1. Introduction 583

§ 2. Van der Pol's method 585

§ 3. Justification of Van der Pol's method 593
  1. The justification of Van der Pol's method for transient processes — 2. Justification of Van der Pol's method for steady-state oscillations

§ 4. Application of Van der Pol's method 601
  1. The valve generator with soft operating conditions — 2. The valve generator whose characteristic is represented by a polynomial of the fifth degree — 3. Self-oscillations in a valve generator with a two-mesh RC circuit

§ 5. Poincaré's method of perturbations 613
  1. The procedure in Poincaré's method — 2. Poincaré's method for almost linear systems

§ 6. Application of Poincaré's method 623
  1. A valve generator with soft self-excitation — 2. The significance of the small parameter μ.

CONTENTS

§ 7. A valve generator with a segmented characteristic   626
     1. A valve generator with a discontinuous ∫ characteristic — 2. A valve oscillator with a segmented characteristic without saturation

§ 8. The effect of grid currents on the performance of a valve oscillator   632

§ 9. The bifurcation or branch theory for a self-oscillating system close to a linear conservative system   635

§ 10. Application of branch theory in the investigation of the modes of operation of a valve oscillator   637
     1. Soft excitation of oscillations — 2. Hard excitation of oscillations

X. Discontinuous Oscillations

§ 1. Introduction   645

§ 2. Small parameters and stability of states of equilibrium   649
     1. Circuit with a voltaic arc — 2. Self-excitation of a multivibrator

§ 3. Small parasitic parameters and discontinuous oscillations   659
     1. The mapping of the "complete" phase space by the paths — 2. Condition for small (parasitic) parameters to be unimportant — 3. Discontinuous oscillations

§ 4. Discontinuous oscillations in systems of the second order   670

§ 5. Multivibrator with one RC circuit   680
     1. Equations of the oscillations — 2. The x, y phase plane for $\mu \rightarrow +0$

§ 6. Mechanical discontinuous oscillations   690

§ 7. Two electrical generators of discontinuous oscillations   696
     1. Circuit with a neon tube — 2. Dynatron generator of discontinuous oscillations   699

§ 8. Frühhauf's circuit   702
     1. "Degenerate" model — 2. The jump postulate — 3. Discontinuous oscillations in the circuit — 4. Including the stray capacitances

§ 9. A multivibrator with an inductance in the anode circuit   712
     1. The equations of "slow" motions — 2. Equations of the multivibrator with stray capacitance C_a. — 3. Discontinuous oscillations of the circuit

§ 10. The "Universal" circuit   724

§ 11. The blocking oscillator   730
     1. The equations of the oscillations — 2. Jumps of voltages and currents — 3. Discontinuous oscillations — 4. Discontinuous self-oscillations of the blocking oscillator

§ 12. Symmetrical multivibrator — 750
   1. The equations of the oscillations — 2. Jumps of the voltages u_1 and u_2 — 3. Discontinuous oscillations of the multivibrator

§ 13. Symmetrical multivibrator (with grid currents) — 758
   1. Equations of the oscillations — 2. Discontinuous oscillations — 3. The point transformation Π — 4. Lamerey's diagram — 5. Self-oscillations of the multivibrator for $E_g \geqslant 0$.

XI. COMMENTS ON MORE RECENT WORKS — 789

APPENDIX: BASIC THEOREMS OF THE THEORY OF DIFFERENTIAL EQUATIONS — 795

REFERENCES — 801

INDEX — 813

PREFACE TO THE SECOND RUSSIAN EDITION

The writer of this Preface is the only one of the three authors of this book who is still alive. Aleksandr Adol'fovich Vitt, who took part in the writing of the first edition of this book equally with the other two authors, but who by an unfortunate mistake was not included on the title page as one of the authors, died in 1937.

Aleksandr Aleksandrovich Andronov died in 1952, i.e. fifteen years after the first edition of the book was published. During those years A.A. Andronov and his disciples continued fruitful work in the field of the non-linear theory of oscillations and in particular of that section of the theory which was expounded in the first edition of this book (autonomous systems with one degree of freedom). Since the publication of the first edition, however, the writer of this preface has taken no part in further development of the problems treated in the first edition. One of the disciples of A.A. Andronov, N. A. Zheleztsov, has undertaken the task of expounding for the second edition of the book the new results achieved (mainly by Andronov's school) in the field of the theory of self-oscillating systems with one degree of freedom. To do this N. A. Zheleztsov has had to rewrite and to enlarge the text of the first edition. E. A. Leontovich-Andronova has also taken part in this work. The parts of the text that have been revised or written anew are indicated by foot-notes at appropriate places.

S. E. Khaikin

NOTE FROM THE ENGLISH EDITOR

THIS translation of the second volume of the second edition of the well-known book *Theory of Oscillations* by Andronov and his collaborators brings to English-speaking readers a full account of the work of these authors. The brilliant translation by Professor Lefschetz of the first edition was a succinct account of the work but was considerably shortened. In so doing some of the flavour and detail of the original work was lost. It is hoped that this translation which, by and large, has not departed too much from the spirit of the Russian edition will be welcomed. There are many practical examples of the most detailed use of the theory, and although the many electrical circuits investigated use thermionic vacuum tubes rather than transistors, the methods demonstrated are of course applicable to all similar types of equations.

INTRODUCTION

In every theoretical investigation of a real physical system we are always forced to simplify and idealize, to a greater or smaller extent, the true properties of the system.

A certain idealization of the problem can never be avoided; in order to construct a mathematical model of the physical system (i.e. in order to write down a set of equations) we must take into account the basic factors governing just those features of the behaviour of the system which are of interest to us at a given time. It is quite unnecessary to try to take into account all its properties without exception. The latter process is not usually feasible and, even if we should succeed in taking into account a substantial part of these properties, we would obtain such a complicated system that its solution would be extremely cumbersome, if not altogether impossible.

Since an idealization of the problem is in any case inevitable, the question arises, first of all, of how far we can go in this direction, i.e. to what extent can we idealize the properties of the system and still obtain satisfactory results? The answer to this question can only be given in the end by experiment. Only the comparison of the answers provided by analysis of our model with the results of the experiment will enable us to judge whether the idealization is legitimate.

This conclusion, of course, is only valid in the case when the theoretical treatment of our idealized scheme (or mathematical model) has been carried out with full rigour[†]. In this case only can we consider a discrepancy between theory and experiment to be indisputable evidence of the inadequacy of the initial idealization and of the necessity of taking into account some new properties of the system, in order to explain the observed phenomena.

We shall see that indications of the validity of an idealization can be obtained not only by comparing the results of the theoretical analysis with experimental data, but also by comparing the results of two different theories, one of which has been developed by using a given idealization and

---

† We shall observe that the term "rigorous theory" does not necessarily mean that this theory provides rigorous quantitative answers to the questions. A rigorous theory may simply give approximate quantitative answers (it can, for example, give an estimate of the amplitude of an oscillatory process by means of inequalities) or may enable us to make qualitative statements (for example, on the existence of a periodic motion).

the other without this idealization. Since the first theory has been developed after neglecting certain facts, while the second, on the contrary, takes into account just these facts, then by comparing the results of the two theories we obtain direct indications on how important these facts are for the solution of the question. We gradually accumulate experience and develop our intuition in this respect and learn to "guess" better and better what is important and what is unimportant to the solution of the problems being considered. Having convinced ourselves by one or other means that a certain fact is of second-order importance in the solution of a given problem, we extend henceforth the result to other analogous problems by neglecting this fact *ab initio*, and then verifying the validity of these simplifications by means of an experiment.

The nature of the idealizations permissible in the analysis of a problem is determined by *the problem in its entirety* and therefore depends not only on the properties of the system considered but also on just which questions we want to answer by our analysis.

Thus, for example, let us consider a system consisting of a small steel sphere falling vertically on to a horizontal steel board. If we are interested in the motion of the sphere as a whole then, generally speaking, we do not make a large error if we assume that the sphere is a material point moving under the action of the force of gravity, the velocity of which instantaneously reverses its sign on reaching the board. If, on the other hand, we are interested in the elastic stresses arising in the sphere at the instant of the impact, then obviously we can no longer consider the sphere as a material point; the sphere must be idealized as an elastic body with given constants characterizing the properties of steel, the nature of the deformations, the time of impact, etc. A similar example can be derived from the theory of electrical systems, where there are cases when the capacitance and inductance can be considered as lumped constants for the calculation of some quantities and as distributed constants for the calculation of others, in the same system.

Thus one and the same idealization can be both "permissible" and "impermissible", or better, either expedient or inexpedient depending on the questions to which we want an answer. An idealization of the properties of a real system i.e. use of a mathematical model, enables us to obtain correct answers to certain questions about the behaviour of the system, but does not, generally speaking, give us the possibility of answering other questions correctly about the behaviour of the same system. This follows from the fact that in constructing a given mathematical model of a real physical system we neglect many of its properties which, while inessential for some processes in the system, can be important or even the deciding factor for others.

The permissibility of an idealization also depends on the quantitative relations which characterize a given problem. For example, we can only neglect friction in a pendulum on condition that the friction is sufficiently small and the time during which we investigate the motion of the pendulum is also not too long. But, when we say "small" or "large", this has only a meaning when we indicate with which other factor the given quantity is small or large. Thus we shall require, in our example, that the damping coefficient be small in comparison with the frequency of the oscillations (i.e. that the logarithmic decrement be small in comparison with unity) and that the time of observation be not too long in comparison with the period of the oscillations. Only when similar inequalities are provided can we consider as exhaustive such quantitative characteristics as "small" or "large".

However, when we set about investigating a problem, it often proves difficult to say in advance with which quantity another quantity should be compared. We then use quantitative relations without indicating these comparisons so that these relations lose their definiteness. Nevertheless, they still retain a certain meaning which is derived from our knowledge of the physical phenomena. Thus, for example, from the point of view of "average human dimensions" observation of a phenomenon for the duration of 1 min is already "not too long". On the other hand a few thousandths of a second is "very short". Therefore we often say that the oscillations of a pendulum are *slowly* attenuated, while the oscillations in an electrical oscillating circuit of high frequency are rapidly attenuated even if the damping ratio of the circuits is very small and close to the damping ratio of the pendulum. Whilst resonant oscillations of mechanical systems are usually considered (at least initially) leaving friction out of account, when investigating the question of the resonant oscillations in an electric oscillatory circuit the ohmic resistance of the circuit is nearly always taken into account from the very beginning. Thus such characteristics as "small" or "large" (without indicating in comparison with what) although apparently devoid of content, still influence us in the choice of the idealizations. In the subsequent analysis these characteristics "small" and "large" assume a well-defined content: it becomes evident by comparison with which quantity a given quantity must be accounted small or large. We shall sometimes begin an analysis with such indetermined assumptions as "large" or "small" without indicating in comparison with what, but the meaning of these statements will always be clarified by the subsequent analysis.

In every physical analysis, and particularly in this exposition the question of which of the properties of a real physical system must be taken into account in constructing a mathematical model, and to what degree of approximation, is an extremely important one. Therefore we must be quite clear about just which type of idealization is to be used in the analysis of oscillatory systems. In fact we shall restrict ourselves throughout the book exclusively to *dynamic models* of real oscillatory systems, i.e. we shall neglect in them fluctuations and all other statistical phenomena[†].

We shall assume correspondingly that the dependent variables, occurring in the equations of a mathematical model, have the physical meaning of quantitative characteristics (true characteristics, not statistical ones) of the state of the system and of one or other processes occurring in it. When we speak about idealizations of real physical systems in the form of dynamic models, then these idealizations are connected in the first place with the number of quantities, determining the state of the system (for examples, co-ordinates and velocities) and, in the second place, with the choice of the laws, connecting these states or the velocities of variation of the states and establishing the relations between them. In these relations, which, in the majority of the cases considered, can be expressed in the form of one or other differential equations, there usually occur a certain number of constant parameters, characterizing the system. For example, for an ordinary electric circuit, in the simplest case the charge and current will serve as the quantities defining the state of the system; the inductance, capacitance and resistance are the constant parameters. The connexion between the quantities, characterizing the state of the system, is determined

---

† The presence of fluctuations in real system must indirectly be taken into account even in the theory of dynamic models of real systems. It is evident that since small random perturbations are inevitable in all physical systems, processes which are possible only in the absence of any random deviations or perturbations whatsoever cannot actually occur in them. Hence there arise the requirements, widely used in the theory of dynamic systems, that the processes represented by a mathematical dynamic model (and corresponding to processes taking place and observed in a real system) be stable both in relation to small variations of the coordinates and velocities, and in relation to small variations of the mathematical model itself. The first requirement leads to the concept of stability of the states of equilibrium of the model and of the processes taking place in it, and the second to the concept of coarseness of dynamic systems.

Statistical models are necessary for the theoretical study of the influence of fluctuations, interferences, etc. on the processes taking place in oscillatory systems. When random processes are taken into account, the motion of the system will be no longer subject to dynamic laws, but to statistical laws. In this connexion questions can arise about the probability of one or other motion, of the more probable motions, and of other probability characteristics of behaviour of the system. The mathematical apparatus for the study of statistical processes in oscillatory systems is provided by the so-called Einstein–Fokker equations [106, 75, 83].

by a differential equation where the constant parameters or combinations of them occur as coefficients.

An important idealization concerns the number of degrees of freedom of the system. In the present book we shall consider mainly those problems which can be solved using a mathematical (dynamic) model of the given system which has one degree of freedom. Any real system, from the point of view of classical physics, will, of course, have not one but a very large number of degrees of freedom.

The concept of number of degrees of freedom was introduced in the theory of oscillations from mechanics, where by number of degrees of freedom is meant the number of coordinates which completely define the space configuration of a mechanical system. In the theory of oscillations, which considers mechanical and other systems, by number of degrees of freedom is meant half the number of the variables which at a given instant of time determine the state of the system completely and uniquely.

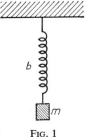

FIG. 1

By considering simplified systems and restricted questions, the concept of system with one degree of freedom can be used. For example, the system consisting of a mass m and a spring (Fig. 1) can be considered as a system with one degree of freedom only if we neglect the mass of the spring (so that the mass of the spring is much smaller than the mass m) and consider the body m as absolutely rigid (meaning the body m is much more rigid than the spring). Moreover, it is evident that we shall restrict our consideration to motions of this body in the vertical direction only. The latter restriction means that the oscillations of the mass m in the vertical direction do not give rise to its swinging as a pendulum (in fact under certain conditions this swinging proves unavoidable). Obviously, in a real system, our assumptions cannot be rigorously observed, for the spring will have a mass, while the body m will have elasticity. Our idealization deprives us of the possibility of answering questions about the motion of separate parts of the body m and the spring.

Let, for example, the spring be initially stretched by a force applied at the point b (Fig. 1) and let this force be removed at the instant of time $t=0$. Such an initial condition is incompatible with the idealization assumed by us, which enabled the whole system to be considered as having one degree of freedom. This example illustrates the general thesis stated above. It is seen that a permissible idealization in relation to the number of quantities determining the state of a system (in particular the number of degrees of freedom of the system) depends not only on the properties of the system itself but also on the nature of the initial conditions which are assigned and on the content of the questions which must be answered; in a word, depends on the nature of the problem formulated by us.

Similarly, an ordinary electric circuit, containing a capacitance, an inductance and a resistance (Fig. 2), can be considered as a system with one degree of freedom only on condition that we leave out of account, for

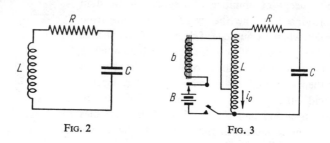

Fig. 2   Fig. 3

example, the capacitance possessed by the separate turns of the induction coil with respect to each other, the shunt conductance, etc. However, notwithstanding this idealization, we shall be able to answer with sufficient accuracy a basic question about the law of variation of the voltage across the plates of the capacitor if the initial conditions are such that the initial voltage across the capacitor and the initial current through the induction coil are assigned. On the contrary, the idealization assumed does not enable us to answer, for example, the question of the law of distribution of the current within the self-inductance coil itself. Nor shall we be able, without making further special assumptions, to solve the question of the variation of the current at the beginning of the process, if at the initial instant of time the current does not flow through all the coil but only through a part of it (connected to a source of direct current (Fig. 3)), since these initial conditions are not compatible with the assumed idealization of the problem. In this case also, we could, just as in the previous one, assign an initial distribution of the current in the coil, not in the form of two conditions for two parts of the coil but in the more general form of a distribution function $i = f(x)$. However, no other distribution except a distribution of current equal in all turns of the coil, is compatible with our idealization of the system. Such arbitrary initial conditions require, generally speaking, that the system be considered to have an infinite number of degrees of freedom.

It is thus seen that, in renouncing considering real systems as distributed systems and considering them with a finite number of degrees of freedom, in particular in our case as a system with one degree of freedom, we must correspondingly restrict the choice of the initial conditions and assign only such initial conditions that are compatible with the idealization. Sometimes, however, there arises a conflict between the initial conditions and the equations of our idealized system, requiring us to modify the nature of the idealization of the system. In several cases, as will be seen below, this conflict can be reconciled without changing the nature of the idealization, but rather by means of a few additional postulates defining the state, compatible with the equation of the model, at which the system arrives in a period of time after the conflict arose.

Similarly the answer to the question whether one or other idealization, connected with the determination of the laws governing the motion of the system, is compatible, depends not only on the properties of the system itself but also on the nature of the problem which has been formulated. This idealization determines the type of equations by which the system is described and leads to the division of systems into "linear" and "non-linear", "conservative" and "non-conservative", "self-oscillatory" and "dissipative", etc.

The question of which law or laws determine the equations of motion of our system must be answered for this choice determines the form of the idealised model. In general, parameters such as resistance, inductance, elasticity, etc., depend on the system state, and on the coordinates, and velocities, and if this has to be taken into account then we inevitably arrive at non-linear differential equations. In order to simplify the problem we must, within certain limits, make a series of simplifying assumptions about the dependence of the parameters on the state of the system.

The simplest and most convenient approach is to assume that the parameters do not in general depend on the state of the system and are constant quantities. Then the mathematical description of the systems considered leads to linear differential equations with constant coefficients, for which there are available standard methods of solution and the investigation of which does not offer any difficulties.

Under which conditions can a satisfactory answer be given to questions about the nature of the motion of a system, if we assume that the parameters of the system do not depend on its state and are constant? This assumption, as we have already seen, does not in general prove to be correct in real physical systems. But we can, in many cases, so choose the regions of variation of the coordinates and velocities that, within them, the parameters of the system will practically (i.e. to a degree of accuracy assigned by us) remain constant. Thus, for example, if the structures of a capacitor and an induction coil are sufficiently rigid and if the largest values achieved by the voltage across the armatures of the capacitor and by the intensity of current in the induction coil are not too large, then, in practice, the capacitance of the capacitor and the inductance of the induction coil can be assumed to be constant in the given region. Similarly, if the current density is sufficiently small, then the resistance of an ordinary metal conductor can be considered to be constant.

We can, in a similar manner, choose such a narrow region of variations of the coordinates and velocities of a mechanical system, that the parameters of the system can be considered to be constant in this region. In the cases which we shall mainly consider, the assumption that the parameters of the system do not depend on the coordinates and velocities amounts to assuming that all forces arising in the system are linear functions of *either* the co-ordinates *or* the velocities *or* the accelerations.

The statement that, within a region of sufficiently small variations of an argument, forces can be considered as linear functions of the coordinates, velocities or accelerations, is derived essentially from mathematical considerations. In fact, if a function can be expanded in the vicinity of a given point into a Taylor series and if, moreover, its first derivative at this point is different from zero, then we can always restrict ourselves for sufficiently small values of the argument to the first term only of the Taylor series, i.e. we can consider the function as linear.

However, these considerations do not give any idea of how large is the region in which a function can be considered as linear. Moreover, cases are possible in real physical systems where the notion of linear forces does not give a correct answer to the question of the motion of the system even in a very narrow (but still physically interesting) region of variations of the co-ordinates and velocities.

The question of the possibility of the "linearization" of a real physical system will be illustrated here by means of the example of a mechanical system subject to friction, for example, a mass m suspended on two springs under the condition that it experiences a certain resistance to motion because of the fluid surrounding it, or else is moved with friction along the surface of some solid body (Fig. 4). The question of the "linearization" of such a system does not raise any difficulty in the case of absence of friction, since the elastic force of a spring is proportional, for small deviations, to the deviation, and the mass of the body can be considered within wide limits to be independent of its velocity. In the presence of friction (we know that a force of friction depends, generally speaking, on velocity) there arises the question of whether we can "linearize" the force of friction, and consider it as a linear function of velocity at least for very small velocities. An answer to this question can only be given by an experiment.

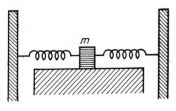

Fig. 4

Completely different force–velocity dependences, generally speaking very complicated ones, are obtained for the case of motion of the body in a gas or liquid or for the case of friction of the body on a solid surface.

In the first case the work done in a displacement essentially depends on the velocity and decreases as the velocity decreases and can be made as small as we like. In the second case, however, of "dry friction", the

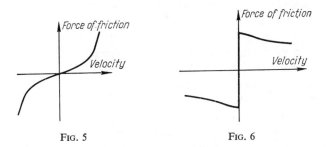

Fig. 5 Fig. 6

work on the contrary depends little on velocity, and however slowly we move the load, some finite and perfectly defined work must be done in a displacement; thus, even for an arbitrarily small velocity, the force of friction has a finite value. We must bear in mind, moreover, that the force of friction is always directed on the side opposite to the velocity and thus it must change its sign, as the velocity passes through zero. Bearing this in mind from the result of our experiments we can establish a connexion between the force of friction and velocity in the region of small velocities. It is evident that in the first case, the case of "fluid friction", the force of friction passes through zero without a jump and changes its sign there (Fig. 5). In the second case, however, as the velocity tends to zero, the force of friction tends from the two sides to finite limits, which are generally speaking different (in particular, for example, of opposite sign but of the same absolute value) and thus undergoes at zero a discontinuous variation (Fig. 6)†. It is evident that in the case of "fluid friction" we can always consider the force of friction to be a linear function of velocity within some, even if small, interval about zero, i.e. we can "linearize" friction and consider the system as linear. In the case, however, of "dry friction" such a

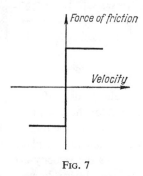

Fig. 7

linearization, even in a region of very small velocities would not reproduce the most typical features of dry friction. Therefore, in the analysis of those problems in which dry friction plays a substantial role, we shall not be able to consider the system as a linear one even if we restrict our considerations to very small values of velocities.

The simplest idealization which can be made in the case of dry friction, i.e. in the case of the friction–velocity dependence represented in Fig. 6, is the assumption made by Coulomb, namely that friction does not depend in value on the velocity. Just as a linear law of friction is the simplest idealization of the cases of fluid friction, Coulomb's law is the simplest idealization of the cases of dry friction. This idealized characteristic of friction is shown in Fig. 7.

It is thus seen that it is not always possible, even within a certain limited region, to consider a system as linear. In many cases, however, it is in fact possible within a limited region, and we can answer the questions of interest to us on the nature and general properties of the motion of the system. The limits of this region are determined by the nature of the relations, existing in real physical systems, between the parameters and the coordinates

† The dependence of the force of friction on velocity in the region of not very small velocity will be dealt with further on in this book. We shall restrict ourselves for the present to the region of very small velocities only.

and velocities, and the nature of the problem which was set; this region, however, where our idealization is applicable, *is always contained within certain limits.*

But if this region is limited, then inevitably the following very important question arises: shall not our system "by itself", by virtue of its properties, go beyond the boundaries of the region in which our idealization is applicable? If this does not happen for initial conditions lying within the region in which the parameters of the system can be assumed to be constant, then our idealized problem enables us to give an answer to a series of questions which can arise in the study of the given system. In order to answer these questions, we can assume the parameters of the system to be independent of the state of the system, and can describe it by means of linear differential equations, i.e. we can consider the system as "linear".

When, however, the system in virtue of its own properties moves beyond the boundaries of the "linear region", it is perfectly evident that questions can arise to which we are not in a position to give an answer by considering the system as linear. The permissibility of one or other idealization, as with regard to the question of the laws connecting the properties of the parameters with the states of the system, depends not only on the properties of the system but also on the nature of the problem which was set, and in particular on the nature of the initial conditions of the problem.

There exist a very large number of very interesting and practically important problems, the answer to which necessitates considering the behaviour of a system beyond the boundaries of the linear region. A number of these arise, for example, in modern radio engineering. As we shall see below, even the theory of the simplest valve oscillator cannot *in principle* be reduced to the investigation of a linear differential equation and requires the study of a non-linear equation; a linear equation, for example, cannot explain the fact that a valve oscillator, independently of the initial conditions, has a tendency to reach determined steady-state conditions. Analogous problems arise in electrical engineering, acoustics, etc.

The consideration of just such questions, *the very nature of which makes the formulation of a non-linear problem inevitable, i.e. forces us to consider the system as non-linear, constitutes the main object of this book.*

A rigorous division of real physical systems into "linear" and "non-linear", "conservative" and "non-conservative", or a division of them according to the number of degrees of freedom, etc. is not possible. Real physical systems are neither linear nor conservative nor can have a finite number of degrees of freedom, since they cannot in general be described with complete accuracy by means of mathematical relations. Therefore any rigorous division, any rigorous classification, cannot be accurately applied to real physical systems. Such a rigorous division may be carried out for abstract schemes only (mathematical models) which are obtained as a result of a certain idealization of the properties of a real physical system.

In particular, the systems which are usually considered in school courses must not be thought of as some kind of special "linear systems" which can be rigorously separated from the other "non-linear systems" such as constitute the object of our analysis. In both cases we often consider one and the same real physical systems, but apply to them different types of idealization. As a result of this idealization, the nature of which is determined not only by the properties of the system itself but also by the content of the problems we have to solve, it is possible to divide systems into linear and non-linear, conservative and non-conservative, to isolate from the number of non-linear non-conservative systems the class of self-oscillatory systems and finally to divide the self-oscillatory systems into continuous (in particular those of the "Thomson type") and "relaxation" ones.

On carrying out this classification we shall always arrive at determined conclusions with respect to the properties of one or other class of systems; it is, however, necessary to bear in mind that these properties characterizing a system are *idealized properties*. Thus, for example, when we speak of the property of a self-oscillatory system of producing oscillations of a constant amplitude for an arbitrarily long time, then this property must obviously be considered as an idealized one. Oscillations in a real self-oscillatory system cannot last for an "arbitrarily long time"; the oscillations of a watch cease when its winding mechanism comes to an end, the oscillations in a valve oscillator cease when the anode battery or the filament battery are discharged. When we speak of oscillations which can last "an arbitrarily long time", then we are ignoring the facts indicated (the finite reserve of energy in the winding mechanism of a watch or in the battery of a valve oscillator). Similarly the statement that every self-oscillatory system, for example a radio transmitter, has a tendency to reach and retain a steady-state condition, i.e. that the "amplitude" and period of the oscillations are constant, has only an approximate meaning. It is easily seen that small external actions, which are always there, and fluctuations, which are unavoidable, will always cause these quantities to vary within certain usually narrow limits. Even more, it is evident that even the concept of periodic motion is also an idealization when referred to a real system. As indicated many times, we underline in any study one or other properties of the real physical system which have a major role in the solution of a given problem and ignore those properties which are of the second order importance.

To which questions shall we try to obtain an answer in the analysis of these non-linear problems?

When we study the behaviour of a dynamic system we are usually interested first of all in the so-called *stationary* motions in the system[†], since just these motions are most typical for the behaviour of a system over long intervals of time.

Which stationary motions are then possible in the systems which we shall consider? (We shall now take into consideration, for the sake of definiteness, dynamic models of mechanical systems only).

They can be first of all states of equilibrium in which velocities and accelerations, determined from the differential equations, which describe the behaviour of the system, reduce to zero. This amounts to saying that no regular force whatsoever, as calculated by the differential equations, acts in the system. But in every physical system there act, besides such regular forces, small irregular forces also, for example of a fluctuation nature. As a consequence of the presence of these forces the system can never be found exactly in a state of equilibrium and accomplishes small motions about the state of equilibrium (Brownian motion). But in the vicinity of a state of equilibrium there already act in the system regular forces also (they are exactly equal to zero in the state of equilibrium only), which can either bring back the system to the state of equilibrium or remove it still farther away. In the first case we shall have stable states of equilibrium and in the second unstable ones. It is clear that in order to study the behaviour of a system we must know not only how to find the states of equilibrium but also how to determine their stability with respect to small variations of the co-ordinates and velocities. Stability is in this case a necessary condition that the system might be found in the vicinity of a given state of equilibrium for an arbitrarily long time.

Further to the number of stationary motions possible in the system, there belong *periodic* motions. Other stationary oscillatory processes, along with periodic processes, are possible in oscillatory systems but in an

---

† A stationary motion is, roughly speaking, a limit motion to which the system tends. When we speak of stationary motions we understand to include in them states of rest also, i.e. we consider a state of rest as a particular case of stationary motion. A precise mathematical definition of stationary motions can be given, by identifying them with the so-called *recurrent* motions of Birkhoff [34, 139, 96]. For systems with one degree of freedom, only states of equilibrium and periodic motions can be recurrent motions. For more general systems recurrent motions can be more complicated motions, for example, quasi-periodic motions.

autonomous† oscillatory system with one degree of freedom, as will be clarified in a more detailed analysis, there can only exist the simplest type of stationary oscillatory motion, namely a periodic process.

Not all possible stationary motions can exist in a real physical system. A necessary condition for a given stationary process to last for an arbitrarily long time is that the system, in the presence of the unavoidable random perturbations, shall accomplish motions sufficiently close to the given stationary motion, and shall not move away from it to any noticeable extent. This requirement is completely analogous to the same requirement in relation to states of equilibrium: in order that a periodic process may last an arbitrarily long time it must be stable with respect to small variations of the coordinates and velocities. Thus we shall have not only to find the periodic processes possible in the system, but also to discuss their stability with respect to small deviations. In precisely the same manner we shall be concerned with the dependence of the motions of the system on the parameters occurring in the equations and which can assume one or other fixed values. The analysis of this dependence enables us to give an answer to a series of basic questions connected with the onset of oscillations, the stopping of oscillations, etc., etc.

In order that stationary processes may exist in a real system a long time, they must be stable not only with respect to small variations of the coordinates and velocities, but also with respect to small variations of the form itself of the differential equations describing the system. These small variations of the form of the differential equations reflect corresponding small variations of the properties of the system which is described by these equations. Thus, since on the one hand we shall never be able to describe with absolute accuracy a real system by means of a mathematical apparatus, while on the other hand no real physical system remains absolutely unvaried over the time of the processes occurring in it, then we have always to allow for the possibility of small variations of the form of the differential equations which describe a physical system‡.

---

† We shall call autonomous such systems as are described by equations which do not contain time explicitly. Therefore, we shall assume, in the analysis of autonomous systems, that external actions do not depend on time.

‡ These small variations of a system or small variations of the form of differential equations shall be assumed at first to be such as not to vary the order of the initial differential equation (or, which is the same, not to vary the number of differential equations of the first order, if we are considering systems of the first order only). This means, in the language of physics, that the small variations being considered for the system are such as not to force us to reject the idealization connected with the number of degrees of freedom.

If, in the analysis of one or other concrete problems, we ascribe to the parameters well-determined fixed values, then this has a meaning only on condition that small variations of the parameters do not substantially alter the nature of the motions and that the behaviour of the ideal model preserves the features in which we are interested. Those aspects, however, of the behaviour of the model which are not preserved under a small variation of the form of the differential equations and of the values of the parameters, are of no physical interest, since they do not reflect properties of a real physical system. Systems which are such as not to vary in their essen-

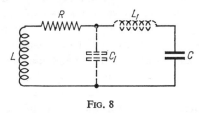

Fig. 8

tial features for a small variation of the form of the differential equations, we shall call "*coarse*" systems, and they serve as useful theoretical models of real physical systems. However, we shall impose a restriction on the small variations of the system, namely that the number of degrees of freedom, and so the order of the equation, shall not increase for these variations. This is a far-reaching restriction, for, from a certain point of view which can be justified physically, we might also consider as "a small variation of the form" of the equation an increase of the order of the differential equation, provided that the coefficients of the new higher-order differential coefficients be sufficiently small. But a "small variation of the form" of the differential equation, consisting in an increase of the order of the equation, is the result of taking into account some new degrees of freedom of the system, i.e. taking into account some of its "parasitic" parameters. Thus, for example, in the case of the electrical circuit, represented in Fig. 8, by taking into account only the inductance L, the capacitance C and the ohmic resistance R of the induction coil, i.e. the basic ("non-parasitic") parameters, we shall obtain a differential equation of the second order. If we take into account, in addition, the total "parasitic" capacitance between the turns of the coil, i.e. the capacitance C_1, and the "parasitic" inductance of the leads, i.e. the inductance L_1, we shall obtain, instead of an equation of the second order, an equation of the fourth order. But since C_1 and L_1 are small quantities, this equation of the fourth order may be simply considered as a small variation of the form of the original differential equation of the second order.

These "small variations" of the form of a differential equation leading to an increase of the order of the equation might be extended further and further at will, for by striving towards a more complete description of the system, we would always arrive at a larger and larger number of degrees of freedom.

Having solved the question of the stability of a stationary motion we ought, strictly speaking, to verify in addition that the given motion does not vanish or lose its stability as the order of the differential equation increases. In fact, if it were shown that a state of equilibrium, stable in the case when only the basic parameters are taken into account, should lose its stability as a consequence of the influence of a small parasitic parameter, which increases the order of the equation, then this would mean that this state of equilibrium is effectively unstable. Therefore the requirement of stability of a state of equilibrium in relation to such variations of the equation is quite natural. It is not difficult to show that it is impossible to construct such an ideal model of a dynamic system (i.e. to define such a class of differential equations), for which a state of equilibrium would always remain stable, even if in the equations of the system there should occur terms with higher-order differential coefficients having small arbitrary analytical coefficients, but different from zero. It is impossible to lay down a general requirement for ideal models of dynamic systems on the invariability of the character of stationary motions as there appear new degrees of freedom (this would be analogous to the requirement of coarseness for small variations of dynamic systems not connected with the appearance of new degrees of freedom). We can only take into account the influence of new degrees of freedom by keeping our attention on the specific characteristics of the new systems. We shall encounter states, the "stability" of which does not depend, essentially, on the properties of the real system, but on the fact that we leave out of account some degree of freedom.

But we cannot help being "naive", for, otherwise, we should have to verify that all possible small parasitic parameters, increasing the order of the equation, shall not disturb the stability of a given state. However, we can never carry out this verification exhaustively, since the number of such parasitic parameters in every system is very large. In addition, as will be shown, it may happen that these parameters act in different directions, so that, in order to verify their influence we have not only to assume the presence of these parameters but also to know the quantitative relations between them. Therefore the validity of the answer to the question of stability of one or other state in a real system, as well as of any other result of a theoretical analysis (inevitably connected with a mathematical model of this system) can be verified only by experiment.

The form in which we shall attempt to obtain answers to the questions of interest will be different in different cases. One could obtain the answers to all questions if the functions which characterize the states of the system and the variations of these states were known. These functions, which we shall have to study in order to determine the behaviour of the system (for example, the dependence of the intensity of current or the voltage upon time), are defined by means of the differential equations which describe a given system and cannot be defined by other means. Only for a very small class of cases, for example for linear equations with constant coefficients, is it possible to reduce the problem of finding such functions to another simpler one, for example to the solution of algebraic equations or to finding the integrals (quadrature) of functions occurring in the differential equations.

It is necessary, therefore, to know how to derive directly from the differential equations themselves knowledge about the character and form of the functions which are determined by these equations.

For example, there arises in the first instance the following problem: to determine the more typical, or qualitative, features of these functions by means of the geometrical construction of the so-called integral curves. We shall call this qualitative integration of the equations. If we are able to integrate qualitatively the differential equation considered, then we obtain a qualitative picture of the possible physical processes in the system model. A large number of questions, having a fundamental practical interest, bear just this qualitative character, i.e. the question of the presence of stable states of equilibrium, the question of the existence of stable periodic processes, the question of a soft or hard mode of excitation, etc.

The qualitative integration makes the quantitative integration also substantially easier or, more precisely, makes it easier to solve the quantitative questions which arise in the physics of oscillations. The theory of oscillations is not concerned in the end with the numerical values of the functions at one or other particular instant of time; it is mainly concerned, rather, with the quantitative characteristics which determine the behaviour of this function over significant intervals of time, e.g., in the case of a periodic function, its period, the values of the coefficients of the expansion in a Fourier series, the spectrum of functions represented by means of Fourier integrals, etc.

However, in order to find these quantities, the theory of oscillations has often to determine, as an intermediate step, numerical values of functions for one or other particular values of the independent variable. The usual approximate methods of numerical integration (e.g. the method of the isoclines, the Runge–Kutta method etc.), which can be used to obtain answers to such questions, also operate directly with the differential equations themselves. The knowledge of a qualitative picture for a given differential equation enables us to employ with greater efficacy and reliability these numerical approximate methods, and to combine them judiciously.

Below we shall have to acquaint the reader with the mathematical apparatus which is needed to investigate functions determined by non-linear differential equations. Since we are restricting ourselves here to the analysis of systems with one degree of freedom, these will be functions determined either by one differential equation of an order not greater than two or by not more than two differential equations of the first order.

In order to make the mastering of this mathematical apparatus easier, we shall begin with an outline of well-known ordinary linear problems using the language, and partly with the aid of the methods, which thereafter in a fully developed form we shall use for the solution of much more complicated non-linear problems.

THEORY OF
OSCILLATORS

THEORY OF
OSCILLATORS

CHAPTER I

LINEAR SYSTEMS

§1. A LINEAR SYSTEM WITHOUT FRICTION (HARMONIC OSCILLATOR)

We shall begin our analysis with the simplest type of oscillatory system; one in which the motion is described by a linear differential equation of the form

$$\ddot{x} + \omega_0^2 x = 0 \tag{1.1}$$

and which in physics is called a *harmonic oscillator*.

An example of such a system is a body of mass m which moves horizontally along a rod under the action of two springs (Fig. 9). In order that

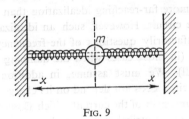

FIG. 9

the analysis of this system may lead to the case of interest to us, we shall make the following simplifying assumptions†. We shall assume firstly, that the spring force on the body is proportional to its displacement x with respect to its position of equilibrium. This assumption is verified in practice, to a reasonable degree of accuracy, for sufficiently small displacements only. We shall assume secondly that the system's motion is not subject to friction either caused by the air and supporting rod or due to internal friction of the springs. This second assumption about the absence of friction is obviously verified in real physical systems to a still smaller degree of accuracy. With these assumptions the motion of such a system

---

† We shall not recall, here or in the sequel, more about the other simplifying assumptions which have been discussed in the Introduction.

is represented by the equation

$$m\ddot{x} + kx = 0, \qquad (1.2)$$

where k is the spring constant. Putting $k/m = \omega_0^2$ we obtain equation (1.1).

An oscillating circuit consisting of a capacitance C and an inductance L (Fig. 10) is an analogous electrical system; for the sake of brevity we shall call such circuits "Thomsonian" circuits. In order to arrive at the case of a linear system without friction, we must of course idealize the properties of this circuit. We must assume firstly that no loss of energy occurs in the system, i.e. that the connecting leads do not have resistance, that energy is not dissipated in the dielectric, and that there is no radiation of electromagnetic energy. These assumptions are never absolutely true in real circuits as is confirmed by the fact that there always occurs a more or less strong but noticeable damping of the oscillations. On idealizing the circuit as a system without losses of energy, we can no longer reproduce this typical feature of all physical systems and in this sense the assumption of the absence of losses of energy is a much more far-reaching idealization than the assumption of the linearity of the circuit. However, such an idealization enables us to answer fairly satisfactorily questions of the frequency and form of the natural oscillations (in those cases when the damping of the oscillations is sufficiently small). We must assume, in addition, that the capacitance C of the capacitor does not depend on its charge nor the inductance L of the coil on the intensity of the current which flows through it. Under these assumptions our electrical system is also governed by an equation of the type (1.1); where if we denote by q the charge of the capacitor we obtain:

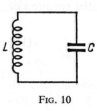

FIG. 10

$$L\ddot{q} + \frac{q}{C} = 0. \qquad (1.3)$$

By writing $1/LC = \omega_0^2$, we arrive again at the equation of the harmonic oscillator (1.1).

We shall recall here the characteristic properties of a harmonic oscillator. The general solution of the differential equation (1.1) has the well-known form

$$x = A \cos \omega_0 t + B \sin \omega_0 t, \qquad (1.4)$$

where A and B are integration constants determined by the initial conditions. If for $t=0$, $x=x_0$ and $\dot{x}=\dot{x}_0$, then

$$x = x_0 \cos \omega_0 t + \frac{\dot{x}_0}{\omega_0} \sin \omega_0 t; \quad \dot{x} = -x_0 \omega_0 \sin \omega_0 t + \dot{x}_0 \cos \omega_0 t. \quad (1.5)$$

This same solution can also be written in the form

$$x = K \cos(\omega_0 t + \alpha); \quad \dot{x} = -K\omega_0 \sin(\omega_0 t + \alpha), \quad (1.6)$$

where

$$K = +\sqrt{A^2+B^2} = +\sqrt{x_0^2 + \frac{\dot{x}_0^2}{\omega_0^2}} \quad \text{and} \quad \text{tg } \alpha = -\frac{B}{A} = -\frac{\dot{x}_0}{\omega_0 x_0} \Big\}$$

$$\left(\cos \alpha = \frac{x_0}{K}, \quad \sin \alpha = -\frac{\dot{x}_0}{\omega_0 K}\right). \quad (1.7)$$

It is seen that the dependence of displacement or charge on time (the oscillogram of the oscillations) is the familiar sinusoid (Fig. 11). Such a "sinusoidal" or harmonic oscillation is characterized by three quantities: K, the maximum deviation or *amplitude* of the oscillations, ω_0, the num-

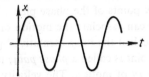

Fig. 11

ber of oscillations in 2π seconds or the angular *frequency*, and α, the so-called *initial phase* of the oscillations which plays a very important role when we are concerned with several simultaneous processes. Since the choice of the phase of the oscillation completely determines the initial instant from which time is measured, then we cannot choose it arbitrarily if the initial instant of time is already assigned by some other process. However, the phase of the oscillations does not play any physical role when we are concerned with one "isolated" process only. An oscillatory motion does not arise, when $x_0=0$ and $\dot{x}=0$, for then the oscillator at the initial instant is in a state of equilibrium and remains so. The amplitude and phase of a harmonic oscillatory motion are determined by the initial conditions, but the angular frequency does not depend on the initial conditions being determined by the parameters of the oscillatory system.

The formulae (1.5) or (1.6) and (1.7) give an exact quantitative description of the motions in the system defined by equation (1.1). They enable us to determine "the future from the present", i.e. enable us to calculate the values of x and $\dot{x}$ for every instant of time t, if they are known at the instant of time $t=0$.

§2. The concept of the phase plane. Representation on the phase plane of the totality of motions of a harmonic oscillator

1. The phase plane

We shall put $\dot{x}=y$ and shall study the motion of a harmonic oscillator by representing this motion on the x, y plane, where x and y are orthogonal cartesian coordinates. To each state of our system, i.e. to each pair of values of the coordinate x and velocity y, there corresponds a point on the x, y plane. Conversely, to each point on the x, y plane there corresponds one and only one state of the system. The x, y plane is called the plane of the states or, otherwise, *phase plane;* it represents the totality of all possible states of our system. To each new state of the system there correspond always new points of the phase plane. Thus, to a variation of state of the system we can associate the motion of a certain point on the phase plane, which is called the "representative" point. A path followed by the representative point is called a *phase path;* it must not be confused with the actual trajectory of motion. The velocity of such representative point is called the *phase velocity;* again this must not be confused with the actual velocity. A curve which is described by the representative point over the whole time of its motion (from $t=-\infty$ to $t=+\infty$) will be called a complete *phase path*[†].

Knowing the solution of the differential equation of a harmonic oscillator (1.1), the equation of a path on the phase plane is easily found. And,

---

† The method of representing the state of a system with n degrees of freedom by assigning one point in a $2n$-dimensional space has already been used in physics for a long time. This $2n$-dimensional space of the states (phases) of a system was given the name of phase space. Hence the terms "phase space" and, in particular, "phase plane" were introduced in the theory of oscillations.

The phase plane was first used for the study of the dynamics of oscillatory systems by Léauté [172], who investigated the operation of a certain automatic control equipment by constructing on the phase plane of this equipment the integral curves and limit cycles (without giving to them this name; he was apparently not aware of the work by Poincaré published a little earlier [108], where the limit cycles first appeared in mathematical literature). Afterwards the remarkable works by Léauté were, unfortunately, almost completely forgotten.

in fact, the equations

$$x = K\cos(\omega_0 t + \alpha); \qquad y = -K\omega_0 \sin(\omega_0 t + \alpha) \qquad (1.6)$$

are the parametric equations of a phase path; eliminating t from these equations, we shall find the coordinate equation of a path:

$$\frac{x^2}{K^2} + \frac{y^2}{K^2\omega_0^2} = 1. \qquad (1.8)$$

It is easily seen that as the parameter K varies this is the equation of a family of similar (i.e. with a constant ratio of the axes) ellipses, such that through each point of the phase plane there passes one and only one ellipse[†], corresponding to a given value of K, i.e. to a given class of initial conditions with one and the same initial value of the total energy of the system. In this case all the x, y plane is filled with ellipses, enclosing each other, except the point $x=0$, $y=0$; the ellipse "passing" through this point degenerates into a point (Fig. 12).

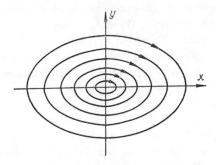

Fig. 12

All these ellipses represent paths of motion of the representative point. Let us see how the representative point will move along one of these ellipses. It is easily seen that, for the direction chosen by us of the coordinate axes, the motion of the representative point along any of the paths will always occur clockwise, since in the upper semi-plane $\dot{x}=y>0$ and x increases with time, while in the lower semi-plane $\dot{x}=y<0$ and hence x decreases with time.

[†] It can happen in other more complicated examples that, on eliminating the time t from the parametric equations of a phase path, we shall obtain the coordinate equation not of one path but of *several at the same time*.

In order to find the value of the phase velocity we shall introduce, the radius-vector

$$\mathbf{r} = \mathbf{i}x + \mathbf{j}y.$$

In this case the phase velocity is represented in the form:

$$\mathbf{v} = \frac{d\mathbf{r}}{dt} = \mathbf{i}\dot{x} + \mathbf{j}\dot{y}$$

or according to (1.6) in the form:

$$\mathbf{v} = \mathbf{i}\{-K\omega_0 \sin(\omega_0 t + \alpha)\} + \mathbf{j}\{-K\omega_0^2 \cos(\omega_0 t + \alpha)\}. \tag{1.9}$$

It is easy to see that the phase velocity, with the exception of the case $K=0$, never reduces to zero, since the sine and cosine never reduce to zero at the same time.

We have investigated the character of the phase plane and have found that to periodic motions occurring in the system there correspond on the phase plane closed paths of the representative point—in our case ellipses —along which the representative point moves with a phase velocity which does not reduce to zero (Fig. 12) and accomplishes a whole circuit in $T_0 = 2\pi/\omega_0$ units of time. To a state of equilibrium of the oscillator there corresponds on the phase plane a phase path degenerating into a point.

Let us assume now that we do not know the character of the motions in the system but have, by some method, come to know the character of the phase paths and the values of the phase velocities. Can we, using this knowledge, derive conclusions with regard to the motions corresponding to these curves? As we shall see later, the general character of the motion, its qualitative features, are already apparent in the character of the phase paths. The phase plane as mapped out by the phase paths gives an easily interpreted "portrait" of a dynamic system.

We have obtained the portrait on the phase plane for the case of a harmonic oscillator by proceeding from an available solution (1.6) of the equation of the oscillator. We can, however, without using this solution, derive from equation (1.1) conclusions about the motion of the representative point on the phase plane. It is just this second approach which offers interest, since it enables us to derive these conclusions without any knowledge of the analytical expressions of the solutions of the initial equation and, consequently, is also applicable in those cases when such analytical expressions as (1.6) cannot be found.

2. Equation not involving time

In order that we may directly arrive at the phase-plane portrait from the initial equation (1.1) without actually integrating this equation, we shall proceed in the following manner. We replace the initial equation of the second order with two equivalent equations of the first order:

$$\frac{dx}{dt} = y; \quad \frac{dy}{dt} = -\omega_0^2 x. \quad (1.10)$$

Dividing one of these equations by the other, we obtain the differential equation

$$\frac{dy}{dx} = -\omega_0^2 \frac{x}{y}. \quad (1.11)$$

These equations determine the so-called *integral curves*, namely such curves that the tangents at every point of them have a slope (the angular coefficient dy/dx) which can be calculated from equation (1.11). It is seen that whilst the dependence of x on t is expressed by the differential equation of the second order (1.1), the dependence of y on x is expressed by a differential equation of the first order. On integrating equation (1.11) we can obtain the equation of the integral curves no longer in a differential but in a finite form. In the given simple case the integral curves, as is easily seen, coincide with the phase paths. However, we shall have in the sequel to distinguish between integral curves and phase paths, since it can happen that one integral curve consists not of one but of several phase paths.

3. Singular points. Centre

Equation (1.11) determines directly at each point of the plane a single tangent to the corresponding integral curve, with the exception of the point $x=0$, $y=0$, where the direction of the tangent becomes indeterminate. As is known from the general theory of differential equations, through those points, for which the conditions of Cauchy's theorem[†] are verified (these include the condition that the differential equation should assign a determined direction of the tangent to the integral curve), there passes one and only one integral curve; on the contrary, for the points in which the direction of the tangent becomes indeterminate and at which consequently the conditions of Cauchy's theorem are not verified, we can

---

[†] To Cauchy's theorem and its significance for the study of the behaviour of the integral curves we shall return again in the sequel (see also Appendix I)

no longer affirm (on the basis of this theorem) that through them there passes one and only one integral curve. Such points where the direction of the tangent is indeterminate are called *singular points* of the given differential equation. However, Cauchy's theorem does not entitle us to affirm that through a singular point there pass either more than one or less than one integral curve (i.e. either no curve or many). But, for the simplest singular points of the first order with which we shall be mainly concerned, this converse thesis proves to be correct.

A differential equation can have, generally speaking, many singular points. In our case there is only one singular point at $x=0$, $y=0$. There exist various types of singular points, differing in the character of the behaviour of the integral curves in the vicinity of the point. In our case through the singular point there passes no integral curve. Such an isolated singular point, in the vicinity of which the integral curves are closed with no singularities, is called a *centre*. For example, the integral curves might be a set of concentric ovals surrounding the singular point. We shall encounter other types of the simplest singular points in later analysis. We shall merely note now that to different types of integral curves there correspond various types of motions of the system and the classification of the singular points is directly connected with the behaviour of the system in the vicinity of the singular point.

4. Isoclines

The equation (1.11) determines a field of tangents on the phase plane. A graphical description of this field is easily obtained, if we construct the family of isoclines[†], which in the case given will be simply straight lines passing through the origin of the coordinates (Fig. 13). Actually, let us find all the points of the phase plane, where the slope of the integral curves is equal to $\varkappa$. Then the equation of this isocline will be, according to (1.11):

$$-\omega_0^2 \frac{x}{y} = \varkappa \quad \text{or} \quad y = \sigma x$$

where

$$\sigma = -\frac{\omega_0^2}{\varkappa}. \tag{1.12}$$

It is easily seen (by giving to σ various values for fixed ω_0^2), that the field

---

[†] An isocline is the locus of the points at which the tangents to all integral curves have the same slope.

investigated consists of line elements, symmetrically disposed with respect to the x and y axes and gradually changing their direction (as the slope σ of the isocline varies) from the horizontal (along the y axis, where $\varkappa=0$) to the vertical (along the x axis where $\varkappa=\infty$).

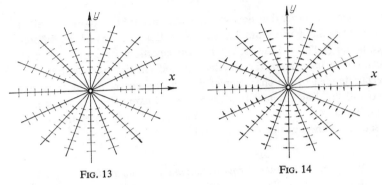

FIG. 13   FIG. 14

Equation (1.11) however, does not give an answer to the question as to the direction and velocity with which the representative point will move on the phase plane. On the other hand, the equations (1.10) determine the phase velocity both in magnitude and direction; actually

$$\mathbf{v} = \mathbf{i}\dot{x}+\mathbf{j}\dot{y} = \mathbf{i}y+\mathbf{j}(-\omega_0^2 x). \qquad (1.13)$$

If we take into account the direction also, then it is expedient to consider instead of the field of line elements (Fig. 13) the vector field (Fig. 14) which characterizes not only the slope of the tangent to the integral curve at a given point, but also the direction of motion along the phase path.

As we have already shown, the phase velocity $v = \sqrt{y^2+\omega_0^4 x^2}$ reduces to zero at the origin of the coordinates only, i.e. at the singular point only.

It is readily seen, by looking at Figs. 13 and 14, that in the case considered the isocline method enables us to obtain at once a notion of the character of the paths on the phase plane. The use of the isocline method in the simple case considered, when the initial differential equation (1.11) permits separation of the variables and is easily integrated, will hardly offer any advantage. In fact, by integrating the equation

$$x\,dx+\frac{1}{\omega_0^2}y\,dy = 0,$$

we obtain

$$\frac{x^2}{2}+\frac{y^2}{2\omega_0^2} = C,$$

or putting $2C=K^2$, we find, as was to be expected, the equation of a family of ellipses as before:

$$\frac{x^2}{K^2}+\frac{y^2}{K^2\omega_0^2} = 1.$$

We must not forget that now we have obtained it by a completely different method without using our knowledge of the solutions of the differential equation (1.1). In those cases when the equation corresponding to (1.11) cannot be integrated, the isocline method enables us to obtain a fairly accurate idea of the character of the integral curves on the phase plane, notwithstanding the fact that an analytical expression for these integral curves cannot be found. In these more complicated cases the use of the isocline method can be of considerable advantage.

5. *State of equilibrium and periodic motion*

We shall now try to derive results in a sense opposite to that of the results arrived at the beginning of this section, when, knowing the motion (knowing the dependence of x on t), we looked for the character of the phase portrait. We shall in fact see what can be said about the motion, if the character of the integral curves on the phase plane and the expression for the phase velocity are known.

We state, first of all, that in our case, all the phase paths (except the path $x=0$, $y=0$, which degenerates into a point) correspond to periodic motions. In fact, all these closed paths are ellipses. If our representative point moves along a closed curve and if it returns after a certain time, having completed a "circuit", to the same point of the phase plane, and has the same position and the same velocity, then the subsequent motion will coincide fully with the preceding one and the process will repeat itself.

It is easily seen that the "recurrence time" or, the period of the motion, is finite. In fact, the length of our ellipse is finite; the phase velocity, on the other hand, never approaches zero during the motion along the ellipse (since it is equal to zero at the origin of the coordinates only and no ellipse passes through the origin of the coordinates). Therefore the representative point moves along an entire ellipse in a finite time and so the period of the process is finite.

We state, in the second place, that the degenerate path or singular point, $x=0$, $y=0$, corresponds to a state of equilibrium. In fact the phase velocity for the point $x=0$, $y=0$ is equal to zero; if the representative point is found at the initial instant at the origin of the coordinates it will remain there, provided that no random deviations remove the representative point away from the point $x=0$, $y=0$.

In general, to states of equilibrium there correspond such points of the phase plane, for which simultaneously $dx/dt=0$ and $dy/dt=0$. This is easily understood from physical considerations. For example, for the mechanical case, $dx/dt=0$ means that the velocity is equal to zero and $dy/dt=0$ indicates that the acceleration, and so the force, is equal to zero.

Generally speaking, to states of equilibrium of a dynamic system there correspond on the phase plane singular points of the equation of the integral curves and, conversely, singular points correspond to states of equilibrium†.

Thus, while we do not know yet the possible motions from the quantitative point of view, still we do know the qualitative character of the possible motions. The results of the qualitative investigation of a linear system without friction (the harmonic oscillator) can be formulated thus: *such a system accomplishes, for any initial conditions, periodic motions about the state of equilibrium $x=0$, $y=0$, with the exception of the case when the initial conditions correspond exactly to the state of equilibrium.*

§3. STABILITY OF A STATE OF EQUILIBRIUM

We can visualize intuitively the meaning of the words "stability of a state of equilibrium". However, this intuitive notion is certainly inadequate and must be replaced by a rigorous concept which we shall be able to use in the sequel.

We shall begin our analysis with the simplest example: let us imagine an ideal pendulum without friction (Fig. 15). It is evident that two states of equilibrium of the pendulum are possible: (1) when we put it at the lowest point a without imparting to it an initial velocity, and (2) when we put it, again without imparting to it an initial velocity, at the uppermost point b. It is also evident that the lower state of equilibrium is stable and the upper one unstable. In fact if the pendulum is found at the point b,

Fig. 15

---

† Consider a dynamic system represented by the equations $dx/dt = P(x, y)$, $dy/dt = Q(x, y)$. If $P(x, y)$ and $Q(x, y)$ have a common factor, which reduces to zero at some points, then there can be states of equilibrium which are not singular points of the equation of the integral curves $dy/dx = Q(x, y)/P(x, y)$. Also if $P(x, y)$ and $Q(x, y)$ have a common factor, going to infinity at singular points of the equation of the integral curves, then these singular points cannot correspond to states of equilibrium.

then an arbitrarily small impulse is sufficient for the pendulum to move with increasing velocity away from the point *b*. A pendulum resting at the point *a* will behave differently. On receiving an impulse it begins to move with decreasing velocity, the distance moved away from the point *a* being the smaller, the smaller the impulse, and then will return back and oscillate about the point *a*. For a sufficiently small impulse, the pendulum does not leave an arbitrarily given region about the point *a* and its velocity does not exceed an arbitrarily assigned value.

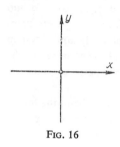

Fig. 16

Proceeding from this example we shall attempt to give a definition of the stability of a state of equilibrium, using to this end the notion of the phase plane. Let the system considered be found in a state of equilibrium. Then the representative point on the phase plane is found in a state of rest at one of the singular points of the equation of the integral curves. If now we remove our system from the state of equilibrium, by imparting to it, for example, a certain impulse†, then the representative point is displaced from the singular point and will begin to move on the phase plane. Let us draw the representative point in black, and leave the singular point white (Fig. 16). We can then characterize a stable state of equilibrium in this manner: if for a sufficient small initial displacement the black point never moves far from the white one, then the white point is a stable state of equilibrium‡.

It is clear that this definition is insufficient. In the first place, shall we call the white point stable if the black point does not move far away for initial displacements in single directions while it does move far away when we displace it, however little, in other directions? It is evident that such a white point will not be stable; it is, so to speak, only "conditionally" stable, if a certain class of displacements is not allowed. We must require then that the black point shall not move far away from the white one as a result of a sufficiently small displacement in any direction.

In the second place, and this is the most important consideration, the terms "does not move far away"; "remains in the vicinity" etc. are

---

† In the theory of stability one usually considers "instantaneous" impulses, the role of which amounts to an instantaneous displacement of the representative point on the phase plane, in other words, to an instantaneous variation of the initial conditions. This is obviously an idealization of real impulses.

‡ This same condition is often formulated thus: a state of equilibrium is stable if a sufficiently small perturbation remains always small.

not sufficiently well-defined. It is clear that the concepts "near" and "far" depend on the concrete physical conditions of the problem. Therefore the words "far" and "near" do not mean anything else but that the black point does or does not leave a certain assigned region surrounding the white point, this region being larger or smaller depending on the conditions of the problem.

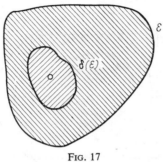

Fig. 17

Thus we shall finally formulate the following definition (Fig. 17): *a state of equilibrium is stable if, for any assigned region of possible deviations from the state of equilibrium (region ε), we can indicate a region δ(ε), containing the state of equilibrium and having the property that no motion, beginning within δ, ever reaches the boundary of the region ε. On the contrary, a state of equilibrium is unstable if we can indicate such a region of deviations from the state of equilibrium (region ε) for which there is no region δ(ε), containing the state of equilibrium and having the property that a motion, beginning within δ, never reaches the boundary of the region ε.*

These definitions are connected with the notion of the phase plane of the system considered. However, a definition of stability can also be formulated without having recourse to this idea.

We can also translate this definition of stability in the language of mathematical inequalities, by denoting by $x(t)$ and $y(t)$ the motion of the black point after a displacement, and assuming for the sake of simplicity that the region ε of permissible deviations is a square (Fig. 18). We shall then obtain the following formulation of our definition: a state of equilibrium $x=\bar{x}$, $y=0$ is called stable if, having previously assigned an arbitrarily small $\varepsilon(\varepsilon > 0)$, it is possible to find such $\delta(\varepsilon)$ that, if for $t=0$

$$|x(0)-\bar{x}| < \delta \quad \text{and} \quad |y(0)| < \delta,$$

then for $0 < t < \infty$

$$|x(t)-\bar{x}| < \varepsilon \quad \text{and} \quad |y(t)| < \varepsilon$$

We shall call the type of stability thus defined *stability in the sense of Liapunov* and shall have just this in mind when we speak simply of stability. Below we shall encounter other definitions of stability and shall be in a position to appreciate the importance of the works by Liapunov [84] on stability.

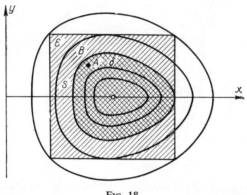

FIG. 18

We shall pass now to the analysis of the stability of a state of equilibrium of a harmonic oscillator. This analysis will enable us to visualize, somewhat intuitively, the necessity of the two regions ε and δ in the definition of stability.

It is easily seen that a singular point of the centre type corresponds to a stable state of equilibrium. Let an arbitrary small region ε be assigned, for example a square one (shaded by slant lines in Fig. 18). Let us choose, of the closed curves containing the singular point, that closed curve S which is tangent to the given square and lies entirely within it. We shall note, incidentally, that this is always possible independently of whether the closed integral curves in the immediate vicinity of a singular point have the form of ellipses or not. For the presence of such a curve it is only necessary that there exists a continuum of closed curves, not having singularities, contained in each other and gathering about a point, which is just what we have in the case of a centre. The region within the curve S (cross-shaded) will be the region $\delta(\varepsilon)$ since, if the initial position of the black point is within this region (point A), then it will never leave the square ε but will accomplish a periodic motion about the state of equilibrium. We could of course have chosen as the region δ any other region contained within the curve S, for example the region within a square,

§ 4. LINEAR OSCILLATOR IN THE PRESENCE OF FRICTION

lying with all its points within the curve S except the vertices which can lie on the curve S†. We can thus affirm that *a state of equilibrium of the centre type is a stable state of equilibrium.*

In order to give an answer to questions in which friction plays an essential role, we must drop one of the ideal features of our harmonic oscillator, namely the absence of friction, while retaining the remaining idealization. We shall assume that the frictional force is proportional to velocity. This assumption also represents an idealization and is found in satisfactory agreement with experiment when we are concerned with liquid friction or air friction for sufficiently small velocities. Any other friction law would destroy the linearity of the oscillator, whereas we are restricting our consideration for the present to linear systems only.

The equation of motion with the assumption made about friction will be:

$$m\ddot{x} + b\dot{x} + kx = 0, \qquad (1.14)$$

where b is the friction coefficient. An electrical analogue of such a mechanical system is a "Thomsonian circuit" with ohmic resistance. Such a circuit obeys the equation

$$L\ddot{q} + R\dot{q} + \frac{q}{C} = 0, \qquad (1.15)$$

where q is the charge on the capacitor and L, R and C are, as usual, the inductance, resistance and capacitance.

Introducing the notation $b/m = 2h$, $k/m = \omega_0^2$ (or correspondingly $R/L = 2h$, $1/LC = \omega_0^2$) we shall obtain the equations (1.14) and (1.15) in the usual form

$$\ddot{x} + 2h\dot{x} + \omega_0^2 x = 0. \qquad (1.16)$$

The solution of this equation is‡:

$$x = Ae^{\lambda_1 t} + Be^{\lambda_2 t}, \qquad (1.17)$$

where λ_1 and λ_2 are the roots of the quadratic equation:

$$\lambda^2 + 2h\lambda + \omega_0^2 = 0. \qquad (1.18)$$

---

† Clearly we cannot choose as the region $\delta(\varepsilon)$ the region ε itself, since, for all initial positions of the black point within the region ε but not within δ, for example at the point B (Fig. 18), the point will certainly leave the region ε.
‡ Excluding the particular case $h^2 = \omega_0^2$.

As is well known, for $h^2 > \omega_0^2$ these roots are real and for $h^2 < \omega_0^2$ are complex. Accordingly, depending on the sign of $h^2 - \omega_0^2$, we shall obtain two types of solutions and two different processes: for $h^2 < \omega_0^2$, a damped oscillatory process and for $h^2 > \omega_0^2$ a damped aperiodic process.

1. Damped oscillatory process

For a sufficiently small friction, when $h^2 < \omega_0^2$, the roots of the characteristic equation (1.18) have the values

$$\lambda_{1,2} = -h \pm j\omega,$$

where

$$\omega = +\sqrt{\omega_0^2 - h^2}, \quad j = \sqrt{-1}, \tag{1.19}$$

and we obtain for the general solution of the equation (1.16):

$$x = e^{-ht}(A\cos\omega t + B\sin\omega t), \tag{1.20}$$

where A and B are determined by the initial conditions. And precisely, if for $t=0$, $x=x_0$, $\dot{x}=\dot{x}_0$, then

$$\begin{aligned} x &= e^{-ht}\left\{x_0\cos\omega t + \frac{\dot{x}_0 + hx_0}{\omega}\sin\omega t\right\} \\ \dot{x} &= e^{-ht}\left\{\dot{x}_0\cos\omega t - \frac{\omega_0^2 x_0 + h\dot{x}_0}{\omega}\sin\omega t\right\}. \end{aligned} \tag{1.21}$$

The solution of (1.20) can be also written in the form:

$$x = Ke^{-ht}\cos(\omega t + \alpha),$$

$$K = +\sqrt{A^2 + B^2} = +\sqrt{x_0^2 + \left(\frac{\dot{x}_0 + hx_0}{\omega}\right)^2},$$

where

$$\begin{aligned} \tan\alpha &= -\frac{B}{A} = -\frac{\dot{x}_0 + hx_0}{\omega x_0} \\ \left(\cos\alpha = \frac{x_0}{K}, \quad \sin\alpha\right. &= \left.-\frac{\dot{x}_0 + hx_0}{\omega K}\right) \end{aligned} \tag{1.22}$$

The expression for the velocity can also be written in a like manner

$$\dot{x} = -K\omega_0 e^{-ht}\sin(\omega t + \alpha + \vartheta), \tag{1.23}$$

where ϑ is determined by the relations

$$\left.\begin{aligned} h &= \omega_0 \sin\vartheta, \\ \omega &= \omega_0 \cos\vartheta. \end{aligned}\right\} \tag{1.24}$$

Formulae (1.22) and (1.23) define one of the types of damped oscillatory motion in which the damping of the amplitude obeys an exponential law.

The functions $x(t)$, as well as the function $\dot{x}(t)$, *are not periodic functions*. In fact, as is known, we call periodic functions such functions for which a certain quantity τ can be found such that

$$f(t+\tau) = f(t)$$

for *any* value of the argument t. The minimum value of τ is called the period of the function $f(t)$. The functions (1.22) and (1.23) do not satisfy this definition, since for them the condition given is not satisfied for

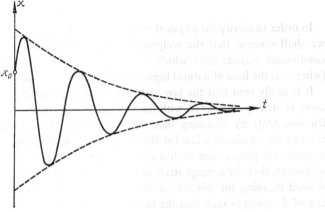

Fig. 19

arbitrary values of the argument t. Therefore we cannot, in this case, speak of a period in the strict sense of this word. However, the interval of time between two successive passages of the system through the position of equilibrium (in the same direction) or between two successive maximum deviations (on one and the same side) is constant and equal to $T=2\pi/\omega$. We shall call this interval of time "conditional period" of a damped oscillatory process. The dependence of the coordinate upon time has the form shown in Fig. 19†.

The rate of damping of the oscillatory process considered can be defined in terms of the quantity h, the so-called damping coefficient. The numerical value of h depends on the choice of the units of time. It is possible, however,

† We shall note that all extrema (both maxima and minima) are not found at the mid-points of the time intervals between corresponding zeros, but are displaced on the left by the quantity ϑ/ω, where ϑ is determined by the formulae (1.24).

to give for such a process another attenuation index which is independent of the choice of the units of measurement.

We shall take the ratio of two successive extrema directed on the same side, for example two successive maxima:

$$\frac{x'_{\max}}{x''_{\max}} = e^{hT} = e^{\frac{2\pi h}{\omega}}.$$

The logarithm of this quantity, is called the logarithmic decrement d, where

$$d = hT = \frac{2\pi h}{\omega}. \tag{1.25}$$

In order to clarify the physical meaning of the logarithmic decrement d, we shall observe that the reciprocal quantity $1/d$ gives the number of conditional periods after which the amplitude is decreased by e times (where e is the base of natural logarithms)[†].

It is easily seen that the law obtained for the damping of the oscillations is strictly connected with the idealization assumed for the law of friction. Only by assuming that the frictional force is proportional to velocity do we obtain a law for the decay of the oscillations according to a geometric progression with a common ratio equal to e^{-d}. It is clear, moreover, that the concept itself of logarithmic decrement has only a well-defined meaning for this law of attenuation and loses its meaning if the law of damping is such that the ratio of two successive maxima does not remain constant. Thus, the concept of logarithmic decrement is applicable to linear systems only unless *ad hoc* conventions are introduced. The determination of the logarithmic decrement can be arrived at from the curve shown in Fig. 19, or from the time dependence of the extrema as obtained from this curve and plotted on a linear–logarithmic scale (instead of the deviations, the maximum deviations are taken as the ordinates). In this latter case the dependence takes the form of a straight line, the slope of which yields directly the damping coefficient h which, multiplied by the conditional period T, gives the logarithmic decrement d (in practice, however, instead of this direct method, the more convenient method of determination of the damping from the resonance curve is more often used).

It is seen that a damped oscillatory motion is characterized by four quantities: the conditional period T (or correspondingly the conditional

---

[†] For example if $d = 0.02$, then this means that after fifty conditional periods the amplitude decreases by e times, i.e. is approximately one-third of its initial value.

angular frequency ω), the logarithmic decrement d, the amplitude K and the phase α.

The conditional period and the logarithmic decrement of the oscillations are determined by the properties of the system; the phase and amplitude, however, remain arbitrary and are determined by the initial conditions.

2. Representation of a damped oscillatory process on the phase plane

We shall pass on, now, to investigating the phase plane of the system considered, i.e. to constructing its "portrait", representing the totality of all possible motions.

Since we know the solution of the differential equation (1.6), we can find the equation of the family of phase paths. According to (1.22) the parametric equations of the paths on the phase plane x, y have the form

$$\left.\begin{array}{l} x = Ke^{-ht}\cos(\omega t+\alpha), \\ y = \dot{x} = -Ke^{-ht}[h\cos(\omega t+\alpha)+\omega\sin(\omega t+\alpha)]. \end{array}\right\} \quad (1.26)$$

We shall show that this is a family of spirals having an asymptotic point at the origin of the coordinates.

We shall employ to this end a linear transformation of coordinates, a method to which we shall have recourse repeatedly in the sequel. We shall pass from the variables x, y to the variables

$$u = \omega x, \quad v = y+hx, \qquad (1.27)$$

which we shall interpret as Cartesian coordinates on *another* plane (the so-called "active" interpretation of a transformation of coordinates[†]). It is apparent that, if we denote ωK by C_1, then

$$u = C_1 e^{-ht}\cos(\omega t+\alpha), \quad v = -C_1 e^{-ht}\sin(\omega t+\alpha).$$

[†] The "active" interpretation of a transformation of coordinates consists in considering the transformation $u = u(x, y)$, $v = v(x, y)$ as the law of a certain point-to-point transformation of the x, y plane into another plane with an orthogonal (cartesian) system of coordinates u, v and of the corresponding deformation of figures.

This deformation of the figures reduces, in our case of a linear and homogeneous transformation (1.27), to a simple rotation and to uniform reduction or increase in length along the two so-called principal axes. It is easily verified that each straight line on the x, y plane passing through the origin of the coordinates is transformed by the relations (1.27) into a new straight line also passing through the origin of the coordinates, the distances of the corresponding points on these straight lines from the origins of the coordinates (in the x, y and u, v planes respectively) being proportional to each other.

The equations of the phase paths on the u, v plane reduce to an even simpler form in the polar coordinates ϱ, φ ($u = \varrho \cos \varphi$, $v = \varrho \sin \varphi$):

$$\varrho = C_1 e^{-ht}, \quad \varphi = -(\omega t + \alpha),$$

or on eliminating time

$$\varrho = C e^{\frac{h}{\omega} \varphi} \tag{1.28}$$

(here $C = C_1 e^{h\alpha/\omega}$ is a new arbitrary constant).

Thus the paths on the u, v plane will be a family of *logarithmic spirals* with an asymptotic point at the origin of the coordinates (Fig. 20). In

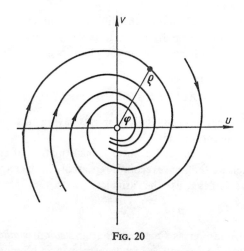

Fig. 20

this connexion, since φ decreases with time and $\varrho \to 0$ as $t \to +\infty$, the representative point, moving along spirals in the u, v plane, approaches asymptotically the origin of the coordinates.

Let us return to the x, y plane. By observing that

$$\varrho^2 = u^2 + v^2 = \omega^2 x^2 + (y + hx)^2 = y^2 + 2hxy + \omega_0^2 x^2,$$

$$\varphi = \arctan \frac{v}{u} = \arctan \frac{y + hx}{\omega x},$$

we obtain from (1.28) the coordinate equation (the equation not involving time) of the phase paths.

$$y^2 + 2hxy + \omega_0^2 x^2 = C^2 e^{2 \frac{h}{\omega} \arctan \frac{y+hx}{\omega x}}. \tag{1.29}$$

Since the deformation of the phase path connected with the inverse transformation (from u, v to x, y) cannot modify their qualitative character, we can state that the family of phase paths (1.29) on the x, y plane is also a family of spirals with an asymptotic point at the origin of the coordinates.

The following can be noted with reference to the character of these spirals. For small h/ω, i.e. small logarithmic decrements, the logarithmic spiral (1.28) lies, over the interval of time of one rotation, close to the corresponding circle $u^2+v^2=$const. By the linear relation (1.27) this circle is transformed into the ellipse $y^2+2hxy+\omega_0^2 x^2=$const. Hence we

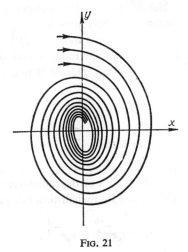

Fig. 21

can conclude that for small h/ω the spiral (1.29) under investigation lies, over the interval of time of one turn, close to the ellipse

$$y^2+2hxy+\omega_0^2 x^2 = \text{const.} \qquad (1.30)$$

(with a value of the constant chosen in a suitable manner).

The family of spirals we are investigating — the family of phase paths on the x, y plane — is shown in Fig. 21. The representative point, moving along an arbitrary spiral, will approach asymptotically (for $t \to +\infty$) the origin of the coordinates, which is a state of equilibrium. The radius-vector of the representative point will decrease at every turn.

To calculate the magnitude of this decrease draw on the x, y plane an arbitrary straight line passing through the origin of the coordinates and denote by $r_0, r_1, r_2, \ldots$, the distances from the origin of the coordinates of the points of intersection of a certain spiral (1.29) with the straight line

(Fig. 22). The straight line we have drawn, together with the points of intersection, is transformed by the relations (1.27) into a straight line, again passing through the origin of the coordinates, where, as we have shown above,

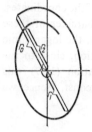

Fig. 22

$$r_0 : r_1 : r_2 : \ldots = \varrho_0 : \varrho_1 : \varrho_2 : \ldots$$

Here, by $\varrho_0, \varrho_1, \varrho_2 \ldots$, we denote the distances on the u, v plane of the transformed points of intersection from the origin of the coordinates.

Hence it follows that to each half-turn of the radius-vector r of the representative point moving on the x, y plane along a spiral (1.29) there also corresponds a half-turn of the radius-vector ϱ on the u, v plane (the angle φ decreasing by π over an interval of time equal to $\pi/\omega = T/2$). We have from (1.28):

$$\varrho_1 = \varrho_0 e^{-\frac{h\pi}{\omega}},$$

$$\varrho_2 = \varrho_0 e^{-2\frac{h\pi}{\omega}}, \ldots$$

Since the distances $r_0, r_1, r_2 \ldots$, and $\varrho_0, \varrho_1, \varrho_2 \ldots$, are proportional to each other, then, clearly, the length of the radius-vector of the representative point on the x, y plane will after a half-turn be equal to

$$r_1 = r_0 e^{-\frac{\pi h}{\omega}} = r_0 e^{-\frac{d}{2}}, \qquad (1.31)$$

after a complete turn, to

$$r_2 = r_0 e^{-\frac{2\pi h}{\omega}} = r_0 e^{-d}$$

and, after n turns, to

$$r_{2n} = r_0 e^{-nd}. \qquad (1.32)$$

It is seen that the decrease of the radius-vector obeys the exponential law, found previously, with the logarithmic decrement d equal to hT.

We have thus established the character of the phase paths. It can be shown further that through each point of the phase plane there passes one and only one spiral, corresponding to a determined value of the constant C or, in other words, corresponding to the initial conditions. The whole plane is filled with spirals winding within each other and along which the representative point approaches the origin of the coordinates asymptotically (for $t \to +\infty$). The only exception is the state of equilibrium, the point $x=0, y=0$, which must be considered as a distinct phase path. As the representative point moves along a spiral the phase velocity never reduces to zero but gradually decreases with each turn, the time of each turn remaining constant and equal to $T=2\pi/\omega$. The phase velocity is always equal to zero for the "motion" represented by the path $x=0$, $y=0$.

3. Direct investigation of the differential equation

We have investigated the character of the motions on the phase plane for the case of a linear oscillator in the presence of friction proportional to velocity, and have established that with small damping ($h^2<\omega_0^2$) there corresponds a motion of the representative point along a phase path of a spiral form, having an asymptotic point at the origin of the coordinates. In this case the origin of the coordinates itself is a state of equilibrium. However, this picture on the phase plane was obtained proceeding from the solution (1.20) found previously. We could have obtained the same picture directly from (1.16) without knowing (1.20).

Let us change, as we have already done, the initial equation of the second order (1.16) into two equivalent equations of the first order

$$\frac{dx}{dt} = y; \quad \frac{dy}{dt} = -2hy - \omega_0^2 x. \tag{1.33}$$

On dividing one equation by the other, we shall obtain the differential equation of the integral curves in the form

$$\frac{dy}{dx} = \frac{-2hy - \omega_0^2 x}{y}. \tag{1.34}$$

It is easily seen that this equation, in a similar manner to equation (1.11), determines on the phase plane a certain field of tangents and, together with the equation (1.33), a vector field with the only singular point $x=0$, $y=0$.

The nature of this field is easily investigated approximately by means of the isoclines. The equation of the isocline at the points of the integral curves which have the slope $\varkappa$, is

$$\frac{-2hy - \omega_0^2 x}{y} = \varkappa \quad \text{or} \quad y = \sigma x,$$

where

$$\sigma = -\frac{\omega_0^2}{\varkappa + 2h}, \tag{1.35}$$

i.e., the isoclines are again straight lines passing through the origin of the coordinates. Having assigned, for example, a sufficiently large number of values of $\varkappa$ (for fixed h and ω_0 which are determined by the system), we shall obtain a family of isoclines and by means of them shall be able to construct the vector field to the required degree of accuracy[†].

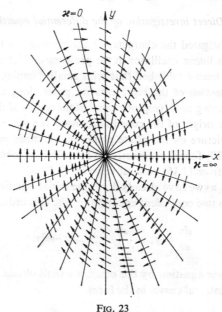

FIG. 23

In Fig. 23 there is shown such a vector field, constructed by means of several isoclines, and the character of the integral curves can already be anticipated from this sketch.

The equation (1.34) obtained after eliminating time permits integration since it belongs to the class of homogeneous equations.

On integrating it according to the usual rules (by means of the substitution $z = y/x$), we shall obtain for our case ($h^2 < \omega_0^2$) the equation of the integral curves

$$y^2 + 2hxy + \omega_0^2 x^2 = Ce^{2\frac{h}{\omega}\arctan\frac{y+hx}{\omega x}},$$

which has already been examined[‡]. We have now obtained this equation

[†] We observe that the isocline method is not only a method of approximate numerical integration but also a method by means of which it is possible to demonstrate rigorously various statements with regard to the integral curves.

[‡] The positive constant of integration C^2 in the expression (1.29) is denoted here by C.

by other means, without knowing the solution of equation (1.16). The expression of the phase velocity **v** is found from the equations (1.13) and (1.33)

$$\mathbf{v} = \mathbf{i} \cdot y + \mathbf{j}(-2hy - \omega_0^2 x)$$

and

$$|\mathbf{v}|^2 = \omega_0^4 x^2 + 4h\omega_0^2 xy + (1+4h^2)y^2. \tag{1.36}$$

Thus this type of approach enables us to see, directly, almost without any calculations, that the phase velocity nowhere reduces to zero, except at the origin of the coordinates $x=0$, $y=0$, but decreases as the representative point approaches the origin.

What can be said of the character of the motions in our system, if the character of the integral curves on the phase plane and the expression for the phase velocity are known?

We can first of all state that all phase paths correspond to oscillatory damped motions, tending to the position of equilibrium (with the exception of the "motion" along the path $x=0$, $y=0$), and that all these paths are spirals. Obviously as the representative point moves along a spiral, the displacement and velocity of the system repeatedly pass through zero, then the spirals are representative of an oscillatory process. In addition, the radius-vector of the representative point decreases after each rotation. This means the process is damped and the maximal values of x and $\dot{x}$ decrease at each turn. It is clear also that the singular point $x=0$, $y=0$ corresponds to a state of equilibrium.

The results obtained from the analysis of the character of the motions on the phase plane can be formulated thus: *given any initial conditions, our system accomplishes damped oscillatory motions around the position of equilibrium $x=0$, $y=0$, except the one case when the initial conditions correspond exactly to the state of equilibrium.*

In the case considered we have only one singular point of the system of integral curves, being an asymptotic point for all integral curves. A singular point, which is the asymptotic point of all integral curves, having the form of spirals enclosed in each other, is called a *focus*.

We shall now elucidate the question of whether this singular point of the focus type is stable. Bearing in mind that the representative point will approach, along any of the integral curves, the singular point, it is easily verified that the condition of stability that we have formulated above is satisfied in this case. In fact we can always choose such a region δ (doubly shaded in Fig. 24) so that the representative point will not leave the boundaries of the region ε (simply shaded). Therefore in this case the state of equilibrium is stable and the singular point is a *stable focus*. The stability of a singular point of the focus type is clearly related to whether the integral curves are winding or unwinding with respect to the direction of motion of the representative point. Since the direction of motion is uniquely determined by the choice of the coordinates (the point must move clockwise), then the stability of the singular point in the case considered is

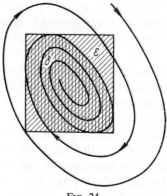

Fig. 24

unambiguously established. Conversely, should the spiral unwind (the direction being reckoned by the same criterion as above) then the singular point would be unstable. As is easily verified, for example, from the equation (1.28), winding of the integral curves is caused by the fact that $h>0$, since only in this case will the radius-vector decrease for a clock-wise motion.

Thus a singular point of the focus type can be either stable or unstable (in contrast to a singular point of the centre type which is always stable). In this example the focus is stable since $h>0$. The physical meaning of this condition of stability is clear: friction must be positive, opposing motion and involving dissipation of energy. Such a frictional force cannot cause instability and if the position of equilibrium in the system was stable in the absence of friction (in the harmonic oscillator), then it will remain stable in the presence of positive friction also. Later in our analysis we shall encounter unstable singular points of the focus type.

The stable focus considered here has a "stronger" stability than the centre considered in the previous section. In fact, in the case of the stable focus there will be satisfied not only the condition of stability in the sense of Liapunov, but also a more stringent requirement. More precisely, for any initial deviations the system returns, after a sufficiently long period of time, arbitrarily close to the position of equilibrium. Such a stability, for which initial deviations not only fail to increase but, on the contrary, decrease, will be called *absolute stability*. In the case considered by us of a linear oscillator, the focus is *absolutely stable*.

4. Damped aperiodic process

We shall now consider the case when the roots of the characteristic equation are real, i. e. when $h^2 > \omega_0^2$. In this case, denoting

$$q = +\sqrt{h^2 - \omega_0^2}, \qquad (1.37)$$

we shall obtain the roots of the characteristic equation:

$$\lambda_1 = -h+q = -q_1, \quad \lambda_2 = -h-q = -q_2 \qquad (1.38)$$

($q_2 > q_1 > 0$). Therefore the general solution of the equation (1.16) can be written in the form

$$\left.\begin{array}{l} x = Ae^{-q_1 t} + Be^{-q_2 t} \\ \dot{x} = -q_1 A e^{-q_1 t} - q_2 B e^{-q_2 t}. \end{array}\right\} \qquad (1.39)$$

Here A and B are determined by the initial conditions. More precisely, if for $t=0$, $x=x_0$ and $\dot{x}=\dot{x}_0$, then

$$x = \frac{\dot{x}_0 + q_2 x_0}{q_2 - q_1} e^{-q_1 t} + \frac{\dot{x}_0 + q_1 x_0}{q_1 - q_2} e^{-q_2 t}. \qquad (1.40)†$$

It will firstly be apparent that, for all initial conditions, the motion is damped, since $q_1 > 0$ and $q_2 > 0$ and thus for $t \to +\infty$, $x(t) \to 0$. To illustrate in greater detail the character of the damping, we shall find t_1 and t_2, the

---

† It is often convenient to write the solution of equation (1.16) for $h^2 > \omega_0^2$ in terms of hyperbolic functions: the general solution in the form

$$x = e^{-ht}(A \cosh qt + B \sinh qt)$$

and the solution satisfying the initial conditions $x=x_0$, $\dot{x}=\dot{x}_0$ for $t=0$, in the form

$$x = e^{-ht}\left\{x_0 \cosh qt + \frac{\dot{x}_0 + h x_0}{q} \sinh qt\right\},$$

$$\dot{x} = e^{-ht}\left\{\dot{x}_0 \cosh qt - \frac{\omega_0^2 x_0 + h \dot{x}_0}{q} \sinh qt\right\}.$$

The latter expressions are obtained from (1.21) by replacing the trigonometric functions by the corresponding hyperbolic functions, and ω by q.

instants of time for which respectively x and $\dot{x}$ reduce to zero. Using (1.40), we find the following equations for the determination of t_1 and t_2:

$$e^{(q_2-q_1)t_1} = \frac{\dot{x}_0+q_1 x_0}{\dot{x}_0+q_2 x_0} = 1 - \frac{x_0(q_2-q_1)}{\dot{x}_0+q_2 x_0}, \tag{1.41}$$

$$e^{(q_2-q_1)t_2} = \frac{q_2(\dot{x}_0+q_1 x_0)}{q_1(\dot{x}_0+q_2 x_0)} = 1 + \frac{\dot{x}_0(q_2-q_1)}{q_1(\dot{x}_0+q_2 x_0)} \tag{1.42}$$

It is seen at once from these equations that each of them has not more than one root. Thus oscillatory damping is impossible and we are dealing with a so-called aperiodic process.

Let us ascertain under which condition the equation which determines t_2 has no positive root. In this case the motion is *monotonically* damped, tending asymptotically to zero. This will occur, as is seen from the expression for t_2 (1.42), if

$$\frac{\dot{x}_0}{\dot{x}_0+q_2 x_0} < 0. \tag{1.43}$$

In Fig. 25 there is shown the region of the initial values which satisfy this inequality (the region II). For the remaining initial conditions $\dot{x}_0/(\dot{x}_0+q_2 x_0) > 0$ the equation which determines t_2 has a positive root. This means that the displacement does not decrease monotonically, but at first increases in absolute value and only after reaching a certain maximum will begin to decrease, tending asymptotically to zero.

Here we have to distinguish two cases, depending on whether, for the initial condition considered, the equation which determines t_1 has a positive root or not. If there is no such root, then, during the time of motion $(0 < t < \infty)$, the displacement retains its sign; the system is moved far from the position of equilibrium, reaches a certain maximum deviation and then monotonically approaches the position of equilibrium but does not pass through it. According to (1.41), this takes place if

$$\frac{x_0}{\dot{x}_0+q_2 x_0} > 0. \tag{1.44}$$

The regions of the initial values leading to motions of such a type are marked in Fig. 25 by the figure *I*.

If the equation determining t_1 has a positive root, then the system at first approaches the position of equilibrium, passes through the position of equilibrium at the instant $t=t_1$, then at the instant $t=t_2$ reaches a certain maximum deviation in a direction opposite to that of the initial deviation

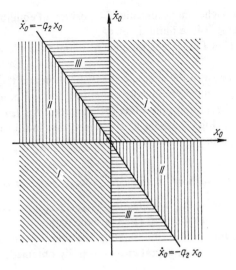

Fig. 25

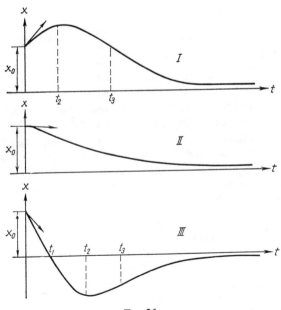

Fig. 26

and finally approaches monotonically the position of equilibrium without, however, reaching it in a finite time†.

The region *III* of Fig. 25 corresponds to initial values leading to such a type of motion.

The relation between the character of the motion and the initial conditions can be represented graphically in another form also, by showing the dependence of the motion upon time for all three cases *I*, *II* and *III*. This has been done in Fig. 26 where it is assumed that in all cases the initial displacement $x_0 > 0$.

5. *Representation of an aperiodic process on the phase plane*

Let us pass now to the investigation of the phase paths on the phase plane x, y ($y = \dot{x}$). The equations (1.39) are the parametric equations of the phase paths in our example. Eliminating the time t from them‡, the coordinate equation of the integral curves is easily obtained:

$$(y+q_1x)^{q_1} = C(y+q_2x)^{q_2}. \tag{1.45}$$

In order to investigate this family of curves we shall again use a linear transformation of coordinates

$$y+q_1x = v; \qquad y+q_2x = u.$$

After this transformation the equation (1.45) takes in the new variables the simple form:

$$v = Cu^a, \quad \text{where} \quad a = \frac{q_2}{q_1} > 1$$

Interpreting u and v as rectangular coordinates, we can say that after the transformation we have obtained a family of "parabolae", where, since $a > 1$: (i) all integral curves, except the curve corresponding to $C = \infty$ are tangential to the horizontal axis at the origin since $dv/du = Cau^{a-1}$, therefore $(dv/du_{u=0}) = 0$; (ii) the integral curves for $C=0$ and $C=\infty$ degenerate into straight lines: for $C_1=0$ we have $v=0$, i.e. the u axis, for $C_1=\infty$ we have $u=0$, i.e. the v axis; (iii) the integral curves are convex

---

† It is easily seen from the equations (1.41) and (1.42) that $\exp[(q_2-q_1)(t_2-t_1)] = q_2/q_1$ and, hence necessarily $t_2 > t_1$.

‡ This can be done, for example, in the following manner. Solving the equations (1.39) with respect to Ae^{-q_1t}, and Be^{-q_2t}, we shall obtain:

$$y+q_1x = (q_1-q_2)Be^{-q_2t}, \qquad y+q_2x = (q_2-q_1)Ae^{-q_1t}.$$

Raising the first expression to the power q_1 and the second to the power q_2 and dividing one of the relations obtained by the other we shall obtain (1.45).

towards the u axis† and their ordinates increase monotonically in absolute value as u increases. The family of parabolae is shown in Fig. 27.

Let us return now to the x, y plane. To the v axis on the u, v plane there corresponds the straight line $y+q_2 x = 0$ on the x, y plane; to the u axis the straight line $y+q_1 x = 0$. The remaining curves of the family

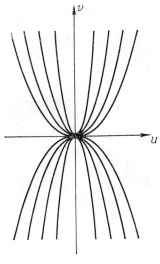

Fig. 27

(1.45) on the x, y plane represent deformed parabolae, tangential to the straight line $y = -q_1 x$ (the u axis). In order to represent this family of curves we must take into account the following additional facts: (i) the curves of the family have horizontal tangents at the points of intersection with the straight line $y = -[q_1 q_2/(q_1+q_2)]x$ ($q_1 q_2/(q_1+q_2) < q_1$); (ii) the curves of the family have vertical tangents at the points of intersection with the x axis; (iii) the slope of those curves which intersect the x axis monotonically increases on the section from the state of equilibrium to the x axis, and varies from $-q_1$ to $+\infty$; (iv) the curves of the family have unlimited parabolic branches, with axes parallel to the straight line $y = -q_2 x$ (as the representative point moves to infinity the slope of the curves $dy/dx \to -q_2$). This family of curves is shown in Fig. 28.

In a manner similar to that of the previous example we can arrive at the results obtained without integrating the differential equation (1.16) but

† Since $v''/v = a(a-1)/u^2$.

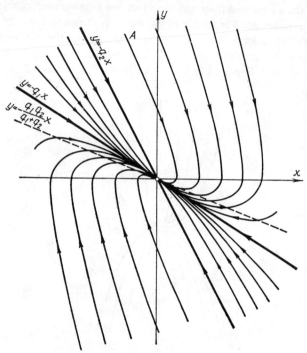

Fig. 28

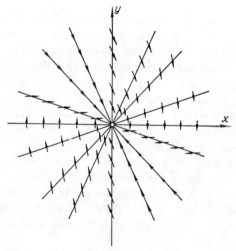

Fig. 29

replacing this second-order equation by two equivalent equations of the first order and eliminating time from them. We shall obtain the same equation of the integral curves:

$$\frac{dy}{dx} = \frac{-2hy - \omega_0^2 x}{v}. \tag{1.34}$$

The only singular point of this family of curves is the point $x=0$, $y=0$, corresponding to a state of equilibrium of the system. The isoclines will be straight lines, defined by the equations (1.35). But since in the case considered $h^2 > \omega_0^2$, then the position of the isoclines will be somewhat different (Fig. 29). In the case considered there are two integral curves which are straight lines, passing through the origin of the coordinates. To find them, we shall substitute the equation of a straight line in the equation of the integral curves (1.34). We shall obtain then for the angular coefficient β of a straight-line integral curve the equation $\beta^2 + 2h\beta + \omega_0^2 = 0$, coinciding with the characteristic equation (1.18). This has in our case the real roots: $\beta_1 = -q_1$, and $\beta_2 = -q_2$, i.e. the straight lines $y = -q_1 x$ and $y = -q_2 x$ will be integral curves. This rules out at once the existence of spiral phase paths circling the origin of the coordinates and, hence, the existence of damped oscillatory motions.

Just as for the integration of (1.34) we shall obtain by the substitution $z = y/x$ (since $h^2 > \omega_0^2$) a result different from the previous case, namely, the equation of a family of integral curves of a "parabolic type":

$$y^2 + 2hxy + \omega_0^2 x^2 = C \left[\frac{\frac{y}{x} + h - \sqrt{h^2 - \omega_0^2}}{\frac{y}{x} + h + \sqrt{h^2 - \omega_0^2}} \right]^{\frac{h}{\sqrt{h^2 - \omega_0^2}}}$$

or

$$(y + q_1 x)^{q_1} = C_1 (y + q_2 x)^{q_2}, \tag{1.45}$$

where $q_1 = h - \sqrt{h^2 - \omega_0^2}$ and $q_2 = h + \sqrt{h^2 - \omega_0^2}$, i.e. the same equation which we obtained above by eliminating t from the solutions of the differential equation.

The direction of motion of the representative point is determined by means of the same reasoning as in the previous cases, namely from the condition that for $y = \dot{x} > 0$ the value of x must increase. Since the slope of the tangent to an integral curve changes its sign only once at the intersection with the x axis, then it is seen at once that the representative point will move along the integral curves in the directions indicated in Fig. 28

by arrows. The velocity of motion of the representative point, just as in the previous cases, reduces to zero only at the origin where at the same time $\dot{x}=0$ and $\dot{y}=0$, i.e. at the singular point of the differential equation.

As discussed already we shall distinguish between *integral curves* and *phase paths*, since one integral curve can correspond to several substantially different motions or phase paths. For example, in the case considered, having assigned a determined value of the constant C, we have not yet fixed a unique path, since in this case each integral curve passes through a singular point and consequently consists of three phase paths (two of which correspond to motions asymptotically tending to the state of equilibrium, and the third is the state of equilibrium itself). Thus all integral curves pass through the singular point. A singular point such that integral curves pass through it, just as the family of parabolae $y = Cx^\alpha$ ($\alpha > 0$) passes through the origin of the coordinates, is called a *node*. It is easily seen that the state of equilibrium, which in this case is a singular point, a node, is stable in the sense of Liapunov since the representative point moves along all integral curves towards the origin of the coordinates. A stable state of equilibrium which corresponds to a singular point of the node type will be called by us a *stable node*. As we shall verify in the sequel, a node can be also unstable, for which it is sufficient that h be negative. As in the case of a focus, the physical meaning of this is that if the state of equilibrium in a system without friction and with one degree of freedom is stable, then the addition of positive friction cannot disturb the stability (even more than that, positive friction makes the position of equilibrium absolutely stable).

Let us consider in somewhat greater detail the physical features of the three types of aperiodic motions represented in Fig. 26. First of all, if the initial velocity and the initial deviation are of the same sign (i.e. if the representative point lies in the region I in Fig. 25), then the system will at first move away from the position of equilibrium, its velocity being gradually decreased (the initial kinetic energy being spent to increase the potential energy and to overcome friction). When the velocity reduces to zero (the point t_2), the system will begin to move back towards the position of equilibrium, the velocity at first increasing, since the restoring force is larger than friction. But as the motion proceeds the frictional force increases (since velocity increases) and the restoring force decreases (since the system approaches the position of equilibrium) and, consequently, starting from a certain instant (the point t_3 in Fig. 26, I), the velocity, having attained at this instant a maximum, will begin to decrease again. The system will approach asymptotically the position of equilibrium.

The other case when the initial velocity and the initial deviation are of different signs, i.e. the initial impulse is directed in a sense opposite to the initial deviation, leads to two different types of motion *(II and III)*. If the initial impulse is small as compared with the initial deviation, then the system owing to the presence of large friction cannot pass through the position of equilibrium and will asymptotically approach the position of equilibrium (the curve *II)*. If, however, the initial velocity is sufficiently large, then the system will pass at a certain moment t_1 through the position of equilibrium (the curve *III*) and after that will still have a certain velocity, directed away from the position of equilibrium, i.e. on the same side as the deviation of the system. Then a motion of the type *I* already considered is obtained; the system reaches a certain maximum deviation and then asymptotically approaches the position of equilibrium. Thus a motion of the type *III* differs from a motion of the type *I* in its first part only (up to the point t_1). However, after the point t_1 a motion of the type *III* is similar to a motion of the type *I*. On the other hand, a motion of the type *I* is similar, after the point t_2, to a motion of the type *II*. And in fact the motion of the representative point along certain phase curves passing through all three regions *I, II* and *III* (for example along the curve marked by the letter *A* in Fig. 28) will belong either to *III* or to *I* or to the type *II*, depending on in which region the representative point lies at the initial instant.

The limit case (when $h^2 = \omega_0^2$) will not be considered in detail. We shall restrict ourselves to some observations, since this case (just as any other case when the relation between the parameters of the system is rigorously fixed) cannot be exactly realized in practice in a physical system and has a value only as a boundary between two different types of damped processes, namely the oscillatory and the aperiodic ones. In the case $h^2 = \omega_0^2$, as is well-known, the solution of the initial differential equation (1.16) must be looked for in the form

$$x = (A+Bt)e^{-qt}.$$

We can, of course, dispense with the search for a solution of the differential equation of the second order, and pass on to an equation of the first order, determining the phase curves (1.34). We shall obtain, in this case also, a family of integral curves of the parabolic type and a stable singular point of the node type, so that, from the point of view of the behaviour of the integral curves and the type of the singular point, this limit case is to be attributed to the case $h^2 > \omega_0^2$ and not to the case $h^2 < \omega_0^2$. The case $h^2 = \omega_0^2$, though having no physical meaning, still presents a certain analytical interest, since it is often convenient to choose the attenuation of the system so that h^2 be as close as possible to ω_0^2. Thus on the one hand we remove from the system oscillations which would be inevitable for h^2 much smaller than ω_0^2 and, on the other hand, a maximum velocity of aperiodic return of the system to zero is obtained (larger than for larger values of h). Just such conditions are those most advantageous for some measuring devices, for example, for galvanometers. However, for an arbitrarily small variation of the parameters of the system, this limit case will change into one of the two other cases considered earlier. Therefore it offers no physical interest and does not reflect typical features of a real physical system. We must, however, bear in mind that dividing systems into oscillatory and aperiodic, which in the case of a linear system can still be done with full mathematical rigour, has, practically speaking, no major physical content, since for large h the system loses its more typical "oscillatory features" even before h^2 has attained the value ω_0^2. Actually, if h^2 is only a little smaller than ω_0^2 then damping in the system is very large and already the second maximum following the initial deviation can be almost unnoticeable in practice. Under such conditions the phenomenon of resonance, one of the most typical phenomena in non-autonomous oscillatory systems, can in a similar manner become unnoticeable.

We shall observe, incidentally, that for certain non-linear systems (for example, systems with "constant", "Coulomb-type" friction or "square-law" friction) the division into oscillatory and aperiodic systems becomes altogether meaningless.

§ 5. OSCILLATOR WITH SMALL MASS[†]

1. Linear systems with half degree of freedom

In our analysis above of a linear oscillator in the presence of friction, we have assumed that all three parameters of the oscillator — the mass (or inductance), the friction coefficient (or resistance) and the coefficient of elasticity (or reciprocal value of the capacitance) — are of equal importance and affect appreciably the properties and behaviour of the system. In the cases when friction is small we can leave the influence of friction on the motion of the system out of account altogether, and will still be in a position to answer certain questions for which friction is a second-

---

[†] The Subsections 1 and 2 have been revised and the Subsections 3 and 4 completely rewritten by N. A. Zheleztsov

OSCILLATOR WITH SMALL MASS

order factor. If, however, friction is large[†], another case may be met when a negligible second-order factor proves to be, *because of its smallness*, one of the two other "oscillatory" parameters of the system: the mass or the coefficient of elasticity.

We shall consider the motion of a body *of small mass* in a medium offering a strong resistance under the action of a spring (this case is the one of greatest relevance to the analysis later of the so-called "relaxation" oscillations). In addition to the assumptions made when formulating the problem of a linear oscillator with friction, we shall neglect now the mass of the moving body. Then the equation of motion is written in the form of a differential equation of the first order

$$b\dot{x} + kx = 0 \tag{1.47}$$

(here, just as previously, x is the displacement with respect to the position of equilibrium and k and b are the positive coefficients of elasticity and friction). We arrive thus at a system with half a degree of freedom. To determine uniquely the state of such a system, the knowledge of one quantity is sufficient (for example the x coordinate) instead of the two necessary for the determination of the state of a system with one degree of freedom. Correspondingly, for systems with half a degree of freedom the phase space is unidimensional and is not a plane but a *line*.

The solution of equation (1.47) has, as is wellknown, the form

$$x = Ae^{-\frac{k}{b}t}$$

or, if we introduce the initial condition $x = x_0$ for $t = 0$[‡],

$$x = x_0 e^{-\frac{k}{b}t}. \tag{1.48}$$

Clearly, $x = 0$ is a state of equilibrium; for all other initial conditions ($x_0 \neq 0$) an oscillator without mass accomplishes an aperiodic damped motion, approaching (for $t \to +\infty$) the state of equilibrium.

[†] We use the terms "small" and "large" without indicating in comparison with what. As was observed in the Introduction, these statements do not in such a form have a great significance. However, it will become clear from the following analysis in comparison with what the friction and resistance must be large.

[‡] We cannot now, within the limits of the present idealization, give an initial value to the velocity $\dot{x}_0$ arbitrarily, independently of the value of x_0, since the values of the velocity $\dot{x}$ and the coordinate x are uniquely connected between each other by the equation (1.47), which we consider to be valid at any instant of time (for the instant $t=0$ we obtain: $\dot{x}_0 = -(k/b)x_0$).

We shall obtain the same picture if we consider the motion of the representative point along the phase line—the straight line x (Fig. 30). The origin of the coordinates is a state of equilibrium; the representative point moves away from other states in a direction towards the state of equilibrium (since on its right $\dot{x} < 0$ and on its left $\dot{x} > 0$).

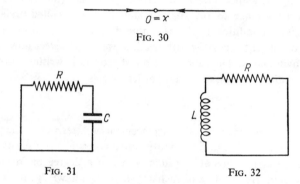

Fig. 30

Fig. 31   Fig. 32

Electric circuits consisting of resistance and capacitance (*RC*-circuit, Fig. 31) or of resistance and inductance (*RL*-circuit, Fig. 32) will also be systems with half a degree of freedom. These also are clearly idealized systems, at which we arrive from suitable real electric circuits, neglecting, in particular, small (parasitic) inductances or capacitances of one or other elements forming the circuits. The equations of motion for such circuits can be written in the form

$$R\dot{q}+\frac{q}{C} = 0 \tag{1.49}$$

for the *RC*-circuit (q being the charge of the capacitor) and

$$L\frac{di}{dt}+Ri = 0 \tag{1.50}$$

for the *RL*-circuit (i being the intensity of current in the circuit).

Their solution will clearly be

$$q = q_0 e^{-\frac{t}{RC}}, \quad i = i_0 e^{-\frac{R}{L}t}.$$

There remains the question of the "lawfulness" or expediency of the representation of a physical system as a system with a half-degree of freedom. (1.47), (1.49) and (1.50) and their solution describe the motions of these real physical systems. We are only dealing here, obviously, with those motions of the physical systems which begin from states *compatible* (to a certain degree of accuracy) with the equations of motion of the corresponding systems with half a degree of freedom†. The answer to this question can be obtained by comparing the results, obtained from the solution of the equations (1.47), (1.49) and (1.50), with experimental data. This comparison indicates the convenience, the "lawfulness" of employing systems with a half-degree of freedom to represent the motions of corresponding physical systems.

We shall now show analytically that taking into account the small mass of an oscillator does not give us anything essentially new, i.e. that the mass, if it is sufficiently small, is not an essential parameter in the problem considered. Let us take into account the small mass of an oscillator and compare the solution of the more "complete" equation of the oscillator with a small mass

$$m\ddot{x} + b\dot{x} + kx = 0, \quad (1.14)$$

where m is small but different from zero, with the solution of the equation of the first order (1.47). For given initial conditions $t = 0, x = x_0, \dot{x} = \dot{x}_0$, we have, according to (1.40) a solution in the form

$$x = x_0 \left[\frac{q_2}{q_2 - q_1} e^{-q_1 t} - \frac{q_1}{q_2 - q_1} e^{-q_2 t} \right] + \frac{\dot{x}_0}{q_2 - q_1} \left[e^{-q_1 t} - e^{-q_2 t} \right], \quad (1.51)$$

where

$$q_1 = \frac{b}{2m} - \sqrt{\frac{b^2}{4m^2} - \frac{k}{m}}, \quad q_2 = \frac{b}{2m} + \sqrt{\frac{b^2}{4m^2} - \frac{k}{m}}.$$

To make the comparison easier, we shall replace the rigorous solution (1.51) of the equation (1.14) with an approximate solution $x_1(t)$ such that the difference between $x(t)$ and $x_1(t)$ and that between their derivatives $\dot{x}(t)$ and $\dot{x}_1(t)$ may be made arbitrarily small (uniformly with respect to t) by choosing a value of m sufficiently small.

Using the expansion of a square root

$$\sqrt{\frac{b^2}{4m^2} - \frac{k}{m}} = \frac{b}{2m} \sqrt{1 - \frac{4km}{b^2}} = \frac{b}{2m} \left(1 - \frac{2km}{b^2} + \ldots \right),$$

we obtain without difficulty

$$x_1(t) = x_0 \left[e^{-\frac{k}{b}t} - \frac{mk}{b^2} e^{-\frac{b}{m}t} \right] + \dot{x}_0 \frac{m}{b} \left[e^{-\frac{k}{b}t} - e^{-\frac{b}{m}t} \right]. \quad (1.52)$$

† Any given idealized system can only be used, as was shown in the Introduction, for analysing those motions of a real physical system, which start from states compatible with the equations of this idealized system.

It can be shown that this solution is, an approximation to the rigorous solution, in the sense that, however small we choose an ε, it is always possible to find such a small m that

$$|x_1(t)-x(t)| < \varepsilon, \quad |\dot{x}_1(t)-\dot{x}(t)| < \varepsilon$$

for all values of t in the interval $0 \leq t \leq +\infty$[†].

Let us compare now (1.48) and (1.52). Denoting the solution of the equation of the first order by $\bar{x}$ and assuming that the same initial values of the displacement for the solutions of the complete equation[‡] and of the equation of the first order we have

$$x_1(t)-\bar{x}(t) = \frac{m}{b}\left\{-\left(\dot{x}_0+\frac{k}{b}x_0\right)e^{-\frac{b}{m}t}+\dot{x}_0 e^{-\frac{k}{b}t}\right\} \quad (1.53)$$

and for the velocities

$$\dot{x}_1(t)-\dot{\bar{x}}(t) = \left(\dot{x}_0+\frac{k}{b}x_0\right)e^{-\frac{b}{m}t} - \frac{mk}{b^2}\dot{x}_0 e^{-\frac{k}{b}t}. \quad (1.54)$$

Since we are now considering only those motions which start from states compatible (to a certain degree of accuracy) with the equation (1.47), i.e. for which $\dot{x}_0+(k/b)x_0$ is equal or close to zero, then, as is seen directly from the relations (1.53) and (1.54), the differences $x_1(t)-\bar{x}(t)$ and $\dot{x}_1(t)-\dot{\bar{x}}(t)$, and hence also the differences $x(t)-\bar{x}(t)$ and $\dot{x}(t)-\dot{\bar{x}}(t)$ can be made as small as we choose by choosing a sufficiently small m and, moreover, for all $0 \leq t < +\infty$. The condition of closeness of the solutions (1.48) and (1.51) clearly implies the validity of the following inequality

$$\frac{mk}{b^2} \ll 1 \quad \text{or} \quad m \ll \frac{b^2}{k}.$$

In other words, if the initial state of the system is compatible with the equation of the first order (1.47) (or is close to a state compatible with this equation), then the latter is sufficiently accurate (the more accurate, the smaller the mass) to represent the motion of an oscillator with a small mass. Allowing for the mass gives, in this case, only a small quantitative

† Note that these inequalities cannot be replaced by inequalities of the type

$$\left|\frac{x_1(t)}{x(t)}-1\right| < \varepsilon, \quad \left|\frac{\dot{x}_1(t)}{\dot{x}(t)}-1\right| < \varepsilon, \quad (\alpha)$$

valid if m is sufficiently small for all values of t. However, over any given interval of values of t, large as we choose, we can make the inequalities (α) be satisfied, by choosing a sufficiently small m.

‡ By "complete system", "complete equation", we shall mean here, for the sake of brevity, an oscillator the mass of which is taken into account, and its equation.

correction, without adding anything essentially new; the mass of an oscillator, provided that it is sufficiently small, proves to be an inessential parameter and the representation of an oscillator with small mass ($m \ll b^2/k$) as a system with a half-degree of freedom (as a system *without mass*) proves to be quite adequate.

2. *Initial conditions and their relations to the idealization*

We shall now consider the case when the initial state of an oscillator with small mass (given x_0 and $\dot{x}_0$) is not compatible with the equation of the first order (1.47), i.e. when $\dot{x}_0 \neq -(k/b)x_0$ and hence $\dot{x}_0 + (k/b)x_0$ is not small. Clearly we cannot expect in this case that the first-order equation will adequately represent *the whole* process of motion of such an oscillator, since this equation is admittedly inapplicable at the initial instant of time. The study of such motions of an oscillator with small mass (the mass can be as small as we choose), must be carried out by using the equation of the second order (1.14) which is compatible with the initial conditions.

To investigate the characteristics of the motions of an oscillator with small mass we shall compare the solution of the equation (1.14) in its approximate form (1.52) with the solution of the equation of the first order. Returning to (1.53) we see that, as before, the difference $x_1(t) - \bar{x}(t)$ and hence also $x(t) - \bar{x}(t)$ can be made as small as we choose for all $0 \leqslant t < +\infty$ provided that we choose a sufficiently small m, notwithstanding the fact that now $\dot{x}_0 + (k/b)x_0$ is not small. It will be readily noticed, however, that a different situation arises for the velocities. In fact, according to (1.54), the difference $\dot{x}_1(t) - \dot{\bar{x}}(t)$ for a small fixed m and for small values of t (for $t \ll m/b$) is close to $\dot{x}_0 + (k/b)x_0$ (this is quite natural since $\dot{x}(0) = \dot{x}_0$ and $\dot{\bar{x}}(0) = -(k/b)x_0$). This quantity does not depend on m and we cannot make it small by choosing a small m. However, on investigating the structure of expression (1.54) and bearing in mind the rapid decrease of $e^{-(b/m)t}$ for a fixed $t > 0$ and a decreasing m, we arrive at the following conclusion: it is always possible, by choosing a sufficiently small m, to achieve for all t, starting from a certain arbitrarily small but well-defined instant $\tau > 0$ (for all $\tau \leqslant t < +\infty$), that the inequality

$$|\dot{x}_1(t) - \dot{\bar{x}}(t)| < \varepsilon \quad \text{or} \quad |\dot{x}(t) - \dot{\bar{x}}(t)| < \varepsilon$$

be satisfied (here, as before, ε is a small arbitrary positive quantity given in advance).

Thus, during the initial stage of the motion (for $0 \leqslant t \leqslant \tau$) the velocity of an oscillator with small mass varies very rapidly (the more rapidly, the smaller the mass) from the initial value $\dot{x}_0$ to values close to those obtained from the solution of equation (1.47). The variation of the coordinate during this interval of time τ tends obviously to zero together with τ (or, what amounts to the same thing, together with m)[†]. It is perfectly clear that the motion of an oscillator with small mass, during this stage of the motion with rapid variations of velocity and, consequently, with large accelerations, cannot be represented by the first-order equation (1.47) since the mass proves to be an important factor (the term $m\ddot{x}$ is not small in comparison with the other terms of equation (1.14). Only after that has the oscillator arrived in a time τ at a state close to one compatible with equation (1.47) (and this means incidentally that the term $m\ddot{x}$ has become very small). The velocity of the oscillator ceases to vary rapidly and its motion will be represented by first-order equation (1.47) (the more accurately, the smaller mk/b^2).

To illustrate what has been said we shall consider the motion of an oscillator with small mass for the following initial values: for $t = 0$, $x = x_0$ and $\dot{x} = 0$ (these initial conditions are clearly incompatible with equation (1.47)). While $\dot{x}$ is very small the term $b\dot{x}$ remains unimportant, and, as follows from the complete equation (1.14), the acceleration is approximately determined by the expression

$$\ddot{x} \approx -\frac{k}{m} x,$$

and since m is very small the acceleration in the system is very large, i.e. the velocity increases extremely rapidly. At the same time the friction force increases and a larger and larger part of the force of the spring is used to overcome it. As a consequence of this the acceleration of the system becomes smaller and smaller and in the end the term $m\ddot{x}$ ceases to be an important factor. The subsequent motion of the system can now be satisfactorily described by the first-order equation (1.47). By this time the velocity acquires a value which is related to the displacement by the equation (1.47), since, as the term $m\ddot{x}$ vanishes, an approximate equality between the terms kx and $(-b\dot{x})$ is established. This rapid transition from a state not compatible with equation (1.47) to a state which is compatible has been followed analytically, using the complete equation of the second-order (1.14) and its solution (1.52).

[†] The duration τ of this initial stage of the motion, in the course of which there occurs a rapid variation of velocity, coincides in order of magnitude with m/b: in a time m/b, the first main term in the expression (1.54) decreases by e times ($e \approx 2.7$) and in a time $5m/b$ by approximately 150 times.

3. Conditions for a jump

As we have seen, in the transition to a state compatible with a first-order equation, the velocity of the system varies very rapidly, while the coordinate of the system remains almost unvaried. However, if the transition itself is accomplished sufficiently rapidly, its details will often be of no interest to us. We can consider this rapid transition as an instantaneous jump and restrict ourselves to determining the final state only, into which the system "jumps" and starting from which the behaviour of the system is determined by the first-order equation (1.47). We can therefore consider the system as having no mass, but must employ another method of analysis of the entire process. We must *add* to the first-order differential equation a *jump condition* which will replace the previous consideration of a brief initial stage of the motion, and determine the state at which the system arrives by this rapid "instantaneous" transition, and from which state the first order equation is valid. This jump condition, which is essentially a different manner of taking into account small parameters (in our case, the small mass of the oscillator) affecting substantially the initial stage of the motion, is formulated either on the basis of an analysis of the system allowing for these small important parameters (this is the regular method) or on the basis of one or other *additional* physical considerations or experimental data[†].

Clearly the jump condition can be formulated in the following manner. If the initial state of the system *(given x_0 and $\dot{x}_0$)* does not satisfy the first-order equation (1.47), then the system passes with a jump to a state compatible with this equation, the velocity of the system x varying *at the time of the jump instantaneously, while the coordinate x remains unchanged*. After such a jump there begins the continuous motion of the system determined by the equation (1.47). We shall observe that here, in formulating the jump condition, we have been guided essentially by the results of the analysis of the system as carried out by means of the second-order equation (1.14) and our postulate is only a simplified formulation of these results.

[†] A similar method, namely the introduction of postulates which replace a more detailed analysis of one or other processes, is often employed. For example, in considering the collision problem in mechanics, the actual process of collision of bodies is often ignored in the analysis and this is replaced by the concept of an "instantaneous" collision, by adding certain postulates which enable one, without considering in any detail the process of collision, to establish the states in which the bodies will be found immediately after the collision.

The jump condition can also be obtained from the consideration of the mapping out of the phase plane of the "complete" system by the phase paths *in the limit case* $m \to 0$ (Fig. 33). Denoting, as is usual, $\dot{x} = y$, we write the equations of motion of the "complete" system in the form

$$\dot{x} = y, \quad \dot{y} = -\frac{kx+by}{m}. \qquad (1.55)$$

On the x, y phase plane a phase line of the system with half a degree of freedom ($m = 0$) is the straight line

$$kx + by = 0. \qquad (1.56)$$

Clearly at any point (x, y) of the phase plane not on this straight line $(kx+by \neq 0)$ $\dot{y} \to \infty$ for $m \to 0$ (while $\dot{x}$ remains finite), i.e. everywhere out-

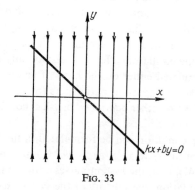

Fig. 33

side the straight line (1.56) there occur rapid, in the limit jump-wise, variations of the state of the system (the velocity y varies with a jump). Further, according to (1.55)

$$\frac{dy}{dx} = -\frac{kx+by}{my}.$$

Therefore outside the straight line $kx + by = 0$ for $m \to 0$ $dy/dx \to \infty$ and the phase paths are vertical straight lines (x=const). Along them, the representative point moves *with a jump* (with a phase velocity tending to infinity for $m \to 0$, x remaining constant during the jump) and arrives at the phase line of the system with half a degree of freedom, namely at the straight line (1.56), since above this line $kx+by > 0$ and $\dot{y} \to -\infty$ for $m \to 0$ and below it $\dot{y} \to +\infty$. Since *all* phase paths of rapid jump-wise motions arrive at the straight line $kx+by = 0$, then the subsequent motion of the representative point occurs along this line towards the state of equilibrium. Below we shall often use similar methods for obtaining the jump conditions in the analysis of "relaxation" oscillations.

Let us illustrate graphically the meaning of the jump conditions introduced here. Since the quantity which varies with a jump is the velocity, we shall compare the diagram of velocity versus time for the case $m \neq 0$ (the second-order equation) with the same diagram for $m = 0$ (the first order equation plus the jump condition).

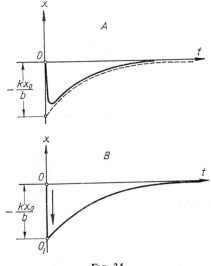

Fig. 34

At the initial instant x and $\dot{x}$ may be assigned as we choose. Let, for example, for $t=0$, $x=x_0 (x_0 > 0)$, $\dot{x}=0$. It is easily proved that the dependence of velocity upon time, following a second-order equation, has the form shown in Fig. 34, A (in constructing the curve, m was assumed to be much less than b^2/k). If, however, we use the first-order equation, then the initial value $x=x_0$ automatically gives the initial value $\dot{x}=-(k/b)x_0$ and the subsequent variation of velocity with time is given by Fig. 34, B. The jump which reconciles the "conflict" between the initial conditions $x=x_0$, $\dot{x}=0$ and the first-order differential equation, is represented in Fig. 34, B by the segment OO_1.

The similarity of Figs 34, A and B is readily seen. Its physical meaning has been clarified in subsection 2.

4. *Other examples*

Let us consider now the oscillations of an RC or RL circuit, beginning from states which do not satisfy the corresponding first-order equation:

$$R\dot{q} + \frac{q}{C} = 0 \tag{1.49}$$

or

$$L\frac{di}{dt} + Ri = 0. \tag{1.50}$$

To analyse such oscillations, we must either pass to other, more "complete" idealizations of the corresponding real electrical circuits, taking into account the important small parameters needed[†], or else add to the equations (1.49) and (1.50) appropriate jump conditions.

Let at the initial instant of time $t=0$ such initial values of the charge q_0 and current $\dot{q}_0$ be assigned in the RC circuit and such values of the current i_0 and its derivative $(di/dt)_0$ in the RL circuits, that the first-order equations for these circuits are *not* satisfied (for example the initial states $q_0 \neq 0$, $\dot{q}_0 = 0$ and $i_0 \neq 0$, $(di/dt)_0 = 0$, which can be assigned by closing the switch in Figs. 35 and 36). To obtain systems compatible with the initial

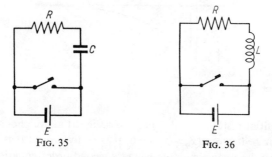

Fig. 35   Fig. 36

conditions given, we shall take into account in the case of the RC circuit the small inductance of the resistor and of the connecting wire, and in the case of the RL circuit the small capacitance of the induction coil, the resistor and the connecting wire. Representing these small parasitic inductance and capacitance as lumped parameters, we arrive at systems the

---

† Which small parameters are important and must be taken into account, depends on the initial state assigned in the real system. In any case, the idealized model obtained as a result of allowing for these small parameters must be compatible with the given initial state.

diagrams of which are shown in Fig. 37 and Fig. 38 (there L_0 and C_0 are small parasitic inductance and capacitance). The equations of the oscillations are now written in the form

$$L_0 \ddot{q} + R\dot{q} + \frac{q}{C} = 0$$

for the circuit of Fig. 37 and

$$C_0 L R \frac{d^2 i}{dt^2} + L \frac{di}{dt} + Ri = 0$$

for the circuit of Fig. 38, i.e. in the form of linear equations of the second order with a small positive coefficient for the higher-order derivative, in complete analogy with the equation (1.14) for the motion of an oscillator with small mass.

We can affirm, on the basis of this analogy, that in the initial stage of the motion in the RC circuit (for a small inductance L_0) there will occur rapid variations of the intensity of current $i = \dot{q}$ (during this time the charge q of the capacitor remains practically unvaried) and in the RL circuit (for small capacitance C_0) rapid variations of di/dt, or of the e.m.f. of the self-inductance (now the current i remains practically unvaried). As a result of the rapid variations of current (in the first case) and of the e.m.f. of the self-inductance (in the second case) the systems arrive in a small interval of time (the duration of which coincides in order of magnitude with L/R

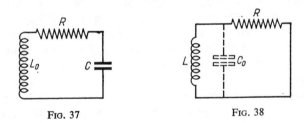

Fig. 37   Fig. 38

or $C_0 R$) at states which are nearly compatible with the first-order equations (1.49) or (1.50). The subsequent motions are satisfactorily represented by the first-order equations (the more accurately, the smaller L_0/CR^2 in the first case and $C_0 R^2/L$ in the second case, with respect to unity‡).

‡ On the basis of the same analogy, we can affirm that the same small parasitic capacitance and inductance represent unessential second-order parameters for oscillations in the circuits when the initial conditions are compatible with the corresponding first-order equations.

If we are not interested in the details of these rapid variations, we can leave out of account the small inductance L_0 in the RC circuit and the small capacitance C_0 in the RL circuit and introduce, instead of the detailed analysis of the initial stage of the motion, suitable jump conditions.

We must admit for the RC circuit jumps of the current i leaving unvaried the charge q of the capacitor and for the RL circuit jumps of the e.m.f. of the self-inductance (or di/dt) leaving unvaried the current i.

Had we admitted instantaneous variations of the intensity of current in the circuit with the self-inductance, i.e. had we assumed that at certain instants $d^2q/dt^2 = \infty$, then we should have admitted the appearance of an infinitely large self-induction e.m.f. Ld^2q/dt^2 across the terminals of the self-induction coil. Similarly, had we admitted instantaneous variations of charge on the plates of the capacitor, then this would have forced us to admit the appearance of infinitely large currents in the circuit (since if q varies with a jump, then $dq/dt = i = \infty$). Both these types of variations are incompatible with the postulates established by us on the character of the jump†.

We shall observe that in all three examples considered we have been dealing with *conservative* jumps, i.e. with such jumps for which the energy of the system did not vary. In fact, in the case of the oscillator without mass, all the energy of the system consisted of the potential energy of the spring and was equal to $kx^2/2$. During the jump the coordinate x remained constant and therefore the energy did not vary either. In just the same manner, in the RC circuit the energy of the system consisted of the energy of the electric field in the capacitor (the energy was equal to $q^2/2C$), and in the RL circuit of the energy of the magnetic field in the self-induction coil ($=Li^2/2$) and, since in the jump the charge q of the capacitor in the first case and the current i in the second do not vary, the energy also remains unvaried.

We must not think, however, that conservativeness is an indispensable condition, valid for all jumps. In mechanics, in the analysis of collisions, we have often to use the notion of non-conservative collisions (in the collision the kinetic energy of colliding bodies "instantaneously" decreases). Similar jumps, for which the energy of the system varies, will be met below (in the theory of the watch and of the valve oscillator with a ∫ characteris-

---

† Clearly the jump conditions formulated above can be obtained from the postulate of the *finiteness* of the currents and voltages through and across separate elements of the circuits. Of course, this postulate is not a consequence of the first-order equations, but is an additional physical hypothesis.

tic). We shall give now only one example of a system with non-conservative jumps.

Let us consider the circuit shown in Fig. 39. The state of the circuit obtaining immediately after closure of the switch (the current in the resistor and the voltage across the capacitor C_1 are equal to zero and the voltage across the capacitor C_2 is equal to E) are clearly incompatible with the equation (1.49) for an RC circuit with capacitance $C = C_1 + C_2$. Neglecting the resistance and the inductance of the switch (in the short-circuit state) and of the conductors connecting the capacitors C_1 and C_2, we must admit that after closure of the switch infinite currents flow through the wires connecting these capacitors, as a result of which the voltages across the capacitors C_1 and C_2 and also the current through the resistor vary with a jump. At the end of this "instantaneous" jump the voltages across the two capacitors must become the same (we shall denote this voltage by v_0) and the current through the resistor must be equal to v_0/R. To determine v_0 we shall note that during the instantaneous redistribution of charge on the capacitors the total charge of the capacitors must not vary, since the currents through the resistor R are always finite. Thus

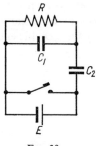

FIG. 39

$$C_1 \cdot 0 + C_2 \cdot E = (C_1 + C_2)v_0,$$

and therefore

$$v_0 = \frac{C_2 E}{C_1 + C_2}.$$

After such a jump of the current a continuous motion will begin, determined clearly by the equation (1.49) (with a capacitance $C = C_1 + C_2$). As is easily calculated, the energy of the system *decreases* in such a jump. In fact, let us compare the energy of the system up to the time of the jump $C_2 E^2/2$ with the energy of the system after the jump $(C_1 + C_2)v_0^2/2$. Clearly

$$\frac{1}{2}(C_1 + C_2)v_0^2 = \frac{C_2}{C_1 + C_2} \cdot \frac{C_2 E^2}{2} < \frac{C_2 E^2}{2}.$$

We have considered just now a jump in the system on the basis of the assumption (additional with respect to equation (1.49)) of the conservation of the sum of the charges of the capacitors during the jump. The same can also be done by considering a more "complete" system, which now permits the given initial conditions. This can be, for example, a system in which account is taken of the small resistance R_1 of the conductors connecting the capacitors (Fig. 40). We leave it to the reader to carry out this analysis.

The examples given here have helped us clarify to a sufficient extent all that has been said with respect to systems the motions of which are represented by linear differential equations of the second order with small positive coefficients of the second derivative.

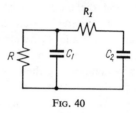

Fig. 40

As we have seen, during the initial stage of the motion there can occur in such systems (under suitable initial conditions) rapid variations of states, after which the motion is described fairly satisfactorily by the corresponding equations of the first order. These rapid variations of states, during which one or other small parameters play an essential role, can only be analysed by taking into account the latter, and so solving corresponding equations of the second order. If, however, we are not interested in the details of this initial very brief stage of the motion, we can replace this analysis of an equation of the second order by the assumption that a state compatible with the equation of the first order is established instantaneously with a jump. In this connexion we must introduce a new postulate (the jump condition) which must determine the state at which the system arrives as a result of the jump and starting from which the motion of the system is represented by the corresponding equation of the first order.

This notion of jump-wise variations of the states of a system will be widely used later in the study of systems with "relaxation" oscillations (see Chapter X).

§ 6. Linear systems with "negative friction"

In the ordinary systems with friction, examples of which have been considered above, the coefficient $h = b/2m$ (or in the electric systems $h = R/2L$) is always a positive quantity, since friction always hinders motion and $b > 0$ (just in the same manner, also $R > 0$). A positive coefficient of friction and a positive resistance mean that to overcome frictional forces (or resistances in an electric circuit), energy is spent. In fact, if in the equation of motion

$$m\frac{d^2x}{dt^2}+b\frac{dx}{dt}+kx = 0 \tag{1.14}$$

we multiply all terms by dx/dt and then take the integral over a certain interval of time from 0 to τ, we shall obtain

$$m\int_0^\tau \frac{d^2x}{dt^2}\cdot\frac{dx}{dt}dt + \int_0^\tau b\left(\frac{dx}{dt}\right)^2 dt + \int_0^\tau kx\frac{dx}{dt}dt = 0.$$

On carrying out the integration we have

$$\left|\frac{m}{2}\left(\frac{dx}{dt}\right)^2\right|_0^\tau + \left|\frac{kx^2}{2}\right|_0^\tau = -\int_0^\tau b\left(\frac{dx}{dt}\right)^2 dt. \tag{1.57}$$

The terms on the left-hand side express the variation of kinetic and potential energy of the system during the time from 0 to τ; their sum clearly determines the variation of the total energy of the system over this interval of time. If $b > 0$, the integral on the right-hand side is positive and the variation of energy is negative, i.e. the energy of the system decreases. This decrease of energy is caused by losses of energy owing to friction.

If b and hence h were negative, then the energy of the system would increase and "friction", in this case, would be a source of energy. It is clear that in a system having no energy sources of its own this is impossible, and b and h together are always positive. However, if a system possesses its own reservoir of energy, then, generally speaking, it can be admitted that $h < 0$ and that the energy of the system increases at the expense of "friction" or "resistance". Certainly this would no longer be friction or resistance in the usual sense. But since this is characterized by the same term of the differential equation as is ordinary friction, namely by the term containing dx/dt, we shall, also in the case of a negative h, employ the term "friction" or "resistance" and shall speak of "negative friction" and "negative resistance".

1. Mechanical example

A very simple example of a mechanical system in which "friction" is negative in a certain region is the structure shown in Fig. 41. On a belt moving uniformly with velocity v_0 there lies a mass m fixed by the springs k_1 and k_2. The friction force exerted by the belt on the load is certainly

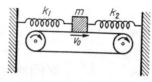

Fig. 41

a very complicated function of the relative velocity of the belt and the body. If we denote the displacement of the load by x and its velocity by $\dot{x}$, then the frictional force acting on the mass m, being a function of the relative velocity $v = v_0 - \dot{x}$, can be written thus $F(v_0 - \dot{x})$. If we denote the coefficient of elasticity by k and consider as proportional to the first power of velocity all remaining frictional forces acting in this system (for example, the resistance of the air or the internal friction of the springs), then the equation of motion of the mass m is written thus

$$m\ddot{x} + b\dot{x} + kx = F(v_0 - \dot{x}), \tag{1.58}$$

where $F(v) = F(v_0 - \dot{x})$ is a function characterizing the dependence of the frictional force on the relative velocity v. Without specifying the form of the function $F(v_0 - x)$ we can restrict our analysis to a region in which $|\dot{x}| \ll v_0$ (for example, by choosing a sufficiently large v_0). We can expand in this region the function F in a series about the values of v_0 and consider only one term of the series. Then $F(v_0 - x) = F(v_0) - \dot{x} F'(v_0) + \ldots$, and, within the limits of this restriction, the equation of motion assumes the form

$$m\ddot{x} + [b + F'(v_0)]\dot{x} + kx = F(v_0) \tag{1.59}$$

The constant term, occurring in the right-hand side, only causes a displacement of the position of equilibrium by the quantity $F(v_0)/k$ in the direction of motion of the ribbon. The sign and amplitude of the velocity coefficient $(b + F'(v_0))$ depend on the form of the characteristic of friction; the quantity $F'(v_0)$ represents the slope of the friction characteristic at the point v_0 and in the case of a decreasing characteristic of friction $F'(v_0) < 0$. If the characteristic of friction decreases in the region of v_0 sufficiently sharply, then $b + F'(v_0) < 0$ and equation (1.59) describes a system with "negative friction". In practice this case is fairly easily set up since the characteristics of friction of solid surfaces usually have the form shown in Fig. 42, and have at the beginning (for small velocities) almost always a more or less significant segment of negative slope. In this region, our struc-

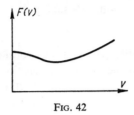

Fig. 42

ture will represent a linear system with "negative friction". We must, however, bear in mind that we have arrived at a linear system with negative friction by restricting ourselves to the region where $|\dot{x}| \ll v_0$. This restriction, as we shall see below, is one of principle and will play an essential role in the answering of the questions of interest.

Another example of a mechanical system in which "friction" is negative within a certain region is the so-called Froude's pendulum [117, 63, 116]. The mechanism of this pendulum is the following: to a shaft rotating uniformly with angular velocity Ω is suspended with a certain friction an ordinary pendulum (Fig. 43). The equation of motion of this pendulum will differ from the equations of motion of an ordinary pendulum only in that we must take into account in this equation the moment of the force of friction of the rotating shaft acting on the bearing from which the pendulum is suspended. Since the frictional force depends on the relative velocity of the rubbing surfaces, i.e. in our case on the relative angular velocity of the shaft and the pendulum $(\Omega - \dot{\varphi})$, then the moment of the frictional force can be written thus:

$$F(\Omega - \dot{\varphi}).$$

Fig. 43

By taking into account, together with the friction of the pendulum on the shaft, air friction, and assuming that this is proportional to the velocity $\dot{\varphi}$, we shall obtain the equation of motion of the pendulum in the following form:

$$I\ddot{\varphi} + b\dot{\varphi} + mgl \sin \varphi = F(\Omega - \dot{\varphi}). \qquad (1.60)$$

The states of equilibrium $\varphi=\varphi_0$, $\dot\varphi=0$ are clearly determined by the equation

$$mgl \sin \varphi_0 = F(\Omega)$$

Let us consider the motion of the pendulum near the lower state of equilibrium (for this $\cos \varphi_0 > 0$). We shall put

$$\varphi = \varphi_0 + \psi$$

where ψ is a small quantity (we shall also consider the velocity $\dot\varphi = \dot\psi$ to be small). Let us expand the non-linear functions $\sin \varphi$ and $F(\Omega - \dot\varphi)$ in power series with respect to ψ and $\dot\psi$. Restricting ourselves to the linear terms, we shall obtain the linearized equation of the small oscillations of the pendulum in the form

$$I\ddot\psi + [b+F'(\Omega)]\dot\psi + mgl \cos \varphi_0 \cdot \psi = 0 \tag{1.61}$$

If $F'(\Omega) < 0$ and is larger in absolute value than b, then the coefficient of $\dot\varphi$ will be negative. In a certain region of values of Ω, where the characteristic of friction decreases sufficiently rapidly, we can, for a sufficiently small b, attain a situation where $b+F'(\Omega)$ remains negative and shall obtain an equation, analogous to the equation of the usual system with friction

$$\ddot x + 2h\dot x + \omega_0^2 x = 0,$$

but differing in that the coefficient h will be negative. It is seen, therefore, that, for an appropriate choice of v_0 in the first system and of Ω in the second one can realize in practice a mechanical system, which in a certain bounded region can be considered as a linear system with negative friction[†].

2. *Electrical example*

An electrical system, the "resistance" of which is negative in a certain region, is also quite feasible. An example of such a system is a valve generator, i.e. a circuit including an electronic valve, an oscillatory circuit and "feedback". For the sake of definiteness, we shall consider the simplest circuit of an oscillator with inductive feedback and an oscillating circuit in the grid circuit (Fig. 44)[‡], neglecting the grid current. For the chosen

---

† Another example of a mechanical system with "negative friction" is an oscillatory system containing a single-phase asynchronous motor [44].

‡ A similar analysis can be carried out for an oscillator with inductive feedback and the oscillating circuit in the anode circuit. We shall not consider other types of valve oscillators, since their analysis either gives nothing new in principle, or leads to differential equations of the third order, i.e. to systems with one and a half degrees of freedom and thus exceeds the limits of this book.

positive directions of the current and of the capacitor voltage we can write for the oscillatory circuit the following equations:

$$i = -C\frac{dv}{dt}, \quad Ri = v - L\frac{di}{dt} - M\frac{di_a}{dt}$$

($-M\,di_a/dt$ represents the feedback e.m.f. induced in the oscillating circuit by the action on it of the anode current, flowing through the coil L_a) or

$$LC\frac{d^2v}{dt^2} + RC\frac{dv}{dt} + v = M\frac{di_a}{dt}. \tag{1.62}$$

Neglecting the anode reaction, i.e. assuming the anode current i_a to depend on the grid voltage $u_g = v$ only (this is sufficiently well observed for triodes

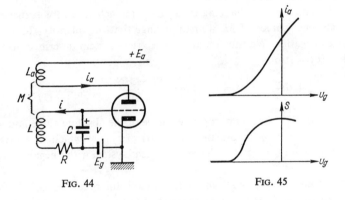

Fig. 44 Fig. 45

with large amplification factors or even better for pentodes), we have clearly

$$\frac{di_a}{dt} = \frac{di_a}{du_g} \cdot \frac{dv}{dt} = S(v)\frac{dv}{dt},$$

where $S = di_a/du_g = S(u_g)$ is the slope of the characteristic of the valve, depending obviously on the grid voltage u_g. A typical valve characteristic and also the dependence of the slope S of the characteristic upon u_g are shown in Fig. 45.

Substituting expression (1.63) in (1.62), we shall obtain

$$LC\frac{d^2v}{dt^2} + [RC - MS(v)]\frac{dv}{dt} + v = 0 \tag{1.64}$$

i.e. the non-linear equations for the oscillations of a valve generator, which we shall analyse in detail later. Here we shall consider only *small* oscillations in the circuit of the oscillator in the vicinity of the state of equilibrium $v=0$. Restricting ourselves to a sufficiently small region of variations of the voltage v, we shall assume S to be constant: so that $S(v)=S_0$ (the slope of the characteristic at the operating point). Then we shall obtain, for such small oscillations, the *linear* equation

$$LC\frac{d^2v}{dt^2}+[RC-MS_0]\frac{dv}{dt}+v = 0. \qquad (1.65)$$

The sign of the coefficient of mutual induction M (for the chosen positive directions of the currents i and i_a) is determined by the relative disposition of the turns of the coils L and L_a. We shall assume that $M > 0$, i.e. that the coils L and L_a are so connected that the currents i and i_a, flowing in the directions indicated in Fig. 44 by arrows ($i > 0$, $i_a > 0$), give rise in the coil L to magnetic fluxes enhancing each other. In such cases, for sufficiently large absolute values of M, we can arrange that the quantity $RC-MS_0$ becomes negative. We thus obtain an electrical system described also by the linear equation

$$\frac{d^2v}{dt^2}+2h\frac{dv}{dt}+\omega_0^2 v = 0 \quad \left(h = \frac{RC-MS_0}{2LC}, \quad \omega_0^2 = \frac{1}{LC}\right),$$

where $h<0$. Thus by an appropriate choice of the absolute value and sign of M we can realize in practice an electrical system which in a certain bounded region can be considered as a linear system with "negative resistance".

All systems considered lead us to a linear differential equation of the form $\ddot{x}+2h\dot{x}+\omega_0^2 x = 0$, where in contrast to the previous cases the coefficient h of this equation is negative (ω_0^2 is positive as before). To investigate the behaviour of these systems we can employ the methods which have been developed above. However, since these methods are applicable independently of the sign of h, we shall not repeat here all the derivations but will use the results obtained in § 4 for the case $h > 0$.

3. Portrait on the phase plane

In order to establish the character of the integral curves on the phase plane in the case $h<0$, it is sufficient in both the cases considered of a linear oscillator (small and large friction) to see how the portrait established earlier is modified when the sign of h varies.

For the case $h^2 < \omega_0^2$, i.e. for the case of not too large "negative friction" we shall again obtain a family of spirals, determined by the equation

$$y^2 + 2hxy + \omega_0^2 x^2 = C^2 e^{2\frac{h}{\omega} \arctan \frac{y+hx}{\omega x}}, \quad (1.29)$$

or in polar coordinates in the u, v plane

$$\varrho = C e^{\frac{h}{\omega} \varphi} \quad (1.28)$$

where, just as before,

$$\omega = +\sqrt{\omega_0^2 - h^2} \quad \text{and} \quad \varphi = -(\omega t + \alpha).$$

However, since now $h < 0$, then, as φ decreases (φ decreases as t increases) ϱ will increase (Fig. 46), i.e. the direction of the spirals is changed into

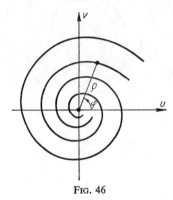

Fig. 46

the opposite one with respect to that obtained for systems with a positive h. Consequently, moving along an integral curve, the representative point will move away from the state of equilibrium (the singular point $x=0$, $y=0$). The singular point is, also in this case, the asymptotic point of a family of spirals winding within each other, i.e. is a singular point of the focus type (Fig. 47).

The velocity of motion of the representative point on the phase plane reduces as in the previous case to zero at the origin of the coordinates only and increases together with the distance of the representative point from the origin. Since, moreover, this velocity is always directed along an integral curve in a direction pointing away from the origin then, for an arbitrary non-zero initial deviation of the system from the state of equilibrium, the system will move eventually as far away as one chooses from this sole state of equilibrium. Owing to this, we cannot indicate such a region $\delta(\varepsilon)$

that, when the representative point is found in it at the initial instant of time, it shall never overstep the boundaries of an assigned region ε. Consequently the only position of equilibrium is in this case unstable: the singular point is an *unstable focus*. It is perfectly clear that the instability

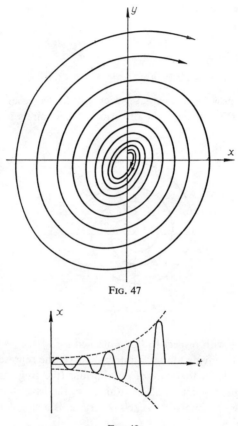

Fig. 47

Fig. 48

of this focus is caused by the fact that $h < 0$. Obviously in the case $h < 0$ and $h^2 < \omega_0^2$ the motion of the system is also an oscillatory process, just as for a small positive h, but the process is no longer a damped but a reinforcing one. The maximum deviations of the system increase with time (Fig. 48) and the dependence of the deviations on time is determined by an expression of the form $x = Ke^{-ht} \cos(\omega t + \alpha)$, where $h < 0$. The law of increase of the maxima is a geometrical progression with the common ratio $e^{-hT} = e^{-d}$, where, since $h<0$, then $d<0$ and $e^{-d}>1$. The quantity $d_1 = -d$ bears in this case the name of *logarithmic increment* of the oscillations. What has been said above with respect to the decrement, applies entirely to the increment. Thus in particular the concept of logarithmic increment is applicable to linear systems only.

Thus, as far as we restrict ourselves to a linear treatment of the system, we obtain an oscillatory process increasing without limits.

We can, in the same manner, analyse the character of the behaviour of the integral curves for the case of a large "negative friction": $h<0$, $h^2 > \omega_0^2$, when the family of integral curves is determined by the equation (1.45).

$$(y+q_1 x)^{q_1} = C_1 (y+q_2 x)^{q_2},$$

where

$$q_1 = h - \sqrt{h^2 - \omega_0^2}$$

and

$$q_2 = h + \sqrt{h^2 - \omega_0^2}.$$

Since $h < -\omega_0 < 0$, then $q_1 < q_2 < 0$, which fact results in a variation of the position of the integral curves $y+q_1 x=0$ and $y+q_2 x=0$; both these

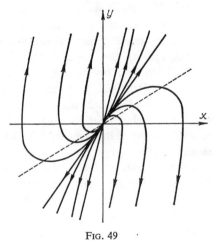

FIG. 49

straight lines will pass in this case through the first and third quadrants since x and y are of the same sign (Fig. 49).

We again obtain a family of integral curves of the "parabolic" type, all curves passing through the unique singular point, situated at the origin of the coordinates. This is a singular point of the node type.

By determining the direction of motion of the representative point on the phase plane, we easily verify that these motions occur in the directions indicated in the Figure by arrows, and, consequently, the representative point, in its motion along any one of the integral curves, tends to move away from the state of equilibrium. On the velocity of motion of the representative point we could repeat all that has been said above. Consequently, however small the initial deviation of the system from the state of equilibrium ($x=0$, $y=0$), the system will move eventually sufficiently far away from the state of equilibrium and thus this state of equilibrium is unstable. Again we shall not be able to indicate such a finite region of initial values $\delta(\varepsilon)$ that the representative point shall not leave an assigned region ε.

The singular point considered is an *unstable node*, the instability being again caused by the fact that $h<0$. We have obtained an aperiodic process, building up according to the law: $x=Ae^{\lambda_1 t}+Be^{\lambda_2 t}$, where λ_1 and λ_2 are positive. Thus, as long as we consider the system as linear, this build-up in it will last indefinitely.

Considering the system as a linear one does not lead to any stable stationary states. It cannot remain in a region close to the state of equilibrium, i.e. deviations in the linear system must increase continuously. But in describing the mechanical and electrical systems which lead us to these cases, and in order to arrive at linear equations, we had to restrict ourselves to considering regions sufficiently close to the state of equilibrium (small x and small y). Therefore, on the one hand, we have to restrict ourselves to considering regions sufficiently close to the state of equilibrium while, on the other hand, considering the motion of the system in these regions leads us to conclude that the system does not remain in this region but inevitably oversteps its boundaries. In other words, a linear treatment enables one to represent correctly the behaviour of the phase paths only in a certain bounded region of the phase plane near the position of equilibrium. However, all phase paths overstep this bounded region. To investigate the behaviour of the system further, we must evidently take into account certain facts which so far we have left out of account and consider now the system as a non-linear one.

It is seen that in the case considered *a linear treatment cannot in principle provide an answer to many questions about the behaviour of the system*, for example to the question of which motions the system will accomplish in the course of a sufficiently long interval of time.

4. Behaviour of the system for a variation of the feedback

We shall sum up here the results obtained, restricting ourselves to the results for a valve generator only; but in the cases of a load on a moving belt and Froude's pendulum the results will be completely analogous.

As long as the feedback is sufficiently small[†] (we assume that the direction of the turns of the coils is such that $M > 0$), we have in the circuit either an aperiodic damping or damped oscillations, depending on whether h^2 is larger or smaller than ω_0^2. If the oscillating circuit itself has such a large resistance that aperiodic damping occurs in it, then choosing a sufficiently large feedback we can arrange that this feedback will "compensate" a large part of the resistance of the oscillating circuit, i.e. $h = (RC - MS_0)/2L$ will be a small positive quantity. Then, in the case of not too large initial deviations (such that the system does not overstep the boundaries of the linear region) an oscillatory damping will occur and not an aperiodic one. By increasing the feedback, we shall pass through a position when $RC - MS_0 = 0$, and into the region where $RC - MS_0 < 0$, and the state of equilibrium is unstable (since $h < 0$).

There will be no longer damping but reinforcement of the oscillations. The larger the absolute value of h, the larger will be the phase velocity of the spiral on the phase plane and the more rapidly will these spirals unwind and grow larger. Finally, for a further increase of the feedback, the system will pass through a position in which $h^2 = \omega_0^2$ and pass into a region where $h^2 > \omega_0^2$ (where h is now negative). In this region we shall again obtain an aperiodic process, but no longer a damped one (as for a large positive h) but a reinforcing one. The rate of increase of the process, determined by the roots of the characteristic equation λ_1 and λ_2 will be the larger, the larger $|h|$ and, hence, the larger the feedback.

Thus, simply by varying the value of h, characterizing the "resistance" of the system (from large positive values to large negative values of h), we can make the system pass successively through five different regions corresponding to various types of motions and states of equilibrium, and

---

[†] In the mechanical systems considered there is no element analogous to the variable feedback. Therefore, in order to vary the regime, some other parameter must be varied, for example the slope of the characteristic of friction. Also, in a valve oscillator we could vary, instead of the value of the feedback, the slope of the characteristic of the valve at the working point, i.e. the value of S_0. Owing to the absence of a feedback in the mechanical systems considered there is no complete analogy between these systems and an ordinary valve oscillator. An electrical analogy of Froude's pendulum can be found in the so-called dynatron oscillator, in which there is no feedback and self-excitation occurs as a consequence of working in a decreasing section of the characteristic of the valve (see Section 7 of this chapter).

more precisely: a stable node, a stable focus, a centre, an unstable focus and an unstable node. In the following section we shall meet one more type of equilibrium, which cannot be attained by a variation of the friction damping in the system.

However, not all of these five types of equilibrium offer the same physical interest. To all the states of equilibrium, except the centre ($h=0$), there correspond finite regions of the values of the parameters of the system, in particular of the parameter M. In other words, to values of M, comprised within given finite limits, can correspond any state of equilibrium except the centre, while to the latter state of equilibrium there corresponds exactly one single critical value M_{crit} obtained from the relation $MS_0 - RC = 0$. If this condition is disturbed, however little, then the system will pass either into the region $h > 0$ or in the region $h < 0$. This means that a state of equilibrium of the centre type is unstable with respect to small variations of the parameters of the system. Since small variations of the parameters of the system can never be avoided, then states unstable with respect to them are not representative of the behaviour of a real physical system. Therefore, a state of equilibrium of the centre type has a physical meaning only as a boundary between two other states, a stable and an unstable focus, just as the case $h^2 = \omega_0^2$ has a meaning only as a boundary between a focus and a node. However, as has already been indicated, we must not attribute to these boundaries a too strict physical meaning. The transition from one type of motion to another occurs, in real systems, gradually and the physical boundary between oscillatory and aperiodic damping is not too sharp, since, as the damping increases, the system loses its oscillatory properties not abruptly but gradually. In other words, in real systems we are not in a position to distinguish a "strong" focus, i.e. a focus with very large h (when h^2 is only a little smaller than ω_0^2) from a "weak" node, i.e. a node for which h^2 is only a little larger than ω_0^2. In the same way we cannot distinguish a very weak damping from a very weak reinforcement, since, in order to observe the difference between these two processes we should have to wait an extremely long time.

We have verified above that by a suitable choice of the voltage and the value of the feedback we can not only achieve a decrease of the damping of the natural oscillations in a system but also arrange that these oscillations become of increasing amplitude. The physical meaning of this phenomenon is quite clear. A decrease of the damping of a system is caused evidently by the fact that, instead of a fraction of the energy being dissipated in the oscillating circuit, a certain amount of energy arrives from an external source (in our case from the anode battery) which partially compensates the losses of energy in the circuit and thus reduces the damping in the system. The stronger the feedback, the more energy arrives from the battery during one period and the larger is the fraction of the losses being compensated and the weaker is the damping of the oscillations. As the feedback is further increased, the energy arriving in the circuit can prove larger than the losses of energy and then the energy in the circuit will increase, and a reinforcement of the oscillations in the circuit will occur.

For a further increase of the energy arriving in the circuit there even occurs an aperiodic reinforcement. How long the reinforcement of the oscillations will last, whether it will stop and exactly when, we cannot say, as long as our analysis is restricted to the linear approach.

In investigating linear equations we are also deprived of the possibility of saying anything about the ultimate process which will be established in the system after a long interval of time, and even whether a periodic process is possible in a given system. We can only affirm that in linear systems a periodic process is impossible. To answer the question of the further behaviour of a real system after it has overstepped the boundaries of the region to which we had restricted our considerations, it is evidently necessary to consider the system as non-linear. Such a non-linear approach is the object of our further analysis. Here we shall only point out that the absence of oscillatory motions in the vicinity of a position of equilibrium by no means indicates the impossibility of oscillatory motions in the given system in general. In particular, if in the vicinity of the position of equilibrium there occurs an aperiodic reinforcement (unstable node), this does not mean that an oscillatory process cannot be established in the system at a later stage. Also we shall see that for the case of a singular point of the node type the existence of a periodic process (non-damped oscillations) is entirely possible.

Let us return, however, to the question of the energy compensating the losses in the system. The picture which one obtains from this point of view is the same both for electrical and for mechanical systems. In the case of an oscillator energy arrives in the circuit from the anode battery and the electron valve is merely the mechanism which regulates in the required manner the inflow of energy in the circuit. In mechanical systems to which all our conclusions can be applied, the source of energy is the motor which drives the belt or the shaft, and transfer of this energy into the oscillating system is caused by the appropriate form of the friction characteristic. More precisely, the form of the friction characteristic is such that the belt or shaft helps the body in its motion in the same direction more than it prevents the opposite motion. Had we chosen in the oscillator such a connection of the coils as to correspond to a negative feedback ($M<0$), or, in the mechanical models, had we fixed the working point not on the decreasing but on the increasing section of the characteristic of friction, then the energy from the battery or the motor would not have been transferred to the oscillating system but, on the contrary, a part of the energy of the oscillators would have been dissipated by the auxiliary mechanism (in the valve at the anode and in the bearing to overcome friction). The damping of the oscillations in the system would not only fail to decrease, but on the contrary, would increase, if the direction of the feedback were the wrong one.

We shall observe, in conclusion (although these questions will not be considered in this book) that in the case of action of an external force on a system with feedback (for example, on a regenerative receiver) it is still possible to obtain an answer to certain questions using a linear idealization. For example, in the case $h<0$, (e.g. the case of an under-excited regenerator and for weak signals), and an external action which does not remove the system from the region in which it can be considered linear, it can be assumed that the feedback only reduces the damping of the system without modifying the linear properties of the system. For sufficiently large signals, however, this statement is no longer correct.

§ 7. LINEAR SYSTEM WITH REPULSIVE FORCE

Thus far we have considered linear systems in which there acts a quasi-elastic force, i.e. a force *attracting* the system to the position of equilibrium and proportional to the displacement of the system. In all cases considered the nature of friction varied, but the force remained an attractive one. However, we shall often encounter systems (and from the point of view of the theory of oscillations these systems are of considerable interest) in which there acts a force which does not attract the system to a position of equilibrium but, on the contrary, *repels* it away from a position of equilibrium, the magnitude of this repulsive force increasing as the displacement of the system increases. In considering these systems there arises first of all the question of the dependence of the repulsive force on the displacement. As we shall see below when considering certain particular examples (and also as follows from general concepts on the expansion of an arbitrary function in series) in a region of sufficiently small deviations we can assume that the repulsive force is proportional to the displacement. We arrive, under such an assumption, to linear systems in which there acts not an attractive but a repulsive force. The behaviour of these systems differs essentially from the behaviour of the linear systems considered above.

As a first example of a linear system with a repulsive force, we shall consider the behaviour of a mathematical pendulum in the immediate vicinity of the upper (unstable) position of equilibrium. We shall firstly assume, for the sake of simplicity, that friction in the pendulum is absent. In this case, if the angle φ is measured from the upper position of equilibrium (Fig. 50) the equation of motion of the pendulum is written:

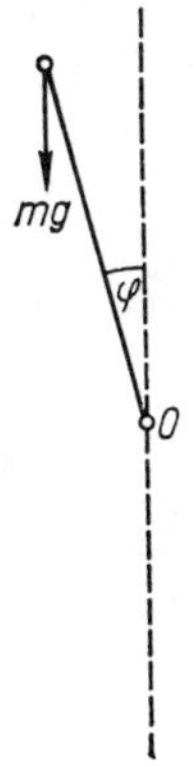

$$ml^2\ddot{\varphi} = mgl\sin\varphi. \qquad (1.66)$$

Restricting ourselves to considering a region sufficiently near to the position of equilibrium, we can replace $\sin\varphi$ by φ. Then the equation takes the form

$$\ddot{\varphi} - \frac{g}{l}\varphi = 0. \qquad (1.67)$$

We have again obtained a linear equation of the second order. This equation, like the equation obtained for the region close to the lower position of equilibrium, does not, evidently, describe motions of the pendulum for all angles φ and is only applicable for sufficiently small values of φ.

Fig. 50

1. Portrait on the phase plane

The equation of our example can be written in a general form:

$$\ddot{x} - nx = 0 \qquad (1.68)$$

where $n > 0$. To investigate the behaviour of a system described by this equation we might choose any of the methods used for the preceding problems, namely, either find the solution of equation (1.68) and then consider the solution found $x = f(t)$ and $\dot{x} = f'(t)$ as parametric equations of the integral curves, or, without integrating equation (1.68), eliminate the time from it and then integrate and consider the equation obtained as the equation of the integral curves. We shall use here this second method. Putting $y = \dot{x}$, we can replace this equation of the second order by two equations of the first order

$$\dot{x} = y, \quad \dot{y} = nx; \qquad (1.69)$$

and eliminating time, we shall obtain only one equation of the first order, connecting x with y:

$$\frac{dy}{dx} = n\frac{x}{y}. \tag{1.70}$$

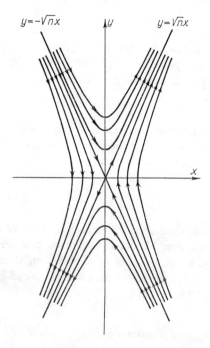

Fig. 51

The state of equilibrium in this system (determined by the condition $dx/dt=0$ and $dy/dt=0$) is the point $x=0$, $y=0$. The isocline $\varkappa=0$ ($dy/dx=0$) is the axis of the ordinates ($x=0$), and the isocline $\varkappa=\infty$ ($dy/dx=\infty$) is the axis of the abscissae ($y=0$). In order to determine exactly the form of the paths of the representative point on the phase plane, we must integrate equation (1.70). The variables are separable and integration gives

$$y^2 - nx^2 = C. \tag{1.71}$$

This is the equation of a family of equilateral hyperbolae referred to their principal axes. For $C=0$ we obtain the two asymptotes of this family: $y=-\sqrt{n}x$ and $y=+\sqrt{n}x$, which pass through the origin of the coordinates. The origin of the coordinates is the only singular point of the family of integral curves. All the remaining integral curves are hyperbolae which do not pass through the origin of the coordinates (Fig. 51). Such a singular point, through which there pass only two integral curves which are asymptotes to all the remaining integral curves, is called a singular point of the *saddle* type.

What conclusions can we derive from the portrait obtained on the phase plane? First of all, bearing in mind that for a positive velocity the coordinate of the system must increase, and for a negative one must decrease, we can indicate by arrows in all four quadrants the directions of motion of the representative point on the phase plane as in Fig. 51. It is easily verified by considering the direction of motion of the representative point, that, wherever the representative point is found at the initial instant (excluding the singular point and the points on the asymptote $y = -\sqrt{n}x$ passing through the second and fourth quadrant) it will always move away in the end from the state of equilibrium, its motion being never oscillatory but aperiodic.

The phase velocity also reduces to zero only at the singular point. Thus, even if the representative point moves at first along one of the integral curves in a direction towards the singular point (the case of motions in the second and fourth quadrants) still it will in the end move as far away as one chooses from the position of equilibrium, except when in motion along the asymptote $y = -\sqrt{n}x$. Therefore, the state of equilibrium is unstable since we cannot choose a region $\delta(\varepsilon)$ such that the representative point, being in this region at the initial instant, shall not in the end overstep the boundaries of a given region ε. It is evident that a singular point of the saddle type is always unstable, and this instability is connected with the very nature of the singular point and the character of the integral curves and not with the direction of motion of the representative point along an integral curve (even if the direction of motion changes the singular point would still be unstable).

As far as the motions along the asymptote $y = -\sqrt{n}x$ are concerned, they represent a special case when the system can only approach the state of equilibrium. For this motion the representative point will approach the origin of the coordinates with a velocity tending to zero, but does not reach the origin of the coordinates in a finite interval of time. This case, the so-called *limitation* motion will be considered later in detail. However, the possibility of such a motion, directed towards the state of unstable equilibrium, is evident from elementary considerations. In fact, for any initial deviation of the pendulum from its upper state of equilibrium it is always possible to choose an initial velocity such that the kinetic energy of the pendulum at the initial instant be exactly equal to the work it must accomplish in order to reach the state of equilibrium. But, as we shall see later, even if we were able to impart with absolute precision such a chosen initial velocity, the pendulum would reach the state of equilibrium only after an infinitely large interval of time.

However, this special case of motion towards the position of equilibrium does not infringe the statement that in this example this state of equilibrium is unstable. In fact, for any initial conditions, different from those especially chosen to correspond exactly to the asymptote $y = -\sqrt{n}x$, the system will always move away from the state of equilibrium. This motion along the asymptote can never be exactly realized in practice, since it corresponds to one initial state and not to a finite region of initial states. Such an initial state (or better, such a "line of initial states") does not form a finite region of initial states and cannot be given with absolute precision in the system. In other words, if we assume that all initial states are equiprobable, the probability of such an initial state which corresponds to a motion towards the singular point is equal to zero. Therefore, any real motion in the system will remove the system away from the state of equilibrium.

Let us now consider the cases when, together with a repulsive force, there exists friction, the friction being either positive or negative. We are led to the first of these cases when considering the pendulum close to the upper position of equilibrium in the presence of a frictional force proportional to velocity. The equation describing the motion of the system will have in this case the form

$$\ddot{\varphi} + 2h\dot{\varphi} - n\varphi = 0, \qquad (1.72)$$

where $n = g/l$ and $h > 0$. We have already met the second case, i.e. the case of $h < 0$ in the equation (1.72), when we considered Froude's pendulum, also in a region close to the upper state of equilibrium.

2. An electrical system

We arrive at the same equation (1.72) when we consider under appropriate assumptions the so-called dynatron oscillator (Fig. 52), a circuit which can sustain self-oscillations owing to the presence in the anode characteristic of a tetrode $i_a = \varphi(u)$ (Fig 53) of a decreasing section (a section for which $di_a/du < 0$)†. Kirchhoff's laws give, for the circuit considered,

$$i = i_a + C\frac{du}{dt}, \quad L\frac{di}{dt} + Ri = E_a - u,$$

or, after eliminating the current i

$$LC\frac{d^2u}{dt^2} + \left[RC + L\frac{di_a}{du}\right]\frac{du}{dt} + u + Ri_a = E_a. \quad (1.73)$$

We have, for states of equilibrium $du/dt = 0$, $d^2u/dt^2 = 0$ and, hence

$$u + Ri_a = E_a. \quad (1.74)$$

By solving the equation obtained compatibly with the equation of the characteristic of the tetrode $i_a = \varphi(u)$ (a graphical solution is given in Fig. 53) we shall find the state of equilibrium of the electrical circuit and it is

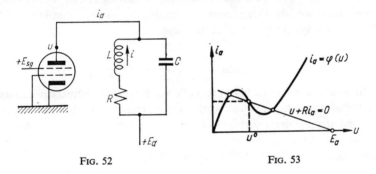

FIG. 52   FIG. 53

clear that, for a given tetrode characteristic, depending on R and E_a, there are either one or three states of equilibrium.

Let us suppose that R and E_a are such that there is a state of equilibrium $(u = u^0, i_a = i_a^0)$ situated on the *decreasing* section of the characteristic $(\varphi'(u^0) < 0)$. Restricting ourselves to the region of small oscillations about this state of equilibrium

$$u = u^0 + v,$$

† As is well known, the anode characteristic of an electronic valve is the relation of the anode current i_a to the anode voltage u for constant voltages at the other electrodes. The anode characteristic of a tetrode has (for sufficiently large screen-grid voltages E_s) a decreasing section owing to the so-called *dynatron effect* occurring in the tetrode for a certain range of anode voltage.

where v is sufficiently small, we can assume the characteristic of the tetrode to be linear

$$i_a = i_a^0 - S_0 v,$$

where $S_0 = -\varphi'(u^0)$ is the absolute value of the slope of the anode characteristic of the tetrode at the working point. We shall obtain for this region of small oscillations the following linear differential equation of the second order:

$$LC\frac{d^2v}{dt^2} + [RC - LS_0]\frac{dv}{dt} + (1 - RS_0)v = 0. \quad (1.76)$$

Just as in the previous cases, our linear equation is applicable for describing oscillations only in a certain bounded region in which the anode voltage u is sufficiently close to the value u^0 (i.e. v is sufficiently small).

If $RS_0 < 1$, then we obtain the "usual" differential equation of the second order, describing a system with an "attractive" force and positive or negative "friction" depending on the sign of the expression $RC - LS_0$. If, however, $RS_0 > 1$ (if the resistance R of the oscillating circuit is sufficiently large) then we obtain an equation, analogous to the equation (1.72) describing a system with "repulsive" force[†].

3. Singular point of the saddle type

Both cases considered so far, that of a pendulum (an ordinary pendulum or Froude's pendulum) close to the upper state of equilibrium and that of a dynatron generator close to a state of equilibrium on the decreasing section of the characteristic (for $RS_0 > 1$) have led us under suitable simplifications to linear differential equations of the form

$$\ddot{x} + 2h\dot{x} - nx = 0 \quad (1.77)$$

where $n > 0$ (for the pendulum $n = g/l$ and for a dynatron oscillator $n = (RS_0 - 1)/LC$ while h can have any sign.

The characteristic equation for the differential equation (1.77) has the form

$$\lambda^2 + 2h\lambda - n = 0$$

and has, independently of the sign of the coefficient h, real roots but of different signs:

$$\lambda_{1,2} = -h \pm \sqrt{h^2 + n}$$

[†] We shall see that in this case the dynatron oscillator has in addition to the one considered, two more states of equilibrium which are stable nodes or foci.

(below we shall denote the positive root by q_1 and the negative one by $-q_2$; q_1 and $q_2 > 0$). Therefore, the general solution for the equation (1.77) can be written in the form

$$x = Ae^{q_1 t} + Be^{-q_2 t}. \qquad (1.78)$$

To find the integral curves on the x, y plane ($y = \dot{x}$ just as before) we shall eliminate the time from the first-order equations

$$\dot{x} = y, \quad \dot{y} = nx - 2hy, \qquad (1.79)$$

equivalent to the equation (1.77), by dividing the second equation by the first:

$$\frac{dy}{dx} = -2h + n\frac{x}{y} \qquad (1.80)$$

As before, the only singular point (the only state of equilibrium) is the origin of the coordinates ($x=0$, $y=0$). We obtain, for the isocline corresponding to a slope of the integral curves $dy/dx = \varkappa$, the equation $-2h + n(x/y) = \varkappa$ or

$$y = \frac{n}{\varkappa + 2h} x. \qquad (1.81)$$

In particular the isocline $\varkappa = 0$ (i.e. $dy/dx = 0$) is the straight line

$$y = \frac{n}{2h} x,$$

and the isocline $\varkappa = \infty$ ($dy/dx = \infty$) the axis of the abscissae ($y=0$). In this case, as also in the case of a singular point of the node type, there are two integral straight lines passing through the singular point, — the straight lines $y = q_1 x$ and $y = -q_2 x$. To determine the form of the other integral curves we can, as before, integrate equation (1.80) by means of a substitution and obtain

$$(y - q_1 x)^{-q_1} = C(y + q_2 x)^{q_2} \qquad (1.82)$$

i.e. an equation which determines a family of curves of the *hyperbolic type*, with asymptotes $y = q_1 x$ and $y = -q_2 x$, which evidently pass through different quadrants. The families of integral curves are shown in Fig. 54 (for $h < 0$) and Fig. 55 (for $h > 0$).

To clarify the form of the integral curves, we can, just as in the case considered earlier of a singular point of the node type, introduce the new variables

$$u = y - q_1 x, \quad v = y + q_2 x$$

and transform the equation (1.82) into

$$v = \frac{C}{u^\alpha},$$

where $\alpha = q_1/q_2 > 0$. This equation determines on the u, v plane a family of curves of the hyperbolic type, the asymptotes of which are the coordinate axes (Fig. 56). Therefore equation (1.82) also determines on the xy plane a family of curves of the hyperbolic type with asymptotes $y = q_1 x$

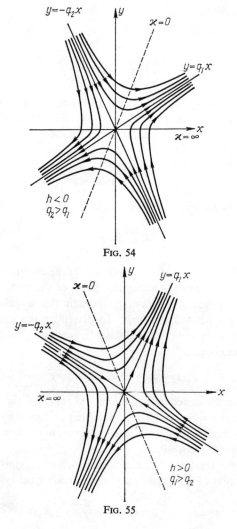

Fig. 54

Fig. 55

and $y = -q_2 x$ which are the straight lines corresponding to the u and v axes.

Thus we have seen that the presence of a resistance, either positive or negative, does not essentially alter the portraits on the phase plane of a

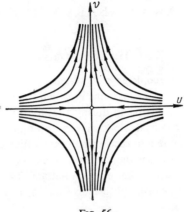

Fig. 56

system with a repulsive force. The singular point is as before a *saddle point*; it is always unstable and all motions in the system will in the end move the system away from the state of equilibrium. After a sufficiently long interval of time has elapsed, the system moves away as far as one chooses and, therefore, will actually overstep the boundaries of the region to which we have restricted our considerations and in which the system can be considered to be linear. In general in all cases of instability of the state of equilibrium, we can, by means of a linear idealization, only describe the behaviour of the system in a certain interval of time and even then only for small initial deviations and provided the system has not had time to overstep the boundaries of the "linear region".

We shall conclude the analysis of linear systems with an observation very important for the later exposition. None of the phase plane portraits considered for various linear systems, except the harmonic oscillator without friction (i.e. except a conservative linear system), has given rise to closed integral curves on the phase plane. All integral curves had branches moving away to infinity. But to periodic processes there must correspond on the phase plane closed integral curves. We can therefore derive from our analysis of linear systems the following important conclusion: *in linear non-conservative systems periodic processes are in general impossible.*

CHAPTER II

NON-LINEAR CONSERVATIVE SYSTEMS

§ 1. Introduction

The analysis of macroscopic physical systems always leads us to consider non-conservative systems, i.e. systems in which the total energy does not remain constant but is dissipated during the motion. In many cases, however, this process of energy dissipation is so slow and affects the character of the motion of the system to such a small extent that a series of questions of interest can be answered without a need to take into account this effect and thus we assume that the sum of potential and kinetic energy remains constant. We arrive, as a result of this idealization, at the notion of conservative systems.

On the other hand, energy is dissipated so rapidly in other systems, that we can no longer neglect this fact and must consider the system as a non-conservative one in order to be able to answer (to a given degree of accuracy) the same questions which could be answered, in the first case, by assuming the system to be conservative. As has been pointed out repeatedly, we classify systems as conservative or non-conservative as a result of an idealization of the properties of real physical systems, the nature of the idealization permissible in each case depending not only on the properties of the system but also on the nature of the questions which are of interest. Thus, for example, to solve the problem of the motion of a pendulum subject to a very small friction (a pendulum suspended on very sharp knife-edges and situated in an enclosure from which air has been removed) for not too-long intervals of time (for example, a hundred periods), we can, with very good accuracy, assume the pendulum to be a conservative system. If, however, we are interested in the motions of the pendulum in the course of a very long interval of time, then, considering the system as a conservative one would no longer be valid. Notwithstanding the fact that energy is dissipated very slowly, it will dissipate so much in a sufficiently long interval of time that the energy remaining in the system will be noticeably less than that possessed by the system at the initial instant of time. In just the same manner the motion of the planets, for example the Earth, can be considered as a conservative motion if the interval of time of interest to us is not too long. For very long intervals involving geological epochs we must take into account in the analysis of Earth motion the so-called tidal friction and can no longer consider the system to be conservative.

The acceptability of a conservative model depends both on the nature of the problem and the properties of the system. The question posed above about the motion of the pendulum during an interval of time equal to a hundred periods cannot be answered at all with a conservative model if the pendulum moves in a medium with large resistance. In this case, even in the period of one swing it will dissipate a considerable fraction of its initial energy and for an interval of time equal to a hundred periods, the sum of kinetic and potential energy of the pendulum cannot be considered even approximately constant.

The consideration of conservative systems not only answers a number of questions but is of special interest here for the following reasons. In the first place we shall extend our understanding of those ideas (the phase plane, the singular points, periodic motions, stability, dependence of a dynamic system on a parameter) which are needed for the analysis of the basic problems of the theory of self-oscillating systems. In the second place, we shall be able to study certain self-oscillating systems only in so far as they are similar to conservative systems.

§ 2. The simplest conservative system

Let us consider the simplest autonomous conservative system with one degree of freedom: the motion of a material point on a straight line under the action of a force depending on distance only. The position of the material point is determined by assigning one number: the abscissa x. The mechanical *state* of the system is determined by the value of x and the velocity $y = \dot{x}$. For the sake of simplicity assume the mass to be equal to unity. The equation of motion of such a system can be written as a second-order equation:

$$\ddot{x} = f(x), \qquad (2.1)$$

where $f(x)$ is a force. In the form of two differential equations of the first order. Equation (2.1) becomes

$$\frac{dx}{dt} = y \quad \text{and} \quad \frac{dy}{dt} = f(x). \qquad (2.2)$$

Unless stated otherwise we shall assume that $f(x)$ is an analytic function over the whole x axis ($-\infty < x < +\infty$), i.e. $f(x)$ is a holomorphic function at each point x†.

The differential equation determining the integral curves on the phase plane is now

$$\frac{dy}{dx} = \frac{f(x)}{y} \quad \text{or} \quad \frac{dy}{dx} = \varphi(x, y), \qquad (2.3)$$

The phase *velocity* of motion v of the representative point can be expressed as

$$v = \frac{ds}{dt} = \sqrt{\left(\frac{dx}{dt}\right)^2 + \left(\frac{dy}{dt}\right)^2} = \sqrt{y^2 + [f(x)]^2}.$$

The velocity of variation of position must be carefully distinguished from the velocity v of the representative point on the phase plane. The first velocity is equal to the ordinate and the second

$$\frac{ds}{dt} = \sqrt{y^2 + [f(x)]^2} = y\sqrt{1 + \left(\frac{dy}{dx}\right)^2} \qquad (2.4)$$

is equal to the length of the normal to the integral curve at a chosen point. From the expression (2.4) it follows that at each point of the phase plane the representative point has a finite non-zero velocity except in states of equilibrium (singular points) at which we have

$$y = 0 \quad \text{and} \quad f(x) = 0.$$

From these conditions all states of equilibrium are situated on the phase plane on the x axis, their abscissae satisfying the equation $f(x) = 0$.

At a point (x_0, y_0) on the phase plane it is necessary to know whether it is always possible to find an integral curve passing through the given point and whether such a curve will be unique. Equation (2.3) determines at each point of the phase plane a unique direction of the tangent, with the exception of the singular points where $y = 0$ and $f(x) = 0$. It will be shown that through each non-singular point there passes one and only one integral curve. We know that such a curve exists and will be unique if the conditions of Cauchy's theorem are satisfied‡. We have considered y as a function of x and had to deal with the equation $dy/dx = f(x)/y = \varphi(x, y)$.

---

† We shall use the following terminology: we shall call the function $f(x)$ analytic in a given region of values of x if it is holomorphic at each point of this region, i.e. if in the vicinity of each point it can be expanded in a power series with a radius of convergence different from zero.

‡ For a formulation of Cauchy's theorem on the existence and uniqueness of the solution of a differential equation (or system of differential equations) see Appendix I.

In this case $d\varphi/dy = -f(x)/y^2$ so $y=0$ is the locus of the points on the phase plane where Cauchy's conditions fail. Now we shall consider x as a function of y. Then the differential equation (2.3) must be written in the form: $dx/dy = y/f(x) = \psi(x,y)$. In this case $d\psi/dx = -yf'(x)/[f(x)]^2$. When $f(x)=0$, there is failure of the continuity conditions and Cauchy's theorem for this equation is not satisfied. The different results obtained by different approaches do not conflict since Cauchy's conditions are only sufficient but not necessary for uniqueness. Hence, we can affirm that through each point of the phase plane there passes one and only one integral curve with the possible exception of the singular points where at the same time $y=0$ and $f(x)=0$. At these points, as we shall see later, for the case of a conservative system, the integral curves either intersect each other and have different tangents, or degenerate into isolated points and have no tangents at all. The phase velocity of the representative point

$$\mathbf{v} = \mathbf{i}y + \mathbf{j}f(x) \tag{2.6}$$

is uniquely determined everywhere and is zero at a singular point only. Hence, by virtue of the continuity assumption, it follows that in the vicinity of a singular point the phase velocity decreases.

Let the conditions of Cauchy's theorem be satisfied for the system of equations (2.2) in a certain region which, assuming $f(x)$ to be analytic for all values of x, is the whole plane. It follows for the dynamic system being considered that the past and the future are uniquely determined by the present, since the value of the initial conditions uniquely determines the motion or the solution of the system (2.2).

Does this still hold true for a motion along integral curves *which intersect each other* at a singular point? We shall show that this is so. The representative point which is found initially at a point on the phase plane which is not a singular point for the equation (2.3), can approach the singular point as t increases without limits in an asymptotic manner only.

A description of the portrait on the phase plane can be arrived at either by means of one equation (2.3) enabling us to investigate the integral curves or by the system of equations (2.2) enabling us to investigate the phase paths. In the second case we obtain effectively the equations of the same integral curves but in the parametrical form $x=x(t)$, $y=y(t)$. This is the law of motion of the representative point along an integral curve on the phase plane. The difference between these two methods of representation of one and the same family of curves is particularly clear in the following example. Let $x=x_0$, $y=y_0$ be the coordinates[†] of a singular

---

† According to equation (2.3), x_0 is a root of the equation $f(x)=0$, and $y_0=0$.

point of equation (2.3), i.e. the coordinates of a point at which the conditions of Cauchy's theorem for the one equation (2.3) fail; then $x = x_0$, $y = y_0$ will be in this case a point at which the conditions of Cauchy's theorem for the system of equations (2.2) are satisfied.

It is easily verified by means of a direct substitution that the system $x = x_0$, $y = y_0$ is a solution of the system of equations (2.2) and is a state of equilibrium. Note that in this case the solution of system (2.2) does not depend on t, so by assigning the initial values $t = t_0$, $x = x_0$, $y = y_0$ we shall obtain for any t_0 a solution in the form $x = x_0$, $y = y_0$.

Let us consider the representative point to move along an integral curve passing through a singular point and to be directed towards the singular point. The velocity of its motion decreases and tends to zero as we approach arbitrarily close to the state of equilibrium. The question of whether the representative point can in a finite time reach the state of equilibrium has been answered but can be considered in a different way. Let us assume that the representative point moving according to the law $x = x(t)$, $y = y(t)$ is found outside a state of equilibrium at the instant of time $t = t_0$ and reaches the state of equilibrium with coordinates $x = x_0$, $y = y_0$ at a certain determined instant of time $t_1 (t_1 > t_0)$, i.e. that $x_0 = x(t_1)$ and $y_0 = y(t_1)$. But then we would obtain two solutions satisfying one and the same initial conditions (for $t = t_1$, $x = x_0$, $y = y_0$), namely $x = x_0$, $y = y_0$ and $x = x(t)$, $y = y(t)$. This is impossible, since at the point x_0, y_0, as we have just observed, the conditions of Cauchy's theorem for the system of equations (2.2) are satisfied.

We shall note that in the sequel we shall encounter systems of equations (similar to (2.2) or of a more general type) for which the conditions of Cauchy's theorem fail at certain points of the phase plane. These derive from dynamical models of real physical systems such that the right-hand sides of these equations of motion are discontinuous (for example, oscillating systems subject to dry friction of the Coulomb type). Our statement on the past being determined by the present is untrue for such models. In these cases too we can, generally speaking, no longer affirm that the system does not reach a state of equilibrium in a finite time. We shall also note that in such cases the singular points of one equation (similar to (2.3)) do not always correspond to states of equilibrium.

§. 3. INVESTIGATION OF THE PHASE PLANE NEAR STATES OF EQUILIBRIUM

If we know the totality of the integral curves on the phase plane for a dynamic system, we are able to comprehend at a glance the whole picture of the possible motions for various initial conditions. The investigation of these integral curves for a conservative system is made very much easier by the fact that equation (2.3) can be easily integrated since the variables are separated. The integral obtained has the form

$$\frac{y^2}{2} + V(x) = h \tag{2.7}$$

where $V(x)$ is such that $V'(x) = -f(x)$ and h is a constant of integration. This equation describes for this case the law of conservation of energy. In fact $y^2/2 = mx^2/2$ is the kinetic energy, $V(x) = \int_0^x f(x)\,dx$ is the work done by the forces acting in the system, or the potential energy of the system, and h is the so-called energy constant depending on the initial conditions.

If, on the other hand, we assign h, then to one and the same value of h there corresponds an infinity of states (x, y) of the system, namely a whole curve $y = \Phi(x)$ on the x, y plane (which can have a series of isolated branches), called the *equi-energy curve*. The representative point will move along one of the branches of this curve if the total energy of the motion is equal to h. It can happen that, having given h, we do not find real values of x and y which satisfy equation (2.7). This means that the energy of our system cannot have this value for any real motion of the system.

We shall assume in the following analysis that $f(x)$ and hence also $V(x)$ are analytic functions over the whole range of x. (A few examples where this does not hold will be considered later).

It will be advantageous to carry out the investigation on the assumption that the function $V(x)$ is given. We shall observe that the values $x = \bar{x}_1, \ldots, x = \bar{x}_i$ (the abscissae of the singular points) where $f(x)$ reduces to zero are points where $V'(x)$ also reduces to zero. Hence these values correspond to extremal values of the potential energy $V(x)$, i.e. to either a minimum, or a maximum, or an inflexion point with horizontal tangent. A classification of the singular points of equation (2.3) can be made in terms of the extremal properties of the potential energy at the singular points. First, a few general observations regarding the type of the integral curves on the phase plane.

(i) The equation (2.7) is not altered if we replace y by $-y$. Hence all curves of this family are symmetric with respect to the x axis.

(ii) The locus of the points where the tangents to the integral curves are vertical is, as is easily seen from (2.2), the x axis except possibly at the singular points.

(iii) The locus of the points where the tangents to the integral curves are horizontal straight lines parallel to the y axis, the equations of which

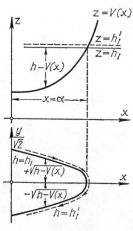

Fig. 57

are of the type $x = \bar{x}_i$, where $\bar{x}_i$ are the roots of the equation $f(x) = 0$, with the possible exception of the points of intersection of these straight lines with the x axis, which also are singular points.

A simple method can be used to construct the integral curves on the phase plane, if the potential energy $V(x)$ is given. We shall use to this end an auxiliary "energy–balance plane" with rectangular axes x and z on which we plot the potential energy $z = V(x)$. Since

$$\frac{y^2}{2} = h - V(x),$$

then, if h, the total energy, is given, the kinetic energy will be represented by the difference of h and $V(x)$. If the kinetic energy is negative then the corresponding motion is impossible.

In Fig. 57 there is shown a section of the energy–balance diagram for a particular form of the curve $z = V(x)$. In order to obtain the integral curve on the phase plane†, which is represented immediately below the energy-balance diagram, we must take the square roots of the differences $h - V(x)$ and then plot them on the phase plane above and below the x axis. It should be remembered in constructing the curve that all the integral curves on the phase plane intersect the x axis with a vertical tangent, provided that they do not intersect it at a singular point‡.

The direction and velocity of motion along the segment of curve thus obtained is easily found by means of the considerations repeatedly employed. On varying h a little, we shall obtain another curve slightly displaced on the phase plane.

This method will now be used to give a picture of the integral curves on the phase plane in the vicinity of states of equilibrium. We shall begin with the case when the state of equilibrium corresponds to a minimum of the potential energy.

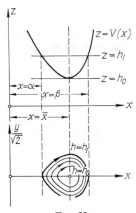

Fig. 58

Let a minimum of potential energy occur when $x = \bar{x}$ and let $V(\bar{x}) = h_0$. The energy-balance diagram in the vicinity of $x = \bar{x}$ will have the form shown in Fig. 58. The integral curve for $h = h_0$ degenerates into an isolated point with coordinates $x = \bar{x}$, $y = 0$. For a value $h = h_1$ close to $h_0 (h_1 > h_0)$ we shall have a closed integral curve. The direction of motion along this closed curve is easily found in the usual manner.

† To simplify the construction, we shall take for the phase plane a scale somewhat different along the ordinate axis, by plotting in the ordinates $y/\sqrt{2}$ instead of y.

‡ The abscissa α of the point of intersection of an integral curve with the x axis is clearly determined by the equation $V(x) = h$. In the case considered in the figure the point $x = \alpha$ is not a singular point, since $f(\alpha) \neq 0$.

As the representative point moves along this closed integral curve, the actual velocity of displacement of the mass reduces to zero twice: for $x=\alpha$ and $x=\beta$ (Fig. 58). However, the phase-velocity of the representative point is never equal to zero, since our curve does not pass through a singular point. The representative point, moving along a closed curve, will return to its initial position after a finite interval of time. Hence it follows that we are dealing with a periodic motion. It is easily seen that intermediate values of $h(h_0 < h < h_1)$ again give closed integral curves which also correspond to periodic motions.

We obtain on the phase plane a whole continuum of closed curves enclosed in each other and encircling the degenerate integral curve $x=\bar{x}$, $y=0$. A singular point of a differential equation with behaviour of the neighbouring integral curves of this type was met when considering linear conservative systems. Such a point, it will be recalled, is termed a centre.

A singular point of the centre type corresponds to a stable state of equilibrium. We shall find now the analytical conditions for the presence of such a singular point and approximate equations of the closed curves in its vicinity.

The expansions in series of $f(x)$ and $V(x)$ about a singular point with coordinate $\bar{x}$ have the form

$$f(x) = a_1(x-\bar{x}) + \frac{a_2}{1 \cdot 2}(x-\bar{x})^2 + \frac{a_3}{1 \cdot 2 \cdot 3}(x-\bar{x})^3 + \ldots, \qquad (2.8)$$

$$V(x) = h_0 - \left\{ \frac{a_1}{1 \cdot 2}(x-\bar{x})^2 + \frac{a_2}{1 \cdot 2 \cdot 3}(x-\bar{x})^3 + \frac{a_3}{1 \cdot 2 \cdot 3 \cdot 4}(x-\bar{x})^4 + \ldots \right\}, \qquad (2.9)$$

where $a_1 = f'(\bar{x}) = -V''(\bar{x})$, $a_2 = f''(\bar{x}) = -V'''(\bar{x})$ etc. Let us transfer the origin of the coordinates to this singular point by putting $x = \bar{x} + \xi$, $y = 0 + \eta$ and let us substitute in the equation of the family investigated (2.7) the expression $V(\bar{x}+\xi)$ in the form of a series. Then the equation of the family of curves can be written thus:

$$\frac{\eta^2}{2} + h_0 - \left\{ \frac{a_1 \xi^2}{1 \cdot 2} + \ldots + \frac{a_k \xi^{k+1}}{1 \cdot 2 \ldots (k+1)} + \ldots \right\} = h. \qquad (2.10)$$

Consider first the case $a_1 \neq 0$. Then in the energy–balance diagram the straight line $z = h_0$ has a contact of the first order with the curve $V(x)$ at the point $x = \bar{x}$. Since $V(x)$ has a minimum for $x = \bar{x}$, then $V''(\bar{x}) > 0$ and

$a_1 < 0$. The curve (2.10) for $h = h_0$ has an isolated singular point at the point $\xi = 0, \eta = 0$.

For a sufficiently small $\alpha = h - h_0$ ($\alpha > 0$) we obtain closed curves, similar to ellipses, since they can be described approximately by the equation

$$\frac{\xi^2}{a^2} + \frac{\eta^2}{b^2} = 1, \tag{2.11}$$

where $b^2 = 2\alpha$ and $a^2 = 2\alpha/|a_1|$.

A motion represented on the phase plane by an ellipse is a harmonic motion. Thus for sufficiently small initial deviations the motion will be close to a harmonic one. As the initial deviations increase, the motion will differ more and more from a harmonic one, and so the period also will vary and be dependent on the value of the initial deviations.

If a_k is the first non-zero coefficient of the expansion (2.8), then, since $a_1 = 0$, $a_2 = 0$, ..., $a_{k-1} = 0$, on the energy–balance diagram the straight line $z = h_0$ has a contact of the k-th order with the curve of the potential energy at the point $x = \bar{x}$. Since $V(x)$ has a minimum for $x = \bar{x}$, then k is necessarily odd and $a_k < 0$. The curve (2.10) has again an isolated point for $h = h_0$ and for a sufficiently small $\alpha = h - h_0 (h > h_0)$ we shall obtain closed integral curves of the type:

$$\frac{\eta^2}{2} + \frac{|a_k|\xi^{k+1}}{1 \cdot 2 \ldots k(k+1)} = \alpha. \tag{2.12}$$

The closed curves around the singular point will not, even in the immediate vicinity of it, resemble ellipses and the corresponding motions will no longer be close to harmonic ones, even for very small deviations.

However, the general topological picture of the motions on the phase plane is not altered: each singular point corresponding to a minimum of potential energy is encircled by a continuum of closed curves, enclosed in each other and corresponding to periodic motions.

We shall consider now the case when a state of equilibrium corresponds to a maximum of the potential energy. The energy–balance diagram is represented in the upper part of Fig. 59 and the phase plane is shown below. We shall obtain on the phase plane for the value $h = h_0$ four branches of a curve with a common point. We shall number these branches I, II, III and IV and shall call them the "arms" of the singular point considered. The character of the arms in the vicinity of a singular point is easily investigated analytically. For values of h close to h_0 ($h_1 > h_0$ and $h_2 > h_0$) we shall have segments of integral curves similar to branches of hyperbolae

(Fig. 59). By varying h between h_1 and h_2 we shall obtain a continuum of intermediate curves.

We shall first consider the motion along the arms (Fig. 59). The representative point, having arrived on the arms II and IV in the vicinity of the state of equilibrium, approaches it asymptotically; if on

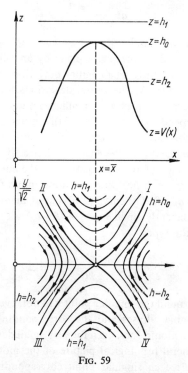

Fig. 59

the other hand it arrives on the arms I and III it moves away from the state of equilibrium. Observe that by replacing t with $-t$, the arms interchange their roles. The motions corresponding to the remaining integral curves have the property that, if the representative point arrives on any of these curves in the vicinity of a state of equilibrium, it will move sufficiently far away from this state of equilibrium in a finite time.

A singular point of a differential equation with such a type of behaviour of the neighbouring integral curves has already been met when considering systems with a repulsive force; it is a singular point of the *saddle* type.

As we have seen when considering a linear system with a repulsive force, a singular point of the saddle type always corresponds to an unstable state of equilibrium. We shall find now analytical conditions for the existence of such a singular point and approximate equations for the integral curves in the immediate vicinity of a state of equilibrium. Proceeding just as in the case of a centre we shall arrive again to the equation (2.10).

$$\frac{\eta^2}{2} + h_0 - \left\{ \frac{a_1 \xi^2}{1 \cdot 2} + \cdots + \frac{a_k \xi^{k+1}}{1 \cdot 2 \cdots (k+1)} + \cdots \right\} = h.$$

We shall begin again with the case of a simple contact between the straight line $z = h_0$ and the curve $z = V(x)$ at the point $x = \bar{x}$, i.e. with the case when $a_1 \neq 0$. Since $V(x)$ has a maximum for $x = \bar{x}$, then $V''(\bar{x}) < 0$

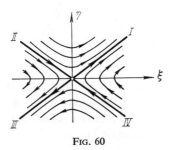

FIG. 60

and $a_1 > 0$. Putting $h = h_0$ we shall obtain the equation of the arms It is easily seen that the origin of the coordinates ($\xi = 0$, $\eta = 0$) is a nodal point of the arms, the equation of the tangents to the arms at this node having the form

$$\eta = +\sqrt{a_1}\xi \quad \text{and} \quad \eta = -\sqrt{a_1}\xi. \qquad (2.13)$$

For small values of $h - h_0 = \alpha$, the family of curves which are close to the singular point behave similarly to hyperbolae determined by the equations

$$\frac{\eta^2}{2\alpha} - \frac{a_1 \xi^2}{2\alpha} = 1. \qquad (2.14)$$

The form of the arms and the character of the integral curves in the immediate vicinity of the singular point is shown in Fig. 60. We have already investigated in the preceding chapter the character of the motion of the representative point along this family of hyperbolae. It is clear that these results are approximately true also for a motion along integral curves in the vicinity of a singular point in this case. As we move away from the singular point the results obtained for a linear system describe the motions less and less accurately.

In the case when the straight line $z = h_0$ and the curve $z = V(x)$ have a contact of the k-th order, $a_1 = 0$, $a_2 = 0$ etc. and only a certain $a_k \neq 0$. Since for $x = \bar{x}$, $V(x)$ has a maximum, then k is necessarily odd and $a_k > 0$. Putting $h = h_0$ we shall again obtain the equation of the arms. It is easily seen that the origin of the coordinates ($\xi = 0$, $\eta = 0$) is the point of mutual contact of the arms (Fig. 61) which in the vicinity of the

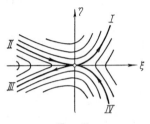

Fig. 61

singular point are close to the curve $\eta^2 = a_k/[1.2\ldots(k+1)]\xi^{k+1}$ and have as common tangent the ξ axis. For small values of α ($\alpha = h - h_0$) we shall obtain the neighbouring integral curves, which in the vicinity of the singular point behave similarly to the curves

$$\frac{\eta^2}{2} - \frac{a_k \xi^{k+1}}{1 \cdot 2 \ldots (k+1)} = \alpha. \tag{2.15}$$

The integral curves differ from hyperbolae even in the vicinity of the singular point and the motions investigated can no longer be described, even approximately, in terms of the picture which was obtained for a linear system with repulsive force. However, as in the preceding cases, the general topological picture of the motions on the phase plane is the same and, therefore, is completely determined by the fact that we are dealing with a maximum of potential energy.

Let us consider now the third and last case when to a state of equilibrium there corresponds on the curve of the potential energy a point of inflexion with horizontal tangent.

The energy-balance diagram and the aspect of the phase plane are shown in Fig. 62. The construction of the integral curves on the phase plane presents no difficulties for all values of h, except the value $h = h_0$ which gives two branches of a curve with the common point $x = \bar{x}, y = 0$. To establish the character of these two arms in the vicinity of a singular point presents some difficulty and to do this an analytical approach is required. Before doing this, which is carried out as in the previous two cases, observe that since we are dealing with a point of inflexion, then necessarily $a_1 = 0$ (since $a_1 = -V''(\bar{x})$) and the first coefficient a_k differing from zero corresponds to an even k. In this case, equation (2.15) takes the form

$$\frac{\eta^2}{2} - \left[\frac{a_k \xi^{k+1}}{1 \cdot 2 \ldots (k+1)} + \frac{a_{k+1} \xi^{k+2}}{1 \cdot 2 \ldots (k+2)} + \cdots \right] = h - h_0. \quad (2.16)$$

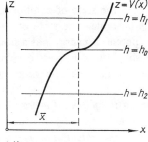

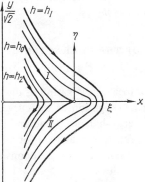

Fig. 62

The curve passing through the singular point will be obtained by putting $h = h_0$. It is easily seen that this curve has at the point $\xi = 0$, $\eta = 0$ a turning point of the first kind. If the representative point arrives on the arm *I* it will tend asymptotically to the state of equilibrium, while if it arrives on the arm *II* it will move away from the state of equilibrium. Clearly the state of equilibrium, just as in the case of a saddle point, is unstable, since the representative point which was found at the initial instant in a finite region $\delta(\varepsilon)$ will necessarily overstep, after a sufficiently long interval of time has elapsed, the boundaries of a finite region ε. The motion along the arm *I* towards the state of equilibrium does not contradict our statement on the instability of states of equilibrium there, just as was the case for analogous motions near a saddle point.

We have considered the three possible cases of extremal values of the potential energy of a system and have related them to the type of singular points and the question of stability of the states of equilibrium[†]. We have verified that in the case of minimum potential energy the state of equilibrium is a singular point of the centre type and is stable; if the potential energy has a maximum, then the state of equilibrium is a singular point of the saddle type and is unstable. The state of equilibrium is unstable also in the case when the potential energy has a point of inflexion. Thus, for this example of the simplest conservative system, two basic theorems on stability can be formulated: first Lagrange's[‡] theorem which states:

If in a state of equilibrium the potential energy is a minimum, then the state of equilibrium is stable, and, secondly, Liapunov's converse theorem:

If in a state of equilibrium the potential energy is not a minimum, then the state of equilibrium is unstable.

§ 4. INVESTIGATION OF THE CHARACTER OF THE MOTIONS ON THE WHOLE PHASE PLANE

Let us pass now from a local investigation of the motions in the vicinity of singular points to an investigation of the curves on the whole plane. We shall again use the energy–balance plane and shall proceed from the assumption that $V(x)$ is a function analytical for all values of x. Later on, when we shall pass to the examples, we shall consider a number of cases when $V(x)$ admits discontinuities.

[†] It is clear that each singular point of the differential equation (2.3) is a singular point in the sense used in differential geometry for the integral curve

$$\frac{y^2}{2} + V(x) = h_0.$$

To a state of equilibrium with a minimum potential energy there corresponds an isolated singular point, to one with maximum potential energy a nodal point (i.e. a point of intersection of the curve with itself) or a point of self-contact, topologically equivalent to a nodal point; finally to a state of equilibrium in which the potential energy has an inflexion point there corresponds a turning point of the first kind.

[‡] This theorem is called sometimes Lejeune-Dirichlet's theorem from the name of the mathematician who first proved it rigorously. This theorem is also true for conservative systems with many degrees of freedom.

Thus we shall assume that on the x, z plane we are given the curve $z = V(x)$, which satisfies the conditions indicated[†], and a straight line $z = h$. We shall construct on the phase plane the totality of all motions which are characterized by the given energy constant. The following basic cases can be met:

(i) The straight line $z = h$ nowhere intersects the curve $z = V(x)$. If in this case the points of the curve $z = V(x)$ lie above the points of the straight line $z = h$, then on the whole phase plane there exist no motions with such

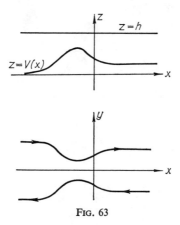

Fig. 63

total energy, since the velocities of such motions would be imaginary. If, however, the straight line $z = h$ lies above the curve $z = V(x)$, then on the phase plane we shall have two branches of phase path symmetrically situated with respect to the x axis (Fig. 63). The representative point, having begun to move from any place on either the upper or lower branch, will continue to move away to infinity. If we replace t by $-t$, i.e. if we make "time to flow in the opposite direction", then the character of the motion of the representative point is not disturbed and only the direction of motion is varied. We shall call such motions (such phase paths), for which the representative point moves for any initial position away to infinity, run-away motions (run-away paths). The motions considered are run-away motions both for $t \to +\infty$ and for $t \to -\infty$. It is easily seen that for values of h close to the one chosen above we shall obtain the same picture and shall have perfectly analogous phase paths.

---

† To simplify the analysis we shall assume that $V(x)$ does not admit points of inflexion at which the tangent is parallel to the x axis.

(ii) The straight line $z = h$ intersects the curve $z = V(x)$ without being tangent to it anywhere (Fig. 64). For the values of x for which $V(x) > h$, there are no phase paths. For the remaining values of x, however, there exist two types of phase paths: they are either branches moving away to infinity (the number of which is not greater than two), or closed branches (the number of which can be arbitrary). The branches moving away to infinity correspond again to run-away motions both for $t \to +\infty$ and for $t \to -\infty$. The closed curves correspond to periodic motions.

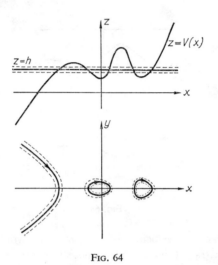

Fig. 64

(iii) The straight line $z = h$ is tangent to the curve $z = V(x)$. Then all phase curves can be divided into the following classes:

(a) *Isolated points*, in the vicinity of which (for a given h) there are no branches of phase curves. These are stable states of equilibrium which we have already discussed. If we vary h, then for an increase of h we shall obtain a closed curve enclosing the isolated point considered, and for a decrease of h we shall not obtain real branches of a curve in the vicinity of the isolated point.

(b) *Isolated finite sections of phase curves*. They can be of two types: either they are simply closed curves corresponding to periodic motions and which we have already discussed or they are phase curves belonging to a number of the so-called *separatrices*, i.e. to the curves passing through singular points. These points of self-intersection, or singular points of the saddle type correspond to the points on the x, z diagram where the straight line $z = h$ is tangent to maxima of the curve $z = V(x)$ (Fig. 65). The separatrices we are discussing now consist of one (in the case of degeneration) or generally speaking, several "links". Each link represents a separate phase path (if it is a terminal one) or consists of two phase paths (if it is an intermediate one)†. The motion along any of these paths is asymptotic towards a state of equilibrium. Such motions are called *limitation* motions. We have already met one example of such a motion when considering a pendulum which is found in the upper position of equilibrium. The motions considered here are limitation motions both

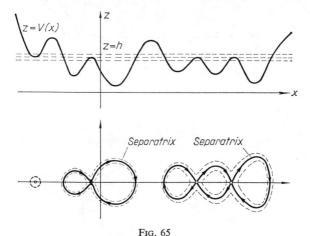

Fig. 65

for $t \to +\infty$ and for $t \to -\infty$. The separatrices are, in a certain sense, exceptional integral curves since to them there correspond points where the straight line $z = h$ is tangent to the curve $z = V(x)$ on the energy-balance plane. Knowledge of them is extremely important for establishing the general picture of integral curves on the phase plane.

As h varies, the character of neighbouring curves will depend essentially on whether we increase or decrease h. For an increase of h we shall obtain an integral curve enclosing the whole separatrix investigated (the whole "chain" of limitation paths). For a decrease of h we shall obtain closed integral curves inside each link (Fig. 65). There follows the important *role of the separatrices as "dividing" curves which separate regions filled with paths of different types.*

---

† The singular points also are separate paths—they correspond to states of equilibrium.

(c) *Infinite sections of phase curves.* There are in this case a number of possible types of curves. In the first place there can be run-away paths of the type which we have already considered in (ii). In the second place there can be a separatrix in the form of an infinite link stretching in one or the other direction. An essential new type of path will be paths which are run-away paths for $t \to +\infty$ and are limitation paths for $t \to -\infty$ or vice versa (Fig. 66). Such paths will also be called separatrices, since on them

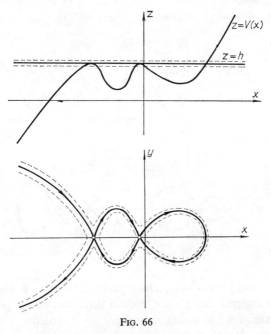

Fig. 66

there are necessarily singular points to which there correspond points of tangency of the straight line $z = h$ with the curve $z = V(x)$ and since, which is very important, the character of neighbouring curves varies substantially depending on whether h increases or decreases.

We shall observe that to the separatrices can be attributed sometimes motions which prove to be run-away motions both for $t \to +\infty$ and for $t \to -\infty$. This can only occur for this example, when the straight line $z = h$ is an asymptote of the curve $z = V(x)$, since then we can obtain a substantial variation of the character of the phase path for a variation of h.

Such an example is illustrated in Fig. 67. For a decrease of h the run-away path turns into a periodic one.

Thus, summarizing the results obtained, we shall give a list of the possible motions:

(i) States of equilibrium;
(ii) Periodic motions;
(iii) Double limitation motions (both for $t \to +\infty$ and for $t \to -\infty$);
(iv) Double run-away motions (both for $t \to +\infty$ and for $t \to -\infty$);
(v) Limitation–run-away motions (limitation motions for $t \to +\infty$ and run-away motions for $t \to -\infty$ or vice versa).

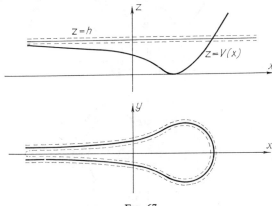

Fig. 67

It can be shown [163] (and we shall return to this) that, for conservative systems, nearly all motions are either periodic motions or doubly run-away motions. If all initial motions on the phase plane are assumed equiprobable, then the probability of occurrence of initial conditions corresponding to motions of the type (i), (iii) and (v) is equal to zero. However, the phase paths corresponding to these motions play an important role on the phase plane; they are separatrices, i.e. curves which separate paths of different types from each other on the phase plane.

Before concluding this chapter we must investigate the distribution law of the singular points on the straight line $y=0$, on which they must fall in this example, and the mutual relationships of singular points and closed phase paths. Both the first and the second problems were solved by Poincaré for the general case of a non-conservative system and we shall give his solution later. For the particular case here an answer to these questions can be obtained from elementary notions. Let us answer the first question. Clearly the maxima and minima of the curve $z - V(x)$ alternate with each other. Hence it follows that singular points of the saddle type and of the centre type also alternate with each other on the axis of the abscisae of the phase plane.

To answer the second question about the mutual relations between closed phase paths and singular points we shall also have recourse to the energy–balance plane (Fig. 68).

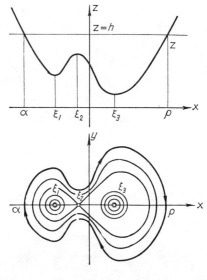

Fig. 68

Consider a closed curve on the phase plane. Then to the points α and ϱ there correspond on the energy–balance plane points at which the straight line $z = h$ intersects the curve $z = V(x)$. Let the function $\Phi(x) = h - V(x)$. For our case $\Phi(\alpha) = 0$, $\Phi(\varrho) = 0$ and $\Phi(x) > 0$ for $\alpha < x < \varrho$. Therefore on the basis of Rolle's theorem we can affirm that there exists such a value $x = \xi \, (\alpha < \xi < \varrho)$ for which $\Phi'(\xi) = 0$ or, equivalently, $V'(\xi) = 0$. We have thus shown that inside a closed phase path there is bound to be at least one singular point or, in other words, that a periodic motion necessarily occurs around a position of equilibrium. From geometrical considerations if this singular point is unique, then it corresponds to a minimum of potential energy and is a singular point of the centre type; if, however, there are several such singular points, then centre and saddle points will always alternate with each other, the number of centre points always being one more than the number of saddle points. We can formulate the following theorem; in the case of a conservative system, *inside a closed phase path there is bound to be an odd number of singular points, the number of centre points being one more than the number of saddle points.*

In concluding this section we shall consider an ordinary pendulum (with one degree of freedom) neglecting frictional forces and without restricting ourselves to small angles of deviation from the vertical. This

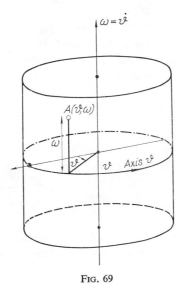

Fig. 69

conservative system oversteps to some extent the limits of our treatment above, since we cannot take a plane as our phase surface. In fact the position of the pendulum is determined by an *angle* (we shall denote it by ϑ) and values of ϑ differing by 2π define one and the same position. Therefore, if we take as the phase surface of the pendulum the usual plane with cartesian coordinates ϑ and $\dot\vartheta$, then the points of this plane $(\vartheta+2k\pi, \dot\vartheta)$, where k is an arbitrary integer, would correspond to the same state as the point $(\vartheta, \dot\vartheta)$, i.e. the requirement for a one-to-one continuous correspondence between the states of the system and the points of its phase surface would fail. This requirement will be satisfied if we take as the phase surface of the pendulum not a plane but a cylinder (Fig. 69)[†].

---

† It is very convenient to represent the phase paths of a pendulum and similar systems not on a cylinder but on the development of a cylinder on a plane in the form of a stripe of width 2π. In this case, however, we must bear in mind that one line of cut

A cylindrical phase surface of a pendulum is clearly connected with the presence of two different types of motions of the pendulum: motions with no revolution round the axis and motions with such revolutions.

The equation of the pendulum can be written in the form

$$I\frac{d^2\vartheta}{dt^2} + mgl \sin \vartheta = 0, \tag{2.17}$$

where I is the moment of inertia, l is the distance from the centre of gravity to the point of suspension and $P = mg$ is the weight of the pendulum (the angle is measured with reference to the downwards vertical). The equation (2.17) can be reduced to a system of two equations of the first order:

$$\frac{d\vartheta}{dt} = \omega, \quad \frac{d\omega}{dt} = -\frac{mgl}{I}\sin\vartheta. \tag{2.18}$$

To obtain the differential equation of the integral curves on the phase cylinder (or on its development) divide the second equation (2.18) by the first one:

$$\frac{d\omega}{d\vartheta} = -\frac{mgl}{I} \cdot \frac{\sin\vartheta}{\omega}. \tag{2.19}$$

On integrating this equation we shall obtain the energy integral (or, in other words, the equation of the family of integral curves of the equation (2.19)):

$$\frac{1}{2}I\omega^2 - mgl \cos\vartheta = h \, (= \text{const}). \tag{2.20}$$

To construct the integral curves use the method indicated in § 3. Having plotted on the auxiliary ϑ, z plane the curve

$$z = V(\vartheta) = -mgl \cos\vartheta \tag{2.21}$$

and having situated the development of the phase cylinder below it, it is easy to construct on the latter the family of integral curves, making use of the fact that, according to (2.20)

$$\omega = \pm\sqrt{\frac{2}{I}} \sqrt{h - V(\vartheta)}.$$

---

of the cylinder is represented on its development by two (boundary) straight lines, and therefore, on using the development of a cylinder as the phase surface we must consider the points of these straight lines (the points of these straight lines having the same values of ϑ) as corresponding to identical states of the system.

Such a construction is given in Fig. 70. The singular point (0,0) is a centre point (to it there corresponds the constant of integration $h = -mgl$). It is surrounded by a continuum of closed phase paths, for which $-mgl < h < +mgl$. These phase paths clearly correspond to periodic oscillations of the pendulum about the lower position of equilibrium with *no turn* round the axis. For a constant of integration $h = +mgl$ an integral curve is obtained

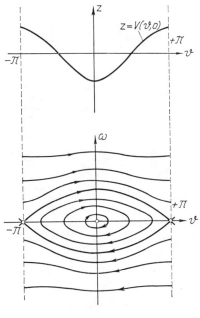

Fig. 70

which passes through a saddle point $(\pm\pi, 0)$, i.e. consists of the saddle point and of its separatrices (to the first there corresponds the upper unstable position of equilibrium, and to the latter, limitation motions of the pendulum for which the pendulum asymptotically approaches, for $t \to +\infty$, the upper position of equilibrium). For $h > +mgl$ we obtain paths situated outside the separatrices and going round the cylinder. Since for each such path the values of ω for $\vartheta \to +\pi$ and for $\vartheta \to -\pi$ coincide, then we can affirm that these paths also are closed (they correspond to periodic rotating motions of the pendulum). By "sticking" together the two boundaries $\vartheta = \pm\pi$ of the development of the cylinder, we shall obtain the phase portrait of the pendulum (Fig. 71). Thus all phase paths of a conservative pendulum, except the singular points (the centre and saddle points), and the saddle separatrices, are closed.

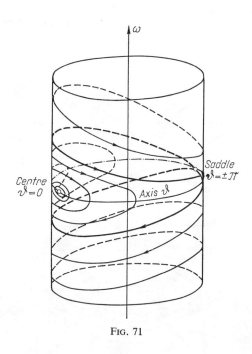

Fig. 71

§ 5. Dependence of the behaviour of the simplest conservative system upon a parameter

We have already stated that a conservative system represents an exceptional system in so far as there exists for it an energy integral. In other words, if we vary in an arbitrary manner the form of the equations of motion then, generally speaking, these equations will no longer satisfy the condition of conservativeness.

However, we shall consider here such variations of the parameters which characterize our system, that the system remains a conservative one. We shall assume that we have one variable parameter only and that only the potential energy of the system depends on this parameter.

The problem is to discover how the aspect of the phase plane varies for a variation of the parameter. We shall not touch upon the important question of how an arbitrary *given* motion will behave for *a sufficiently slow variation of the parameter*[†].

The basic elements which determine the qualitative picture of the integral curves for a conservative system are the singular points and the separatrices. If we know the form of the separatrices (singular points of the saddle type are points of mutual intersection of separatrices) and the relative position of the separatrices and states of equilibrium of the centre type, we can reproduce in a general way the whole portrait of the integral curves.

As the parameter is varied, the integral curves will vary. If we assume that the potential energy is an analytic function of the parameter, then these variations will occur continuously. The general form of the integral curves will undergo *quantitative* variations only, and only for certain special so-called "bifurcation" values of the parameter shall we have *qualitative* variations of the character of the integral curves. The bifurcation values of the parameter will be, in this case, the values of the parameter for which a variation of the number or character of the singular points and separatrices occurs.

More generally, we can give the following definition which is not connected with the conservativeness of the system: a value of the parameter $\lambda = \lambda_0$ will be called by us ordinary if such a finite ε ($\varepsilon > 0$) exists that for all λ satisfying the condition $|\lambda - \lambda_0| < \varepsilon$ we have the same topologic structure in the mapping-out of the phase plane by the integral curves. The other values of the parameters for which this condition is not satisfied will be called *bifurcation or branch* values.

We shall outline in fair detail the theory, developed by Poincaré [182, 183] on the dependence of the states of equilibrium upon a parameter, since we shall need it for the investigation of self-oscillating systems. We shall only illustrate by means of examples other bifurcation cases connected with the dependence of the separatrices upon a parameter.

Let us assume that the potential energy of the system (2.1), and so the force also, is a function of a parameter λ[‡]. The positions of equilibrium ($x = \bar{x}$) are characterized by the fact that for them the force is equal to zero, i.e.

$$f(\bar{x}, \lambda) = 0. \tag{2.22}$$

By solving this equation with respect to $\bar{x}$, the positions of equilibrium of

---

[†] The answer to this question happens to be particularly difficult for a conservative system; in this case the question is studied by the so-called theory of adiabatic invariants.

[‡] We are assuming that $f(x, \lambda)$ is an analytic function of x on the whole range of x and an analytic function of λ for a certain range of values of λ.

this conservative system can be found for some or other value of the parameter λ and the variations of the positions of equilibrium for a variation of λ can be examined.

The dependence of the positions of equilibrium on the parameter can be illustrated graphically by the so-called *bifurcation diagram*, i.e. the curve $f(\bar{x}, \lambda) = 0$ constructed on the λ, $\bar{x}$ plane. Let, for example, the curve have the form shown in Fig. 72. The straight line $\lambda = \lambda_0$ parallel to the axis of the ordinates intersects the curve $f(\bar{x}, \lambda) = 0$ at three points. This clearly indicates that for the given value of the parameter $\lambda = \lambda_0$ the system has three positions of equilibrium $\bar{x} = \bar{x}_1$, $\bar{x} = \bar{x}_2$ and $\bar{x} = \bar{x}_3$.

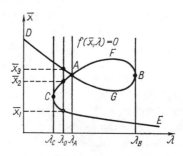

Fig. 72

As λ decreases the positions of equilibrium $\bar{x} = \bar{x}_1$ and $\bar{x} = \bar{x}_2$ approach each other, they merge with each other for $\lambda = \lambda_C$ and then disappear (for $\lambda < \lambda_C$ there is only one position of equilibrium: $\bar{x} = \bar{x}_3$). The value of the parameter $\lambda = \lambda_C$ is thus a bifurcation value. Also the values $\lambda = \lambda_A$ and $\lambda = \lambda_B$, for which there is also a variation in number of equilibrium positions of the system, will be bifurcation values.

On differentiating equation (2.22) with respect to λ we have

$$\frac{df}{dx} \cdot \frac{d\bar{x}}{d\lambda} + \frac{df}{d\lambda} = 0$$

or

$$\frac{d\bar{x}}{d\lambda} = -\frac{f'_\lambda(\bar{x}, \lambda)}{f'_x(\bar{x}, \lambda)}. \qquad (2.23)$$

Hence it follows that in a neighbourhood of a point of the curve $f(\bar{x}, \lambda) = 0$ for which $f'_x(\bar{x}, \lambda) \neq 0$, $\bar{x}$ is a continuous differentiable function of λ.

Therefore, if for a certain value of the parameter $\lambda = \lambda_0$ the system of equations

$$\left.\begin{array}{l} f(x, \lambda) = 0, \\ f'_x(x, \lambda) = 0 \end{array}\right\} \tag{2.24}$$

has no real solutions for x, we can affirm that in a sufficiently small neighbourhood of this value of the parameter $\lambda = \lambda_0$ the abscissae $\bar{x}$ of *all* positions of equilibrium are continuous differentiable functions of the parameter λ and their number cannot vary. Therefore such a value $\lambda = \lambda_0$ is not a bifurcation point (in the sense that as λ passes through the value $\lambda = \lambda_0$ no variation of the number of states of equilibrium occurs).

Let now, at a certain point $(\lambda, \bar{x})$ of the curve $f(\bar{x}, \lambda) = 0$, $f'_x(\bar{x}, \lambda)$ be also zero. If $f'_x(\bar{x}, \lambda) = 0$ and $f'_\lambda(\bar{x}, \lambda) \neq 0$, then the curve has at this point a vertical tangent and when λ passes (in a suitable direction) through the value corresponding to this point, two real roots for $\bar{x}$ merge with each other and then become complex[†]. This is a bifurcation point at which there occurs a variation of the number of the states of equilibrium (the points B and C in Fig. 72). If, however, at the point $(\lambda, \bar{x})$ of the curve $f(\bar{x}, \lambda) = 0$ both $f'_x(\bar{x}, \lambda)$ and $f'_\lambda(x, \lambda)$ vanish, then we are dealing with a "singular" point (in the sense of differential geometry) of this curve. This point (the point A in Fig. 72) will also be a bifurcation point, since for a value of λ corresponding to this point the number of states of equilibrium is always different from that for adjacent values of this parameter.

Thus the points of the curve $f(\bar{x}, \lambda) = 0$ for which $f'_x(\bar{x}, \lambda) = 0$ are bifurcation points and the corresponding values of λ bifurcation values. In addition to these, the parameter λ will have bifurcation values at those values for which the curve $f(\bar{x}, \lambda) = 0$ goes to infinity (this will take place if the curve has unlimited branches with vertical asymptotes).

To each position of equilibrium $x = \bar{x}$ there corresponds a given state of equilibrium $(x = \bar{x}, \dot{x} = 0)$ and a certain singular point on the phase plane. The nature of the singular points, or, which is the same, the stability of the states of equilibrium, is determined by the sign of the derivative $f'_x(\bar{x}, \lambda) = -V''_{xx}(\bar{x}, \lambda)$. Then for

$$f'_x(\bar{x}, \lambda) > 0 \tag{2.25}$$

(minimum potential energy) the state of equilibrium is stable (of the centre type) and for

$$f'_x(\bar{x}, \lambda) < 0 \tag{2.26}$$

[†] We exclude from our considerations the case when the curve $f(\bar{x}, \lambda) = 0$ has at this point a point of inflexion. To exclude this case it is sufficient, for example, to assume that at this point $f''_x(\bar{x}, \lambda) \neq 0$.

(maximum potential energy) the state of equilibrium is a saddle point and is unstable.

It is not difficult to give, following Poincaré, a simple rule for a rapid determination of the stability of a state of equilibrium by means of the bifurcation diagram. Let us mark off (by shading it) the regions of the λ, $\bar{x}$ plane where $f(\bar{x}, \lambda) > 0$ (the curve $f(\bar{x}, \lambda) = 0$ will be clearly their boundary). If a given point $(\lambda, \bar{x})$ lies *above* a shaded region, it will correspond to a stable state of equilibrium[†]. In fact, in the vicinity of this point

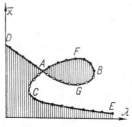

Fig. 73

the function $f(\bar{x}, \lambda)$ decreases as $\bar{x}$ increases (for λ fixed) from positive values, inside the shaded region to zero on the curve $f(\bar{x}, \lambda) = 0$. Consequently $f'_x(\bar{x}, \lambda) < 0$ and this corresponds to a singular point of the centre type and to *stability* of the state of equilibrium. If, however, the point of the curve $f(\bar{x}, \lambda) = 0$ lies *below* a shaded area, then it corresponds to *instability* of the state of equilibrium, since similar arguments show that for it the inequality $f'_x(\bar{x}, \lambda) < 0$ takes place. Following this rule, we find at once that, for example, in Fig. 73 the points of the segments of curve DA, AFB and CE (shown as a thick line with points) correspond to stable states of equilibrium while the points of the segments AGB and AC (shown as a thin line with small circles) correspond to unstable states of equilibrium.

If we now move on the bifurcation diagram along the curve $f(\bar{x}, \lambda) = 0$, the nature of the state of equilibrium, i.e. its stability or instability, will be retained until we arrive at a bifurcation point. It is easily seen that if we continue to move further along the curve, following the direction of the tangent (i.e. ensuring that the tangent rotates continuously), then at the bifurcation point a stable state of equilibrium is changed into an unstable one and vice versa. In Fig. 73 such a change of stability occurs at the points A, B and C.

[†] We are assuming the usual directions of the coordinate axes: the $\bar{x}$ axis is directed upwards and the λ axis to the right.

Thus for a variation of the parameter λ, states of equilibrium can appear or vanish, in a finite region of the phase plane, in pairs only, in which connexion (and this is *a differentiating feature of conservative systems*) a state of equilibrium can vary its stability, for example change from stable to unstable, *only after merging with other states of equilibrium*.

From the point of view of the change of stability, the states of equilibrium of conservative systems form a *closed* system, the behaviour of which for a variation of a parameter can be studied independently of the behaviour of the separatrices.

The values of the parameter for which states of equilibrium merge with one another or go to infinity belong obviously to bifurcation values of the parameter, but, generally speaking, they do not exhaust all bifurcation values, since there can be important variations in the character of the separatrices without a corresponding variation of the number and character of the states of equilibrium. With respect to bifurcation values of this second type we shall make no general statements, but examine them in concrete examples. We shall illustrate in these examples all we have said above in relation to bifurcation values, where there occurs a variation in the character of the states of equilibrium.

1. Motion of a point mass along a circle which rotates about a vertical axis

Let us consider the motion of a mass m along a circle of radius a when this circle rotates about its vertical diameter with constant angular velocity

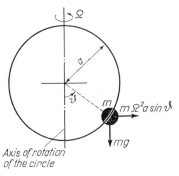

Fig. 74

Ω (Fig. 74). A pendulum oscillating on a rotating platform can serve as an example of such a conservative system.

The position of the point of mass m will be determined by the angle ϑ in a system of coordinates attached to the rotating circle. To write down the equation of motion in a rotating non-inertial system of coordinates using Newton's second law, it is necessary to introduce the inertial forces, in our case the centrifugal force. The moment of the gravitational force with respect to the centre of the circle is equal to $-mga \sin \vartheta$, the centrifugal force is equal to $m\Omega^2 a \sin \vartheta$ and its moment is equal to $+m\Omega^2 a^2 \sin \vartheta \cos \vartheta$. Therefore, neglecting frictional forces, we shall obtain the tfollowing equation of motion of the system considered:

$$I \frac{d^2\vartheta}{dt^2} = m\Omega^2 a^2 \sin \vartheta \cos \vartheta - mga \sin \vartheta, \qquad (2.27)$$

where $I = ma^2$ is the moment of inertia of the material point (with respect o the centre of the circle). If we introduce the dimensionless parameter

$$\lambda = \frac{g}{\Omega^2 a}$$

and a new non-dimensional time

$$t_{\text{new}} = \Omega t$$

(below, differentiation with respect to the new time is indicated by a dot), then equation (2.27) is reduced to the following form, containing a single parameter,

$$\dot{\vartheta} = \omega, \quad \dot{\omega} = (\cos \vartheta - \lambda) \sin \vartheta. \qquad (2.28)$$

To illustrate, using this conservative system (2.28), the qualitative variation of the character of the separatrices for a variation of the parameter and without a variation of the number of singular points, we shall assume that the parameter λ can assume an arbitrary value $-\infty < \lambda < +\infty$ notwithstanding the fact that for the physical system considered the values $\lambda \leqslant 0$ have no physical meaning[†].

Since the position of the material point is uniquely determined by the angle ϑ, we shall again take a *cylinder* as the phase surface of the system considered and represent the phase paths on the development of this cylinder. The equation of the integral curves will be obtained by dividing one of the equations (2.28) by the other:

$$\frac{d\omega}{d\vartheta} = \frac{(\cos \vartheta - \lambda) \sin \vartheta}{\omega}. \qquad (2.29)$$

† Note that the value $\lambda = 1$ is obtained for $\Omega = (g/a)^{1/2}$, i.e. when the angular velocity of rotation of the circle coincides with the angular frequency of small oscillations of the mass m about the lower position of equilibrium for $\Omega = 0$.

The energy integral will be

$$\omega^2 - (\sin^2 \vartheta + 2\lambda \cos \vartheta) = h \tag{2.30}$$

(it is seen at once from (2.30) that the integral curves are symmetric with respect to ϑ and ω).

The positions of equilibrium are determined by the equation

$$f(\vartheta, \lambda) \equiv (\cos \vartheta - \lambda) \sin \vartheta = 0. \tag{2.31}$$

Clearly, for any λ, the system has the positions of equilibrium $\vartheta = 0$ and $\vartheta = \pm \pi$. In addition, for $|\lambda| < 1$ there exist two more positions of equilibrium $\vartheta = +\vartheta_0$ and $\vartheta = -\vartheta_0$, where $\vartheta_0 = \cos^{-1} \lambda$. Figure 75 shows the

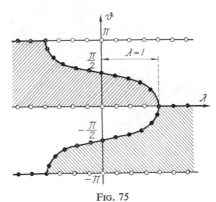

Fig. 75

bifurcation diagram for the positions of equilibrium (the shaded area and the symbols in this figure have the same meaning as in the previous example). Thus for $\lambda > +1$ the system has two singular points: a centre point ($\vartheta = 0$, $\omega = 0$) and a saddle point ($\vartheta = \pm \pi$, $\omega = 0$). For $-1 < \lambda < +1$ it has four singular points: two centre points ($\vartheta = \pm \vartheta_0$, $\omega = 0$) and two saddle points ($\vartheta = 0$, $\omega = 0$) and ($\vartheta = \pm \pi$, $\omega = 0$). Finally, for $\lambda < -1$, it has again two singular points: a centre point ($\vartheta = \pm \pi$, $\omega = 0$) and a saddle point ($\vartheta = 0$, $\omega = 0$).

To determine the separatrices we shall make use of the fact that each separatrix passes through a corresponding singular point of the saddle type at which the constant h of the energy integral can be easily evaluated. The equation of the one of them passing through the saddle point ($\vartheta = \pm \pi$, $\omega = 0$) (this point is a saddle point for $\lambda > -1$), has the form

$$\omega^2 = \sin^2 \vartheta + 2\lambda (\cos \vartheta + 1). \tag{A}$$

The equation of the second, passing through the point (0,0), which is a saddle point for $\lambda < 1$, will be

$$\omega^2 = \sin^2\vartheta + 2\lambda(\cos\vartheta - 1). \tag{B}$$

Both these separatrices, having the form of a "figure of eight", are shown

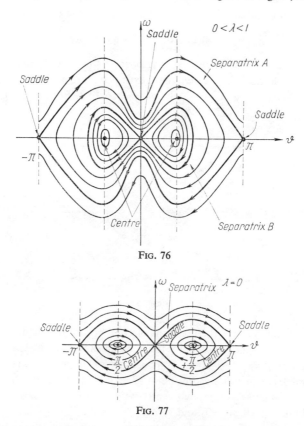

Fig. 76

Fig. 77

in Fig. 76 for the case $0 < \lambda < 1$. For $\lambda = 0$ the two separatrices merge with each other and we obtain the picture shown in Fig. 77. For $-1 < \lambda < 0$ we obtain the same picture as for $0 < \lambda < +1$ but shifted by π along the ϑ axis (Fig. 78). In the case $0 < \lambda < 1$ (Fig. 76), within the outward separatrix (the separatrix A) there are three regions of periodic motions, two simply connected ones (where closed phase paths encircle one of the centres) and one doubly connected (where closed phase paths encircle centre points, the saddle point $\vartheta = 0$, $\omega = 0$ and the separatrix B). The phase paths situated

DEPENDENCE OF THE BEHAVIOUR UPON A PARAMETER

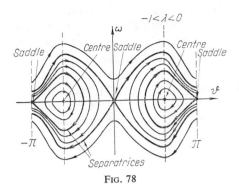

Fig. 78

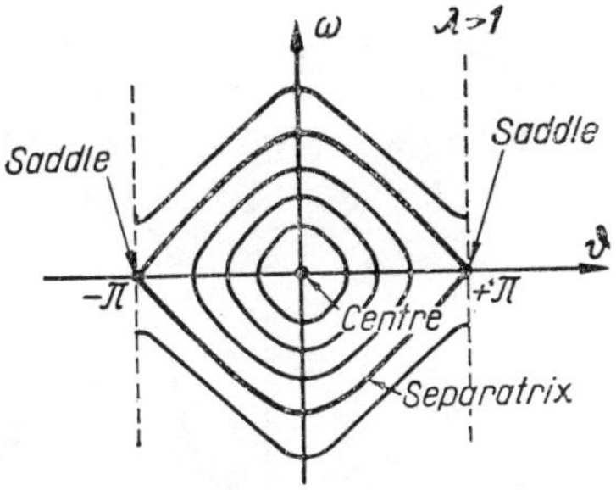

Fig. 79

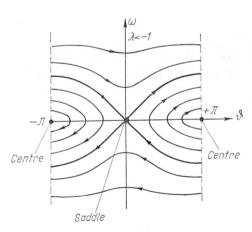

Fig. 80

outside the outward separatrix are always closed and go round the cylinder (this takes place for any λ); they correspond, clearly, to periodic motions of the mass running round the whole circle. Since for $\lambda=0$ the separatrices merge with each other, then for this value of λ, the doubly connected region does not exist. The qualitative topological picture of the phase curves is modified and hence $\lambda=0$ is a bifurcation value. In a similar manner, since for $|\lambda|>1$ a new picture of the integral curve is obtained (Figs. 79 and 80), also the values $\lambda=+1$ and $\lambda=-1$ are bifurcation values of the parameter λ.

2. *Motion of a material point along a parabola rotating about its vertical axis*

We shall consider, as a second example, the following problem.

Let us assume that a material point of mass m can move freely along a parabola determined by the equation $x^2 = 2pz$ and rotating with constant angular velocity Ω about the z axis (Fig. 81). A model for this problem is

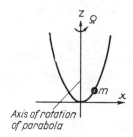

Fig. 81

the well-known demonstration model of a heavy sphere in a cup having the form of a paraboloid of revolution. To construct the equations of motion of the point we could proceed just as in the previous problem, namely, by introducing the inertial forces (i. e. again the centrifugal force) and writing the equation expressing Newton's second law for motions in the x, z plane. We shall proceed, however, somewhat differently, in order to recall to the reader Lagrange's equations of the second kind, which we shall need soon.

Lagrange's equations of the second kind have the form

$$\frac{d}{dt}\left(\frac{\partial L}{\partial \dot{q}_i}\right) - \frac{\partial L}{\partial q_i} = 0, \tag{2.32}$$

where L is the Lagrangean function which for the usual cases of mechanics represents the difference between the kinetic and potential energy, i.e.

$$L = T - V. \tag{2.33}$$

The potential energy of the system is the energy of the material point in the field of the gravitational force, i. e.

$$V = mgz. \tag{2.34}$$

The kinetic energy is made up of the energy of rotation of the body about the vertical axis and the energy of the motion in the x, z plane (since the directions of these motions are orthogonal to each other). Consequently

$$T = \frac{m\Omega^2 x^2}{2} + \frac{m}{2}(\dot{x}^2 + \dot{z}^2) \tag{2.35}$$

Replacing $\dot{z}$ by $x\dot{x}/p$ (from the equation of the parabola) and constructing the Lagrangean function (2.33), we shall obtain:

$$L = \frac{m}{2}\left(1 + \frac{x^2}{p^2}\right)\dot{x}^2 - \frac{m}{2}\lambda x^2, \tag{2.36}$$

where $\lambda = (g/p) - \Omega^2$, and Lagrange's equation is written thus:

$$m\left(1 + \frac{x^2}{p^2}\right)\ddot{x} + m\frac{\dot{x}^2}{p^2}x + m\lambda x = 0$$

or

$$\ddot{x} = -\frac{\left(\lambda + \frac{\dot{x}^2}{p^2}\right)x}{\left(1 + \frac{x^2}{p^2}\right)}.$$

Putting $\dot{x} = y$, we have

$$\frac{dx}{dt} = y; \quad \frac{dy}{dt} = -\frac{\left(\lambda + \frac{y^2}{p^2}\right)x}{\left(1 + \frac{x^2}{p^2}\right)}$$

and, dividing one by the other,

$$\frac{dy}{dx} = -\frac{\left(\lambda + \frac{y^2}{p^2}\right)x}{\left(1 + \frac{x^2}{p^2}\right)y}.$$

The first integral of equation (2.32), the so-called energy integral, has the form: $(\partial L/\partial \dot{x})\dot{x} - L = $ const. (The validity of this is easily verified by a direct substitution). As can be seen from the expression for T and V, the energy integral has the following form:

$$\frac{m}{2}\left[\left(1+\frac{x^2}{p^2}\right)y^2 + \lambda x^2\right] = \text{const.}$$

The equation $f(x, \lambda) = 0$ is thus: $m\lambda x = 0$; therefore $\partial f/\partial x = m\lambda$ and thus $\lambda = 0$ is a bifurcation value of the parameter. For various values of λ the following types of motions and states of equilibrium are obtained:

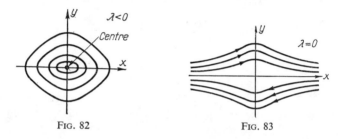

Fig. 82   Fig. 83

(i) $\lambda > 0$ ($\Omega^2 < g/p$): one stable state of equilibrium of the centre type $x = y = 0$. The form of the integral curves on the phase plane (closed curves, enclosed in each other) is shown in Fig. 82. In this case the material point will accomplish oscillations about the state of equilibrium $x=0$, $y=0$.

(ii) $\lambda = 0$ ($\Omega^2 = g/p$): an infinite number of states of equilibrium corresponding to the straight line $y=0$. The form of the integral curves on the phase plane is shown in Fig. 83. The material point will either rest at any place of the parabola or move monotonically in the direction of the initial velocity imparted to it. For t tending to infinity, the velocity tends to zero. A maximum of velocity is obtained at the vertex of the parabola.

(iii) $\lambda < 0$ ($\Omega^2 > g/p$): one unstable state of equilibrium $x=0, y=0$ of the saddle type. It is seen at once from the energy integral that the straight lines $y = \pm(-\lambda)^{1/2}p$ satisfy the equation of motion and are therefore integral curves. These integral "curves" correspond to such motions of the material point along the rotating parabola for which the projection of the velocity of the point on the x axis remains constant. The general form of the integral curves for this case is shown in Fig. 84. If the initial velocity is sufficiently large (larger than $(-\lambda)^{1/2}p$), then the character of the motion is the same as in the case $\lambda = 0$. For smaller initial velocities, the point either moves monotonically on one side and has a minimum value of velocity at the vertex, or, without reaching the vertex, turns back. These two latter types of motion are separated by two integral curves passing through the singular point; the representative point can move along one of them towards the state of equilibrium, approaching it asymptotically.

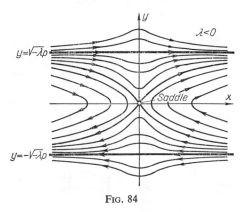

Fig. 84

3. Motion of a conductor carrying a current

We shall consider one last example: an infinite rectilinear conductor, along which there flows an electric current of magnitude I, attracts a conductor AB of length l and mass m along which there flows a current i. In addition the conductor AB is attracted by the spring C (Fig. 85). We shall take as the origin of the x axis the position A_0B_0 of the conductor AB for which the spring is not deformed and shall denote by a the coordinate of the conductor carrying the current I. We shall assume that the conductors are always parallel to each other and that the current is derived from the ends of the conductor AB by means of conducting wires which are perpendicular to the current I. Then the interaction force of the conductors can be taken to be equal to

$$f_1 = \frac{2I \cdot il}{d},$$

where $d = a-x$ (here all quantities are expressed in units of the CGS-system). Assuming the force exerted by the spring to be equal to kx then the total force acting on the conductor AB is

$$f(x, \lambda) = -kx + \frac{2Iil}{a-x} = k\left(\frac{\lambda}{a-x} - x\right), \qquad (2.37)$$

where $\lambda = 2Iil/k$. The equation, connecting the parameter λ and the coordinate of the position of equilibrium $\bar{x}$, has the form

$$f(x, \lambda) \equiv k\left(\frac{\lambda}{a-x} - x\right) = 0$$

or

$$x^2 - ax + \lambda = 0.$$

The bifurcation or branch diagram is shown in Fig. 86. The equation

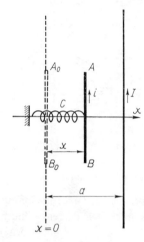

Fig. 85

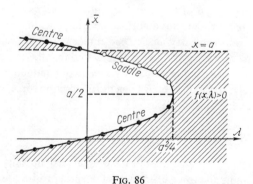

Fig. 86

$f(x, \lambda) = 0$ has a multiple root for $\lambda = a^2/4$. This means that for $x = a/2$ and $\lambda = a^2/4$, not only the function $f(x, \lambda)$ itself but also its derivative

$f'_x(x, \lambda)$ reduce to zero. Consequently, $\lambda = a^2/4$ is a bifurcation value of the parameter. The equations of motion have the form

$$\frac{dx}{dt} = y, \quad \frac{dy}{dt} = -\frac{k}{m}\left(x - \frac{\lambda}{a-x}\right) = \frac{k}{m}\left\{\frac{x^2 - ax + \lambda}{a-x}\right\},$$

hence we obtain

$$\frac{dy}{dx} = \frac{k}{m}\frac{x^2 - ax + \lambda}{(a-x)y}. \tag{2.38}$$

For this system there exists, in addition to the singular points, a "singular"

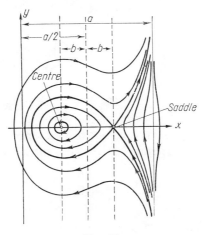

Fig. 87

straight line $x = a$, on which the force $f(x, \lambda)$ goes to infinity[†]. The energy integral has the form

$$\frac{my^2}{2} + \frac{1}{2}kx^2 + k\lambda \ln|a-x| = C. \tag{2.39}$$

(i) Let us first of all consider the case $\lambda < a^2/4$ (Fig. 87). There are two singular points in this case, one of the singular points being a centre point (for which $x = (a/2) - b$ where $b = [(a^2/4) - \lambda]^{1/2}$) and the other a saddle point (for it $x = (a/2) + b$). For both singular points $y = 0$. The tangents

[†] It is evident that we have to exclude from our analysis both the straight line $x = a$ and a small neighbourhood of it, since they correspond to states of the system in which the conductor AB is found at a place occupied by the conductor carrying the current I, and which consequently are not realized in practice in a physical system.

to the integral curves are vertical on the axis $y=0$ (and on the straight line $x=a$, but this case has been excluded from the analysis) and are horizontal on the vertical straight lines, passing through either of the two singular points. The singular straight line $x=a$ is an integral curve and at the same time an asymptote of the remaining integral curves. The equation of the separatrix will be obtained by substituting in the energy integral

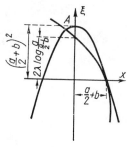

Fig. 88

$y=0$ and $x=(a/2)+b$ (i.e. the condition that the separatrix passes through the saddle point) and determining from this the energy constant C we have

$$C_0 = \frac{k}{2}\left(\frac{a}{2}+b\right)^2 + k\lambda \ln\left(\frac{a}{2}-b\right),$$

and hence the equation of the separatrix has the form

$$\frac{my^2}{2} + \frac{k}{2}\left[x^2 - \left(\frac{a}{2}-b\right)^2\right] + k\lambda \ln\frac{a-x}{\frac{a}{2}-b} = 0.$$

The second root of this equation for $y=0$, i.e. the coordinate of the point of intersection of the separatrix with the x axis, can be found by means of a graphical construction shown in Fig. 88. To do this, we shall plot the two curves

$$\xi = \left(\frac{a}{2}+b\right)^2 - x^2 \quad \text{and} \quad \xi = 2\lambda \ln\frac{a-x}{\frac{a}{2}-b}$$

or $x = a - [(a/2)-b]e^{\xi/2\lambda}$, and find the second point of intersection A of these curves in addition to the point $x=(a/2)+b$, $y=0$. From the phase portrait (Fig. 87) the following conclusions can be drawn: the segment of conductor AB will accomplish oscillations if the initial conditions are such that the representative point is found at the initial instant inside the loop of the separatrix. In particular, for an initial velocity equal to zero, the conductor AB will oscillate if its deviation from the position of equilibrium is not too large.

(ii) Let us consider now the second case $\lambda > a^2/4$. In such a case the equation $f(x, \lambda) = 0$ has no real roots and the system has no singular points

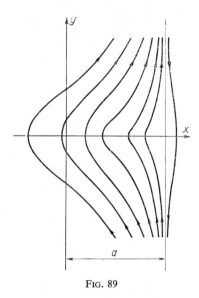

FIG. 89

(states of equilibrium). The behaviour of the integral curves for this case is shown in Fig. 89. For any initial conditions, the conductor AB eventually approaches the straight line x (the position of the long conductor) with a velocity increasing without limits. Oscillating motions are clearly impossible in this case.

(iii) The value $\lambda = a^2/4$ corresponds to a third case, intermediate between the first and the second one. It is readily seen that in the first case, as λ increases, the two singular points draw together and, for $\lambda = a^2/4$, merge with each other. This process of drawing together of the singular points is illustrated in Fig. 90. It is evident that for $\lambda = a^2/4$ there is one singular point only (Fig. 91) of the type corresponding to the case when the potential energy of the system has an inflexion point. Thus this type of singular point can be considered as the result of merging of a centre and

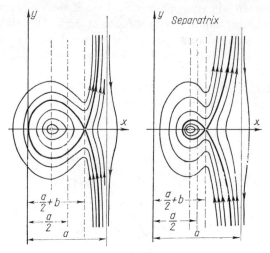

Fig. 90

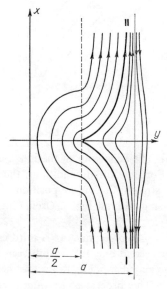

Fig. 91

a saddle point. Such a singular point corresponds to an unstable state of equilibrium. Periodic motions are impossible in this third case also. For all initial conditions the conductor moves with a velocity increasing without limits towards the infinite conductor. The arms *I* and *II* passing through the singular point delimit two types of motions, differing from each other in that, for motions of the first type (at the initial instant the system is found in the region bounded by the straight line $x = a$ and the arms *I* and *II*) the conductor *AB* (Fig. 85) moves towards the straight line $x=a$ without passing through the position of equilibrium. For the second type of motion (at the initial instant the system is found outside the region bounded by the arms *I* and *II* and the straight line $x = a$) the conductor *AB* always passes through the position of equilibrium.

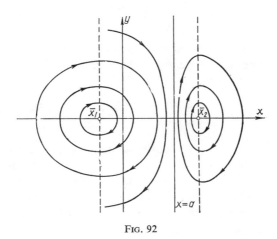

Fig. 92

(iv) Let us consider finally, the last case $\lambda < 0$ (the variation of the sign can be achieved by reversing the direction of one of the currents *i* or *I*). In this case there always exist two real roots of the equation $f(x, \lambda) = 0$. One of these two roots $\bar{x}_{1,2} = (a/4) \pm [(a^2/4 - \lambda)]^{1/2}$ is always negative and the other is larger than *a*. Both states of equilibrium ($\bar{x}_1 < 0$ and $\bar{x}_2 > a$) are centres and are stable. The remaining integral curves are closed and encircle either the first or the second state of equilibrium, the line which separates these two types of closed curves being the "singular" line $x=a$ (Fig. 92). Thus in the case $\lambda < 0$ all motions of the conductor *AB are* oscillating (periodic).

The examples given clarify to a sufficient extent the question of the dependence of the character of the motions in a conservative system upon a parameter and we shall now pass to further questions arising in the analysis of conservative systems.

§ 6. THE EQUATIONS OF MOTION

Up to now we have only considered the simplest conservative systems. We shall pass now to more complicated ones.

To construct the equations of motion of complicated conservative systems it is advantageous to use Lagrange's equations of the second kind. Denoting by $L(q,\dot{q})$ a certain function (let it be a single-valued function of the coordinate q and the velocity $\dot{q}$) which we shall call the Lagrangean function, we have Lagrange's equation in this form:

$$\frac{d}{dt}\left(\frac{\partial L}{\partial \dot{q}}\right) - \frac{\partial L}{\partial q} = 0. \tag{2.40}$$

The equation is invariant with respect to any transformation of the coordinate q. This means, in other words, that, putting $q = f(\varphi)$ we shall again obtain an equation of the type (2.40), i.e.

$$\frac{d}{dt}\left(\frac{\partial L}{\partial \dot{\varphi}}\right) - \frac{\partial L}{\partial \varphi} = 0.$$

This invariance property of Lagrange's equation offers a great advantage, since it allows the possibility of writing at once the equations of motion for any system of coordinates we may choose, if the Lagrangean function of the system is known. For the usual conservative mechanical systems (on condition that the reference system is inertial) the Lagrangean function represents the difference between kinetic and potential energy. In a similar manner in the simplest electrical system the Lagrangean function represents the difference between magnetic and electric energy, if we choose as the generalized coordinates the integrals of the independent circuit currents $q = \int i\,dt$ (in circuits which comprise capacitors the q's are evidently the charges on these capacitors). There is particular advantage in using Lagrange's equations for constructing the equations of motion of electromagnetic systems[†].

It must be observed, however, that the Lagrangean function cannot always be represented as the difference of two energies. In such cases it is not always possible to indicate in advance a "physical" rule for constructing the Lagrangean function. Then it is only possible, by suitably choosing the function L, to reduce in a purely analytical manner the equations of

---

[†] The equations of motion of electric and electromagnetic systems, written in the form of Lagrange's equations of the second kind, are often called Lagrange-Maxwell's equations.

motion to the required form. It is known that in the case of an autonomous conservative system, it is possible to write for Lagrange's equations, the so-called "energy integral" which is expressed thus:

$$\dot{q}\frac{\partial L}{\partial \dot{q}} - L = h. \tag{2.41}$$

It is easily verified by simple differentiation that the time derivative of the left-hand side of this equation reduces to zero owing to Lagrange's equation. However, the expression (2.41) does not always denote the energy of the system in the physical meaning of this word. Introducing together with the coordinate q a second variable $p = \delta L/\delta \dot{q}$, the moment or impulse, and constructing the function

$$H = p\dot{q} - L = H(p, q), \tag{2.42}$$

the so-called Hamilton's function, we can reduce the equation of motion (2.40) to two differential equations of the first order:

$$\frac{dq}{dt} = \frac{\partial H}{\partial p}; \quad \frac{dp}{dt} = -\frac{\partial H}{\partial q}, \tag{2.43}$$

which bear the name of *Hamilton's equations*. The Hamiltonian form of the equations of motion offers considerable advantage for the analyses of a series of problems in mathematics, astronomy and physics. A number of methods of integration of the equations of motion are associated with this form.

Hamilton's equations are invariant not only with respect to transformations of the variables but also with respect to the so-called *canonical* transformations, which play an important role in the study of conservative systems with many degrees of freedom.

We shall observe that the "energy integral" for Hamilton's equations can be written at once

$$H(p, q) = h = \text{const.} \tag{2.44}$$

There are two examples which illustrate the application of Lagrange's and Hamilton's equations:

1. Oscillating circuit with iron core

Let us consider as a first example of a non-linear conservative system an electrical circuit which comprises an inductive coil containing an iron core [197] (Fig. 93). In order that we may consider the system as a conservative one, we must neglect the resistance of the circuit and the hysteresis loss. If we assume that the whole magnetic flux Φ passes through all w turns of the coil, then on the basis of Kirchhoff's law we obtain for the intensity of current i in the circuit the following equation

$$\frac{1}{C}\int i\,dt + w\frac{d\Phi}{dt} = 0, \qquad (2.45)$$

where Φ is a certain function of i, non-linear owing to the presence of an iron core in the coil. The approximate form of the function $\Phi(i)$ for an iron core is shown in Fig. 94.

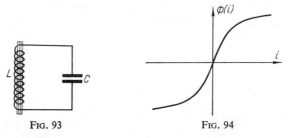

Fig. 93   Fig. 94

Equation (2.45) can be easily reduced to Lagrange's form. To do so we shall replace i with $\dot{q}$, where q is the charge on the plates of the capacitor, and introduce the notation

$$L = L(q, \dot{q}) = w\int \Phi(\dot{q})\,d\dot{q} - \frac{q^2}{2C}. \qquad (2.46)$$

In this case

$$\frac{\partial L}{\partial \dot{q}} = w\Phi(\dot{q}) \quad \text{and} \quad \frac{\partial L}{\partial q} = -\frac{q}{C},$$

and equation (2.45) takes Lagrange's form

$$\frac{d}{dt}\left(\frac{\partial L}{\partial \dot{q}}\right) - \frac{\partial L}{\partial q} = 0.$$

We can write, for Lagrange's equation, the energy integral

$$\dot{q}\frac{\partial L}{\partial \dot{q}} - L = h.$$

In this case this energy integral has the form

$$h = w\Phi(\dot{q})\dot{q} - w\int \Phi(\dot{q})\,d\dot{q} + \frac{q^2}{2C} = \text{const.} \qquad (2.47)$$

Here h does actually represent the total energy of the system. In fact, the electrostatic energy in a capacitor is $V = q^2/2C$, and the magnetic energy in the coil is defined as the work done against the self-induction e.m.f., i.e. can be expressed as

$$T = w \int \frac{d\Phi(i)}{dt} \dot{q} \, dt = w \int \dot{q} \, d\Phi(\dot{q}) \tag{2.48}$$

or as a result of integration by parts

$$T = w\Phi(\dot{q})\dot{q} - w \int \Phi(\dot{q}) \, d\dot{q}. \tag{2.49}$$

Consequently $h = T+V$. However, in this case, $L \neq T-V$ and we have an example of the fact that the Lagrangean function is not always equal to the difference between kinetic and potential energy.

Introducing the new variable $p = \partial L/\partial \dot{q} = w\Phi(\dot{q})$ we can reduce our equation to Hamilton's type. Hamilton's equation is now

$$H(p, q) = \int \Psi(p) \, dp + \frac{q^2}{2C}, \tag{2.50}$$

where $\Psi(p)$ is the function obtained by solving the expression $p = w\Phi(\dot{q})$ with respect to $\dot{q}$. The character of the function $\Phi(\dot{q})$, as is seen from the curve of Fig. 94, is such that the transformations $p = w\Phi(\dot{q})$ are continuous and single-valued in both directions. Hamilton's equations will be written

$$\frac{dp}{dt} = -\frac{\partial H}{\partial q} = -\frac{q}{C}; \quad \frac{dq}{dt} = \frac{\partial H}{\partial p} = \Psi(p).$$

The manner of behaviour of the integral curves on the phase plane is determined by the energy integral, which using (2.47)–(2.49) can be written in the form

$$w \int \frac{\partial \Phi}{\partial \dot{q}} \dot{q} \, d\dot{q} + \frac{q^2}{2C} = \text{const.} \tag{2.51}$$

This expression is analogous to that which we have obtained in the analysis of the examples of conservative systems in § 5, except for the difference that q and $\dot{q}$ appear to have been interchanged. We can, therefore, make the same statements in relation to the character of the integral curves as we have made for the simplest conservative systems. The expression under the sign of integral is always greater than zero and therefore $\int [\partial \Phi(\dot{q})/\partial \dot{q}] \dot{q} \, d\dot{q}$ is a positive function, the derivative of which reduces to zero at the point $\dot{q}=0$ only. Therefore, $\dot{q}=0$ corresponds to a minimum of energy and the singular point $q=0$, $\dot{q}=0$ is a centre point; it corresponds to a stable state of equilibrium. All integral curves are closed curves, contained in each other and encircling the singular point. More precisely we shall be able to determine the character of the integral curves on assigning a known analytical expression for the function $\Phi(i)$. In the absence of superimposed magnetization this function is fairly well approximated by the expression

$$\Phi(i) = A \arctan \frac{wi}{S} + B \frac{wi}{S},$$

where A, B and S are positive constants. Using this expression we obtain

$$\frac{\partial \Phi}{\partial \dot{q}} = \frac{Aw}{S} \frac{1}{1 + \frac{w^2 \dot{q}^2}{S^2}} + B \frac{w}{S}$$

and then

$$w \int \frac{\partial \Phi}{\partial \dot{q}} \dot{q} \, d\dot{q} = \frac{Aw^2}{S} \int \frac{\dot{q} \, d\dot{q}}{1 + \frac{w^2 \dot{q}^2}{S^2}} + B \frac{w^2}{S} \int \dot{q} \, d\dot{q}.$$

The first integral is evaluated by means of the substitution $\dot{q}^2 = z$ and we obtain finally

$$\frac{AS}{2} \ln \left(\frac{\dot{q}^2}{2} + \frac{S^2}{2w^2} \right) + \frac{Bw^2}{2S} \dot{q}^2 + \frac{q^2}{2C} = \text{const.}$$

This equation determines a family of curves of elliptical type. In Fig. 95 there is shown a family of these curves plotted for certain particular values of the parameters.

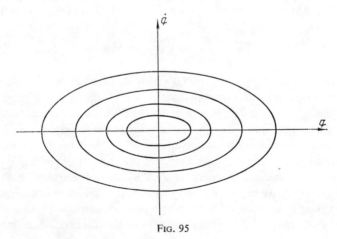

Fig. 95

2. Oscillating circuit having a Rochelle salt capacitor

We shall consider as the second example of a non-linear conservative system an oscillating circuit with a capacitor in which the dielectric is Rochelle salt (Fig. 96), which has electric properties analogous to the magnetic properties of iron. A non-linear dependence between the electric

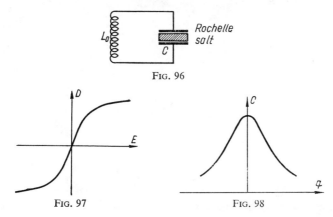

Fig. 96

Fig. 97 Fig. 98

induction D and the field intensity E is typical of Rochelle salt (Fig. 97), as a consequence of which the capacitance proves to be a function of the charge or the voltage. We call the ratio of the charge on the plates of the capacitor to the potential difference produced by this charge, the capacitance $C(q)$ of such a capacitor. The dependence of the capacitance $C(q)$, thus defined, upon the value of the charge on the plates of the capacitor is shown approximately in Fig. 98.

Neglecting the ohmic resistance and the loss due to hysteresis, we have, owing to the fact that C is a function of q, a non-linear conservative system. According to Kirchhoff's law we can write for the circuit[†]

$$L_0 \frac{d^2q}{dt^2} + \frac{q}{C(q)} = 0. \tag{2.54}$$

[†] Note that the capacitance could have been defined in a different manner also, for example as $C_1(q) = dq/du$, where u is the potential difference. In this case the differential equation of a circuit, containing a capacitor with a Rochelle dielectric, would have taken the form

$$L_0 \ddot{q} + \int \frac{dq}{C_1 q} = 0. \tag{2.54a}$$

We can use indifferently either of equations (2.54) and (2.54a), since both give one and the same dependence of $\ddot{q}$ upon q. Apparently, the second definition $C_1 = dq/du$ is the most advantageous for a non-autonomous system.

This equation can also be reduced easily to Lagrange's form. Let us introduce the following function of the state of the system:

$$L = L(q, \dot{q}) = \frac{L_0 \dot{q}^2}{2} - \int \frac{q\,dq}{C(q)}. \tag{2.55}$$

In this case

$$\frac{\partial L}{\partial \dot{q}} = L_0 \dot{q}, \quad \frac{\partial L}{\partial q} = -\frac{q}{C}$$

and equation (2.55) can be written in Lagrange's form

$$\frac{d}{dt}\left(\frac{\partial L}{\partial q}\right) - \frac{\partial L}{\partial \dot{q}} = 0.$$

The energy integral is

$$\frac{L_0 \dot{q}^2}{2} + \int \frac{q}{C(q)}\,dq = h = \text{const.} \tag{2.56}$$

It is easily seen that h is the total energy of the system, since the energy of the charge of the capacitor is equal to the work of the current which charges the capacitor

$$V = \int \frac{q}{C(q)} \dot{q}\,dt = \int \frac{q\,dq}{C(q)}. \tag{2.57}$$

In addition, however, in contrast to the preceding one, the Lagrangean function $L = T - V$, i.e. is equal to the difference between the magnetic and electrostatic energy of the system. Equation (2.55) can be easily reduced, by means of the substitution $p = \partial L / \partial \dot{q} = L_0 \dot{q}$, to Hamilton's form in a similar manner as was done in the preceding example.

Equation (2.56) is the equation of the family of integral curves on the phase plane q, $\dot{q}$. Since the function $\int q\,dq/C(q)$ has a minimum for $q=0$, then $q = 0$, $\dot{q} = 0$ is a singular point of the centre type, corresponding to a stable state of equilibrium.

In order to determine more precisely the form of the integral curves we must define in some manner or other the form of the function $C(q)$. In the general case, if together with the variable voltage across the plates of the capacitor there exists a certain constant voltage (by analogy with superimposed magnetization we shall call this constant voltage "superimposed electrification"), then the capacitance of the capacitor will no longer vary equally on both sides of the point $\bar{q}=0$. Assuming this, we can approximate to the dependence between C and q in a certain bounded region of values of q by means of the following expression:

$$C(q) = \frac{C_0}{1+C_1q+C_2q^2}$$

(a graph of this function $C(q)$ is shown in Fig. 99). Substituting the expression for $C(q)$ in expression (2.56), we shall obtain

$$\frac{L_0\dot{q}^2}{2}+\frac{q^2}{2C_0}+\frac{C_1q^3}{2C_0}+\frac{C_2q^4}{2C_0} = \text{const.} \qquad (2.58)$$

This equation determines a family of closed curves, contained in each other (Fig. 100). The lack of symmetry of these curves with respect to the i

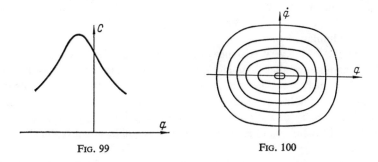

Fig. 99   Fig. 100

axis is caused by the presence of the term $C_1q^3/2C_0$ in the equation of the family. However, this term appears as a result of superimposed electrification. In the absence of this, $C(q)=C(-q)$ and the asymmetry of the integral curves vanishes. We shall obtain a family of curves of an elliptical type and only those of these curves for which the term q^4 is important will differ noticeably from ellipses.

§7. General properties of conservative systems

In the theory of oscillations of conservative systems with one degree of freedom, we are interested primarily in the stationary states, namely the states of equilibrium and periodic motions. All remaining motions, as we have verified in the analysis of the simplest conservative systems, either go to infinity or tend to a state of equilibrium of the saddle type (limitation motions). We have already considered in detail the states of equilibrium in the simplest conservative systems. We must now examine in greater detail the character of the periodic motions possible in the simplest conservative systems.

1. Periodic motions and their stability

Periodic motions in conservative systems are characterized first of all by the property that they do not occur in isolation. If for $h = h_0$ we have a closed path on the phase plane, i.e. a periodic motion, then, as we have seen, this closed curve is surrounded by a continuum of near-by paths, obtained for neighbouring values of h. Periodic motions occur as continua of periodic motions and fill whole regions of the phase plane, one closed path round another. This means, physically, that if one periodic motion is possible, then an infinite number of them are possible, and the maximum swings and maximum values of the velocities can vary continuously, depending on the initial conditions, within certain finite or infinite limits.

The fact of the very existence of periodic motions is not enough and we must enquire whether these motions are stable. Therefore we must formulate rigorously the concept of stability of motion as we have done for the concept of stability of positions of equilibrium. We shall take the definition of stability of a motion given by Liapunov and which corresponds fully to the definition of stability of states of equilibrium, given in Chapter I, § 3.

To periodic motions there correspond motions of the representative point along a given closed phase path. We shall enclose this point in a small region ε which moves together with the representative point. *If for a given region ε which can be as small as we choose, we can indicate such a region $\delta(\varepsilon)$ that every representative point situated at the initial instant in this region $\delta(\varepsilon)$ never oversteps the boundaries of the region ε then the motion considered is stable in the sense of Liapunov.* More intuitively we can formulate this condition of stability in the following manner. Let the motion be subjected to a small perturbation such as an instantaneous jump in an arbitrary direction. Then the representative point will be displaced and will then continue its motion along another path. Let us indicate that after this jump the representative point is "blackened" (Fig. 101). Then the initial perturbed motion whose stability we are investigating, i.e. the motion which would occur if there had been no jump, will be represented by the motion of a white representative point. The motion after the jump, the perturbed one, is represented by the motion of a black representative point. If the black point, which is found at the initial instant (i.e. immediately after the jump) sufficiently close to the white one, always remains sufficiently close to the latter, then the representative white point is stable in the sense of Liapunov[†].

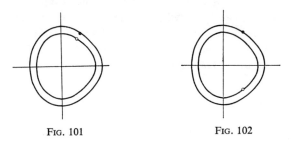

Fig. 101   Fig. 102

It is easily seen that, generally speaking, a motion in a conservative system is *unstable* in the sense of Liapunov, since in the general case the period of rotation of the representative point along different integral curves is different. As a result of this the black and the white points, however small the initial separation, will depart more and more from each other and after a certain number of periods we obtain the picture shown in Fig. 102. Then, however, they will again begin to approach each other. Still, for an arbitrarily small (but different from zero) initial distance, the distance between them will not always be less than a given number. The distance between the black and the white points will not increase in comparison with the initial distance in the special case when the black and white points move along the same path, i.e. when the perturbation is such that the representative point jumps back on the same path (we shall observe incidentally that this special type of perturbation can be realized in practice only for a well-determined relation between the variation of the coordinate and the variation of velocity). However, this case does not contradict our statement on the instability of motion, since we have been discussing a region $\delta(\varepsilon)$ whereas a segment of path does not represent such a region.

Periodic motions in a conservative system will be stable in the sense of Liapunov only when there is *isochronism*, i.e. when the period of rotation is one and the same for different paths. However, even in this case, we shall have no absolutely stable closed paths, i.e. no such paths to which the representative point will again tend asymptotically after a sufficiently

[†] See also the definition of stability of a periodic motion in the sense of Liapunov in Chapter V, Sections 6, 7, where there are outlined analytical methods for investigating stability. These, however, are only suitable for non-conservative systems.

small perturbation. This last type of path is, generally speaking, impossible in conservative systems with one degree of freedom. We shall only meet them in the analysis of non-conservative systems. Although, as we have just seen, periodic motions in conservative systems are unstable in the sense of Liapunov, they still possess a certain type of stability. More precisely, a sufficiently close path will always lie, in its entirety, in the immediate vicinity of the chosen one. Such a type of stability bears the name of *orbital* stability; this stability plays an essential role in the general theory of the behaviour of integral curves.

2. Single-valued analytic integral and conservativeness

So far we have considered such conservative systems for which Hamilton's equations are valid. At the same time, from the point of view of the character of the phase plane or phase surface, and also therefore from the point of view of the character of the possible motions in the system, it would be natural to include in conservative systems certain systems, for which Hamilton's equations are not valid. We shall give, therefore, a more general definition of conservative systems and shall establish certain properties of conservative systems which derive from this definition.

To each dynamic system there corresponds a certain phase surface, uniquely and fully determined topologically, with a grid of phase paths situated on it, such that to each point of the phase surface there corresponds a fully determined state of the system and vice versa; this correspondence is mutually continuous and single-valued. We shall consider as *a necessary attribute of a conservative system the existence of a single-valued integral of the form*

$$F(u, v) = C, \tag{2.59}$$

where u and v are the coordinates which determine the position of the point on the phase surface. We shall assume, to avoid superfluous discussions, that the function $F(u, v)$ is a *single-valued analytic function;* but, according to the nature of the problem, it cannot be identically equal to a constant quantity. Considering C as a third coordinate measured along the normal to the phase surface, we can interpret equation (2.59) as the equation of a certain new surface constructed above the phase surface. The surface constructed in this manner has the property that lines of equal level (the level being measured along the axis C) are integral curves. In the case when the phase surface is a plane, the lines of equal level, i.e. the integral curves, represent intersections of the surface $F(u, v) = C$ with the plane parallel to the phase plane and determined by the equation $C=C_0$, where C is the coordinate and C_0 is a constant (Fig. 103).

Knowing one such surface, we can construct an infinite number of them. In fact we are interested exclusively in the lines of equal level themselves while their relative height does not interest us at all. Hence, we can vary according to an arbitrary law the "scale" of the C axis, by compressing or

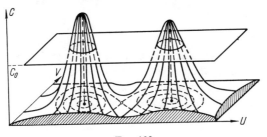

Fig. 103

stretching it, or different parts of it in an arbitrary manner. We shall obtain new surfaces, all of them having the property that lines of equal level are integral curves. In the language of analysis this indicates the evident fact, that if $F(u, v) = C$ is an integral of a certain equation, then $\Phi[F(u, v)] = C$ will also be an integral of this equation.

The singular points of the equilevel curves correspond to singular points of the system of integral curves. Thus isolated points of the equilevel curves correspond to a centre, nodal points to a saddle and cuspidal points to singular points obtained from the merging of a centre with a saddle point. The differential equation of the integral curves, as follows from equation (2.59), has the form

$$\frac{dv}{du} = -\frac{\frac{\partial F}{\partial u}}{\frac{\partial F}{\partial v}}. \qquad (2.60)$$

Singular points correspond to those values of u and v, for which $\partial F/\partial u$ and $\partial F/\partial v$ reduce simultaneously to zero. It can happen that $\partial F/\partial u$ and $\partial F/\partial v$ reduce simultaneously to zero not only at isolated points but also along a certain analytical curve. We shall show that this curve is necessarily an integral curve, i.e. that the points of this curve satisfy an equation $F(u, v) = $ const. Let us assume that the curve we are discussing is given in the parametric form

$$u = u(s), \quad v = v(s).$$

Then
$$\frac{dF}{ds} = \frac{\partial F}{\partial u}\frac{du}{ds} + \frac{\partial F}{\partial v}\frac{dv}{ds}$$
or, since $\partial F/\partial u = 0$ and $\partial F/\partial v = 0$, then
$$\frac{dF}{ds} = 0.$$
hence
$$F = \text{const},$$

i.e. $F(u, v)$ retains a constant value along the curve. It is easily seen that such a case occurs if the corresponding curve of equal slope consists of points at which the tangent plane is parallel to the phase surface as, for example, when the surface $F(u, v) = C$ has the form of a crater the edges of which are situated at a constant level (Fig. 104). No singular point can

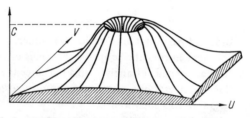

Fig. 104

be of such a type that through it there passes an infinite number of integral curves, which fill completely a certain part of the plane, since in this case all curves would have to be of one level. Owing to the analyticity of $F(u, v)$ this function would be everywhere constant which contradicts the condition assumed. Hence we conclude that singular points in conservative systems cannot be nodes or foci. It can be shown by a completely analogous reasoning that in a conservative system there cannot be a closed integral curve around which other integral curves spiral. Finally, it can be affirmed that if there exists one closed path, then there must exist a whole continuum of such curves, completely filling a portion of the plane. This follows directly from the fact that the integral curves represent constant-level lines of the continuous surface $F(u, v) = C$. Therefore there cannot exist one isolated closed path, since if one constant-level line on a continuous surface is closed then all nearby constant-level lines would also be closed.

Let us investigate now the motions along these paths as a function of time. Since equation (2.60) is the result of eliminating time from the equations of motion, then, in order to return to the equations of motion we must bear in mind that, together with the elimination of time, a certain function $S(u, v) = 1/Q(u, v)$ occurring in both equations, may have been eliminated. Therefore the equations of motions can be written in the general form thus:

$$\left.\begin{aligned}\frac{du}{dt} &= \frac{\partial F}{\partial v}\frac{1}{Q(u,v)} = U(u, v), \\ \frac{dv}{dt} &= -\frac{\partial F}{\partial u}\frac{1}{Q(u,v)} = V(u, v).\end{aligned}\right\} \quad (2.61)$$

These more general equations of a conservative system bear the name of Pfaff's equations. We shall assume for $S(u, v)$ that this is a singlevalued analytic function on the whole u, v plane and that it does not reduce to zero for any values of u and v.

We could make more general assumptions on the function $S(u, v) = 1/Q(u, v)$; for example we could allow this function to reduce to zero or lose its holomorphicity along isolated curves. The corresponding equations are met with in practice fairly often as ideal models of real systems and these models (for example, when the isolated curves mentioned above coincide with phase paths) undoubtedly deserve to be attributed to the class of conservative systems. However, we shall not carry out the investigation and classification of such pathological cases here and shall restrict ourselves to only a few remarks concerning terminology, and shall consider one example (No. 6 of this section).

It is easily seen that in the particular case

$$Q(u, v) = 1$$

we obtain equations of Hamilton's type

$$\frac{dv}{dt} = -\frac{\partial H}{\partial u}, \quad \frac{du}{dt} = \frac{\partial H}{\partial v}.$$

Here, according to a notation generally adopted, F is denoted by H. Hamilton's equations, as we have seen, have a single-valued integral $H = \text{const}$, usually but not always representing the energy integral.

Equations (2.61) are equivalent to the equation

$$V(u, v)\, du - U(u, v)\, dv = 0,$$

which, as is known, always admits an integrating factor. Therefore, any dynamic system described by two differential equations of the first order can be reduced formally to the type (2.61). However, not all systems described by these equations are conservative. The reason for this is that in the case when a conservative system is described by equations of the type (2.61), the functions F and Q are subject to prescribed conditions (single-valuedness, analyticity, etc.). When Hamilton's equations are considered in classical mechanics, then H occurring in them is the energy, and therefore these conditions are automatically verified.

We shall observe that if a dynamic system is defined by differential equations of the general form

$$\frac{du}{dt} = U; \qquad \frac{dv}{dt} = V,$$

then no general methods exist which would enable us to establish whether the system described by these equations is conservative or not. Often the non-conservativeness of a system can be established at once, for example, by showing the existence of absolutely stable or unstable states of equilibrium. In general, however, we can establish the conservative nature of the integral curves only by finding by some method a single-valued integral of the system.

3. Conservative systems and variational principle

A feature of conservative equations is their variational behaviour.

As is known, Hamilton's equations can be obtained by means of Hamilton's variational principle

$$\delta \int_0^{t_1} L \, dt = \delta \int_0^{t_1} (p\dot{q} - H) \, dt = 0. \qquad (2.62)$$

By making use of the fact that δq reduces to zero for $t=0$ and $t=t_1$, expression (2.62) can be brought to the form

$$\int_0^{t_1} \left\{ \left(\dot{q} - \frac{\partial H}{\partial p} \right) \delta p + \left(\dot{p} + \frac{\partial H}{\partial q} \right) \delta q \right\} dt = 0$$

hence, by virtue of the "basic lemma" of variational calculus, we obtain Hamilton's equations

$$\dot{q} = \frac{\partial H}{\partial p}, \quad \dot{p} = -\frac{\partial H}{\partial p}.$$

We shall consider now a more general variational principle and assume that the integrand in the integral being varied is a linear combination of the general form

$$\delta \int_0^{t_1} \{X\dot{x}+Y\dot{y}+F\}\, dt = 0,$$

where X, Y and F are single-valued analytic functions of x and y only.

In this more general case, the variational equations or the equations of motions have the form

$$Q(x, y)\frac{dy}{dt} = \frac{\partial F}{\partial x} \quad \text{and} \quad Q(x, y)\frac{dx}{dt} = -\frac{\partial F}{\partial y},$$

where $Q(x, q) = X'_y - Y'_x$.

These are Pfaff's equations, the well-known and most general form of equations which describe a conservative system.

4. Integral invariant

We shall introduce now the concept of integral invariant. Let us consider first the corresponding problem in a general form, independently of conservativeness. We shall then employ the results obtained for conservative systems.

Let a certain dynamic system be defined by equations of the general type

$$\dot{x} = P(x, y), \quad \dot{y} = Q(x, y). \tag{2.63}$$

We shall interpret the representative points on the phase plane as particles of a certain two-dimensional "liquid" and the phase paths as lines of current of a stationary flow of this "liquid" on the phase path, assuming that nowhere are there sources or sinks of "liquid". Let $\varrho(x, y)$ be the "density" of this imaginary liquid. Let us consider the set of representative points (i.e. the totality of "liquid particles") which filled at the instant of time $t=0$ a certain region $G(0)$ on the phase plane.

The "mass" of the "liquid film" considered is evidently expressed by the integral

$$I(0) = \iint_{G(0)} \varrho(x_0, y_0)\, dx_0\, dy_0$$

(x_0, y_0 are the coordinates of the representative points at $t=0$). Our "liquid" flows on the phase plane, following lines of current defined by the equations of motion (2.63) or by their solution

$$x = x(t; x_0, y_0), \quad y = y(t; x_0, y_0) \tag{2.64}$$

(since x_0, y_0 are the initial values of the coordinates of the representative points, then, evidently, $x(0; x_0, y_0) = x_0$ and $y(0; x_0, y_0) = y_0$). Along these paths there will move the liquid "particles" which fill at the instant $t=0$ the "volume" $G(0)$. Let us denote by $G(t)$ the region which will be filled by this set of "particles" at the instant of time t. The "liquid mass" in this new "volume" will be

$$I(t) = \iint\limits_{G(t)} \varrho(x, y) \, dx \, dy \qquad (2.65)$$

and must be equal to $I(0)$ if our interpretation of the motion of the representative points on the phase plane as a stationary flow of certain "liquid" with density $\varrho(x, y)$ and without sources or sinks is correct, since for the "liquid" the law of conservation of "mass" must be satisfied. More precisely, such an interpretation of the motion of the representative points is only possible in the case when such a function $\varrho(x, y)$ (the "density" of the liquid) can be chosen so that the "mass of liquid" remains unvaried during the motion. In this case the equation of motion (2.63) admits a two-dimensional positive *integral invariant*. Thus the expression (2.65) is an integral invariant (the function $\varrho(x, y)$ is called the phase density of the integral invariant[†]) if for any initial region $G(0)$, $I(t) \equiv I(0)$ or, which amounts to the same,

$$\frac{d}{dt} \iint\limits_{G(t)} \varrho(x, y) \, dx \, dy \equiv 0$$

for any region of integration $G(t)$.

Let us find the condition that the function $\varrho(x, y)$ must satisfy in order that expression (2.65) be an integral invariant of the equations (2.63). The basic difficulty in differentiating the integral (2.65) with respect to time is that the region $G(t)$, over which integration is carried out, varies with time. In order to surmount this difficulty we shall change, under the sign of integral, from the variables x, y to the variables x_0, y_0 by means of the Jacobian[‡]

[†] Below we shall assume the function $\varrho(x, y)$ to be positive definite

$$0 \leq \varrho(x, y) < M,$$

where M is a certain constant number. In addition, this function must not be identically equal to zero in any finite region.

[‡] We shall prove that the Jacobian is not equal to zero (in this case only will the transformation of variables introduced be a one-to-one transformation). Differentiating $D(t; x_0, y_0)$ with respect to time, we shall obtain

$$\frac{\partial D}{\partial t} = \begin{vmatrix} \frac{\partial \dot{x}}{\partial x_0} & \frac{\partial y}{\partial x_0} \\ \frac{\partial \dot{x}}{\partial y_0} & \frac{\partial y}{\partial y_0} \end{vmatrix} + \begin{vmatrix} \frac{\partial x}{\partial x_0} & \frac{\partial \dot{y}}{\partial x_0} \\ \frac{\partial x}{\partial y_0} & \frac{\partial \dot{y}}{\partial y_0} \end{vmatrix} = D\{P'_x(x, y) + Q'_y(x, y)\},$$

since, considering $\dot{x}$ and $\dot{y}$ as functions of x and y according to the equations of motion

GENERAL PROPERTIES OF CONSERVATIVE SYSTEMS

$$D = \frac{\partial(x, y)}{\partial(x_0, y_0)} = \begin{vmatrix} \dfrac{\partial x}{\partial x_0} & \dfrac{\partial y}{\partial x_0} \\ \dfrac{\partial x}{\partial y_0} & \dfrac{\partial y}{\partial y_0} \end{vmatrix} = D(t; x_0, y_0) \neq 0. \quad (2.67)$$

After passing to the new variables x_0, y_0, we have

$$I(t) = \iint\limits_{G(0)} \varrho(x, y) \frac{\partial(x, y)}{\partial(x_0, y_0)} \, dx_0 \, dy_0, \quad (2.68)$$

where now, by x and y we shall understand the functions $x(t; x_0, y_0)$ and $y(t; x_0, y_0)$, i.e. the solutions of the differential equations (2.63). Also

$$\frac{dI(t)}{dt} = \iint\limits_{G(0)} \frac{\partial}{\partial t} [\varrho D] \, dx_0 \, dy_0,$$

since the region of integration does not now depend on time. Since $dI(t)/dt$ must be identically equal to zero for *any* region of integration $G(O)$, the expression under sign of integral must also be identically equal to zero (for any x_0, y_0), i.e.[†]

$$\frac{\partial}{\partial t} \{\varrho(x, y) \cdot D(t; x_0, y_0)\} \equiv 0. \quad (2.69)$$

Since

$$\frac{\partial D}{\partial t} = D\{P'_x + Q'_y\}$$

(2.63) and x and y as functions of t; x_0; y_0; according to the solution (2.64) of these equations we have

$$\frac{\partial \dot{x}}{\partial x_0} = \frac{\partial P(x, y)}{\partial x_0} = P'_x(x, y) \frac{\partial x}{\partial x_0} + P'_y(x, y) \frac{\partial y}{\partial x_0}$$

and analogous expressions for $\partial \dot{x}/\partial y_0$, $\partial \dot{y}/\partial x_0$ and $\partial \dot{y}/\partial y_0$. Integrating with respect to time (for assigned x_0, y_0) we obtain

$$D(t; x_0, y_0) = D(0; x_0, y_0) e^{\int_0^t \{P'_x(x, y) + Q'_y(x, y)\} \, dt}$$

where $x = x(t; x_0, y_0)$ and $y = y(t; x_0, y_0)$. Therefore

$$D(0; x_0, y_0) = \frac{\partial(x_0, y_0)}{\partial(x_0, y_0)} = 1,$$

$$D(t; x_0, y_0) = e^{\int_0^t \{P'_x(x, y) + Q'_y(x, y)\} \, dt} \neq 0.$$

[†] We write the derivatives with respect to time as partial derivatives, since x, y and $D(t; x_0, y_0)$ depend not only on time but also on x_0, y_0.

then

$$\frac{\partial}{\partial t}[\varrho D] = D\frac{\partial \varrho}{\partial t} + \varrho\frac{\partial D}{\partial t} = D\left\{\frac{\partial \varrho}{\partial x}P + \frac{\partial \varrho}{\partial y}Q + \varrho\frac{\partial P}{\partial x} + \varrho\frac{\partial Q}{\partial y}\right\} =$$
$$= D\left\{\frac{\partial}{\partial x}(\varrho P) + \frac{\partial}{\partial y}(\varrho Q)\right\}.$$

and the condition (2.69), since $D \neq 0$, reduces to the condition

$$\frac{\partial}{\partial x}(\varrho P) + \frac{\partial}{\partial y}(\varrho Q) \equiv 0 \qquad (2.70)$$

for any x and y.

It is easily shown that Hamilton's equations always admit an integral invariant with constant phase density (which without detracting from generality can always be put equal to unity). In fact, in the case of Hamilton's equation, putting $x = q$, $y = p$ and $\varrho = 1$, the condition (2.70) can be reduced to the condition

$$\frac{\partial}{\partial q}\left\{\frac{\partial H}{\partial p}\right\} + \frac{\partial}{\partial p}\left\{-\frac{\partial H}{\partial q}\right\} = 0, \qquad (2.71)$$

which is satisfied identically by virtue of the interchangeability of the operations of differentiation.

Thus the phase area (the "two-dimensional phase volume") is an integral invariant for Hamilton's equation. This statement, first proved by Liouville, bears the name of *Liouville's theorem*.

To understand Liouville's rather abstract theorem, we shall consider examples in which the invariance of the phase area is easily established directly.

Example I. Harmonic motion:

$$\frac{dp}{dt} = -q, \quad \frac{dq}{dt} = p,$$

$$p = a\cos(t+\varphi), \quad q = a\sin(t+\varphi).$$

It is easily seen that during a certain time each radius-vector

$$\mathbf{r} = \mathbf{i}a\sin(t+\varphi) + \mathbf{j}a\cos(t+\varphi),$$

characterizing a state of the system will be rotated by one and the same angle. Any figure will simply rotate without changing form and consequently without changing area (Fig. 105).

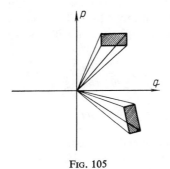

FIG. 105

Example II. Motion under the action of a constant force:

$$\frac{dp}{dt} = -q, \quad \frac{dq}{dt} = p,$$

$$p = p_0 - gt, \quad q = q_0 + p_0 t - \frac{gt^2}{2}.$$

If at the instant $t=0$ we isolate on the phase plane the square of vertices (q_0, p_0), (q_0+a, p_0) (q_0, p_0+a) and (q_0+a, p_0+a) then as t increases this square will be distorted more and more (Fig. 106) but the area of the

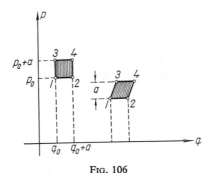

FIG. 106

figure will remain constant, since the sides parallel to the q axis, i.e. the ones connecting the points of equal initial velocity p_0 will move parallel to themselves and in addition the distance between them and their length will remain equal to *a*. We shall obtain, instead of a square of side *a*, a parallelogram of base *a* and height *a*, i.e. of the same area as the square.

If we use the phase plane, not with the variables p and q, but with the variables q and $\dot{q}$, i.e. if we proceed not from Hamilton's equations but from Lagrange's equations, then Liouville's theorem will no longer apply. However, generally speaking, we shall still have an integral invariant. In fact

$$\iint_G dp\,dq = \iint_{G^*} \begin{vmatrix} \dfrac{\partial p}{\partial \dot{q}} & \dfrac{\partial p}{\partial q} \\ \dfrac{\partial q}{\partial \dot{q}} & \dfrac{\partial q}{\partial q} \end{vmatrix} dq\,d\dot{q} = \iint_{G^*} \dfrac{\partial^2 L}{\partial \dot{q}^2} dq\,d\dot{q}.$$

Thus with the variables q and $\dot{q}$ the phase area is no longer constant but is equal to $\partial^2 L/\partial \dot{q}^2$. Therefore, in order that Lagrange's equations may admit an integral invariant, it is sufficient that $\partial^2 L/\partial \dot{q}^2$ be finite and of constant sign, for example positive. In practical cases this condition is usually satisfied.

The more general equations of conservative systems, i.e. Pfaff's equations (2.61), also have an integral invariant and in fact the integral invariant with phase density $Q(u, v)$ is

$$I = \iint_{G(t)} Q(u, v)\,du\,dv,$$

since the condition for this expression to be an integral invariant of equations (2.61)

$$\frac{\partial}{\partial u}(QU) + \frac{\partial}{\partial v}(QV) = \frac{\partial}{\partial u}\left\{\frac{\partial F}{\partial v}\right\} + \frac{\partial}{\partial v}\left\{-\frac{\partial F}{\partial u}\right\} \equiv 0$$

is satisfied identically by virtue of the interchangeability of differentiation. It is easily seen that the expression $Q\Phi(F)$ where Φ is any function and F is the left-hand side integral of the conservative system (2.59), can be used to transform the integral invariant in the form of a phase density. In fact $\Phi(F)$ is a constant of the motion; it is therefore quite clear that, if $\iint Q\,du\,dv$ is an integral invariant, then $\iint Q\,\Phi(F)\,du\,dv$ will also be an integral invariant. It can be shown that this is the general form of an integral invariant. In other words, the ratio of two different expressions for phase density integral invariants, equated to a constant quantity, is always an integral of the system.

Let us turn now to the physical interpretation of the representative points as "particles of a two-dimensional liquid" and of their motions as a stationary flow of such a "liquid" (without sources or sinks). As has already been shown at the beginning of this paragraph, such an interpretation is only possible in the case of existence of an integral invariant; its phase density $\varrho(x, y)$ can be taken as the "liquid density" and the integral invariant itself will express the law of conservation of the "liquid mass".

Let us consider a "liquid" current comprised between two sufficiently close phase paths, i.e. a strip of current (Fig. 107) which is analogous to a tube of current in hydrodynamics. By virtue of the law of conservation of the "liquid mass" the "liquid" current through one cross-section of this strip (for example through the cross-section l_1 must be equal to the current through any other cross-section of the same strip of current (for example, through the cross-section l_2). If we denote by w_1 and w_2 the phase velocities on these cross-sections†, i.e. the velocity of flow of the "liquid" on these cross-sections of the strip of current, then, evidently‡,

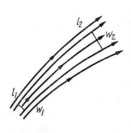

Fig. 107

$$\varrho_1 w_1 l_1 = \varrho_2 w_2 l_2,$$

where ϱ_1 and ϱ_2 are the densities of the "liquid" in the first and second cross-sections of the strip of current.

Thus, if we know the phase paths and the phase densities we can determine the relative distribution of phase velocities along the paths, i.e. we can determine the phase velocity at any point of a given phase path, if this is known for any one point of this path.

From the existence of an integral invariant with phase density, limited and of determined sign, there follows once again the impossibility of the existence in conservative systems of states of equilibrium of the node or focus type and of closed phase paths which are approached asymptotically by nearby phase paths (i. e. limit cycles). In fact, assuming the contrary to be true, we shall have on the phase plane "strips of current", the cross-

† The cross-sections of the "strip of current" must be sufficiently small, so that on each cross-section the phase velocities can be assumed to be the same.

‡ It is easily seen that the liquid current through any closed contour is equal to zero. In fact, the flow of liquid into a closed contour is determined, as is well known, by the integral

$$\oint_\Gamma \varrho(\dot{y}\, dx - \dot{x}\, dy) = \oint_\Gamma \varrho Q\, dx - \varrho P\, dy = \iint_S \left[\frac{\partial}{\partial x}(\varrho P) + \frac{\partial}{\partial y}(\varrho Q) \right] dx\, dy$$

and is equal to zero by virtue of condition (2.70) (the last integral is obtained by using Green's first formula; S denotes the two-dimensional region lying inside the contour Γ).

sections of which will decrease without limits as these "strips of current" approach states of equilibrium of the node or focus types or a limit cycle. However, phase velocities will remain finite (and in approaching states of equilibrium will even tend to zero) and consequently, as we approach states of equilibrium or limit cycles, the phase density must increase without limits, which is impossible.

5. *Basic properties of conservative systems*

Let us consider now in somewhat greater detail the motions which are possible in a conservative system. We shall begin with the positions of equilibrium. The positions of equilibrium are determined by putting equal to zero the right-hand sides of the equations (2.61):

$$\frac{\partial F}{\partial u} = 0 \quad \text{and} \quad \frac{\partial F}{\partial v} = 0.$$

These positions of equilibrium either correspond to singular points of the system or form lines of equilibrium (in the case of existence of common factors of $\partial F/\partial u$ and $\partial F/\partial v$), which then, as we have seen, necessarily coincide with integral curves.

We have seen that singular points cannot be points to which there converge an infinite number of paths, filling continuously a region of the plane and, therefore, positions of equilibrium cannot be absolutely stable.

Closed paths correspond to periodic solutions: we have already seen that if there is just one such periodic solution then other motions cannot spiral about and towards it (and also cannot spiral away from it).

In other words, in a conservative system there cannot exist absolutely orbitally stable paths either. If in a conservative system there is one closed path, then there must exist an infinite number of them, continuously filling a certain region of the phase plane, these closed paths being enclosed in one another. The physical meaning of this is that if one periodic motion is possible then an infinite number of them is possible, the maximum possible swings and maximum possible velocities being of arbitrary values within limits depending on the initial conditions. It is easily seen that the period of the oscillations are, generally speaking, different for different maximum swings and so depend on the initial conditions. Systems admitting isochronous oscillations i.e. oscillations the period of which does not depend on the maximum swing, are an exceptional case; the case already considered in Chapter I of the harmonic oscillator is an example. In the case when the phase surface is topologically equivalent to a plane, inside closed paths there are bound to be one or several singular points (if such singular point is only one, this is necessarily a centre). Oscillations in the system are only accomplished about one or several positions of equilibrium, some of which must necessarily be stable. If, however, for example, the phase surface is a cylinder, then there can be closed paths which do not surround singular points and paths which go round the cylinder; in such systems there can occur periodic motions along closed curves which do not surround singular points. As an example we can cite the rotation of an undamped pendulum with a large initial velocity. Further, there are possible integral curves with one or several singular points; the first correspond to doubly limitation motions, i.e. to motions which for t tending to $+\infty$ and t tending to $-\infty$ tend to one and the same position of equilibrium. The second ones correspond to limitation motions which for $t \to +\infty$ tend to one position of equilibrium and for $t \to -\infty$ to another. There are also possible limitation–run-away motions which, for t tending to infinity from one direction, tend to a position of equilibrium and, for t tending to infinity from the other direction, move away to infinity and finally doubly run-away motions which move away to infinity for both ways of tending to infinity by t.

The following terminology will be used. If the equations of motion of a system (defined by two autonomous equations of the first order) admit a single-valued analytic integral, then we shall say that the structure of the integral curves on the phase plane for such a system has a *conservative character*. A system having a single-valued analytic integral will be referred to as a *conservative* system if it has an integral invariant, satisfying the following requirement: (i) the region of integration $G(t_0)$ can be chosen arbitrarily, provided that it is not intersected by certain isolated curves, (ii) for a further variation of t, $G(t)$ does not tend to zero, remaining in a finite part of the phase plane.

In conclusion we shall point out one more property, which has been mentioned before, namely the instability of conservative systems in relation to a variation of the form of the differential equations. It can be shown that the slightest variation of the form of a differential equation will, generally speaking, substantially modify the whole picture on the phase plane and destroy the conservativeness of the system. To illustrate this thesis, which will be formulated rigorously later and established for the general case, the following example can be adduced. The equation of a harmonic oscillator $\ddot{x} + \omega_0^2 x = 0$ may be considered as a particular case of the equation of a linear oscillator

$$\ddot{x} + h\dot{x} + \omega_0^2 x = 0.$$

For $h=0$ we obtain a conservative system having one singular point (a centre point) and integral curves in the form of a family of ellipses enclosed in one another. For $h \neq 0$, but arbitrarily small and equivalent to an arbitrarily small variation of the form of the differential equation, the system will no longer be conservative for the singular point becomes a focus and the closed integral curves vanish and spirals appear.

Thus a conservative system is a very special case of a dynamic system, a case which is only realized in practice for critical values of certain system parameters (and therefore hardly realizable in practice). A variation of these parameters usually alters the form of the differential equations and gives rise to failure of the conservative property[†].

6. Example. Simultaneous existence of two species

The examples considered so far were either mechanical or electrical systems for which the question of conservativeness is answered at once by physical arguments. However, there are possible cases where simple arguments for answering the question of the conservativeness of the system can no longer be applied. A necessary criterion of conservativeness is the differentiating feature cited in the preceding section for the existence of a single-valued analytic integral of the form $F(u, v) = C$. An example of such a system for which the question of conservativeness cannot be answered in advance comes from biology and is due to Volterra [175, 199, 45]. Consider the simultaneous existence of two species of animals (for example two types of fishes). The first species feeds upon the products of the medium, products which we shall assume to be always present in sufficient quantity. Fishes of the second species feed upon fishes of the first type only. The number of individuals of each species is, of course, an integral number and, consequently, can vary only by jumps, but in order to be able to apply the methods of differential calculus, we shall consider them as continuous functions of time. Let us denote the number of individuals of the first species by N_1 and that of the second species by N_2. We shall assume that if the first species lived alone, then the number of its individuals would increase continuously at a rate of increase proportional to the number of individuals present; then we can write

$$\frac{dN_1}{dt} = \varepsilon_1 N_1,$$

[†] We may recall that in Section 5 we considered specially chosen variations of the parameters of the system which did not destroy the conservativeness of the system.

7] GENERAL PROPERTIES OF CONSERVATIVE SYSTEMS

ε_1 being greater than zero. This coefficient ε_1 depends on the mortality and the birth rate. If the second species lived alone, it would progressively become extinct, since it would have nothing on which to feed. We can therefore write for the second species

$$\frac{dN_2}{dt} = -\varepsilon_2 N_2.$$

If the two species live together, then the coefficient characterizing the rate of increase of the first species will be the smaller, the larger N_2, since fishes of the first species are eaten by fishes of the second species. We shall make the simplest assumption that the coefficient of increase ε_1 decreases in magnitude proportionally to N_2; in a similar manner we shall assume that the coefficient of decrease of the second species ε_2, by virtue of the presence of the first species (the presence of food), varies in magnitude in proportion to N_1. Under these assumptions we obtain the following system of differential equations:

$$\frac{dN_1}{dt} = N_1(\varepsilon_1 - \gamma_1 N_2); \qquad \frac{dN_2}{dt} = -N_2(\varepsilon_2 - \gamma_2 N_1) \qquad (2.72)$$

ε_1, ε_2, γ_1 and γ_2 being all greater than zero[†]. By multiplying the first equation by γ_2, the second by γ_1 and adding we shall obtain

$$\gamma_2 \frac{dN_1}{dt} + \gamma_1 \frac{dN_2}{dt} = \varepsilon_1 \gamma_2 N_1 - \varepsilon_2 \gamma_1 N_2;$$

then multiplying the first by ε_2/N_1 and the second by ε_1/N_2 and adding we have

$$\varepsilon_2 \frac{1}{N_1} \frac{dN_1}{dt} + \varepsilon_1 \frac{1}{N_2} \frac{dN_2}{dt} = -\varepsilon_2 \gamma_1 N_2 + \varepsilon_1 \gamma_2 N_1.$$

Consequently

$$\gamma_2 \frac{dN_1}{dt} + \gamma_1 \frac{dN_2}{dt} - \varepsilon_2 \frac{d \ln N_1}{dt} - \varepsilon_1 \frac{d \ln N_2}{dt} = 0.$$

This last equation can be integrated and we have the single-valued integral

$$\gamma_2 N_1 + \gamma_1 N_2 - \varepsilon_2 \ln N_1 - \varepsilon_1 \ln N_2 = \text{const.}$$

[†] We shall observe that equations of the type (2.72) are also obtained (under suitable simplifying conditions) as a result of certain problems of the kinetics of chemical processes; see, for example, [123].

We can write this integral in the following form:

$$F(N_1, N_2) = e^{-\gamma_2 N_1} e^{-\gamma_1 N_2} N_1^{\varepsilon_2} N_2^{\varepsilon_1} = \text{const}.$$

It is easily verified that the expression

$$\iint \frac{dN_1\, dN_2}{N_1 N_2}$$

will be an integral invariant. On the basis of this we conclude that the system considered is conservative. Let us pass now to investigating the form of the integral curves. To this end we shall rewrite equation (2.73) in the following form

$$N_1^{-\varepsilon_2} e^{\gamma_2 N_1} = C N_2^{\varepsilon_1} e^{-\gamma_1 N_2},$$

and shall construct the curves

$$Y = N_1^{-\varepsilon_2} e^{\gamma_2 N_1}; \qquad X = N_2^{\varepsilon_1} e^{-\gamma_1 N_2},$$

hence the required path is determined by the relation

$$Y = CX.$$

We shall take two mutually perpendicular straight lines and shall situate on them the axes OX and ON_1 and the axes OY and ON_2 respectively as is shown in Fig. 108. In the second and fourth quadrant we shall plot

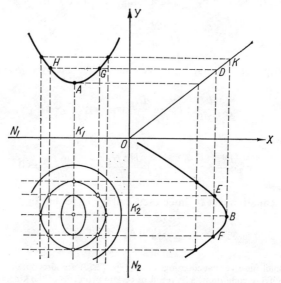

Fig. 108

respectively the curves X and Y. The form of these curves is easily determined from the following table:

N_1	0	$k_1 = \dfrac{\varepsilon_2}{\gamma_2}$	$+\infty$		N_2	0	$k_2 = \dfrac{\varepsilon_1}{\gamma_1}$	$+\infty$
$\dfrac{dY}{dN_1}$	$-$	0	$+$		$\dfrac{dX}{dN_2}$	$+$	0	$-$
Y	$+\infty \searrow$	min	$\nearrow +\infty$		X	0	$\nearrow$ max $\searrow$	0

since

$$\frac{dY}{dN_1} = Y\left(-\frac{\varepsilon_2}{N_1}+\gamma_2\right) \quad \text{and} \quad \frac{dX}{dN_2} = X\left(\frac{\varepsilon_1}{N_2}-\gamma_1\right).$$

In the first quadrant we shall draw the line $Y=CX$. Let us take a point on the straight line OK, for example the point D. Draw through it two straight lines, one parallel to the OX axis and one to the OY axis. Let E, F, G and H be the points of intersection of these straight lines with the curves X and Y. Through the points E and F draw two straight lines parallel to the OX axis, and through the points H and G draw two straight lines parallel to the OX axis. The points of intersections of these straight lines will belong to the integral curve $Y=CX$. The locus of such points

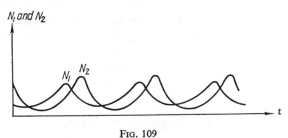

Fig. 109

when the points D slide along the straight line OK is just the required integral curve. It is easily seen that the integral curves are all closed, except one corresponding to the coordinate axis. The state of equilibrium is a singular point of the centre type with coordinates

$$N_1 = \frac{\varepsilon_2}{\gamma_2} \quad \text{and} \quad N_2 = \frac{\varepsilon_1}{\gamma_1}.$$

Thus, in the case investigated the variation of the number of both species follows a periodic law. In Fig. 109 there are given the dependences of N_1 and N_2 upon time.

CHAPTER III

NON-CONSERVATIVE SYSTEMS

Two classes of systems have been considered so far: firstly systems which are non-conservative but linear and we have verified that periodic motions are in general impossible for this class of systems; secondly we have considered systems which are conservative (linear or non-linear) and have verified that periodic motions are possible in these systems, but that there are possibly an infinite number of them, their amplitudes being entirely determined by the initial conditions. On the other hand, as we have already repeatedly indicated, we are mainly interested in those periodic motions which have their amplitudes determined by the properties of the system itself. In addition we are interested, in the first instance, in systems such that the character of their motion does not vary substantially for small sufficiently general variations of the systems themselves. Conservative systems, as has been indicated in § 7, Chapter II, do not satisfy this requirement. We shall further see that only non-conservative non-linear systems are adequate mathematical models of real physical systems which are such as to provide answers to questions concerning the physics of oscillations. In the present chapter we shall encounter examples of two basic types of such non-linear non-conservative systems: *dissipative* systems and *self-oscillating* systems.

§ 1. DISSIPATIVE SYSTEMS

The case considered here differs from the conservative system considered earlier by containing forces which do not admit a potential. But by introducing "generalized forces" Lagrange's equation for this system may be written

$$\frac{d}{dt}\left(\frac{\partial L}{\partial \dot{q}}\right) - \frac{\partial L}{\partial q} - \Phi = 0, \qquad (3.1)$$

where the generalized force Φ is a certain function of q and $\dot{q}$[†]. In the par-

---

[†] A generalized force is defined by the relation: $\delta A = \Phi \, \delta q$, where δA is the work of non-potential forces on the system for a small variation of the coordinate q (for a virtual variation in the sense of analytical mechanics). For example, the work done in the

ticular case of "linear friction" or ohmic resistance, Φ is a linear function of velocity: $\Phi = -b\dot{q}$, and Lagrange's equation takes the form

$$\frac{d}{dt}\left(\frac{\partial L}{\partial \dot{q}}\right) - \frac{\partial L}{\partial q} + b\dot{q} = 0.$$

If the non-conservative forces are frictional they must oppose the motion. Consequently the following condition is always satisfied:

$$\Phi\dot{q} \leqslant 0, \tag{3.2}$$

where the equality to zero can never be satisfied identically except in the case when $\dot{q} \equiv 0$, i.e. when the system is found in a state of rest. On multiplying (3.1) by $\dot{q}$ we obtain the energy–balance equation:

$$\frac{dW}{dt} - \Phi\dot{q} = 0, \tag{3.3}$$

where $W = \dot{q}(\partial L/\partial \dot{q}) - L$ (W = const. is an integral of equation (3.1) for $\Phi = 0$). For ordinary systems W is the total energy and by (3.3) and (3.2), this energy *always decreases* during motion ($\dot{q} \neq 0$). If the energy W cannot tend to $-\infty$, then we can affirm that it has a limiting value W_0, while $\Phi\dot{q}$ and hence $\dot{q}$ tend to zero[†]. We shall call such systems *dissipative*. Periodic motions in dissipative systems are clearly impossible, since the energy of the system always decreases during the motion.

As an example of a dissipative system consider the large deviations of an ordinary pendulum in the presence of a frictional force. Assume that $\Phi = -b\dot{q}$ and $b > 0$. The Lagrangean function for the pendulum is

$$L = \frac{I\dot{\varphi}^2}{2} + mgl(\cos\varphi - 1),$$

and Lagrange's equation is

$$I\ddot{\varphi} + b\dot{\varphi} + mgl\sin\varphi = 0. \tag{3.4}$$

The portrait on the development of the phase cylinder is determined by

$$\frac{d\omega}{d\varphi} = -\frac{b\omega + mgl\sin\varphi}{I\omega} \tag{3.5}$$

where $\omega = d\varphi/dt$. The singular points are clearly ($\varphi=0$, $\omega=0$) and ($\varphi = +\pi$, $\omega = 0$). The point (0,0) corresponds to a stable state of

---

resistance R of an electrical system by the passage of the charge δq is equal to $U_R \delta q$ where $U_R = R\dot{q}$ is the potential difference across the resistance; thus, in this case $\delta A = -R\dot{q}\,\delta q$ and $\Phi = -R\dot{q}$. Similarly for forces of viscous linear friction $\Phi = -b\dot{q}(b>0)$.

[†] Generally speaking W_0 depends on the initial conditions.

equilibrium. It is either a stable node (for $b^2 > 4Imgl$) or a stable focus (for $b^2 < 4Imgl$). The saddle point ($\pm\pi$, 0) corresponds to an unstable state of equilibrium.

The tangents to the integral curves are vertical at points on the φ axis ($\omega = 0$) and horizontal at points on the curve $\omega = -(mgl/b)\sin\varphi$. For $b^2 < 4Imgl$ and $b^2 > 4Imgl$, the phase portraits are shown in Figs. 110 and 111 respectively. On joining the development of the cylinder along the

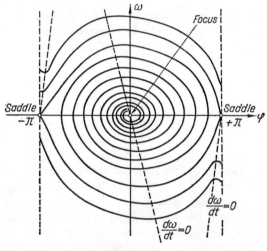

Fig. 110

lines $\varphi = \pm\pi$ we shall obtain the phase cylinder mapped out by the phase paths. These figures show that there are no periodic motions and that for almost all initial conditions (except the initial conditions corresponding to the states of equilibrium and to the stable arms of the saddle points) the system tends to stable equilibrium.

If condition (3.2) does not hold, then the system is no longer dissipative; energy increases at the expense of the "forces of friction" since $\Phi\dot{q} > 0$ implies $dW/dt > 0$. We have encountered such cases in the valve generator and Froude's pendulum. We can no longer assert that periodic motions are impossible. If $\Phi = -b\dot{q}^2$, where $b > 0$ then it is evident that such a "force of friction" resists the motion for $\dot{q} > 0$ and assists it for $\dot{q} < 0$. Here the energy-balance equation

$$\frac{dW}{dt} + b\dot{q}^3 = 0.$$

DISSIPATIVE SYSTEMS

It may be easily shown from this equation that for the usual mechanical and electrical problems a "force of friction" of the type $\Phi = -b\dot{q}^2$ does not destroy the conservativeness of the system and that a continuum of periodic motions is possible with amplitudes depending on the initial conditions.

To explain this further, consider an oscillator described by

$$2\ddot{x} + \dot{x}^2 + x = 0.$$

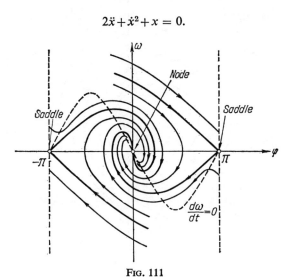

Fig. 111

Putting $\dot{x} = y$ yields the equation of the integral curves

$$\frac{dy}{dx} = -\frac{x + y^2}{2y} \tag{3.7}$$

or $2y\,dy/dx = -x - y^2$ and $d(y^2)/dx + y^2 = -x$. Integrating the latter gives

$$y^2 = Ce^{-x} + 1 - x$$

or

$$(y^2 + x - 1)e^x = C, \tag{3.8}$$

where C is constant.

The family of integral curves (3.8) is shown in Fig. 112. The value $C = -1$ corresponds to the singular point $(0,0)$ of equation (3.7). For $0 > C > -1$ we obtain closed curves encircling the origin of the coordinates and contained in each other. For $C \geqslant 0$ the curves (3.8) have infinite branches (to the value $C = 0$ there corresponds the parabola $y^2 = 1 - x$ which is a separatrix between the closed curves and the curves with infinite branches.

If the representative point is found initially inside the separatrix, then the motion is periodic but not sinusoidal with an amplitude entirely determined by the initial conditions and so of the type met with in conservative systems.

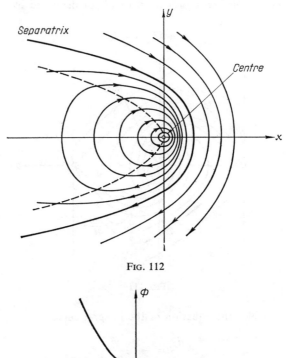

Fig. 112

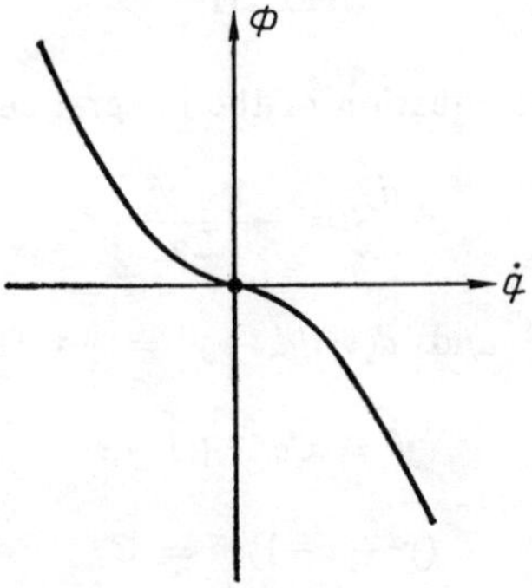

Fig. 113

It is easy to pass on from this case to an *oscillator* with a force of friction proportional to the square of velocity (Fig. 113) and expressed by the relation

$$\Phi = -b\left(\frac{dq}{dt}\right)^2 \operatorname{sgn}\left(\frac{dq}{dt}\right)$$

($b > 0$). The equation of motion

$$m\frac{d^2q}{dt^2} + b\left(\frac{dq}{dt}\right)^2 \operatorname{sgn}\left(\frac{dq}{dt}\right) + kq = 0$$

is reduced by a change of variables to the equation

$$2\ddot{x} + \dot{x}^2 \operatorname{sgn} \dot{x} + x = 0, \tag{3.9}$$

or

$$\left.\begin{aligned} \dot{x} &= y, \\ 2\dot{y} &= -x - y^2 \operatorname{sgn} y. \end{aligned}\right\} \tag{3.10}$$

The only state of equilibrium is the origin (0,0). Further, if the curve $y = f(x)$ is a phase path, then the curve $-y = f(-x)$ is also a phase path. Therefore, by retaining above the x axis the picture shown in Fig. 112 and constructing in the lower half-plane paths symmetrical (with respect to the origin) to the paths in the upper half-plane, we shall obtain the phase portrait of an oscillator with square-law friction (Fig. 114).

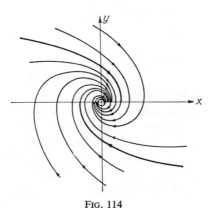

Fig. 114

The state of equilibrium (0,0) is stable and all remaining spiral-like phase paths approach it asymptotically for $t \to +\infty$†.

§2. Oscillator with Coulomb friction

Another example of a dissipative system is an oscillator with "dry" friction (Fig. 115). In the absence of friction the system is a harmonic oscillator. A linear law such as is assumed for viscous friction is totally

---

† In Chapter VIII, Section 9 we shall show that the phase paths have this behaviour, by reducing the problem to a certain point-transformation of one straight line into another.

inadequate to represent the characteristics of "dry" friction between solid unlubricated surfaces. The basic features of these characteristics are reproduced sufficiently well, at least for small velocities, by assuming a "constant" or Coulomb friction. This "constant" friction is constant in amplitude but not in direction since the direction of the force of friction is always opposite to the direction of velocity. The dependence of a Coulomb friction f on the velocity v can be represented by the diagram shown in Fig. 116. Note that for $v=0$ the value of f can assume, depending

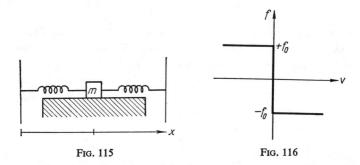

Fig. 115   Fig. 116

on the values of the other forces acting on the system, any values between $+f_0$ and $-f_0$. The mass m is acted upon not only by the force of dry friction but also by the force of tension of the springs $-kx$. It remains at rest ($\dot{x}=0$) if the tension force of the springs does not exceed f_0 in absolute value. Thus, all the positions of the oscillator with coordinates $-f_0/k \leqslant x \leqslant f_0/k$ can be positions of rest. If, however, $|kx| > f_0$ then the oscillator is in motion. When the oscillator is in motion the force of Coulomb friction is $f = +f_0$ for $\dot{x} < 0$ and $f = -f_0$ for $\dot{x} > 0$.

Thus, we can describe the Coulomb force of friction by the equations

$$f = \begin{cases} +f_0 & \text{for} \quad \dot{x} < 0, \\ +f_0 & \text{for} \quad \dot{x} = 0 \quad \text{and} \quad kx > f_0, \\ +kx & \text{for} \quad \dot{x} = 0 \quad \text{and} \quad |kx| \leqslant f_0, \\ -f_0 & \text{for} \quad \dot{x} = 0 \quad \text{and} \quad kx < -f_0 \\ -f_0 & \text{for} \quad \dot{x} > 0 \end{cases} \quad (3.11)$$

(a Coulomb force of friction is thus a non-linear discontinuous function and depends not only on the velocity $\dot{x}$ but also on the coordinate x of the oscillator).

The non-linear differential equation of motion of the oscillator

$$m\ddot{x} = -kx + f \quad (3.12)$$

can be written in the form of two *separate linear* equations, one of which is valid for a motion towards the left:

$$m\ddot{x}+kx = +f_0 \quad (\dot{x} < 0) \qquad (3.12a)$$

and the other for a motion towards the right:

$$m\ddot{x}+kx = -f_0 \quad (\dot{x} > 0). \qquad (3.12b)$$

Suppose that, at the initial instant of time, $\dot{x}<0$. The motion of the system is described by (3.12a). The velocity decreases until at $t=t_1$ the system reaches $x=x_1$ and the velocity reduces to zero. Then the velocity changes its sign and the system will move in the opposite direction[†]. The opposite motion now is described by the second equation (3.12b), where now the initial conditions are the coordinate and velocity $(x_1, 0)$ which the system possessed at the end of the previous stage of the motion. This process continues until the body finally remains at rest. Let $k/m = \omega_0^2$ and $f_0/m = a\omega_0^2$ where $a = f_0/k$. The equations of the motion are

$$\ddot{x}+\omega_0^2 x = \begin{cases} +a\omega_0^2 & \text{for} \quad \dot{x} < 0, \\ -a\omega_0^2 & \text{for} \quad \dot{x} > 0. \end{cases} \qquad (3.13)$$

Let $\xi_1 = x-a$, when $\dot{x}<0$ and let $\xi_2 = x+a$, when $\dot{x}>0$, then $\ddot{\xi}_1+\omega_0^2\xi_1 = 0$ (for $\xi_1 < 0$) and $\ddot{\xi}_2+\omega_0^2\xi_2 = 0$ (for $\xi_2>0$) but the variables are referred to different origins. Hence the motion of the system is obtained by combining two halves of harmonic oscillations centred on two different positions of equilibrium at distances $+a$ and $-a$ from the point $x=0$. Changing from one mode to another takes place at the instant when the velocity of the system reduces to zero while the coordinate is different from zero.

To find displacement x as a function of t proceed as follows (Fig. 117). Let the initial position be x_{01} with zero initial velocity $\dot{x}_{01}$. If x_{01} is positive, the velocity will at first be negative with a position of equilibrium displaced by $+a$, (in Fig 117 by a above the time axis). Finally the system reaches the maximum downwards deviation x_{02}, where $|x_{02}| = |x_{01}|-2a$.

Then for $\dot{x}>0$ the second equation becomes valid and consequently, there will be a part oscillation with a position of equilibrium displaced by $-a$, i.e. by the quantity a below the time axis. At the end of this half-

[†] The body, of course, may also remain at rest. Whether it will or not stay at rest depends on whether the maximum value of the force of friction f_0 is larger or smaller than the elastic force $|kx_1|$.

oscillation the system reaches the maximum deviation x_{03} (above the t axis in Fig. 117) where $|x_{03}| = |x_{02}| - 2a = |x_{01}| - 4a$. The maximum displacement reached decreases each time in absolute value by $2a$, and successive maxima form a decreasing arithmetical progression with constant difference equal to $4a$. It is clear that this progression consists of a finite

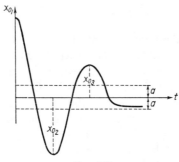

Fig. 117

number of terms and the motion ceases after the lapse of a finite number of oscillations. In fact, when the maximum displacement decreases to a value smaller than $3a$, for example, to the value x_{03} in Fig. 117, then the subsequent motion takes the system into the region enclosed between the straight lines $+a$ and $-a$, and comes to rest. It then remains at rest since in this region $|kx| \leq f_0$.

Thus, the successive maxima decrease not according to a geometrical progression as in a viscously-damped linear oscillator but according to an arithmetical progression, and the term "logarithmic decrement" has no meaning. The interval of time between two maxima in the same sense does not depend, in the case of constant friction, on the value of the force of friction and equals the period of the harmonic oscillator[†]. But, as is easily verified by examining Fig. 117, the time interval between a maximum and the following zero value is larger than that between the zero value and the following minimum. This difference is the more noticeable, the smaller the maximum. Finally there is another difference between systems with linear and those with constant friction in that the division into oscillatory and aperiodic systems loses in general its meaning, since for an arbitrary friction it is always possible to choose a sufficiently large initial deviation so

† Note that, in the case of constant friction, the intervals of time between zero values of the coordinate, corresponding to motion of the system on one and the same side (during the oscillatory stage) is no longer the same and cannot therefore serve as the basis of the definition of "conditional period".

that the system performs a number of oscillations before it stops. The physical meaning of this property of systems with constant friction becomes particularly clear when considering the energy balance.

Starting at x_{01} with zero initial velocity, the initial energy is the potential energy $v_1 = kx_{01}^2/2$. The work A spent in overcoming the force of friction does not depend on the velocity but only depends on the path (since the force of friction is constant), so that during the first half "conditional"-period this work amounts to

$$A_1 = (|x_{01}| + |x_{02}|)f_0,$$

while the potential energy at $x = x_{02}$ is

$$V_2 = \frac{kx_{02}^2}{2};$$

Since $V_1 - V_2 = A_1$, then

$$\frac{k}{2}(x_{01}^2 - x_{02}^2) = (|x_{01}| + |x_{02}|)f_0,$$

$$|x_{01}| - |x_{02}| = \frac{2f_0}{k} = 2a$$

or

$$A_1 = 2f_0(|x_{01}| + a).$$

A_1 increases linearly, while V_1 increases according to a square law. Consequently, for a large enough x_{01} the reserve of energy in the system at the end of the first half conditional-period will be large and it will oscillate at the start.

Consider now the motion on the phase plane. Put $dx/dt = y$ and so:

$$\frac{dy}{dx} = \frac{-\omega_0^2(x-a)}{y} \quad \text{for} \quad y < 0,$$

$$\frac{dy}{dx} = \frac{-\omega_0^2(x+a)}{y} \quad \text{for} \quad y > 0;$$

whose integrals are

$$\frac{(x-a)^2}{R_1^2} + \frac{y^2}{R_1^2\omega_0^2} = 1 \quad \text{for} \quad y < 0, \tag{3.14a}$$

$$\frac{(x+a)^2}{R_2^2} + \frac{y^2}{R_2^2\omega_0^2} = 1 \quad \text{for} \quad y > 0, \tag{3.14b}$$

where R_1 and R_2 are constants of integration. The equations (3.14a) and (3.14b) define a family of "semi-ellipses" the centres of which are displaced successively to the right (3.14a) and to the left (3.14b) by a. By "matching" the initial conditions and taking into account the direction of motion on the phase plane, it is easy to construct the phase paths as shown in Fig. 118. All phase paths are spirals formed from semi-ellipses and motion

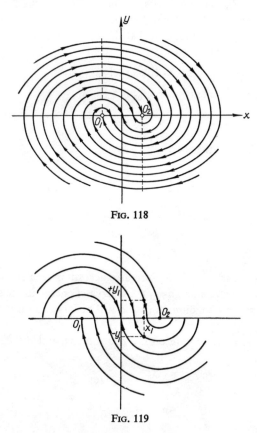

Fig. 118

Fig. 119

is along them to the segment O_1O_2 which is the locus of the states of equilibrium. Thus, in general, free oscillations are of decreasing amplitude and stop after a finite number of swings, which depends on the initial conditions. In the particular case when the initial condition corresponds to a point on the segment O_1O_2 the system remains at rest. On this segment, however, the system possesses a certain special type of "instability". Let the system be initially at rest at the point $+x_1$ and give it an initial velocity $\pm y_1 (y_1 > 0)$. Then, as shown in Fig. 119, the representative point moves from

(x_1, y_1) to a point on O_1O_2 further away from the origin, or from $(x_1, -y_1)$ to a point on O_1O_2 nearer the origin. However, the upper semi-ellipses approach O_1O_2 more steeply than the lower ones, and consequently, for the $-y_1$ jump the system gets nearer in the end to the "true" position of equilibrium (the origin of the coordinates) and it is moved away from this position of equilibrium by the $+y_1$ jump. Hence, if jumps act both in the one and the other direction equally often (for example this may be achieved by subjecting the system to an artificial systematic shaking) then the net effect is to move the system to a region near the "true" position of equilibrium.

The presence of a whole region of positions of equilibrium (the so-called "stagnation") and the approach to the "true" position of equilibrium as a result of impulses are observed to a smaller or greater extent in all measuring and indicating instruments in which there is dry friction. It is clear that "fluid" friction cannot play such a role; so in certain measuring systems and indicating devices a very ingenious method is used to change dry friction into fluid friction. A motion backwards and forwards along the bearing is provided and then the component of the force of friction between the axis and the bearing in the *direction of rotation* is proportional to the velocity of rotation (if the velocity is sufficiently small) and the device behaves as a system having fluid and not dry friction (sometimes called a Brown arrangement).

§ 3. Valve Oscillator with a ∫ Characteristic

The method used in the preceding section is not limited to dissipative systems. This method of replacing a non-linear equation by several linear ones with "matching" of the initial conditions will be applied to two examples of some value. They enable us to approach the theory of periodic processes in non-linear systems. Consider a valve oscillator with the oscillating circuit in the anode circuit and inductive feedback (Fig. 120). Neglecting the grid current and employing the notation indicated in the figure, we have

$$Ri = -v - L\frac{di}{dt}, \quad i = i_a + C\frac{dv}{dt},$$

and after eliminating v:

$$LC\frac{d^2i}{dt^2} + RC\frac{di}{dt} + i = i_a. \qquad (3.15)$$

Here $i_a = i_a(u_g)$ is the anode current depending only on the voltage at the grid

$$u_g = -M\frac{di}{dt}. \tag{3.16}$$

We shall assume that the characteristic of the valve $i_a = i_a(u_g)$ has a saturation current I_s and rises steeply to this value, and the working point lies on this sloping section of the characteristic. Such a characteristic is shown in Fig. 121 as a dotted line. If the amplitude of the voltage oscilla-

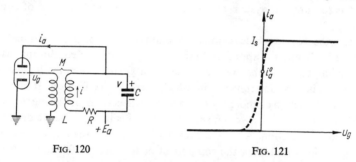

Fig. 120   Fig. 121

tions on the grid are so large that for the greatest part of the time the anode current i_a is either equal to zero (the valve is cut-off) or to the saturation current I_s, then we can sufficiently well represent the properties of such a valve by the idealized ∫ characteristic, shown in the same Fig. 121 by a continuous line:

$$i_a = \begin{cases} 0 & \text{for} \quad u_g < 0, \\ I_s & \text{for} \quad u_g > 0. \end{cases} \tag{3.17}$$

We shall assume that for $u_g = 0$, $i_a = i_a^0$. As we shall see later, the amplitude of the oscillations of the grid voltage will be the larger the smaller the damping in of the oscillating circuit. Hence it can be concluded that our idealization of the characteristic of the valve (the so-called ∫ characteristic) is of physical significance in the case of sufficiently small damping in the resonant circuit and a sufficiently strong feedback coupling to the grid.

The coils are arranged so that $M<0$ (as we shall see this leads to self-oscillations). The equation (3.15) for the current in the oscillating circuit can be reduced to

$$\ddot{x}+2h\dot{x}+\omega_0^2 x = \begin{cases} 0 & \text{for} \quad \ddot{x} < 0, \\ \omega_0^2 & \text{for} \quad \ddot{x} > 0, \end{cases} \tag{3.18}$$

where

$$x = \frac{i}{I_s}, \quad \omega_0^2 = \frac{1}{LC} \quad \text{and} \quad 2h = \frac{R}{L}.$$

The equation (3.18) has a discontinuous right-hand side, since the anode current varies with a jump when x, and so u_g, passes through zero. As a consequence of this we must, in addition to the equation (3.18), determine how the system behaves for a passage of $\dot{x}$ through zero. The physical requirement that the voltages and currents are bounded implies that x and $\dot{x}$ are *continuous* everywhere and in particular at $\dot{x}=0$†. Therefore, as in the previous case, there are two modes of action subject to different differential equations, and the initial conditions of one mode are the final conditions of the other.

Each of the equations (3.18) determines a damped "half-oscillation" (we shall assume the damping to be small).

The one for which $\dot{x}>0$ governs a "half-oscillation" about a position of equilibrium is displaced by one unit in a direction *opposite* to that associated with the corresponding "half-oscillation" in the case of dry friction. It follows from this that for sufficiently small initial displacements and sufficiently small linear damping the swings of the oscillations increase in succession rather than decrease as was the case for dry friction (Fig. 122).

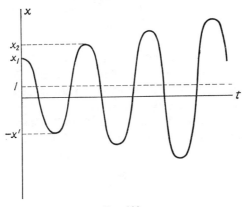

Fig. 122

It will be shown that this build-up of oscillations does not last and that undamped oscillations with a constant "amplitude" are finally established.

This build-up of periodic oscillations in a valve generator can be followed on the phase plane x, y ($y=\dot{x}$). It is clear that the phase paths in the lower

† In fact, if x (or the current i) varied with a jump then the induction e.m.f. and the grid voltage which are proportional to di/dt would be infinitely large. This is impossible and hence x is a continuous function of time. Jumps of the capacitor voltage v are similarly impossible. But $L\,di/dt = -v - Ri$ and therefore di/dt (or $\dot{x}$) by virtue of the continuity of v and i will also be continuous.

half-plane ($y<0$) coincide with the phase paths (spirals) of a damped linear oscillator (Chapter I, Section 4) and in the upper half-plane ($y>0$) with the paths but for a similar oscillator with the state of equilibrium displaced to the point (1,0). These "half-turns" of spirals will form the entire phase portrait, which are *continuous* curves. The only state of equilibrium, and of course a stable one, is the point (x_0, 0) where $x_0 = i_a^0/I_s$. The general aspect of the phase plane of a valve generator with a discontinuous characteristic is shown in Fig. 123.

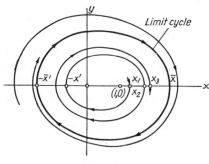

Fig. 123

Consider an arbitrary phase path reaching the lower half-plane at a point x_1 on the positive x axis (Fig. 123). After passing through the lower half-plane it intersects the negative x axis at $x = x'$ reaches the upper half-plane and returns to the positive x axis at $x = x_2$.

In the lower half-plane this path is a "half-turn" of a spiral for an oscillator with damping and with a state of equilibrium at the point (0,0). Then from (1.31)

$$x' = x_1 e^{-\frac{d}{2}},$$

where $d = hT = 2\pi h/\sqrt{\omega_0^2 - h^2}$ is the logarithmic decrement. In the upper half-plane this path is also a "half-turn" of the same spiral but for an oscillator with the state of equilibrium displaced on the right by one unit. Therefore, according to the same relationship

$$x_2 - 1 = (x' + 1)e^{-\frac{d}{2}},$$

Eliminating x', we have

$$x_2 = 1 + e^{-\frac{d}{2}} + x_1 e^{-d}. \tag{3.19}$$

This is a *sequence function* relating consecutive points of intersection of a phase path with the positive x axis.

There is one point, the fixed point, which is transformed into itself so that $x_1 = x_2$. This defines a *closed* phase path. Substituting in (3.19) $x_1 = \bar{x}$, $x_2 = \bar{x}$ we shall obtain for the fixed point

$$\bar{x} = 1 + e^{-\frac{d}{2}} + \bar{x} e^{-d},$$

or

$$\bar{x} = \frac{1 + e^{-\frac{d}{2}}}{1 - e^{-d}} = \frac{1}{1 - e^{-\frac{d}{2}}} > 1. \quad (3.20)$$

Thus, there is a unique *closed* phase path, corresponding to periodic undamped oscillations in the generator. However, it must be shown that these undamped oscillations can actually be generated. Firstly, it must be known under what initial conditions the periodic motion is established, and whether it is established when the initial values of x and $\dot{x}$ are sufficiently small. Secondly, whether the periodic motion is *stable* with respect to arbitrarily small variations of the initial conditions. Consider the graph of the sequence function (3.19), the so-called "Lamerey's diagram" (Fig. 124). It is evident that the graph of the sequence function (3.19) is a

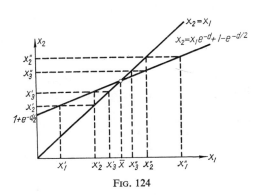

Fig. 124

straight line with slope e^{-d} intercepting the x_2 axis at $1 + e^{-d/2}$. The fixed point $\bar{x}$ must lie on the straight line $x_1 = x_2$ and consequently is the point of intersection of this straight line with the graph of the sequence function.

Let us take an arbitrary phase path different from a closed one and consider the sequences of points $x_1', x_2', x_3', \ldots,$ and $x_1'', x_2'', x_3'' \ldots,$ (Fig. 124).

In one sequence the initial point $x_1' < \bar{x}$ and for the other $x_1'' > \bar{x}$. As seen from Fig. 124 the points of both sequences x_1', x_2', x_3', ..., and x_1'', x_2'', x_3'', ..., approach the fixed point $\bar{x}$†. This means that phase paths, either outside or inside the closed phase path, approach it asymptotically as $t \to \infty$. Such an isolated closed phase path, to which all neighbouring paths tend and which corresponds to a periodic mode of operation in the system, will be called a *limit cycle*.

Thus, whatever the initial conditions, undamped oscillations are established and these undamped oscillations are stable‡. The "amplitude" of these oscillations is determined by the properties of the system and not by the initial conditions. Such oscillations will be called *self-oscillations* and the systems in which self-oscillations are possible *self-oscillating systems*††.

The amplitude of the self-oscillations‡‡ is

$$x^* = \frac{1}{2}(\bar{x} + \bar{x}') = \frac{1}{2}\bar{x}(1 + e^{-d/2}) = \frac{1}{2}\frac{1 + e^{-d/2}}{1 - e^{-d/2}} = \frac{1}{2}\operatorname{cth}\frac{d}{4},$$

and so the current in the oscillating circuit is

$$I = I_s \cdot x^* = \frac{I_s}{2}\operatorname{cth}\frac{d}{4}. \tag{3.21}$$

For $d \ll 1$, expanding $e^{-d/2}$ in a power series and restricting ourselves to the principal terms the amplitude of the steady-state oscillations of current is approximately

$$I = \frac{2I_s}{d}.$$

The "period" of these stationary oscillations happens to coincide with the "period" of the damped oscillations of the linear resonant circuit but

---

† It is easily shown analytically that the sequence x_1, x_2, x_3, ..., for any values of x_1 has for its limit point the fixed point $\bar{x}$ of the transformation. In fact, as is easily seen,

$$x_n = \frac{1 - e^{-(n-1)d}}{1 - e^{-d/2}} + x_1 e^{-(n-1)d}$$

and hence, $x_n \to 1/(1 - e^{-d/2}) = \bar{x}$ for $n \to \infty$ and arbitrary values of x_1. This also follows from Königs' theorem which will be considered in detail in Chapter V, Section 7.

‡ It is easily shown that this periodic motion is stable in the sense of Liapunov. The reasoning, given in the text, only shows the absolute orbital stability of the motion. In the sequel we shall treat in detail the analysis of stability of periodic motions in non-conservative systems (see Section 6, Chapter V).

†† In Chapter V we shall give for the case of one degree of freedom a mathematical definition of the concepts of "self-oscillation" and "self-oscillating" system.

‡‡ By amplitude of periodic oscillations is meant half the difference between the maximum and minimum values of the oscillation.

generally speaking, in other self-oscillations, the "period" is far from coincident with the "period" of the damped oscillations of the linear circuit which is part of the self-oscillating system.

The energy needed for maintaining non-damped oscillations is provided by the anode battery, but the anode current only flows for half of each period when the valve is switched on. Thus the mean power furnished by the battery is

$$W_A = \frac{I_s E_a}{2},$$

where E_a is the voltage applied by the battery. Since i is a known function of time it is possible to calculate the power dissipated in the oscillatory circuit, but as will be shown below, for sufficiently small dampings the oscillations are close to sinusoidal. In this case the amplitude is about $2I_s/d$ and the voltage amplitude in the oscillatory circuit is $V \approx I\omega L$. The mean power dissipated in the oscillating circuit over a period is therefore

$$W = \frac{RI^2}{2} = \frac{RI}{2} \cdot \frac{2I_s}{d} = \frac{RII_s}{\frac{\pi R}{\omega L}} = \frac{I_s \omega L I}{\pi} = \frac{V I_s}{\pi},$$

and the efficiency of the generator is

$$\eta = \frac{W}{W_A} = \frac{I_s V}{\pi} \cdot \frac{2}{I_s E_a} = \frac{2}{\pi} \frac{V_A}{E_a}.$$

The amplitude of the voltage V across the oscillating circuit must be less than the voltage E of the anode battery, otherwise at certain instants of time, the valve would have to conduct with zero or even negative anode voltages, which is not possible. Consequently, the efficiency of the generator under the above assumptions cannot exceed $2/\pi$ i.e. 64%.

The efficiency of a generator can be greater than 64% if the working point on the characteristic is displaced to the left, i.e. if the anode current is switched on not for a zero applied grid voltage but for a certain positive voltage.

Fig. 125 shows the variation of $x = x(t)$ and the function representing this periodic process in the period $T = 2\pi/\omega$ can be written:

$$\left.\begin{aligned}
x &= 1 - \frac{1}{1 - e^{-\frac{\pi h}{\omega}}} e^{-ht} \frac{\cos(\omega t - \vartheta)}{\cos \vartheta} \quad \text{for} \quad 0 \leqslant t \leqslant \frac{\pi}{\omega}, \\
x &= -\frac{e^{\frac{\pi h}{\omega}}}{1 - e^{-\frac{\pi h}{\omega}}} e^{-ht} \frac{\cos(\omega t - \vartheta)}{\cos \vartheta} \quad \text{for} \quad \frac{\pi}{\omega} \leqslant t \leqslant \frac{2\pi}{\omega},
\end{aligned}\right\} \quad (3.22)$$

where

$$\omega = +\sqrt{\omega_0^2 - h^2}, \quad \vartheta = \arctan\frac{h}{\omega}.$$

Here the constants of integration are chosen in order that $x(0) = x(2\pi/\omega) = -\bar{x}'$ and that for both expressions at $t = \pi/\omega$ $x = \bar{x}$, $\dot{x}(0) = \dot{x}(\pi/\omega) = \dot{x}(2\pi/\omega) = 0$, $\dot{x} > 0$ for $0 < t < \pi/\omega$ and $\dot{x} < 0$ for $\pi/\omega < t < 2\pi/\omega$. Using

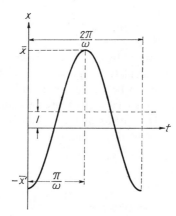

Fig. 125

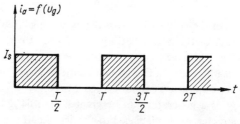

Fig. 126

these $x(t)$ can be expanded in a Fourier series but it is easier to note that $i_a = F(t)$ is a periodic sequence of rectangular pulses (Fig. 126) in which during one half-period, when $u_g > 0$, $i_a(u_g) = I_s$ and during another half-period, when $u_g < 0$, $i_a(u_g) = 0$. We can therefore represent the function $f_1(\dot{x}) = i_a(-M\dot{x})/I_s = i_a(u_g)/I_s$ occurring in the equation

$$\ddot{x} + 2h\dot{x} + \omega_0^2 x = \omega_0^2 f_1(\dot{x}) \tag{3.23}$$

not as a function of $\dot{x}$ but as an "external force" $f(t)$ which is acting on the linear oscillating circuit, is periodic, is given as a function of time, and the frequency of which is not arbitrary but coincides with the "conditional frequency" of damped oscillations in the system, i.e. with the frequency $\omega = \sqrt{\omega_0^2 - h^2}$. Thus the problem is reduced to investigating the action of an external force on an oscillating circuit. If $f(t)$ is of the form

$$f(t) = \begin{cases} 1 & \text{for} \quad 0 < t < \dfrac{\pi}{\omega}, \\ 0 & \text{for} \quad \dfrac{\pi}{\omega} < t < \dfrac{2\pi}{\omega}, \end{cases}$$

its expansion in a Fourier series is

$$f(t) = \frac{1}{2} + \frac{2}{\pi} \sum_{k=0}^{\infty} \frac{1}{2k+1} \sin(2k+1)\omega t. \tag{3.24}$$

Since we are dealing here with a linear problem, for which the superposition principle holds, then the complete solution for $x(t)$ can be written in the form

$$x(t) = \frac{a_0}{2} + \sum_{1}^{\infty} (a_s \cos s\omega t + b_s \sin s\omega t). \tag{3.25}$$

and so

$$\dot{x} = \sum_{1}^{\infty} (-a_s s\omega \sin s\omega t + b_s s\omega \cos s\omega t),$$

$$\ddot{x} = \sum_{1}^{\infty} (-a_s s^2 \omega^2 \cos s\omega t - b_s s^2 \omega^2 \sin s\omega t).$$

Substituting the values of x, $\dot{x}$ and $\ddot{x}$ in equation (3.23) and equating to zero the coefficients of the sines and cosines we shall find that all Fourier coefficients of *even* order are equal to zero (except $a_0 = 1$) and that the Fourier coefficients of odd order are determined by the equations

$$-\omega^2 s^2 a_s + 2h\omega s b_s + \omega_0^2 a_s = 0,$$

$$-\omega^2 s^2 b_s - 2h\omega s a_s + \omega_0^2 b_s = \frac{2\omega_0^2}{\pi s}$$

where $s = 2k+1$ ($k = 0, 1, 2, \ldots$).

Solving these equations we find the expressions required for the odd Fourier coefficients:

$$a_{2k+1} = -\frac{2\omega_0^2}{\pi(2k+1)} \frac{2h\omega(2k+1)}{[\omega_0^2 - \omega^2(2k+1)^2]^2 + 4h^2\omega^2(2k+1)^2},$$

$$b_{2k+1} = \frac{2\omega_0^2}{\pi(2k+1)} \frac{\omega_0^2 - \omega^2(2k+1)^2}{[\omega_0^2 - \omega^2(2k+1)^2]^2 + 4h^2\omega^2(2k+1)^2},$$

and for the squares of the amplitudes of the corresponding harmonics

$$R_{2k+1}^2 = a_{2k+1}^2 + b_{2k+1}^2 =$$
$$= \frac{4\omega_0^4}{\pi^2(2k+1)^2} \frac{1}{[\omega_0^2 - \omega^2(2k+1)^2]^2 + 4h^2\omega^2(2k+1)^2} \quad (3.26)$$

As a measure of the departure from a sine wave of $x(t)$ we shall use the harmonic coefficient defined by

$$\varkappa^2 = \frac{\sum\limits_{s=2}^{\infty}(a_s^2 + b_s^2)}{a_1^2 + b_1^2}.$$

The value of this coefficient for small values of h is easily estimated from (3.26). The square of the amplitude of the fundamental of $x(t)$ $a_1^2 + b_1^2$, increases without limits for $h \to 0$, but the remaining terms of the expansion have frequencies far removed from resonance and the sum of the squares of their amplitudes tends to a finite positive limit as $h \to 0$. Thus for a sufficiently small h the periodic self-oscillations are arbitrarily close in form to sinusoidal ones.

It is wellknown in the theory of resonant systems that a markedly non-sinusoidal external force can maintain almost sinusoidal oscillations in a linearly damped harmonic oscillator. We can say therefore that in the problem of a generator with a ∫-characteristic we are dealing, for a sufficiently small h, with *self-resonance*, i.e. with resonance under the action of a force generated by the motion of the system itself.

In the valve oscillator with a ∫-characteristic the mode of excitation is said to be "soft" because oscillations build up for *any* initial conditions. If, however, the valve characteristic is biassed so that the vertical part of the characteristic does not pass through the point $u_g = 0$, then an impulse is required to initiate oscillations and the excitation mode is "hard". A biassed ∫-characteristic can serve as a satisfactory model for two cases: firstly, when the variable grid voltages greatly exceed the saturation voltage of the valve; and secondly, when the working point is displaced either towards the region of the saturation current or to the region where the anode current is equal to zero. In this case the behaviour of the valve generator will be determined by equations of the form

$$\left.\begin{array}{ll}\ddot{x} + 2h\dot{x} + \omega_0^2 x = \omega_0^2 & \text{for} \quad \dot{x} > b, \\ \ddot{x} + 2h\dot{x} + \omega_0^2 x = 0 & \text{for} \quad \dot{x} < b,\end{array}\right\} \quad (3.27)$$

where, for the case of the bias shown in Fig. 127 the quantity $b > 0$ (as before, damping is assumed to be small). The change from one mode of operation to the other now takes place not for $\dot{x}=0$ but for $\dot{x}=b$. The phase plane is modified (Fig. 128), and is now divided not along the straight line $y=0$ but along the straight line $y=b$ and the upper half-plane must be displaced on the right by one unit. From continuity considerations there

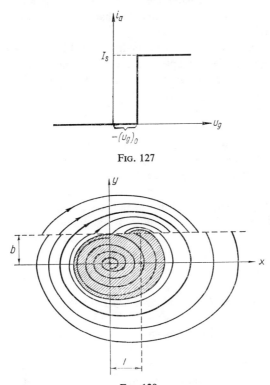

Fig. 127

Fig. 128

follows the necessary existence of one closed path consisting of two "half-spirals". To this path tend all other paths, except those which lie inside the shaded region formed by the limit spiral and the straight line $y=b$. Starting from all the initial states situated inside this region the system tends to a stable state of equilibrium (a stable focus). It appears from Fig. 128 that a periodic process is generated only when the initial voltage or intensity of current in the oscillating circuit is sufficiently large[†].

[†] This problem will be considered in greater detail in Section 4, Chapter VIII.

§ 4. THEORY OF THE CLOCK. MODEL WITH IMPULSES

A clock is an oscillating system which maintains oscillations whose amplitude is independent of the initial conditions. To start the clock a large initial impulse is usually needed. If the initial impulse is too small, then the clock comes to rest again. We shall consider an idealized model of the clock.

Any clock mechanism can be broadly divided into three parts: (1) an oscillating system, for example, a pendulum, a balancewheel, etc., (2) a source of energy such as a weight or spring, (3) a trigger mechanism, connecting the oscillating structure with the energy source. For fixed positions of the oscillating system (referred to, for the sake of brevity, as the pendulum) the trigger mechanism acts and energy is given to the pendulum during a short impulse. In a good clock the impulse is of very short duration. The trigger mechanism usually acts twice during a period close to the position of equilibrium where the velocity is greatest. It is important to note that the instant of time when the trigger mechanism begins to act is entirely determined by the position of the pendulum. In addition, the manner of its action and the magnitude of the impulse also depend on the state of the pendulum. Consequently, all forces which arise in the mechanism depend only on the positions and velocities of the separate parts of the system, and not on the time. Thus the clock is an *autonomous* system.

To simplify the discussion, we shall assume that the pendulum receives *an instantaneous impulse at the instant of its passage through the position of equilibrium* once per period and that this results in an instantaneous increase in its velocity. Two more simple assumptions might prove appropriate, either that the change in velocity is constant, or that the change in kinetic energy is constant. If v_0 and v_1 are the velocities before and after the impulse then the assumptions are equivalent to saying that either $v_1 - v_0 = $ const. or that $(mv_1^2/2) - (mv_0^2/2) = $ const. These assumptions do not of course exhaust all possible types of impulses but in the case of a driving mechanism with weights, the weight is lowered the same distance at each impulse (so doing the same work). The second assumption, that the pendulum receives equal amounts of energy, is quite natural[†]. The first assumption implies that the lower the velocity of the system before the impulse the less energy it receives during the impulse, which is perhaps less natural but not altogether impossible, so both these hypotheses will be discussed.

[†] This second assumption is the one usually introduced in the theory of the clock. See, for example [133].

In addition to the assumption about the impulse, it will be assumed firstly that the magnitude of the force of friction is proportional to velocity ("linear friction"), and secondly that the magnitude of the force of friction is independent of velocity ("constant friction"). These assumptions lead to quite different results.

1. *The clock with linear friction*

We shall begin by considering a constant-momentum impulse, received only once in a period†. This can be investigated by a method similar to that used for the analysis of the valve generator with a ∫ (discontinuous) characteristic. If the logarithmic decrement is d (the damping is assumed to be small) and the increment of velocity during an impulse is denoted by a, then for an initial velocity y_1 (at an instant immediately following the final impulse) the velocity after one period will be

$$\left. \begin{array}{l} \text{immediately before the impulse } y' = y_1 e^{-d}, \\ \text{immediately after the impulse } \quad y_2 = y_1 e^{-d} + a. \end{array} \right\} \quad (3.28)$$

To be periodic it is necessary that $y_2 = y_1 = \bar{y}$, where $\bar{y}$ is the stationary amplitude. Therefore

$$\bar{y} = \frac{a}{1 - e^{-d}}.$$

As was shown for the valve generator, this stationary amplitude is stable and no matter how small the value of y_1, the oscillations will grow.

Therefore this model of a clock is self-exciting since oscillations are built up in it for arbitrary small initial conditions. The portrait on the phase plane is shown in Fig. 129. The representative point, after following a spiral to the positive y axis, makes a jump upwards by an amount a and then continues its motion along another spiral. It is clear from considerations of continuity that by virtue of the jump one of the paths of the representative point along one of the spirals proves to be a *closed* one corresponding to a periodic motion. Our idealized model possesses the property of self-excitation even for an arbitrarily small initial impulse, but real clocks need a certain initial impulse of finite magnitude to start and so this model must be rejected.

Now assume the linearity of friction but adopt the second law of impulse, i.e. assume

$$y_2^2 - y'^2 = h^2 = \text{const}$$

† One impulse per period occurs for example in the trigger mechanism used in chronometers.

In this case

immediately before the impulse $y' = y_1 e^{-d}$,

immediately after the impulse $y_2 = \sqrt{y'^2 + h^2} = \sqrt{y_1^2 e^{-2d} + h^2}$.

The stationary amplitude $\bar{y}$ is determined from the condition

$$y_2 = y_1 = \bar{y} \quad \text{and} \quad \bar{y}^2(1 - e^{-2d}) = h^2 \quad \text{or} \quad \bar{y} = \frac{h}{\sqrt{1 - e^{-2d}}}.$$

Self-excitation of oscillations will occur as before in this case and the portrait on the phase plane is only modified in that the jumps a along the

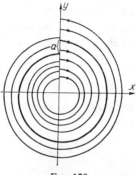

Fig. 129

y axis will no longer be constant in value but will be functions of y' (i.e. of the velocity which precedes the impulse):

$$a = \sqrt{y'^2 + h^2} - y'.$$

Therefore, as the velocity preceding the impulse increases, the increments of velocity will decrease. It is again clear from considerations of continuity that a spiral must exist, motion along which leads to a closed path or oscillation.

A similar analysis can be carried out when the trigger mechanism delivers impulses to the oscillating system twice in a period, i.e. for each passage of the system through the position of equilibrium†. The picture on the phase plane for this case is shown in Fig. 130. The phase paths consist of "half-turns" of the spirals of a damped oscillator and have

---

† Two impulses per period occur in the majority of trigger mechanisms and, in particular, in the anchor escapement used in pocket watches.

jumps on the y axis corresponding to the instantaneous impulses delivered to the pendulum by the trigger mechanism.

Let us indicate by v the values of the velocities *immediately after* the impulses. It is evident that the velocity v_2 after an impulse is uniquely determined by the velocity v_1 of the system after the previous impulse (Fig. 130). For the two assumptions introduced above for the impulse laws we shall obtain the following sequence functions connecting v_1 and v_2:

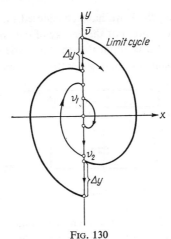

Fig. 130

$$v_2 = v_1 e^{-\frac{d}{2}} + a, \qquad (3.29)$$

if the impulse law is $\Delta y = a = $ const. for all impulses, and

$$v_2^2 = v_1^2 e^{-d} + h^2, \qquad (3.30)$$

if the impulse is $\Delta(y^2) = h^2 = $ const.

In both cases there is a single fixed point

$$\bar{v} = \frac{a}{1 - e^{-d/2}}$$

for the first impulse law, and

$$\bar{v} = \frac{h}{\sqrt{1 - e^{-d}}}$$

for the second. In both cases the system has a unique periodic motion which, as is easily shown by constructing the graphs of the sequence functions, is stable and is established for any initial conditions. The graphs of the sequence functions (Lamerey's diagrams) have a form similar to that of Fig. 124 (in the second case this graph should be plotted for v^2 instead of v). Again, a model with linear friction does not explain the need for an initial finite impulse to start the clock and must therefore be rejected. It is necessary to assume that the clock is a self-oscillating system with dry friction.

2. Valve generator with a discontinuous $\int$ characteristic

In the theory of the clock we have considered impulses which instantaneously vary the momentum or the energy of the system. It is natural to ask whether such an idea is applicable to the analysis of electronic valve oscillators. Let us assume that in a generator with an oscillating circuit in

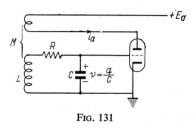

Fig. 131

the grid circuit (Fig. 131) there are such large sinusoidal oscillations that the grid voltage goes far, both into the region where the anode current is zero and into the saturation region. Then the grid voltage v (Fig. 131) changes its sign twice in a period. When v passes through zero in the positive direction, the anode current changes extremely rapidly from zero to I_s. During the switching on of the anode current the inductive e.m.f. $M di_a/dt$ increases very rapidly from the zero (when $i_a = 0$), to a certain very large value and then, when i_a approaches I_s, decreases at approximately the same rate down to zero. When v passes through zero in the negative direction, the anode current varies extremely rapidly from value I_s to zero. At the same time the inductive e.m.f. varies very rapidly from the zero to a certain very large negative value, and then again very rapidly goes to zero. Thus there is, in the oscillating circuit, an e.m.f. of short duration but very large value. If we assume that the current transition occurs with a jump (i.e. if we assume a $\int$ characteristic) then we shall have a voltage impulse which is infinitely large. There is a complete analogy with the case of the clock when the oscillating mass is acted upon at appropriate instants of time by an instantaneous impulse, communicating to this mass a fixed momentum.

In mechanics the action of a force $f(t)$, differing from zero during a sufficiently small interval of time τ, can be considered as an instantaneous impulse, causing a sudden change in momentum. Similarly if the induction e.m.f. $\mathscr{E} = M di_a/dt$ is different from zero during a sufficiently small interval of time then it gives rise to a rapid (for $\tau \to 0$, instantaneous) variation of the induction flux by an amount

$$\Delta(L\dot{q}) = \int_t^{t+\tau} \mathscr{E}\, dt = \int_t^{t+\tau} M \frac{di_a}{dt} dt = M[i_a(t+\tau) - i_a(t)] = M\Delta i_a,$$

At the instant of a jump of the induction flux the coordinate q itself (the charge of the capacitor) does not vary. The equation of the valve generator with a resonant circuit in the grid path (1.62) is

$$LC\ddot{v} + RC\dot{v} + v = M\frac{di_a}{dt},$$

Integrate with respect to time over the interval from t to $t+\tau$, during which the anode current varies from 0 to I_s. We shall then obtain

$$LC[\dot{v}(t+\tau) - \dot{v}(t)] + RC[v(t+\tau) - v(t)] + \int_t^{t+\tau} v\, dt = \int_t^{t+\tau} M\frac{di_a}{dt} dt.$$

In the limit, for a $\int$ characteristic, the grid voltage $v(t+\tau) \to v(t)$ as τ tends to zero. Also

$$\int_t^{t+\tau} v\, dt \to 0$$

and

$$LC[\dot{v}(t+\tau) - \dot{v}(t)] \to \int_t^{t+\tau} M\frac{di_a}{dt} dt = M\Delta i_a$$

or $$\Delta(L\dot{q}) \to M\Delta i_a.$$

Obviously as the grid voltage passes through zero in a positive direction ($q=0$, $\dot{q}>0$). $\Delta i_a = I_s$ and $\Delta(L\dot{q}) = MI_s$, whilst for a passage of v through zero in an opposite direction ($q=0$, $\dot{q}<0$) $\Delta(L\dot{q}) = -MI_s$. Then the equations of the system are

$$\left.\begin{aligned} \ddot{q} + 2h\dot{q} + \omega_0^2 q &= 0, \quad (q \neq 0), \\ \Delta\dot{q} &= \frac{M}{L} I_s \quad (q=0,\ \dot{q}>0), \\ \Delta\dot{q} &= -\frac{M}{L} I_s \quad (q=0,\ \dot{q}<0) \end{aligned}\right\} \quad (3.31)$$

with the additional condition that q varies continuously. Thus the oscillogram of any motion consists of arcs of damped sinusoids

$$q = Ae^{-h(t-t_0)} \cos\left[\omega(t-t_0)+\varphi\right],$$

beginning and ending on the time axis. At the join of two arcs there exists a difference of slopes determined by the jump.

The analysis of the system (3.31) is completely analogous to that carried out for the clock in the case of two impulses per period and for an impulse law of constant momentum per impulse.

3. Model of the clock with Coulomb friction

We have already considered the motion of an oscillator with Coulomb friction (Section 3, Chapter III), and with suitable units the equation of motion is

$$\left.\begin{array}{l}\ddot{x}+x = -f_0 \quad \text{for} \quad \dot{x} > 0, \\ \ddot{x}+x = +f_0 \quad \text{for} \quad \dot{x} < 0, \end{array}\right\} \tag{3.32}$$

where f_0 represents the force of friction. The phase paths will be spirals formed by segments of semi-ellipses (actually by semicircles because of the units chosen) the centre of the semicircles in the upper half-plane being the point $(-f_0, 0)$ and in the lower half-plane the point $(+f_0, 0)$. To simplify the analysis assume that an impulse occurs at $x = -f_0$ rather than $x = 0$. First, assume that

$$mv_1 - mv_0 = \text{const},$$

or, using the notation of the phase plane

$$\Delta y = a. \tag{3.33}$$

Let the velocity of the balance-wheel or pendulum immediately after an impulse be equal to v_1; the point A_1 representative of this state is at $(-f_0, v_1)$. The representative point will move along a circle with centre $(-f_0, 0)$ and with radius $R_1 = v_1$ (Fig. 132). On reaching the positive x axis, the representative point will fall on the *segment of rest* $(-f_0 \leqslant x \leqslant +f_0, y=0)$, consisting of the states of equilibrium (provided $R_1 \leqslant 2f_0$), or else (for $R_1 > 2f_0$) will pass into the lower half-plane and move along the semicircle with centre at the point $(+f_0, 0)$ the radius of which is $R_2 = R_1 - 2f_0 = v_1 - 2f_0$. If $R_2 \leqslant 2f_0$, the representative point will again fall on the segment of rest (but this time from below). Only for $R_2 > 2f_0$, i.e. for $v_1 > 4f_0$, will the representative point pass into the upper half-plane and move there along a quarter of a circle with centre at $(-f_0, 0)$ and of radius $R_3 = R_2 - 2f_0 = v_1 - 4f_0$, until it reaches "the axis of the impulses" ($x = -f_0$, $y > 0$) at the point $y = R_1$:

$$v_1' = v_1 - 4f_0. \tag{3.34}$$

Here the system receives an impulse as a result of which the velocity y instantaneously increases by a, and the representative point passes with a jump to the point $A_2(-f_0, v_2)$ where

$$v_2 = v_1' + a = v_1 - 4f_0 + a \quad (v_1 > 4f_0). \tag{3.35}$$

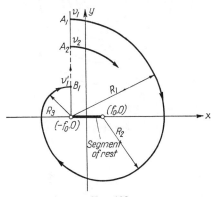

Fig. 132

The relation (3.35) is the required sequence function. Clearly, the sequence of the velocities $v_1, v_2, v_3, v_4, \ldots$ following successive impulses is an arithmetical progression the general term of which is

$$v_n = v_1 + (n-1)(a - 4f_0).$$

It is easily seen that the character of the possible motions depends on the sign of $a - 4f_0$.

Case I. $a - 4f_0 < 0$. In this case (Fig. 133), whatever the initial conditions, the oscillations will be damped and the representative point will make a finite number of swings and reach the state of equilibrium, $-f_0 < x < f_0$.

Case II. $a - 4f_0 > 0$. In this case if at the initial instant the representative point lies inside the region $a_1 b_1 c_1$ (Fig. 134), then the system will reach in a finite time a state of equilibrium. If, however, the initial conditions correspond to points lying outside this region or on its boundary, the oscillations will increase without limit.

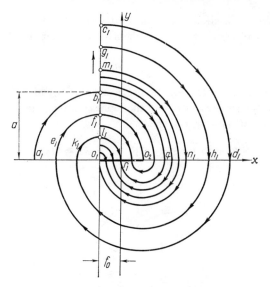

Fig. 133

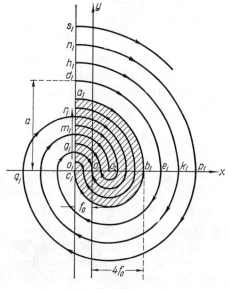

Fig. 134

Case III. $a - 4f_0 = 0$. In this ideal case if the initial conditions lie within the region $a_1 b_1 c_1$ (Fig. 135), then the representative point will arrive at a state of equilibrium without having completed one revolution. If, however, the initial conditions lie outside this region, then all motions are periodic with an amplitude which is determined by the initial conditions.

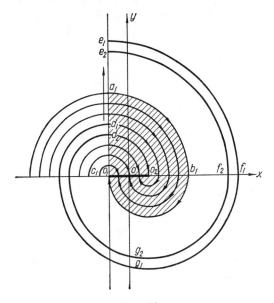

Fig. 135

We are dealing with a continuum of periodic motions and the system is unstable with respect to small variations of a parameter, which is typical of a conservative system. Any change in f_0 will change this case to either case I or II, i.e. to an essentially different portrait. Thus the assumption of constant friction, $mv_1 - mv_2 = \text{const.}$ and the law of impulse is unable to reproduce one of the most essential features of a real clock, namely that there is only one periodic motion with a well-determined amplitude, independent of the initial conditions. It is sufficient, however, to vary the assumption about the law of impulses in order to obtain a satisfactory model. We assume that

$$\Delta \left(\frac{mv^2}{2} \right) = \text{const.} \quad \text{or} \quad \Delta(y^2) = h \tag{3.36}$$

where h is a constant. Hence, as follows from the expression (3.36), the jump $a = (y^2+h^2)^{1/2} - y$. The sequence function for the velocities of the balance-wheel after an impulse will be

$$v_2^2 = (v_1 - 4f_0)^2 + h^2 \quad (v_1 > 4f_0) \tag{3.37}$$

since $v_1' = v_1 - 4f_0$ (equ. 3.34). The representative point returns to the "semi-axis of the impulses" $x = -f_0$, $y > 0$ for $v_1 > 4f_0$ only. The graph of the sequence function (3.37) is shown in Fig. 136. The intersection of

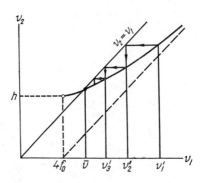

Fig. 136

this hyperbola with the straight line $v_1 = v_2$, if it exists, will give us $\bar{v}$ which is the velocity after an impulse for a *periodic* motion. It is evident that such a point exists for

$$h > 4f_0, \tag{3.38}$$

and is unique. We have

$$\bar{v}^2 = (\bar{v} - 4f_0)^2 + h^2,$$

hence

$$\bar{v} = \frac{h^2}{8f_0} + 2f_0$$

provided that $\bar{v} > 4f_0$, which is true if $h > 4f_0$. The amplitude of the periodic oscillation is

$$\bar{x} = \bar{v} - f_0 = \frac{h^2}{8f_0} + f_0.$$

If the condition (3.38) is satisfied then a *unique* periodic process with a fixed amplitude is possible in the system. It corresponds in the phase plane to a closed path formed by parts of circles and a segment of length a on the line $x = -f_0$. It can be shown, either by constructing "Lamerey's ladder"

(Fig. 136) or by making use of Königs's Theorem (see Section 7, Chapter V), that the fixed point $\bar{v}$ is *stable* and that neighbouring motions steadily approach the periodic motion.

It follows that the limiting periodic motion is stable in the sense of Liapunov. The picture on the phase plane (Fig. 137) shows that this model possesses the two most typical features of a clock system: (1) the presence of a unique periodic process of fixed amplitude and (2) the necessity of an initial impulse of a certain magnitude to start the oscillation. The larger

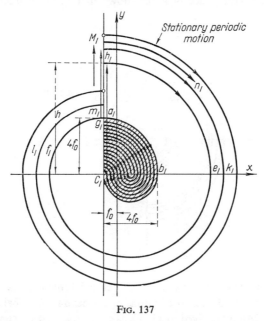

Fig. 137

the constant friction, the larger must be the initial impulse. The assumption itself of a constant friction does not of course include all the properties of the system and only reflects the most typical feature of this system. Of course in a clock there are both linear and Coulomb types of friction present; the friction of the trigger mechanism is better represented by means of Coulomb friction and the resistance of the air to the motion of the pendulum by linear friction. Introducing linear friction would not yield anything new, except that parts of spirals would have to be drawn on the phase plane instead of parts of circles. But constant friction does involve the existence of a new property; the absence of self-excitation and the necessity of an initial impulse to start a periodic process.

We shall obtain the same results by considering a model of a clock with two accelerating impulses at $rx = -f_0$, $y > 0$ and at $x = +f_0$, $y < 0$ during each oscillation. Again we have to assume that each impulse increases the kinetic energy of the pendulum or by a constant amount, obtaining the sequence function

$$v_2^2 = (v_1 - 2f_0)^2 + h^2,$$

where v_1 and v_2 are velocities of the pendulum before and after the impulse. The portrait on the phase plane as shown in Fig. 138 is easily obtained

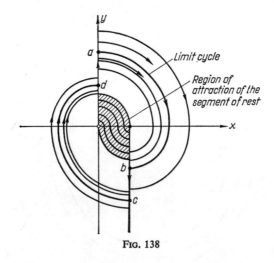

Fig. 138

and shows the stable limit cycle as the closed broken curve *abcda* corresponding to periodic self-oscillations of the pendulum or balance wheel.

Some conclusions arrived at above can be explained by simple energy considerations. Note that for a linear friction, the energy dissipated during a period is proportional to the square of the amplitude and, for a constant friction, it is a linear function of the amplitude. Also for an impulse according to the law $v_1 - v_2 = \Delta v_0 = $ const. the energy increases by

$$\frac{m}{2}(v_0 + \Delta v_0)^2 - \frac{m v_0^2}{2} = \frac{m}{2}(2v_0 \Delta v_0 + \Delta v_0^2),$$

which is a linear function of the amplitude. For an impulse $(mv_1^2/2) - (mv_0^2/2) = $ const. the energy of the system increases by a constant quantity. The main results now become clear. Any periodic process is only possible when the energy has the same value at the beginning and end of a period. In the first case ("linear friction" and impulse law $v_1 - v_0 =$ const) the energy losses increase in proportion to the square of the amplitude and the energy entering the system is a linear function of the amplitude. It is

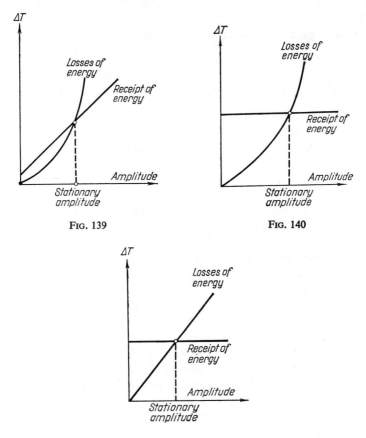

Fig. 139

Fig. 140

Fig. 141

clear that an energy balance is achieved for one and only one amplitude and so only one stationary amplitude exists (Fig. 139). In the second case (linear friction and impulse law $v_1^2 - v_0^2 =$ const) the losses are proportional to the square of the amplitude and the energy entering the system is a constant quantity. Again there exists only one stationary amplitude for which an energy balance takes place (Fig. 140). In the third case (constant friction and impulse law $v_1 - v_0 =$ const) both the losses and the energy entering the system are linear functions of the amplitude. Therefore, either there is no stationary amplitude or infinitely many.

Finally, in the fourth case (constant friction and impulse law, $v_1^2 - v_0^2 = $ const) the energy losses are a linear function of the amplitude, while the energy entering the system has a constant value and again there is only possible one stationary amplitude. (Fig. 141).

§ 5. Theory of the clock.
Model of a "recoil escapement" without impulses[†]

In the preceding section we have considered a few simple models of clock mechanisms, which enabled us to clarify certain basic properties such as the existence of a unique periodic motion and the necessity of an initial jump of a finite value for the excitation of these oscillations. To establish this hard mode of self-excitation we had to assume dry friction in the clock. These models, however, are only coarse models and cannot reproduce other important quantitative characteristics or explain the dependence of the period of oscillations on the forces of the driving mechanism and friction[‡].

In order to relate the rate or frequency of the clock to these parameters a more detailed analysis is required taking into account the main features of the type of clock mechanism and in particular of the trigger device employed in it[††]. Below we shall consider under certain simplifying assumptions the dynamics of the clock provided with the so-called "recoil escapement"[‡‡]. A schematic diagram of this escapement is shown in Fig. 142. The escape wheel, connected by a system of gears to the driving mechanism, meshes with the ends *(pallets)* of the anchor, which is on the same axis as the balance-wheel or pendulum of the clock. The driving mechanism produces (via the escape wheel and the anchor) a moment M which is applied to the balance-wheel and depends on the relative position of the escape wheel

---

[†] Written by N. A. Zheleztsov.

[‡] The force of the driving mechanism and the force of friction are the parameters of the clock which vary the most. For example, the force of a spring winding mechanism decreases as the spring unwinds, while the forces of friction depend in a noticeable measure on the position of the clock. This leads to a certain variation of the period of the oscillations of the clock.

[††] A considerable contribution to the development of the dynamics of the clock was made in recent years in the works by N. N. Bautin [22–25, 27, 28].

[‡‡] Such an escapement, also called "anchor escapement", finds application in wall clocks, alarm clocks, etc. The "barrel" escapement, which was used in the earliest constructions of clocks, in particular in Huygens's clocks, has similar dynamics to the trigger type mentioned above.

and the anchor. Figure 142(a) shows the middle position of the anchor of the balance-wheel ($\varphi=0$), for which the escape wheel coming into contact with the right-hand pallet P_1 of the anchor by means of the tooth A_1 turns the balance-wheel in a direction opposite to the rotation of the clock hand (we shall take this direction as the positive φ direction). This accelerated motion of the balance-wheel will last until the balance-wheel is rotated through an angle φ_0 (Fig. 142(b)) and the tooth A_1 escapes from the

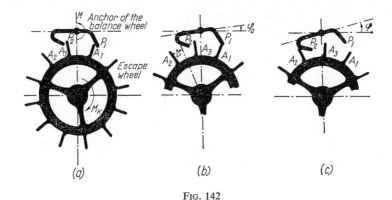

Fig. 142

pallet P_1 (the angle $2\varphi_0$ is called the "angle of lift" of the balance-wheel). After the tooth A_1 has escaped from the pallet P_1 the balance-wheel will continue to move in the positive φ direction, while the escape wheel, having become free from the balance-wheel rapidly accelerates (the acceleration being the larger, the smaller the moment of inertia of the escape wheel). In this motion the tooth A_2 and the left-hand pallet P_2 of the anchor move towards each other† and after a certain small interval of time the tooth A_2 will deliver a counter-impulse to the balance-wheel via the pallet P_2. After tooth A_2 has made contact with pallet P_2 the escape wheel will generate a negative moment on the axis of the anchor, thus retarding the motion of the balance-wheel. The balance-wheel, however, can still rotate because of its inertia by a certain positive angle thus compelling the escape wheel to make a certain "recoil" (hence the name of the escapement). Only then will motion in the opposite direction begin, being speeded up by the moment applied by the escape wheel via pallet P_2 (Fig. 142(c)). The balance-wheel passes through the middle position ($\varphi=0$) and as the

† This is the so-called "fall of the escape wheel". The angle Δ of Fig. 142 (b) is called the angle of fall of the escape wheel.

angle $\varphi = -\varphi_0$ is reached the tooth A_2 escapes from pallet P_2. After the "fall of the escape wheel" the tooth A_3 comes into contact with the pallet P_1, and the motion of the balance-wheel in the negative φ direction is retarded, and then there begins again motion in the positive direction ($\dot{\varphi} > 0$), etc. Thus the escape wheel impels the balance-wheel twice during its oscillation and, doing positive work, compensates the energy loss due to friction.

The analysis of clocks with such a trigger device is fairly complicated [16, 22]. To simplify the problem, we choose a model with *one* degree of freedom, on the assumption that *the moment of inertia of the escape wheel I_k is equal to zero*[†]. The "fall of the escape wheel" after each tooth escapes from a pallet will be thus *instantaneous*, and one of the pallets, either the right-hand one P_1 or the left-hand one P_2, will always be found in contact with a tooth of the escape wheel.

The kinematics of the escapement, are such that for $\varphi \leqslant -\varphi_0$, only the right-hand pallet P_1 is in contact with a tooth of the escape wheel, and for $\varphi \geqslant +\varphi_0$, only the left-hand pallet P_2, and for $-\varphi_0 < \varphi < +\varphi_0$ any of the pallets. For a contact of a tooth of the escape wheel with the pallet P_1 the moment M applied to the axis of the balance-wheel by the escape wheel is greater than zero, and for a contact with the pallet P_2 the moment $M < 0$. Therefore the moment developed by the escape wheel on the axis of the balance-wheel $M = M(\varphi)$ over the interval $-\varphi_0 < \varphi < +\varphi_0$ is a *twovalued* function of the angle of rotation of the balance-wheel φ. $M > 0$ or $M < 0$ according to which of the pallets is in contact with a tooth of the escape wheel.

This imposes certain limitations on the shape of the phase plane trajectories for the dynamic model of this clock. In fact we cannot use the usual plane with Cartesian coordinates φ and $\dot{\varphi}$[‡], since assigning a point $(\varphi, \dot{\varphi})$ where $-\varphi_0 < \varphi < +\varphi_0$ does not uniquely determine the forces acting in the system and therefore does not uniquely determine the state of the system. To correspond with the *two-valuedness* of the moment M upon φ we can use as the phase surface a "plane with superposition" (Fig. 143), consisting of two half-planes superimposed: (*I*) $\varphi < \varphi_0$ and (*II*) $\varphi > -\varphi_0$. The first of them corresponds to contact of a tooth of the escape

[†] If we reduce the moment of inertia of the escape wheel (together with the whole system of gears and the driving mechanism) to the axis of the balance-wheel, then in many clock mechanisms it will amount to a few per cent of the moment of inertia I of the balance wheel. Our assumption that $I_k = 0$ will be fairly well satisfied for such clocks.

[‡] We shall ignore the fact that owing to construction reasons the angles of rotation are limited and shall assume that φ can take any value.

wheel with the right-hand pallet P_1, the second one with the left-hand pallet P_2. The points of this *two-sheet* phase surface and the states of the system have a *one-to-one correspondence*. In this connexion the passage of the representative point from the sheet (*I*) to the sheet (*II*) occurs for $\varphi = +\varphi_0$ and the reverse passage for $\varphi = -\varphi_0$. As the representative point passes from one sheet to another, its abscissa remains unvaried. Owing to our assumption that the moment of inertia of the escape wheel I_k is equal

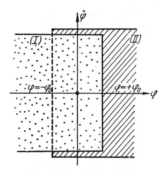

Fig. 143

to zero, the ordinate $\dot\varphi$ (the velocity of the balance-wheel) will also remain constant despite the presence of counter-impulses delivered to the pallets by the teeth of the escape wheel at the end of the "fall" of the escape wheel[†].

To simplify further the analysis of the dynamics of this clock model we shall make the following assumptions about the forces acting on the balance-wheel of the clock. First of all, assume that the moment M applied to the balance-wheel by the escapement is constant in absolute value, i.e. $M = +M_0$ when the pallet P_1 is in contact with a tooth of the escape wheel and $M = -M_0$ for a contact with the pallet P_2 (Fig. 144)[‡]. Secondly we shall represent the forces of friction as Coulomb-friction forces and denote

---

[†] The passage of the representative point from one sheet to the other corresponds to the process of disengagement of the escape wheel from the balance-wheel, i.e. its "fall" (a rotation by an angle Δ) and the counter-impulse of the corresponding pallet. The escape wheel, by virtue of the assumption about the inertia $I_k = 0$ does not have any moment of momentum and, hence, for a counter-impulse of one of its teeth against either pallet (for simplicity we shall assume the impact to be absolutely inelastic) will not modify the velocity of the balance-wheel $\dot\varphi$.

[‡] The form of the function $M = M(\varphi)$ for a contact of a tooth of the escape wheel with a given pallet is determined by the profile of the pallet, and the profile can be so chosen that M be constant.

the maximum moment of the friction by f_0, assuming it to be independent of the angle or rotation of the balance-wheel[†].

On the basis of the dynamic model for a clock with "recoil" escapement, we shall examine two types of clock mechanisms. One of them has a balance-wheel without a natural period (the centre of gravity of such a balance-wheel lies on its axis of rotation and there is no spring to bring it to the middle position $\varphi=0$). This type of clock mechanism does not possess

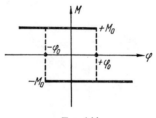

Fig. 144

good stability and is therefore used in those cases when a high stability of the period of the oscillations is not needed (it is used, for example, in the automatic trigger of photo equipment, in the ringing device of alarm clocks etc.) [16, 25, 67].

A second type of clock mechanism has a balance-wheel with a natural period (a balance-wheel with spring or a pendulum) which, when the escapement is disconnected, can perform damped oscillations. This second type, having a sufficiently good stability of motion, is used in some types of clocks such as wall clocks[‡].

1. Model of clock with a balance-wheel without natural period

The dynamic equations of this model can be put in the form

$$I\frac{d^2\varphi}{dt^2} = f\left(\varphi, \frac{d\varphi}{dt}\right) + M(\varphi),$$

[†] The forces of dry friction in the oscillating system of a clock arise in two places: in the bearings of the axis of the balance-wheel and in the sliding of a tooth of the escape wheel along one of the pallets of the anchor of the balance-wheel. The latter forces of friction are the basic ones in the majority of clocks and are evidently proportional, for a given coefficient of friction, to the force with which a tooth is pressed against the pallet, and so to the force of the driving mechanism.

[‡] The dynamics of early clocks without a pendulum, often called pre-Galileian clocks, is similar to the dynamics of the first type of clock mechanism. The dynamics of the second type of clock is close to the dynamics of Huygens's clocks [128].

THEORY OF THE CLOCK. MODEL WITHOUT IMPULSES

where I is the moment of inertia of the balance wheel, $M = M(\varphi)$ is the moment produced by the escapement on the axis of the balance-wheel, and $f(\varphi, d\varphi/dt)$ is the moment of the forces of Coulomb friction. During motion of the balance-wheel ($d\varphi/dt \neq 0$)

$$f\left(\varphi, \frac{d\varphi}{dt}\right) = -f_0 \operatorname{sgn} \frac{d\varphi}{dt}$$

and the equation of motion takes the form

$$I \frac{d^2\varphi}{dt^2} = -f_0 \operatorname{sgn} \frac{d\varphi}{dt} \pm M_0 \tag{3.39}$$

(the upper sign applying for a contact of a tooth of the escape wheel with the pallet P_1, the lower one for a contact with the pallet P_2).

Let us introduce the new non-dimensional variables

$$x = \frac{\varphi}{\varphi_0}, \quad t_{\text{new}} = +\sqrt{\frac{M_0}{I \cdot \varphi_0}}\, t;$$

then equation (3.39) reduces to the following form:

$$\left.\begin{array}{l} \dot{x} = y, \\ \dot{y} = -F \operatorname{sgn} y - (-1)^n \end{array}\right\} \quad (y \neq 0), \tag{3.41}$$

where

$$F = \frac{f_0}{M_0},$$

n being the ordinal number of the pallet which is found in contact with a tooth of the escape wheel (a dot indicating differentiation with respect to the new non-dimensional time).

In the case of $M_0 \leqslant f_0$ (i.e. $F \geqslant 1$) and of an oscillator at rest ($y = 0$) the moment of the escapement cannot overcome the forces of dry friction, therefore $\dot{y} = 0$ (i.e. $d^2\varphi/dt^2 = 0$) and any state $(x, 0)$ is a state of equilibrium. In this case no periodic motions are possible and any motion ends with arrival at one of the states of equilibrium.

Therefore we shall now assume that $M_0 > f_0$ and so $F < 1$. In this case the system has no states of equilibrium.

By dividing the second of the equations (3.41) by the first one, we obtain for the phase paths on the sheet (I) ($x < +1$; the right-hand pallet P_1 is in contact with a tooth of the escape wheel and $M = +M_0$) the following equation

$$\frac{dy}{dx} = \frac{1 - F \operatorname{sgn} y}{y};$$

and after integrating

$$\frac{y^2}{2} - (1+F)x = \text{const} \qquad (3.42a)$$

on the lower half-plane of the sheet ($y<0$), and

$$\frac{y^2}{2} - (1-F)x = \text{const} \qquad (3.42b)$$

on the upper one ($y>0$). Thus the phase paths on the sheet (I) consist of two parabolae (3.42a) and (3.42b), and the representative point moves to the left on the lower half of the sheet, since there $\dot{x} = y < 0$, and to the right on the upper one (Fig. 145). All phase paths on the sheet (I) reach its boundary on the semiaxis $x = +1, y > 0$.

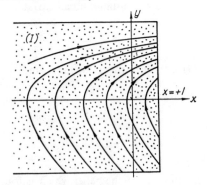

Fig. 145

The phase paths on the sheet (II) are *symmetrical* (with respect to the origin of the coordinates) with the paths on the sheet (I), since the equations (3.41) for the phase paths on the sheet (II) (the half-plane $x > -1$) reduce to the equations for the paths on the sheet (I) for a change of the variables x and y into $-x$ and $-y$. To clarify the possible motions of the balance-wheel we shall draw two axes: (v) where $x = -1, y = -v < 0$ and (v') where $x = +1, y = v' > 0$ and shall consider the sequence of the points of intersection with them of an arbitrary phase path, i.e. the sequence v, v_1, v_2, v_3 (Fig. 146)[†]. Let the representative point pass at the point $(-1, -v)$ from the sheet (II) to the sheet (I). It will move along the

† It is evident that the points of these axes correspond to states of the system for which, after the disengagement of one of the pallets, the other pallet comes into contact with a tooth of the escape wheel; v and v' are absolute values of the velocity of the balance-wheel in these states.

parabola (3.42a) and reach the axis of the abscissae at the point $(-\xi, 0)$, ξ being evidently determined by the equation

$$\frac{v^2}{2} + (1+F) = (1+F)\xi,$$

or

$$v^2 = 2(1+F)(\xi - 1). \tag{3.43a}$$

Then the representative point moves on the upper half of the sheet (I) and reaches the boundary of this sheet at the point $(+1, v_1)$, where $v_1 > 0$ and is determined by the relation

$$(1-F)\xi = \frac{v_1^2}{2} - (1-F),$$

or

$$v_1^2 = 2(1-F)(\xi + 1). \tag{3.43b}$$

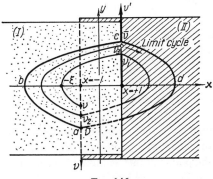

Fig. 146

Thus the phase paths on the sheet (I) establish between the points of the axes (v) and (v') a certain one-to-one continuous correspondence or, in other words, generate a point transformation of the axis (v) into the axis (v') represented by the sequence function (3.43a) and (3.43b) (the sequence function is written in a parametric form; the parameter being ξ, the maximum deviation of the balance-wheel)†. Afterwards the representative point passes onto the sheet (II) and, moving along the corresponding phase path (for which there is a symmetric one on the sheet (I)), reaches the semi-axis (v) at a certain point $(-1, -v_2)$. Owing to the symmetry of the phase paths

† Of course, the parameter ξ is easily eliminated, and the sequence function can be written in an explicit form. However, in many cases it is difficult to obtain the sequence function written in its explicit form, while it is comparatively easy to obtain it in a parametric form (see Chapter VIII).

on the sheets (*I*) and (*II*), v_2 is determined from v_1 by the same sequence function in the relations (3.43a) and (3.43b). In other words the point transformation of the axis (v') into the axis (v) coincides with the point transformation of the axis (v) into the axis (v') and therefore we shall speak below of a single point transformation of the axes (v) and (v') into each other.

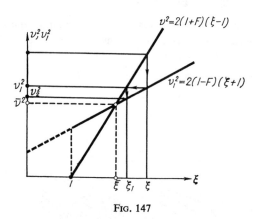

Fig. 147

Consequently each subsequent velocity in the sequence of the balance-wheel velocities $v, v_1, v_2, v_3, \ldots$ (at the instants of change of the contacting pallet), is determined from the preceding one by this sequence function. The fixed point $\bar{v}$ for which $v = v_1 = \bar{v}$ corresponds, clearly, to a *symmetric limit cycle*, being the points of intersection of this limit cycle with the axes (v) and (v'). We have for the fixed point

$$(1+F)(\bar{\xi}-1) = (1-F)(\bar{\xi}+1),$$

and therefore the amplitude of the self-oscillations of the balance-wheel is

$$\bar{\xi} = \frac{1}{F} \tag{3.44}$$

and

$$\bar{v}^2 = 2\frac{1-F^2}{F}. \tag{3.45}$$

In order to study the stability of the limit cycle construct on a single diagram the curves $v^2 = v^2(\xi)$ and $v_1^2 = v_1^2(\xi)$ obtaining the two straight lines shown in Fig. 147. Their point of intersection is the fixed point. If v is chosen (Fig. 147) then ξ can be determined from the straight line (3.43a)

Then from the straight line (3.43b) determine v_1; from v_1 as a new initial point, ξ_1 can be found and then v_2, etc. "Lamerey's ladder" thus constructed tends to the fixed point owing to the fact that the straight line $v^2 = 2(1+F)(\xi-1)$ has a steeper slope than the straight line $v_1^2 = 2(1-F)(\xi+1)$. The sequence $v, v_1, v_2, v_3, \ldots$ tends to $\bar{v}$ for any $v > \bar{v}$, or for $v < \bar{v}$. This shows the stability of the unique periodic motion of the balance-wheel and it shows that this motion will result for any initial conditions.

The amplitude of the self-oscillations of the balance-wheel is given by the formula (3.44) or in ordinary angle units

$$\bar{\varphi} = \varphi_0 \bar{v} = \varphi_0 \frac{M_0}{f_0}.$$

The limit cycle (it is shown in Fig. 146) consists of four arcs of a parabola on each of which the acceleration $\ddot{y}$ of the balance-wheel is constant. The acceleration $\ddot{y}$ is equal to $1+F$ on the arc of parabola ab and therefore the time taken by the representative point to move along this arc of the limit cycle is equal to

$$\tau_1 = \frac{\bar{v}}{1+F};$$

similarly along the arc bc $\ddot{y} = 1-F$ and the time of transit is equal to

$$\tau_2 = \frac{\bar{v}}{1-F}.$$

Therefore the period of the self-oscillations (in units of the non-dimensional time) is equal to

$$\tau = 2(\tau_1 + \tau_2) = \frac{4\sqrt{2}}{\sqrt{F(1-F^2)}} \qquad (3.46)$$

or in ordinary units

$$T = 4\sqrt{\frac{2\varphi_0 I}{M_0}} \cdot \frac{1}{\sqrt{F(1-F^2)}}. \qquad (3.47)$$

Thus the period of the self-oscillations of the balance-wheel depends both on the force of the driving mechanism and on the force of friction. The moment M_0 developed by the escapement is proportional to the force of the driving mechanism. Also the pressure exerted by the teeth of the escape wheel is proportional to the force of the drive on the pallets and therefore, approximately, so is the frictional moment f_0 acting on the balance-wheel. We can therefore assume, to this degree of accuracy, that the coefficient F, and so the amplitude of the self-oscillations $\bar{\xi}$ or $\bar{\varphi}$, do not depend on the force of the driving mechanism and are essentially determined by the coefficient of friction of a tooth of the escape wheel on a pallet. The period of the self-oscillations depends on both M_0 and F (graphs of the dependence of T on M_0 and F are given in Fig. 148 and Fig. 149). As

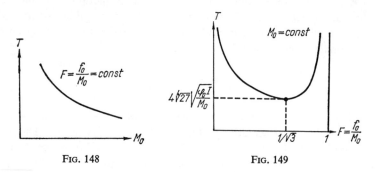

FIG. 148   FIG. 149

a quantitative measure of the stability of motion of the clock against a variation of driving force and of the coefficient of friction the following quantities can be evaluated:

$$S_M = \frac{1}{\dfrac{M_0}{T}\left|\dfrac{\partial T}{\partial M_0}\right|_{F=\text{const}}}, \quad S_f = \frac{1}{\dfrac{F}{T}\left|\dfrac{\partial T}{\partial F}\right|_{M_0=\text{const}}}.$$

They are the ratios of the percentage variations of the period to the percentage variation of one or other parameters. Proceeding from the formulae (3.47),

$$S_M = 2 \quad \text{and} \quad S_f = 2\frac{1-F^2}{|1-3F^2|}. \tag{3.48}$$

Maximum stability of the motion of the clock for a variation of the coefficient of friction ($S_f = \infty$ or $\partial T/\partial F = 0$) is obtained for $F = 3^{-1/2}$, but the stability of motion for a variation of the force of the driving mechanism is always small ($S_M = 2$).

2. Model of clock with a balance-wheel having a natural period

The balance-wheel is now acted upon not only by the force of friction and the forces exerted by the escapement but also by an elastic restoring force.

The equation of motion is now

$$I\frac{d^2\varphi}{dt^2}+k\varphi = f\left(\varphi, \frac{d\varphi}{dt}\right)+M(\varphi),$$

where the notation of the preceding sub-section has been retained and $k\varphi$ is the moment of the spring of the balance-wheel[†]. When the balance-wheel is *in motion* ($d\varphi/dt \neq 0$) and if the new variables

$$x = \frac{\varphi}{\varphi_0}, \quad t_{\text{new}} = \omega_0 t = \sqrt{\frac{k}{I}}\,t$$

and the non-dimensional parameter of the clock

$$\lambda = \frac{M_0}{k\varphi_0} \quad \text{and} \quad r = \frac{f_0}{k\varphi_0},$$

are introduced, then the equations above can be reduced to two differential equations of the first order:

$$\left.\begin{aligned}\dot{x} &= y,\\ \dot{y} &= -x - r\,\text{sgn}\,y - \lambda(-1)^n,\end{aligned}\right\} \qquad (3.49)$$

As before, sheet (*I*), corresponding to a contact of the right-hand pallet P_1 with a tooth of the escape wheel, is the half-plane $x < +1$, and the sheet (*II*), corresponds to a contact of the left-hand pallet P_2, and is the half-plane $x > -1$.

The phase paths on sheet (*II*) are symmetrical with the paths on the sheet (*I*) with respect to the origin of the coordinates. Since friction is of dry Coulomb type, equilibrium occurs at each state at which the motion ceases ($d\varphi/dt = 0$ or $y=0$), and the sum of the moments of the forces of the spring and the escapement does not exceed the maximum moment of the force of friction of rest, i.e.

$$|M_0 - k\varphi| < f_0 \quad \text{or} \quad |x - \lambda| < r.$$

Clearly, three cases can arise according to the values of the parameters: (a) if $\lambda + r \leq 1$, then on the sheet (*I*) there is a segment $0_1^+, 0_1^-$, ($y=0$, $\lambda - r \leq x \leq \lambda + r$), consisting of states of equilibrium (Fig. 150); (b) if $\lambda + r > 1$, but $\lambda - r \leq 1$, then the points of the segment ($y=0$, $\lambda - r \leq x \leq 1$) will be states of equilibrium, and finally (c) if $\lambda - r > 1$ there are no states of equilibrium.

By integrating (3.49) for the sheet (*I*) (where $n=1$), it is easily verified that the phase paths will be arcs of the semicircles:

$$y^2 + [x - (\lambda + r)]^2 = \text{const} \qquad (3.50\text{a})$$

[†] This equation is valid for a clock with a pendulum having a small angular swing.

in the lower half of the sheet, with a centre at the point 0_1^+ $(\lambda-r, 0)$. Figure 150 shows the phase paths on the sheet (I) for the case $\lambda+r<1$. The phase paths beginning in the shaded region arrive after a finite time at the segment $0_1^+ \, 0_1^-$. All remaining phase paths reach the boundary of the sheet on the axis

$$x = +1, \quad y > 0. \tag{350b}$$

To explain the possible motions of the clock, again draw on the phase plane two axes: (v), $x=-1$, $y=-v$ $(v>0)$ and (v'), $x=+1$, $y=v'>0$ and consider the point transformation between them which the phase paths

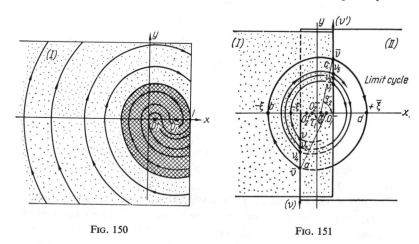

Fig. 150   Fig. 151

determine. Suppose the representative point moves from the sheet (II) on to the sheet (I) at the point $(-1, -v)$ of the axis v (Fig. 151). It will arrive at the axis of the abscissae at the point $(-\xi, 0)$ where $\xi>1$ and is given by

$$v^2 = [\xi+\lambda+r]^2-[1+\lambda+r] = \xi^2+2(\lambda+r)\xi-1-2(\lambda+r). \tag{3.51a}$$

If $-\xi<\lambda-r$ then the representative point crosses the axis of the abscissae and moves to the upper half of the sheet (I) along the semicircle (3.50b):

$$y^2+[x-(\lambda-r)]^2 = [\xi+\lambda-r]^2$$

and will either reach the semi-axis (v') at the point $(+1, v_1)$, determined by the equation

$$v_1^2 = [\xi+\lambda-r]^2-[1-\lambda+r]^2 = \xi^2+2(\lambda-r)\xi-1+2(\lambda-r), \tag{3.51b}$$

or will arrive at the segment of rest (stagnation) at one of the states of equilibrium. The latter takes place when

$$\xi+\lambda-r < 1-\lambda+r \quad \text{or} \quad \xi < \xi_1 = 1-2(\lambda-r).$$

The relations (3.51a) and (3.51b) are the sequence functions written in parametric form. The sequence function for the point transformation of the axis (v') into the axis (v), as established by the phase paths on sheet (II), has the same form because of the symmetry. This sequence function determines the sequence of points of intersection of any phase path with the axes (v) and (v') i.e. the sequence $v, v_1, v_2, v_3, \ldots$ The fixed point $\bar{v}$ where $v = v_1 = \bar{v}$ corresponds to a symmetric limit cycle (Fig. 151).

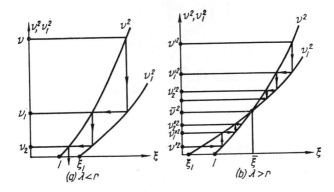

Fig. 152

To find the limit cycle and to assess its stability we can construct Lamerey's diagram (Fig. 152). Having constructed on it the curves (3.51a) and (3.51b) (the first constructed for $\xi > 1$ only and the second for $\xi > \xi_1 = 1-2(\lambda-r)$), it is easy to find the fixed point as the point of intersection of these curves (in Fig. 152, v^2 and v_1^2 have been plotted and so the curves (3.51a) and (3.51b) are parabolae). Evidently, if $\xi_1 > 1$, for which $\lambda < r$, then the curves (3.51a) (3.51b) do not intersect and the sequence of numbers $v, v_1, v_2, \ldots$, will be monotonically decreasing so that for any initial conditions the system arrives at a state of equilibrium. In this case there will be no self-oscillations (Fig. 152(a)).

If, however, $\xi_1 < 1$, for which

$$\lambda > r, \tag{3.53}$$

then the curves (3.51a) and (3.51b) have a single point of intersection, which is a stable fixed point (Fig. 152(b). There is then a single *stable limit cycle*, corresponding to self-oscillations of the clock (the limit cycle for $\lambda + r < 1$ and $\lambda > r$ is shown in Fig. 151).

Depending on the values of the parameters λ and r ($\lambda > r$) we have either soft or hard excitation. If $r < \lambda < 1 + r$, then there is not only a stable

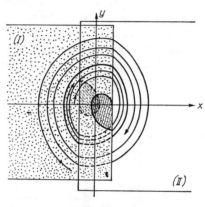

Fig. 153

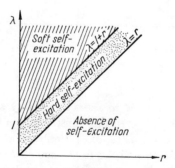

Fig. 154

limit cycle but also segments of stable states of equilibrium (on each sheet) and self-excitation cannot occur for initial conditions inside the shaded area of Fig. 153. If, however, $\lambda > 1 + r$, then there are no states of equilibrium and *all* phase paths approach the limit cycle as $t \to \infty$ and there is soft self-excitation. Fig. 154 shows the regions of various modes of operation of the clock. Now, periodic motion of the oscillating system of the clock exists for $\lambda > r$ only or, which is the same, for $M_0 > f_0$. Equating

v^2 to v_1^2 in the expressions (3.51a) and (3.51b) we obtain for the amplitude of oscillation

$$\bar{\xi} = \frac{\lambda}{r}, \qquad (3.54)$$

or in units of angle†.

$$\bar{\varphi} = \varphi_0 \bar{\xi} = \varphi_0 \frac{M_0}{f_0}$$

The period is found by noting that the representative points move along the phase paths with an angular velocity equal to unity‡. Therefore the time of transit (in units of non-dimensional time) along one of the arcs of a semicircle which form the limit cycle, is equal to the angle subtended at its centre by the arc, and the period is thus

$$\tau = 2(\pi + \tau_1 - \tau_2),$$

where τ_1 and $\pi - \tau_2$ are the subtended angles of the arcs ab and bc of the limit cycle (Fig. 151). Clearly τ_1 and τ_2 satisfy the inequalities

$$0 < \tau_1 < \frac{\pi}{2} \quad \text{and} \quad 0 < \tau_2 < \pi,$$

and are determined by the relations

$$\cos \tau_1 = \frac{1+\lambda+r}{\bar{\xi}+\lambda+r} \quad \text{and} \quad \cos \tau_2 = \frac{1-\lambda+r}{\bar{\xi}+\lambda-r}. \qquad (3.55)$$

Since $(1+\lambda+r)/(\bar{\xi}+\lambda+r) > (1-\lambda+r)/(\bar{\xi}+\lambda-r)$, then $\tau_1 < \tau_2$ and the period of the self-oscillations is

$$\tau = 2\pi - 2(\tau_2 - \tau_1) < 2\pi. \qquad (3.56)$$

---

† We obtained the same expression for the amplitude of the self-oscillations in the case of the clock with a balance-wheel without a natural period (see (3.44)). This follows from the fact that the moment of the spring of the balance-wheel $k\varphi$ is conservative. Since the work of the escapement during a time equal to the period of the self-oscillations is equal to $4M_0 \varphi_0$ while the work of the Coulomb forces of friction during the same interval of time is equal to $4f_0\bar{\varphi}$, the energy-balance equation can be written in the form $4M_0\varphi_0 = 4f_0\bar{\varphi}$, independently of whether the balance-wheel has a spring or not, since the work of the spring during a self-oscillation period is equal to zero. We shall obtain, from this energy-balance equation, for both types of clocks: $\bar{\varphi} = \varphi_0 \frac{M_0}{f_0}$.

‡ In fact, according to the equations (3.49) the square of the phase velocity is equal to $x^2 + \dot{y}^2 = y^2 + [x(-1)^n \lambda + r\, sgn y]^2 = R^2$, therefore the angular velocity is equal to unity.

In ordinary units the period is

$$T = \frac{\tau}{\omega_0} = \tau\sqrt{\frac{I}{k}} < 2\pi\sqrt{\frac{I}{k}}. \qquad (3.57)$$

It is always less than the period of the free oscillations of the balance-wheel or pendulum.

The period T of the self-oscillations depends on the parameters λ (i.e. on the force of the driving mechanism) and $\bar{\xi} = \lambda/r$ (i.e. the friction). The most interesting case in practice is for small values of λ and $r (r < \lambda \ll 1)$[†]. For a given $\bar{\xi}$, and for λ and r both tending to zero, τ_1 and τ_2 tend to $\tau^0 = \cos^{-1}(1/\bar{\xi})$ and τ tends to 2π. For both λ and $r \ll 1$, the following approximate formulae hold:

$$[-\sin \tau^0](\tau_1 - \tau^0) = \frac{1+\lambda+r}{\bar{\xi}+\lambda+r} - \frac{1}{\bar{\xi}} = \frac{(\bar{\xi}-1)(\lambda+r)}{\bar{\xi}(\bar{\xi}+\lambda+r)}$$

and, neglecting $\lambda + r$ in comparison with $\bar{\xi}$,

$$\tau_1 \approx \tau^0 - \frac{(\bar{\xi}-1)(\lambda+r)}{\bar{\xi}^2 \sin \tau^0};$$

and similarly

$$\tau_2 \approx \tau^0 + \frac{(\bar{\xi}+1)(\lambda-r)}{\bar{\xi}^2 \sin \tau^0}$$

and, consequently,

$$\tau \approx 2\pi - 4\lambda \frac{\sqrt{\bar{\xi}^2-1}}{\bar{\xi}^2} = 2\pi - 4r\sqrt{1-\frac{r^2}{\lambda^2}}, \qquad (3.57)$$

since $\sin \tau^0 = (1+\bar{\xi}^{-2})^{\frac{1}{2}}$. A graph of τ as a function of $r/\lambda = 1/\bar{\xi}$ (for a constant λ) is given in Fig. 155 $\left(\partial\tau/\partial r = 0 \text{ for } r/\lambda = 2^{-\frac{1}{2}}\right)$. By considering τ as a function of λ and $r/\lambda = 1/\bar{\xi}$, the following expressions are easily obtained for the stability of motion under variations of the force of the

---

[†] Just as in the preceding sub-section, we shall assume that the maximum moment of the friction of rest f_0 is proportional to the force with which the teeth of the escape wheel are pressed against the pallets of the balance-wheel or, otherwise, to the force of the driving mechanism. Then the ratio $\lambda/r = \bar{\xi}$ will not depend on the force of the driving mechanism but will be determined by the coefficient of friction between the surfaces of the tooth of the escape wheel and of the anchor of the balance-wheel.

driving mechanism and of the coefficient of friction

$$S_M = \cfrac{1}{\cfrac{\lambda}{2\pi}\left|\cfrac{\partial \tau}{\partial \lambda}\right|_{\frac{r}{\lambda}=\text{const}}} = \cfrac{\pi \bar{\xi}^2}{2\lambda \sqrt{\bar{\xi}^2-1}} = \cfrac{\pi}{2r\sqrt{1-\cfrac{r^2}{\lambda^2}}}$$

and

$$S_f = \cfrac{1}{\cfrac{r}{2\pi}\left|\cfrac{\partial \tau}{\partial r}\right|_{\lambda=\text{const}}} = \cfrac{2\pi\sqrt{1-\cfrac{r^2}{\lambda^2}}}{r\left(1-2\cfrac{r^2}{\lambda^2}\right)}.$$

(3.58)

The stability of motion of the clock is the better, the smaller r and λ. Furthermore, the stability of motion of the clock with a "balance-wheel having a natural period" can be made considerably better than the stability of the clock with a balance-wheel without a natural period [23].

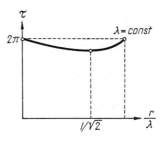

Fig. 155

§ 6. Properties of the simplest self-oscillating systems

The particular examples of the clock and of the valve generator (with a characteristic consisting of rectilinear segments) considered in the preceding sections, have basic features which place them in a special class, the class of *self-oscillating* systems. The common feature is their ability to perform *self-oscillations* which do not depend, generally speaking, on the initial conditions but are determined by the properties of the system itself. Examples of this class are: the electric bell, generators of saw-tooth and relaxation oscillations, the oscillating electric arc and wind and string instruments, etc. Self-oscillations can also arise in the front suspension of motor cars (the so-called phenomenon of "shimmy" of the wheels) [124, 54–56].

Another typical feature of self-oscillating systems is that there occurs a compensation of the energy losses at the expense of a certain source of energy. There is bound to exist such a source of energy and in an autonomous system (not acted upon by forces depending explicitly on time) the energy source must produce a force which itself is not a given function of time. For example, the anode battery gives a certain constant voltage independent of time, but the energy supplied by the battery will vary periodically in the presence of oscillations. Thus a self-oscillating system is an apparatus which *produces a periodic process at the expense of a non-periodic source of energy*. From this viewpoint a steam engine is a self-oscillating system.

§ 7. Preliminary discussion of nearly sinusoidal self-oscillations

Self-oscillatory systems with one degree of freedom have the equation[†]

$$\ddot{x}+\omega_0^2 x = F(x, \dot{x})-2h\dot{x} = f(x, \dot{x}). \tag{3.59}$$

We arrive at an equation of this type if an oscillatory circuit with linear damping forms part of the system. In an ordinary valve generator $\omega_0^2 = 1/LC$, $2h = R/L$ and $F(x, \dot{x})$ is, in suitable units,[‡] the e.m.f. acting on the oscillating circuit due to feedback coupling. The work done by the force compensates the losses of energy in the resistance and a periodic process becomes possible.

A *basic* problem for the theory of non-linear oscillations in autonomous systems with one degree of freedom is to determine for a given function $f(x, \dot{x})$ whether stable self-oscillations are possible and, if so, to find approximately the characteristics (amplitude, period, form) of such oscillations. Most of the material expounded later is essentially connected with this basic problem. It is worthwhile, however, to make a preliminary non-rigorous investigation of an important class of self-oscillations, the so-called *nearly-sinusoidal self-oscillations*[††]. Suppose there exists a periodic solu-

[†] In a number of very simple cases $F(x, \dot{x})$ does not depend on x, so that instead of (3.59) we have
$$\ddot{x}+\omega_0^2 x = \psi(\dot{x}).$$

[‡] The dimension of $F(x, \dot{x})$ may not coincide with the dimension of the electromotive force.

[††] Although self-oscillations differ substantially in their physical nature (in the character of the acting forces) from the oscillations of conservative systems, nevertheless the *form* of steady-state self-oscillations may differ little from the form of the oscillations of a conservative system. [*continued on next page*]

tion of (3.59) corresponding to a closed phase path on the phase plane which is outside a circle of fixed radius R_0. Then if $f(x, \dot{x})$ is sufficiently small everywhere outside this circle the periodic process will be close to a sinusoidal one†. On the other hand the requirement for the function $f(x, \dot{x})$

In particular, in a series of cases very important in practice the form of self-oscillations differs very little from the form of the oscillations of a linear harmonic oscillator (in the sense of having a very small klirr factor). If we look, for example, at the oscillogram of an oscillator with a ⌐ characteristic in the case of a small h, we shall not be able to distinguish it from the oscillogram of a harmonic oscillator.

† To clarify this statement we shall give a proof. Consider the system Equation

$$\frac{dx}{dt} = y;$$
$$\frac{dy}{dt} = -x + \varphi(x, y).$$

(3.59) is easily reduced, by an appropriate change of variables, to such a system.

Suppose this system has a periodic motion the phase path of which lies outside a circle of radius R_0, and further that $|\varphi(x, y)|$ be less than εR_0 outside this circle, ε being less than $\frac{1}{2}$. In polar coordinates

$$\dot{\varphi} = -1 + \frac{x\varphi(x, y)}{r^2};$$
$$\dot{r} = +\frac{y\varphi(x, y)}{r}.$$

Outside the circle of radius R_0 we have

$$\left|\frac{x\varphi(x, y)}{r^2}\right| < \frac{|\varphi(x, y)|}{R_0} < \varepsilon < \frac{1}{2}; \qquad \left|\frac{y\varphi(x, y)}{r}\right| < |\varphi(x, y)| < \varepsilon R_0.$$

Let us give an estimate of the "correction" to the period of a harmonic oscillator

$$\int_0^{2\pi} \dot{\varphi}(t) \, dt = -2\pi + \alpha, \quad \text{where} \quad |\alpha| < 2\pi\varepsilon,$$

$$\int_0^{\tau} \dot{\varphi}(t) \, dt = -2\pi,$$

where τ is the required period. Hence $\left|\int_{2\pi}^{\tau} \dot{\varphi}(t) \, dt\right| < 2\pi\varepsilon$, which gives the correction to the period $|\tau - 2\pi| < 4\pi\varepsilon$.

The maximum variation of the radius vector during a period is given by

$$\Delta r < \int_0^{\tau} |\dot{r}_{\max}| \, dt < \int_0^{\tau} \varepsilon R_0 \, dt < \varepsilon R_0 (2\pi + 4\pi\varepsilon).$$

Hence it follows that the closed path corresponding to the periodic solution lies on the phase plane between two concentric circles the difference between the radii of which is less than $R_0(2\pi\varepsilon + 4\pi\varepsilon^2)$. It is evident that if we know before-hand that the path of a periodic motion lies between two circles of radii R_0 and $R_1 (R_0 < R_1)$, then it is sufficient for us to require $\varphi(x, y)$ to be small only in the region between the two circles.

to be small is not altogether necessary. Self-oscillations are possible, having a form arbitrarily close to a sinusoidal one, although the function $f(x, \dot{x})$ assumes arbitrarily large values at certain instants of the motion. We have been concerned with such systems in the theory of the clock and in the theory of the generator with a ∫ shaped characteristic, since if we are concerned with an instantaneous transmission of a finite momentum, then this can take place only as a result of the action of an infinitely large force.

Let us recall certain elementary propositions about resonance. Resonance in a linearly damped oscillator occurs when an external periodic force sets up a motion in the oscillator which is close to one of its natural oscillations; close in the sense that the period of this motion is nearly that of a natural oscillation and the harmonic factor is sufficiently small. Consider an external periodic force $\Phi(t)$, with period $2\pi/\omega$, acting on a harmonic oscillator with linear damping, whose natural frequency is also ω. Then

$$\ddot{x} + \omega^2 x = \Phi(t) - 2h\dot{x}. \tag{3.60}$$

Put $\Phi(t)$ in the form

$$\Phi(t) = P \cos \omega t + Q \sin \omega t + G(t), \tag{3.60a}$$

having isolated the resonant terms[†]. There exists a natural oscillation

$$x_1(t) = \frac{P \sin \omega t - Q \cos \omega t}{2h\omega}, \tag{I}$$

for which the resonant terms of the external force are compensated by the force of friction. It is easily seen that for a sufficiently small h (if $P^2 + Q^2 \neq 0$) a periodic motion is sustained under the action of the force $\Phi(t)$, which is arbitrarily close to the natural oscillation (I) in the sense that for this motion the natural oscillation (I) will swamp to an arbitrary extent the remaining terms of the expansion in the Fourier series, and that the harmonic factor will be arbitrarily small. To prove this, denote the difference between the exact solution of the equation (3.60) and the natural oscillations $x_1(t)$ by $z(t)$ so that $x(t) = x_1(t) + z(t)$. Clearly, $z(t)$ is generated by the non-resonant terms of $G(t)$ and satisfies the equation

$$\ddot{z} + \omega^2 z = G(t) - 2h\dot{z},$$

[†] i.e. having chosen constants P and Q so that

$$\int_0^{\frac{2\pi}{\omega}} G(t) \cos \omega t \, dt = 0, \quad \int_0^{\frac{2\pi}{\omega}} G(t) \sin \omega t \, dt = 0.$$

where
$$G(t) = \frac{P_0}{2} + \sum_{n=2}^{\infty} (P_n \cos n\omega t + Q_n \sin n\omega t)$$

If by $z(t)$ we understand the "forced" solution of this equation, i.e. if
$$z(t) = \frac{C_0}{2} + \sum_{n=2}^{\infty} C_n \cos(n\omega t + \varphi_n)$$
where
$$C_n = \frac{\sqrt{P_n^2 + Q_n^2}}{\sqrt{(n^2-1)^2\omega^4 + 4h^2 n^2 \omega^2}} \leqslant \frac{\sqrt{P_n^2 + Q_n^2}}{\omega^2},$$

we can write harmonic factor or coefficient in the form
$$\varkappa^2 = \frac{\sum_{n=2}^{\infty} C_n^2}{\frac{P^2 + Q^2}{4h^2\omega^2}},$$

or, since
$$\sum_{n=2}^{\infty} C_n^2 < \frac{1}{\omega^4}\left[\frac{P_0^2}{2} + P^2 + Q^2 + \sum_{n=2}^{\infty}(P_n^2 + Q_n^2)\right] = \frac{1}{\pi\omega^2}\int_0^{\frac{2\pi}{\omega}} G^2(t)\,dt,$$

$$\varkappa^2 \leqslant \frac{\int_0^{\frac{2\pi}{\omega}} G^2(t)\,dt}{P^2 + Q^2} \cdot \frac{4h^2}{\pi\omega}.$$

Thus the condition for the harmonic factor to be small takes the form
$$\int_0^{\frac{2\pi}{\omega}} \frac{G^2(t)\,dt}{\pi\omega} \ll \frac{P^2 + Q^2}{4h^2}.$$

If $\Phi(t)$ is given for a sufficiently small h, the harmonic factor can be made sufficiently small, whatever the spectrum of $\Phi(t)$ and if $P^2 + Q^2 \neq 0$.

We are interested in the case, not of an external force (forced oscillations) but where the system itself generates the force acting on it. The equation of motion is
$$\ddot{x} + \omega_0^2 x = F(x, \dot{x}) - 2h\dot{x}. \tag{3.59}$$

If the periodic motion of this system is $x = \varphi(t)$, $\dot{x} = \dot{\varphi}(t)$, then it is clear that this solution satisfies the equation

$$\ddot{x}+\omega_0^2 x = F[\varphi(t), \dot{\varphi}(t)] - 2h\dot{x}, \qquad (3.61)$$

which is the equation of a system under a force depending explicitly on time[†]. Thus self-oscillations can be considered as forced oscillations due to a force determined by the form of the self-oscillations themselves. If the function of time $F[\varphi(t), \dot{\varphi}(t)]$ satisfies the resonance conditions and if its period is sufficiently close to $2\pi/\omega_0$, then it is convenient to speak of *self-resonance*[‡].

Note that the form of (3.61) is not unique. If often proves expedient to write this equation as

$$\ddot{x}+\omega^2 x = F(\varphi, \dot{\varphi}) + (\omega^2-\omega_0^2)\varphi - 2h\dot{x}, \qquad (3.62)$$

where ω is the frequency of the self-oscillations, and thus to consider the action of the variable external force

$$F_1(\varphi, \ddot{\varphi}) = F(\varphi, \ddot{\varphi}) + (\omega^2-\omega_0^2)\varphi$$

acting on the linear oscillator with a different ("corrected") frequency.

It may be that in the form (3.61) the conditions of resonance will not be satisfied, whereas in the form (3.62), with an appropriate choice of ω, they will be satisfied.

We shall show, using the notion of self-resonance and having *postulated* the existence of a periodic nearly-sinusoidal solution for equation (3.59), that approximate expressions for the amplitude of the fundamental waveform and for its frequency.

Assume that a periodic solution of the equation (3.59) is close as a harmonic small factor to the sinusoidal oscillation

$$x_0(t) = A \cos \omega t; \qquad \dot{x}_0(t) = -A\omega \sin \omega t,$$

where A and ω are so far undetermined constants. Substitute in equation (3.62) the zero-order approximation $x_0(t) \doteq A \cos \omega t$ instead of the exact solution for φ and again consider the self-oscillations as forced oscillations.

[†] We shall observe that the equation (3.61) is satisfied by the periodic motion only and, generally speaking, *is not satisfied* by other motions determined by the equation (3.59). Hence it follows that, proceeding from the solutions of this non-autonomous system, it is impossible to consider questions of stability.

[‡] Using the notion of self-resonance, it can be concluded that if the function $F[x(t), \dot{x}(t)]$ in the equation (3.59) *considered as a function of time* does not depend on the type of the oscillations in the oscillating circuit (for example, on the magnitude of the swing), then decreasing damping of the oscillating circuit will lead to more nearly-sinusoidal oscillations, provided the period also tends to $2\pi/\omega_0$.

We obtain
$$\ddot{x}+\omega^2 x = F_1(A\cos\omega t, -A\omega\sin\omega t)-2h\dot{x}. \qquad (3.63)$$

Expanding $F_1(A\cos\omega t, -A\omega\sin\omega t)$ in a Fourier series, we have (see (3.60a))
$$F_1(A\cos\omega t, -A\omega\sin\omega t) = P(A)\cos\omega t + Q(A)\sin\omega t + G(A, t),$$
where
$$P(A) = \frac{\omega}{\pi}\int_0^{\frac{2\pi}{\omega}} F_1(A\cos\omega t, -A\omega\sin\omega t)\cos\omega t\, dt,$$

$$Q(A) = \frac{\omega}{\pi}\int_0^{\frac{2\pi}{\omega}} F_1(A\cos\omega t, -A\omega\sin\omega t)\sin\omega t\, dt.$$

The forced solution of equation (3.63) has the form
$$x_1(t) = \frac{P(A)\sin\omega t - Q(A)\cos\omega t}{2h\omega} + z_1(A, t),$$

where $z_1(A, t)$ are the terms resulting from the non-resonant term $G(A, t)$. If we assume that ω, $P(A)$ and $Q(A)$ are given, then there exists a fixed natural oscillation
$$\frac{P(A)\sin\omega t - Q(A)\cos\omega t}{2h\omega},$$
for which the resonant terms of the external force are compensated by the force of friction developed by this natural oscillation. Hence this natural oscillation can be identified with the natural oscillation $x_0(t)=A\cos\omega t$. This gives at once two equations[†]
$$P(A) = 0, \quad Q(A)+2h\omega A = 0, \qquad (3.64)$$
which "select" the A and ω.

[†] These equations are the equations obtained by equating to zero the coefficients of the "resonant terms". Consider the equation
$$\ddot{x}+\omega^2 x = F(\varphi, \dot{\varphi})+(\omega^2-\omega_0^2)x-2h\dot{\varphi} = f_1(\varphi, \dot{\varphi}).$$
Assuming the existence of oscillations, close to the sinusoidal oscillation $x=A\cos\omega t$, we obtain the following problem belonging to the theory of the forced oscillations of a harmonic oscillator *without friction;*
$$\ddot{x}+\omega^2 x = P(A)\cos\omega t + [Q(A)+2h\omega A]\sin\omega t + G(A, t).$$
Absence of an unlimited build-up of oscillations occurs only when the coefficients of the resonant terms of the external force are equal to zero. This observation leads again to the equations (3.64).

Even so, the amplitude and frequency obtained from the equations (3.63) are not, generally speaking, the amplitude of the fundamental waveform and the frequency of the *exact* periodic solution (even if, as we have assumed, such an exact solution actually exists and has a small harmonic factor†), since in passing to the "forced" problem we substituted $A \cos \omega t$ for the exact solution.

A better approximation to the amplitude of the fundamental waveform and the frequency of the exact solution may be expected if the "first approximation"‡ replaces the "zero" approximation in the "forced" problem

$$x_1(t) = A \cos \omega t + z_1(A, t).$$

In a similar manner we can obtain (instead of (3.64)) new and modified conditions for the determination of A and ω, and find the "second approximation"

$$x_2(t) = A \cos \omega t + z_2(A, t).$$

Such a formation of successive "approximations" can be continued without limits, but to substantiate this method and to prove the existence of a periodic solution needs a special mathematical analysis. We shall take this up again when we study Poincaré's quantitative methods.

The assumption that self-oscillations are close to sinusoidal ones is widely used. Many approximate quantitative methods for the analysis of valve generators such as Barkhausen–Moeller's method (the method of the "average slope" or "quasilinear method") [18, 136, 178, 73, 74, 29] or Van der Pol's method (186, 90], are based on this assumption. Also Poincaré's methods [184, 185] are convenient used in cases where the oscillations are close to sinusoidal††.

To conclude and to illustrate the idea of self-resonance, we shall carry out the evaluation of the period and amplitude of the self-oscillations of

---

† We must underline that the presence of real solutions of the equations (3.64) does not by itself imply the existence of periodic solutions of the differential equation (3.59).

‡ Note that, if the "first approximation" represents with sufficient accuracy the required periodic motion, which by our assumption is close to a sinusoidal one, then the condition for smallness of the harmonic factor must be satisfied. If this condition is not satisfied, then, generally speaking, we can no longer state whether the A and ω obtained from the equations (3.64) will represent with sufficient accuracy the waveform in our solution or what the harmonic factor will be in subsequent "approximations".

†† Often oscillators are far removed from being sinusoidal either accidentaly or because special conditions of operation have been chosen. These include oscillations in many kinds of multivibrators and generators of saw-tooth voltages, etc.

the clock with recoil escapement, and a balance-wheel with a natural period, i.e. the clock considered in Sub-section 2 of Section 5.

The equation of motion can be written in the form (3.49)

$$\ddot{x} + \omega^2 x = F_1(x, \dot{x}) + (\omega^2 - 1)x = F(x, \dot{x}),$$

where $F_1(x, \dot{x}) = -r\,\text{sgn}\,\dot{x} - (-1)^n \lambda$ is the sum of the reduced moments of the forces of dry friction and of the escapement.

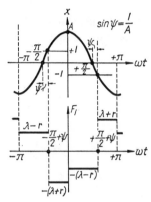

Fig. 156

Assume that the periodic solution of this equation is close to a sinusoidal one

$$x = A \cos \omega t$$

(this occurs, of course for $r < \lambda \ll 1$). Such an oscillation, and the form of the function F_1, are shown in Fig. 156. On evaluating the first Fourier coefficients for the function $F[x(t), \dot{x}(t)]$

$$P(A) = \frac{4}{\pi}\left[r - \frac{\lambda}{A}\right]$$

$$Q(A) = (\omega^2 - 1)A - \frac{4\lambda}{\pi}\sqrt{1 - \frac{1}{A^2}},$$

we obtain the following equations (according to (3.64)) for the amplitude A and the frequency ω of the periodic solution

$$r - \frac{\lambda}{A} = 0, \quad (\omega^2 - 1)A - \frac{4\lambda}{\pi}\sqrt{1 - \frac{1}{A^2}} = 0.$$

Hence

$$A = \frac{\lambda}{r}, \quad \omega^2 = 1 + \frac{4\lambda}{\pi A}\sqrt{1 - \frac{1}{A^2}} = 1 + \frac{4r}{\pi}\sqrt{1 - \frac{r^2}{\lambda^2}}.$$

Recalling that the self-oscillations of the clock are close to sinusoidal ones for r and $\lambda \ll 1$ only, we obtain

$$\omega \approx 1 + \frac{2r}{\pi}\sqrt{1 - \frac{r^2}{\lambda^2}} \quad \text{and} \quad \tau = \frac{2\pi}{\omega} \approx 2\pi - 4r\sqrt{1 - \frac{r^2}{\lambda^2}}.$$

These relations coincide with the formulae (3.54) and (3.57) which were the result of a rigorous analysis of the same problem.

CHAPTER IV

DYNAMIC SYSTEMS WITH A FIRST ORDER DIFFERENTIAL EQUATION†

WE shall proceed now to a systematic exposition of the theory of non-linear systems and of the methods of investigation and solution of non-linear differential equations, particular attention being given to qualitative integration, the importance of which has already been noted.

The most general case which we shall investigate is a system described by one non-linear differential equation of the second order or, alternatively, by two differential equations of the first order. First, however, consider the simpler case of non-linear systems with half a degree of freedom described by one non-linear differential equation of the first order systems

$$\frac{dx}{dt} = f(x). \tag{4.1}$$

A dynamic model of this kind is only an approximation to some real problem, but we shall suppose the right-hand side of (4.1) is an analytic function over the whole x axis with the possible exception of a finite number of points.

The general theory has for its ultimate object the discovery of the form of the function $x(t)$; the portrait on the uni-dimensional phase "space", having only an auxiliary although important role.

§ 1. THEOREMS OF EXISTENCE AND UNIQUENESS

Consider the t, x plane. The solutions of our equation $x = \varphi(t)$ are curves on this plane. These will be called integral curves (but should not be confused with the integral curves on the phase plane.)

Let $x = x_0$, $t = t_0$ be a given initial point. If the conditions of Cauchy's theorem‡ are satisfied for the equation (4.1) (for example, if the function

---

† Section 6 (Sub-section 2) of this chapter has been revised, and Section 5 (Sub-sections 2–4, 6), Section 6 (Sub-section 1) and Section 7 have been completely re-written by N. A. Zheleztsov.

‡ See Appendix I.

$f(x)$ is analytic over a certain interval containing x_0) then there is a unique solution of (4.1) that satisfies these initial conditions. Thus a unique integral curve passes through the point (t_0, x_0). This integral curve can be extended until x reaches a value for which $f(x)$ is not holomorphic. If the function $f(x)$ is analytic over the whole range of x then the solution can be extended until x reaches infinity†. If, however, x never goes to infinity, then the solution is useful from $t = -\infty$ to $t + \infty$

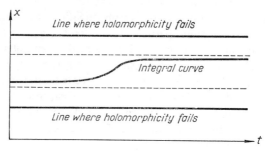

Fig. 157

Even when there exist singular points where the conditions of holomorphicity fail, cases are possible where the solutions are valid from $t = -\infty$ to $t = +\infty$ In these cases a solutions run, for example, between two straight lines parallel to the t axis whose ordinates are singular points of the function $f(x)$ (Fig. 157).

Summarizing, the following can be stated. The whole (t, x) plane can be divided into strips parallel to the t axis, such that the ordinates of the boundaries are singular points of the function $f(x)$. In each such strip, there passes through any point a unique integral curve. These curves are analytic and do not intersect one another within a strip.

As yet, nothing can be inferred about what happens on the boundaries of these strips. Boundaries may be crossed continuously by an integral curve or there may be a discontinuity.

Consider an example having a physical interest, when Cauchy's conditions are not satisfied; the fall of mass m with acceleration g with zero initial velocity.

† Note that this can happen at a finite time. Then the solution is valid (in the sense indicated here) right up to this instant. A simple example is provided by the equation $dx/dt = 1 + x^2$.

From the law of conservation of the energy, we have

$$\frac{mv^2}{2} = mg(x-x_0),$$

hence, taking the root with the positive sign for motion in one direction, we obtain

$$\frac{dx}{dt} = +\sqrt{2g(x-x_0)}. \tag{4.2}$$

Let us find the solution of this equation corresponding to the initial conditions $t=t_0$, $x=x_0$. It is easily seen that for this value of x the function $f(x) = [2g(x-x_0)]^{\frac{1}{2}}$ is non-holomorphic, since the derivative $f'(x)$ becomes infinite at $x=x_0$ and, along the straight line $x=x_0$ the conditions of Cauchy's theorem are not satisfied.

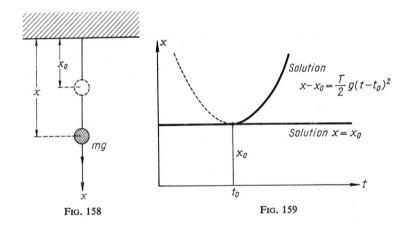

FIG. 158   FIG. 159

We can conclude that on this straight line the solutions may not be unique or perhaps even non-existent.

There, this question can be settled by direct integration, giving from Equation (4.2) the solution

$$x-x_0 = \frac{1}{2} g(t-t_0)^2.$$

Only the branch to the right-hand side of the axis of symmetry need be considered since taking the positive root implies $dx/dt > 0$.

In addition the equation has one more solution satisfying the same initial conditions,
$$x = x_0.$$
This solution can be obtained by the usual rules for the envelope of the family of parabolae $x-x_0 = g(t-t_0)^2/2$ with a variable parameter t_0. Thus it is seen (Fig. 159) that through each point of the straight line $x=x_0$ there pass not one but two integral curves, i.e. the conditions of uniqueness of the solution fail. The physical meaning of this multiple solution is easily shown. It follows from the fact that we have proceeded not from Newton's law of motion $md^2x/dt^2 = f$, but from the law of conservation of energy. From the point of view of the law of conservation of energy, the body can, under the given initial conditions, either fall with constant acceleration or remain in a state of rest. This illustrates that even for the case of one degree of freedom, the law of conservation of the energy is insufficient to establish the law of motion.

§ 2. Qualitative character of the curves on the t, x plane depending on the form of the function $f(x)$

We assume that $f(x)$ is an analytic function for all values of x, and that the equation $f(x)=0$ has no real roots. Then dx/dt can only have one sign, and all solutions are monotonic functions, either increasing or decreasing from $t = -\infty$ to $t = +\infty$. If, however, $f(x)=0$ has real roots $x = x_1, x = x_2,...$, $x = x_k$, they are, clearly, states of equilibrium. The corresponding integral curves on the t, x plane are straight lines, parallel to the t axis and dividing the x, t plane into strips. Since integral curves cannot intersect (by virtue of Cauchy's theorem) then each must be entirely contained in one of these strips and so will be monotonic, since $f(x)$ does not change its sign within a strip. Moreover, it is easily seen that if an integral curve is contained in a strip between two straight lines parallel to the t axis ($x=x_i$ and $x=x_{i+1}$), then it will approach one of these straight lines for $t \to +\infty$ and to the other for $t \to -\infty$. If, however, an integral curve is contained in a region bounded on one side only by such a straight line, then this integral curve goes to infinity either for a finite value of t or for $t = \pm \infty$; on the other side it will tend to the boundary line.

Thus, knowing $f(x)$, the qualitative character of the curves on the t, x plane is easily established.

It is clear that these curves, provided that $f(x)$ is an analytic function, cannot be periodic, since they are monotonic. This observation will prove important later.

§ 3. MOTION ON THE PHASE LINE

Consider now the representation of the motions in a uni-dimensional phase space, which in this case is the x axis. (Fig. 160).

For a given x, the representative point has a velocity $f(x)$,

$$\frac{dx}{dt} = f(x). \tag{4.1}$$

We shall assume, as before, that on the whole x axis except, possibly, at a finite number of points, $f(x)$ is an analytic function. Then, by virtue of Cauchy's theorem, the motion of the representative point is determined by

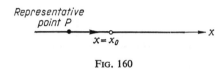

Fig. 160

the initial conditions (t_0, x_0), until the point reaches the boundary of the region of analyticity. The character of the motion of the representative point on the phase line does not depend on the instant at which the motion began, since the equation of motion does not depend explicitly on time. This follows from the fact that each individual path on the phase line corresponds not to one motion but to an infinite number of motions, beginning at different times.

For two points A and B, situated on the same path, there corresponds a finite interval of time during which the representative point moves from A to B. It should be noted that the representative point moving along the path cannot reach a point of equilibrium determined by the equation $f(x)=0$ in a finite interval of time. In fact, if the representative point, moving according to the law $x=\varphi(t)$, could reach a point of equilibrium at $x=x_0$, at a certain finite $t=t_0$, then there would be two *different* solutions for the differential equation (the first $x=\varphi(t)$ and the second $x=x_0$) that assume the same value for $t=t_0$, which in fact contradicts Cauchy's theorem. The path of the representative point, which tends asymptotically to a state of equilibrium without reaching it in a finite time, will be either a segment or half a straight line with one end at the point $x=x_0$ (Fig. 160).

The point $x=x_0$ itself does not belong to the path but is a path in its own right; an important distinction to make.

We shall formulate now for the straight phase line a theorem on the continuity of the dependence of the solution upon the initial conditions[†]. To do this, consider the motion of two representative points, $P_1\{(x=x_1(t)\}$ and $P_2\{(x=x_2(t)\}$, that started to move at the same instant $t=t_0$, and follow them during a certain finite interval of time T, during which P_1 does not leave the analytic region. Then the theorem on the continuity of the dependence of the solution upon the initial conditions reads: for any T and any $\varepsilon (\varepsilon > 0)$ it is always possible to find a positive δ dependent on T and ε, such that

$$|x_1(t)-x_2(t)| < \varepsilon \quad \text{for} \quad t_0 \leqslant t \leqslant t_0+T,$$

if

$$|x_1(t_0)-x_2(t_0)| < \delta.$$

Now assume that $f(x)$ is an analytic function over the whole x axis. If the equation $f(x)=0$ has no real roots, then all motions have one and the same path coinciding with the entire straight phase line. If, however, $f(x)$ has the real roots $x=x_1$, $x=x_2$, $x=x_3$,..., $x=x_k$, then there can be paths of various types:

(a) states of equilibrium;
(b) intervals between two roots;
(c) intervals between one of the roots and infinity (half straight lines).

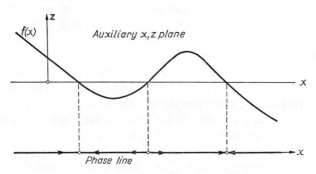

Fig. 161

On each path the motion takes place in a determined direction, since the sign of $f(x)$ does not vary over a path. If $f(x)>0$, the representative point moves towards the right; if $f(x)<0$ the representative point moves towards the left. The points where $f(x)=0$ correspond to states of equilib-

[†] We give here a somewhat different formulation of this theorem from the one given in Appendix I, namely a formulation which is suitable for a uni-dimensional phase space.

rium. Knowing the form of the curve $z = f(x)$ and using these arguments, we can divide the phase (straight) line into paths and indicate the direction of motion of the representative point along the paths[†]. An example of such a construction is shown in Fig. 161. This gives a clear picture of the possible motions of a dynamic system described by one differential equation of the first order. Knowing the states of equilibrium and their stability will establish a qualitative picture of the possible motions. In particular,

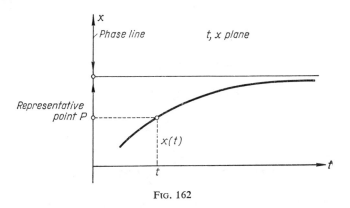

Fig. 162

when $f(x)$ is analytic over the whole straight line, periodic motions are impossible. The behaviour of the integral curves on the t, x plane can also be established. Consider the phase line coincident with the x axis of the (t, x) plane, and let the representative point move on the phase line.

On the t, x plane construct the point of abscissa t and ordinate equal to the displacement of the representative point along the x axis at a given instant t. The abscissa of this point is time and therefore varies. The ordinate, generally speaking, also varies, since the representative point moves. Consequently, the point on the t, x plane will move describing a certain curve. This curve will be an integral curve (Fig. 162).

§4. Stability of the States of Equilibrium

We have already given a definition of stability of a state of equilibrium in the sense of Liapunov. To repeat, a state of equilibrium $x = x_0$ will be stable in the sense of Liapunov if, having assigned an arbitrarily small

---

[†] The direction of motion of the representative point is indicated on the straight phase line with arrows.

positive ε, it is always possible to find such a δ that

$$|x(t)-x_0| < \varepsilon \quad \text{for} \quad t_0 \leqslant t < +\infty, \quad \text{if} \quad |x(t_0)-x_0| < \delta.$$

Liapunov gives a rule for investigating the stability of states of equilibrium. We shall first formulate the rule itself and then give its justification. Consider small deviations from the state of equilibrium $x=x_0$, and put $x=x_0+\xi$. $f(x)$ is still an analytic function. Replacing the variable x by the variable ξ in

$$\frac{dx}{dt} = f(x), \tag{4.1}$$

we obtain

$$\frac{d\xi}{dt} = f(x_0+\xi) = f(x_0)+f'(x_0)\xi+\frac{1}{1\cdot 2}f''(x_0)\xi^2+\ldots \tag{4.2}$$

and since $f(x_0)=0$, the equation (4.1) reduces to

$$\frac{d\xi}{dt} = a_1\xi+a_2\xi^2+a_3\xi^3+\ldots, \tag{4.3}$$

where

$$a_1 = f'(x_0); \qquad a_2 = \frac{1}{2}f''(x_0) \quad \text{etc.}$$

Liapunov's rule consists in neglecting the non-linear terms of equation (4.3). We then obtain a *linear* equation

$$\frac{d\xi}{dt} = a_1\xi, \tag{4.4}$$

which is the equation of the first approximation. The integral of (4.4) is readily found

$$\xi = ce^{\lambda t}, \quad \text{where} \quad \lambda = a_1 = f'(x_0).$$

Liapunov states that if $\lambda<0$, the equilibrium state $x=x_0$ is stable, and if $\lambda>0$, the state of equilibrium is unstable.

If $\lambda=0$, then equation (4.4) is inadequate to determine the stability.

In the simple case considered it is very easy to justify this rule. On multiplying both sides of the equation (4.3) by ξ we have

$$\frac{1}{2}\frac{d(\xi^2)}{dt} = a_1\xi^2+a_2\xi^3+\ldots = F(\xi). \tag{4.5}$$

Write $F(\xi)$ in a Taylor's expansion, noting that

$$F(0) = 0, \quad F'(0) = 0, \quad F''(0) = 2a_1,$$

$$F(\xi) = \frac{\xi^2}{1\cdot 2}F''(\vartheta\xi) \quad (\text{where } 0 < \vartheta < 1),$$

and put $\varrho = \xi^2/2$; then the equation (4.5) becomes

$$\frac{d\varrho}{dt} = \frac{\xi^2}{1 \cdot 2} F''(\vartheta\xi). \tag{4.6}$$

If $F''(0) < 0$, i.e. if $a_1 < 0$, then, by virtue of the continuity of the function $F''(\xi)$, $F''(\vartheta\xi) < 0$ for sufficiently small values of $|\xi|$. Hence it follows, according to (4.6), that $d\varrho/dt < 0$ for the same values of $|\xi|$. If $\varrho = \xi^2/2$ decreases then $|\xi|$ decreases. It follows that the condition $a_1 = f'(x_0) < 0$ is sufficient for the stability of the state of equilibrium at $x = x_0$, since there always exists about $x = x_0$ a region of initial values from which the representative point will asymptotically approach the state of equilibrium. In

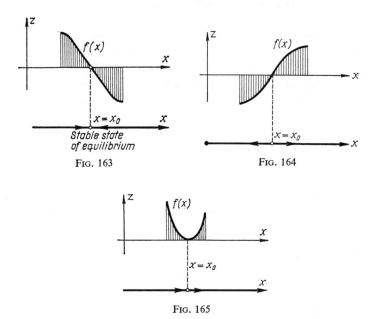

Fig. 163   Fig. 164

Fig. 165

exactly the same way it can be shown that for $a_1 = f'(x_0) > 0$ the state of equilibrium is unstable. The case when $a_1 = 0$ needs special investigation.

Thus Liapunov's rule is justified, since the result of the investigation of the stability of the state of equilibrium by means of the complete non-linear equation

$$\frac{d\xi}{dt} = a_1\xi + a_2\xi^2 + a_3\xi^3 + \ldots$$

agrees with the investigation of the stability by means of the linear equation

$$\frac{d\xi}{dt} = a_1\xi,$$

provided that $a_1 \neq 0$. In this example with an equation of the first order it is easy to investigate stability directly from the properties of the function $f(x)$ near to the state of equilibrium $x=x_0$. Since $f(x)=0$, three essentially different cases occur, and are illustrated in Figs. 163, 164 and 165.

(1) $f(x)$ changes its sign near $x=x_0$ from positive to negative as x increases (Fig. 163). Hence $f'(x_0)<0$ and x_0 is stable.

(2) $f(x)$ changes its sign near $x=x_0$ from negative to positive as x increases (Fig. 164). Hence $f'(x_0)>0$ and there is unstable point at $x=x_0$.

(3) $f(x)$ does *not* change its sign in the vicinity of the state of equilibrium $x=x_0$ as x increases (Fig. 165). This means that a representative point, situated sufficiently close to the position of equilibrium on one side of it, will approach it, and one situated on the other side will move away from it. It is clear that the state of equilibrium proves unstable in the sense of Liapunov, for there is instability on one side and stability on the other.

In this case $f'(x_0) = 0$.

§ 5. Dependence of the character of the motions on a parameter

In every real system, the motion is subjected to the influence of a series of factors. Small variations of these factors are unavoidable in every real system, and must always be taken into account. We can do this in our problem by making the right-hand side of the differential equation depend on a certain parameter λ

$$\frac{dx}{dt} = f(x, \lambda); \tag{4.7}$$

and study the variations in the solutions under variations of λ.

A state of equilibrium is given by the equation

$$f(x, \lambda) = 0. \tag{4.8}$$

This equation determines on the λ, x plane a certain curve (Fig. 166) which expresses the dependence of the coordinates of the states of equilibrium upon the parameter λ.

Now by Liapunov's rule a state of equilibrium $x=\bar{x}$ is stable if

$$f'(\bar{x}, \lambda) < 0, \tag{4.9}$$

and is unstable if
$$f'(\bar{x}, \lambda) > 0. \tag{4.10}$$

Thus the theory of the dependence of the states of equilibrium of a dynamic system with an equation of the first order on a parameter *is an exact copy* of the theory of the dependence of the states of equilibrium

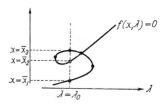

Fig. 166

of the simplest conservative system with one degree of freedom upon a parameter. Just as before we shall be concerned with the bifurcation values of the parameter, with the change of stability, etc. This is illustrated by examples.

1. *Voltaic arc in a circuit with resistance and self-induction*

This circuit leads to a non-linear differential equation of the first order, if we take into account only the elements of the system shown in Fig. 167. The non-linearity of this equation is caused by the fact that the arc is a

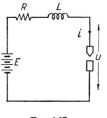

Fig. 167

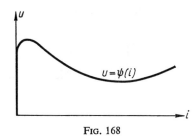

Fig. 168

conductor not obeying Ohm's law. The relation between the arc voltage and the current is given graphically by the so-called *static* characteristic of the arc, $i = \varphi(u)$ or $u = \psi(i)$, where u is the voltage and i the current (Fig. 168).

The circuit has the following differential equation:

$$L\frac{di}{dt} + Ri + \psi(i) = E,$$

or

$$\frac{di}{dt} = f(i) = \frac{E - Ri - \psi(i)}{L}. \qquad (4.11)$$

The states of equilibrium $i = I$ are determined by $f(i) = 0$, i.e. by the equation

$$E - Ri - \psi(i) = 0. \qquad (4.12)$$

In order to find the roots of this equation we plot the arc characteristic $u = \psi(i)$ and the line $u = E - Ri$. Points of intersection are the values of current I in the states of equilibrium (Fig. 169). On the same graph

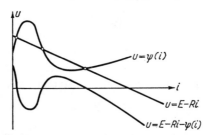

Fig. 169

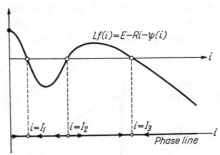

Fig. 170

plot the curve $u = (E - Ri) - \psi(i)$ which for some scale represents the function $f(i)$. Knowing $f(i)$, the paths on the straight phase line (Fig. 170)[†]

[†] Since the current i has been chosen as the coordinate of the system (this determines uniquely u and di/dt), the phase line will be the i line. The u straight line cannot serve as the phase line, since the current i is not a single-valued function of the potential difference u across the arc and therefore assigning u is not sufficient to determine uniquely the state of the system.

can be constructed. In this example there exist three states of equilibrium: $i=I_1$, $i=I_2$ and $i=I_3$. The first and last of which, as follows from the stability criteria given above, are stable, and the middle one unstable.

Now let E be the variable parameter, and construct on the E, I plane the curve $f(i, E) = 0$ or
$$E - RI - \psi(I) = 0$$

(Fig. 171). This curve, as can be seen from the diagram, has two branch points and hence two branch values of the parameter E: $E = E_1$ and $E = E_2$.

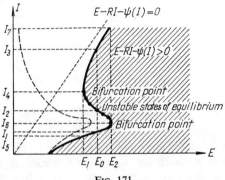

Fig. 171

The value E_2 corresponds to such a large battery voltage (for a given R) that the states of equilibrium I_1 and I_2 merge with each other and disappear, so that for a further increase of E there remains only one stable state of equilibrium $i = I_3$, corresponding to a large current. The value E_1 corresponds to such a small value of the battery voltage (for a given R) that the states of equilibrium $i = I_2$ and $i = I_3$ merge with each other and disappear and for a further decrease of E we have only one stable state of equilibrium I_1, corresponding to a very small current. The diagram shows that if we vary E slowly and continuously we shall have abrupt transitions of the system at the branch points from one state of equilibrium into another. The arc current will increase, according to equation (4.11), from I_6 to I_7 (for $E = E_2$) and will drop from I_4 to I_5 (for $E = E_1$). The picture showing the dependence of the steady current I on the voltage E has a hysteretic character (Fig. 172). Similarly we can construct the analogous diagram for a fixed E and a variable R.

We have considered the case of sufficiently small resistance in the circuit of the arc, such that $R < |\psi'|_{\max}$, where $|\psi'|_{\max}$ is the maximum absolute value of the slope of the arc characteristic $u = \psi(i)$ on its decreasing section.

If, however,
$$R > |\psi'|_{\max}, \tag{4.13}$$
then for all values of E there is only *one* state of equilibrium and that is *stable* (Fig. 173). Stability is independent of L, even for arbitrarily small

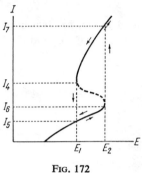

Fig. 172

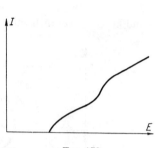

Fig. 173

values of L. This result is definitely contradicted by experiment and the condition (4.13) is not sufficient to ensure the stability of the state of equilibrium on the decreasing section of the characteristic. This emphasises the fact that, in describing this system with a first order equation (with half a degree of freedom), we have neglected some parameters that are essential to the case when $R > |\psi'|_{\max}$. We shall consider in Chapter V, § 5 the problem of a voltaic arc in a circuit with resistance and self-inductance and also capacitance[†].

2. Dynatron circuit with resistance and capacitance

As a second example of an electric circuit with half a degree of freedom we shall consider the circuit shown in Fig. 174.

The equation of such a system (taking into account only the elements shown in Fig. 174[‡]) is

$$RC\frac{du}{dt} + u + Ri = E. \tag{4.14}$$

---

† As we shall see, the inertia of the ionic processes in the arc can be replaced approximately by a certain "equivalent" self-inductance connected in series with the arc.

‡ If the anode circuit of the tetrode does not have an actual capacitor, the capacitance C will represent the small parasitic capacitance of the anode to the other electrodes and the parasitic capacitance of the resistor R.

$i = \varphi(u)$ is the anode current of the tetrode and is a non-linear single-valued function of the anode voltage u. The graph of this function as pointed out in Chapter I, § 7, has a section with a negative slope (Fig. 175).

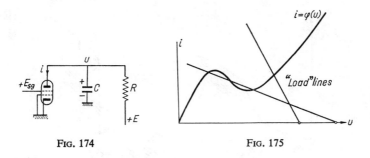

Fig. 174   Fig. 175

The states of equilibrium $u = U$ are determined by

$$E - u - Ri = 0 \qquad (4.15)$$

and may be found graphically as the points of intersection of the characteristic $i = \varphi(u)$ and the "load" line $E - u - Ri = 0$ (Fig. 175). It is evident that for a given anode characteristic $i = \varphi(u)$ there are either one or three states of equilibrium depending on the values of E and R. If we take as a variable parameter the battery voltage E, while R remains fixed, we can plot on the E, U plane the curve (4.15) relating the states of equilibrium U to the parameter E (Fig. 175). Two cases are possible. If the resistance R is sufficiently small so that $R|\varphi'|_{\max} < 1$, where $|\varphi'|_{\max}$ is the maximum absolute value of the negative slope of the characteristic, then for any values of E there is one state of equilibrium (Fig. 176a). If, however, $R|\varphi'|_{\max} > 1$, then, over a certain range of voltages E, there are three states of equilibrium (Fig. 176b). In the latter case the values E_1 and E_2 are branch values.

Shown shaded is the region in which

$$E - u - Ri > 0$$

i.e. the region where $du/dt > 0$. We can easily determine the stability of the states of equilibrium, for the points of the curve (4.15) which lie *above* this region correspond to stable states of equilibrium, and the points *below* it to unstable ones. Thus in the case when there is one state of equilibrium, it is always stable. If, however, there are three states of equilibrium, then the extreme ones ($u = U_1$ and $u = U_3$ in Fig. 176b) are stable and the middle one ($u=U_2$) is unstable. The phase lines in the presence of one and three states of equilibrium are shown in Fig. 176c. Whatever the initial conditions the system approaches one of the states of equilibrium.

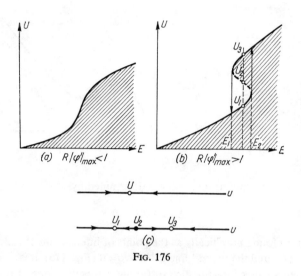

Fig. 176

3. Valve relay (bi-stable trigger circuit)

In the first order equation is obtained when analysing the valve relay which is shown in Fig. 177, C_a represents the small parasitic capacitance of the anode of V_1 †

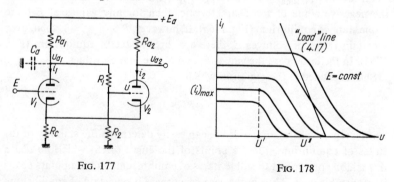

Fig. 177　　　　　　　　　　Fig. 178

† We shall not consider other parasitic capacitances or parameters. The parasitic capacitance of the grid of the valve V_2 is usually compensated by connecting a suitable capacitor in parallel with the resistor R_1.

From Kirchoff's laws we obtain an equation for the anode voltage of V_1

$$C_a \frac{du_{a1}}{dt} + i_1 + \frac{u_{a1} - E_a}{R_{a1}} + \frac{u_{a1}}{R_1 + R_2} = 0 \qquad (4.16)$$

where grid current, if any, is neglected. We can assume that the anode currents of the valves are functions of the voltage E at the grid of valve V_1 and u at the grid of V_2. The voltages u and E as well as u_{a1} are measured with respect to the earth terminal of the circuit, and E will be the variable parameter.

Fig. 178 shows a family of valve characteristics relating anode current i_1 of valve V_1 to the voltage u, for various constant values of E (and when V_1 and V_2 are connected as shown). For sufficiently small values of $u (u \leqslant U')$ the valve V_2 is cut off ($i_2 = 0$) and the anode current i_1 of the valve V_1 is independent of $u (i_1 = i_{1\,\text{max}} = \text{const})$. For $u > U'$ the anode current of the valve V_2 is not zero and an increase of u causes i_2 to increase with the cathode voltage u_c. Hence i *decreases* until finally, for $u = U''$ the valve V_1 is cut off. For $u > U''$ the valve V_1 does not conduct; $i_1 = 0$. The voltages U' and U'', for which the valves V_1 and V_2 are respectively cut off, and the maximum current $i_{1\,\text{max}}$ depend on the value of E. We shall also assume that the maximum negative slope of the valve characteristic S_0 does not depend on E.

The grid voltage of the valve V_2 is equal to

$$u = \beta u_{a1},$$

where $\beta = R_2/(R_1 + R_2)$ is the transmission factor of the voltage divider formed by R_1 and R_2†. Eliminating the variable u_{a1} from the equation (4.16) we obtain the following equation for the voltage u:

$$\frac{C_a}{\beta} \frac{du}{dt} = \frac{E_a}{R_{a1}} - i_1(u, E) - \frac{u}{\beta}\left(\frac{1}{R_{a1}} + \frac{1}{R_1 + R_2}\right). \qquad (4.16a)$$

The states of equilibrium are determined by the equation

$$\frac{E_a}{R_{a1}} - i_1 - \frac{u}{\beta}\left(\frac{1}{R_{a1}} + \frac{1}{R_1 + R_2}\right) = 0 \qquad (4.17)$$

and may be found graphically as the points of intersection of the valve characteristic $i_1 = i_1(u, E)$ with the "load" line (4.17). There are either one or three states of equilibrium.

† We recall that we are neglecting the parasitic capacitance of the grid junction of the valve V_2. If the parasitic capacitance of this junction is equal to C_2, then the expression for the transmission factor of the voltage divider as given in the text remains valid if the resistor R_1 is shunted by a capacitance equal to $R_2 C_2/R_1$.

If the slope of the load line (4.17) $(1/\beta)\left(\dfrac{1}{R_{a_1}} + \dfrac{1}{R_1+R_2}\right)$ is larger than S_0, so that

$$\beta \frac{S_0 R_{a1}}{1 + \dfrac{R_{a1}}{R_1+R_2}} < 1,$$

then for each E there is a unique stable state of equilibrium. In this case (see Fig. 178) the equilibrium value of the voltage u on the grid of the valve V_2 decreases as E increases, so that the output voltage U_{a2} at the anode of the valve V_2 depends upon E as shown in Fig. 179. The output voltage

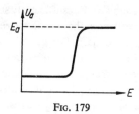

Fig. 179

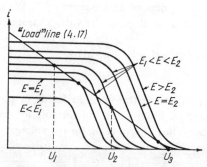

Fig. 180

U_{a2} is a continuous function of the input voltage E and the circuit acts as a voltage amplifier with large amplification, due to the presence in the circuit of a *positive* feedback.

If, however,

$$\beta \frac{S_0 R_{a1}}{1 + \dfrac{R_{a1}}{R_1+R_2}} > 1,$$

i.e. if the slope of the straight line (4.17) is less than S_0 (Fig. 180), then over the range $E_1 < E < E_2$ the system has three states of equilibrium U_1, U_2 and

U_3, two of which (U_1 and U_3) are stable and one (U_2) is unstable[†]. In this case, therefore, the output voltage U_{a2} is related to E by an S-shaped curve (Fig. 181), the dotted section corresponding to unstable states of equilibrium. The voltages $E=E_1$ and $E=E_2$ for which the states of equilibrium merge together are branch values. Strictly speaking, the characteristic of the circuit shown in Fig. 181 is a *static* one, relating the *equilibrium* values of the voltage U_{a2} to various fixed values of E. However, if we vary the

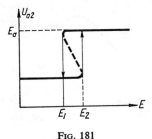

Fig. 181

input voltage E, sufficiently slowly (and continuously), then at the branch points ($E=E_2$, $\dot{E}>0$ and $E=E_1$, $\dot{E}<0$) there will be abrupt changes of the output voltage, according to the equation (4.16a).

Thus the circuit works as a valve *relay*, having rapid transitions from one state of equilibrium into another, and also having "hysteresis".

4. Motion of a hydroplane

We shall consider now a few examples of mechanical systems satisfactorily described by a differential equation of the first order. As a first example we shall consider the rectilinear motion of a hydroplane (leaving pitching and rolling out of account). The equation of its motion may be written according to Newton's second law in the form

$$m\frac{dv}{dt} = T(v) - W(v), \qquad (4.18)$$

[†] In fact, the linearized equation, valid in the vicinity of a state of equilibrium $u=U$ has the form

$$\frac{C_a}{\beta}\frac{d\xi}{dt} = -A\cdot\xi.$$

where

$$\xi = u-U \quad \text{and} \quad A = \frac{1}{\beta}\left(\frac{1}{R_{a1}} + \frac{1}{R_1+R_2}\right) + \left(\frac{di_1}{du}\right)_{u=U}.$$

For the states of equilibrium $u=U_1$ and $u=U_3$, $A>0$ and for $u=U_2$ $A<0$. Therefore the first two states of equilibrium are stable and the last is unstable.

where v is the velocity of the ship, m is its mass and T and W are respectively the thrust (the tractive force) of the screw and the resistance of the ship. T and W are functions of the velocity v, the force of thrust of the screw decreasing monotonically as the velocity increases, while the resistance over a certain range of velocities has a negative slope (Fig. 182)[†].

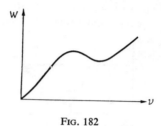

Fig. 182

The "states of equilibrium" $v = V = $const. are determined, clearly, by

$$T(v) = W(v).$$

Typical cases are shown in Fig. 183a and b (in the same figure there are also shown the corresponding phase lines divided into phase paths). It is

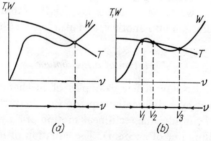

Fig. 183

easily seen that if there is only one condition of uniform motion (Fig. 183a), this motion is stable, whilst in the presence of three conditions of uniform motion (Fig. 183b) the motions with velocities $v = V_1$ and $v = V_3$ are stable, and the motion with the intermediate velocity $v = V_2$ is unstable.

[†] For sufficiently small velocities of motion the hydroplane "floats", its weight being balanced by hydrostatic forces. As the velocity increases, the resistance to motion increases and a larger and larger fraction of the weight of the craft is balanced by the hydrodynamic lift forces. The craft rises and the wetted area decreases, and over a certain range of velocities the resistance decreases for an increase in speed. For higher velocities the ship "planes" and the resistance again increases with increase of velocity.

This latter case can occur when the thrust $T=T(v)$ is almost independent of V, as is often the situation when towing models of hydroplanes in test tents.

It is clear that, whatever the initial conditions, the hydroplane reaches a stable condition of uniform motion.

5. Single-phase induction motor

As a second example we shall consider the problem of the rotation of the rotor of a single-phase induction motor. This type of motor is not made in large sizes and is used when the starting torque need be small;

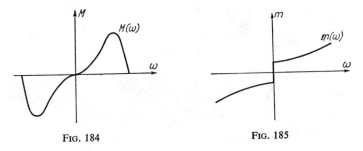

Fig. 184   Fig. 185

they are often used, for example, to drive small fans. The variation of torque with angular shaft velocity is shown in Fig. 184. The rotation of the motor is opposed by bearing friction and the air resistance of the fan, and together they can be represented approximately by the graph of friction moment versus velocity shown in Fig. 185.

Denoting the torque by $M(\omega)$, the moment of the forces of friction by $m(\omega)$ and the moment of inertia by I, we can write the equation of motion as

$$I\frac{d\omega}{dt} = M(\omega) - m(\omega).$$

The "states of equilibrium" are the states of rest, $\omega = 0$ and certain speeds of uniform rotation given by the equation

$$M(\omega) - m(\omega) = 0.$$

To find the roots of this equation construct two auxiliary curves $z = M(\omega)$ and $z = m(\omega)$ and find their points of intersection (Fig. 186). Then, as before, we plot the function $f(\omega) = [M(\omega) - m(\omega)]/I$ and mark the paths on the phase line (Fig. 187).

In this case there are stable states of equilibrium: $\omega=\omega_0$, $\omega=\omega_2$ and $\omega=\omega'_2$ and two unstable ones $\omega=\omega_1$ and $\omega=\omega'_1$. The stability of the state $\omega = \omega_0 = 0$, corresponding to complete rest, shows the rotor does not start by itself but has to be speeded up in some manner beyond the states ω_1 or ω_2 after which it accelerates to the normal angular velocity ω_2. The

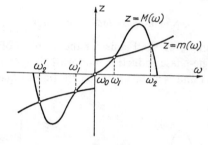

Fig. 186

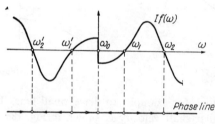

Fig. 187

motor can rotate in both directions (two stable states of equilibrium ω_2 and ω'_2) and the direction of the steady-state rotation depends only on the direction in which it is started. Special methods are employed sometimes (additional coils, poles, etc.) which introduce an asymmetry in the graph in Fig. 187 and a certain initial torque appears so that the motor starts by itself.

6. Frictional speed regulator

Let us consider one more mechanical system, described under certain simplifying assumptions by an equation of the first order: the frictional regulator, the construction of which is shown diagrammatically in Fig. 188. Such frictional regulators are employed in a number of astronomical instruments, telegraph equipment, gramophones etc. for stabilizing a velocity of rotation. Their action is based on the fact that as the velocity of rotation of the regulator increases, the spheres of the regulator diverge and at a certain angle $\vartheta=\vartheta_0$ the braking blocks rub against the limit ring. The resulting braking moment will increase with the velocity of rotation and tend to hold the system at a uniform velocity which proves to be nearly independent of the moment M applied to the regulator by the associated mechanism.

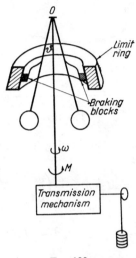

Fig. 188

Assuming the regulator to be ideally rigid and that during the process of regulation $\vartheta \equiv \vartheta_0$ we could obtain a dynamic model described by a differential equation of the first order. However, in order to obtain an expression for the pressure of the braking blocks against the limit ring we will not assume that $\vartheta \equiv \vartheta_0$ and will consider the regulator as a system with two degrees of freedom (with the generalized coordinates φ and ϑ). Lagrange's function for such a system is thus

$$L = \frac{1}{2} I(\vartheta)\omega^2 + \frac{1}{2} J_\vartheta \dot\vartheta^2 - V(\vartheta),$$

where $I(\vartheta)$ is the moment of inertia of the regulator with respect to its axis, J_ϑ is the moment of inertia of the spheres of the regulator with respect to the point 0, and $V(\vartheta)$ is the potential energy of the regulator. We shall neglect all forces of friction except the forces of dry friction at the braking blocks on the limit ring. The moment m of these forces about the axis will be assumed to be proportional to the moment N of the forces which press the braking blocks against the limit ring (the moment being referred to the point 0), so that

$$m = N\varphi(\omega),$$

where $\varphi(\omega)$ is a function determined by the properties of the rubbing sur-

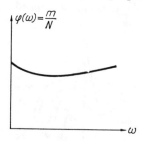

Fig. 189

faces of the braking blocks and the ring (see Fig. 189). Then Lagrange's equations of motion are

$$\frac{d}{dt}\left(\frac{\partial L}{\partial \omega}\right) = M - m, \quad \frac{d}{dt}\left(\frac{\partial L}{\partial \dot{\vartheta}}\right) - \frac{\partial L}{\partial \vartheta} = -N$$

or

$$I(\vartheta)\dot{\omega} + I'(\vartheta)\omega\dot{\vartheta} = M - m, \quad J_\vartheta \ddot{\vartheta} - \frac{1}{2}I'(\vartheta)\omega^2 + V'(\vartheta) = -N. \quad (4.19)$$

Putting $\vartheta \equiv \vartheta_0$ we obtain, firstly, the expression for the moment N

$$N = \frac{1}{2}I'(\vartheta_0)\omega^2 - V'(\vartheta_0)$$

or

$$N = a(\omega^2 - \omega_0^2), \quad (4.20)$$

where $a = I'(\vartheta_0)/2$, and $\omega_0^2 = 2V'(\vartheta_0)/I'(\vartheta_0)$. And secondly, we obtain the equation of motion of the first order

$$I(\vartheta_0)\dot{\omega} = M - a\varphi(\omega)(\omega^2 - \omega_0^2). \quad (4.21)$$

The states of equilibrium $\omega = \Omega$ are clearly determined by the equation

$$a\varphi(\Omega)(\Omega^2 - \omega_0^2) = M. \quad (4.22)$$

The velocity of uniform rotation Ω depends on the moment M applied to the axis of the regulator by the connected mechanism and is given by equation (4.22). The derivative† $d\Omega/dM$ is

$$\frac{d\Omega}{dM} = \frac{1}{2a\Omega\varphi(\Omega) + a\varphi'(\Omega)(\Omega^2 - \omega_0^2)}. \quad (4.23)$$

† In all the expressions (4.20) – (4.23) and below we shall assume that $\omega > \omega_0$. Only in this case $N > 0$ and the braking blocks are pressed against the limit ring.

The stability of the conditions of uniform rotation is thus determined by an equation of the first approximation

$$I(\vartheta_0)\frac{d(\Delta\omega)}{dt} = -a\left\{2\Omega\varphi(\Omega)+\varphi'(\Omega)(\Omega^2-\omega_0^2)\right\}$$

or

$$I(\vartheta_0)\frac{d(\Delta\omega)}{dt} + \frac{\Delta\omega}{\left(\dfrac{d\Omega}{dM}\right)} = 0. \qquad (4.24)$$

Clearly the conditions of uniform rotation are stable if $d\Omega/dM > 0$ and unstable if $d\Omega/dM < 0$. Stability depends on the function $\varphi(\omega)$ and especially on the value of the negative slope of the friction characteristic and on the value of the parameter ω_0[†]. The static characteristic of the

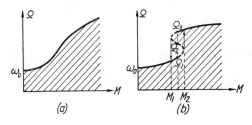

FIG. 190

regulator $\Omega = \Omega(M)$ is either a single-valued monotonic function (Fig. 190a) being stable for every value of the moment M, or else has an S-shaped form (Fig. 190b). In the latter case, for $M_1 < M < M_2$, the system has three conditions of uniform rotation with velocities Ω_1, Ω_2 and Ω_3, two of which ($\Omega = \Omega_1$ and $\Omega = \Omega_3$) are stable and one ($\Omega = \Omega_2$) is unstable. The values $M = M_1$ and $M = M_2$ are branch values. As the moment M passes through these values there is a transition from one condition of uniform rotation to another[‡].

[†] ω_0 may be varied within certain limits by displacing the limit ring.

[‡] The dynamic model of the friction regulator has no periodic oscillations and its motions cease on arrival at a stable condition of uniform rotation. On the other hand, under certain conditions real friction regulators have no stable condition of uniform rotation and in them there arise self-oscillations [132, 9]. To explain this self-excited oscillation it is necessary to relax the assumption of an absolutely rigid regulator and allow for a large but finite rigidity of the flat springs on which the braking blocks are fixed. This leads to a dynamic model with one and a half degrees of freedom (described by a differential equation of the third order). This analysis is outside the scope of this book.

§ 6. Periodic motions

The dynamic systems with equations of the first order so far considered have had only states of equilibrium, as stationary states, and periodic motions did not exist. This is because the systems were governed by the equation in

$$\frac{dx}{dt} = f(x) \tag{4.1}$$

having a single-valued right-hand side. In fact, periodic motion becomes possible in systems of the first order only when the right-hand side of the equation (4.1), i.e. the function $f(x)$, is *multi-valued* over at least a certain range of x. Consider a harmonic oscillator with *given total energy h*. Its equation is

$$\frac{1}{2} m\dot{x}^2 + \frac{1}{2} kx^2 = h \quad (= \text{const.})$$

or, after reducing it to the form (4.1)

$$\dot{x} = \pm \sqrt{\frac{2}{m}} \sqrt{h - \frac{kx^2}{2}},$$

which has the periodic solution

$$x = \sqrt{\frac{2h}{k}} \cos(\omega_0 t + \alpha),$$

where $\omega_0 = (k/m)^{\frac{1}{2}}$ and α is an arbitrary constant. We cannot use for the phase line of this first-order system the straight segment $-A \leqslant x \leqslant +A$, where $A = +(2h/k)^{\frac{1}{2}}$ is a given amplitude of the oscillations, since x does not determine uniquely the velocity $\dot{x}$ of the system. But we can take as the phase line any simple closed curve, for example, a circle (Fig. 191). To each value of x there correspond *two* points on the circle, and thus the possibility of establishing a one-to-one continuous correspondence between the points of this circle and the states of the harmonic oscillator. We can assume, for example, that on the upper semicircle $\dot{x} = +(2/m)^{\frac{1}{2}} [h-(kx^2/2)]^{\frac{1}{2}}$ and on the lower one $\dot{x} = -(2/m)^{\frac{1}{2}} [h-(kx^2/2)]^{\frac{1}{2}}$; then a point of the circle, determines x and $\dot{x}$ unequivocally.

This situation proves to be common to all dynamic systems of the first order: periodic motions are only possible in systems whose phase lines have closed sections. Therefore *the multi-valuedness of the right-hand side of the equation (4.1) over a certain interval of x is a necessary condition for the existence of periodic solutions.*

Below we shall consider two examples of physical systems, described by equations of first order with a *double-valued* right-hand side.

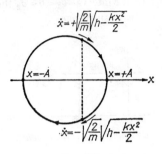

Fig. 191

1. Two-position temperature regulator

The first example will be the two-position (relay) temperature regulator, as shown in Fig. 192. The temperature θ of the oven (measured relative to the surrounding medium) obeys the following heat-balance equation

$$C\frac{d\theta}{dt} = W - K\theta, \qquad (4.25)$$

where C is the thermal capacity of the oven, W is the power supplied to the oven by the heater, and $K\theta$ is the thermal power lost by the oven to the

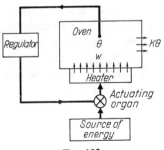

Fig. 192

surrounding medium. The temperature of the oven is measured by a thermocouple (or by some other thermometer) which, via the regulator and an actuating element, controls the power supplied to the oven.

We shall consider a *two-position* regulator of the "on–off" type, which according to the temperature of the oven θ, either disconnects the heater ($W=0$) or supplies it with a fixed power ($W=W_0$). The characteristic of such a regulator is shown in Fig. 193. It always shows hysteresis: if the switching-on occurs at the temperature θ_1, then its switching off occurs at a temperature $\theta_2 > \theta_1$. In the interval $\theta_1 < \theta < \theta_2$ the characteristic of the regulator $W = W(\theta)$ is a two-valued one.

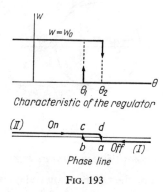

Fig. 193

Corresponding to this two-valued characteristic is the phase line of the system (the oven with the regulator) which will be a line with "superposition" (Fig. 193) consisting of two half lines, one of which (*I*) ($\theta > \theta_1$) corresponds to the heater cut-off, and the other (*II*) ($\theta < \theta_2$) to the heater switched on. Note that the passage of the representative point from the half-straight line (*I*) to the half-straight line (*II*) occurs at the point *c* only (for $\theta = \theta_1$) and the reverse passage at the point *a* (for $\theta = \theta_2$). The phase line has the closed section *a b c d a* and the phase path corresponding to periodic motion can only be the closed curve *a b c d a* (since the phase line does not admit other closed phase paths).

The solution of equation (4.25) for $W=0$ (when the representative point is found on the half-straight line (*I*)) is:

$$\theta = A e^{-\frac{K}{C} t}, \qquad (4.26)$$

and, for $W = W_0$ when the representative point is found on the half-straight line *II*), is

$$\theta = \frac{W_0}{K} + B e^{-\frac{K}{C} t}. \qquad (4.27)$$

For $\theta_\infty = W_0/K < \theta_2$ (θ_∞ being the ultimate temperature reached with the heater permanently switched on) the system has a single state of equilibrium θ_0

$$\theta_0 = \frac{W_0}{K}$$

on the half-straight line (*II*). As is easily verified, this state of equilibrium will be established for arbitrary initial conditions. The equilibrium tem-

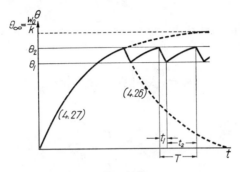

Fig. 194

perature θ_0 will depend both on the power W_0 of the heater and on the rate of heat loss.

Now consider the condition

$$\theta_\infty = \frac{W_0}{K} > \theta_2. \tag{4.28}$$

In this case, over the *whole* half-straight line (*II*) $d\theta/dt > 0$ and over the half-straight line (*I*) $d\theta/dt < 0$. Therefore on the phase line there are no states of equilibrium and the representative point *after the first switching of the regulator* will describe the *closed* phase path *a b c d a*. A *self-oscillating mode of operation* will be established. The oven temperature oscillates within the limits θ_1 and θ_2. For such a self-oscillating mode the oven temperature does not cross the limits of this region, which can be made quite narrow for quite wide variations in the power W_0 and the heat loss rate K.

An oscillogram of temperature oscillations is shown in Fig. 194 and comprises segments of the exponential curves (4.26) and (4.27) suitably placed to make the temperature θ a continuous function of time. The periodic oscillations have a form markedly different from that of harmonic oscillations, and are saw-tooth shaped relaxation oscillations.

The interval t_1 during which the heater is switched off and the temperature falls from θ_2 to θ_1 is determined by (4.26), so

hence
$$\theta_1 = \theta_2 e^{-\frac{K}{C}t_1},$$

$$t_1 = \frac{C}{K} \ln \frac{\theta_2}{\theta_1}.$$

Similarly the interval t_2 during which the heater is switched on and the temperature increases from θ_1 to θ_2 is determined by (4.27), so

$$\theta_2 = \frac{W_0}{K} + \left(\theta_1 - \frac{W_0}{K}\right) e^{-\frac{K}{C}t_2}$$

or

$$t_2 = \frac{C}{K} \ln \frac{\frac{W_0}{K} - \theta_1}{\frac{W_0}{K} - \theta_2}.$$

Thus the period of the self-oscillations is

$$T = t_1 + t_2 = \frac{C}{K} \left\{ \ln \frac{\theta_2}{\theta_1} + \ln \frac{\frac{W_0}{K} - \theta_1}{\frac{W_0}{K} - \theta_2} \right\}$$

or

$$T = \frac{C}{K} \ln \frac{\frac{\theta_\infty}{\theta_1} - 1}{\frac{\theta_\infty}{\theta_2} - 1}. \tag{4.29}$$

Obviously the closer θ_1 and θ_2 are to each other, the smaller the period and the more frequent the switching on and off of the heater[†].

[†] We have assumed that the oven temperature is the same throughout the *whole* volume of the oven. This is valid only for sufficiently slow temperature oscillations. If the finite velocity of heat transfer between different parts of the oven is taken into account, then the system has an equation of a higher order, or is even a distributed system. The temperature in the self-oscillating mode will now exceed somewhat the switching limits of the regulator, and the period will not tend to zero as $\theta_1 \to \theta_2$. This is in full agreement with experimental data.

2. Oscillations in a circuit with a neon tube

As a second example of a dynamic system with half a degree of freedom, the oscillations of which are described by a differential equation of the first order (4.1) with a two-valued right-hand side, consider the circuit of a relaxation oscillator with a neon tube (Fig. 195)[†]. Such a circuit has been analysed by Van der Pol, Friedlaender and others [152, 153, 188, 146, 143].

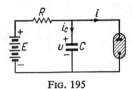

Fig. 195

In analysing this circuit, we shall neglect stray inductance or capacitance in elements of the circuit, and also assume that the current i through the neon tube is a function of the voltage u across it and is determined by the *static* characteristic if the tube $i = \varphi(u)$[‡]. Thus we select as our model a dynamic system with half a degree of freedom, described by a non-linear differential equation of the first order

$$C\frac{du}{dt} + i = \frac{E-u}{R}$$

or

$$\frac{du}{dt} = \frac{1}{RC}\{E - u - R\varphi(u)\} = f(u). \qquad (4.30)$$

The current i and the voltage u is not determined by Ohm's law, but is expressed by the non-linear relation $i = \varphi(u)$, having a hysteretic character. The most typical features of a neon tube of immediate use are the following. For small voltages the tube is non-conducting; and the tube "fires" and current flows only at a certain ignition voltage U_1. Then a certain current I_1 is established at once. For $u > U_1$, the current increases according to an almost linear law. If the voltage is now decreased to the value U_1 the

---

[†] Note that the analysis of a thyratron generator of saw-tooth voltages is analogous to the analysis of a circuit with a neon tube.

[‡] We are assuming that at each instant of time the values of i and u do not differ from the *static* values, which is true only for sufficiently slow processes, when the rate of change of u is considerably less than the rate at which the gas discharge forms in the neon tube. The time to initiate the gas discharge is of the order of hundreds of microseconds. Therefore, the results in the text will only be valid for frequencies which do not exceed about kc/s.

discharge in the tube is not quenched. If u is decreased still further the current decreases gradually and eventually the tube is quenched abruptly at a certain cut-off voltage U_2, when the current is I_2, $U_2 < U_1$ and $I_2 < I_1$. All these essential features of the neon-tube generator are shown in Fig. 196a.

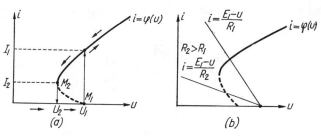

Fig. 196

Of course, this circuit possesses a small parasitic self-inductance and the current through the neon-tube or the circuit cannot vary instantaneously. However, since we are neglecting these factors, we shall assume that the current through a neon-tube varies instantaneously from 0 to I_1 at firing and from I_2 to 0 at quenching, as is indicated by the arrows in Fig. 196a†. The dotted section $M_1 M_2$ of the static characteristic is not followed under static conditions of operation, owing to its instability. The characteristic shown in Fig. 196a coincides, by and large, with those measured experimentally.

The states of equilibrium of the system are determined from the condition $f(u) = 0$, so from (4.30)

$$\frac{E-u}{R} = \varphi(u). \tag{4.31}$$

As usual, plot the curve $i = \varphi(u)$ and the straight line $i = (E-u)/R$ and find their points of intersection. Let $E > U_1$ always, so that the battery voltage is larger than the firing voltage, then there only exists one point of intersection whose position depends on the values of the parameters E and R (Fig. 196b). The stability of this state of equilibrium $u = U$ is determined by the sign of $f'(u)$. It is easily seen that if the state of equilibrium lies on the upper section of the characteristic then it is stable, and

---

† Thus we are assuming that, during these rapid (jump-wise) variations of the current in a neon-tube, the current i is no longer determined by the static characteristic $i = \varphi(u)$ and the system does not obey the equation (4.30).

is unstable if it lies on the lower section[†]. Therefore, for each $E > U_1$, we can, by increasing R, pass from a stable state of equilibrium to an unstable one. The larger E, the larger must be the critical resistance R_{crit}, for which the point of intersection passes on to the lower part of the characteristic where equilibrium is unstable.

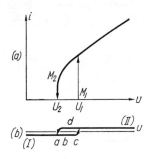

FIG. 197

We restrict the initial states of the point (u, i) representing the state of the neon tube to be either *on the section $i = 0$* or *on the upper section of the static characteristic*[‡], since the neon-tube never operates on the *lower section*.

The current i in the tube is a *two-valued* function of the voltage u over the interval $U_2 < u < U_1$ and single-valued outside this interval (Fig. 197a). Thus, the phase line will be a *line with superposition* (Fig. 197b) consisting of two half-straight lines (*I*) $u < U_1$ and (*II*) $u < U_2$, the first of which corresponds to the tube quenched and the second to the tube fired. The representative point passes from one half-straight line to the other (the firing and the quenching of the neon tube) for $u = U_1$ and $u = U_2$, at the end points of these half-straight lines. Since the phase line admits the single closed phase path *abcda*, only a single periodic process is possible.

It is easy to follow the motion of the system by means of this phase line. In the case when R is sufficiently small and the position of equilibrium is

[†] In fact, it is found that a state of equilibrium on the lower section is stable if the resistance R is sufficiently large and the capacitance C sufficiently small (see Section 5, Chapter V). The circuit is stabilized by the inertia of the gas discharge, i.e. by the finite rate at which the current grows.

[‡] If we are interested in oscillations starting from other initial states such as those which lie on the lower part of the characteristic then our dynamic model would be unsuitable. We must extend this model and take into account small parasitic parameters such as the inertia of the gas discharge, and consider, instead of the static characteristic of the neon valve the differential equation which represents the dynamics of the gas discharge (see Section 7, Sub-section 1, Chapter X).

stable (it lies on the half-straight line *II*), then, when the capacitor is connected in parallel with the tube, this will flash (its current increases) and thereafter the voltage across the tube and the current through it will begin to decrease. The rate of change of *u* will be determined by the equation (4.30), but will be in any case finite. The decrease in the voltage *u* will last until a stable state of equilibrium $u=U$ is reached (Fig. 198) (strictly speaking, the system will approach asymptotically this state of equilibrium).

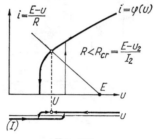

Fig. 198

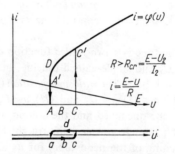

Fig. 199

If, however, *R* is so large that on the half-straight line *II* there are no stable states of equilibrium (in this case the state of equilibrium lies on the lower section of the characteristic and is unstable), then $du/dt < 0$ on the half-straight line (*II*) and $du/dt > 0$ on the half-straight line (*I*). A *periodic process* is established in the circuit after the first firing or quenching, being represented on the phase line by the closed phase path *abcda*, and on the *u*, *i* plane by the closed curve $ABCC'DA'A$ (Fig. 199). The section *abc* of this closed phase path corresponds to the process of charging of the capacitor *C* via the resistor *R*. and the section *cda* to the discharging of the capacitor via the fired neon tube.

The oscillograms are shown in Fig. 200. Until the periodic process begins, the form of these curves depends on the initial conditions. Thus, for example, if the circuit is switched on by closing the switch B_2 in Fig. 201, the switch B_1 having been closed in advance then $u_0 = E$ and the character

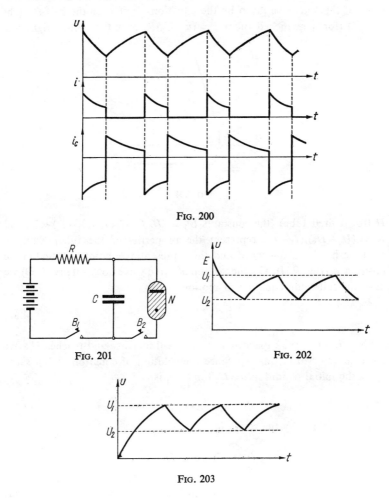

Fig. 200

Fig. 201

Fig. 202

Fig. 203

of the process will be that shown in Fig. 202. If, however, the switch B_2 has been closed in advance and the circuit is switched on by closing the switch B_1 then $u_0 = 0$ (Fig. 201) and the early form of the process will be different, as in Fig. 203.

In order to determine quantitative characteristics (the period, the amplitude and the form of the oscillograms), we need to know the non-linear function $i = \varphi(u)$. It is convenient to use a piece-wise linear function $\varphi(u)$, as an approximation to the real curve (Fig. 204). Let the upper stable sections of this curve be given by the equations $i=0$ over the half-straight line (I) (for a quenched tube) $i = (u-U_0)R_i$ over the half-straight line

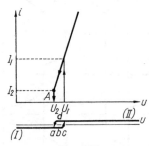

Fig. 204

II (for a fired tube) (the constants $U_0 = (U_2 I_1 - U_1 I_2)/(I_1 - I_2) < U_2$ and $R_i = (U_1 - U_2)/(I_1 - I_2)$, represent the properties of the tube). Suppose the tube has been quenched and the representative point is found at the position a (Fig. 204). On the section abc (the valve does not fire, $i=0$) we have the following equation of motion:

$$RC\frac{du}{dt} = E - u. \tag{4.32}$$

Since $E > U_1$, the capacitor voltage u will increase and reaches U after a certain interval of time τ_1. Since the solution of equation (4.32), satisfying the initial condition $u = U_2$ for $t = 0$, is

$$u = E - (E - U_2)e^{-\frac{t}{RC}}, \tag{4.33}$$

τ_1 is determined by the relation

$$U_1 = E - (E - U_2)e^{-\frac{\tau_1}{RC}}$$

or

$$\tau_1 = RC \ln \frac{E - U_2}{E - U_1}. \tag{4.34}$$

After the valve is fired the equation of motion takes the form

$$C\frac{du}{dt} = \frac{E - u}{R} - \frac{u - U_0}{R_i}.$$

Put $\varrho = RR_i/(R+R_i)$, then this equation reduces to

$$C\frac{du}{dt} = \frac{E}{R} + \frac{U_0}{R_i} - \frac{u}{\varrho}. \tag{4.35}$$

Let $U_2/\varrho > (E/R) + (U_0/R_i)$, which takes place if

$$R > R_{\mathrm{cr}} = \frac{E-U_2}{I_2}. \tag{4.36}$$

Then, for $U_2 < u < U_1$, $du/dt < 0$, i.e. the capacitor voltage will decrease, since for $R > R_c$ the current of discharge of the capacitor via the fired tube is always larger than the charging current via the resistor R. After a further time τ_2 the capacitor voltage reaches U_2 and the tube is quenched. The solution of the equation (4.35) for $u = U_1$ at $t = 0$ is

$$\frac{u}{\varrho} = \frac{E}{R} + \frac{U_0}{R_i} + \left(\frac{U_1}{\varrho} - \frac{E}{R} - \frac{U_0}{R_i}\right) e^{-\frac{t}{\varrho C}}. \tag{4.37}$$

Putting $u = U_2$ and $t = \tau_2$ and solving with respect to τ_2 we obtain

$$\tau_2 = \varrho C \ln \frac{\dfrac{U_1}{\varrho} - \dfrac{E}{R} - \dfrac{U_0}{R_i}}{\dfrac{U_2}{\varrho} - \dfrac{E}{R} - \dfrac{U_0}{R_i}} \tag{4.38}$$

where $(U_1 - U_0)/R_i = I_1$ and $(U_2 - U_0)/R_i = I_2$. The period of the self-oscillations is equal to

$$\tau = \tau_1 + \tau_2. \tag{4.39}$$

The oscillogram of the capacitor voltage consists of segments of exponential curves (see the equations (4.33) and (4.37)) and its form is markedly

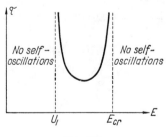

Fig. 205

different from that of a sinusoid. The period of the self-oscillations is proportional to the capacitance C. The dependence of the period τ upon the battery voltage E, other parameters being constant, is shown qualitatively in Fig. 205. Self-oscillations occur for $U_1 < E < E_{crit} = U_2 + RI_2$ only. For E approaching either U_1 or E_{crit}, $\tau \to +\infty$ (in the first case this is due to τ_1 tending to infinity, in the second one to τ_2 tending to ∞).

In conclusion, we must emphasize that this dynamic system of the first order reproduces satisfactorily the processes taking place in a relaxation oscillator with a neon tube only if the capacitance C is sufficiently large to prevent high frequency oscillations.

§7. Multivibrator with one RC circuit

Another example of an electrical self-oscillating system is the RC-coupled multivibrator shown in Fig. 206. Such a multivibrator generates a periodic sequence of quasirectangular voltage pulses.

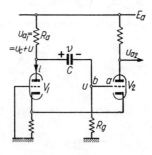

Fig. 206

The equations of the multivibrator, taking into account the circuit elements shown in Fig. 206 and neglecting the grid current of the valve V_2, are

$$i + C\frac{dv}{dt} + \frac{u+v-E_a}{R_a} = 0,$$
$$C\frac{dv}{dt} = \frac{u}{R_g}.$$
(4.40)

Neglecting the anode conductance, we assume the anode currents to be single-valued functions of the grid voltage u of the valve V_2. In particular, the dependence of the anode current i of the valve V_1 upon this voltage is given by the characteristic $i = \varphi(u)$, shown in Fig. 207. Below, in order to simplify the analysis, we shall assume the point of maximum negative slope is at the middle of the descending section, where $u = 0$. We shall denote by S_0 the maximum absolute value of the slope of the descending section where $S_0 = -\varphi'(0)$, and so $|\varphi'(u)| \leq S_0$†.

Eliminating v, the voltage across the capacitor C, from the equations (4.40), we obtain a differential equation of the first order for the voltage u on the grid of the valve V_2.

$$C(R_a+R_g)\left[1+\frac{R_a R_g}{R_a+R_g}\varphi'(u)\right]\frac{du}{dt}+u = 0. \tag{4.41}$$

Since the current i is a *single-valued* function of the voltage u, assigning u determines du/dt uniquely and so determines the state of the dynamic

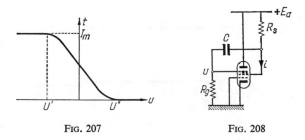

Fig. 207 Fig. 208

model of the multivibrator. We shall take, therefore, as the phase line of the system the straight line u.

The only state of equilibrium is the state $u = 0$. Its stability is determined, clearly, by the following linearized equation (the equation of the first approximation)

$$C(R_a+R_g)[1-K]\frac{du}{dt}+u = 0, \tag{4.42}$$

where

$$K = \frac{S_0 R_a R_g}{R_a+R_g}$$

is the amplification between the points a and b in Fig. 206, the point a being disconnected from b.

If $K<1$, the state of equilibrium $u=0$ is stable and is reached for any initial conditions, since, by virtue of the inequality $|\varphi'(u)| \leq S_0$, $du/dt>0$ for $u<0$ and $du/dt<0$ for $u>0$. The division of the phase straight line

† The dynamics of a single valve multivibrator or "transitron" (Fig. 208) is similar to the dynamics of the circuit being considered. The dependence of the current i of the screen grid upon the voltage u of the third grid is given by a characteristic similar to the one shown in Fig. 207.

into phase paths for this case of a non-excited multivibrator is shown in Fig. 209a.

A different picture is obtained for $K>1$ (Fig. 209b) when the state of equilibrium $u=0$ is unstable. Let U_1 and U_2 be the values of u for which the coefficient of du/dt in (4.41) reduces to zero. Clearly, U_1 and U_2 are determined by the equation

$$\frac{R_a R_g}{R_a + R_g} \varphi'(u) = -1$$

and $U' < U_1 < 0 < U_2 < U''$. For a passage through $u = U_1$ or $u = U_2$, du/dt changes its sign. Therefore $du/dt > 0$ for $u < U_1$, $du/dt < 0$ for $U_1 < u < 0$, $du/dt > 0$ for $0 < u < U_2$ and, finally, $du/dt < 0$ for $u > U_2$.

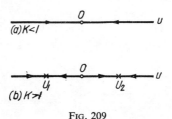

Fig. 209

Hence, whatever the initial conditions, the representative point arrives either at the point $u = U_1$ or at the point $u = U_2$, which, however, are not states of equilibrium although there are no phase paths leaving them.

The dynamic model of a multivibrator used here takes *certain properties only* of a real multivibrator into account and is *unsatisfactory* because it does not reproduce the oscillations in a real multivibrator.

The fact is, *we have left out of account certain essential factors which influence in a radical manner the laws of the oscillating processes in the multivibrator*, at least during certain stages of the motion.

It turns out that the parasitic capacitances C_a and C_g of the anode of the valve V_1 and the grid of V_2 and the capacitance C_c of the common cathode connection are essential in determining the behaviour of the oscillations during certain stages of the motion.

When we take them into account we arrive at a dynamic model of the second order (with 1 degree of freedom) which represents sufficiently well an actual multivibrator. Such a dynamic model of a multivibrator will be considered in Chapter VIII, Section 5 and Chapter X, Section 4.

Another approach consists in "correcting" the dynamic model of the first order by introducing certain *additional postulates* about the transition from the states $u = U_1$ and $u = U_2$. The equation (4.41) is replaced during certain stages of the oscillations. These additional postulates follow from certain additional physical considerations. This method will be used in Chapter X in the analysis of a series of oscillating systems with "discontinuous" oscillations[†].

Here, however, we shall continue to use a dynamic model of the first order, with the addition of *postulates about the jumps of the voltage u* at the grid of the valve V_2. It is well-known that for $K \gg 1$, multivibrator self-oscillations have a "discontinuous" character: comparatively slow variations of the voltage u are periodically replaced by very rapid ones. The rates of the latter are determined by the rates of charging and discharging the parasitic capacitances of the circuit, the most important of them being the capacitances C_a and C_g. For sufficiently small parasitic capacitances, we can assume these rapid variations of voltage to be almost instantaneous. The equation (4.41) is manifestly unsuitable for describing the motion of the system after it has arrived at the state $u = U_1$ or at the state $u = U_2$. We shall assume therefore, that the system moves away from these states by a sudden transition, thus reaching a state where (4.41) is again applicable. To determine the states into which the system jumps, it is merely necessary to assume that infinite voltages and currents cannot exist. Then the charging current $C\, dv/dt$ of the capacitor is always finite. Thus, when u changes abruptly, the voltage v across the capacitor C does not vary, since otherwise $dv/dt = \infty$, which is impossible. The continuity of voltage across the capacitor C during the jump in u is sufficient to determine uniquely the states at which the system arrives as a result of a jump[‡].

On eliminating $C\, dv/dt$ from the equations (4.40) we obtain v as a function of the voltage u,

$$v = F(u) = E_a - R_a \varphi(u) - \left(1 + \frac{R_a}{R_g}\right) u, \qquad (4.43)$$

[†] Discontinuous oscillations is the name given to such oscillations for which there are both relatively slow variations and very rapid ones of the state of the system. The multivibrator is a typical representative of a generator of discontinuous oscillations.

[‡] In other problems, for example in the problem of the oscillations of two multivibrators connected with each other [37], the condition of continuity of the voltages across the capacitors during the jumps is not sufficient to determine uniquely the states of the system after the jumps.

If, on the other hand, the "law of the jumps" is obtained by a limit analysis of a "more complete" dynamic model (taking into account essential parasitic parameters but assuming them to tend to zero), then the states of the system after the jumps are always determined uniquely (see Chapter X).

A limit analysis of a "more complete" dynamic model of a multivibrator (see Chapter X, Section 4) shows that the voltage v does actually remain constant during the jumps, and that the jumps begin not only for $u = U_1$ and $u = U_2$ but also for arbitrary values of u in the interval $U_1 < U < U_2$.

valid of course, for the states of the multivibrator for which the equations (4.41) or (4.40) are satisfied). v is a single valued and continuous function of u, and its graph for $K>1$ is shown in Fig. 210(a). The states of the multivibrator immediately before a jump ($u=U_1$ or $u=U_2$) and after a jump ($u=U_1'$ or $u=U_2'$) are such that (4.41) and (4.43) are valid for them. Also

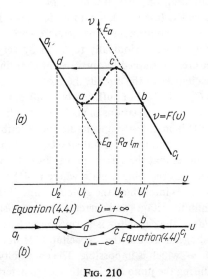

Fig. 210

v does not vary during the jump so that the state of the multivibrator ($u=U_j'$) immediately after the jump from the state $U_j (j=1,2)$ is determined by the equation

$$F(U_j) = F(U_j')$$

or

$$R_a \varphi(U_j') + \left(1 + \frac{R_a}{R_g}\right) U_j' = R_a \varphi(U_j) + \left(1 + \frac{R_a}{R_g}\right) U_j. \qquad (4.44)$$

The graphical solution of this equation is shown in Fig. 210(a).

Thus the oscillations in the multivibrator prove to be periodic and consist of slow variations of the voltage u from U_1' to U_2 and from U_2' to U_1 obeying the equations (4.41), and of jump-wise variations from U_1 to U_1' and from U_2 to U_2' determined by the jump condition. This periodic motion corresponds in Fig. 210(a) to the closed curve abcda (the sections bc and da corresponding to the "slow" variations and the sections *ab* and *cd* to the jump-wise variations of the voltage *u*). Oscillograms of the voltages u, v and u_{a2} are shown in Fig. 211. The oscillations of the voltage v across the capacitor C are continuous and are "saw-tooth" shaped, while the oscillations of the anode voltage u_{a2} of the valve V_2 are nearly "rectangular".

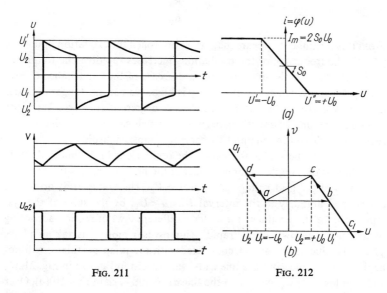

FIG. 211   FIG. 212

To determine the period we have to integrate the equation (4.41). Now

$$dt = -C(R_a+R_g)\left[1+\frac{R_a R_g}{R_a+R_g}\varphi'(u)\right]\frac{du}{u}, \qquad (4.45)$$

over the interval from $u=U_2'$ to $u=U_1$ and from $u=U_1'$ to $u=U_2$. In the intervals over which the integration is to be carried out, i.e. in the region where $i=0$ and in the region where $i=I_m=2S_0U_0$, the characteristic of Fig. 212(*a*) represents sufficiently well the properties of a real valve pair (the two valves with a common cathode resistor). In these intervals we have $\varphi'(0)=0$ and the equation (4.41)

$$C(R_a+R_g)\frac{du}{dt}+u = 0$$

is linear and easily integrated. As a result a very simple formula for the period is obtained, which is highly typical of processes of this nature,[†]

$$T = 2C(R_a+R_g) \ln(2K-1), \qquad (4.46)$$

where, as before

$$K = \frac{S_0 R_a}{1+\dfrac{R_a}{R_g}}.$$

From this formula, it appears that, as we approach the excitation boundary ($K \to 1$), the frequency of the oscillations increases rapidly. The frequency of the oscillations also increases for a decrease of the capacitance C. But, strictly speaking, when the frequency is high, we can no longer consider the multivibrator as a system with half a degree of freedom. We must allow for the small parasitic capacitances which cause the oscillations to cease to be discontinuous and to become almost sinusoidal.

We have thus been able to analyse the oscillations in the multivibrator, by adding to the simple dynamic model of the first order a postulate about the jumps of the grid voltage of the valve V_2. The voltage u is no longer determined uniquely in the interval $U_2' < u < U_1$, by the state of the system, since for these values of u we have different laws of motion according to whether a "slow" or a "rapid" (jump-wise) motion is taking place. The phase line of the model, completed with the jump postulate, will not be the straight line u but the line with "superposition" shown in Fig. 210(b) and topologically equivalent to the line $a_1 ab$ and $c_1 cd$ in Fig. 210(a). Over the sections $a_1 a$ and $c_1 c$ the motion is determined by the equation (4.41) and the jumps from a to b and from c to d, shown with thick lines, are determined by the postulate about the jumps of $u(du/dt = +\infty$ over the ab section and $du/dt = -\infty$ over the cd section). The phase line, just as in the other examples discussed in Section 7 of this chapter, admits a closed phase path, which corresponds to discontinuous periodic oscillations of the multivibrator.

[†] The dependence of the voltage v across the capacitor C upon the voltage u at the grid of the valve V_2, during the slow variation is shown in Fig. 212 (b). According to the equation (4.44) for the section bc we have

$$\left(1+\frac{R_a}{R_g}\right) U_1' = R_a I_m - \left(1+\frac{R_a}{R_g}\right) U_0, \quad \text{i.e.} \quad U_1' = U_0(2K-1).$$

Therefore, the duration of the "slow" variation along the section bc is equal to

$$C(R_a+R_g)\ln(2K-1).$$

By virtue of the symmetry of the characteristic of the valve-pair, the duration of the "slow" motion along the section da will be the same. Therefore for the total period the formula (4.46) is again obtained.

CHAPTER V

DYNAMIC SYSTEMS OF THE SECOND ORDER†

We shall consider in this chapter autonomous dynamic systems of the second order (with 1 degree of freedom) whose motion may be described by two differential equations of the first order

$$\frac{dx}{dt} = P(x, y), \quad \frac{dy}{dt} = Q(x, y). \tag{5.1}$$

Such systems are the most general case of the systems forming the object of our analysis, and we have already met many examples in which they have arisen.

We must emphasize that, for a mathematical study of these systems, the equations (5.1) are not sufficient: we need to know the phase space of the system, the points of which are in a *one-to-one continuous correspondence* with the states of the system‡.

However, the nature of the phase space, like the differential equations, must be derived from the physical problem. If, for example, we know that our system returns to a previous state when x varies by 2π, then this suggests that perhaps we need a cylindrical phase space. The differential equations by themselves do not determine the character of all possible motions of the system and their possible phase paths in the space until this space has been chosen. To clarify this, consider the simplest linear system

$$\frac{dx}{dt} = a, \quad \frac{dy}{dt} = b.$$

If x and y are ordinary cartesian coordinates of the phase plane, then the phase paths are straight lines. On the phase plane we have a continuum of open trajectories. If, however, x and y are orthogonal curvilinear coordinates on a torus (for example, x is the azimuth in the meridian plane and

† Sections 5 and 12 have been revised and Section 1, Section 3 (Sub-section 1), Section 7 (Sub-section 2 and 3) and Sections 9 and 11 have been written anew by N.A. Zheleztsov.

‡ Of course, in dynamic systems of the second order (with 1 degree of freedom) the phase space is two-dimensional, i.e. is a certain surface, since the state of the system is completely determined by assigning a number-pair x, y.

y is a polar angle with vertex on the axis of the torus) then the phase paths for the same system of differential equations form either a continuum of closed curves (if a and b are commensurable) or a continuum of paths which cover the surface of the torus (if a and b are incommensurable). In the first case the actual motion is periodic and in the second, quasi-periodic. This shows the importance of knowing the order of connexion of the phase space before examining the phase paths.

In this chapter we shall restrict ourselves to the most important case, when the phase surface is an ordinary plane. Later on, in Chapter VII, we shall meet examples from mechanics of cylindrical phase surfaces, and in Chapter VIII we shall consider a few systems with a phase surface having many sheets.

§ 1. Phase paths and integral curves on the phase plane

Consider a system of two autonomous differential equations of the first order

$$\frac{dx}{dt} = P(x, y), \quad \frac{dy}{dt} = Q(x, y), \tag{5.1}$$

describing a dynamic system of the second order[†], on the assumption that between the states of this dynamic system and the points x, y of the phase plane there is a one-to-one continuous correspondence. The functions $P(x, y)$ and $Q(x, y)$ will be assumed to be analytic over the whole phase plane.[‡]

The conditions of Cauchy's theorem on the existence and uniqueness of the solution of a system of differential equations (see Appendix I) are satisfied for the equations (5.1), and there exists therefore a unique system of functions: $x = x(t)$ and $y = y(t)$ satisfying both (5.1) and the given initial conditions $x = x_0$, $y = y_0$ at $t = t_0$. Since the solution depends on the initial conditions, it is sometimes convenient to write such a solution in the form

$$x = \varphi(t - t_0; x_0, y_0), \quad y = \psi(t - t_0; x_0, y_0)^{\dagger\dagger}, \tag{5.2}$$

where φ and ψ are analytic functions in all three arguments t, x_0, and y_0.

[†] If there is one equation of the second order $\ddot{x} = f(x, \dot{x})$ then by the substitution $y = \dot{x}$ we are always able to reduce it to the form $\dot{x} = y$, $\dot{y} = f(x, y)$.

[‡] The requirement of analyticity of the functions $P(x, y)$ and $Q(x, y)$ has only been introduced for the sake of a certain simplification of the demonstrations and can be replaced by the less stringent requirement that these functions have continuous partial derivatives of suitable orders (in a number of cases, of the first order).

[††] This way of writing down the solution is only possible for autonomous systems. If

Each solution (5.2) (with given x_0, y_0, t_0) may be considered as a parametric equation of a certain curve on the plane x, y along which the representative point moves as t varies. Such curves are called *phase paths*.

On the other hand, the solution (5.2) may also be considered as the equation of a curve in the x, y, t space, i.e. of an integral curve of (5.1). Thus each phase path is the projection on the phase plane of a certain integral curve in the x, y, t space†. In addition, because equations (5.1) are auto-

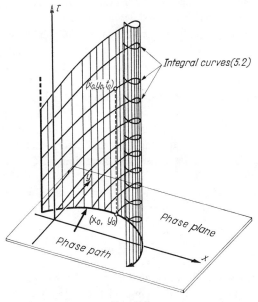

Fig. 213

nomous, all the integral curves (5.2) with the same x_0, y_0 but different t_0 form in the x, y, t space a cylindrical surface with generators parallel to the t axis. Hence, their projections on the phase plane are one and the same phase path (Fig. 213). In other words each phase path corresponds to all

---

$x = \varphi(t; x_0, y_0)$, $y = \psi(t; x_0, y_0)$ be the solution of the equations (5.1) satisfying the initial conditions: $x = x_0$, $y = y_0$ at $t = 0$, then it is evident that the functions φ and ψ are such that $\varphi(0; x_0, y_0) \equiv x_0$ and $\psi(0; x_0, y_0) \equiv y_0$. Since the equations (5.1) are autonomous (their right-hand sides, the functions P and Q, do not depend explicitly on the time t), then the system of functions (5.2) will also be a solution, and in fact (by virtue of Cauchy's theorem) the *only solution satisfying the initial conditions:* $x = x_0$, $y = y_0$ at $t = t_0$.

† The two other projections of the integral curve (5.2) on the planes x, t and y, t are, clearly, the ordinary oscillograms of the variations of x and y for some motion of the system.

motions of the dynamic system which pass through the same states and differing from each other only in the origin of time.

Since the conditions of Cauchy's existence theorem are satisfied by (5.1), then through each point of the x, y, t space there passes a unique integral curve of (5.1) i.e. the integral curves in the x, y, t space cannot intersect each other. Because equations (5.1) are autonomous this can also be said about the phase paths; they also cannot intersect each other since *through each point of the phase plane there passes a unique phase path*[†].

To illustrate what has been stated, let us consider the following example. If at a certain point $(\bar{x}, \bar{y})$ the functions $P(x, y)$ and $Q(x, y)$ reduce to zero, then the equations (5.1) have the solution: $x \equiv \bar{x}$, $y \equiv \bar{y}$; the phase path corresponding to this state of equilibrium consists of one isolated point. By virtue of the property mentioned above the representative point cannot, by moving along any other paths, reach this point in finite time. Similarly, if the representative point is not on a limit cycle, it will not reach this limit cycle in any finite time. Therefore, periodic oscillations or a state of equilibrium are reached in dynamic systems described by the equations (5.1), where right-hand sides satisfy the conditions of Cauchy's theorem, only asymptotically (as $t \to +\infty$).

If we divide one of the equations (5.1) by the other, we eliminate time and shall obtain an equation of the first order:

$$\frac{dy}{dx} = \frac{Q(x, y)}{P(x, y)}, \qquad (5.3)$$

which in many cases can be integrated more easily than (5.1). The solution of this equation $y = y(x; C)$, or in implicit form $F(x, y) = C$, where C is a constant, is a family of integral curves such that the slope of a tangent is determined by the equation (5.3)[‡]. It can be shown by applying Cauchy's

---

† In fact, should two phase paths pass through a certain point (x^*, y^*), then through each point of the straight line $x = x^*$, $y = y^*$ in the x, y, t space there would pass two different integral curves of the equations (5.1), which contradicts Cauchy's theorem.

Note that the integral curves of a *non-autonomous* system $\dot{x} = P(x, y, t), \dot{y} = Q(x, y, t)$ do not intersect each other, just as before, provided that the conditions of Cauchy's theorem are satisfied, but their projections on the x, y plane will, generally speaking, intersect each other.

‡ Now, by integral curves we mean the integral curves of the equation (5.3) only.

Also observe that two curves, each a solution of the differential equation (5.3) and forming the analytic continuation of one another, are usually referred to as one integral curve, even if such a curve passes through a point where this equation loses its meaning. For example, we obtain automatically solutions corresponding to integral curves of this kind, in cases where the integration of the equation (5.3) can be reduced to quadratures.

theorem to the equation (5.3) that, because functions $P(x, y)$ and $Q(x, y)$ are analytic, only one integral curve of the equation (5.3) passes through a point of the x, y plane except possibly at the singular points where the equation loses its meaning. For our system only the points at which $P(x, y) = 0$ and $Q(x, y) = 0$ will be singular points and so states of equilibrium of the system (5.1)†. At these points the integral curves intersect each other.

Obviously each phase path is an integral curve or part of an integral curve, while an integral curve or an arc of integral curve not passing through a singular point is necessarily a phase path. On the other hand, an integral curve passing through a singular point always consists of a number of phase paths. Nevertheless, by integrating the simpler equation (5.3) and finding its integral curves, we also map out the phase plane by the paths: the phase paths will be: (*a*) the singular points (the states of equilibrium); (*b*) the integral curves not passing through singular points; and (*c*) the arcs of integral curves comprised between two singular points or between a singular point and infinity. But, the equation (5.3) gives no indications whatever about the direction of the representative point along the phase paths, since time was eliminated from it. The direction of motion of the representative point is determined from the equation (5.1).

§2. Linear systems of the general type

We shall consider first the simplest dynamic systems of the form (5.1), namely those represented by a system of two linear equations

$$\frac{dx}{dt} = ax+by; \quad \frac{dy}{dt} = cx+dy, \tag{5.4}$$

where a, b, c, d are constants.

As is well known, the general solution of (5.4) has the form‡

$$\left.\begin{array}{l} x = C_1 e^{\lambda_1 t} + C_2 e^{\lambda_2 t}, \\ y = C_1 \varkappa_1 e^{\lambda_1 t} + C_2 \varkappa_2 e^{\lambda_2 t}, \end{array}\right\} \tag{5.5}$$

where λ_1 and λ_2 are the roots of the characteristic equation

$$\lambda^2 - \lambda(a+d) + (ad-bc) = 0, \tag{5.6}$$

† The proof is completely analogous to that carried out for the simplest conservative systems (see Chapter II, Section 2).
‡ We are assuming that both roots have real parts different from zero and that there are no multiple roots.

and the so-called distribution coefficients $\varkappa_1$ and $\varkappa_2$ are determined by the relations

$$\left. \begin{array}{l} a-\lambda_k+b\varkappa_k = 0, \\ c+(d-\lambda_k)\varkappa_k = 0 \end{array} \right\} \quad (5.7)$$

(the latter relations form a compatible system of equations, since λ_k are the roots of the characteristic equation) and so

$$\varkappa_1 = \frac{\lambda_1-a}{b} = \frac{c}{\lambda_1-d}, \quad \varkappa_2 = \frac{\lambda_2-a}{b} = \frac{c}{\lambda_2-d}. \quad (5.8)$$

Also

$$\varkappa_1+\varkappa_2 = \frac{\lambda_1-a}{b}+\frac{\lambda_2-a}{b} = \frac{d-a}{b} \quad \text{and} \quad \varkappa_1\varkappa_2 = \frac{\lambda_1-a}{b} \cdot \frac{\lambda_2-a}{b} = -\frac{c}{b}$$

and, hence, $\varkappa_1$ and $\varkappa_2$ are the roots of equation

$$b\varkappa^2+(a-d)\varkappa-c = 0. \quad (5.9)$$

We shall not discuss the time dependence of the solutions but pass at once to the analysis of possible paths on the phase plane.

To this end, as in Chapter I for the particular case $a = 0$, it is convenient to introduce the linear homogeneous transformation

$$\xi = \alpha x+\beta y, \quad \eta = \gamma x+\delta y. \quad (5.10)$$

We shall reduce the system (5.4) to the canonical form

$$\frac{d\xi}{dt} = \lambda_1\xi, \quad \frac{d\eta}{dt} = \lambda_2\eta, \quad (5.11)$$

where λ_1 and λ_2 are as yet unknown constants.

This is always possible for the assumptions made about the nature of the roots of the equation (5.6). Differentiating (5.10), we have

$$\frac{d\xi}{dt} = \alpha\frac{dx}{dt}+\beta\frac{dy}{dt}; \quad \frac{d\eta}{dt} = \gamma\frac{dx}{dt}+\delta\frac{dy}{dt}.$$

Replacing dx/dt and dy/dt by expressions from (5.4), we arrive at the relations

$$\lambda_1(\alpha x+\beta y) = \alpha(ax+by)+\beta(cx+dy),$$
$$\lambda_2(\gamma x+\delta y) = \gamma(ax+by)+\delta(cx+dy).$$

Equating coefficients of x and y we obtain four equations, linear and homogeneous in α, β, γ and δ:

$$\left. \begin{array}{ll} \alpha(a-\lambda_1)+\beta c = 0, & \gamma(a-\lambda_2)+\delta c = 0, \\ \alpha b+\beta(d-\lambda_1) = 0, & \gamma b+\delta(d-\lambda_2) = 0. \end{array} \right\}. \quad (5.12)$$

These equations have no solutions for α, β, γ and δ not identically zero, except in the case when λ_1 and λ_2 are roots of the equation

$$\lambda^2 - (a+d)\lambda + (ad-bc) = 0, \tag{5.6}$$

i.e. are roots of the characteristic equation. The first pair of equations (5.12) only determines the ratio α/β, the second γ/δ. Since the roots of the characteristic equation are assumed unequal then these ratios are not equal to each other and, hence, α, β, γ and δ may be so chosen that the determinant

$$\begin{vmatrix} \alpha & \beta \\ \gamma & \delta \end{vmatrix} \neq 0.$$

Hence, the equations (5.10) can be solved for x and y, and thus (5.10) is a one-to-one transformation. Therefore in the non-degenerate case it is always possible to transform the original system into the canonical form.

Let us consider the different cases that can arise.

1. *The roots λ_1 and λ_2 are real and of the same sign.* Then the coefficients of the transformation are real and we have a transformation of the real x, y plane into the real ξ, η plane. Our problem consists in investigating the transformed ξ, η phase plane, where the canonical system is valid.

$$\frac{d\xi}{dt} = \lambda_1 \xi, \quad \frac{d\eta}{dt} = \lambda_2 \eta, \tag{5.11}$$

and then interpreting the results in the x, y plane.

Dividing one canonical equation by the other

$$\frac{d\eta}{d\xi} = \frac{\lambda_2}{\lambda_1} \frac{\eta}{\xi}. \tag{5.13}$$

Integrating

$$\eta = C |\xi|^a, \quad \text{where} \quad a = \frac{\lambda_2}{\lambda_1}. \tag{5.14}$$

Let λ_2 be the root such that $|\lambda_2| > |\lambda_1|$.

Then, since λ_1 and λ_2 are of the same sign, $a > 1$ and the integral curves are of the parabolic type (Fig. 214). All integral curves, except the η axis which corresponds to $C = \infty$, are tangent at the origin to the ξ axis, which also is an integral curve of the equation (5.13). The origin is a singular point, and it is a node.

The directions of motions on the phase plane are easily found.

If λ_1 and λ_2 are negative, then by (5.11), $|\xi|$ and $|\eta|$ decrease with time. As t increases the representative point approaches the origin without ever reaching it, since this would contradict Cauchy's theorem which, for (5.11), is valid over the whole ξ, η plane. The origin is a *stable node*. If λ_1 and λ_2 are positive then $|\xi|$ and $|\eta|$ increase with time and the representative point moves away from the origin, which is an *unstable node*.

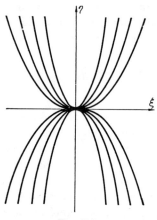

Fig. 214

Let us now return to the x, y plane. As we know, the general character of the phase-portrait near the state of equilibrium is not different but on this plane tangents to the integral curves at the origin no longer coincide with the coordinate axes. It is of interest to establish their directions. Since on the ξ, η plane the tangents are the axes $\xi=0$ and $\eta=0$, it suffices to establish which curves on the x, y plane correspond to the straight lines $\xi=0$ and $\eta=0$ on the ξ, η plane. Equations (5.10) show that the ξ axis (line $\eta=0$) corresponds to the line

$$\gamma x + \delta y = 0, \quad \text{or} \quad y = -\frac{\gamma}{\delta}x, \tag{5.15}$$

passing through the origin with slope

$$\varkappa_1 = -\frac{\gamma}{\delta} = \frac{c}{a-\lambda_2} = \frac{d-\lambda_2}{b}.$$

Similarly the axis η (line $\xi=0$) corresponds to the line

$$\alpha x + \beta y = 0, \quad \text{or} \quad y = -\frac{\alpha}{\beta}x, \tag{5.16}$$

passing through the origin with slope

$$\varkappa_2 = -\frac{\alpha}{\beta} = \frac{c}{a-\lambda_1} = \frac{d-\lambda_1}{b}.$$

These slopes coincide with the distribution coefficients $\varkappa_1$ and $\varkappa_2$ determined by (5.7) or (5.8) and so are the roots of the equation (5.9).

The lines $y=\varkappa_1$ and $y=\varkappa_2$ are on the one hand integral curves for the equation $dy/dx = (cx+dy)/(ax+by)$ (just as the straight lines $\xi=0$ and $\eta=0$ are integral curves for the equation $d\eta/d\xi = a\eta/\xi$†), and on the other hand, the first of them is a tangent to all integral curves but one—the straight line $y = \varkappa_2 x$‡. It is now easy to indicate the behaviour of the

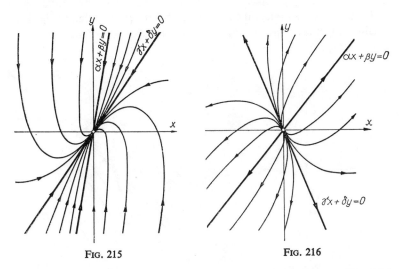

Fig. 215   Fig. 216

phase paths about a stable node (Fig. 215) or an unstable one (Fig. 216).
2. *The roots λ_1 and λ_2 are real but of different signs.* The transformation from the x, y coordinates to the ξ, η coordinates is again real. As before

---

† Each of the lines $y = \varkappa_1 x$ or $y = \varkappa_2 x$ is not one path but consists of *three* paths (two motions towards the state of equilibrium or away from the state of equilibrium, and the state of equilibrium itself).

‡ The direction $\varkappa_1$ for the tangent to the continuum of integral curves is determined at the node, by (5.7) or (5.8) from the root of the characteristic equation λ_1 with smallest modulus. If the directions of the tangents to the integral curves at the node are determined without previously solving the characteristic equation, but as roots of the equation (5.9), then the direction of the tangent to the continuum of integral curves corresponds obviously to the root for which the expression $|a+b\varkappa|$ has the smallest value, since according to (5.7) $\lambda_k = a+b\varkappa_k$.

the canonical system is

$$\frac{d\xi}{dt} = \lambda_1 \xi; \qquad \frac{d\eta}{dt} = \lambda_2 \eta, \tag{5.11}$$

but now λ_1 and λ_2 are of different signs.

The equation of the curves on the new phase plane is

$$\frac{d\eta}{d\xi} = -a\frac{\eta}{\xi}, \quad \text{where} \quad a = \left|\frac{\lambda_2}{\lambda_1}\right|. \tag{5.17}$$

Integrating, then

$$\eta = C|\xi|^{-a}. \tag{5.18}$$

This equation determines a family of curves of the hyperbolic type, having the axes as asymptotes. The coordinate axes are integral curves again and will be the only integral curves passing through the origin† (Fig. 217).

Obviously, the origin is again a singular point, but now of the type known as a saddle point.

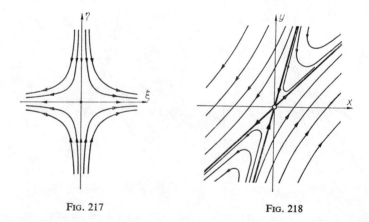

Fig. 217   Fig. 218

Let $\lambda_1 > 0$ and $\lambda_2 < 0$. Then the representative point will move away from the origin of the coordinates when it is on the ξ axis, and, when it is found on the η axis, will approach nearer and nearer the origin without ever reaching it. The directions of motions along the remaining phase paths can easily be obtained by considerations of continuity (Fig. 217). As we know, a saddle point is unstable. Again, on the x, y plane, the same qualitative picture of the phase portrait near the origin is retained (Fig. 218), and the

---

† Each of these integral curves, passing through the origin, consists of three phase paths: two motions towards the state of equilibrium (or away from it) and the state of equilibrium itself.

slopes of the lines passing through the singular point (the separatrices of the saddle point) are given by the equation

$$b\varkappa^2 + (a-d)\varkappa - c = 0.$$

A border-line case between a node and a saddle point occurs when one of the roots of the characteristic equation (5.6) (say λ_1) reduces to zero. This happens when $ad - bc = 0$. In this case the coefficients of the right-hand sides of equations (5.4) are proportional to each other ($a/c = b/d$) and

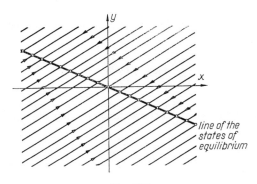

Fig. 219

the states of equilibrium are all the points of the line $ax + by = 0$. The remaining integral curves form a family of parallel straight lines with slope $\varkappa_2 = c/d$, along which the representative point either approaches the states of equilibrium or moves away from them, according to the sign of the second root of the characteristic equation $\lambda_2 = a + d$ (Fig. 219)[†].

3. *λ_1 and λ_2 are complex conjugates.* It is easily seen that for x and y real then ξ and η are complex conjugates. However, by introducing an intermediate transformation, a final real linear homogeneous transformation can be obtained. Put

$$\left.\begin{aligned} \lambda_1 &= a_1 + jb_1, & \xi &= u + jv, \\ \lambda_2 &= a_1 - jb_1, & \eta &= u - jv, \end{aligned}\right\} \quad (5.19)$$

where a_1, b_1, u and v are real quantities. Then it can be shown that the transformation of x, y into u, v is, under our assumptions, real, linear, homogeneous and has a non-zero determinant.

† Fig. 219 shows the case when $a + d > 0$ and the states of equilibrium are stable.

Using (5.19) we have

$$\frac{du}{dt} + j\frac{dv}{dt} = (a_1 + jb_1)(u + jv),$$

$$\frac{du}{dt} - j\frac{dv}{dt} = (a_1 - jb_1)(u - jv),$$

hence

$$\frac{du}{dt} = a_1 u - b_1 v; \qquad \frac{dv}{dt} = a_1 v + b_1 u. \tag{5.20}$$

Consider first of all the form of the integral curves on the (u, v) phase plane. The differential equation of these curves

$$\frac{dv}{du} = \frac{a_1 v + b_1 u}{a_1 u - b_1 v} \tag{5.21}$$

is integrated more easily in polar coordinates. Putting $u = r\cos\varphi$, $v = r\sin\varphi$, we have

$$\frac{dr}{d\varphi} = \frac{a_1}{b_1} r,$$

and therefore

$$r = Ce^{\frac{a}{b}\varphi}. \tag{5.22}$$

On the u, v phase plane the curves are a family of logarithmic spirals, each of which has an asymptotic point at the origin. The origin is a singular point of the focus type (Fig. 220).

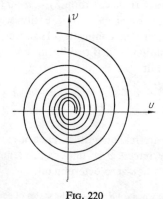

Fig. 220

LINEAR SYSTEMS OF THE GENERAL TYPE

Let us establish the character of motion of the representative point along the phase paths. Multiply the first of the equations (5.20) by u and the second by v and add

$$\frac{1}{2}\frac{d\varrho}{dt} = a_1\varrho, \quad \text{where} \quad \varrho = u^2+v^2. \tag{5.23}$$

Thus, for $a_1 < 0$ ($a_1 =$ Re) the representative point approaches the origin continuously without ever reaching it, and, thus, for $a_1 < 0$ the origin is a *stable focus*.

If, however, $a_1 > 0$ then the representative point moves continuously away from the origin which is now an *unstable focus*.

In passing from the u, v plane to the original x, y plane the spirals will remain spirals, but somewhat deformed (Fig. 221).

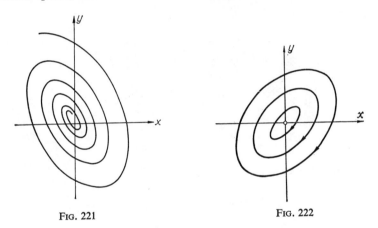

FIG. 221   FIG. 222

For $a_1 = 0$ the phase paths on the u, v plane will be the circles $u^2+v^2 = $ const, to which there correspond on the x, y plane the ellipses

$$by^2+(a-d)xy-cx^2 = \text{const.}$$

$a_1 = 0$ occurs for $a+d = 0$, and the singular point is a *centre* (Fig. 222). In the general non-degenerate linear system being considered here ($ad-bc \neq 0$) there can be six types of states of equilibrium, depending on the nature of the roots of the characteristic equation:

(1) Stable node (λ_1 and λ_2 are real and negative);
(2) Unstable node (λ_1 and λ_2 are real and positive);
(3) Saddle point (λ_1 and λ_2 are real and of different signs);

(4) Stable focus (λ_1 and λ_2 are complex and $Re\ \lambda < 0$);
(5) Unstable focus (λ_1 and λ_2 are complex and $Re\ \lambda > 0$);
(6) Centre (λ_1 and λ_2 are imaginary).

The first five types of states of equilibrium are "coarse" ones: their character does not change for sufficiently small variations of the right-hand sides of the equations (5.4).

The relation between the states of equilibrium and the roots of the characteristic equation may be shown graphically, as in Fig. 223.

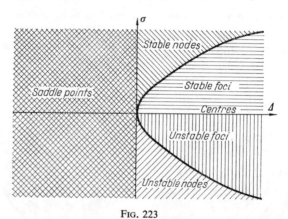

Fig. 223

Let us introduce the notation

$$\sigma = -(a+d), \quad \Delta = \begin{vmatrix} a & b \\ c & d \end{vmatrix}.$$

Then the characteristic equation can be written as

$$\lambda^2 + \sigma\lambda + \Delta = 0. \tag{5.24}$$

On the plane with rectangular coordinates σ and Δ mark out the regions corresponding to the various states of equilibrium. States of equilibrium are stable if λ_1 and λ_2 have negative real parts. A necessary and sufficient condition for this is that $\sigma > 0$, $\Delta > 0$. On the diagram this condition corresponds to points situated in the first quadrant. The singular point will be of the focus type if λ_1 and λ_2 are complex. This condition corresponds to the points for which $\sigma^2 - 4\Delta < 0$, and so lie between the branches of the parabola, $\sigma^2 = 4\Delta$. The points of the axis $\sigma = 0$, $\Delta > 0$ correspond to states of equilibrium of the centre type. Similarly λ_1 and λ_2 will be real but of different signs and the singular point a saddle point, if $\Delta < 0$, etc. If the coefficients of the linear system a, b, c, d depend on some parameter, then, σ and Δ also depend on it and a curve can be plotted on the (σ, Δ) plane which passes from one region into another for certain branch values of the parameter. It will be useful to note that in the case of equal roots, $\sigma^2 - 4\Delta = 0$ and this corresponds to the boundary between nodes and foci on the stability chart (Fig. 223).

§ 3. EXAMPLES OF LINEAR SYSTEMS

To illustrate the preceding observations consider two circuits which under suitable simplifying conditions are described by linear differential equations and in which any of the states of equilibrium can be obtained by varying certain parameters.

EXAMPLE 1. Small oscillations of a dynatron generator. We have already considered this circuit in Chapter I (§ 7, Sub-section 2) as an example of a system with a repulsive force (for $RS_0 > 1$). We shall consider now small oscillations near the state of equilibrium and when the working point lies on the section of the tetrode characteristic with negative slope. This circuit (see equation (1.76)) has the following linear equation:

$$LC\frac{d^2u}{dt^2} + [RC - LS_0]\frac{du}{dt} + [1 - RS_0]u = 0$$

or, if we introduce the non-dimensional time $t_{new} = \omega_0 t$, where $\omega_0 = (LC)^{-\frac{1}{2}}$ and the non-dimensional parameters $r = \omega_0 RC$ and $s = \omega_0 LS_0$,

$$\ddot{u} + (r-s)\dot{u} + (1-rs)u = 0 \tag{5.25}$$

(here a dot denotes differentiation with respect to the new non-dimensional time).

The roots of the characteristic equation,

$$\lambda^2 + (r-s)\lambda + (1-rs) = 0, \tag{5.26}$$

and so the states of equilibrium depend on the parameters r and s. The stability diagram on the r, s plane is shown in Fig. 224.

For $rs > 1$, i.e. above the hyperbola $rs = 1$, the roots λ_1, λ_2 of (5.26) are real and of different signs, and a state of equilibrium is a saddle point. The roots of the characteristic equation are complex for

$$(r-s)^2 < 4(1-rs) \quad \text{or} \quad (r+s)^2 < 4$$

i.e. below the straight line $r+s=2$, a state of equilibrium is a focus. Between the line $r+s=2$ and the hyperbola $rs=1$ the state of equilibrium is a node. The stability of a node or focus, as we have seen, is determined by the sign of the coefficient of λ in the characteristic equation:

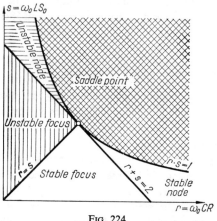

Fig. 224

namely, for $r>s$ the node or focus is stable and for $r<s$ unstable. Thus, the segment of the straight line $r=s$ up to the intersection with the hyperbola $rs=1$ and the segment of hyperbola on the right of this point of intersection form the boundary of the region of stability of the generator. If the state of equilibrium is unstable, then the dynatron generator will move away from the neighbourhood of this state of equilibrium. However, the use of a linear equation does not enable us to state any more about the final operation in the generator.

EXAMPLE 2. "Universal" circuit. A second example is the so-called universal circuit investigated by Khaikin [125], shown in Fig. 225 or in

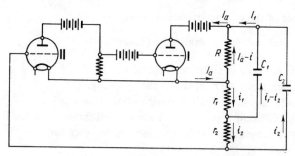

Fig. 225

equivalent form (Fig. 226), provided of course, it is suitably idealized and "linearized". More precisely we shall assume that the characteristics of valves are linear, which is true only for small variations of the grid voltages. Therefore, linearization prevents an analysis over the whole region of inputs.

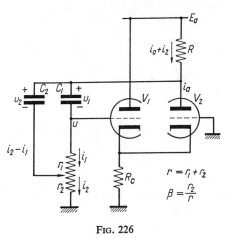

Fig. 226

We shall neglect, as usual, the grid currents and the anode conductance. Proceeding from Kirchhoff's equations we shall obtain (with the notation of Fig. 226) the following equations:

$$\left. \begin{array}{l} r_1 i_1 = u_2 - u_1, \quad R(i_a + i_2) + u_2 + r_2 i_2 = E_a, \\ C_1 \dfrac{du_1}{dt} = i_1, \quad C_2 \dfrac{du_2}{dt} = i_2 - i_1, \end{array} \right\} \quad (5.27)$$

where, in the linear approximation (close to state of equilibrium $i_1 = i_2 = 0$, $u = 0$)

$$i_a = i_{a0} - Su = i_{a0} - S(r_1 i_1 + r_2 i_2).$$

S is the modulus of the negative slope of the anode characteristic of V_2 when the valves V_1 and V_2 are coupled with a common cathode resistance R_c, measured at the working point (a state of equilibrium). Differentiating the first two equations with respect to time and using the last two, as well as the expression for the anode current of the valve V_2, we obtain two equations of the first order for the currents i_1 and i_2:

$$\left.\begin{aligned}\frac{di_1}{dt} &= \frac{-\left(\frac{1}{C_1}+\frac{1}{C_2}\right)i_1+\frac{1}{C_2}i_2}{r_1}, \\ \frac{di_2}{dt} &= \frac{\left[\frac{1}{C_2}-RS\left(\frac{1}{C_1}+\frac{1}{C_2}\right)\right]i_1+(RS-1)\frac{1}{C_2}i_2}{R+r_2(1-RS)}\end{aligned}\right\} \quad (5.28)$$

or, if $k = RS \geqslant 0$, $r = r_1 + r_2$ and $\beta = r_2/r$ ($0 \leqslant \beta \leqslant 1$)

$$\left.\begin{aligned}\frac{di_1}{dt} &= \frac{-\left(\frac{1}{C_1}+\frac{1}{C_2}\right)i_1+\frac{1}{C_2}i_2}{(1-\beta)r}, \\ \frac{di_2}{dt} &= \frac{\left[\frac{1}{C_2}(1-k)-\frac{1}{C_1}k\right]i_1+(k-1)\frac{1}{C_2}i_2}{R-\beta r(k-1)}\end{aligned}\right\}. \quad (5.29)$$

To determine the nature of the singular point at $(i_1 = i_2 = 0)$, consider the characteristic equation of (5.29), which is

$$C_1 C_2 (1-\beta) r [R - \beta r(k-1)] \lambda^2 + [R(C_1+C_2)-(k-1)r(C_1+\beta C_2)]\lambda + 1 = 0. \quad (5.30)$$

The nature of its roots and the nature of the singular point, depend on the four non-dimensional parameters k, β, R/r and C_2/C_1. By choosing various values it is possible to obtain all the types of the singular points discussed above. Here we shall assume that only k and β are variable parameters.

The plane of the parameters k and β can be divided into regions, each of which corresponds to a special singular point (Fig. 227). For $k=0$, we obtain two real negative roots, and so the singular point is a stable node[†]. This might have been anticipated since for $k=0$ the valve-pair plays no role at all, and only damped aperiodic motions can occur. These correspond to a stable node. Next, for

$$k > 1 + \frac{R}{r\beta} \quad (5.31)$$

[†] In fact, for $k=0$, the coefficients of λ^2 and λ are positive, as is the discriminant of the equation

$$[R(C_1+C_2)+r(C_1+\beta C_2)]^2 - 4C_1 C_2 (1-\beta) r [R+\beta r] =$$
$$= [C_1(R+r)-C_2(R+\beta r)]^2 + 4C_1 C_2 [R+\beta r]^2 > 0.$$

the coefficient of λ^2 is negative and so the singular point is a saddle. The points lying below the hyperbola $k = 1+(R/r\beta)$ correspond to either a node or a focus. In this case the stability of the singular point is determined

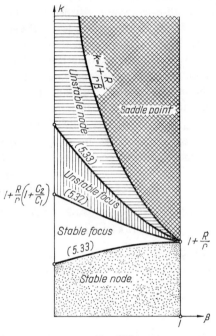

Fig. 227

by the sign of the coefficient of λ. This coefficient reduces to zero on the hyperbola

$$k = 1 + \frac{R}{r}\frac{C_1+C_2}{C_1+\beta C_2}, \qquad (5.32)$$

and is positive below it and negative above it. Since $0 \leqslant \beta \leqslant 1$,

$$\frac{1}{\beta} \geqslant \frac{C_1+C_2}{C_1+\beta C_2}$$

and the hyperbola (5.32) lies below the hyperbola $k = 1+(R/r\beta)$. It is, therefore, the boundary self-excitation of the circuit.

The boundary, which divides the regions of real and complex roots (and so the node and the focus) is determined by the condition

$$[R(C_1+C_2)-(k-1)r(C_1+\beta C_2)]^2 - 4C_1C_2(1-\beta)r[R-\beta r(k-1)] = 0 \quad (5.33).$$

The curve defined by (5.33) on the plane (k,β) has two branches, one of which (the boundary between the unstable nodes and unstable focuses) passes between the hyperbola (5.32) and $k = 1+(R/\beta r)$, and the other below the hyperbola (5.32), but above the axis $k = 0$.

If the condition of self-excitation is satisfied and the singular point is unstable, then we can assert that the system leaves the state of equilibrium and so determines the character of this motion, but cannot go further than this statement, since we have restricted ourselves to linear equations. The analysis of the non-linear equations of the "universal" circuit (see Chapter X, Section 10) shows that, when the conditions of self-excitation are satisfied, self-oscillations are established in the circuit which are continuous oscillations for $k < k_{\text{crit}} = 1+(R/r\beta)$ (or, which is the same, for $\beta < \beta_{\text{crit}} = R/\sqrt{r}\,(k-1)$) and discontinuous ones for $k > k_{\text{crit}}$ (or for $\beta > \beta_{\text{crit}}$[†]). It also appears to be more correct to call the region $k > 1+(R/r\beta)$ on the diagram of Fig. 227, a region of the "rapid" motions (jumps), removing the system away from the state of equilibrium, rather than a "saddle" region.

§ 4. STATES OF EQUILIBRIUM AND THEIR STABILITY

Let us return now from the particular case of a linear system to the general case of a dynamic system described by two differential equations of the first order:

$$\frac{dx}{dt} = P(x, y), \quad \frac{dy}{dt} = Q(x, y). \tag{5.1}$$

In the states of equilibrium the phase velocity is zero so that we must find the points of intersection on the phase plane of the curves

$$P(x, y) = 0, \quad Q(x, y) = 0. \tag{5.34}$$

These points will be *singular* points of the differential equation of the first order, determining the integral curves

$$\frac{dy}{dx} = \frac{Q(x, y)}{P(x, y)}. \tag{5.3}$$

In this sense, the states of equilibrium are singular points of this family of integral curves.

A state of equilibrium is called stable in the sense of Liapunov if, having assigned about the state of equilibrium an *arbitrary* region ε, it is

[†] The fact that both continuous and discontinuous self-oscillations are possible in the circuit, is the reason for its name.

always possible to find a corresponding region $\delta(\varepsilon)$ such that when the representative point is found in the region $\delta(\varepsilon)$ (for $t=t_0$) it will *never* (for $t>t_0$) leave the region ε. A state of equilibrium is called unstable if there exists such a region ε about the state of equilibrium that, for it, we cannot choose a region $\delta(\varepsilon)$ possessing the property indicated above. Poincaré [185] and Liapunov [84] have given an analytic method for investigating the stability of states of equilibrium. We shall outline this method and shall give its justification.

We are interested in the stability of the state of equilibrium (x_0, y_0), a point of intersection of the curves $P(x, y) = 0$ and $Q(x, y) = 0$. It is convenient therefore to introduce new independent variables ξ and η defined as the displacements from the position of equilibrium (on the phase plane)

$$x = x_0 + \xi, \quad y = y_0 + \eta. \tag{5.35}$$

By our assumption $P(x, y)$ and $Q(x, y)$ are analytic functions and can be expanded about (x_0, y_0) so that from (5.1) we have†

$$\left. \begin{aligned} \frac{d\xi}{dt} &= a\xi + b\eta + [p_{11}\xi^2 + 2p_{12}\xi\eta + p_{22}\eta^2 + \cdots], \\ \frac{d\eta}{dt} &= c\xi + d\eta + [q_{11}\xi^2 + 2q_{12}\xi\eta + q_{22}\eta^2 + \cdots], \end{aligned} \right\} \tag{5.36}$$

where

$$a = P'_x(x_0, y_0), \quad b = P'_y(x_0, y_0),$$
$$c = Q'_x(x_0, y_0), \quad d = Q'_y(x_0, y_0)$$

etc.

The method established by Liapunov for investigating stability reduces to the following. Ignore in (5.34) the non-linear terms, and obtain then a system of *linear* equations with constant coefficients, the so-called system of equations of the first approximation

$$\frac{d\xi}{dt} = a\xi + b\eta, \quad \frac{d\eta}{dt} = c\xi + d\eta. \tag{5.37}$$

The solution of this system of equations will be readily written, as soon as we know the roots of the characteristic equation

$$\begin{vmatrix} a-\lambda & b \\ c & d-\lambda \end{vmatrix} = 0.$$

---

† We assume these expansions have linear terms in ξ and η so that the singular points are simple.

Liapunov has shown that, in the case when both roots of this equation have real parts different from zero, then an investigation of the equations of the first approximation always gives the correct answer to questions of stability near a state of equilibrium in the system (5.1). More precisely, if both roots have a negative real part and if, therefore, all solutions of the equations of the first approximation are damped, then the state of equilibrium will be stable; if, however, both roots have a positive real part, i.e. if the system of equations of the first approximation has solutions that increase with time, then the state of equilibrium is unstable.

To demonstrate these propositions by Liapunov, consider separately the case of real values of λ and the case of complex values of λ.

1. *The case of real roots of the characteristic equation*

By means of the linear homogeneous transformation

$$u = \alpha\xi + \beta\eta, \quad v = \gamma\xi + \delta\eta \tag{5.38}$$

we can reduce the system of equations of the first approximation to the so-called canonical form

$$\frac{du}{dt} = \lambda_1 u;$$

$$\frac{dv}{dt} = \lambda_2 v,$$

where λ_1 and λ_2 are just the roots of the characteristic equation. Let us apply the same transformation to the system (5.1). We shall obtain

$$\left.\begin{aligned}\frac{du}{dt} &= \lambda_1 u + (\bar{p}_{11}u^2 + 2\bar{p}_{12}uv + \bar{p}_{22}v^2) + \ldots, \\ \frac{dv}{dt} &= \lambda_2 v + (\bar{q}_{11}u^2 + 2\bar{q}_{12}uv + \bar{q}_{22}v^2) + \ldots\end{aligned}\right\} \tag{5.39}$$

Multiply the first equation by u, the second by v and add, then

$$\frac{1}{2}\frac{d\varrho}{dt} = \lambda_1 u^2 + \lambda_2 v^2 + \ldots = \Phi(u, v), \tag{5.40}$$

where $\varrho = u^2 + v^2$.

Let us consider separately three cases: λ_1 and λ_2 both negative, λ_1 and λ_2 both positive, λ_1 and λ_2 of different signs.

1. If λ_1 and λ_2 are both negative, then the curve $\Phi(u, v) = 0$ has an isolated point at the origin, and the surface $z = \Phi(u, v)$ has a maximum at

the origin. Hence it follows that there exists a region S near the origin in which $\Phi(u, v) < 0$. At the origin $\Phi(0, 0) = 0$. The presence of such a region readily enables us to determine the stability of the state of equilibrium.

Let a region ε be assigned about the origin. We shall choose for δ the region inside a circle around the origin, lying entirely both in the region ε and in the region S (Fig. 228). If initially the representative point is somewhere within the region $\delta(\varepsilon)$, then it will never leave this region and, therefore, never reach the boundary of the region ε, for $d\rho/dt < 0$ for all points of the region[†].

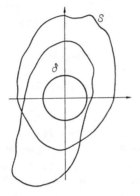

Fig. 228

Furthermore, since $\varrho = u^2 + v^2$ decreases monotonically from any initial value $\varrho = \varrho_0$ as time increases, then, for $t \to \infty$, ϱ tends either to zero or to a certain limit $\varrho_1(\varrho_1 > 0)$. But the limit different from zero must be discarded, since if $\varrho_0 \geqslant \varrho \geqslant \varrho_1$ for a finite velocity $|d\varrho/dt| > \gamma > 0$, and ϱ would decrease after an unlimited time by an arbitrarily large quantity and could not remain positive. It is clear that these propositions hold in terms of the ξ, η plane.

Each circle on the u, v plane lying entirely inside the region S is a "cycle without contact" (Poincaré's terminology), since all integral curves intersect it (for negative values of λ_1 and λ_2 the curves cross it from the outside) and none is tangent to it. We can plot a whole family of such circles each containing the next one and tending to the origin. Since a circle on the u, v plane corresponds to an ellipse on the ξ, η plane, then the state of equilibrium on the ξ, η plane may be encircled by a family of ellipses each

---

[†] Except the point $u = v = 0$. However, when the representative point is found at the origin of the coordinates, it will remain at rest there.

containing the next one, tending to the origin and which are cycles without contact (Fig. 229). If the representative point intersects the largest of the cycles without contact then it is bound to intersect all remaining ones, and tend asymptotically to the singular point.

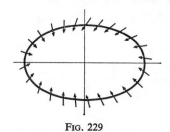

Fig. 229

2. If λ_1 and λ_2 are both positive, then the curve $\Phi(u, v) = 0$ still has an isolated point at the origin, but the surface $z = \Phi(u, v)$ will now have a minimum at the origin. Hence it follows that there exists a region S near the origin in which $\Phi(u, v) > 0$ (except $\Phi(0, 0) = 0$).

In this case the state of equilibrium is unstable, and it can be proved as follows. Displace the representative point at $t = t_0$ to any point of the region δ, except the origin. Since in S, $\Phi(u, v) = d\varrho/dt > 0$, then the representative point will move away from the origin monotonically as time increases; this can be untrue only if the representative point leaves the region S. Denote by ϱ_0 the value of u^2+v^2 at $t=t_0$ and by ϱ_e the value of u^2+v^2 at the boundary of the region ε. It is evident that in the ring between the circles $\varrho=\varrho_0$ and $\varrho=\varrho_e$, $\Phi(u, v) = d\varrho/dt$ has a certain *positive* lower limit. Therefore the representative point will move away from the origin and reach the boundary of ε in a finite time. The state of equilibrium is unstable in the sense of Liapunov.

Just as in the previous case, all qualitative statements remain valid when we pass to the ξ, η plane. In this case also, there exists on the ξ, η plane a family of concentric ellipses that are cycles without contact. A representative point situated sufficiently close to the state of equilibrium is bound to move away from it intersecting all the cycles without contact.

3. If λ_1 and λ_2 have the same sign, then the curve $\Phi(u, v) = 0$ has a branch point at the origin and the surface $z = \Phi(u, v)$ has an extremum of the saddle type at the origin. Therefore about the origin there are alternate regions in which $\Phi(u, v) > 0$ and in which $\Phi(u, v) < 0$, the boundary of separation being the curve $\Phi(u, v) = 0$, with a simple branch point at the origin (Fig. 231).

Otherwise about the origin there exists a circle with a radius different from zero which intersects the curve $\Phi(u, v) = 0$ four times. Let us call the region inside this circle the S region; this S region is divided by the curve $\Phi(u, v) = 0$ into four internal regions in such a manner that in two of them

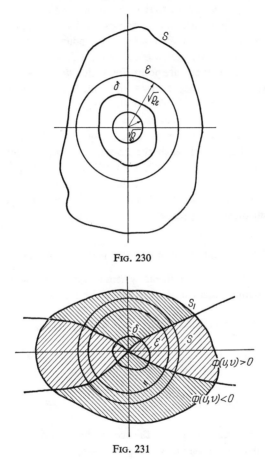

Fig. 230

Fig. 231

$\Phi(u, v) > 0$ and in the other two $\Phi(u, v) < 0$. We shall prove that now the state of equilibrium is unstable. Differentiating $d\varrho/dt$ once more and replacing du/dt and dv/dt by their values from the differential equation we obtain

$$\frac{1}{4}\frac{d^2\varrho}{dt^2} = \lambda_1^2 u^2 + \lambda_2^2 v^2 + \ldots = \Phi_1(u, v).$$

The surface $z = \Phi_1(u, v)$, as is easily verified, has a minimum at the origin. Therefore, there exists about the origin a region S_1, inside which $\Phi_1(u, v) > 0$ (at the origin $\Phi_1(0, 0) = 0$). And so $d^2\varrho/dt^2 > 0$ inside S_1. Take as ε a region bounded by a circle lying entirely both in the region S and in the region S_1†. We prove that it is *impossible* to choose a region δ, containing the origin, such that a representative point, initially at any point of the region δ, can never reach the boundary of the region ε.

To do this, assume that such a region δ did exist. Since it must contain the origin, then in it there are points for which $\Phi(u, v) > 0$. Displace the representative point at $t = t_0$ to any such point. Since for $t = t_0$, $\Phi(u, v) = d\varrho/dt > 0$ and since in S_1 $d^2\varrho/dt^2 > 0$ (the region ε has been chosen inside the region S_1 and the region δ cannot have parts lying outside ε), then the representative point will move away from the origin with increasing velocity and will reach the boundary of the region ε in a finite time. We have arrived, thus, at a contradiction. The required region δ cannot be chosen. The state of equilibrium is unstable in the sense of Liapunov. It is evident that the same applies to the corresponding state of equilibrium on the ξ, η plane.

2. The characteristic equation with complex roots

In this case, as we know, a linear system can be reduced by means of a real linear homogeneous transformation to the form

$$\frac{du_1}{dt} = a_1 u_1 - b_1 v_1, \quad \frac{dv_1}{dt} = a_1 v_1 + b_1 u_1,$$

where $\lambda_1 = a_1 + jb$ and $\lambda_2 = a_1 - jb$. Apply the same transformation to the non-linear system and we obtain a non-linear system

$$\frac{du_1}{dt} = a_1 u_1 - b_1 v_1 + \ldots; \quad \frac{dv_1}{dt} = a_1 v_1 + b_1 u_1 + \ldots \quad (5.41)$$

Multiply the first equation by u_1, the second by v_1 and add, then we obtain the following expression, where $\varrho = u_1^2 + v_1^2$.

$$\frac{1}{2} \frac{d\varrho}{dt} = a_1(u_1^2 + v_1^2) + \ldots = \psi(u_1, v_1).$$

Since there are no terms less than the second and higher orders then $\psi(u_1, v_1)$ has a maximum or a minimum at the origin according to the

† We can use, in particular, for the region ε, the region S which can always be so chosen as to be entirely situated within the region S_1.

sign of a_1. Repeating exactly the procedures carried out in the case of real roots having the same sign, we shall find that in the case $a_1 < 0$ the state of equilibrium is stable in the sense of Liapunov and even asymptotically stable, while in the case $a_1 > 0$ the state of equilibrium is unstable in the sense of Liapunov. In both cases sufficiently small circles in the vicinity of the origin will serve as the cycles without contact. In passing on to the ξ, η plane, this family of circles is transformed into a family of ellipses without contact, which are crossed either from without or from within according to the sign of a_1.

We have thus justified Liapunov's method of ignoring the non-linear terms, for the case when the characteristic roots are not equal and have non-zero real parts. The restriction that there are no equal roots is unessential—we have introduced it for the sole purpose of simplifying the proof. The restriction that the real parts are different from zero in both roots is, however, an essential one. Assuming that the equation considered is of a general form, it cannot be removed. Thus, Liapunov's theorem on the stability of the states of equilibrium can be formulated as follows: *if the real parts of the roots of the characteristic equation are negative, then the state of equilibrium is stable; if one or both real parts are positive, then the state of equilibrium is unstable.*

If the real parts of both roots of the characteristic equation are equal to zero or if one root is equal to zero and the other is negative, then the equations of the first approximation do not give an answer to the question of the stability of the state of equilibrium.

Thus the stability of a state of equilibrium of the system (5.1) is completely determined by the corresponding equations of the first approximation (5.37) in the case when both roots of the characteristic equation have real parts different from zero. It may be shown (we shall not do this here) that in this case the equations of the first approximation determine not only the stability of the state of equilibrium but also the character of the phase paths in a sufficiently small neighbourhood of the state of equilibrium. Moreover, the states of equilibrium (the singular points) for which the real parts of both roots of the characteristic equation are different from zero, are *coarse:* their character, i.e. the character of the phase paths in a nearby region, is preserved for sufficiently small variations of the functions $P(x, y)$ and $Q(x, y)$ and of their first-order derivatives (see Chapter VI, Section 4 for more details). Thus, in exactly the same manner as in § 2, we have here five types of coarse states of equilibrium: stable node, unstable node, stable focus, unstable focus and saddle point. To investigate the character of coarse states of equilibrium it is convenient to use the diagram shown in Fig. 223. Now we have

$$\sigma = -[P'_x(x_0, y_0) + Q'_y(x_0, y_0)]$$

and

$$\Delta = \begin{vmatrix} P'_x(x_0, y_0) & Q'_x(x_0, y_0) \\ P'_y(x_0, y_0) & Q'_y(x_0, y_0) \end{vmatrix}. \quad (5.42)$$

Coarse states of equilibrium correspond to all points lying outside the axis $\Delta = 0$ and the semi-axis $\sigma = 0$, $\Delta > 0$. In the case of a node and a saddle point, as we know, the integral curves reach the singular point

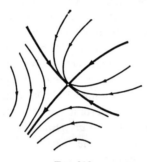

Fig. 232

along two directions, which can be determined from the corresponding linear equations. Using the results of Section 2, the following equations determine the slopes $\varkappa$ of this direction:

$$P'_y(x_0, y_0)\varkappa^2 + \{P'_x(x_0, y_0) - Q'_y(x_0, y_0)\}\varkappa - Q'_x(x_0, y_0) = 0.$$

The points on the axis $\Delta = 0$ and the semi-axis $\sigma = 0$, $\Delta > 0$ correspond to *non-coarse* states of equilibrium, whose character is varied by arbitrarily small variations of the functions $P(x, y)$ and $Q(x, y)$ and their derivatives. Thus their character and stability is not determined by the linearized equations (5.37). The points of the axis $\sigma = 0$, $\Delta > 0$ can correspond to a centre or an unstable focus or a stable focus. The points of the axis $\Delta = 0$ correspond to *multiple* singular points the simplest of which (a point of the saddle-node type) is shown in Fig. 232[†].

[†] Multiple singular points or singular points for which $\Delta = 0$, are, evidently, points of *contact* of the curves $P(x, y) = 0$ and $Q(x, y) = 0$. Owing to this, for arbitrarily small variations of the functions $P(x, y)$ and $Q(x, y)$ a multiple singular point can be split into two or more singular points. Singular points for which $\Delta \neq 0$ bear the name of *simple* singular points, and their number cannot vary for sufficiently small variations of the functions $P(x, y)$ and $Q(x, y)$.

§ 5. Example: States of Equilibrium in the Circuit of a Voltaic Arc

As an example illustrating the application of Liapunov's methods, we shall consider the equilibrium of a Voltaic arc connected in series with an inductance and shunted by a capacitance (Fig. 233). This circuit is a modified version of the arc generator (Chapter IV, Section 5). Neglecting again

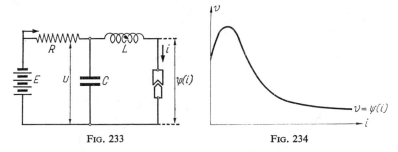

Fig. 233   Fig. 234

the inertia of the ionic processes in the arc, we easily obtain by means of Kirchhoff's laws the following equations (the notation of Fig. 233 is used)

$$L\frac{di}{dt} = u - \psi(i), \\ C\frac{du}{dt} = \frac{E-u}{R} - i, \quad \quad (5.43)$$

$\psi(i) = v$ is the voltage across the arc and is a single-valued function of the arc current i (Fig. 234).

The equilibrium is determined from $du/dt = 0$ and $di/dt = 0$, or

$$u = E - Ri, \quad u = \psi(i). \quad \quad (5.44)$$

The points of intersection of the curves are states of equilibrium. Depending on the value of E and R, there can be either one (Fig. 235) or three (Fig. 236) singular points. Following Liapunov's method, substitute $u = u_0 + \eta$ and $i = i_0 + \xi$ in (5.43) where (u_0, i_0) is one state of equilibrium. Further, expanding the arc characteristic $\psi(i_0 + \xi)$ in a series $\psi(i_0 + \xi) = \psi(i_0) + \xi\psi'(i_0) + \ldots$ and retaining only the first term we obtain, in view of (5.44), two linear equations for ξ and η.

$$\frac{d\eta}{dt} = -\frac{\eta}{RC} - \frac{\xi}{C}, \\ \frac{d\xi}{dt} = \frac{\eta}{L} - \frac{\varrho\xi}{L}, \quad \quad (5.45)$$

where $\varrho = \psi'(i_0)$ is the slope of the arc characteristic at the point corresponding to the equilibrium state (u_0, i_0). The arc resistance ϱ is a variable quantity which for certain values of i_0 can assume negative values; however, in using this concept, we must keep in mind the proviso made when we first introduced the term "negative resistance" (Chapter I, Section 6).

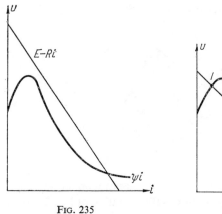
Fig. 235

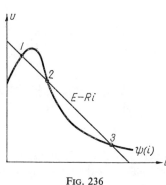
Fig. 236

The characteristic equation is

$$\begin{vmatrix} -\dfrac{\varrho}{L} - \lambda & \dfrac{1}{L} \\ -\dfrac{1}{C} & -\dfrac{1}{RC} - \lambda \end{vmatrix} = 0,$$

or

$$\lambda^2 + \lambda\left(\frac{1}{RC} + \frac{\varrho}{L}\right) + \frac{1}{LC}\left(\frac{\varrho}{R} + 1\right) = 0. \tag{5.46}$$

The nature of the roots of this equation depends on the values of four parameters: R, C, L and ϱ. In order to establish the nature of these roots for all possible values of the parameters, we can construct three stability diagrams on the R, ϱ; L, ϱ and C, ϱ planes. We must bear in mind that L, C and R can assume positive values only, whereas ϱ can assume both positive and negative values.

To construct the R, ϱ diagram, first write the condition for complex roots:

$$[L - RC\varrho]^2 - [2R\sqrt{LC}]^2 < 0. \tag{5.47}$$

The left-hand side is the product of two factors which vanish separately for
$$L - RC\varrho + 2R\sqrt{LC} = 0;$$
and
$$L - RC\varrho - 2R\sqrt{LC} = 0.$$

Each of these equations determines a hyperbola; one asymptote being the

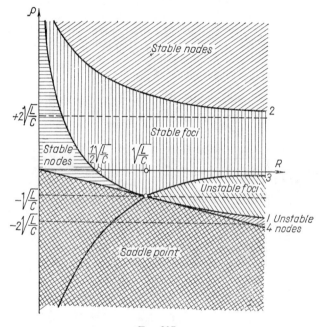

Fig. 237

ϱ axis for both curves, and the other: the line $\varrho = -2(L/C)^{\frac{1}{2}}$ for the first curve and the line $\varrho = +2(L/C)^{\frac{1}{2}}$ for the second curve.

The "curvilinear wedge" formed by the two hyperbolae 1 and 2 (Fig. 237) represents the region of complex roots. The boundary of the region of roots with positive real parts (stable nodes and foci), is the hyperbola 3, $L + RC\varrho = 0$, situated in the fourth quadrant and intersecting the hyperbola 1 at the point $R = (L/C)^{\frac{1}{2}}$ and $\varrho = -(L/C)^{\frac{1}{2}}$. It is evident that all nodes and foci lying above this hyperbola are stable and those lying below it are unstable. Finally the boundary of the region of saddle points is the straight line 4, with equation $R + \varrho = 0$, since for $R + \varrho < 0$ the roots of the equation (5.46) will always be of different signs. Thus below the straight line $\varrho = -R$ is the region of saddle points. The complete stability diagram for R and ϱ is shown in Fig. 237.

As is seen from this diagram, for $\varrho > 0$ there exist only stable singular points. These points will be foci if the arc resistance ϱ is not too large and

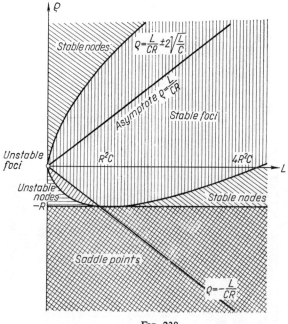

Fig. 238

if the resistance of the load, which shunts the circuit, is not too small. For $\varrho < 0$ the states of equilibrium can be stable only if $|\varrho|$ is not too large and if, on the other hand, R is neither too small nor too large. For $\varrho < 0$ three types of instability are possible: an unstable node, an unstable focus and a saddle point. Moreover a focus (a stable or an unstable one, depending on the sign of ϱ) is obtained for $|\varrho| < 2(LC)^{\frac{1}{2}}$ if R is sufficiently large. In general, for $|\varrho| < 0$ and $|\varrho| < 2(L/C)^{\frac{1}{2}}$, it is possible by varying R to obtain any singular point. If, however, $\varrho < 0$ and $|\varrho| > 2(L/C)^{\frac{1}{2}}$ only unstable singular points are possible, either as saddle points or as unstable nodes, depending on the value of R.

Similar stability diagrams can be constructed for the L, ϱ (Fig. 238) and for C, ϱ (Fig. 239) planes. For both diagrams the boundary of the region of complex roots is expressed by the equation

$$L^2+(RC\varrho)^2-2RCL\varrho-4R^2CL = 0$$

so that

$$\varrho = \frac{L}{CR} \pm 2\sqrt{\frac{L}{C}}.$$

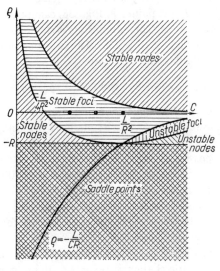

Fig. 239

On the L, ϱ diagram this boundary is a single curve with an asymptote $\varrho = L/CR$, a vertical tangent at the point $L=0, \varrho=0$ and horizontal tangent at the point $L = R^2C, \varrho=-R$. On the C, ϱ diagram this boundary is split into two curves of hyperbolic type with asymptotes $C=0$ and $\varrho=0$. The boundary of the region of stability of nodes and foci is determined by the equation $RC\varrho = -L$, and is a hyperbola on the C, ϱ diagram and a line on the L, ϱ diagram. The boundary of the region of saddle points is given by the equation

$$\varrho = -R.$$

We obtain finally the stability diagrams shown in Fig. 238 and Fig. 239. These diagrams agree completely with the first one shown in Fig. 237 and they enable us to assess the nature of the singular points for arbitrary values of the parameters R, C, L and ϱ[†].

It is clear that a state of equilibrium lying where $\varrho > 0$ (for example, the point 1 in Fig. 236) is always stable and by knowing L, C, R, and ϱ we establish at once whether there is a focus or a node. If, however, the state of equilibrium lies where $\varrho < 0$ and the slope of the load line $u = E - Ri$ is smaller than the slope of the arc characteristic, i.e. $|\varrho| > R$ (see point 2 in Fig. 236) then this state of equilibrium is a saddle point and is unstable, for any value of L/C. Finally a state of equilibrium, lying where $\varrho < 0$ but for which $|\varrho| < R$ (the point 3 in Fig. 236) cannot be a saddle point but is either a focus or a node. This equilibrium is stable for small C (Fig. 239), and is unstable for small L (Fig. 238). The two conditions for the stability of a state of equilibrium where $\varrho < 0$ are

$$|\varrho| < R \quad \text{and} \quad L > |\varrho| RC,$$

and reduce, for $L \neq 0$, to a single condition: $|\varrho| < R$, when $C = 0$ (Section 6, Chapter IV). However, since all circuits have a certain, however small, capacitance, the stability of a state of equilibrium where $\varrho < 0$ requires that the circuit possess a certain, not too small inductance.

In the analysis of the stability of the states of equilibrium in a circuit with a Voltaic arc we have used the *static* characteristic of the arc, which, strictly speaking, only applies to steady-state equilibrium processes in the arc. Our analysis, therefore, will only be adequate for sufficiently slow oscillations. If, however, L and C are small and the frequency is high, then the inertia of the ionic processes in the arc plays an important role and we cannot use the static characteristic of the arc but must use instead the dynamic (differential) equations which reproduce, to some degree of accuracy, the *dynamics* of the arc discharge. It is found that the inertia of the arc discharge is a *stabilizing* factor, sufficient to produce equilibrium with a small capacitance C also without any external inductance.

The simplest differential equation of the first order which reproduces to some extent the dynamics of the processes in an arc at a state of equilibrium (v_0, i_0) is

$$\tau \left(\frac{d\xi}{dt} - \frac{v_0}{i_0} \frac{d\eta}{dt} \right) + \xi - \varrho\eta = 0 \qquad (5.48)$$

where $\xi = v - v_0$, $\eta = i - i_0$ and v is the voltage across the terminals of the arc [200, 51]. The time constant τ characterizing mostly thermal inertia of

---

[†] More useful and practical parameters are the combinations L/CR and $\sqrt{L/C}$, in addition to ϱ. [Ed.].

the electrodes is of the order of 10^{-3}–10^{-4} sec. We obtain as limit cases from the equation (5.48) both the linearized static characteristic $\xi = \varrho\eta$ putting the derivatives equal to zero, and the dynamic characteristic for high frequencies $\xi = (v_0/i_0)\eta$ (the thermal state of the arc has no time to vary) if we assume the derivatives to be so large that the terms $\xi - \varrho\eta$ in the equation can be neglected.

For an arc circuit without inductance but with capacitance, we have, in addition to the equation (5.48),

$$C\frac{d\xi}{dt} = -\frac{\xi}{R} - \eta. \tag{5.49}$$

The characteristic equation for the system (5.48) and (5.49) is

$$C\tau\frac{v_0}{i_0}\lambda^2 + \left\{\tau\left(1+\frac{v_0}{Ri_0}\right) + \varrho C\right\}\lambda + \left(1+\frac{\varrho}{R}\right) = 0, \tag{5.50}$$

and hence, a state of equilibrium on the descending section of the static characteristic ($\varrho < 0$) will be stable, if

$$|\varrho| < R \quad \text{and} \quad \tau\left(1+\frac{v_0}{Ri_0}\right) > |\varrho|C.$$

These conditions are satisfied for sufficiently large resistances R and for sufficiently small capacitances C. Thus the circuit of an arc with small capacitance (for example, with $C < \tau/|\varrho|$) will have a stable state of equilibrium where $\varrho < 0$ and without any inductance in its circuit, provided that $|\varrho| < R$. This conclusion is found in qualitative agreement with experimental data.

§ 6. Limit Cycles and Self-Oscillations

Let us examine periodic motions, which, as we know, can occur in systems described by

$$\frac{dx}{dt} = P(x, y), \quad \frac{dy}{dt} = Q(x, y). \tag{5.1}$$

If $T(T > 0)$ is the smallest number for which, for all t,

$$x(t+T) = x(t),$$
$$y(t+T) = y(t),$$

then the motion $x = x(t)$, $y = y(t)$ is periodic with period T. As we know, to a periodic motion there corresponds a closed phase path, and conversely to each closed path there correspond an infinite set of periodic motions differing from each other by the choice of the origin of time. We have already met closed phase paths in the analysis of conservative systems, where they always formed a whole continuum of concentric ovals (for example, paths around a centre). In certain examples of self-oscillating

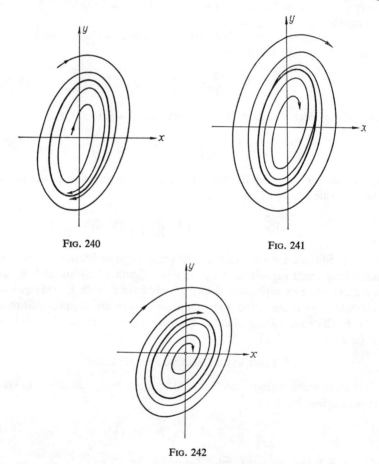

Fig. 240

Fig. 241

Fig. 242

systems (see Chapter III, Sections 3–5) we found an isolated closed curve towards which its adjoining paths approached from both sides in a spiral fashion. Such isolated closed paths are *limit cycles*.

We shall call a limit cycle *orbitally stable* if there exists a region (ε) on the phase plane which contains the limit cycle, such that all phase paths starting in the neighbourhood (ε) approach the limit cycle asymptotically as $t \to +\infty$; conversely if there exists at least one phase path in ε that does not approach the limit cycle for $t \to +\infty$, then the limit cycle is orbitally *unstable*.

A stable limit cycle is shown in Fig. 240, and unstable limit cycles are shown in Fig. 241 and Fig. 242. The unstable cycles shown in Fig. 242 in which all paths on one side approach them and on the other side move away from them for $t \to +\infty$, are sometimes called "semi-stable" or double, because for a suitable variation of a parameter of the system, such cycles usually split into two, one of which is stable and the other is unstable.

In addition to the stability of a limit cycle as a path, the definition of which has been given above, hence the name *orbital stability*, we can also speak of stability in the sense of Liapunov of the periodic motion corresponding to a limit cycle.

More precisely, a periodic motion $x = \varphi(t)$, $y = \psi(t)$ of period T, is stable in the sense of Liapunov if for every $\varepsilon > 0$, a positive $\delta(\varepsilon)$ can be found such that for any other motion $x = x(t)$, $y = y(t)$ satisfying the conditions

$$|x(t_0) - \varphi(t_0)| < \delta \quad \text{and} \quad |y(t_0) - \psi(t_0)| < \delta,$$

the inequalities

$$|x(t) - \varphi(t)| < \varepsilon \quad \text{and} \quad |y(t) - \psi(t)| < \varepsilon$$

are satisfied for all $t > t_0$. Below we use mainly the concept of orbital stability.

The orbital stability of a limit cycle (just as for the stability in the sense of Liapunov of the corresponding periodic motions) is determined by the sign of its *characteristic exponent*

$$h = \frac{1}{T} \int_0^T \left\{ P'_x[\varphi(t), \psi(t)] + Q'_y[\varphi(t), \psi(t)] \right\} dt,$$

where $x = \varphi(t)$, $y = \psi(t)$ is an arbitrary periodic motion on the limit cycle. Then *a limit cycle is stable if $h < 0$ and unstable for $h > 0$* (to the value $h = 0$ there correspond both stable and unstable limit cycles).

To investigate the stability of a periodic motion $x = \varphi(t)$, $y = \psi(t)$ the equations can be linearized as was done for the case of singular points. Put $x = \varphi(t) + \xi$ and $y = \psi(t) + \eta$; substitute these expressions in the equations (5.1); expand the functions $P(\varphi + \xi, \psi + \eta)$ and $Q(\varphi + \xi, \psi + \eta)$ in power series with respect to ξ and η and neglect the non-linear terms, then we obtain linear equations of the first approximation for the perturbations ξ and η

$$\frac{d\xi}{dt} = P'_x[\varphi(t),\ \psi(t)]\xi + P'_y[\varphi(t),\ \psi(t)]\eta,$$

$$\frac{d\eta}{dt} = Q'_x[\varphi(t),\ \psi(t)]\xi + Q'_y[\varphi(t),\ \psi(t)]\eta.$$

This is a system of linear differential equations with periodic coefficient of period T (since P'_x, P'_y, Q'_x, Q'_y are periodic functions of φ and ψ with period T). The general form of its solution is

$$\xi = C_1 f_{11}(t) e^{h_1 t} + C_2 f_{12}(t) e^{h_2 t},$$

$$\eta = C_1 f_{21}(t) e^{h_1 t} + C_2 f_{22}(t) e^{h_2 t},$$

where f_{jk} are certain periodic functions of period T. The solutions for ξ and η depend upon the characteristic exponents h_1 and h_2. The sign of their real parts determine whether these solutions are increasing or decreasing. Because the system of equations (5.1) is autonomous it so happens that one of the characteristic exponents is equal to zero and the other is equal to h [185]. As stated just now the sign of this exponent determines whether the motion is stable [8].

Before considering the proof of this stability condition let us examine, for later use, the physical interpretation of limit cycles. If we require that in real physical systems the qualitative character of possible motions is preserved for arbitrary small variations of the system then, as will be seen, we exclude the existence of non-isolated closed curves.

Thus the characteristic exponent is bound to be different from zero and the orbital stability of a limit cycle implies the stability in the sense of Liapunov of all periodic motions corresponding to it. Also for such motions the period and the "amplitude"[†] are ultimately independent of the initial conditions.

We have already studied certain equations of the type (5.1), when we examined certain examples of self-oscillatory systems, and they were shown to have limit cycles with a negative characteristic exponent, and that stationary periodic processes were actually represented by these limit cycles. We conclude (and it can be proved) that *the presence of limit cycles in the phase portrait of a dynamic system described by (5.1) is a necessary*

---

†More precisely we should say: "the period and the whole spectrum of the amplitudes obtained by expanding the periodic motion in a Fourier series".

and sufficient condition for the possibility (under suitable initial conditions) of self-oscillations in the system [3.5]

An unstable limit cycle having a positive characteristic exponent can, of course, appear in the phase portrait of "coarse" systems, but it does not correspond to a real periodic process; it only has the role of a "watershed" on both sides of which the paths have a different behaviour. For example, the existence of an unstable cycle explains the "hard" mode of excitation for which small initial deviations in the system are damped, and large ones are reinforced.

§7. POINT TRANSFORMATIONS AND LIMIT CYCLES

As we have seen in Chapter III, Sections 3–5 one of the methods for finding limit cycles and determining their stability is to find the law of a certain point transformation, and evaluate the corresponding sequence function.

1. Sequence function and point transformation

The sequence function concept was introduced by Poincaré. Draw on the phase plane

$$\frac{dx}{dt} = P(x, y), \quad \frac{dy}{dt} = Q(x, y) \tag{5.1}$$

through non-singular points a line (or arc) AB such that the phase paths of the system (5.1) intersect it without being tangent to it[†].

Let Q be a point on L (the arc AB) at a distance s from A, and let C be a phase path passing through Q at time $t = t_0$. It may happen that for any value of $t > t_0$ the path C does not intersect L again and then the point Q "does not possess a consecutive point on L". However, it may happen that C does intersect L again for $t > t_0$. Let $\bar{t}$ be the first value of $t > t_0$ for which C intersects L, and $\overline{Q}$ the point of intersection at t. Then the point Q "has a consecutive point $\overline{Q}$ on L" (Fig. 243).

It is easily shown, from the theorem on the continuity of the dependence on the initial conditions, that if Q has a consecutive point then all points of L sufficiently close to Q have consecutive points. Clearly $\bar{s}$ is a function of s, and this function

$$\bar{s} = f(s) \tag{5.52}$$

is called a *sequence function* (law of a certain *point transformation*) establishing a single-valued correspondence between the points of L (or part of it)

---

† This may be called a *line segment* or *arc without contact*.

and their consecutive points on L. It is geometrically evident that a sequence function is found when paths intersect having the character of spirals or closed paths, and that if a certain value $s = s_0$ corresponds to a closed

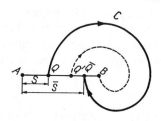

Fig. 243

curve, then $\bar{s} = f(s_0) = s_0$ and point Q and its consecutive point $\bar{Q}$ coincide. Such points are known as fixed points. Poincaré found properties of the function $\bar{s} = f(s)$, given here without proof.

Property I.

If a point Q_0, at $s = s_0$, has a consecutive point on L, then the function $\bar{s} = f(s)$ is a holomorphic function of s at the point $s = s_0$.

Property II.

The differential coefficient $d\bar{s}/ds$ is always positive.

The first property is a consequence of the theorem that the solutions of the system (5.1) with analytic right-hand sides are analytic functions of the initial conditions while the second property is a consequence of Cauchy's theorem (that phase paths cannot intersect each other).

Suppose a point Q_0 at $s = s_0$ has a consecutive point (not coinciding with the ends A or B of the segment L). Let it move along L from the point Q_0 in any direction then it may arrive at a point $s = s'$ such that all points of L within the interval $s_0 < s < s'$ (or $s' < s < s_0$) will have consecutive points, while the point Q' at $s = s'$ will have no consecutive point on the segment L. It can be shown that a path passing through the point Q' will end at a singular point without intersecting L again. If the singular point is simple, this point can only be a saddle point[†].

[†] This point cannot be a node, nor a focus. In fact, let us assume that a path, passing through Q', ends (without intersecting L again) at a node or a focus. Then, as is easily shown, all paths passing through the points of L, corresponding to values of s near s',

It can happen that points corresponding to $s>s'$ again have consecutive points. Thus the sequence function exists for $s<s'$ and for $s>s'$. For $s=s'$ the sequence function is not defined (Fig. 244 and Fig. 245).

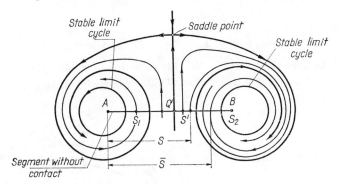

Fig. 244

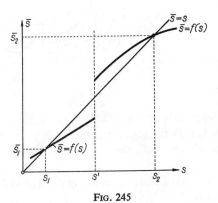

Fig. 245

2. Stability of the fixed point. Koenigs's theorem

If we know the sequence function of a certain arc L then we can find the fixed points s^* and the limit cycles by solving

$$f(s^*) = s^*.$$

Doing this graphically produces (Fig. 246) the so-called Lamerey's diagram. For the purpose of determining the stability of the limit cycle, we

---

but sufficiently close to s', would also end at this singular point, without intersecting L again. However, it would follow from this that the points corresponding to values of s, smaller than s', have no consecutive points, which contradicts our assumption.

need only consider the sequence of the points of intersection with the segment L, of the phase paths that lie in a neighbourhood of the limit cycle, i.e. the sequence

$$s, \ s_1, \ s_2, \ \ldots, \ s_n, \ s_{n+1}, \ \ldots, \tag{S}$$

where

$$s_1 = f(s), \ s_2 = f(s_1), \ \ldots, \ s_{n+1} = f(s_n)$$

If the limit cycle is stable, then (by definition of stability) there exists a neighbourhood (ε) of it such that all phase paths with initial points in this

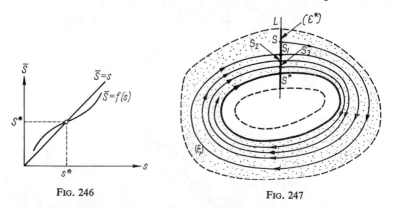

Fig. 246   Fig. 247

neighbourhood approach asymptotically the limit cycle as $t \to +\infty$. This also means that on the segment L there exists a neighbourhood (ε^*) to the fixed point s^* (Fig. 247), such that all s, belonging to (ε^*), $s_n \to s^*$ as $n \to +\infty$.

A fixed point is *stable* if there exists a neighbourhood (ε^*) to this point, so that all sequences

$$s, \ s_1, \ s_2, \ \ldots, \ s_n, \ s_{n+1}, \ \ldots$$

with initial point s in (ε^*) converge to this fixed point. Therefore to a stable limit cycle there corresponds a stable fixed point, and conversely. The definition of an unstable fixed point follows immediately. It corresponds to an unstable limit cycle.

Conditions for the stability of the fixed point s^* of a point transformation expressed by the sequence function $\bar{s} = f(s)$, and so of the corresponding limit cycle are stated in *Koenigs's theorem* [168, 169][†]:

[†] We give here a general formulation of Koenigs's theorem, suitable also for the case when $d\bar{s}/ds < 0$, which can occur for dynamic systems (5.1) with non-analytic right-hand sides or with a phase surface different from the ordinary plane.

7] POINT TRANSFORMATIONS AND LIMIT CYCLES

The fixed point s^* of a point transformation $\bar{s} = f(s)$ is stable, if

$$\left| \frac{d\bar{s}}{ds} \right|_{s=s^*} < 1, \tag{5.53a}$$

and unstable, if

$$\left| \frac{d\bar{s}}{ds} \right|_{s=s^*} > 1. \tag{5.53b}$$

To prove Koenigs's theorem let

$$\xi = s - s^*, \quad \bar{\xi} = \bar{s} - s^*$$

Then to the sequence of points $s, s_1, s_2, \ldots, s_n, s_{n+1}, \ldots$, there will correspond the sequence of positive numbers

$$|\xi|, \quad |\xi_1|, \quad |\xi_2|, \ldots, \quad |\xi_n|, \quad |\xi_{n+1}|, \ldots,$$

where $\xi_n = s_n - s^*$.

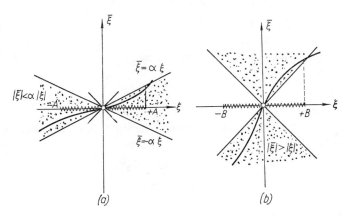

Fig. 248

If $|d\bar{s}/ds|_{s=s^*} < 1$ then on the segment L there exists a neighbourhood of the fixed point $|\xi| \leqslant A$ (Fig. 248a) such that for all points on it, except $\xi = 0$.

$$|\bar{\xi}| < \alpha |\xi| \tag{5.54}$$

where α is a positive number *smaller than unity*. Therefore each sequence of the positive numbers

$$|\xi|, \quad |\xi_1|, \quad |\xi_2|, \ldots$$

is a monotonic decreasing sequence with a lower boundary and, therefore, by Cauchy's well-known theorem on the convergence of such numerical sequences, tends to a certain limit, which, however, cannot be different from zero†. Thus, when condition (5.53a) is satisfied, any sequence of points $s, s_1, s_2, \ldots$, with initial points in the neighbourhood $s^* - A \leq s \leq s^* + A$ converges to s^* and, therefore, the fixed point s^* is stable.

It, however, the condition (5.53b) is satisfied, then there exists such a neighbourhood $|\xi| \leq B$, for the points of which $|\bar{\xi}| > |\xi|$ (Fig. 248b). Therefore any sequence of numbers $|\xi|, |\xi_1|, |\xi_2|, \ldots$, cannot converge to the limit $\xi = 0$, and the sequences $s, s_1, s_2, \ldots$ (with initial points $s^* - A \leq s \leq s^* + A$) cannot converge to s^*. Therefore, in this case, the fixed point will be unstable. Thus we have proved Koenigs's theorem‡. This theorem does not determine the stability of the fixed point if $|d\bar{s}/ds| = 1$ (an additional investigation is required, since the stability is determined by the signs of a higher-order derivative of the sequence function).

3. A condition of stability of the limit cycle

We shall prove, using Koenig's theorem, that the limit cycle is stable if the characteristic exponent $h < 0$.

Let C_0 be a limit cycle of (5.1) described by

$$x = \varphi(t), \quad y = \psi(t)$$

where φ and ψ are periodic with period T.

Let us introduce in a neighbourhood of this limit cycle a new curvilinear system of coordinates u, v (Fig. 249), by putting

$$\left. \begin{array}{l} x = \varphi(u) - v\psi'(u), \\ y = \psi(u) + v\varphi'(u). \end{array} \right\} \quad (5.55)$$

The straight lines $u = $ const. are orthogonal to the limit cycle and the curves $v = $ const. are closed curves (the curve $v = 0$ coincides with the limit cycle C_0). From the Jacobian of the transformation which is always positive on or near the limit cycle, it can be shown that each point of the plane (in this region) corresponds to a single pair of numbers (u, v).

---

† In fact, if this limit were different from zero and equal to a $(a > 0)$, then for all $n |\xi_n| > a > 0$ and, by the condition (5.54),

$$|\xi_n| - |\xi_{n+1}| > \left(\frac{1}{\alpha} - 1\right) |\xi_{n+1}| > \left(\frac{1}{\alpha} - 1\right) a,$$

which contradicts Cauchy's criterion for the limit of a numerical sequence.

‡ Since $d\bar{s}/ds = f'(s) > 0$, the condition of stability of the fixed point will be the inequality $f'(s_0) > 1$ and the condition of instability the inequality $f'(s_0) < 1$.

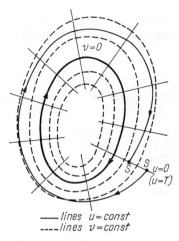

Fig. 249

In the annular region near C_0 $|v| \leqslant A$ and the equations (5.1) are

$$[\varphi' - v\psi'']\frac{du}{dt} - \psi'\frac{dv}{dt} = P(\varphi - v\psi, \psi + v\varphi'),$$

$$[\psi' + v\varphi'']\frac{du}{dt} + \varphi'\frac{dv}{dt} = Q(\varphi - v\psi', \psi + v\varphi').$$

Then

$$\frac{du}{dt} = \frac{P(\varphi - v\psi', \psi + v\varphi')\varphi' + Q(\varphi - v\psi', \psi + v\varphi')\psi'}{D},$$

$$\frac{dv}{dt} = \frac{-P(\varphi - v\psi', \psi + o\varphi')[\psi' + v\varphi''] + Q(\varphi - v\psi', \psi + v\varphi')[\varphi' - v\psi]}{D}$$

or, dividing one equation by the other

$$\frac{dv}{du} = \frac{-P(\varphi - v\psi', \psi + v\varphi')[\psi' + v\varphi''] + Q(\varphi - v\psi', \psi + v\varphi')[\varphi' - v\psi'']}{P(\varphi - v\psi', \psi + v\varphi')\varphi' + Q(\varphi - v\psi', \psi + v\varphi')\psi'}.$$

(5.56)

Bearing in mind the identities

$$P(\varphi, \psi) = \varphi', \quad Q(\varphi, \psi) \equiv \psi' \qquad (5.57)$$

it is easily verified that the denominator of the right-hand side of the equation (5.56) does not reduce to zero for $v = 0$ for nor in a certain neighbourhood of the limit cycle $v = 0$†. In addition, the right-hand side of this equation is clearly a periodic function of u with period T.

† In this neighbourhood the equation (5.56) has no singular points and, therefore, each integral curve consists of one phase path.

Let us take as the arc without contact L a segment of the normal $u=0$ (clearly, this same segment corresponds to $u=nT$, where n is an integer), and denote by

$$v = \Phi(u, s) \tag{5.58}$$

the solution of the equation (5.56) satisfying the initial condition: $v=s$ for $u=0$, i.e. point M on L. By virtue of the theorem on the dependence of the solutions of the equations (5.1) or the equations (5.56) upon the initial conditions, each phase path intersecting L at $t=t_0$ sufficiently close to the point of intersection N_0 of the limit cycle will again intersect this segment for t close to t_0+T_0. Therefore the consecutive point of intersection of the path (5.58) with L is determined, clearly, by the relation

$$\bar{v} = \bar{s} = \Phi(T, s) = f(s). \tag{5.59}$$

This sequence function exists in a certain neighbourhood of point M_0, which is, of course, the fixed point at $v=s=0$.

The stability of the fixed point M_0 (and, hence, also the stability of the limit cycle C_0) is determined by the quantity, $f'(0)$. Knowing the functions $P(x, y)$ and $Q(x, y)$ it is possible to find the value of $f'(0)$. The denominator of the right-hand side of the equation (5.56) does not reduce to zero near the limit cycle (for $|v| \leq A$). Therefore, in this neighbourhood, the right-hand side of the equation (5.56) is an analytic function and can be represented in the form of a power series with respect to v;

$$\frac{dv}{du} = A_1(u)v + A_2(u)v^2 + \ldots \tag{5.56a}$$

(the coefficients $A_1, A_2, \ldots$ of the series are periodic functions of u with period T). By employing the identities $P'_x\varphi' + P'_y\psi' \equiv \varphi''$ and $Q'_x\varphi' + Q'_y\psi' \equiv \psi''$ (obtained from identity (5.57)), it is easily calculated that

$$A_1(u) = P'_x + Q'_y - \frac{d}{du} \ln(\varphi'^2 + \psi'^2).$$

On the other hand, since the solutions of equations with analytic right-hand sides are analytic functions of the initial conditions (see Appendix I), then the solution (5.58) is an analytic function of s and can be expanded in a power series with respect to s

$$v = \Phi(u, s) = a_1(u)s + a_2(u)s^2 + \ldots$$

(the constant term is equal to zero, since to the value $s=0$ there corresponds the limit cycle $v \equiv 0$). To find the functions $a_i(u)$ we substitute this series in the equation (3.56a) and equate the coefficients of equal powers of s. Then we obtain

$$a_1'(u)s + a_2'(u)s^2 + \ldots \equiv A_1(u)[a_1(u)s + a_2(u)s^2 + \ldots] + \\ + A_2(u)[a_1(u)s + a_2(u)s^2 + \ldots]^2 + \ldots$$

and

$$a_1' = A_1(u)a_1,$$
$$a_2' = A_1(u)a_2 + A_2(u)a_1^2, \quad \text{etc.}$$

Integrating these equations with the initial conditions

$$a_1(0) = +1 \quad \text{and} \quad a_i(0) = 0 \quad (i = 2, 3, \ldots,)$$

(the latter conditions are obtained from the evident identity: $\Phi(0, s) \equiv s$), we can find the coefficients in the expansion of the function $\Phi(u, s)$. In particular

$$\ln a_1(u) = \int_0^u A_1(t)\, dt = \int_0^u (P_x' + Q_y')\, dt - \ln \frac{[\varphi'(u)]^2 + [\psi'(u)]^2}{[\varphi'(0)]^2 + [\psi'(0)]^2}$$

and, hence,

$$f'(0) = a_1(T) = e^{\int_0^T (P_x' + Q_y')\, dt}$$

since functions φ and ψ, and thus their derivatives, are periodic functions with period T.

Thus the limit cycle C_0 is *stable*, if its characteristic exponent

$$h = \frac{1}{T} \int_0^T [P_x'(\varphi, \psi) + Q_y'(\varphi, \psi)]\, dt < 0,$$

for then $0 < f'(0) < 1$. And *unstable*, if

$$h > 0$$

for then $f'(0) > 1$.

§ 8. Poincaré's indices

Before analysing some specific dynamic systems of the second order, it is worthwhile outlining certain general theorems which enable us to obtain some, often very incomplete, information about the phase portrait of a system.

We shall outline, in the first place, the general laws formulated by Poincaré [108] for the simultaneous existence of singular points and closed paths.

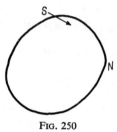

Fig. 250

Consider a phase plane defined as usual by the equations (5.1) where $P(x, y)$ and $Q(x, y)$ will be assumed to be analytic over the whole phase plane.

Consider a point S on a *simple closed curve N not passing through states of equilibrium*, and draw through S a vector (P, Q) which is tangent to the phase path through this point (Fig. 250). If we move the point S along the curve N, the vector will rotate continuously. When S has traversed the closed curve N and has returned to the initial position, then the vector (ϱ, Q) will have rotated by an angle $2\pi j$, where j is an integer. The direction of rotation of the vector will be positive if it coincides with the sense in which the point S goes round N; to be definite, let the point S describe the curve N in an anti-clockwise direction and j can be either a positive or a negative number or be equal to zero. The integer j, in a certain sense is independent *of the form of the closed curve N*. In fact if N changes continuously without crossing any singular points, the angle by which the vector rotates can also only vary continuously, and since it is an integer times 2π it remains constant. Therefore, all other closed curves, provided that they contain the same singular points as the curve N, will yield the same number j. The integer j bears the name of *index* of the closed curve N with respect to the vector field (P, Q).

Let us encircle with N a certain state of equilibrium or point. Then the Poincaré index is determined only by the nature of the singular point, and the index of such a closed curve may be referred to the singular point itself and we may call it Poincaré's index of the singular point.

It is easily verified by direct examination (Fig. 251) that Poincaré's indices for a centre, a node and a focus are all +1 and that Poincaré's index for a saddle-point is −1.

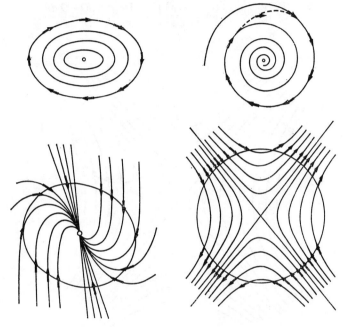

Fig. 251

The validity of the following statements can also be verified by direct examination:

(1) the index of a closed curve, not surrounding any singular points, is equal to zero (Fig. 252);

(2) the index of a closed curve surrounding a number of singular points is equal to the sum of the indices of these points;

(3) the index of a closed path is +1 (see Fig. 251, the case of a centre), since then the direction of the vector always coincides with the direction of the tangent to the curve N;

(4) the index of a closed curve along which the vectors are directed either all inwards or all outwards is +1 (see Fig. 251, the case of a saddle

point†).

These statements were obtained essentially by examining single examples and making use of certain considerations of continuity based on geometrical intuition, but they may be proved analytically. It is easily seen that the index of a closed curve N can be expressed by the curvilinear integral

$$j = \frac{1}{2\pi} \oint_N d\left\{\arctan \frac{Q(x,y)}{P(x,y)}\right\} = \frac{1}{2\pi} \oint_N \frac{P\,dQ - Q\,dP}{Q^2 + P^2}.$$

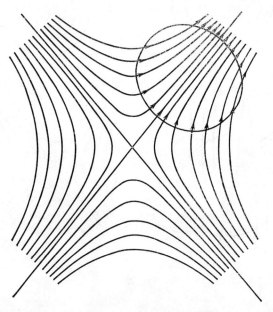

Fig. 252

This is a curvilinear integral of an exact differential; hence, if inside the region bounded by N the corresponding integrands and their derivatives are continuous, then the integral is equal to zero. This proves rigorously our first statement that the index of a closed curve N, inside which there ar no singular points, is equal to zero‡, since our assumptions about

---

† The index does not take into account the direction of motion along the phase paths; for example a stable node and an unstable one have both the index 1.

‡ The converse cannot be affirmed, since there can be singular points of higher order (for which $\Delta = 0$) with an index equal to zero.

$P(x, y)$ and $Q(x, y)$ imply the continuity of the integrands and their derivatives except at singular points. Let us calculate Poincaré's index for a singular point. We shall assume that for this point $\varDelta = ad - bc \neq 0$ (see equation 5.4) and is thus simple. Assume that the singular point is at the origin so that

$$\frac{dx}{dt} = ax + by + P_2(x, y), \quad \frac{dy}{dt} = cx + dy + Q_2(x, y),$$

where P_2 and Q_2 are power series beginning with terms of at least the second order in x and y.

We shall first prove that in evaluating the index of a simple singular point we can neglect the terms of higher orders, such as P_2 and Q_2. Since the index does not depend on the shape of the curve, then we can take for the curve N a circle of sufficiently small radius ϱ ($\varrho > 0$).

In polar coordinates $x = \varrho \cos \varphi$, $y = \varrho \sin \varphi$, the curvilinear integral becomes an ordinary definite integral

$$j = I(\varrho) =$$
$$= \frac{1}{2\pi} \int_0^{2\pi} \frac{(a\cos\varphi + b\sin\varphi)d(c\cos\varphi + d\sin\varphi) - (c\cos\varphi + d\sin\varphi)d(a\cos\varphi + b\sin\varphi) + \varrho F(\varrho, \varphi)\, d\varphi}{(a\cos\varphi + b\sin\varphi)^2 + (c\cos\varphi + d\sin\varphi)^2 + \varrho G(\varrho, \varphi)},$$

where $F(\varrho, \varphi)$ and $G(\varrho, \varphi)$ are power series with respect to ϱ with coefficients which are periodic functions of φ.

The definite integral $I(\varrho)$ is a continuous function of ϱ for sufficiently small values of φ (since $\varDelta \neq 0$). Therefore $\lim_{\varrho \to 0} I(\varrho) = I(0)$. We know, on the other hand that the curvilinear integral does not depend upon ϱ for sufficiently small values of ϱ. Hence it follows that, for sufficiently small values of ϱ, $I(\varrho) = I(0)$, and

$$j = I(0) =$$
$$= \frac{1}{2\pi} \int_0^{2\pi} \frac{(a\cos\varphi + b\sin\varphi)d(c\cos\varphi + d\sin\varphi) - (c\cos\varphi + d\sin\varphi)d(a\cos\varphi + b\sin\varphi)}{(a\cos\varphi + b\sin\varphi)^2 + (c\cos\varphi + d\sin\varphi)^2}.$$

It is thus proved that in evaluating Poincaré's index for a simple singular point (with $\varDelta \neq 0$) the non-linear terms can be neglected. To evaluate $I(0)$ it is expedient to return to ordinary coordinates and write the expression for j as

$$j = I(0) = \oint_N \frac{(ax+by)d(cx+dy) - (cx+dy)d(ax+by)}{(ax+by)^2 + (cx+dy)^2},$$

where N is an arbitrary simple closed curve surrounding the origin. Now choose for N the ellipse Γ

$$(ax+by)^2+(cx+dy)^2 = 1;$$

then, as simple calculations show,

$$j = I(0) = \frac{\Delta}{2\pi} \oint_\Gamma (x\,dy - y\,dx),$$

where

$$\Delta = \begin{vmatrix} a & b \\ c & d \end{vmatrix}$$

or as is well known:

$$j = I(0) = \frac{\Delta}{\pi} S,$$

where S is the area of the ellipse. Since $S = \pi/|\Delta|$, then

$$j = \frac{\Delta}{|\Delta|}.$$

It follows at once that Poincaré's index for a node, a focus and a centre is equal to $+1$ and for a saddle-point is equal to -1. Poincaré's index for a singular point of higher order can be different from ± 1, since $\Delta = 0$. For example, for a singular point of the saddle-node type $j = 0$ (see Fig. 253). A proof of this follows directly from the basic properties of a curvilinear integral.

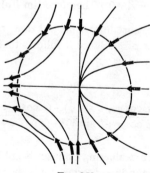

Fig. 253

COROLLARY 1. Inside a closed phase path there is at least one singular point, since the index is ± 1, while the index of any closed oval inside which there is no singular point is zero.

COROLLARY 2. If inside a closed phase path there is one singular point, then it must be a node or a focus.

COROLLARY 3. If inside a closed phase path there are only simple singular points then the number of such singular points must be odd, the number of saddle-points being one less than the number of nodes and foci.

It also follows that if the point at infinity is absolutely stable or absolutely unstable, then the sum of the indices of all singular points which are found at a finite distance is equal to $+1$.

§9. SYSTEMS WITHOUT CLOSED PATHS

The difficulties which arise in investigating specific examples of models are very great and it is necessary to have recourse to various methods of numerical integration. There are cases, however, when the investigation can be fairly simply carried out using the general theory. One such case (possibly the most important one) is when we are able to prove in some manner that there are no closed phase paths on the phase plane.

A number of criteria give sufficient conditions for the absence of closed phase paths, but they all fail to give a regular method for proving their absence in the general system (5.1). However, they are of a definite practical interest. Again we assume $P(x, y)$ and $Q(x, y)$ to be analytic over the whole phase plane of the system (5.1).

Bendixson's criterion [137] reads: *if on a certain singly-connected region on the phase plane the expression $(\partial P/\partial x)+(\partial Q/\partial y)$ is of constant sign, then in this region there are no closed contours formed entirely by phase paths of the dynamic system (5.1)*†

Now Green's theorem states

$$\iint \left(\frac{\partial P}{\partial x}+\frac{\partial Q}{\partial y}\right) dx\, dy = \oint (P\, dy - Q\, dx).$$

If the contour integral is taken along a curve consisting entirely of phase paths, then by virtue of the equations (5.1) it is equal to zero and, therefore, the double integral is also equal to zero. Hence, the expression $(\partial P/\partial x)+(\partial Q/\partial y)$ is bound to change its sign somewhere within the contour taken. Our statement is thus proved.

A well-known generalization of Bendixson's criterion is *Dulac's criterion* [148, 108]: *if a continuous function $B(x, y)$ with continuous derivatives*

---

† The criterion remains valid when $P'_x+Q'_y$ reduces to zero at separate points or on certain curves in this region.

exists, such that in a certain single-connected region on the phase plane the expression $[\partial(BP)\partial x]+[\partial(BQ)\partial y]$ is of constant sign, then in this region there exist no closed contours consisting entirely of phase paths of the system (5.1). The proof is similar to that of Bendixson's criterion, and we omit it.

We shall present now criteria concerned with the much less stringent requirement that there are no closed single paths, or limit cycles, and hence no periodic solutions of the system (5.1). A number of criteria could be given on the basis of the theory of the indices outlined in § 8; but we mention only the more important. In the sequel we encounter yet more criteria based on the properties of the so-called "curve of the contacts".

1. If no singular points exist in the system, then closed phase paths cannot exist.

2. If one singular point only exists its index not being equal to $+1$ (for example, a saddle-point), then there cannot be closed phase paths or limit cycles.

3. If the system possesses several singular points, the sum of the indices of any combination of them being different from $+1$, then closed phase paths do not exist.

4. If a system has simple singular points only, and through all points with indices $+1$ there pass integral curves receding to infinity, then such a system has no closed phase paths.

To illustrate the various criteria we shall consider a few physical systems.

EXAMPLE 1. *Symmetrical valve relay* (trigger). As a first example we shall consider the manner of operation of the symmetrical valve relay or trigger circuit shown in Fig. 254 (the capacitances C_a and C_g represent small parasitic capacitances). Under certain conditions this circuit has two stable states of equilibrium and can be flipped from one state of equilibrium into the other by applying a suitable voltage impulse to an appropriate terminal. It has a fairly wide application in electronic equipment; for example in cathode-ray oscillographs as the triggering circuit for the time-base circuit, in counters of electrical pulses, and in electronic digital computers (with transistors instead of valves).

We shall consider a simplified circuit diagram (Fig. 225), obtained from the complete scheme (Fig. 254) on the assumption that $CR_1 = C_g R_2$. Then the anode to grid transmission coefficient β is constant[†] and

[†] Strictly speaking, the equations of the voltage divider are differential equations. For example, for the divider which transmits the anode-voltage oscillations of the left-hand valve to the grid of the right-hand one (Fig. 254):

$$i = C\frac{d(u_{a1}-u_2)}{dt}+\frac{u_{a1}-u_2}{R_1} = C_g\frac{du_2}{dt}+\frac{u_2-E_g}{R_2},$$

9] SYSTEMS WITHOUT CLOSED PATHS 307

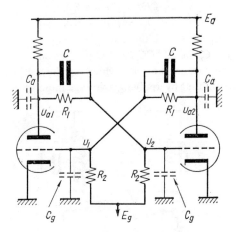

Fig. 254

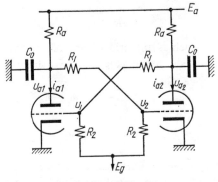

Fig. 255

where i is the current flowing through the divider, hence

$$\frac{R_1 R_2}{R_1+R_2}(C+C_g)\frac{du_2}{dt}+u_2 = \frac{R_1 R_2}{R_1+R_2} C \frac{du_{a1}}{dt}+\frac{R_2}{R_1+R_2}(u_{a1}-E_g)+E_g.$$

However, if $CR_1 = C_g R_2$, then this equation has a solution for all values of u_{a1}.

$$u_2 = E_g + \frac{R_2}{R_1+R_2}(u_{a1}-E_g)+Ae^{-\overline{CR_1}}$$

and, therefore, for $t \gg CR_1$ the response to a step change in u_{a1} is the second relation of (5.60) whatever the past variable values of u_{a1}. The current required by the divider is equal to $i = [CC_g/(C+C_g)]du_{a1}/dt+(u_{a1}-E_g)/(R_1+R_2)$. We can therefore replace the RC divider (on condition that $R_1 C = R_2 C_g$) by a divider consisting only of the resistances R_1 and R_2 shunted by the capacitance $CC_g/(C+C_g)$.

$$u_1 - Eg = \beta(u_{a2} - E_g).$$
$$u_2 - Eg = \beta(u_{a1} - E_g), \tag{5.60}$$

where $\beta = R_2/(R_1+R_2)$ and the equivalent capacitance at the anodes (C_0 in Fig. 255) is $C_0 = C_a + [CC_g/(C+C_g)]$. This simplified circuit enables us to examine the trigger as a relay which is "flipped" from one state of equilibrium into another by applying a voltage impulse to a *non-symmetric*

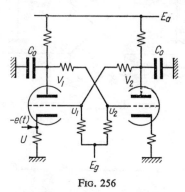

Fig. 256

point of the circuit, such as the grid of one of the valves, or to a small resistance in a cathode circuit (Fig. 256). We have in the notation of Fig. 255, the following equations:

$$C_0 \frac{du_{a1}}{dt} + i_{a1} + \frac{u_{a1}-E_g}{R_1+R_2} + \frac{u_{a1}-E_a}{R_a} = 0,$$
$$C_0 \frac{du_{a2}}{dt} + i_{a2} + \frac{u_{a2}-E_g}{R_1+R_2} + \frac{u_{a2}-E_a}{R_a} = 0,$$

then neglecting the anode conductance, we assume that the anode current of each valve depends only on the voltage of its grid, so that $i_{a1} = f(u_1)$ and

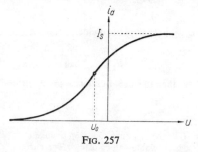

Fig. 257

$i_{a2} = f(u_2)$. We also assume that the valve characteristic $i_a = f(u)$, shown in Fig. 257, possesses the following properties:

(1) $f'(u) \geq 0$, with $0 \leq f(u) \leq I_s$, the saturation current of the valve;

(2) the slope $f'(u)$, has a single maximum and decreases monotonically to zero on each side of this maximum. The equations then reduce to

$$C_0 R \frac{du_1}{dt} = -u_1 - \beta R f(u_2) + E,$$
$$C_0 R \frac{du_2}{dt} = -u_2 - \beta R f(u_1) + E, \qquad (5.61)$$

where

$$R = \frac{R_a(R_1 + R_2)}{R_a + R_1 + R_2}, \quad E = E_g + \beta \frac{R}{R_a}(E_a - E_g) = \frac{R_2 E_a + (R_1 + R_a) E_g}{R_1 + R_2 + R_a},$$

or to

$$\frac{du_2}{du_1} = \frac{u_2 + \beta R f(u_1) - E}{u_1 + \beta R f(u_2) - E}. \qquad (5.62)$$

The states of equilibrium are clearly determined by

$$u_2 + \beta R f(u_1) - E = 0, \qquad (5.63a)$$
$$u_1 + \beta R f(u_2) - E = 0. \qquad (5.63b)$$

and can be considered as the points of intersection of the curves (5.63a) and (5.63b) on the phase plane (note that the first is the isocline of the horizontal tangents and the second is the isocline of the vertical tangents). Obviously for all values of the parameters there exists a "symmetrical" state of equilibrium (U, U), lying on the bisector $u_1 = u_2$, an integral straight line of the equation (5.62). This point is determined by

$$U + \beta R f'(U) - E = 0,$$

which, for the chosen function $f(u)$, has only one solution. Also if the point (a, b) is a state of equilibrium, then so is the point (b, a). Thus the total number of states of equilibrium is always odd. To find the states of equilibrium we must construct the curves (5.63a) and (5.63b) and find their points of intersection, as shown in Fig. 258a for the case of $\beta R f'(U) < 1$ when there is only one symmetrical state of equilibrium, and in Fig. 258b for the case of $\beta R f'(U) > 1$ where there are three states of equilibrium[†].

[†] If $\beta R f'(U)$ is close to unity while the slope $f'(u)$, is not at the maximum then the system can, generally speaking, have even more than three states of equilibrium. For $\beta R f'(U) < 1$ their number can be equal to 5, 9, 13, ..., and for $\beta R f' U > 1$ to 7, 11, ..., depending on the form of the characteristic $f(u)$ on the section from $u = U$ to the point of maximum slope.

Suppose a state of equilibrium is (u_1^0, u_2^0). Now put

$$u_1 = u_1^0 + \xi, \quad u_2 = u_2^0 + \eta;$$

Then, as is easily verified, the equations of the first approximation are

$$C_0 R \frac{d\xi}{dt} = -\xi - \beta R f'(u_2^0)\eta, \quad C_0 R \frac{d\eta}{dt} = -\eta - \beta R f'(u_1^0)\xi,$$

and the characteristic roots are determined by the equation

$$\begin{vmatrix} C_0 R\lambda + 1 & \beta R f'(u_2^0) \\ \beta R f'(u_1^0) & C_0 R\lambda + 1 \end{vmatrix} = 0,$$

or

$$C_0 R \lambda_{1,2} = -1 \pm \beta R \sqrt{f'(u_1^0) f'(u_2^0)}.$$

Bearing in mind that $-\beta R f'(u_1^0)$ is the slope of the tangent to the curve (5.63a) and $-1/\beta f'(u_2^0)$ is the slope of the tangent to the curve (5.63b), we see that the "symmetrical" state of equilibrium is a stable node for

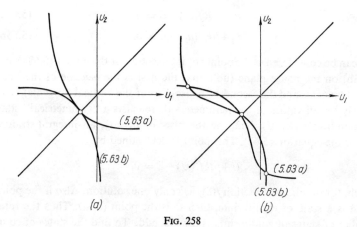

Fig. 258

$\beta R f'(U) < 1$ and an unstable saddle-point for $\beta R f'(U) > 1$. Therefore, this state of equilibrium is stable if it is unique, and unstable if there are three states of equilibrium. In the latter case the remaining two states of equilibrium are stable.

Further, if $P(u_1, u_2)$ and $Q(u_1, u_2)$ denote the right-hand sides of (5.61), then

$$\frac{\partial P}{\partial u_1} + \frac{\partial Q}{\partial u_2} = -2 < 0$$

and according to Bendixson's criterion, there are no closed contours consisting of phase paths, nor any limit-cycles. It is also easily seen that all

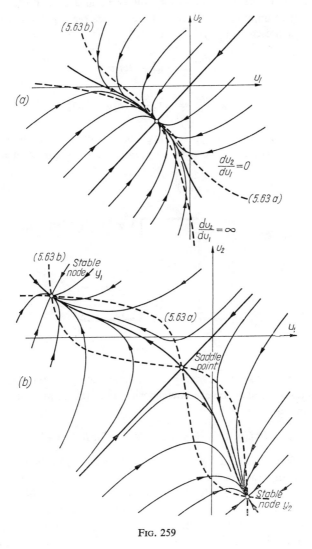

Fig. 259

phase paths are directed from infinity inwards. The phase portrait is shown diagrammatically in Fig. 259a (for $R\beta f'(U) < 1$) and in Fig. 259b (for $R\beta f'(U) > 1$).

Now let $\beta R f'(U) > 1$ and suppose the trigger circuit to be in a state of equilibrium at the node Y_1, when V_1 is cut-off and the valve V_2 conducts. Let us apply to the valve V_1 a short voltage impulse $e(t)$, (for an example of a negative impulse to the cathode resistance see Fig. 256). The equations in the presence of $e(t)$, become

$$C_0 R \frac{du_1}{dt} = -u_1 - \beta R f(u_2) + E,$$
$$C_0 R \frac{du_2}{dt} = -u_2 - \beta R f[u_1 + e(t)]$$
(5.64)

If $e(t)$ is a rectangular impulse (Fig. 260), we can assume the system to be autonomous during the path duration τ, and can construct its phase

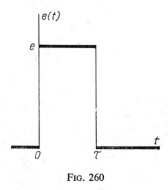

Fig. 260

portrait again. As before, each path terminates at one of the stable nodes. The states of equilibrium are now determined as the points of intersection of the curves

$$u_2 + \beta R f(u_1 + e) - E = 0 \quad \text{and} \quad u_1 + \beta R f(u_2) - E = 0,$$

only the first of which differs from (5.63) by a shift e to the left (if $e > 0$).

Let the amplitude of the impulse e be so large that, during the duration τ, the trigger circuit has only one state of equilibrium Y (a stable node) (Fig. 261). Now immediately after the passage of the leading edge of the pulse (at $t = 0$) the state of the system will be represented on the phase plane (Fig. 261) by the point Y_1, which was a stable node before the arrival of the pulse, but is now and for $0 < t < \tau$ a simple non-singular point. Then by (5.64) the representative point moves along a phase path towards the stable node Y, and if τ is large enough the representative point will cross the bisector

$u_1 = u_2$, before the impulse vanishes. When $e(t) = 0$ it will be in the region of "attraction" of the node Y_2 on the phase plane (Fig. 259(b) and will approach it asymptotically. Thus if the impulse is of sufficiently large

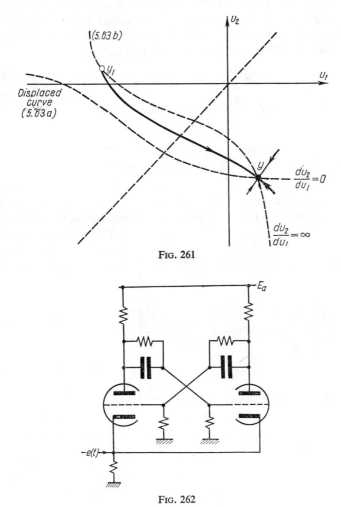

Fig. 261

Fig. 262

amplitude e and duration τ, it will "flip" the trigger from the stable node Y_1 to the stable node Y_2. A second similar impulse has no further effect. A pulse suitably injected elsewhere is needed to change back to the original equilibrium state.

We shall make one more observation having a certain practical interest. This trigger circuit is used as a counter for pulses but these are applied to a *symmetrical* point of the circuit (for example, to a common cathode resistance, see Fig. 262) and then *each* impulse flips the trigger from one state of equilibrium into another. Now the simplified circuit considered here has not got this property. In fact the equations of the simplified circuit for a symmetrical-fed rectangular pulse of amplitude e are

$$C_0 R \frac{du_1}{dt} = -u_1 - \beta R f(u_2+e) + E,$$

$$C_0 R \frac{du_2}{dt} = -u_2 - \beta R f(u_1+e) + E$$

or

$$\frac{du_1}{du_2} = \frac{-u_1 - \beta R f(u_2+e) + E}{-u_2 - \beta R f(u_1+e) + E},$$

which, clearly, has as an integral curve, $u_1 = u_2$. Therefore during the time of the action of the impulse the representative point cannot cross the bisector $u_1 = u_2$ and thus fall into the region of "attraction" of the other node.

This reflects a property of real trigger circuits and in fact in order that the

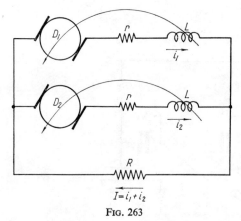

Fig. 263

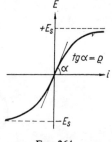

Fig. 264

trigger may work as a counting network, it must have a sufficiently large capacitance C (Fig. 254), invalidating our condition $CR_1 = C_g R_2$.

EXAMPLE 2. *Dynamos working in parallel on a common load.* Let us consider two similar d.c. dynamos with series excitation, connected in parallel and supplying a common load (Fig. 263). Neglecting magnetic hysteresis we write the e.m.f. E of each dynamo as a single-valued function of the current i in its excitation winding: $E = \psi(i)$. We assume that $\psi(i)$ for each dynamo has the form shown in Fig. 264.

Then, (1) $E = \psi(i)$ is a continuous differentiable *odd* function, saturating at E_s and $-E_s$ for large positive and negative values of i.
(2) the derivative $\psi(i) \geqslant 0$ and decreases as $|i|$ increases.

The equations of the system, in the notation of Fig. 263, are

$$L\frac{di_1}{dt} = \psi(i_1) - (r+R)i_1 - Ri_2, \\ L\frac{di_2}{dt} = \psi(i_2) - (r+R)i_2 - Ri_1. \quad (5.65)$$

The equation of the integral curves is

$$\frac{di_2}{di_1} = \frac{\psi(i_2) - (r+R)i_2 - Ri_1}{\psi(i_1) - (r+R)i_1 - Ri_2}. \quad (5.66)$$

The states of equilibrium are determined by the equations

$$\psi(i_1) - (r+R)i_1 - Ri_2 = 0, \quad (5.67a)$$

$$\psi(i_2) - (r+R)i_2 - Ri_1 = 0, \quad (5.67b)$$

and their characters by the roots of the characteristic equation

$$\lambda^2 + \sigma\lambda + \Delta = 0,$$

the coefficients of which, for the state of equilibrium (I_1, I_2), are easily calculated to be

$$\sigma = \frac{1}{L}[2(r+R) - \psi'(I_1) - \psi'(I_2)], \\ \Delta = \frac{1}{L^2}\{[\psi'(I_1) - (r+R)][\psi'(I_2) - (r+R)] - R^2\}. \quad (5.68)$$

The discriminant of the characteristic equation is

$$\frac{\sigma^2}{4} - \Delta = \frac{[\psi'(I_1) - \psi'(I_2)]^2}{4L^2} + \frac{R^2}{L^2} > 0.$$

So the states of equilibrium can only be *nodes* or *saddle* points. A state of equilibrium at the origin of the phase plane, $i_1 = i_2 = 0$, corresponds to non-excited machines. For this state of equilibrium we have $\sigma L = 2(r+R-\varrho)$ and $\Delta L^2 = (\varrho-r)[\varrho-r][\varrho-(r+2R)]$, where $\varrho = \psi'(0) > 0$. Therefore this state is a stable node ($\sigma > 0, \Delta > 0$) for $r > \varrho$, a saddle point ($\Delta < 0$) for $r < \varrho < r+2R$, and an unstable node ($\sigma < 0, \Delta > 0$) for $r+2R < \varrho$.

To find the states of equilibrium A lying on the bisector $i_1 = i_2$ we shall put $i_1 = i_2 = a$ in equations (5.67a) and (5.67b). Then the coordinates of the points A are given by

$$\psi(a) - (r+2R)a = 0. \tag{5.69}$$

These are the required states of equilibrium since under these conditions both machines work correctly, supplying maximum power to the resistance R. From the graphical solution of the equation (5.68), shown in Fig. 265,

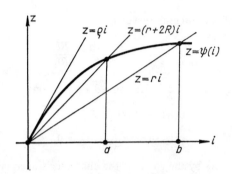

Fig. 265

such states of equilibrium are seen to exist only for $r+2R < \varrho$ and even then there are only two of them: $A(a, a)$ and $A_1(-a, -a)$ where $a < 0$. We have for the A points: $\sigma L = 2[r+R-\psi'(a)]$ and $\Delta L^2 = [\psi'(a)-r][\psi'(a)-(r+2R)]$. Since $\psi'(a) < r+2R$ then these useful states of equilibrium are stable nodes for $r > \psi'(a)$ only and are unstable saddle points for $r < \psi'(a)$.

On the other bisector $i_2 = -i_1$ if $r < \varrho$, then there exist two states of equilibrium: $B(b, -b)$ and $B_1(-b, b)$. These are "harmful" states of equilibrium in which the dynamos feed each other and the current through the load R, $I = i_1 + i_2$, is equal to zero. The coordinate $b(b > 0)$ is determined, clearly, by the equation (see Fig. 265)

$$\psi(b) - rb = 0, \tag{5.70}$$

We have for the B points: $\sigma L = 2[r+R-\psi'(b)] > 0$, $\Delta L^2 = [\psi'(b)-r] \cdot [\psi'(b)-(r+2R)] > 0$, since $\psi'(b) < r$. Therefore the B points, if they exist, are always *stable nodes*.

To find which modes of operation actually occur we construct a "gallery of phase portraits", taking the resistances r and R as variable parameters.

First of all, as follows from the equations (5.65), note that

$$\frac{1}{2}L\frac{d}{dt}(i_1^2+i_2^2) = -R(i_1+i_2)^2-r(i_1^2+i_2^2)+i_1\psi(i_2)+i_2\psi(i_1) < 0$$

at the points of a circle $i_1^2+i_2^2$ of sufficiently large radius. Each such circle is therefore a *cycle without contact* and all phase paths go from infinity into the region inside this cycle without contact (i.e. the point at infinity is absolutely unstable). Therefore the sum of Poincaré's indices for all states of equilibrium is equal to $+1$.

In addition, by virtue of the symmetry of the circuit and the fact that the function $\psi(i)$ is odd, both bisectors $i_2=i_1$ and $i_2=-i_1$ are integral curves of the equation (5.66) and the whole phase portrait is symmetrical with respect to these straight lines; in particular the positions of the state of equilibrium are symmetrical to each other with respect to these lines†. Therefore, we need only consider the behaviour of the phase paths in the quadrant K_1, enclosed between these integral straight lines and containing the positive i_1 axis. The states of equilibrium, as we have already indicated, are points of intersection of the curves (5.67a) and (5.67b)‡ the equations of which can be written in explicit form:

$$i_2 = \varphi(i_1), \tag{5.67a}$$

$$i_1 = \varphi(i_2), \tag{5.67b}$$

where

$$\varphi(i) = \frac{\psi(i)-(r+R)i}{R}. \tag{5.71}$$

Curves (5.67a) and (5.67b) are shown for various values of the parameters r and R in Fig. 266. The continuous line represents the curve (5.67a) for $i_1 \geqslant 0$, and the dotted lines represent the curve (5.67b) as the mirror image (5.67a) in the straight lines $i_2=i_1$, and $i_2=-i_1$ (within the limits of the K_1 quadrant). It is possible to have up to nine singular points.

Case I. $r > \varrho$. In this case (Fig. 266, I) $\varphi'(0) = [\varrho-(r+R)]/R < -1$ and the curve (5.67a) for $i_1>0$ lies entirely in the K_2 quadrant, and its mirror image in $i_2 = -i_1$ lies in K (the curve (5.67b)). There is one state of equi-

---

† Another consequence of the symmetry of the phase portrait with respect to the bisectors $i_2=i_1$ and $i_2=-i_1$ is that all integral curves of the equation (5.66) pass through the states of equilibrium along directions with slopes $\varkappa_{1,2} = \pm 1$.

‡ These curves are isoclines on the phase plane: the first is the isocline $\varkappa = \infty$ and the second is the isocline $\varkappa = 0$.

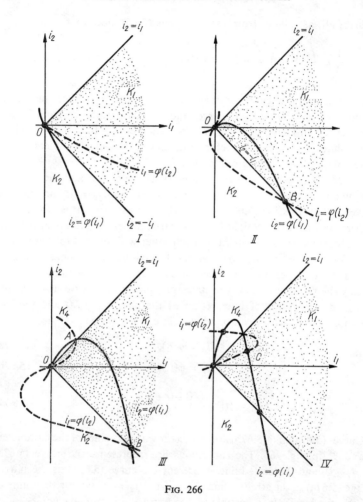

Fig. 266

librium on the phase plane (Fig. 267, *I*) stable at the origin 0. Since there are also no closed phase paths†, then all phase paths approach asympto-

---

† If on the phase plane there were a closed path, then, according to the theory of Poincaré's indices, it would surround the node 0, which is impossible since through it there pass the integral straight lines $i_2 = i_1$ and $i_2 = -i_1$ which go to infinity. For the same reason, there are no closed phase paths for other values of the parameters of the system either (through each node, as we shall see, there pass the integral straight line $i_2 = i_1$ or $i_2 = -i_1$; therefore, a closed phase line, if such a one did exist, could not surround any of the nodes and have a Poincaré's index equal to $+1$).

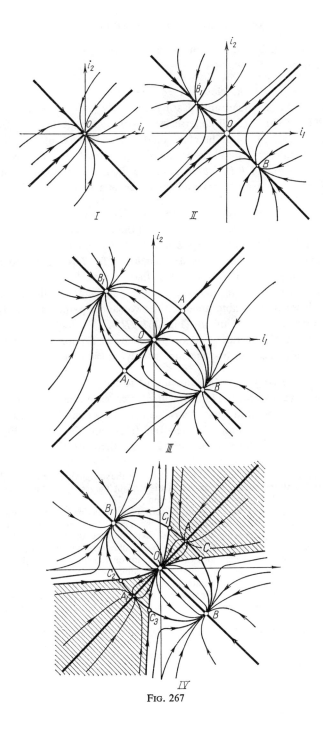

Fig. 267

tically the node 0, whatever the initial conditions, corresponding to both machines not being excited[†].

Case II. $r < \varrho < r+2R$. Now $-1 < \varphi'(0) < +1$ and the curve (5.67a) lies partly in the K_1 quadrant passing for $i_1 = b$ into the K_2 quadrant (Fig. 266, *II*). Curve (5.67b) intersects the curve (5.67a) at the origin 0 and at the point $B(b, -b)$ only. Therefore, on the phase plane (Fig. 267, *II*) there are three states of equilibrium: the saddle point 0 and the two stable nodes $B(b, -b)$ and $B_1(-b, b)$. A node is a "harmful" condition when one machine acts as a generator and the other as a motor.

Case III. $r+2R < \varrho$, $r < \psi'(a)$. Since $r+2R < \varrho$, then $\varphi'(0) > +1$ and the curve (5.67a) is (for small values of i_1) in the K_4 quadrant then, for $i_1 = a$, passes in K_1 and, finally, for $i_1 = b$ into K_2 (Fig. 266, *III*). Correspondingly the curve (5.67b) lies in K_1 for $0 \leqslant i_2 \leqslant a$ and for $i_2 \leqslant b$ only. Assuming $\varphi'(a) = [\psi'(a) - (r+R)]/R > -1$, then these curves do not intersect each other inside K_1 quadrant and there are five states of equilibrium on the phase plane: the unstable node 0, the two saddle points A and A_1, and the two stable nodes B and B_1 (Fig. 267, *III*)[‡].

Again a stable operating condition will be at a point B, whatever the initial conditions, with one machine supplying the other.

Case IV. $r+2R < \varrho$, $r > \psi'(a)$. In contrast to the previous case, $\varphi'(a) < -1$ and the curves (5.67a) and (5.67b) will intersect each other *inside* the quadrant K_1 at one point at least. Below we shall only consider the case when this point of intersection is a *single* one (the point $C(c', c'')$ in Fig. 266, *IV*)[††]. Thus, on the phase plane (Fig. 267, *IV*) there are nine states of equi-

---

[†] All integral curves of the equation (5.66), except the straight line $i_2 = i_1$, pass through the node 0, with the same slope as that of the straight line $i_2 = -i_1$. This is easily verified using the method given in the footnote on page 261.

[‡] The assumptions made above on the form of the function $\psi(i)$ are insufficient to prove the absence of points of intersection of the curve (5.67a) and (5.67b) inside the K_1 quadrant. Generally speaking, for $r < \psi'(a)$, depending on the form of the function $\psi(i)$ there can be any even number of such points of intersection, and correspondingly five, thirteen, twenty-one etc. states of equilibrium on the phase plane, three, seven, eleven ... of which will be nodes and the remaining ones saddle points, since the sum of Poincaré's indices for all states of equilibrium is equal to $+1$.

We observe that now on the phase plane three are closed contours made up of phase paths (for example, the contour ABA_1B_1A in Fig. 267, *III*), although as before there are no closed phase paths.

[††] Generally speaking, for $r > \psi'(a)$ and depending on the form of the function $\psi(i)$ there can be any odd number of points of intersection of the curves (5.67a) and (5.67b) inside the K_1 quadrant and, on the phase plane, nine, seventeen, twenty-five, etc., states of equilibrium.

librium: the unstable node 0, the four stable nodes A, A_1, B and B_1, and four C points: $C(c', c'')$, $C_1(c'', c')$, $C_2(-c', -c'')$ and $C_3(-c'', -c')$. It is easily verified on the basis of the theory of Poincaré's indices that these are saddle points. In fact the sum of Poincaré's indices for all states of equilibrium, as we have already seen, is equal to $+1$; the five known states

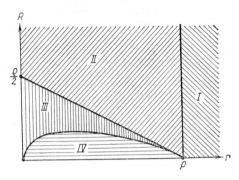

Fig. 268

of equilibrium which are the points 0, A, A_1, B and B_1 belonging to the integral straight lines $i_2 = i_1$ and, or $i_2 = -i_1$, are nodes and the sum of their indices is equal to $+5$, hence the sum of the indices of the four C points must be equal to -4, i.e. the C points must be saddle points. Stable stationary operating conditions of the machines correspond to the stable nodes A, A_1 (when power is supplied by both machines to the external circuit) and B, B_1 (with no external power being supplied). The attainment of one or other operating condition will depend on the initial conditions; if the initial state of the system corresponds to a point of the region bounded by the separatrices shaded in Fig. 267, *IV*, then the machines operate correctly.

Figure 268 shows the branch curves mapping out the (r, R) stability plane into various regions (the numbering of the regions coincides with that of Fig. 267, where corresponding phase portraits are shown). The branch curves are (1) the straight line $r = \varrho$ separating *I* from *II* (2) the straight line $r + 2R = \varrho$ and (3) the curve $r = \psi'(a)$, where a is determined by the equation (5.69). Region *IV* is that in which the correct operating conditions of the machines (with supply power to the external circuit) are stable. The equation of this latter branch curve can be written

$$r = \psi'(a), \quad R = \frac{1}{2}\left[\frac{\psi(a)}{a} - \psi'(a)\right],$$

where a is a parameter which can take all the values from 0 to $+\infty$. It is easily seen that this curve passes through the points $(\varrho, 0)$ and $(0, 0)$ and lies below the straight line $r+2R = \varrho$.

Obviously, the circuit diagram shown in Fig. 263 cannot be accepted as a practical one, since operating conditions, corresponding to a supply

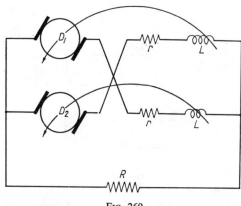

Fig. 269

of power to the external circuit by both machines are only stable for limited values of the parameters (*IV* in Fig. 268). This defect is absent in the circuit with a cross connexion of the excitation windings shown in Fig. 269. In this case, as is easily verified (and we leave it to the reader), operating conditions for which one machine supplies the other cannot be stable, while operating conditions in which both machines supply power to the external circuit exist for $r+2R<\varrho$ and are stable.

EXAMPLE 3. *Oscillator with quadratic terms* [20]. Let us consider an oscillator the equation of which

$$m\frac{d^2x}{dt^2} = -ax+\alpha x^2 - b\frac{dx}{dt} + \beta\left(\frac{dx}{dt}\right)^2$$

contains quadratic terms in the expressions for the force of the spring and for the force of friction. This equation can be written as (where $m=1$)

$$\frac{dx}{dt} = y = P(x,y), \quad \frac{dy}{dt} = -ax-by+\alpha x^2+\beta y^2 = Q(x,y). \quad (5.72)$$

Let us use Dulac's criterion, taking as the multiplier $B(x,y)$ the function $B(x, y) = be^{-2\beta x}$. Since, as is easily seen, $[\partial(BP)/\partial x]+[\partial(BQ)/\partial y] = -b^2 e^{-2\beta x} < 0$ for $b \neq 0$, the system (5.72), according to Dulac's criterion, has no closed phase paths nor even closed contours made up of various phase paths, and, hence, cannot be self-oscillating.

EXAMPLE 4. *One more example of non-self-oscillating system* [26]. We shall prove that the system of equations

$$\left.\begin{array}{l}\dot{x} = x(ax+by+c) = P(x,y),\\ \dot{y} = y(a'x+b'y+c') = Q(x,y).\end{array}\right\} \quad (5.73)$$

which is encountered in the non-linear theory of oscillations[†], does not have limit cycles. This system has as straight integral curves, the axes $x=0$ and $y=0$, and has on them three states of equilibrium, $(0,0)$, $(0,-c'/b')$ and $(-c/a, 0)$. In addition, if $\delta = ab'-a'b \neq 0$, there is one more state of equilibrium (x_0, y_0) not lying on the coordinate axes and determined by the system of equations:

$$ax+by+c = 0, \quad a'x+b'y+c' = 0.$$

Therefore, if the system (5.73) has a closed phase path, then the latter should lie within the limits of one quadrant, containing the point (x_0, y_0), without intersecting any of the straight integral lines, and should surround the state of equilibrium (x_0, y_0)[‡]. This is, however, impossible according to Dulac's criterion. Let us take as the multiplier B the function $B(x,y) = x^{k-1}y^{h-1}$, where k and h are so far undetermined constants. Then

$$\frac{\partial}{\partial x}(BP) + \frac{\partial}{\partial y}(BQ) = x^{k-1}y^{h-1}\{(a+ka+ha')x+(kb+hb'+b')y+kc+hc'\}.$$

Taking as the constants k and h the solution of the system of equations

$$ka+ha'+a = 0, \quad kb+hb'+b' = 0,$$

i.e.

$$k = \frac{b'(a'-a)}{\delta} \quad \text{and} \quad h = \frac{a(b-b')}{\delta},$$

we obtain

$$\frac{\partial}{\partial x}(BP) + \frac{\partial}{\partial y}(BQ) = B(x,y)\frac{b'c(a'-a)+ac'(b-b')}{\delta} \neq 0$$

[†] We are led to this system, in particular, in problems on self-oscillating systems with 2 degrees of freedom using Van der Pol's method [112, 176, 177].

[‡] It is clear that for $\delta=0$, when the state of equilibrium (x_0, y_0) does not exist, there are no closed phase paths.

within the limits of each quadrant of the phase plane, provided that

$$\sigma = b'c(a'-a)+ac'(b-b') \neq 0.$$

Therefore, according to Dulac's theorem the system (5.73), for $\sigma \neq 0$, has no closed phase paths or limit cycles[†].

§ 10. The behaviour of the phase paths near infinity

The behaviour of the phase paths in sufficiently distant regions of the phase plane is very useful in helping to provide a qualitative picture of the phase portrait. This behaviour is sometimes found very easily.

Thus, from equations (5.1)

$$\frac{1}{2}\frac{dr^2}{dt} = P(x,y)x + Q(x,y)y = R(x,y).$$

If outside any sufficiently large circle $R(x, y)$ assumes a fixed sign and retains it for all values of x and y outside the circle, then all sufficiently large circles with centres at the origin serve as cycles without contact. The sign of $R(x, y)$ determines whether the point at infinity is stable $(+)$ or unstable $(-)$. However, generally speaking, such an elementary method does not yield an answer $[R(x, y)$ does not retain a determined sign] and the question requires a special investigation.

Much more information can be found by means of Bendixson's transformation

$$x = \frac{u}{u^2+v^2}; \quad y = \frac{v}{u^2+v^2}; \quad u = \frac{x}{x^2+y^2}; \quad v = \frac{y}{x^2+y^2}, \quad (5.74)$$

which transfers infinitely remote points of the x, y plane to the origin of the u, v plane (Fig. 270).

It is easy to find the true direction along the paths by transforming the system

$$\frac{dx}{dt} = P(x,y), \quad \frac{dy}{dt} = Q(x,y).$$

[†] *If, on the other hand, $\sigma=0$, the system of equations (5.73) is conservative, having integral curves*

$$x^k y^h (ac'x + b'cy + cc') = \text{const.}$$

All the region enclosed between the coordinate axes and the straight line $ac'x + b'cy + cc' = 0$ (they are straight integral lines) is entirely filled with closed phase paths surrounding the state of equilibrium (x_0, y_0), which in this case, is a centre.

However, notwithstanding its seeming simplicity, Bendixson's transformation usually leads to a multiple singular point of a high order at the origin of the u, v plane, since each integral curve of the x, y plane at infinity is transformed into an integral curve reaching or leaving the origin of the u, v plane. The investigation of multiple singular points of high order is usually very complicated, and we can only apply Bendixson's method in a very few cases[†].

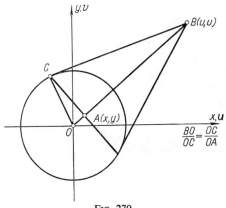

Fig. 270

A more convenient transformation is due to Poincaré [181] which, although more complicated in its conception, leads to simpler calculations. Geometrically it consists of a central projection of the x, y plane on to a sphere of unit radius, which is tangent to the x, y plane at the origin (Fig. 271). A point N on the x, y plane will correspond to only one point N_1 lying on the line $0_1 N$ passing through the centre of the sphere, and so lying on the lower hemisphere. The points at infinite distance on the x, y plane will be transformed into the equator. By definition of our transformation, it is clear that straight lines of the x, y plane are transformed into great circles of the sphere and in particular the straight lines passing through the origin of the coordinates are transformed into great circles perpendicular to the equator. For example, the line PR is transformed into the great circle of the sphere passing through the points $P_1 R_1$. Integral curves of the plane are transformed into corresponding curves of the sphere, and saddle points, nodes and foci retain their character.

---

† Generally speaking, both Bendixson's transformation and Poincaré's transformation (which leads to simpler calculations) are used with advantage only in the cases when $P(x, y)$ and $Q(x, y)$ are linear in x, y.

New singular points, however, appear on the equator. For example, the projections of paths for which $y \to 0$ at an infinite distance from the origin, pass through the points C, D, and those for which $y \to \pm \infty$ pass through the points A, B Thus, singular points on the equator *need not be points of intersection of the curves $P(x, y) = 0$ and $Q(x, y) = 0$*, but are determined by the behaviour of the integral curves at infinity. It follows that

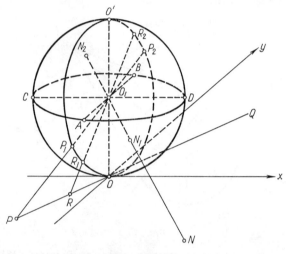

Fig. 271

this transformation is convenient for determining the behaviour of the curves at infinity.

Poincaré used the following formulae of transformation

$$x = \frac{1}{z}, \quad y = \frac{\tau}{z}. \tag{5.75}$$

It is evident that $\tau = y/x$ is the slope of a line joining the origin with the point (x, y). The coordinate lines $z = \text{const.}$ and $\tau = \text{const.}$ transform to the x, y plane as straight lines parallel to the y axis and straight lines passing through the origin respectively. On the sphere they are great circles passing through the diameter AB ($z = \text{const}$) or OO' ($\tau = \text{const}$). It is evident that these circles are not perpendicular to each other. They are, however, perpendicular in a neighbourhood of the equator ($z = 0$) except in small areas in the vicinity of the points A and B. We can construct a plane on which z and τ will serve as rectangular cartesian coordinates. This will be the plane tangent to the sphere at the equator and passing through the point under investigation. The τ axis will be a tangent line lying in the plane of the equator and directed to the side of the positive y axis. The z axis will go vertically downwards. The points of this τ, z plane will be obtained by projecting the points of the sphere from its centre O_1 and it will be clearly convenient to study on this τ, z plane all points at infinity on the x, y plane except the "ends" of the y axis represented on the sphere by the points A and B and their neighbourhood. To study the latter points we need a similar transformation:

$$y = \frac{1}{z}, \quad x = \frac{\tau}{z}. \tag{5.76}$$

Let us now transform the original equation

$$\frac{dx}{dt} = P(x, y), \quad \frac{dy}{dt} = Q(x, y)$$

in terms of Poincaré's coordinates given by $x=1/z$ and $y=\tau/z$. We have

$$dx = -\frac{dz}{z^2}, \quad dy = \frac{z\,d\tau - \tau\,dz}{z^2}$$

and

$$\frac{dz}{dt} = -P\left(\frac{1}{z}, \frac{\tau}{z}\right)z^2, \quad \frac{d\tau}{dt} = -P\left(\frac{1}{z}, \frac{\tau}{z}\right)\tau z + Q\left(\frac{1}{z}, \frac{\tau}{z}\right)z \tag{5.77}$$

or, by eliminating time

$$\frac{dz}{d\tau} = \frac{-z}{\dfrac{Q\left(\dfrac{1}{z}, \dfrac{\tau}{z}\right)}{P\left(\dfrac{1}{z}, \dfrac{\tau}{z}\right)} - \tau}. \tag{5.78}$$

For infinitely remote points, lying close to the "ends" of the y axis, we use the transformation (5.76). In this case the equations (5.77) and (5.78) take respectively the form

$$\left.\begin{aligned}
\frac{dz}{dt} &= -Q\left(\frac{\tau}{z}, \frac{1}{z}\right)z^2, \\
\frac{d\tau}{dt} &= -Q\left(\frac{\tau}{z}, \frac{1}{z}\right)\tau z + P\left(\frac{\tau}{z}, \frac{1}{z}\right)z, \\
\frac{dz}{d\tau} &= \frac{-z}{\dfrac{P\left(\dfrac{\tau}{z}, \dfrac{1}{z}\right)}{Q\left(\dfrac{\tau}{z}, \dfrac{1}{z}\right)} - \tau}.
\end{aligned}\right\} \tag{5.79}$$

It is easily seen from (5.78) that if the identity $Q[(1/z), (\tau/z)] \equiv \tau P[(1/z), (\tau/z)]$ is not obeyed, then the equator, defined by the equation $z=0$, is an integral curve. In the case when $Q=\tau P$, the solution will be $\tau=$const. and so all integral curves intersect the equator at right angles. The singular points lying on the equator are determined by the relations $z=0$, $Q[(1/z), (\tau/z)]/P[(1/z), (\tau/z)]=\tau$. The singular points lying near the "ends" of the y axis must be investigated using the equation (5.79). Such a singular point ($z=0$, $\tau=0$) exists, if the conditions

$$z = 0, \quad \frac{P\left(0, \dfrac{1}{z}\right)}{Q\left(0, \dfrac{1}{z}\right)} = 0 \tag{5.80}$$

are satisfied simultaneously. The investigation of the character and the stability of infinitely remote singular points which are found in this manner is carried out by the usual methods.

EXAMPLE 1. Consider the case of a simple linear oscillator with friction, the differential equation of which has the form

$$\frac{dy}{dt} = -hy - \omega_0^2 x = Q(x,y); \quad \frac{dx}{dt} = y = P(x,y).$$

After carrying out Poincaré's transformation (5.75) we have

$$\frac{dz}{dt} = \tau z, \quad \frac{d\tau}{dt} = -\tau^2 - h\tau - \omega_0^2, \tag{5.81}$$

and the infinitely remote singular points are determined by the relations

$$z = 0, \quad \tau^2 + h\tau + \omega_0^2 = 0,$$

hence

$$\tau = -\frac{h}{2} \pm \sqrt{\frac{h^2}{4} - \omega_0^2}. \tag{5.82}$$

It is easily verified that no singular points lying at the "ends" of the y axis exist. In fact

$$\frac{\dfrac{1}{z}}{-\dfrac{h}{z}} = -\frac{1}{h} \neq 0.$$

Since the equator is an integral curve, two cases are possible: either $h^2/4 < \omega_0^2$ and the equator is a limit cycle which is clearly stable if $h<0$ and unstable if $h>0$; or $h^2/4 > \omega_0^2$ and there are four singular points on the equator situated in pairs diametrically opposite to each other. The slopes of the lines joining pairs are given by the expressions

$$\tau_1 = -\frac{h}{2} + \sqrt{\frac{h^2}{4} - \omega_0^2} < 0;$$

$$\tau_2 = -\frac{h}{2} - \sqrt{\frac{h^2}{4} - \omega_0^2} < 0.$$

To determine the stability of the singular points put

$$z = \xi, \quad \tau = \tau_i + \eta.$$

Substituting these in the equation (5.81) we have

$$\frac{d\xi}{dt} = -(\tau_i+\eta)\xi; \quad \frac{d\eta}{dt} = -(\tau_i+\eta)^2 - h(\tau_i+\eta) - \omega_0^2,$$

or neglecting the terms of higher orders

$$\frac{d\xi}{dt} = -\tau_i\xi; \quad \frac{d\eta}{dt} = -2\tau_i\eta - h\eta. \qquad (5.83)$$

The characteristic exponent is determined by the equation

$$\begin{vmatrix} -\tau_i-\lambda & 0 \\ 0 & -2\tau_i-h-\lambda \end{vmatrix} = 0,$$

hence

$$\lambda_1 = -\tau_i, \quad \lambda_2 = -2\tau_i - h.$$

Thus $\lambda_1 > 0$ and the singular points at infinity are all unstable. For τ_1, $\lambda_2 = -2[(h^2/4) - \omega_0^2]^{\frac{1}{2}}$ and so the point τ_1 is a saddle point, while for τ_2, $\lambda_2 = +2[(h^2/4) - \omega_0^2]^{\frac{1}{2}}$ and point τ_2 is an unstable node. The behaviour of the integral curves at infinity for the case $h^2/4 < \omega_0^2$ is shown in Fig. 272 in an *orthogonal* projection of the sphere on the plane tangent to the sphere at the lower point, whilst Fig. 273 shows the behaviour for the case $h^2/4 > \omega_0^2$.

EXAMPLE 2. An investigation at infinity sometimes answers unequivocally the question of the existence of a limit cycle. Suppose that we know the singular point at infinity to be absolutely unstable. Then, if the only singular point at a finite distance is an unstable node or a focus, there is bound to be at least one stable limit cycle (the rigorous proof of this statement, which is very important for the qualitative investigation of dynamic systems with 1 degree of freedom, will be given in the next chapter).

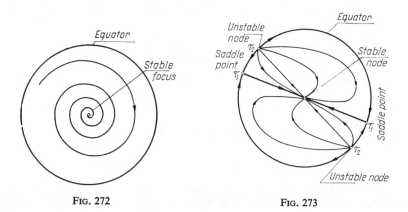

Fig. 272   Fig. 273

We shall show the application of this by proving the existence of a limit cycle for the simplest self-oscillating circuit having a resonant circuit in the grid lead of the valve (see for more details Chapter VII).

Kirchoff's equation for the voltage across the capacitor of the oscillating circuit, as we have seen in Chapter I, Section 6 (see equation (1.64)), has the form

$$CL\frac{d^2u}{dt^2} + [RC - MS(u)]\frac{du}{dt} + u = 0,$$

where $S = S(u)$ is the slope of the characteristic of the valve (we assume that $S \to 0$ for $u \to \infty$). This equation can be reduced to

$$\frac{du}{dt} = y, \quad \frac{dy}{dt} = -\frac{u}{LC} - \frac{1}{LC}[RC - MS(u)]y. \quad (5.84)$$

If we put, according to (5.75), $u = 1/z$ and $y = \tau/z$, we shall obtain

$$\frac{dz}{dt} = -\tau z, \quad \frac{d\tau}{dt} = -\tau^2 - \frac{1}{CL}\left[RC - MS\left(\frac{1}{z}\right)\right]\tau - \frac{1}{CL}. \quad (5.85)$$

This system has the same states of equilibrium as the system (5.81) for $h = R/L$ and $\omega_0^2 = 1/CL$. Hence the behaviour of the phase paths of the initial system at infinity must be the same as in an oscillator with natural frequency $(LC)^{\frac{1}{2}}$ and damping factor R/L. We have investigated such a case just now and we know that the singular points at infinity are stable. Therefore, if the only singular point which is found at a finite distance is unstable, then the equation of the valve generator is bound to have at least one stable limit cycle.

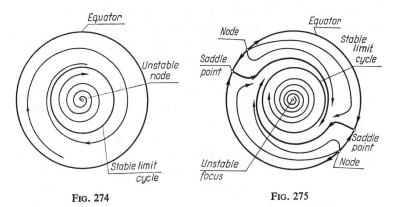

Fig. 274   Fig. 275

A picture of the orthogonal projection of the lower half of Poincaré's sphere on the plane tangent to the sphere at the lower pole is shown in Fig. 274 and Fig. 275 (a semi-stable limit cycle is counted as two) where only the parity of the number of cycles is preserved.

To sum up, we have shown that by investigating the behaviour of the integral curves in distant parts of the plane the equation of the valve generator and other devices can be shown to have at least one limit cycle. Now a question of principle suggests itself: namely: why was a proof necessary, for it is well known that oscillations do occur in valve generators of this type? We have proved no more than that the mathematical model, which corresponds to our idealized generator, sustains a stable periodic process. If it had been proved that our equation had no limit cycle, then this would have meant that some essential fact had been left out of account and that our idealized model was unsuitable. The practical value of any rigorous analysis and, in particular, of the existence of limit cycles, consists in the comparison of the results of the analysis with the experimental data for it enables us to assess the worth of the idealization or mathematical model in use. If agreement is good the model can be used for further analysis or for design; if not it must be changed or modified.

§ 11. Estimating the position of limit cycles[†]

To investigate qualitatively a dynamic system with one degree of freedom, described by the equation

$$\frac{dx}{dt} = P(x, y), \quad \frac{dy}{dt} = Q(x, y), \qquad (5.1)$$

there is no need to find all the phase paths. It suffices only to find the number, nature and the relative position of the singular points and of the limit cycles, as well as the behaviour of the separatrices. The knowledge of these basic paths is sufficient for a qualitative study of a dynamic system of the type (5.1).

The existence of the states of equilibrium and of their character can be found by means of the comparatively simple methods outlined in Sections 2 and 4 of this chapter, but general methods locating limit cycles are, so far, unknown. Apart from a few cases, to be studied later (see Chapters VIII, IX and X), special methods must be devised for each type of problem or recourse must be made to numerical or graphical integration.

Fig. 276

A procedure sometimes effective in proving the existence of and locating limit cycles consists in constructing on the phase plane cycles without contact, on which the velocity vector of the representative point is either directed everywhere outwards from, or everywhere inwards to, the region bounded by this curve.

If the velocity vector of the representative point on a closed curve is directed always to one side of the curve, except a certain number of points where it is tangent, then, clearly, we are dealing at these points with a contact of an even order (Fig. 276). From our point of view this curve is not different in principle from a cycle without contact, and therefore we shall consider such closed curves with isolated points of contact of even order as included in the class of the cycles without contact.

If, inside such a cycle without contact there are no stable singular points and the velocity vector on it is directed everywhere inwards, then there exists at least one stable limit cycle, within the cycle without contact. Similarly if the point at infinity is unstable and there exists a cycle without contact, on which the velocity vector is directed everywhere outwards and outside which there are no stable singular points, then there exists at least one stable limit cycle lying outside the cycle without contact. Similar considera-

[†] Written by N. A. Zheleztsov.

tions can be adduced for investigating unstable limit cycles but, in this case, an opposite direction of the velocity vector is necessary[†].

Suppose now that by means of two cycles without contact we succeed in isolating on the phase plane an annular (doubly connected) region G, not containing within it states of equilibrium. Then if the velocity vector of the representative point on these cycles without contact is never directed outside G, then in this annular region there exists at least one stable limit cycle. If the velocity vector on both cycles without contact is never directed into G, then there exists at least one unstable limit cycle within this annular region. If, finally, the velocity vector of the representative point is directed on one cycle without contact everywhere outside, and on the other everywhere inside the annular region G bounded by them, then in this region there are either no limit cycles or an even number of them, one half of these being unstable[‡].

In certain cases it is possible to prove the uniqueness (or the absence) of a limit cycle in a given annular region, by using *Dulac's criterion for an annular region* [148]: the dynamic system (5.1) cannot have more than one closed phase path (or more than one closed contour made up of paths) in an annular region (G), if in this region the expression

$$\frac{\partial}{\partial x}(BP) + \frac{\partial}{\partial y}(BQ)$$

has a constant sign. $B(x, y)$ is a certain function, continuous and having continuous first derivatives in the region (G). It is evident that in the region (G) there cannot be closed phase paths which can be reduced by a continuous deformation to a point without leaving the region (G), as follows from the criterion due to Dulac.

To prove our criterion, assume that the system (5.1) has two closed phase paths in (G) $abca$ and $a_1 b_1 c_1 a_1$ (Fig. 277). Then, for the closed contour $abcaa_1 c_1 b_1 a_1 a$ $\oint B(P\,dy - Q\,dx) = 0$. However, according to Green's theorem $\oint B(P\,dy - Q\,dx) = \iint_{(\Sigma)} [\partial(BP)/\partial x + \partial(BQ)/\partial y]\,dx\,dy$, where the integration is carried out over the region (Σ) between the closed curves $abca$ $a_1 b_1 c_1 a_1$. Thus the integral $\iint_{(\Sigma)} [\partial(BP)/\partial x + \partial(BQ)/\partial y]\,dx\,dy$ must be equal to zero, which contradicts the fact that the integrand has a constant sign in the region (Σ) which is a part of the region (G). In the region (G) therefore

---

[†] These statements, and also the ones made below, are fairly obvious geometrically. Their rigorous proof is based on the general theory of the behaviour of phase paths.

[‡] The proof of these statements is given in Chapter VI, Section 2 (see Theorem V).

there cannot be more than one closed phase path of the system (5.1). It is evident that the criterion retains its validity, when the expression $[\partial(BP)/\partial x]+[\partial(BQ)/\partial y]$ is of the same sign everywhere in the region (G) except at certain points or curves where it may vanish.

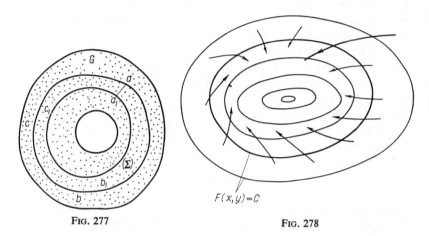

Fig. 277   Fig. 278

In certain problems it is possible to find cycles without contact among the curves belonging to a given family of simple concentric ovals. Let

$$F(x, y) = C \tag{5.86}$$

represent such a family, filling the phase plane.

We shall assume that to each curve of the family (5.86) (Poincaré called this family a topographical system of curves) there corresponds a unique value of C and that a curve with a given C contains in itself all the curves with smaller C). The representative point in its motion along a closed phase path will intersect curves (5.86). For such a motion, clearly,

$$\frac{dC}{dt} = F'_x(x, y)P(x, y) + F'_y(x, y)Q(x, y) = \Phi(x, y),$$

and all the curves of the topographical system on which the function $\Phi(x, y)$ retains the same sign are cycles without contact. Thus, if on a certain curve of the topographic system $\Phi(x, y) \leq 0$, then this curve is a cycle without contact, and all phase paths, intersecting this curve, move into the region inside it (Fig. 278). Similarly, if on a certain curve of the family (5.86) $\Phi(x, y) \geq 0$, then the phase paths intersect this curve, going into the region outside it. It is also clear that, in annular regions, in which the function $\Phi(x, y)$ is of constant sign, there cannot be limit cycles (or closed phase paths). Limit cycles can only exist in the annular regions in which the function $\Phi(x, y)$ is of variable sign.

A somewhat different version of the same method for finding curves without contact is the so-called method of the contact curve due to Poincaré [181, 108]. Poincaré's *contact curve* is a curve at the points of which the phase paths of the system (5.1) are tangent to the curves of a given topographical system (5.86). Its equation is clearly

$$\frac{P}{Q} = -\frac{F'_y}{F'_x}.$$

If the topographical system is so chosen that the curve of contacts is closed, then we can draw the largest and the smallest curve of the topographical system which are tangent to the contact curve. Then all the curves of the topographical system lying outside this largest curve and those inside the smallest curve, are cycles without contact, and the limit cycles, if such exist, are situated in the annular region bounded by these two curves of the topographical system and which contains the contact curve.

EXAMPLE 1. To illustrate what has been stated above, we shall consider two examples of qualitative investigation of dynamic systems. As a first example, consider the equations

$$\frac{du}{dt} = y, \quad \frac{dy}{dt} = -\frac{u}{LC} - \frac{1}{LC}[RC - MS(u)]y, \quad (5.84)$$

describing the oscillations of a valve generator. The slope of the valve characteristic $S(u)$ is an even function monotonically decreasing as $|u|$ increases. We shall take for the topographical system the family of ellipses $CLy^2 + u^2 = A^2$, then

$$\frac{1}{2}\frac{d(A^2)}{dt} = [MS(u) - RC]y^2 = \Phi(u, y).$$

If $MS(0) < RC$ then, over the whole phase plane $\Phi(u, y) \leq 0$ and, hence, $d(A^2)/dt \leq 0$, i.e. all phase paths approach the origin which is the stable state of equilibrium. If, however, $MS(0) > RC$, then the single state of equilibrium $(0, 0)$ is unstable and, moreover, there exists such a segment $|u| \leq |u_0|$ on which $MS(u) - RC \geq 0$. All ellipses with $A \leq u_0$ are therefore cycles without contact, since on them $\Phi(u, y) \geq 0$, and the phase paths intersect them in the region outside the ellipse $LCy^2 + u^2 = u_0^2$. Hence, in the region inside this ellipse there are no limit cycles[†]. At least one stable limit cycle lies outside the ellipse $LCy^2+u^2 = u_0^2$, for there are no states of equilibrium, and infinity is an unstable point (since $S(u) \to 0$ for $u \to \infty$).

EXAMPLE 2. As a second example, we shall give a complete qualitative investigation of the dynamic system [19]

$$\begin{aligned} \frac{dx}{dt} &= ax+by-x(x^2+y^2), \\ \frac{dy}{dt} &= cx+dy-y(x^2+y^2), \end{aligned} \quad (5.87)$$

which arises in the problem of the synchronization of a valve oscillator, when Van der Pol's method is used [190, 7].

First of all the phase portrait is symmetrical with respect to the origin, since the equations (5.87) are invariant for a change of the variable x, y into $-x$, $-y$; secondly the equation of the integral curves

$$\frac{dy}{dx} = \frac{cx+dy-y(x^2+y^2)}{ax+by-x(x^2+y^2)}$$

has the straight lines $y=k_1 x$ and $y=k_2 x_1$ as its integral curves, k_1 and k_2 being the roots of the quadratic equation $bk^2+(a-d)k-c = 0$,

$$k_{1,2} = \frac{d-a \pm \sqrt{(a-d)^2+4bc}}{2b}$$

(on condition, of course, that the discriminant of the equation $\delta = (a-d)^2 + 4bc > 0$). It also follows from the equations (5.87) that the point at infinity is absolutely unstable, i.e. that in distant parts of the phase plane the representative point moves towards the origin.

The singular points satisfy the equations

$$ax+by-x(x^2+y^2) = 0, \quad cx+dy-y(x^2+y^2) = 0.$$

The roots of this system of equations are $x=0$, $y=0$ and

$$x_{1,2} = \pm \sqrt{\frac{a+bk_{1,2}}{1+k_{1,2}^2}}, \quad y_{1,2} = \pm k_{1,2} \sqrt{\frac{a+bk_{1,2}}{1+k_{1,2}^2}}.$$

Thus in a finite part of the phase plane there can be, depending on the parameters of the equations (5.87) either one or three or five states of

---

[†] The absence of limit cycles inside the ellipse $LCy^2+u^2 = u_0^2$ for $MS(0) > RC$ and over the whole phase plane for $MS(0) < RC$ also follows, as is easily seen, from Bendixson's criterion.

equilibrium. The state of equilibrium $(0,0)$ always exists, and its character is determined by the coefficients

$$\sigma = -(a+d) \quad \text{and} \quad \Delta = ad-bc$$

of the characteristic equation $\lambda^2 + \sigma\lambda + \Delta = 0$ (the discriminant is the expression δ introduced previously). Other states of equilibrium, if they exist, lie on the integral straight lines $y = k_1 x$ and $y = k_2 x$ and therefore can only be nodes or saddle points[†].

The following cases are clearly possible. Case I. $\delta > 0$, $\Delta > 0$, $\sigma < 0$. In this case (Fig. 279, I) there are five singular points (states of equilibrium): the unstable node $(0,0)$ and two saddle points and two stable nodes. There are no limit cycles, since through all singular points pass straight integral curves extending to infinity. This follows from the fact that the point at infinity is absolutely unstable and, hence, the sum of Poincaré's indices for all singular points is equal to $+1$. Therefore, of the four points outside the origin, two of them are saddle points and two are stable nodes.

Case II. $\delta > 0$, $\Delta < 0$. Now (Fig. 279, II) the origin is a saddle point and outside the origin there are two stable nodes. As before there are no limit cycles.

Case III. $\delta > 0$, $\Delta > 0$, $\sigma > 0$. On the phase plane (Fig. 279, III) there is a single state of equilibrium, the stable node $(0,0)$, which all phase paths approach asymptotically. Two integral straight lines $y = k_1 x$ and $y = k_2 x$ pass through the node and therefore there are no limit cycles.

Case IV. $\delta < 0$, $\sigma > 0$. The only state of equilibrium is a stable focus $(0,0)$. As will be proved below, there are no limit cycles, and therefore (Fig. 279, IV) all paths approach the origin.

Case V. $\delta < 0$, $\sigma < 0$. In this case the origin is the only state of equilibrium, and is an unstable focus. Since the point at infinity is unstable, there is at least one stable limit cycle. It follows from Dulac's criterion that for $\delta < 0$ there cannot be more than one limit cycle whereas in Case IV there would be an even number of them, if they existed. To prove this take, for

---

† The numerator $a+bk_{1,2}$ under the root sign in the expression for the coordinates of the singular points is a root of the characteristic equation $\lambda^2 + \sigma\lambda + \Delta = 0$. Therefore there are *no* singular points outside the origin if the point $(0, 0)$ is a focus or a stable node; or there are two singular points if $(0, 0)$ is a saddle point, and if the point $(0, 0)$ is an unstable node, then outside the origin there are four singular points.

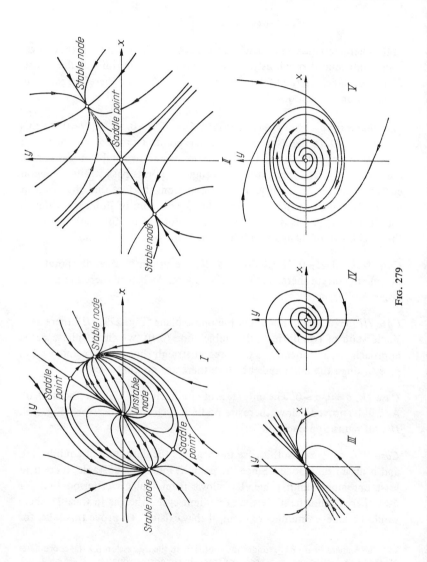

Fig. 279

the multiplier B, the function

$$B(x, y) = \frac{1}{by^2 - cx^2 + (a-d)xy},$$

which, since $\delta = (a-d)^2 + 4bc < 0$, is continuous with its derivatives everywhere except at the origin and, hence is of constant sign. Then, if we denote by $P(x, y)$ and $Q(x, y)$ the right-hand sides of the equations (5.87) we have

$$\frac{\partial}{\partial x}(BP) + \frac{\partial}{\partial y}(BQ) = -2(x^2 + y^2)B(x, y).$$

This expression does not change its sign in the annular region which is obtained by excluding an arbitrarily small neighbourhood around the origin. Thus, there is one limit cycle and the phase portrait is shown in Fig. 279, V.

To determine the boundaries within which the limit cycle is situated in the case V, take for the topographical system the family of circles

$$x^2 + y^2 = R^2.$$

As we have already seen,

$$\frac{1}{2}\frac{d(R^2)}{dt} = ax^2 + (b+c)xy + dy^2 - (x^2 + y^2)^2$$

or, in polar coordinates

$$\frac{1}{2}\frac{d(R^2)}{dt} = \frac{R^2}{2}[a+d+(a-d)\cos 2\varphi + (b+c)\sin 2\varphi] - R^4.$$

It is easily seen that

$$R_1^2 \leqslant a+d+(a-d)\cos 2\varphi + (b+c)\sin 2\varphi \leqslant R_2^2,$$

where

$$R_1^2 = \frac{a+d-\sqrt{(a-d)^2+(b+c)^2}}{2}$$

and

$$R_2^2 = \frac{a+d+\sqrt{(a-d)^2+(b+c)^2}}{2}.$$

Therefore, for $(a+d)^2 > (a-d)^2 + (b+c)^2$ (i.e. $4ad > (b+c)^2$), all circles with radii $R \leqslant R_1$, and $R \geqslant R_2$ are cycles without contact, since for $R \leqslant R_1$, $dR/dt \leqslant 0$ and for $R \geqslant R_2$, $dR/dt \leqslant 0$. A single limit cycle is situated between the circles of radii R_1 and R_2. If, however, $4ad < (b+c)^2$, then the quantity R_1 is imaginary and the cycles without contact will only be the circles with radii $R \geqslant R_2$ (on them, as before, $dR/dt \leqslant 0$) and the limit cycle lies inside the circle of radius R_2.[†]

Thus, depending on the parameters of the equations (5.87) we shall have one of the cases discussed above. The regions of existence of each of them are shown on the plane of the parameters σ and Δ in Fig. 280.

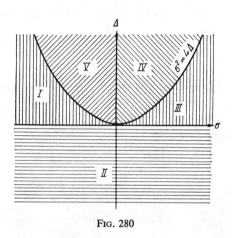

Fig. 280

It is sometimes possible to prove the presence or absence of limit cycles for differential equations of the type (5.1) using considerations specific to the system. Such an analysis, which offers great physical interest, has been given by Liénard [174] for the equation of an electronic oscillator under certain simplifying assumptions about the symmetry of the characteristic.

§ 12. Approximate methods of integration

Often the non-linear differential equations in their general form cannot be directly integrated nor can the phase portrait be constructed in any rigorous manner, and then the simplest available method (sometimes the only one) is that of approximate graphical integration. A major disadvantage of numerical integration is that the results are restricted in generality and make the survey of a problem in its entirety very difficult. However, in cases when the method of approximate graphical integration proves the

---

[†] In the latter case, cycles without contact, outside which there lies the limit cycle, can be found among the ellipses

$$by^2 + (a-d)xy - cx^2 = \text{const.}$$

only possible one, then the best graphical method would appear to be the method of isoclines[†]. The equation of the paths, after eliminating time, is

$$\frac{dy}{dx} = \frac{Q(x, y)}{P(x, y)} = f(x, y), \qquad (5.3)$$

The curves $f(x, y) = C$ are the isoclines where the integral curve phase paths have the slope C. We can construct on the phase plane a family of isoclines and, providing there is a sufficiently dense field of isoclines can

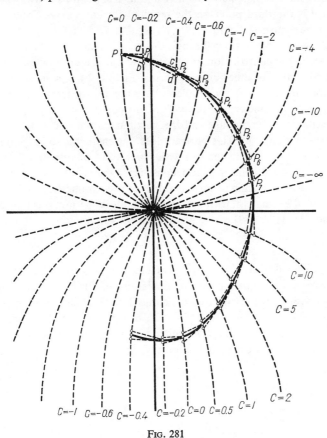

Fig. 281

[†] We are restricting ourselves to a very brief outline of the isocline method since this method is fairly widely employed and its description can be easily found in the literature. See, for example, [110].

construct an approximate phase portrait. Suppose an integral curve passes through an arbitrary point P on the isocline $C=0$. Now draw through P two segments, one in the direction of the tangent corresponding to the isocline $C=0$ and the other in the direction of the tangent corresponding to the adjoining isocline $C=0.2$. Prolong the segments until they meet the latter isocline at the points a and b and take the point P_1 lying half-way between a and b as the next point of our integral curve. From P_1 draw two straight lines at slopes corresponding to the isoclines $C = 0.2$ and $C = 0.4$, intersecting the latter at c and d. The point P_2 halfway between c and d is the third point on the required integral curve. Continue and obtain the sequence of points $P, P_1, P_2, P_3, P_4, \ldots$, which is, approximately, the integral curve passing through the point P. In a similar manner we can extend the construction of this integral curve and draw on the phase plane a number of other integral curves. Finally, by repetition we obtain an approximate but fairly detailed phase portrait in terms of numerical values of the parameters. On the basis of this portrait we can judge whether self-oscillations are possible and the maximum values of x and y during these oscillations, etc. This portrait does not enable us to judge how the behaviour of the system changes with a variation of one or other of its parameters. To do this we must construct a whole "gallery" of phase portraits corresponding to various values of the parameter.

A typical example, illustrating this method, is the investigation due to Van der Pol [188, 189] of the phase plane of the equation

$$\ddot{v} - \varepsilon(1-v^2)\dot{v} + v = 0.$$

This equation is typical of the models of a series of self-oscillatory problems. For example, the equation for the oscillations of a valve generator with a cubic valve characteristic can be reduced to Van der Pol's equation. Van der Pol himself used this equation in the theory of the oscillations in a symmetric multivibrator, in which there is an inductance.

On writing the equation in the form

$$\frac{dv}{dt} = y, \quad \frac{dy}{dt} = -v + \varepsilon(1-v^2)y$$

we obtain finally the equation of the integral curves

$$\frac{dy}{dv} = \varepsilon(1-v^2) - \frac{v}{y}.$$

By giving positive numerical values to the parameter ε and using the method of the isoclines, Van der Pol obtains the "gallery of phase portraits" shown in Fig. 282 (*a*, *b* and *c* correspond respectively to the case of small intermediate and large values of ε). The state of equilibrium $(0, 0)$ is always unstable for $\varepsilon > 0$ (for $0 < \varepsilon < 2$, it is an unstable focus and for $\varepsilon > 2$ an

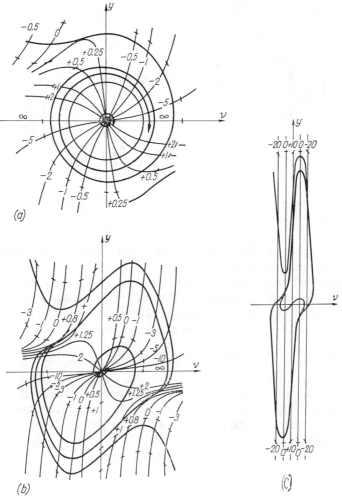

Fig. 282

unstable node). For all $\varepsilon > 0$ there is a single limit cycle and hence, self-oscillations, the mode of excitation being "soft" for any initial conditions. However, the range and the form of these self-oscillations varies with ε.

For small positive ε, the limit cycle is close to a circle (the self-oscillations are nearly sinusoidal) (Fig. 282(a)). As ε increases, the form of the self-oscillations differs from sinusoidal (the limit cycle differs from a circle) (Fig. 282(b) and (c)) until finally for $\varepsilon > 2$ the initial growth of the oscillations becomes aperiodic[†].

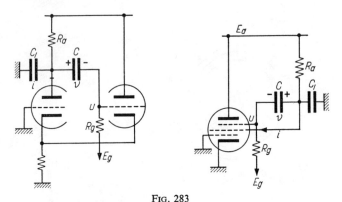

Fig. 283

As a second example we shall construct, by the method of the isoclines, the phase portraits of a valve generator with a two-section RC circuit. The circuit diagrams of two versions of such a generator (one with a twin triode, and one with a pentode under transitron conditions of operation) are shown in Fig. 283. The equations of both circuits, under our usual assumptions and with the notation of Fig. 283, are

$$C\frac{dv}{dt} = \frac{u-E_g}{R_g}; \quad \frac{E_a-(u+v)}{R_a} = i(u) + C\frac{dv}{dt} + C_1\frac{d(u+v)}{dt}, \quad (5.88)$$

where $i = i(u)$ is the characteristic of the twin triode (or of the valve under transitron conditions of operation). In order that these circuits may work as self-oscillatory systems it is essential that this characteristic (Fig. 284) has a section with a negative slope.

We have, clearly, for the only state of equilibrium

$$u = E_g,$$
$$v = E_a - R_a i(E_g) - E_g.$$

[†] We must emphasise that, generally speaking, the form of the self-oscillations is not connected with the character of the singular point situated inside the limit cycle. Therefore, the connexion observed in the case of Van der Pol's equation must not be generalized to all other self-oscillating systems (for example, to a valve generator with a different valve characteristic).

To simplify these introduce the new variables x and y, proportional respectively to the variable components of the grid voltage and of the voltage across the capacitor C,

$$u = E_g + u_0 x, \quad v = E_a - R_a i(E_g) - E_g + \alpha u_0 y,$$

and the dimensionless time $t_{\text{new}} = T t_{\text{old}}$. Let the reduced dimensionless characteristic $\varphi(x)$ be

$$\varphi(x) = \frac{1}{u_0 S}[i(E_g + u_0 x) - i(E_g)],$$

where u_0 and T are certain units of voltage and time, and S is the absolute value of the slope of the characteristic at the "working point", corres-

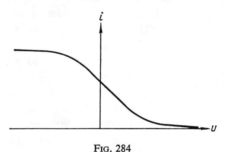

Fig. 284

ponding to the state of equilibrium ($S = |di/du|$ for $u = E_g$). Then the equations (5.88) assume the form

$$\frac{CR_g}{T} \dot{y} = x; \quad \frac{C_1 R_a}{T} \dot{x} = R_a S \varphi(x) - \left[1 + \frac{R_a}{R_g}\left(1 + \frac{C_1}{C}\right)\right] x - \alpha y$$

(the dot denotes differentiation with respect to the new dimensionless time). On choosing

$$\alpha = 1 + \frac{R_a}{R_g}\left(1 + \frac{C_1}{C}\right) \quad \text{and} \quad T = CR_g + (C + C_1)R_a,$$

we shall reduce the equations (5.88) to

$$\dot{y} = x; \quad \mu \dot{x} = -y - x - K\varphi(x) \tag{5.89}$$

with the two dimensionless parameters

$$\mu = \frac{R_a C_1}{R_g C} \frac{1}{\left[1 + \frac{R_a}{R_g}\left(1 + \frac{C_1}{C}\right)\right]^2}, \quad K = \frac{S R_a}{1 + \frac{R_a}{R_g}\left(1 + \frac{C_1}{C}\right)} \tag{5.90}$$

The amplification coefficient K is greater than or equal to zero, and the positive parameter μ, which usually coincides in order of magnitude with C_1/C, does not exceed $\dfrac{1}{4}$.

The only state of equilibrium is at the origin and its characteristic equation is easily seen to be

$$\mu\lambda^2 + (1-K)\lambda + 1 = 0, \qquad (5.91)$$

for $\varphi'(0) = -1$. Therefore, this state of equilibrium is stable for $K < 1$ and unstable for $K > 1$. It is a focus for $(K-1)^2 < 4\mu$ and a node for $(K-1)^2 > 4\mu$. The (μ, K) stability chart is shown in Fig. 285.

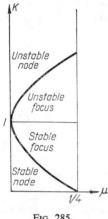

Fig. 285

The point at infinity is always unstable. In fact, for large absolute values of the voltage u we fall on to the horizontal sections of the characteristic, where i or $\varphi(x)$ are constant. Therefore the circuit behaves in distant regions as a linear one with a stable node and so all phase paths come from infinity to the region of finite x and y‡. And if at the origin there is an unstable state of equilibrium (for $K > 1$), then at least one stable limit cycle exists on the phase plane. In the case of a characteristic $\varphi(x)$ with a slope monotonically decreasing in absolute value as we move away from the "working point", this limit cycle will be the only one‡. To find this limit cycle, we use the method of the isoclines.†† From (5.89) the equation of the integral curves is

$$\frac{dy}{dx} = -\frac{\mu x}{y + x + K\varphi(x)}; \qquad (5.92)$$

and the equation of an isocline for an integral curve with slope $\varkappa$ will be

$$y = -\left(1 + \frac{\lambda}{\varkappa}\right) x - K\varphi(x). \qquad (5.93)$$

† Strictly speaking, the point at infinity is unstable in all directions, except perhaps parallel to the y axis. It is easily verified, however, that this direction cannot be stable for the equator of Poincaré's sphere does not contain any singular points.

‡ For $K < 1$, when the origin is a stable state of equilibrium, there are no limit cycles (for example, from Bendixson's criterion) and all phase paths approach the origin asymptotically, as $t \to +\infty$.

†† Later on, in Section 5, Chapter VIII, we shall find the limit cycle for the case of a piece-wise-linear characteristic by the method of the point transformation.

In particular the isocline for $\varkappa = 0$ is the y axis and the isocline for $\varkappa = dy/dx = \infty$ is the curve

$$y = -x - K\varphi(x). \tag{5.93a}$$

The limit cycles and also certain other phase paths constructed by this method are shown in Figs. 286–289. The characteristic used for these plots is

$$\varphi(x) = \begin{cases} +\dfrac{2}{3} & \text{for} \quad x \leqslant -1, \\ -x + \dfrac{x^3}{3} & \text{for} \quad |x| \leqslant 1, \\ -\dfrac{2}{3} & \text{for} \quad x \geqslant +1 \end{cases} \tag{5.94}$$

In the regions $x > +1$ and $x < -1$ the equations (5.89) are linear and have rectilinear phase paths $y = \varkappa^* x + \left(\dfrac{2}{3}\right) K$ (for $x > +1$) and $y = \varkappa^* x - \left(\dfrac{2}{3}\right) K$ (for x -1) where $\varkappa^*$ are the roots of the equation $\varkappa^2 + \varkappa + \mu = 0$. We can thus construct a curve without contact, containing inside itself the limit cycle. This curve without contact (*ABCDEFA* in Figs. 286–288) is made of the phase paths *ABC* and *DEF* and the vertical straight segments *CD* and *FA*.

For $0 < K - 1 \ll \mu^{\frac{1}{2}}$ the oscillations in the circuits are nearly sinusoidal since the limit cycle in Fig. 286 ($\mu = 0\cdot 2$ and $K = 1\cdot 2$) is almost an ellipse. As the "drive" of the circuit increases so that the inequality $K - 1 \ll \mu^{\frac{1}{2}}$ is no longer true the form of the limit cycle changes (Figs. 287 and 288) and the self-oscillations differ more and more from sinusoidal ones[†]. For $\mu \ll 1$ and $\mu \ll K - 1$ the self-oscillations approach the form of discontinuous oscillations (Fig. 289), since the phase velocity of the motion of the representative point outside the curve (5.93a) becomes very large (it tends to infinity for $\mu \to 0$), as follows from the second equation (5.89). Then we obtain the phase portrait of a multivibrator with one RC circuit

---

† This is just why additional elements (thermistors, limit diodes) must necessarily be introduced in RC generators of pure sinusoidal oscillations to limit the amplitude of the self-oscillations and ensure that the inequality $0 < K - 1 \ll \mu^{\frac{1}{2}}$ is always satisfied.

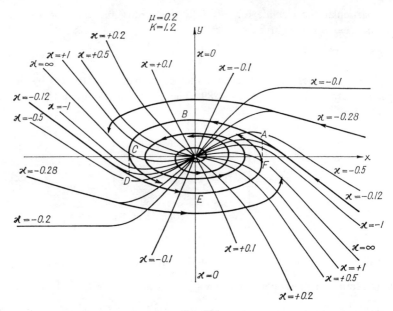

Fig. 286

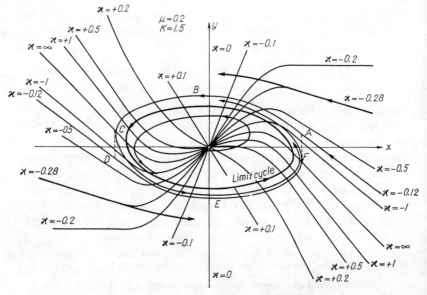

Fig. 287

12] APPROXIMATE METHODS OF INTEGRATION

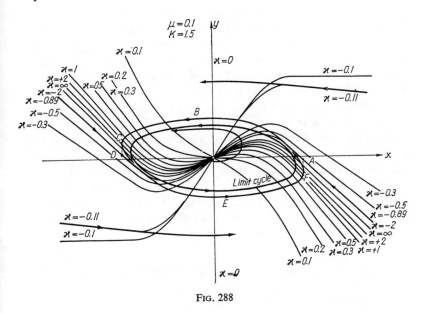

Fig. 288

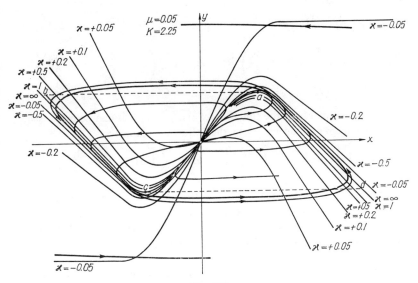

Fig. 289

and a small parasitic capacitance $C_1(C_1 \ll C)$. For small values of μ (i.e. for $C_1 \ll C$) the limit cycle is situated in a small neighbourhood of the curve *abcda* which consists of arcs of the curve (5.93a) and horizontal straight segments, and is the *limit position* (for $\mu \to 0$) of the limit cycle. This will be met again in the analysis of various systems with discontinuous oscillations (see Chapter X).

Sometimes, if we know approximately the period τ of an oscillation, then it suffices to know the values of the coordinate and velocity, separated by intervals of time, of say $\tau/10$, in order to obtain the motion in this process. Such problems — the calculation of the values of functions, determined by given differential equations and given initial conditions — can be solved by means of one of the methods of approximate numerical integration, for example, by Adam's or Runge's methods. The latter method is the simplest and possibly the most suitable and we shall give therefore a brief outline of its application to a problem. Suppose we have[†]

$$\frac{dx}{dt} = P(x, y), \quad \frac{dy}{dt} = Q(x, y), \tag{5.1}$$

and $x = x_0$ and $y = y_0$ at $t = t_0$. We need to find new values of x and y after a small interval of time Δt. To do this we write down the expressions

$$\Delta x_1 = P(x_0, y_0)\Delta t, \qquad \Delta y_1 = Q(x_0, y_0)\Delta t,$$

$$\Delta x_2 = P\left(x_0 + \frac{\Delta x_1}{2}, \ y_0 + \frac{\Delta y_1}{2}\right)\Delta t; \quad \Delta y_2 = Q\left(x_0 + \frac{\Delta x_1}{2}, \ y_0 + \frac{\Delta y_1}{2}\right)\Delta t,$$

$$\Delta x_3 = P\left(x_0 + \frac{\Delta x_2}{2}, \ y_0 + \frac{\Delta y_2}{2}\right)\Delta t; \quad \Delta y_3 = Q\left(x_0 + \frac{\Delta x_2}{2}, \ y_0 + \frac{\Delta y_2}{2}\right)\Delta t,$$

$$\Delta x_4 = P(x_0 + \Delta x_3, \ y_0 + \Delta y_3)\Delta t; \quad \Delta y_4 = Q(x_0 + \Delta x_3, \ y_0 + \Delta y_3)\Delta t.$$

Then the increments of the functions x and y at $t_0 + \Delta t$ can be expressed, to a high degree of accuracy, in the following manner:

$$\Delta x = \frac{1}{3}\left[\Delta x_2 + \Delta x_3 + \frac{\Delta x_1 + \Delta x_4}{2}\right]; \quad \Delta y = \frac{1}{3}\left[\Delta y_2 - \Delta y_3 + \frac{\Delta y_1 + \Delta y_4}{2}\right].$$

Taking $x_1 = x_0 + \Delta x$ and $y_1 = y_0 + \Delta y$ as the new initial values, we can calculate the values of x and y for the instant of time $t_0 + 2\Delta t$ and so on. If the functions P and Q are complicated, this evaluation is very laborious and Adam's method may prove more convenient.

[†] More details on Runge's method and also an exposition of other methods o numerical integration can be found in [110, 76].

CHAPTER VI

FUNDAMENTALS OF THE QUALITATIVE THEORY OF DIFFERENTIAL EQUATIONS OF THE SECOND ORDER[†]

§ 1. Introduction

This chapter has a purely mathematical character. Its object is to make more precise certain concepts used in the previous chapter, and to prove the assumptions that are at the basis of the analysis of the examples of dynamic systems of the second order.

In order that the qualitative theory of differential equations may be used with full confidence in the investigation of non-linear oscillations, we need to be acquainted not only with the results of theory, but also, to a certain extent, with the methods by means of which these results are obtained. In this chapter there are results concerning the general theory of the behaviour of the paths on the phase plane and also some of the proofs, but the reader who wishes to go further should consult some of the classic papers by Poincaré, Bendixson, Liapunov and others given in the references, and some modern text on ordinary differential equations. There are suitable texts by Lefschetz (39), Coddington and Levinson (11), and Bellman (2) amongst others.

We will consider, just as in Chapter V, a system of differential equations of the second order

$$\frac{dx}{dt} = P(x, y), \quad \frac{dy}{dt} = Q(x, y) \tag{6.1}$$

with functions $P(x, y)$ and $Q(x, y)$ *analytic* over the whole x, y phase plane, and having no common factor so that they cannot be written as

$$\left. \begin{array}{l} P(x, y) = P^*(x, y) f(x, y), \\ Q(x, y) = Q^*(x, y) f(x, y). \end{array} \right\} \tag{6.2}$$

[†] Sections 1, 3 and 4 of this chapter have been written by E. A. Leontovich–Andronova.

Under this assumption the curves

$$P(x, y) = 0 \quad \text{and} \quad Q(x, y) = 0$$

can only have a finite number of points of intersection on each finite part of the plane and, hence, the dynamic system (6.1) can only have a *finite number of equilibrium states.*

The first question that naturally arises concerns the types of phase paths possible in dynamic systems of the second order. The paths encountered in earlier examples (see Chapters II, III and V) were either equilibrium states or closed paths or paths approaching equilibrium states or closed paths for $t \to +\infty$ (or for $t \to -\infty$). It was shown by Bendixson that using two general theorems — Cauchy's theorem on the existence and uniqueness of the solution of a system of differential equations and the theorem on the continuity of the dependence of this solution upon the initial conditions (see Appendix I) that exhaustive information can be obtained on the possible character of a separate path [137, 81]. The following section (Section 2) will be devoted to this analysis.

From the examples of previous chapters, it might be expected that to draw a qualitative phase portrait we need know only a finite number of the so-called "singular" paths, such as states of equilibrium, closed paths and separatrices. Section 3 of this chapter is devoted to this topic. A rigorous definition of "singular" and "non-singular" paths is given and it is shown that singular paths divide the whole set of paths into separate regions or cells filled with non-singular paths that have the same behaviour [17, 80, 145].

Sections 4 and 5 of this chapter are devoted to another series of questions. In Section 4 certain general requirements that the system (6.1) must satisfy to correspond to a real physical problem are formulated. Thus, the qualitative phase portrait of such a system must remain unvaried for all sufficiently small variations of the right-hand sides. The systems possessing these properties are called "coarse" or structurally stable. We give in Section 4 a rigorous mathematical definition of coarseness or structural stability, establish necessary and sufficient conditions for a system to be coarse, and the types of "singular" paths and types of cells filled by ordinary paths, which are possible in such a system [17].

In § 5 we consider the dependence of the qualitative phase portrait upon a parameter occurring in the right-hand sides of the system (6.1). With a certain hypothesis of a "general" character it can be assumed for all values of the parameter, except the branch values (see Chapter II, Section 5), that the system is coarse. As the parameter passes through a branch value a transition takes place from one coarse system into another, together with a variation of the qualitative structure. In Section 5 we study this variation of the qualitative structure and, in particular, how limit cycles appear or disappear [10–13].

§ 2. General Theory of the Behaviour of Paths on the Phase Plane. Limit Paths and Their Classification

1. Limit points of half-paths and paths

We shall introduce first of all certain elementary concepts which will be used below.

Let
$$\begin{rcases} x = \varphi(t-t_0); \quad x_0, y_0) = x(t), \\ y = \psi(t-t_0); \quad x_0, y_0) = y(t) \end{rcases} \quad (6.3)$$

be a solution of the system (6.1) and let L be the path corresponding to this solution. The part of the path whose points correspond to $t \geqslant t_0$ will be referred to as the *positive half-path*† and will be denoted by L^+ or $L_{M_0}^+$, where M_0 is the point corresponding to the value $t=t_0$. Similarly the part of the path whose points correspond to $t \leqslant t_0$ will be referred to as the *negative half-path* and will be denoted by L^- or $L_{M_0}^-$.

If for all values of $t \geqslant t_0$ (or $t \leqslant t_0$), for which the solution of (6.1) is defined, the representative point $M[x(t), y(t)]$ remains in a certain limited region of the plane, then all possible values of $t \geqslant t_0$ ($t \leqslant t_0$) correspond to the points of the half-path $L_{M_0}^+$ ($L_{M_0}^-$) there. If the representative point $M[x(t), y(t)]$ remains in a certain limited region of the plane both for $t \geqslant t_0$ and for $t \leqslant t_0$ then, evidently, the solution is defined for all t within $-\infty < t < +\infty$.

Below, we only consider *half-paths and paths that lie entirely in a certain limited region of the plane*. Sometimes, when *all* points of a path are being considered, we shall call it an *entire* or *full* path.

Most important concepts for the sequel are that of a *limit point* of a half-path and that of a *limit path*. A point M^* is called a limit point of L^+ (or L^-) if, for every $\varepsilon > 0$ and any $T > t_0$ (any $T < t_0$), there exists in the ε-neighbourhood of M^* a point of L^+ (L^-) that corresponds to a value of $t > T$ ($t < T$).‡

---

† A half-path is sometimes called a semi-orbit. Ed.
‡ We repeatedly consider points situated at a distance smaller than a certain assigned ε from a given point or from a given path, or, generally, from a given points set K. The totality of all points that are situated at a distance smaller than ε from the points of a given set K will be referred to as the ε-neighbourhood of this set.

From the above definition of limit point† of a half-path it follows at once that, if ξ^*, η^* are the coordinates of a limit point M^* of L^+, then there is a sequence of values of t

$$t_1, \quad t_2, \ldots, t_n, \ldots \qquad (t_n \to +\infty \quad \text{for} \quad n \to +\infty)$$

such that
$$\lim_{n \to +\infty} x(t_n) = \xi^* \quad \text{and} \quad \lim_{n \to +\infty} y(t_n) = \eta^*. \tag{6.4}$$

Conversely, the existence of such a sequence $\{t_n\}$ for which the conditions (6.4) are satisfied, implies that the point $M^*(\xi^*, \eta^*)$ is a limit point of L^+. It is also evident that, if the point M^* is a limit point of L^+ for a given initial position M_0 of the representative point on L^+, then M^* will be a limit point of L^+ for any other choice of the point M_0 on L^+.

The point M^* is called a *limit point of an entire path L*, if M^* is a limit point of either the L^+ or the L^- belonging to L (in the first case M^* is called an ω-limit point and in the second an α-limit point of the path L).

The limit point of a path L can either belong to the path L itself or not. Any state of equilibrium is its own unique limit point (both the ω- and the α-limit point). All points of a closed path are also, clearly, ω- and α-limit points of the path. In fact, the motion corresponding to a closed path L

$$x = x(t), \quad y = y(t)$$

has a period T_0 and each point $M(\xi, \eta)$ of this path corresponds to an infinite number of values of t

$$t_1 = \tau, \quad t_2 = \tau + T_0, \ldots, \quad t_n = \tau + (n+1)T_0, \ldots,$$

and also
$$t_1' = \tau, \quad t_2' = \tau - T_0, \ldots, \quad t_n' = \tau - (n-1)T_0, \ldots$$

A path tending to an equilibrium state (a node or focus or a saddle-point) has this equilibrium state as its unique limit point. For a L^+ (or L^-) winding on to a limit cycle, all points of this limit cycle are clearly limit points. It is evident that in the last two examples the limit point is not a point of the half-path.

---

† The term "limit point" is also used in the theory of sets. In the theory of sets, a point M^* is called a limit point of the set K, if in an arbitrary small neighbourhood of M^* there are points of the set K not coinciding with M^*. These two concepts are not to be confused, so instead of the term "limit point" when discussing sets we shall use the term "cluster point".

2. The first basic theorem on the set of limit points of a half-path

We shall first prove the following theorem, which enables us to introduce the concept of limit path.

THEOREM OF THE LIMIT PATH. *If M^* (ξ^*, η^*) is a limit point of the half-path L^+, then each point of the path L^* passing through M^*, is also a limit point for L^+.*

Let $M'(\xi', \eta')$ be any point on L^* differing from M^*. There are an infinite number of motions possible on L^* differing only in their initial conditions, but for all these motions the transit time τ between the points M^* and M' is always the same. Consider the ε—neighbourhood of the point $M'(\varepsilon > 0)$. Since a solution of (6.1) depends continuously upon the initial conditions, it is always possible to find for any ε, a $\delta > 0$ such that any path passing through a point of the δ—neighbourhood of M^* at $t = \tau^*$ also passes through a point of the ε—neighbourhood of M' at $t = \tau^* + \tau$. The point M^* is a limit point for L^+, so there are an infinite number of points $M_n(x_n, y_n)$ on L^+, corresponding to the infinite sequence $\{t_n\}$, in the δ—neighbourhood of the point M^*. But on L^+ there also exist an infinite number of points $M'_n(x'_n, y'_n)$ that correspond to the sequence $\{t'_n\} = \{t_n + \tau\}$, $n \to \infty$ in the ε—neighbourhood of the point M'. Furthermore, in the case when $\tau < 0$, it is always possible to choose a large $n = n_0$ so that $t'_{n_0} = t_{n_0} + \tau > t_0$, and the points $M'_n(x'_n, y'_n)$ ($n \geqslant n_0$) must belong to L^+. But we can make ε as small as we wish and so the point M' is a limit point for L^+. However, we can take any point of the path L^* as the point M', and therefore all its points are limit points for L^+.

The path L^* will be called the *limit path* for the half-path L^+. It is evident that all the points of L^* will be either points of a region G or points on the boundary of G. When the limit point of the path L is a point on itself, L is called a *self-limit* path. Obviously equilibrium states and closed paths are self-limit paths.

In the theory of sets, as is well-known, a set of points on a plane is called *closed* if it contains all its cluster points. Thus, if a sequence of points belonging to a given closed set K tend to a point N_0, this point N_0 is bound to be a point of the set K. A closed set is called *connected* if it cannot be represented as the sum of two closed sets without common points.

Let K be the set of *all* limit points of a given half-path L^+. The following basic theorem characterizes this set.

FIRST BASIC THEOREM. *The set of the limit points of a given half-path L^+ is a closed connected set and consists of entire paths.*

To prove that the set K is closed (in the sense of set theory), i.e. that every cluster point of the set K belongs to K, let M be a cluster point of the set K. Then, by definition in an arbitrary neighbourhood of M there are points of K, i.e. limit points of L^+. Hence, there are points of the half-path L^+ itself that correspond to arbitrarily large values of t. Thus M is a limit point of the half-path L^+.

To show that the set K is connected first assume that it is non-connected. Hence, being closed, it can be represented as the sum of two closed sets K_1 and K_2 without common points (K_1 and K_2 contain all limit points of L^+). Let ϱ_0 be the minimum distance between points of the sets K_1 and K_2. Suppose $\varepsilon < \varrho_0/3$ then the ε—neighbourhoods of K_1 and K_2 will be without common points. As the points of the sets K_1 and K_2 are limit points for the half-path L^+, then in the ε—neighbourhoods of these points there are bound to be infinite sequences of points of L^+ that correspond to values of t increasing without limit. However, owing to the continuity of the half-path, outside the ε—neighbourhoods of K_1 and K_2 there must also be infinite sets of points of L^+ that correspond to values of t increasing without limit. Since, by hypothesis, the half-path L^+ lies in a bounded region of the plane, these points must have at least one cluster point M_1, and since they correspond to values of t increasing without limit, then, M_1 will be a limit point of the half path L^+. The point M_1 cannot belong to the set K_1 nor to the set K_2 and, hence, L^+ must have limit points that differ from the points of the sets K_1 and K_2, which contradicts the assumption made. Thus the second assertion of the theorem is proved.

The last assertion of the theorem that the set of limit points of the half-path L^+ consists of entire paths, follows at once from the preceding theorem.

Since, from our original assumptions, the number of the equilibrium states in the system is finite in every bounded region of the phase plane, then it follows from the theorem just proved that when there are no points differing from equilibrium states among the limit points of the half-path L^+ this *half-path will have one and only one limit point, i.e. one equilibrium state*. It is also clear that if K is the set of all limit points of a given half-path, then, for an arbitrary small $\varepsilon > 0$, all points of this half-path that correspond to $t > T$, where T depends on ε, will lie in the ε—neighbourhood of the set K.

We have proved the first basic theorem for the case of paths on a phase plane. It is valid, however, also for paths on any phase surface (for example, on a torus) and also in a phase space with n dimensions when the system has n equations of the first order.

3. Auxiliary propositions

Before proceeding to prove the second basic theorem, which indicates which paths can be limit paths, we shall enunciate a series of auxiliary propositions connected with the so-called "segment without contact". Suppose $M_0(x, y)$ is a point on the phase plane, which is not an equilibrium state, and L_0 is the path through M_0. D is a straight line through M_0 that is not tangent at M_0 to the path L_0. It is evident that we can isolate on D a segment that contains the point M_0 and is not tangential at any of its points with any of the paths of the system (6.1). Such a segment is called *a segment without contact*, or a transversal.

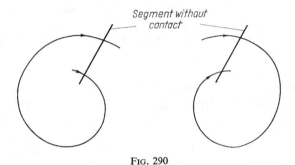

Fig. 290

The following series of propositions about a segment without contact will be needed later. Some of these propositions are quite evident, and will not be proved.

I. The straight line D divides the phase plane into two distinguishable parts. Let the motion $x = x(t)$, $y = y(t)$ be given on the path L_0[†] reaching M_0 at $t = t_0$. Since the straight line D is not tangent to L_0 at M_0 and the right-hand sides of the equations (6.1) are continuous, then we can always find $t_1 < t_0$ and $t_2 > t_0$ such that the motions on the path for $t_1 < t < t_0$ lie entirely on one side of the straight line D, and the motions on the path for $t_0 < t < t_2$ lie entirely on the other side of the straight line.

II. Again, from the continuity of the right-hand sides of the system (6.1), it follows that *all paths intersect a segment without contact in one and the same direction*, as t increases.

[†] In the following propositions we assume that if a path L_0 is given, then the motion along this path, i.e. the solution of the system (6.1) corresponding to this path with a certain choice of the value t_0, is also given.

In particular, if a phase path intersects the segment without contact twice, it can only intersect it as is shown in Fig. 290 and not as shown in Fig. 291.

III. However small we may take $\Delta > 0$, there is always a neighbourhood of the point M_0 such that every path that passes at $t = t_0$ through this neighbourhood, intersects the segment without contact at $t = t_0'$ where $|t_0' - t_0| < \Delta$.

Fig. 291

IV. *Every part of the path that corresponds to values of t inside a finite interval $\alpha \leqslant t \leqslant \beta$ can only have a finite number of points of intersection with any segment without contact.*

This can be proved by a *reductio ad absurdum*. The path L is assumed to have an infinite number of points of intersection with a certain segment without contact l and that all these points correspond to values of t contained between α and β. By the Bolzano–Weierstrass theorem we can choose from the infinite set of values of t corresponding to these points of intersection, a sequence $t_1, t_2, \ldots, t_n, \ldots$, tending, for $n \to +\infty$ to a certain value $\tau(\alpha \leqslant \tau \leqslant \beta)$ and such that the points $M_n(x_n, y_n)$ on L corresponding to t_n tend to the point $M_0(x_0, y_0)$ corresponding to $t = \tau$. This point M_0 must clearly lie on l since the points M_n lying on this segment tend to M_0. However, from proposition I, for values of t sufficiently close to τ, there cannot be points on the path L that are on the segment without contact. But this contradicts the fact that τ is a limit value of t corresponding to the points of intersection of L with l, and so there are values of t arbitrarily close to τ that correspond to points of intersection of L with l. We have arrived at a contradiction and thus the number of points of intersection must be finite.

V. *The points of intersection of a non-closed path L_0, with any segment without contact l that correspond to consecutive values of t are also consecutive, on the segment l.*

It can easily be shown, using proposition II and the fact that the paths do not intersect themselves, that the solutions shown in Fig. 292 are not possible.

The proposition can also be formulated thus: *consecutive points of intersection of a positive half-path with an arbitrary segment without contact l are disposed on the segment l in order of increasing time.*

VI. *A closed path can have only one point of intersection with a segment without contact.*

This can again be proved by a *reductio ad absurdum*, and it is shown that all points of intersection of the closed path L_0 with the segment without contact l must necessarily coincide.

VII. Let us consider a non-closed positive half-path L^+, for which the path L^* (not an equilibrium state) is a limit path. *If a segment without contact is drawn through an arbitrary point M_0 of the path L^*, then on this segment there will be an infinite sequence of points of the half-path L^+ (arranged in order of increasing values of the time t) tending to the point M_0.*

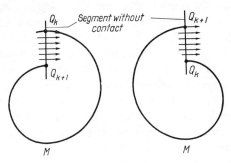

Fig. 292

This proposition is a consequence of the first basic theorem and of the propositions III and V.

VIII. Let $x=x(t)$, $y=y(t)$ be a motion along the path L, the point M_0 of this path corresponding to the value $t=t_0$ and the point M_1 to the value $t=t_1$. Let l be a segment without contact passing through the point M_1. Then, *for arbitrarily small $\varepsilon > 0$ and $\Delta > 0$, there is a $\delta = \delta(\varepsilon, \Delta)$ such that the representative point, which at $t=t_0$ is at a distance smaller than δ from the point M_0 must necessarily intersect the segment without contact l, at $t=t_1'$, where $(t_1-t_1') > \Delta$, remaining during the interval $t=t_0$ to $t=t_1'$, at a distance less than ε from the points of the path L corresponding to values of t between t_0 and t_1.*

This proposition (valid for both $t_1 > t_0$ and $t_1 < t_0$) is a consequence of the theorem on the continuous dependence of the solutions on the initial conditions and of the propositions III.

4. *Second basic theorem on the set of the limit points of a half-path.*

If a half-path L^+ is not closed and has at least one limit path that is not an equilibrium state, the half-path L^+ itself cannot be a limit path.

Let L^*, be a limit path for the half-path L^+. To prove the theorem by a *reductio ad absurdum*, assume that the half-path L^+ is itself a limit path for a certain half-path L_1^+.

A segment without contact l passes through a point P on the path L^*, and so the point P is a limit point for the half-path L^+. Then on the segment

l there will be an infinite number of points of the paths L^+ arranged in order of increasing values of t (proposition VII).

Suppose P_1, P_2 and P_3 are three consecutive points of intersection of L^+ with l. Then the point P_2 will be a limit point for the half-path L_2^+, but from the proposition VII, there must be a sequence of points of the half-path L_1^+, on the segment P_1P_2 or on the segment P_2P_3, tending to the point P_2. This is impossible since the half-path L_1^+ can intersect each of the segments P_1P_2 and P_2P_3 only once, as is easily shown. In fact, let Q be one of the points of intersection of the half-path L_1^+ with the segment P_1P_2. The representative point is at Q when $t=\tau$, and for $t>\tau$, either enters the region lying inside the closed curve $P_1MP_2P_1$ formed by the arc P_1MP_2 of the half-path L^+ and the segment without contact P_1P_2, or leaves this region. If it enters for $t>\tau$ then it will not be able to leave it again, since paths do not intersect each other, and all paths intersect the segment without contact in one and the same direction. Hence, the representative point will not intersect the segment without contact P_1P_2 for $t>\tau$.

A similar argument can be used if the representative point leaves the region bounded by the closed curve $P_1MP_2P_1$ for $t>\tau$. Thus the assumption that the half-path L^+ is a limit path for the half-path L_1^+ leads to a contradiction and the theorem is proved.

It follows from this theorem that a *non-closed path cannot be a self-limit path*.

This theorem reflects features that are characteristic of the plane and need not be true for paths on other phase spaces. It is not true, for example, for paths on a torus or when the phase space is a three-dimensional Euclidean space.

From the second basic theorem it follows that the only types of limit paths are (1) equilibrium states; (2) closed paths; (3) non-closed paths having only states of equilibrium as limit points. There are two more theorems to add which enable us to establish the combinations of these types of limit paths which are possible as the set of all limit points of a half-path.

THIRD THEOREM. *If a half-path L^+ has a closed limit path L_0, then L_0 is the only limit path for L^+.*

If the half-path L^+ is itself closed then all its points are limit points and it cannot have any other limit points. In this case the theorem is evident.

Suppose L^+ is not closed, and on it the corresponding motion is where $x=x(t)$, $y=y(t)$.

The path L_0 corresponds to $x = \bar{x}(t)$, $y = \bar{y}(t)$.

$$\bar{x}(t+h) \equiv \bar{x}(t), \quad \bar{y}(t+h) \equiv \bar{y}(t).$$

P is a point on L_0 at which $t = \tau = \tau+h = \tau+2h, \ldots, l$ a segment without contact through P made entirely inside the ε-neighbourhood of L_0. By proposition VII, there is on l a sequence of points of the half-path $L^+: P_1, P_2, P_3, \ldots, P_n, \ldots$, tending to the point P. The points $P_1, P_2, \ldots$, are arranged on l in order of increasing values of t:

$$t_1, t_2, t_3, \ldots, t_n, \ldots \quad (t_n \to +\infty \quad \text{for} \quad n \to +\infty).$$

By proposition VIII for $0 < \Delta < h/3$ there is a region of small radius $\delta = \delta(\varepsilon, \Delta)$ about P, such that a path passing through this region at $t = \tau$ remains in the ε-neighbourhood of L_0 from $t = \tau$ to $t = T$ intersecting l at $t = T$ where $|T - (\tau+h)| < \Delta$. It follows therefore that each point of intersection of l by L^+ in the region δ about P, lies successively nearer to P and that the part of the half-path L^+ corresponding to values of $t > \tau$ is contained inside the ε-neighbourhood of L_0.

The closed path L_0 must contain all the limit points of the half-path L^+, for if L^+ has a limit point Q not on L_0 but at a distance $d > 0$ from it, then in an arbitrarily small neighbourhood of Q there must be points of L^+ corresponding to arbitrarily large values of t. However, from what has been stated above, however small we take $\varepsilon > 0$, it is always possible to find a $t = \tau_0$ such that all points of L^+ corresponding to $t > \tau_0$ lie entirely inside the ε-neighbourhood of L_0. We can always make ε smaller than $d/2$, so that Q is inside the ε-neighbourhood of L_0. Hence, arbitrarily close to the point Q there cannot be points of L^+ that correspond to arbitrarily large values of t. We have arrived at a contradiction and thus the theorem must be true.

FOURTH THEOREM. *If among the limit points of a half-path there are no equilibrium states then the half-path is either closed, or else is non-closed but has a closed limit path (the Poincaré–Bendixson Theorem).*

This theorem follows immediately from Theorem III and has a further consequence in the following theorem, which is very often used.

FIFTH THEOREM. *Let G be a closed doubly connected (annular) region, that does not contain equilibrium states and is not left by paths as t increases (as t decreases). Then inside such a region G there are bound to be one or more stable (unstable) limit cycles.*

In fact the set of limit points of every non-closed path entering the region G as t increases (as t decreases) lies entirely in this region and hence does not contain singular points. Then, from Theorem IV, this set is a closed path or limit cycle. Thus in the region G there is at least one limit cycle, but there can be more than one. If we assume that among these limit cycles there are no "semi-stable" ones (they are only possible, in "non coarse" systems: see Section 4 of this chapter), then if all paths enter the region G as t increases, there is at least one stable limit cycle, and if all paths leave the region G as t increases, at least one unstable limit cycle.

If there are semi-stable limit cycles in the region G, the validity of the theorem is established by a more complicated reasoning. We use this theorem when there is a region between two cycles without contact into which all paths enter as t increases (as t decreases). The theorem formulated is also valid in the case of the piece-wise–linear systems considered in the chapters VIII and X. We shall use the theorem in these chapters without giving the obvious modifications that are needed to the proof of the theorem.

5. Possible types of half-paths and their limit sets

The theorems above enable us to establish the possible character of the set of the limit points of a half-path that lies entirely in a finite region of the plane. This set can be one of the following types: I. One equilibrium state. II. One closed path. III. The aggregate of equilibrium states and of paths tending to these states of equilibrium both for $t \to +\infty$ and $t \to -\infty$.

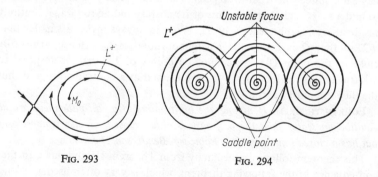

Fig. 293   Fig. 294

It is easily seen that the equilibrium states occurring in the set of limit points of the type III cannot be foci or nodes since every path that reaches a sufficiently small neighbourhood of such a singular point tends to it and cannot have any other limit points. Hence the singular points that can occur in a set of limit points of the type III, if simple, are bound to be saddle points, while the paths differing from equilibrium states must be the separatrices. Knowing the possible types of limit sets, we can state that there are, on a plane, the following possible half-paths: (1) a singular point; (2) a closed path; (3) a half-path tending to a singular point; (4) a half-path tending to a closed path; (5) a half-path tending to a limit set of the type III†.

These half-paths, except the last one, have been repeatedly encountered in preceding examples. A very simple example of type III is shown in Fig. 293, where the half-path L^+ tends to a limit set consisting of a separatrix leaving and returning to the same saddle point. A more complicated type is shown in Fig. 294, where the half-path L^+ (the external one) tends to a limit set consisting of two equilibrium states and four separatrices that tend to these singular points both for $t \to +\infty$ and for $t \to -\infty$.

§ 3. Qualitative features of the phase portrait on the phase plane. Singular paths

1. Topologically invariant properties and topological structure of the phase portrait

The analysis of the particular examples of dynamic systems given in the preceding chapters suggests that the qualitative appearance of the phase plane depends only on certain singular paths. In these examples there was a finite number of singular paths separating the phase plane into regions in which all paths behaved in the same manner. These singular paths were equilibrium states, limit cycles, and the separatrices of the saddle-points. It is natural to ask whether there is always a finite number of such singular paths, how they can be found in the general case, and whether the types so far considered are exhaustive. This section is devoted to clarifying these questions [17, 80].

Let us first recall the concept of topological representation (or transformation), which is a one-to-one continuous point transformation of the plane into itself so that to each point $M(x, y)$ there corresponds a single point $M'(x', y')$ of the same (or of another) plane. Also to any two arbitrarily close points M_1 and M_2 there correspond arbitrarily close points M_1' and M_2'. The reciprocal transformation of a topological transformation is, clearly, also a topological transformation. Any topological transformation is determined by single-valued continuous functions

$$x' = \varphi(x, y), \qquad y' = \psi(x, y),$$

† If the phase surface is not a plane, the types of paths given may not exhaust all possible types.

that can be solved uniquely

$$x = \varphi_1(x', y'), \quad y = \psi_1(x', y'),$$

where φ_1 and ψ_1 are certain continuous functions. It is evident that the form of curves, regions and sets on the plane can vary drastically under the transformation, but certain properties remain unvaried. Thus a closed curve, after an arbitrary topological transformation of the plane into itself, forms another closed curve though it may be very different in shape. A rectilinear segment may, after a topological transformation, correspond to a certain arc, but this arc is bound to be an arc without self-intersection. The properties that remain unvaried for all possible topological transformations are called *topologically invariant properties or topological characteristics*.

Now the dynamic system (6.1) determines a certain family of paths or phase portrait. Under all possible topological transformation of the plane into itself it is evident that the number and the mutual disposition of the closed paths, the equilibrium states, etc. remain unvaried; if an equilibrium state of the system (6.1) had been a saddle point, then its character is preserved. It is easily seen, however, from geometrical considerations, that a focus or a node are topologically identical, as it is always possible to find a topological transformation of the plane into itself for which a node is transformed into a focus or vice versa.

We can now make more precise certain concepts: *the topological structures of the phase portraits determined by two systems of the form (6.1) are called identical if there exists a one-to-one continuous transformation for which the paths of one system are transformed into the path of the other (in this context a path is transformed into a path both in the direct and inverse transformation)*. This definition of identity of two structures is an indirect definition of topological structure. It can be said that *by topological structure of the phase portrait we mean all the properties of the portrait that are invariant for all possible topological transformations of the plane into itself*.

2. Orbitally stable and orbitally unstable (singular) paths

We still assume that the system (6.1) is considered in a limited region G of the plane. Consider a path L lying entirely in a region G, and with a positive half-path L_M^+ beginning at the point M. Note that the ε-neighbourhood of the half-path L_M^+ is bound to contain the ε-neighbourhood of the limit set of this half-path.

We say that *a positive half-path L_M^+ is orbitally stable if, for any $\varepsilon > 0$, there is a $\delta(\varepsilon) > 0$ such that for each path L' passing, for $t = t_0$, through an arbitrary point M' belonging to the δ-neighbourhood of M the half-path $L_{M'}^{'+}$ ($t > t_0$) lies entirely in the ε-neighbourhood of the half-path L_M^+.*

A path L is called orbitally stable for $t \to +\infty$ or ω orbitally stable if each positive half-path detached from it is orbitally stable. It can be shown (and this is geometrically evident) that if a positive half-path of the path L is orbitally stable, then every other positive half-path detached from this path will also be orbitally stable[†].

Half-paths or paths that are not orbitally stable for $t \to +\infty$ are said to be orbitally unstable for $t \to +\infty$, or ω-orbitally unstable. Of course, if a path L is orbitally unstable for $t \to +\infty$ and M is any one of its points, then there is an $\varepsilon_0 > 0$ such that for an arbitrarily small $\delta > 0$ a path L' exists, passing through a point of the δ-neighbourhood of M at $t = t_0$ yet outside the ε_0-neighbourhood of the half-path L for a certain $t > t_0$. Note that the presence of orbitally unstable paths does not contradict the theorem about the continuous dependence of paths (solutions) on the initial conditions, since in this theorem only a finite interval of time t is considered.

What has been said of a positive half-path can be repeated with obvious modifications for a negative half-path. Thus we can speak of a path orbitally stable for $t \to -\infty$ or α-orbitally stable and of a path orbitally unstable for $t \to -\infty$ or α-orbitally unstable. We shall call a path L, that is orbitally stable both for $t \to +\infty$ and for $t \to -\infty$ *orbitally stable* or *non-singular*. Every path that is not orbitally stable will be called *orbitally unstable* or *singular*. Thus, a singular path is bound to be orbitally unstable in at least one "direction", i.e. for $t \to +\infty$ or $t \to -\infty$ or for both $t \to +\infty$ and $t \to -\infty$.

We might recall here (see, for example, Chapter II, Section 7) that a path that is orbitally stable for $t \to +\infty$ need not be stable in the sense of Liapunov for $t \to +\infty$.

The concept of orbital stability and instability of a half-path and of a path characterizes the behaviour of this half-path or path only in relation to the nearby half-paths and paths. It is geometrically evident that every half-path tending to an equilibrium state of the saddle or focus type is orbitally stable, as will be all half-paths tending to limit cycles. In fact the following paths will be clearly orbitally stable or non-singular: paths

---

† A rigorous proof of this geometrically evident fact is not trivial.

tending to nodes or foci for $t \to +\infty$ and for $t \to -\infty$, or tending to a node for $t \to +\infty$ ($t \to -\infty$) and to a limit cycle for $t \to -\infty$ ($t \to +\infty$), and also paths tending to limit cycles both for $t \to +\infty$ and for $t \to -\infty$ (all such paths are orbitally stable both for $t \to +\infty$ and for $t \to -\infty$).

It is easily seen from these examples that when a path is non-singular (orbitally stable) all paths near it behave similarly. This is not the case, however, for "singular" paths. Nodes and foci are orbitally stable either for $t \to +\infty$ or for $t \to -\infty$ but can never be orbitally stable both for $t \to +\infty$ and for $t \to -\infty$; a saddle point is orbitally unstable both for $t \to +\infty$ and for $t \to -\infty$. Stable and unstable limit cycles can be orbitally stable either for $t \to +\infty$ only, or for $t \to -\infty$ only. Half-paths tending to a saddle point (the separatrices of the saddle points) are orbitally unstable. In fact if L_M^+ is a half-path that tends to a saddle point; there is always an $\varepsilon(\delta) > 0$ such that for every $\delta > 0$, half-paths not coinciding with L_M^+ and passing through points of the δ-neighbourhood of the point M are bound to leave (as t increases) the ε-neighbourhood of L_M^+.

3. The possible types of singular and non-singular paths

THEOREM I. *Every path that is a limit path for some path not coinciding with itself is singular or orbitally unstable.*

Let L^* be a path that is a limit path for at least one path L not coinciding with L^* (for the sake of definiteness let L approach L^* for $t \to +\infty$). If L^* is a state of equilibrium or not there are bound to be points on L, such as M, at a distance $\delta > 0$ from the points of the path L^*. These points M could only not exist if the path L were a limit path for L^*. However, this is impossible since L has limit points differing from singular points so by Theorem III of § 2, L cannot be a limit path for any path. If $\varepsilon_0 < \delta$, then the point M will lie outside the ε_0-neighbourhood of L^*. But L^* is a limit path for L, and in the δ-neighbourhood of every point L^* there will be points of L corresponding to values of t larger than that corresponding to the point M. As the point M of the path L lies outside the ε_0-neighbourhood of L^*, then clearly L^* is orbitally unstable for $t \to -\infty$ (α-orbitally unstable), which proves the theorem.

For a half-path L^+, whose set K of limit points are not all equilibrium states the ε-neighbourhood of the limit set K is a part of the ε-neighbourhood of L^+. Then, for any $\varepsilon > 0$, there is a $T(\varepsilon)$ such that the points of L^+ corresponding to $t > T$ lie entirely in the ε-neighbourhood of K. Let P be a point, not an equilibrium state, of the set K, and let l be a segment without contact through P. By proposition VII on l there is a sequence of points of L^+, $P_1, P_2, \ldots, P_n$, corresponding to values of t increasing without limit and tending to the point P. If C_i is a closed curve consisting of the arc $P_i P_{i+1}$ of the half-path L^+ and the part $P_i P_{i+1}$ of the segment l then all such closed curves C_i, for large enough i, will lie entirely in the ε-neighbourhood of the limit set K, which lies either inside all these curves or, outside them (Fig. 295 and Fig. 296). The region G_i is bounded by the closed curve C_i and the limit set K (see the regions shaded in Fig. 295 and Fig. 296), and for any $\varepsilon > 0$, and sufficiently large i, this region G_i

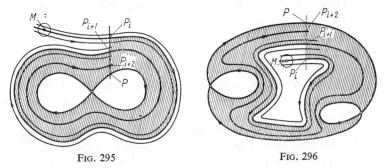

Fig. 295   Fig. 296

is contained entirely in the ε-neighbourhood of K. It is evident that all points of the part $P_{i+1} P$ of the segment l, except P and P_{i+1}, belong to the region G.

THEOREM II. *A non-closed half-path L^+, that has points differing from states of equilibrium among its limit points, is orbitally stable.*

To prove this theorem, it suffices to show that for any $\varepsilon > 0$, all paths passing sufficiently close to a point of the half-path L^+ eventually enter and stay in the ε-neighbourhood of the limit set K, as t increases. From the preliminary remarks for any $\varepsilon > 0$ there is an integer $I(\varepsilon)$ such that, for any $i > I$, the region G_i lies entirely in the ε-neighbourhood of K.

Let M be a point of the half-path L^+ and let $i > I$ be a fixed integer. By proposition VIII, it is always possible to find a small neighbourhood of the point M such that every path passing through this neighbourhood for $t = t_0$ intersects the segment l at $t = T$ arbitrarily close to the point P_{i+1} and between the points P_i and P_{i+2} (see Figs. 295 and 296). But, for values $t > T$, this path will clearly be in the region G_i and cannot leave this region and so intersects either the curve C_i (see proposition V) or the limit set K consisting of entire paths (by Theorem II, Section 2). Thus the theorem is proved.

It follows in particular from this theorem that every half-path approaching a limit cycle is orbitally stable.

To determine when a closed path is orbitally stable or unstable it should be noted that when $P(x, y)$ and $Q(x, y)$ are analytic functions, the following two cases can occur:

(1) either all paths differing from the given closed path L and passing through a sufficiently small neighbourhood of L, are not closed;
(2) or all paths passing through all points sufficiently close to L are closed.

It is evident that the first case occurs when the path L is a limit cycle and the second case arises in a conservative system.

THEOREM III. *A closed path L_0, that is not a limit path for any one non-closed path, is orbitally stable.*

To prove the theorem we shall prove first that all paths passing through points sufficiently close to L_0 are closed. In fact, if among paths arbitrarily close to L_0 there could be non-closed paths, then we would have the case (1) indicated above: i.e. all paths except L_0, that pass through points sufficiently close to L_0, would be non-closed. Then it is easily seen that the path L_0 is bound to be a limit path for a non-closed path, which contradicts the hypothesis. Hence, all paths, passing through the points of a certain sufficiently small neighbourhood of L_0 are closed, and it follows immediately from the continuous dependence on the initial conditions that the paths lie entirely in the ε-neighbourhood of L_0, and thus L_0 is orbitally stable.

To complement these theorems we make some remarks on half-paths that tend to an equilibrium state.

It is evident from examples that such half-paths can be orbitally stable (for example, half-paths tending to a focus or a node) or orbitally unstable (half-paths tending to a saddle point). In such examples the equilibrium state was simple but it can be shown that, even when the equilibrium state is multiple, if the half-path tending towards it is orbitally unstable then it must be a boundary for a certain saddle region. Without giving the proof, we shall discuss this in some detail.

If a half-path L_M^+, tending to the equilibrium state 0, is orbitally unstable, then there is a path that leaves some ε_0-neighbourhood of L, as t increases. Now consider the ε_0-neighbourhood of 0 which is small enough to contain only one equilibrium state, at 0, and no closed path. Let point Q on L_M^+ correspond to $t = \tau$ and lie on a segment without contact l in the ε_0-neighbourhood of 0 (see Fig. 297).

All paths that pass through points sufficiently close to M are bound to intersect (as t increases) the segment l. Suppose that a half-path L'^+

passes through a point Q' on the segment l and, without leaving the ε_0-neighbourhood of 0, tends to the equilibrium state 0. It is easily seen that all paths passing through the part QQ' of l must also tends towards 0.

If the segment l did intersect on both sides of the point Q half-paths tending towards 0 without leaving the ε_0-neighbourhood of 0, then this would clearly contradict the hypothesis. Therefore, through points of the segment l close to Q, and on one side at least, there are bound to be paths that leave the ε_0-neighbourhood of 0 as t increases (Fig. 298). It can be

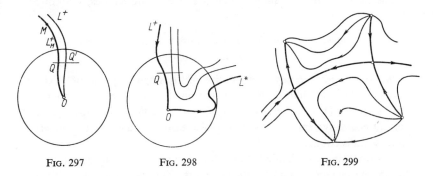

Fig. 297   Fig. 298   Fig. 299

shown that a negative half-path L^{-*} must exist that tends towards 0, bounding, together with the half-path L_M^+, a "saddle region" and having, for a sufficiently small ε_0, points outside the ε_0-neighbourhood of the equilibrium state 0 (see Fig. 298).

We call orbitally unstable half-paths tending to an equilibrium state *the separatrices of this equilibrium state*. Thus every half-path obtained from a non-closed limit path is bound to be a separatrix, but a separatrix need not be a limit path. In this case it is a path separating paths of different behaviour from each other. A simple example is shown in Fig. 299.

Now we can draw exhaustive conclusions as to which half-paths and paths are orbitally unstable. Thus, every orbitally unstable (singular) path belongs to one of the following types:

(1) an equilibrium state[†],
(2) a limit cycle,
(3) a non-closed path, at least one half-path of which is a separatrix of an equilibrium state.

---

† Equilibrium state is orbitally unstable in the case when one path at least tends to it. If, on the other hand, an equilibrium state is a centre then, clearly, it is orbitally stable. However, in all cases we will attribute equilibrium state to singular paths.

The property of a path being singular or non-singular is a topologically invariant property. Thus, the following theorem holds:

THEOREM IV. *If the phase portraits determined by two dynamic systems in a bounded region G are identical, i.e. if there exists a transformation of the plane into itself for which the paths of these systems correspond to each other, then orbitally stable half-paths are transformed into orbitally stable half-paths, and orbitally unstable half-paths are transformed into orbitally unstable ones.*

The proof of this theorem, which presents no difficulties, is omitted.

4. Elementary cell regions filled with non-singular paths having the same behaviour

We shall consider now the totality of all singular paths of a given system (6.1) in a *bounded* region of the plane. It can be shown that for the system (6.1), *the number of singular paths is finite*. For the simplest case of coarse systems this may be established on the basis of the material in the next section†.

The singular paths divide the region G into partial regions whose points lie on *non-singular (orbitally stable)* paths. The boundary of each such partial region are singular paths and points that are on the boundary of the region G. We restrict ourselves to regions whose boundaries do not contain boundary points of the region G, and these regions we call *elementary cells* (or simply *cells*). It is evident that cells consist of entire orbitally stable (i.e. non-singular) paths. In addition, by an argument analogous to that used to prove Theorem I of Section 1, it is shown that the boundary of each cell consists of entire singular paths. The points of one singular path can be boundary points for several cells. Furthermore, due to the fact that the number of singular paths is finite, the number of cells in a region G is also finite.

To consider in greater detail the behaviour of non-singular paths in one cell, we give first certain simple but important, auxiliary propositions.

I. *About every point of an orbitally stable half-path L^+ tending to the equilibrium state 0, there is always a neighbourhood such that all paths passing through it are orbitally stable for $t \to +\infty$ and tend also towards 0 as does L^+.*

To prove this proposition, it is sufficient to prove first that all half-paths that pass through a sufficiently small neighbourhood of any point

---

† The proof that the number of singular paths is finite when $P(x, y)$ and $Q(x, y)$ are general analytic functions, is fairly complicated and outside the scope of this book.

of L^+, owing to the orbital stability of L^+ for $t \to +\infty$, will not leave the ε-neighbourhood of L^+. Hence, the limit set of such paths also lies entirely in the ε-neighbourhood of L^+. This limit set must consist of entire singular paths and, since in the ε-neighbourhood of L^+ there is only one singular path, the equilibrium state 0, then the limit set must consist of the single equilibrium state 0, which proves the proposition I.

II. *About every point of the half-path L^+ which has a limit path differing from an equilibrium state, it is always possible to indicate a neighbourhood such that all paths passing through points of this neighbourhood are orbitally stable for $t \to +\infty$, and for $t \to -\infty$ have the same limit set as has L^+.*

III. *About every point of a closed orbitally stable path a neighbourhood exists such that all orbitally stable paths passing through points of this neighbourhood are closed and each lies inside another.*

Propositions II and III are proved by means of arguments analogous to those used in a proof of the proposition I.

Using these auxiliary propositions, it is possible to prove a number of theorems that completely characterize the behaviour of the paths of one and the same cell.

THEOREM V. *If all paths that belong to one and the same cell are not closed, they have the same ω- and α-limit sets.*

This theorem is almost self-evident, for if there are two paths L and L' belonging to the same cell and having different limit sets for $t \to +\infty$ (or $t \to -\infty$), then these paths can be connected by an arc l lying in the cell. Through all points of the arc l pass orbitally stable paths and the arc l from the path L to the path L' must reach a point M_0 that is either the last point with a path having the same limit set as L or the first point with a path having the same limit set as L', or, lastly, through M_0 there is a path that has a limit set differing from the limit sets of L and L'. None of these possibilities can be true, for by the Propositions I and II of this chapter, all paths passing through points of the arc l sufficiently close to M_0 must have the same limit set as, say, the path through M_0, and there cannot be more than one limit set for all the paths in the cell.

By an argument completely analogous to that sketched out for the last theorem, we can demonstrate the following theorem:

THEOREM VI. *If inside a cell there is at least one closed path, then all paths of this cell are closed, one lying inside another, and between any two paths of this cell there cannot be points not belonging to this cell.*

These theorems give a precise meaning to the assertion made before with the words: "the non-singular paths inside each cell behave in the same manner". It is evident that there is no place for singular paths inside a cell.

5. Simply connected and doubly connected cells

The possible types of distinct cells is of importance so we shall try to classify cells according to the topological structure of their phase portraits. We can either consider a cell by itself, or a cell together with its boundary (consisting of entire singular paths), i.e. a closed cell. It is not proposed to discuss at length the classification of cells, but we will give (without proofs) some basic relevant propositions.

The basic topological characteristic of any region, and of a cell, is its *order of connexion*[†]. For a cell there are only two possibilities as given by the following theorem[‡].

THEOREM VII. *Any cell cannot be more than doubly connected.*

Evidently cells filled with closed paths are always doubly connected, as follows at once from Theorem VI and from the fact that inside a closed path there is always an equilibrium state. Cells filled with non-closed paths can be either simply or doubly connected.

THEOREM VIII. *When a cell filled with non-closed paths is doubly connected, then one of its boundary continua is the α-limit set and the other the ω-limit set for the paths of this cell.*

Thus, in the case of a doubly connected cell filled with non-closed paths, the cell cannot have a boundary point that is not a limit point for the paths of this cell.

Using these theorems it is possible to describe in an exhaustive manner the boundaries that are possible for cells, and to establish the geometrically evident conditions under which two cells, considered with or without their boundaries, have the same topological structure for their phase portraits, but this investigation would be outside the scope of this book. The number of different types of cells (i.e. of cells with a different topological structure) is finite when a cell is considered without its boundary. The number of different types of closed cells (a cell together with its boundary)

[†] The boundary of every region can consist of either one connected arc or boundary continuum i.e. a closed connected set, − or of two, three etc. boundary continua. If the boundary consists of one boundary continuum, then the region is Singly connected; if it consists of two, then the region is called respectively doubly connected, etc. The simplest example of a Singly connected region is the region inside a simple closed curve, of a doubly connected one − the annular region between two simple closed curves. Note that in the case of a doubly connected region, the internal boundary arc can be a separate point. Clearly regions with a different orders of connexion are not topologically identical.

[‡] The proof of this theorem, although simple in its conception, is rather lengthy. It is based on the following auxiliary proposition: on each of the boundary continua of a cell there must be limit points of the paths of this cell.

increases without limits as the number of the equilibrium states of the dynamic system increases. In the case of coarse systems, however, there can only be a finite number of types of closed cells.

However, an exhaustive classification of closed cells in the case of the so-called "coarse systems" will be given in the next section (Section 4). In this section we will only give certain (geometrical) examples of simply connected and doubly connected cells.

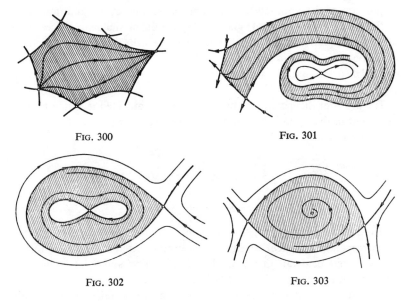

Fig. 300   Fig. 301

Fig. 302   Fig. 303

Examples of simply connected regions are shown in Fig. 300 and 301 (see also Figs. 306 and 309). Examples of doubly connected regions are shown in Figs. 301 and 303 (see also Fig. 305). In these figures the singular paths that occur on the boundaries of the cells are shown with thick lines[†].

In conclusion, without proofs, we add some general considerations about the complete qualitative investigation of a given dynamic system (A) in a region G.

Singular paths divide the region G into partial regions, either cells or partial regions the boundary of which include points of the boundary of

† In the example in Fig. 301 the boundary of the cell has a fairly complicated character. All the points of the figure-of-eight curve are the so-called "unattainable points of the boundary", for no simple arc exists with one end a point of the figure-of-eight curve and the remaining points belonging to the cell.

G. If we know the topological structure of the paths in all these partial regions, and also know the relative position of these regions, then the qualitative investigation of the dynamic system in the region G will be complete.

In order to know the mutual disposition of the partial regions, we must know the disposition of the singular paths and the behaviour of the paths in the cells.

If the type of every equilibrium state is known, together with the mutual disposition of the limit sets (equilibrium states, limit cycles and limit sets of the type III, see Section 1) and the disposition of the separatrices that are not limit paths, then this establishes completely the topological structure of the phase portrait of the paths in the region G. The description of the mutual disposition of the singular paths is called a *diagram* by Leontovich [82].

The proof of this geometrically evident fact is outside the scope of this book, but in the following section (Section 4) we return to this topic when we consider coarse systems.

§ 4. Coarse systems

1. Coarse dynamic systems

The question of what properties dynamic systems (models) must possess to correspond to physical systems, has been considered briefly in the Introduction, and in detail in the works of Andronov and Pontriagin [17], Leontovich and Mayer [80], and Debaggis [145].

In setting out the differential equations we cannot take account of all the factors that influence in some manner or other the behaviour of the physical system. On the other hand, none of the factors taken into account can remain absolutely constant during a motion of the system, so that when we attribute to the parameters perfectly determined values; this has only a meaning on condition that small variations of the parameters do not substantially vary the character of the motion. A certain number of parameters corresponding to physical parameters of the problem occur in the functions P and Q of our system equations, so these functions are never known exactly. Small variations of these parameters must leave unchanged the qualitative structure of the phase portrait. Therefore, if certain qualitative features appear for well-determined quantitative relations between the parameters but vanish for an arbitrarily small variation of the parameters, then it is clear that such qualitative features are not, generally speaking, observed in real systems.

It is natural, therefore, to separate the class of dynamic systems whose topological structure of the phase paths does not vary for small variations of the differential equations. We call such systems *"coarse"* or *structurally stable*, and we will give a rigorous mathematical definition of coarse systems and their basic properties.

Let the given system (*A*)

$$\frac{dx}{dt} = P(x, y), \quad \frac{dy}{dt} = Q(x, y)$$

be considered in a certain bounded region *G* of the plane. We assume that the boundary *C* of the region *G* is a "cycle without contact", i.e. a simple closed curve such that the paths of (*A*) are never tangent to *C*. Curve *C* must be large enough to include all physically useful values of *x* and *y*. This assumption is not necessary but it removes non-essential complications in the analysis. It also restricts the class of dynamic systems without affecting the practically important ones.

Corresponding to (*A*), there is the modified system (*Ã*)

$$\left. \begin{aligned} \frac{dx}{dt} &= P(x, y) + p(x, y), \\ \frac{dy}{dt} &= Q(x, y) + q(x, y), \end{aligned} \right\} \qquad (6.5)$$

where $p(x, y)$ and $q(x, y)$ are small and analytical, and have small partial derivatives. For all sufficiently small $p(x, y)$ and $q(x, y)$, the curve *C* is also a cycle without contact for paths of (*Ã*).

There are certain fundamental theorems on the variations of the solutions of a system of differential equations for small variations of the right-hand sides of these equations, on which the analysis below is based. The first of these theorems, the Theorem IV of Appendix I, can be enunciated in a geometrical form as follows:

Given an arbitrary finite interval of time, it is always possible to construct a system (Ã) so close to a given system (A), and having nearly identical initial points, that the corresponding paths of the systems (A) and (Ã) will differ by an arbitrarily small amount during the chosen interval.

A second theorem is Theorem V of Appendix I, which gives a more precise indication than Theorem IV of the closeness of the solutions of the systems (*A*) and (*Ã*) in the case when the right-hand sides of (*A*) and (*Ã*), and also their partial derivatives, differ by some small amounts.

By this theorem, *if*

$$x = \varphi(t-t_0, x_0, y_0),$$
$$y = \psi(t-t_0, x_0, y_0)$$

are solutions of the system (A) and

$$x = \tilde{\varphi}(t-t_0, \tilde{x}_0, \tilde{y}_0),$$
$$y = \tilde{\psi}(t-t_0, \tilde{x}_0, \tilde{y}_0)$$

are solutions of the system ($\tilde{A}$), then in an arbitrary finite interval of time, the pairs of functions φ and $\tilde{\varphi}$, ψ and $\tilde{\psi}$, $\partial\varphi/\partial x_0$ and $\partial\tilde{\varphi}/\partial \tilde{x}_0$, $\partial\psi/\partial x_0$ and $\partial\tilde{\psi}/\partial \tilde{x}_0$, $\partial\varphi/\partial y_0$ and $\partial\tilde{\varphi}/\partial \tilde{y}_0$, $\partial\psi/\partial y_0$ and $\partial\tilde{\psi}/\partial \tilde{y}_0$ will differ by an arbitrarily small amount, when the right-hand sides of the system ($\tilde{A}$) and their partial derivatives are sufficiently close to the right-hand sides of the system (A) and their partial derivatives, the initial point $\tilde{M}(\tilde{x}_0, \tilde{y}_0)$ being sufficiently close to the point $M(x_0, y_0)$.

By these theorems it appears that for small variations of the right-hand sides every path varies little in a part corresponding to a finite interval of time. It does not follow from this, however, that a path will vary little during an infinite interval of time. Even less, does it mean that the phase portraits of close systems always have the same character[†].

We now define a system with structural stability or coarseness: *a system (A) is called "coarse" (in a region G) if for any $\varepsilon > 0$, there is a $\delta > 0$ such that for all possible analytic functions $p(x, y)$, $q(x, y)$ that satisfy in G the inequalities*

$$\left. \begin{array}{lll} |p(x,y)| < \delta, & |q(x,y)| < \delta, & |p'_x(x,y)| < \delta, \\ |p'_y(x,y)| < \delta, & |q'_x(x,y)| < \delta, & |p'_y(x,y)| < \delta, \end{array} \right\} \quad (6.6)$$

there exists a topological transformation of G into itself, for which each path of the system (A) is transformed into a path of the modified system ($\tilde{A}$) (and conversely), the points that correspond to each other in this transformation being found at distances less than ε.

Two regions, G_1 and G_1^* are ε-close, if there exists a topological transformation for which the corresponding points are found at a distance

---

[†] A very simple example of a system for which the qualitative appearance of the paths varies even for small variations of the right-hand sides, is

$$\frac{dx}{dt} = ax+by; \qquad \frac{dy}{dt} = -bx+ay,$$

for which, for $a=0$ all paths are closed and for arbitrarily small $a \neq 0$ there is no closed path.

less than ε. Suppose that systems (A_1) and $(\tilde{A}_1)$ are defined in G_1 and G_1^* respectively, then the phase portrait in G_1 is ε-identical with that of the closed region G_1^* provided that the topological transformation of G_1 into G_1^* also transforms the paths of (A_1) into the paths of the system (A_1^*).

Let the system (A) be defined in the region G and let G_1 be a closed region contained entirely (together with its boundary) in G. The system (A) is called *coarse* in the region G_1 if for any $\varepsilon > 0$ there is a $\delta > 0$ such that, for any $(\tilde{A})$ satisfying the inequalities (6.6), in the region G, there is found a closed region G_1^* inside G whose mapping by the paths of (A) is ε-identical with the mapping of G_1 by the paths of $(\tilde{A})$. It follows immediately that for a sufficiently small $\varepsilon > 0$[†] and a suitable $\delta > 0$ in the ε-neighbourhood of every equilibrium state of (A) there will be only one equilibrium state of $(\tilde{A})$ of the same character as that of (A); and in the ε-neighbourhood of every limit cycle of (A) only one limit cycle of the system $(\tilde{A})$, etc.

With regard to the necessary and sufficient conditions for the coarseness of a system, we make one very important remark: the limitations imposed by the requirement of coarseness on these dynamic systems have been shown by De Baggis [145] to be such that they isolated the "general case". In other words, non-coarse systems are exceptional systems (see also § 5 of this Chapter).

When, now a system $(\tilde{A})$ is said to be close to the system (A), because of arbitrarily small corrections to the right-hand sides of the system (A) it must be understood that not only are the functions $p(x, y)$, $q(x, y)$ small but also their partial derivatives.

2. Coarse equilibrium states

We establish first any limitations on its equilibrium states which are imposed by requiring a system to be coarse.

The following theorem holds:

THEOREM I. *In a coarse system there cannot be an equilibrium state for which*

$$\Delta \equiv \begin{vmatrix} P'_x(x_0, y_0) & P'_y(x_0, y_0) \\ Q'_x(x_0, y_0) & Q'_y(x_0, y_0) \end{vmatrix} = 0.$$

In fact, if $\Delta = 0$ at a state of equilibrium $0(x_0, y_0)$ then this clearly means

---

[†] More precisely, for an $\varepsilon > 0$ such that in the ε-neighbourhood of any given equilibrium state of (A) other than 0 there are no other equilibrium states and such that in the ε-neighbourhood of any given limit cycle of the system (A) there are no other limit cycles.

that the curves $P(x, y) = 0$, $Q(x, y) = 0$ do not intersect simply but have a contact of a certain order. It is easily shown that in this case there are always analytic functions

$$\tilde{P}(x, y), \quad \tilde{Q}(x, y),$$

arbitrarily close to the functions $P(x, y)$, $Q(x, y)$ such that in an arbitrarily small ε-neighbourhood of $O(x_0, y_0)$ the curves

$$\tilde{P}(x, y) = 0, \quad \tilde{Q}(x, y) = 0$$

have more than one common point. This, however, implies that the system (A) cannot be coarse and the theorem is proved.

But when $\Delta(x_0, y_0) \neq 0$, the isoclines

$$P(x, y) = 0, \quad Q(x, y) = 0$$

have a simple intersection point at $O(x_0, y_0)$. It is easily shown that if we take the functions $\tilde{P}(x, y), \tilde{Q}(x, y)$ sufficiently close to $P(x, y)$, $Q(x, y)$, then the curves $\tilde{P}(x, y) = 0$, $\tilde{Q}(x, y) = 0$ will only have one common point, in the neighbourhood of $O(x_0, y_0)$. It is essential, for this to be true, that the partial desiratives of p and q are small. Of course, we cannot conclude that the condition $\Delta \neq 0$ is sufficient for an equilibrium state to exist in a coarse system.

Let us enumerate the equilibrium states possible when $\Delta \neq 0$. If

$$\sigma = P'_x(x_0, y_0) + Q'_y(x_0, y_0), \quad \Delta = \begin{vmatrix} P'_x(x_0, y_0) & Q'_x(x_0, y_0) \\ P'_y(x_0, y_0) & Q'_y(x_0, y_0) \end{vmatrix},$$

then, as we have seen (see Sections 2 and 4, Chapter V), the following cases are possible:

(1) $\Delta > 0$, $\sigma^2 - 4\Delta > 0$. The roots of the characteristic equation are real and of the same sign. The equilibrium state is a node (stable or unstable depending on the sign of σ).

(2) $\Delta < 0$. The roots of the characteristic equation are real and of different signs. The equilibrium state is a saddle point.

(3) $\Delta > 0$, $\sigma^2 - 4\Delta < 0$, $\sigma \neq 0$. The roots of the characteristic equation are complex conjugate. The equilibrium state is a focus (stable or unstable depending on the sign of σ).

It is easily verified that in (1), (2) and (3) the equilibrium state is "coarse", i.e. can exist in a coarse system.

(4) $\Delta > 0$, $\sigma = 0$. The roots of the characteristic equation are purely imaginary. Now, the character of the equilibrium state has not, in general,

been established (for a linear system the equilibrium state is a centre). This case is considerably more complicated than the cases (1), (2) and (3) and, in fact, the equilibrium state is always "non-coarse", i.e. cannot exist in a coarse system.

To investigate the case (4) and case (3) we assume here that the roots of the characteristic equation are complex conjugate.

Let the equilibrium state 0 be at the origin so that the system (A) is in its canonical form,

$$\begin{aligned} \frac{dx}{dt} &= ax - by + g(x, y), \\ \frac{dy}{dt} &= bx + ay + h(x, y), \end{aligned} \quad (6.7)$$

where $g(x, y), h(x, y)$ are power series expansions beginning with terms at least of the second order, and $\lambda_1 = a + jb$, $\lambda_2 = a - jb$, where $b \neq 0$; for $a \neq 0$ we have case (3) and for $a = 0$ case (4). The functions $g(x,y)$, $h(x,y)$, can also be written

$$g(x, y) = P_2(x, y) + P_3(x, y) + \ldots,$$
$$h(x, y) = Q_2(x, y) + Q_3(x, y) + \ldots,$$

where $P_i(x, y)$ and $Q_i(x, y)$ are homogeneous polynomials of degree i.

In polar coordinates equation (6.7) is

$$\begin{aligned} \frac{dr}{dt} &= \frac{ar^2 + r\cos\theta g(r\cos\vartheta, r\sin\vartheta) + r\sin\theta h(r\cos\vartheta, r\sin\vartheta)}{r} = \\ &= ar + r^2[P_2(\cos\theta, \sin\theta)\cos\theta + Q_2(\cos\theta, \sin\theta)\sin\theta] + \ldots \\ &\quad \ldots + r^i[P_i(\cos\theta, \sin\theta)\cos\theta + Q_i(\cos\theta, \sin\theta)\sin\theta] + \ldots \\ \frac{d\theta}{dt} &= \frac{1}{r^2}[br^2 + g(r\cos\theta, r\sin\theta) r\cos\theta - h(r\cos\theta, r\sin\theta) r\sin\theta] = \\ &= b + r[Q_2(\cos\theta, \sin\theta)\cos\theta - P_2(\cos\theta, \sin\theta)\sin\theta] + \ldots \\ &\quad \ldots + r^{i-1}[Q_i(\cos\theta, \sin\theta)\cos\theta - P_i(\cos\theta, \sin\theta)\sin\theta] + \ldots \end{aligned}$$

(6.8)

As $b \neq 0$, then for all sufficiently small r

$$\frac{d\theta}{dt} \neq 0.$$

the sign of b determining the sign of $d\theta/dt$, and therefore any half line

$$\theta = \text{const}$$

does not meet any paths at points sufficiently close to but differing from the origin.

From equation (6.8) we have

$$\frac{dr}{d\theta} = \frac{ar + r^2[P_2(\cos\theta, \sin\theta)\cos\theta + Q_2(\cos\theta, \sin\theta)\sin\theta] + \cdots}{b + r[Q_2\cos\theta - P_2\sin\theta] + r^2[Q_3\cos\theta - P_3\sin\theta] + \cdots} = R(r, \theta),$$

and since the denominator does not reduce to zero for $r=0$, then we can expand the right-hand side in a power series

$$\frac{dr}{d\theta} = R(r, \theta) = rR_1(\theta) + r^2 R_2(\theta) + \cdots, \qquad (6.9)$$

where the coefficient $R_i(\theta) = R_i(\theta + 2\pi)$ and the series converges for all θ, at least for sufficiently small values of r. It is easily seen that

$$\left.\begin{array}{l} R_1(\theta) = \dfrac{a}{b}, \\[2mm] R_2(\theta) = \dfrac{P_2\cos\theta + Q_2\sin\theta}{b} - \dfrac{a}{b^2}(Q_2\cos\theta - P_2\sin\theta). \end{array}\right\} \quad (6.10)$$

Let

$$r = f(\theta, r_0)$$

be the solution of the differential equation (6.9) such that

$$f(0, r_0) \equiv r_0.$$

It is clear that to every such solution of the equation (6.9) there corresponds a path of the system (A) which intersects the half line $\theta = 0$ at a point $(r_0, 0)$ and, conversely, to each path that intersects the half line $\theta = 0$ sufficiently close to the origin there corresponds a solution $r = f(\theta, r_0)$ where r_0 has some given value. In addition, it can be shown (from Theorem II in Appendix I) that all paths differing from the equilibrium state 0 and passing sufficiently close to 0 must intersect the straight line $\theta = 0$ at points near the origin. Therefore, the solution $r = f(\theta, r_0)$, for all sufficiently small r_0, corresponds to these paths, and since $R(r, \theta)$ is analytic, the function $f(\theta, r_0)$ will be analytic in θ and r_0 (Theorem III, Appendix I) and can be expanded as a power series in r_0. This series is convergent for $0 \leqslant \theta \leqslant \pi_2$ when $r_0 \leqslant \varrho_0$, where $\varrho_0 > 0$ is a certain sufficiently small quantity;

$$r = f(\theta, r_0) = u_1(\theta) r_0 + u_2(\theta) r_0^2 + \cdots \qquad (6.11)$$

Substituting the expression (6.11) in equation (6.9) we have

$$\left(\frac{du_1}{d\theta} r_0 + \frac{du_2}{d\theta} r_0^2 + \cdots\right) = R_1(\theta)(u_1 r_0 + u_2 r_0^2 + \cdots) + \\ + R_2(\theta)(u_1 r_0 + u_2 r_0^2 + \cdots)^2 + \cdots$$

Hence, from this identity we obtain the recurrent differential equations determining the functions $u_i(\theta)$:

$$\left. \begin{aligned} \frac{du_1}{d\theta} &= u_1 R_1 \\ \frac{du_2}{d\theta} &= u_2 R_1 + R_2 u_1^2, \end{aligned} \right\} \quad (6.12)$$

Remembering $f(0, r_0) = r_0$, then clearly

$$u_1(0) = 1, \quad u_i(0) = 0, \quad i = 2, 3, \ldots$$

which, together with the equations (6.12), determine the functions $u_i(\theta)$. In particular

$$u_1(\theta) = e^{\frac{a}{b}\theta},$$

so that in the case (4), when $a=0$, it follows that

$$u_1(\theta) = 1.$$

Since the straight line $\theta=0$ is not tangential to the paths of (A) at any points sufficiently close to the origin 0, then a small segment of this straight line with one end at 0 will be analogous to a segment without contact. If $\theta=2\pi$ then for every given $r_0(0 < r_0 < \varrho_0)$ a value of r corresponds to the "last" intersection of a path with the half line $\theta=0$, and the function $r = f(2\pi, r_0) = u_1(2\pi)r_0 + \ldots$, is completely analogous to the sequence function discussed at Section 7, Chapter V. From this function, we can examine the character of the paths in a neighbourhood of the equilibrium state 0.

To do this introduce the function $\psi(r_0) = f(2\pi, r_0) - r_0 = \alpha_1 r_0 + \alpha_2 r_0^2 + \ldots$, where

$$\alpha_1 = u_1(2\pi) - 1 = e^{2\pi\frac{a}{b}} - 1,$$

$$\alpha_k = u_k(2\pi), \quad k > 1.$$

Obviously, only the values of r_0, for which

$$\Psi(r_0) = f(2\pi, r_0) - r_0 = 0,$$

correspond to closed paths. Note that in case (4) when $a=0$, then $\alpha_1=0$. In addition, the coefficients α_i possess the following properties: if $\alpha_1=0$, then necessarily $\alpha_2=0$, or in general, if $\alpha_1 = \alpha_2 = {'\ldots'} = \alpha_{2n-1} = 0$, then

necessarily $\alpha_{2n}=0$, so the first non-zero coefficient is always of odd order[†].
There are now two possibilities:

(a) At least one coefficient α_j is non-zero
(b) All coefficients α_i are zero 0.

Let α_j be the first of the coefficients differing from zero. Then for all sufficiently small $r_0 > 0$

$$\Psi(r_0) = \alpha_j r_0^j + \ldots$$

is different from zero, and paths passing sufficiently close to 0 are spirals that tend to 0 either for $t \to +\infty$ when $\alpha_j < 0$ and $b > 0$ (and so $\psi(r_0) < 0$ and $d\psi/dt > 0$) and when $\alpha_j > 0$ and $b < 0$ (and so $\psi(r_0) > 0$ and $d\psi/dt < 0$) or for $t \to -\infty$ when $\alpha_j < 0$ and $b < 0$, (and so $\psi(r_0) < 0$ and $d\psi/dt < 0$) and when $\alpha_j > 0$ and $b < 0$, ($\psi(r_0) > 0$ and $d\psi/dt > 0$). The equilibrium state is a focus. This focus can be stable or unstable according to the signs of b and α_j. When $j > 0$ we will call the equilibrium state a *multiple focus of multiplicity j* or a *j-tuple focus*. For $a \neq 0$ then $j = 1$, as we have already seen.

Otherwise when $\alpha_i = 0$, $\psi(r_0) \equiv 0$ and all paths that pass through a sufficiently small neighbourhood of 0 are closed, so that 0 is a *centre*. It can be shown in this case that (A) possesses an analytical integral,

$$x^2 + y^2 + A_3(x, y) + \ldots = C.$$

As a preliminary to showing that in a coarse system there cannot be a multiple focus or a centre, we make the following remarks.

Let us consider the modified system $(\tilde{A})$, sufficiently close to (A) and of the form

$$\begin{aligned} \frac{dx}{dt} &= \tilde{a}x - \tilde{b}y + \tilde{g}(x, y), \\ \frac{dy}{dt} &= \tilde{b}x + \tilde{a}y + \tilde{h}(x, y), \end{aligned} \quad (6.13)$$

In polar coordinates we find an equation analogous to (6.9):

$$\frac{dr}{d\theta} = \tilde{R}(r, \theta) = \tilde{R}_1(\theta)r + \tilde{R}_2(\theta)r^2 + \ldots, \quad (6.14)$$

If

$$r = \tilde{f}(\theta, r_0) = \tilde{u}_1(\theta)r_0 + \tilde{u}_2(\theta)r_0^2 + \ldots$$

---

† If $a = 0$, then in the formulae (6.12) $R_1 = a/b = 0$, and so $du_1/d\theta = 0$, leading to, $u_1 = 1$. The second of the equations (6.12) will then be $du_2/d\theta = R_2(\theta)$, but $R_2(\theta)$ is a homogeneous function in cos $|\theta$ and sin $|\theta$ of the third degree. Integrating the last equation from 0 to 2π then $u_2(2\pi) = 0$ since $u_2(0) = 0$. It can be shown similarly that the first non-zero coefficient α_j is of an odd order (see [84] and also [13]).

is a solution of the equation (6,14), then, the functions $\tilde{u}_i(\theta)$ satisfy (6.12) with $R_i(\theta)$ replaced by $\tilde{R}_i(\theta)$. Therefore

$$\tilde{u}_1(\theta) = e^{\frac{\tilde{a}}{b}\theta},$$

and for system $(\tilde{A})$ there is a corresponding sequence function

$$r = \tilde{f}(2\pi, r_0),$$

and also the function

$$\tilde{\Psi}(r_0) = \tilde{f}(2\pi, r_0) - r_0.$$

By the Theorem V of Appendix I it is easily shown that for any system $(\tilde{A})$ sufficiently close to (A), the function

$$r = \tilde{f}(2\pi, r_0)$$

is also defined for all values of r_0 where $0 < r_0 < \varrho_0$ and is, with its derivative, arbitrarily close to the function $f(2\pi, r_0)$ and its derivative.

We now prove the following theorem:

THEOREM II. *A coarse system cannot have equilibrium states for which*

$$\varDelta > 0, \quad \sigma = P'_x + Q'_y = 0.$$

We begin by assuming the contrary to hold, i.e. that a coarse system (A) has an equilibrium state for which the conditions are valid. If this state is at the origin then the equations for system (A) are

$$\left. \begin{array}{l} \dfrac{dx}{dt} = -by + g(x, y) = P(x, y), \\[2mm] \dfrac{dy}{dt} = bx + h(x, y) = Q(x, y). \end{array} \right\} \qquad (6.15)$$

The two possibilities (a) and (b) first discussed could apply, and so the equilibrium state is either a multiple focus or a centre. The modified system $(\tilde{A})$ is

$$\left. \begin{array}{l} \dfrac{dx}{dt} = \tilde{a}x - by + g(x, y), \\[2mm] \dfrac{dy}{dt} = \tilde{a}y + bx + h(x, y), \end{array} \right\} \qquad (6.16)$$

for which $\tilde{a} \neq 0$ (the sign of $\tilde{a}$ is yet to be chosen).

Let

$$\Psi(r_0) = f(2\pi, r_0) - r_0,$$
$$\tilde{\Psi}(r_0) = \tilde{f}(2\pi, r_0) - r_0$$

be functions constructed respectively for the systems (A) and $(\tilde{A})$ and defined for all $0 < r_0 < \varrho_0$. The two cases (a) and (b) can now be considered separately for (A).

(a) *The equilibrium state $O(0, 0)$ of (A) is a multiple focus.* Let α_{2k+1} be the first non-zero coefficient, and we assume, to be definite, that $b > 0$ and $\alpha_{2k+1} < 0$, so that the multiple focus of (A) is stable (when unstable the analysis is similar). Now, the function $\Psi(r_0)$ has the form

$$\Psi(r_0) = r_0^{2k+1}(\alpha_{2k+1} + \ldots),$$

and there is always a small $r_0' < \varrho_0$ for which

$$\Psi(r_0') < 0.$$

However, there is a modified system $(\tilde{A})$ (see (6,16)) so close to the system (A) that for all

$$0 < r_0 < \varrho_0$$

the corresponding function $\tilde{\Psi}(r_0)$ is arbitrarily close to the function $\Psi(r_0)$ so that

$$\tilde{\Psi}(r_0') < 0.$$

On the other hand, the sign of $\tilde{\Psi}(r_0)$

$$\tilde{\Psi}(r_0) = r_0(\tilde{\alpha}_1 + \ldots)$$

is the same as the sign of $\tilde{\alpha}_1$ for all sufficiently small r_0 less than r_0'.

If we take $\tilde{\alpha} > 0$, then $\tilde{\alpha}_1 = e^{2\pi\tilde{a}/b} - 1 > 0$ and there is an $r_0'' < r_0'$ for which

$$\tilde{\Psi}(r_0'') > 0,$$

and

$$\tilde{\Psi}(r_0'') > 0, \quad \tilde{\Psi}(r_0') < 0.$$

Hence, there is bound to exist an $r_0^*(r_0'' < r_0^* < r_0')$ such that $\tilde{\Psi}(r_0^*) = 0$. This means that through a point $r = r_0^*$ of the half line $\theta = 0$ there is a closed path — a limit cycle — of the system $(\tilde{A})$. It is easily verified that the smaller $\tilde{\alpha}$ the smaller is the neighbourhood around 0 in which the limit cycle lies.

If (A) is coarse then, in a certain small neighbourhood of 0, the phase portraits determined by (A) and $(\tilde{A})$ must be identical. This, however, is clearly impossible, since we can always take a neighbourhood of the point 0 such that in it there is no limit cycle of the system (A), while, from what we just proved, for a sufficiently small $\tilde{\alpha} < 0$ a limit cycle of $(\tilde{A})$ will be in this neighbourhood. The original assumption leads to a contradiction.

(b) *The equilibrium state 0 of (A) is a centre.* For $\bar{a} \neq 0$ the equilibrium state 0 of $(\bar{A})$ is a focus (stable or unstable depending on the sign of $\bar{a}$). Hence, the equilibrium state 0 has a different character for (A) and $(\bar{A})$, and the system (A) cannot be coarse. Thus the theorem is proved.

It follows from Theorems I and II that in a coarse system only simple equilibrium states of the types (1), (2) and (3) are possible. These states of equilibrium are called "coarse" because the phase portraits in near neighbourhoods of the state and the state of a modified close system are topologically identical. In particular, when the point 0 of (A) is a saddle point, the point $\tilde{0}$ of $(\bar{A})$ is also a saddle point, and the separatrices of the saddle point $\tilde{0}$ are displaced a small amount from the separatrices of the saddle point 0 of system (A).

3. Simple and multiple limit cycles. Coarse limit cycles

To establish the conditions for a closed path to exist in a coarse system, we first consider the neighbourhood of an arbitrary closed path that is not necessarily a path of a coarse system. The analysis is analogous to that carried out for a multiple focus and a centre. Thus, let L_0 be a closed path, with

$$x = \varphi(t), \quad y = \psi(t)$$

any periodic motion on L_0 with period τ. l is a segment without contact or transversal drawn through an arbitrary point Q on L_0 and with Q as an internal point. Let s be a parameter on this segment and

$$\bar{s} = f(s)$$

be the sequence function on this segment (see Section 7, Chapter V). The functions $f(s)$ and $\Psi(s) = f(s) - s$ are analytic functions of s (see Section 7, Subsection 3, Chapter V).

If $s = s_0$ corresponds to the point Q on l, through which passes L_0, then, clearly,

$$\Psi(s_0) = f(s_0) - s_0 = 0.$$

If the characteristic exponent h of L_0 is not zero, then, as is known (see Section 7, Chapter V), for $h < 0$ when $d\bar{s}/ds < 1$ and $\Psi'(s_0) < 0$, the path L_0 is a stable limit cycle, and for $h > 0$ when $d\bar{s}/ds > 1$ and, hence $\Psi'(s_0) > 0$, the path L_0 is an unstable limit cycle.

In both these cases $s = s_0$ is a simple root of $\Psi(s) = 0$. Therefore, when $h \neq 0$, the limit cycle is simple.

Let us now examine the case not analysed in Chapter V, § 7, when $h=0$, i.e. $d\bar{s}/ds=1$ and $\Psi'(s_0)=0$. Now $s=s_0$ is a multiple root of $\Psi(s)=0$. There are two alternatives:

(1) At least one of the derivatives of $\Psi(s)$ does not reduce to zero at $s=s_0$, i.e. an integer $k>1$ exists such that

$$\Psi'(s_0) = \ldots = \Psi^{(k-1)}(s_0) = 0; \qquad \Psi^{(k)}(s_0) \neq 0,$$

and therefore

$$\Psi(s) = (s-s_0)^k \left[\Psi^{(k)}(s_0) + (s-s_0)\Psi^{(k+1)}(s_0) + \ldots\right].$$

Hence, there is always a number $d>0$ such that for all $s \neq s_0$ and satisfying

$$|s-s_0| < d,$$

$\Psi(s)$ does not reduce to zero, so that the part of l for which $(s-s_0)<d$ is only intersected by one closed path L_0. This closed path L_0 is called a *multiple limit cycle of order* k.

Consider the case when k is odd, and assume that $\Psi^{(k)}(s_0)<0$. Then, for $s<s_0$

$$\Psi(s) > 0, \quad \text{i.e.} \quad f(s) > s,$$

and for $s>s_0$

$$\Psi(s) < 0, \quad \text{i.e.} \quad f(s) < s.$$

Therefore, each consecutive point on l is closer to the point Q than the preceding one $((t_2)>(t_1))$. L_0 is the only closed path intersecting this part of l, as in Theorem IV in § 2 of this Chapter, each path differing from L_0 but intersecting l sufficiently close to Q tends to the limit cycle L_0 as $t \to +\infty$. The limit cycle L_0 is stable and of odd order.

If $\Psi^{(k)}(s_0)>0$, it can be shown similarly that every path intersecting l sufficiently close to Q tends to the limit cycle L_0 as $t \to -\infty$. The limit cycle L_0 is unstable and of odd order.

Consider now the case when k is even. Then for all $s \neq s_0$ we have either $\Psi(s)>0$ and $f(s)>s$ if $\Psi^{(k)}(s_0)>0$, or $\Psi(s)<0$ and $f(s)<s$ if $\Psi^{(k)}(s_0)<0$. When $\Psi^{(k)}(s_0)>0$, all paths that pass through points of l corresponding to $s<s_0$ tend to L_0 as $t \to +\infty$, and all paths that pass through points of l corresponding to $s>s_0$ tend to L_0 as $t \to -\infty$, and vice versa when $\Psi^{(k)}(s_0)<0$.

It is clear that in this case (even k) the limit cycle L_0 is unstable, but a limit cycle of this type is sometimes termed "semi-stable" (of an even order), the term "unstable" being reserved for a cycle to which adjacent paths tend as $t \to -\infty$.

For $k>1$ the limit cycle is also called a "multiple limit cycle".

(2) All derivatives of $\Psi(s)$ are zero at $s=s_0$,

$$\Psi^{(i)}(s_0) = 0.$$

Then, clearly, since $\Psi(s)$ is analytic

$$\Psi(s) \equiv 0;$$

and the sequence function is simply

$$\bar{s} = s.$$

Therefore all paths passing through points sufficiently close to L_0 are closed.

These results can be represented on Lamerey's diagram, on which the sequence function, $\bar{s}=f(s)$, and the straight line $\bar{s}=s$ are plotted. Closed paths correspond to values of s for which

$$f(s) = s,$$

i.e. the intersection points on the diagram. If the common point is a simple point of intersection, then the corresponding closed path is a limit cycle for which $d\bar{s}/ds \neq 1$. If the common point is a multiple point of contact then the limit cycle will be a multiple limit cycle. In particular when $\bar{s}=f(s)$ coincides with the line $\bar{s}=s$ we have case (2). However, we will now show that a multiple limit cycle for which $k>1$ (the case (1)), or a closed path in whose neighbourhood all paths are closed (the case (2)), cannot exist in a coarse system.

A segment without contact l for the paths of the system (A) will also be a segment without contact for the paths of any modified system $(\tilde{A})$ sufficiently close to (A). In addition, if s_1 and $s_2 (s_1 < s_2)$ correspond to points of l which are not its ends, then it can be shown from the Theorems IV and V of Appendix I that for all $s_1 \leqslant s \leqslant s_2$ we can define on l a sequence function for $(\tilde{A})$

$$\bar{s} = \tilde{f}(s),$$

which with its derivative $\tilde{f}'(s)$ differs arbitrarily little from the function $f(s)$ and $\tilde{f}'(s)$ respectively, provided $(\tilde{A})$ is sufficiently close to (A).

It is natural to expect, from what has been said, that a closed path for which the characteristic exponent $h=0$, cannot exist in a coarse system. In fact let R_0 be the common point of $\bar{s}=f(s)$ and $\bar{s}=s$ corresponding to such a closed path. At R_0 the curve $\bar{s}=f(s)$ is either tangential to the straight line $\bar{s} = s$ (the case (1)) or coincides with the straight line $\bar{s} = s$ (the case (2)). In both these cases, there is a function $\bar{s} = \tilde{f}(s)$ arbitrarily close to $f(s)$ such that $\bar{s} = \tilde{f}(s)$ either intersects $\bar{s} = s_0$ more than once in an arbitrarily small neighbourhood of R_0, or has no point of intersection (the case when $\bar{s} = f(s)$ has tangency of even order with $\bar{s} = s$, see Fig. 313, and also the case when it coincides with the straight line $\bar{s} = s$). If, moreover, we prove that there is a modified system $(\tilde{A})$ arbitrarily close to (A) for which such a function $\tilde{f}(s)$ is the sequence function on the segment l, then, clearly this will mean that for suitably chosen small variations of the right-hand sides of the system (A), the closed path is either split into a number of limit cycles or vanishes (k even or case (2)). Hence the system (A) cannot be coarse. Thus the proof that in a coarse system there are no multiple limit cycles can be carried out by constructing a modified system $(\tilde{A})$ for which the sequence function $\tilde{f}(s)$ exhibits the required properties.

We state first one auxiliary lemma:

LEMMA. *There is a function*

$$z = F(x, y),$$

defined in a region G and having continuous partial derivatives up to at least the second order, such that:

(1) $F(\varphi, \psi) \equiv 0$ *(i.e. the function $z = F(x, y)$ reduces to zero at the points of the path L_0)*; (2) $[F'_x(\varphi, \psi)]^2 + [F'_y(\varphi, \psi)]^2 \neq 0$

This lemma has a very simple geometric meaning. In the x, y, z space the function $z = F(x, y)$ represents a smooth surface that passes through the path L_0 lying in the x, y plane and is not tangent at any point of L_0 with the x, y plane[†].

[†] The proof of the existence of this function $F(x, y)$ is not too difficult. Let us consider the curvilinear system of coordinates (u, v) introduced in Chapter V, Section 7, Subsection 3 (see (5.55)). The curves $v = $ const. are closed curves, the curve $v = 0$ being clearly the closed path L_0. At the points of L_0,

$$D = \begin{vmatrix} \varphi'(u) - v\varphi''(u), & -\psi'(u) \\ \psi'(u) + v\varphi''(u), & \varphi'(u) \end{vmatrix}$$

does not reduce to zero. Therefore, in a neighbourhood of each point of L_0 we can write u as $u = \Phi(x, y)$. It is easily verified that the function $u = \Phi(x, y)$ is a single-valued analytic function defined in a certain neighbourhood of L_0 and that on L_0 it is zero. It is easily shown, in addition, that the function $\Phi(x, y)$ satisfies the conditions of our lemma. The function $\Phi(x, y)$ has been defined only in a certain, generally speaking, small neighbourhood of the path L_0. However, by known theorems on the continuation of a function, it is always possible to find a function $z = F(x, y)$ defined in the whole region in which the system (A) is defined and coinciding with $\Phi(x, y)$ in a certain neighbourhood of the path L_0.

This function $F(x, y)$ enables us to construct a modified system $(\tilde{A})$ possessing the required properties.

The theorem that establishes the conditions to be satisfied in order that a closed path may exist in a coarse system is:

THEOREM III. *In a coarse system there are no closed paths for which*

$$h = \frac{1}{\tau} \int_0^\tau [P'_x(\varphi, \psi) + Q'_y(\varphi, \psi)] \, dt = 0.$$

If a closed path L_0 of (A) having the parametric equations

$$x = \varphi(t), \quad y = \psi(t),$$

satisfies the condition

$$h = 0,$$

then, by the foregoing, either this closed path is a multiple limit cycle of order $k(k>1)$ (the case (1)) and then there exists a neighbourhood of L_0 that does not contain any closed path except L_0, or all paths in a neighbourhood of L_0 are closed. We shall consider first the case (1).

Let us arrange that Q on l and L_0 corresponds to $s=0$. Then if $s = f(s)$ is the sequence function on l and $\psi(s) = f(s) - s$, we have $\psi(0) = 0$ and, in this case, $\psi'(0) = \psi''(0) = \ldots, = \psi^{(k-1)}(0) = 0$ but $\psi^{(k)}(0) \neq 0$. We assume, that $\psi^{(k)} > 0$ (in the case $\psi^{(k)} < 0$ the reasoning is similar). First consider an auxiliary modified system whose right-hand sides are not analytic†

$$\left. \begin{array}{l} \dfrac{dx}{dt} = P(x, y) + \lambda F(x, y) F'_x(x, y) = P^*(x, y), \\[6pt] \dfrac{dy}{dt} = Q(x, y) + \lambda F(x, y) F'_y(x, y) = Q^*(x, y), \end{array} \right\} \qquad (6.17)$$

where λ is a parameter and the function $F(x, y)$ satisfies the preceding lemma, so that the right-hand sides of this system have continuous partial derivatives of the first order. We shall call (6.17) the system (A^*_λ).

Since, by the choice of $F(x,y)$

$$F[\varphi(t), \psi(t)] \equiv 0,$$

† Such systems have not been considered before. However, if the right-hand sides of a system are not analytic but have continuous partial derivatives, then the Theorem I on the existence and uniqueness of the solution and also the Theorem II of Appendix I are satisfied for such a system. If the function $F(x, y)$ having the properties (1) and (2) of the lemma were analytic, then the system (A^*_λ) considered below would also be analytic and the subsequent arguments of this theorem would be considerably simplified. However, a rigorous proof of the existence of an *analytic* function satisfying the conditions (1) and (2) of the lemma is considerably more complicated than the argument given below.

then, clearly,
$$x = \varphi(t), \quad y = \psi(t)$$
is a solution of (A_λ^*), and the path L_0 is also a path of the system (A_λ^*). Evidently, for all sufficiently small values of λ, the system (A_λ^*) will be arbitrarily close to the system (A). We only consider such small values of $\lambda(|\lambda|<\eta)$ for which l remains a segment without contact for the system (A^*) as well as (A). Let
$$\bar{s} = f^*(s, \lambda)$$
be the sequence function for (A_λ^*) on l and $\psi^*(s, \lambda) = f^*(s, \lambda) - s$.

To find $\bar{s} = f^*(s, \lambda)$ since L_0 is a path both of the (A) and (A_λ^*), we can use the same system of curvilinear coordinates u, v (see Section 7, Chapter V) as in the case of the system (A). Let the equation analogous to the equation (5.56) for the system (A_λ^*) be $dv/du = g^*(u, v, \lambda)$ with the solution $v = \Phi^*(u, s, \lambda)$ being equal to s for $u = 0$ (we can arrange l to be a segment on the line $u = 0$). Then the sequence function $f^*(s, \lambda) = \Phi^*(\tau, s, \lambda)$ where τ is the period on L_0. As the right-hand sides of the system (A_λ^*) and $\Phi^*(\tau, s, \lambda)$ are not analytic, the function $\psi^*(s, \lambda) = f^*(s, \lambda) - s$ is also non-analytic and the reasoning carried out in Chapter V, Section 7 (being based on the fact that the functions $g^*(u, v, \lambda)$ and $\Phi^*(u, s, \lambda)$ can be expanded in series) cannot be used here. It is easily shown, however, that the function $g^*(u, v, \lambda)$ must necessarily have continuous partial derivatives of the first order. Hence it follows, by known theorems, that the function $\Phi^*(u, s, \lambda)$ has a continuous derivative with respect to s and this derivative is a solution of the differential equation
$$\frac{d}{du}\frac{\partial \Phi^*}{\partial s} = \frac{\partial g^*}{\partial v}\frac{\partial \Phi^*}{\partial s}.$$

From this equation we obtain, just as in Chapter V, Section 7, Sub-section 3,
$$f^{*\prime}(0) = e^{\int_0^\tau (P_x^{*\prime} + Q_y^{*\prime})\, dt}.$$

By the Theorem V of Appendix I, the function $\Psi^*(s, \lambda)$ and its derivative are arbitrarily close to the function $\Psi(s)$ and its derivative, for sufficiently small values of λ. But L_0 is also a path of the system (A_λ^*) and clearly we have
$$\Psi^*(0, \lambda) = 0.$$

From the expressions for $P^*(x, y, \lambda)$ and $Q^*(x, y, \lambda)$ and also that, by hypothesis
$$h = \frac{1}{\tau}\int_0^\tau (P_x' + Q_y')\, dt = 0,$$
we have

and
$$f^{*\prime}(0, \lambda) = e^{\int_0^\tau (P_x^{*\prime}+Q_y^{*\prime})\,dt} = e^{\lambda \int_0^\tau (F_x^{\prime 2}+F_y^{\prime 2})\,dt} > 0$$

and

$$\Psi^{*\prime}(0, \lambda) = e^{\lambda \int_0^\tau [F_x^{\prime 2}+F_y^{\prime 2}]\,dt} - 1.$$

Thus, the closed path L_0 is a simple limit cycle for the system A_λ^*, stable for $\lambda < 0$ and unstable for $\lambda > 0$.

By hypothesis we have
$$\Psi^{(k)}(0) > 0,$$
and since
$$\Psi(s) = \Psi^{(k)}(0)s^k + \ldots,$$
there is always an $s_1 < 0$ as small as we please, such that
$$\Psi(s_1) > 0.$$

However, for all sufficiently small values of λ, the function $\Psi^*(s, \lambda)$ differs arbitrarily little from the function $\Psi(s)$ and therefore we can always find a λ^* (of any sign) such that

$$\Psi^*(s_1, \lambda^*) > 0. \tag{6.18}$$

However, if we take $\lambda^* < 0$ we find

$$\Psi^{*\prime}(0, \lambda^*) = e^{\lambda^* \int_0^\tau (F_x^{\prime 2}+F_y^{\prime 2})\,dt} - 1 < 0,$$

and so it is always possible to choose an $s_2 > 0$ ($s_2 > s_1$) that

$$\Psi^*(s_2, \lambda^*) < 0. \tag{6.19}$$

Therefore, from (6.18) and (6.19), it follows that in (A_λ^*) there is one more closed path besides L_0 intersecting the segment l, for a certain value of s lying between s_1 and s_2.

Finally, because
$$\Psi^*(0, \lambda^*) = 0, \quad \text{while} \quad \Psi^{*\prime}(0, \lambda^*) < 0,$$

it is always possible to find an $s_3 > 0$ such that $\Psi^*(s_3, \lambda^*) > 0$, i.e. $\Psi^*(s, \lambda)$ reduces to zero once more in the interval $s_3 < s < s_2$ and the system (A_λ^*) has one more closed path (in addition to L_0) that intersects the segment l over the interval.

It is always possible, however, to construct a modified system $(\tilde{A})$, for any fixed λ^*,

$$\frac{dx}{dt} = \tilde{P}(x, y); \quad \frac{dy}{dt} = \tilde{Q}(x, y), \tag{6.20}$$

whose right-hand sides are analytic in x and y, and so close to the system (A^*_λ) that we also have

$$\tilde{\Psi}(s_3) > 0, \quad \tilde{\Psi}(s_2) < 0, \quad \tilde{\Psi}(s_1) > 0,$$

where $\tilde{\Psi}(s)$ is the sequence function of $(\tilde{A})$ similar to $\Psi(s)$. Now, there are bound to exist values $\bar{s}_1$ and $\bar{s}_2$ for which

$$\tilde{\Psi}(\bar{s}_2) = 0 \quad \text{and} \quad \tilde{\Psi}(\bar{s}_1) = 0,$$

and so there are no less than two closed paths in the system $(\tilde{A})$ intersecting l at points corresponding to $\bar{s}_1$ and $\bar{s}_2$. By choosing sufficiently small values of s and λ, and a system $(\tilde{A})$ sufficiently close to (A^*_λ) it is always possible to arrange that $(\tilde{A})$ is arbitrarily close to (A) and its closed paths lie in an arbitrarily small neighbourhood of the path L_0. Hence, the system (A) cannot be "coarse" and thus for Case I (a multiple limit cycle) the theorem is proved.

In Case II, when all the paths are closed, we consider the same auxiliary non-analytic system (A^*_λ), and we have

$$f^{*\prime}(0, \lambda^*) = e^{\lambda * \int\limits_0^\tau (P^{*\prime}_x + Q^{*\prime}_y)\, dt}$$

Therefore, over the interval of s the function $\Psi^*(s, \lambda)$ is not identically equal to zero, and is easily seen that every system $(\tilde{A})$ with analytic right-hand sides, sufficiently close to the system (A^*_λ), has a corresponding function $\tilde{\Psi}(s)$ that is also not identically equal to zero. This means, however, that among the paths of $(\tilde{A})$ which intersect l (in the interval of s) there are both closed and non-closed ones. Since we can find a system $(\tilde{A})$ arbitrarily close to the system (A) and exhibiting this property, then clearly, the system (A) cannot be coarse.

We can now state that when the limit cycle L_0 of (A) is simple ($h \neq 0$, and $f'(0) \neq 1$ and $\Psi'(0) \neq 0$), this limit cycle is "coarse", i.e. can exist in a coarse system. The point of intersection R_0 between $\bar{s} = f(s)$ and $\bar{s} = s$ is a simple point of intersection where the curve $\bar{s} = f(s, \lambda)$ is not tangent to $\bar{s} = s$. Then the curve $\bar{s} = f(s, \lambda)$ corresponding to an arbitrary function $\tilde{f}(s)$ sufficiently close to $f(s)$ will have one and only one common point

R with the straight line $\bar{s}=s$, R being arbitrarily close to R_0†. Clearly, from this it follows that every modified system $(\tilde{A})$ sufficiently close to the system (A), will have one and only one limit cycle $\tilde{L}_0$ arbitrarily close to the limit cycle L_0 of (A). By virtue of the fact that $\tilde{f}'(s)$ differs arbitrarily little from $f'(s)$, this limit cycle $\tilde{L}_0$ will be stable if the limit cycle L_0 is stable and will be unstable if the limit cycle L_0 is unstable.

It can be shown from this that the phase portraits near L_0 of the paths of the system (A) and of the paths of $(\tilde{A})$ are separated by only a small distance.

Note that, both for equilibrium states and for limit cycles, the requirement of coarseness imposes an analytic condition on the system of differential equations. From the topological point of view, for simple and multiple equilibrium states and for simple and multiple limit cycles the mapping of a neighbourhood by the paths can be the same (for example, for a multiple limit cycle of an odd order and for a simple limit cycle).

4. Behaviour of a separatrix of saddle points in coarse systems

The requirement of coarseness imposes a restriction on the character of the separatrices. If a separatrix of the saddle point 0, tending to this saddle point for $t \to +\infty$, also tends to a saddle point (different from 0 or coinciding with 0) for $t \to -\infty$, then we shall say briefly that this separatrix "goes from saddle point to saddle point".

THEOREM IV. *In coarse systems there are no separatrices that go from saddle point to saddle point.*

To prove the theorem we assume that in the coarse system (A) there is a separatrix L going from one saddle point 0 to another saddle point $0'$ (Fig. 229). The proof when L returns to the same saddle point (Fig. 293) is completely analogous.

Consider a modified system $(\tilde{A}_\alpha)$ of the form

$$\frac{dx}{dt} = P - \alpha Q, \quad \frac{dy}{dt} = Q + \alpha P, \tag{6.21}$$

which has the same equilibrium states as the system (A) (and only those), for

$$P - \alpha Q = 0; \quad Q + \alpha P = 0$$

simultaneously, only when

$$P = 0 \quad \text{and} \quad Q = 0.$$

† It will be seen that it is essential that the functions $f(s)$ and $\tilde{f}(s)$ and their derivatives $\tilde{f}'(x)(s)$ and $f'(s)$ are close functions.

The angle φ between the tangents at a point $M(x, y)$ to path of (A) and a path of $(\tilde{A}_\alpha)$ will be given by

$$\tan \varphi = \frac{\dfrac{Q+\alpha P}{P-\alpha Q} - \dfrac{Q}{P}}{1 + \dfrac{Q+\alpha P}{P-\alpha Q} \cdot \dfrac{Q}{P}} = \alpha,$$

and is the same everywhere in the region G. This amounts to saying that the vector field of the system $(\tilde{A}_\alpha)$ is rotated through a constant angle with respect to the vector field of (A).

Therefore, the points 0 and 0′ are also equilibrium states of $(\tilde{A}_\alpha)$ and since by hypothesis the system (A) is coarse, these points 0 and 0′ must be saddle points of the systems $(\tilde{A}_\alpha)$ with a separatrix L_α going from the saddle point 0 to the saddle point 0′. There is always a small $\varepsilon > 0$ such that the ε-neighbourhood of L contains only the equilibrium states 0 and 0′ of the system (A) and does not contain any closed path in its entirety (see the corollaries I and II of the index theory and Section 8, Chapter V), nor contains in its entirety any separatrix of the saddle points 0 and 0′ except L. For all sufficiently small values of α, the separatrix L_α of the system $(\tilde{A}_\alpha)$ will lie entirely in this ε-neighbourhood of L. The separatrices L and L_α may or may not have common points.

We assume first that they have no common points and there is a simple closed curve C_0, consisting of L, L_α and the saddle points 0 and 0′ and lying in the ε-neighbourhood of L. The separatrix L_α of $(\tilde{A}_\alpha)$ is, clearly, an "arc without contact" for the paths of the system (A) (since the field of the system $(\tilde{A}_\alpha)$ is rotated by a constant angle with respect to the field of the system (A)) which intersect L_α from the same direction. Among the paths of the system (A) that intersect L_α let L' be one that is not a separatrix of 0 or 0′. Let us assume, for example, that L' enters C_0 as t increases. As t increases further, L' cannot leave C_0, since it cannot intersect L (L and L' are paths of (A)) nor can it intersect L_α again (since then it would intersect L_α in two directions). Hence, as $t \to +\infty$, L' must tend to a limit set lying entirely in C_0 and thus in the ε-neighbourhood chosen of L. In this neighbourhood, however, there cannot be any limit set (see Subsection 5, Section 2 of this chapter), since it is easily shown that in such a limit set there would enter at least one separatrix of the saddle points 0 and 0′ different from L, and in the ε-neighbourhood of L there is no other separatrix but L_α. This contradicts the hypothesis and thus the theorem is proved for this case.

In the case when L and L_α have common points we need only consider a simple closed curve C_0, consisting of the point 0 and of the paths L

and L_α between the point 0 and their nearest common point or between two adjacent common points, and use the same argument as above. This finally proves the theorem.

5. Necessary and sufficient conditions of coarseness

Combining the results above we can now formulate the necessary conditions for the coarseness (or structural stability) of a system (A) in a region G.

I. The system must have only simple (coarse) equilibrium states, i.e. equilibrium states for which the real parts of the roots of the characteristic equation are different from zero. An alternative statement is that there cannot be equilibrium points $x=x_0$, $y=y_0$ for which

(a)
$$\varDelta = \begin{vmatrix} P'_x(x_0, y_0), & P'_y(x_0, y_0) \\ Q'_x(x_0, y_0), & Q'_y(x_0, y_0) \end{vmatrix} = 0;$$

or (b) for $\varDelta > 0$, $\sigma = [P'_x(x_0, y_0) + Q'_y(x_0, y_0)] = 0$.

II. There must be only simple (coarse) limit cycles, each with a non-zero characteristic exponent h.

III. There cannot be separatrices joining two saddle points.

Thus, in a coarse system, the only possible singular paths are: simple (coarse) equilibrium states; simple (coarse) limit cycles; and separatrices of saddle points that lead to a node, a focus, or a limit cycle; or the boundary of G which is a cycle without contact. An equilibrium point cannot be a centre, and all limit paths are simple equilibrium points or limit cycles.

Thus structural stability or coarseness implies the exclusion of singular paths of the multiple type. The conditions I, II and III enunciated above are necessary conditions for a given system to be coarse, and it can be shown (see De Baggis) that these conditions are also *sufficient*. Thus, the following converse theorem, basic to the theory of coarse systems, is valid.

THEOREM V. *The system*
$$\frac{dx}{dt} = P(x, y), \quad \frac{dy}{dt} = Q(x, y)$$

in a region G bounded by a cycle without contact C is coarse if it has

(1) *only states of equilibrium for which $\varDelta \neq 0$ and, if $\varDelta > 0$, $\sigma \neq 0$;*
(2) *only limit cycles for which $h \neq 0$;*
(3) *only separatrices that do not join saddle points.*

The rigorous proof of this theorem consists in constructing, for every modified system $(\bar{A})$ sufficiently close to the system (A), a topological transformation of the region G into itself, for which the paths of the system $(\bar{A})$ are transformed into paths of the system (A) and points corresponding to each other are found at an arbitrarily small distance from each other.

6. Classification of the paths possible in coarse systems

We assume to be definite that all the paths of (A) crossing the cycle without contact-C which is the boundary of G, enter G as t increases. There are sixteen different types of paths grouped into five basic types in the table below and shown in Fig. 304.

Singular (orbitally unstable) paths

I. States of equilibrium	stable focus (or node)	(1)
	unstable focus (or node)	(2)
	saddle point	(3)
II. Limit cycles	stable	(4)
	unstable	(5)
III. Separatrices	unwinding from an unstable focus or node	(6)
	unwinding from an unstable cycle	(7)
	entering the region G through the boundary cycle without contact	(8)
	tending to a stable focus or node	(9)
	tending to a stable limit cycle	(10)

Non-singular (orbitally stable) paths[†]

IV. Paths tending to a stable focus or node	unwinding from an unstable focus or node	(11)
	unwinding from an unstable cycle	(12)
	entering the region G through the boundary cycle without contact	(13)
V. Paths tending to a stable cycle	unwinding from an unstable focus or node	(14)
	unwinding from an unstable cycle	(15)
	entering the region G through the boundary cycle without contact.	(16)

[†] It can be shown that in coarse systems all non-singular paths are not only orbitally stable but are also stable in the sense of Liapunov both for $t \to +\infty$ and for $t \to -\infty$.
For example, for paths tending to a limit cycle, for $t \to +\infty (t \to -\infty)$, see Section 6, Chapter V.

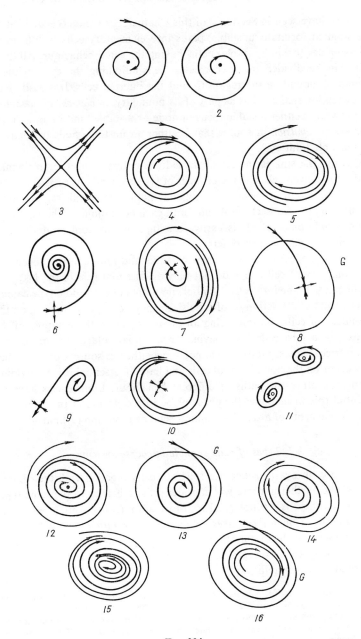

Fig. 304

As we have seen in Section 2 of this Chapter, the region G is divided by the singular (orbitally unstable) paths into elementary cells each filled by non-singular (orbitally stable) paths having the same behaviour. All these cells can be divided into two classes: cells adjoining the cycle without contact C bounding the region G, and the internal cells. It is easily seen that each internal cell has as part of its boundary an *element of attraction* or *sink* that is either a stable focus or node or a stable limit cycle, and one *element of repulsion* or source that is either an unstable node or focus or an unstable limit cycle.

Clearly, the sink is the set of the ω-limit points of every non-singular path of the given cell, and the element of repulsion or source is the set of the α-limit points of every non-singular path of the cell (see Section 3 of this chapter, Subsections 4 and 5). In each cell adjoining the boundary there is only one sink. It is easily seen that in this context the role of different singular paths is different.

Foci (or nodes) serve as sources or sinks and although they occur on the boundary of cells, they do not play an essential role in the mapping of the phase plane into cells. Equilibrium points of the saddle type cannot be elements of attraction or of repulsion; they occur, as do nodes, on the boundary of cells, without being an essential part in the mapping of the phase plane into cells, but having an essential role in generating the separatrices. The separatrices cannot serve either as sources or sinks, but they occur on the boundaries of cells and have an essential role in dividing from each other the paths of different behaviour. Limit cycles have an essential role in mapping the phase plane into cells and at the same time serve as elements of attraction (ω-limit sets) or repulsion (α-limit sets).

7. *Types of cells possible in coarse systems*

Here we always consider the separate cells together with their boundaries, and furthermore we will consider cells to belong to the same type, only when there is a topological transformation (transforming paths into paths) between them *that leaves the direction of rotation unvaried*[†].

It can also be shown that there is only a finite number of types of cells in a coarse system. To begin classifying cells, let us take an arbitrary cell. There are two possibilities:

---

† Topological transformations are divided into two classes: transformations that retain orientation and transformations that change the orientation.
A simple example of a topological transformation that changes the orientation is a specular reflection. Hence two cells related by specular reflection will be considered as belonging to different types.

(a) no saddle point on the boundary; *(b)* a saddle point on the boundary.

Let us consider the first case *(a)*. There is bound to occur a limit cycle on the boundary, since the plane cannot be divided into cells by equilibrium points and the singular paths which form a boundary must be separatrices (but then there would be a saddle point), limit cycles, and equilibrium states. If a limit cycle forms part of the boundary, then again two classes can arise:

(AI) The paths of the cell lie outside the cycle;
(AII) The paths of the cell lie inside the cycle.

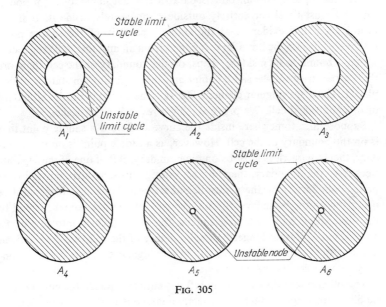

Fig. 305

In class *AI*, since there is no saddle point there must be one more (external) limit cycle. Since it is evident that there are other permissible singular paths on the boundary, then, bearing in mind the direction of rotation and the stability, we obtain four different types of regions: AIa_1, AIa_2, AIa_3 and AIa_4 (Fig. 305, the cases A_1, A_2, A_3, A_4). In class *AII* there can be two alternatives: either there is one more limit cycle inside and we again have the same types as above, or there is a focus (or node) inside and, then, bearing in mind the direction of rotation and the stability, we have two types of cells: $AIIb_1$ and $AIIb_2$ (Fig. 305, the cases A_5 and A_6).

In the second basic case when a saddle point occurs on the boundary, there are two classes again;

(BI) No limit cycle on the boundary;
(BII) A limit cycle on the boundary.

Consider the first, *BI*. As is known, a saddle point has four separatrices: two stable and two unstable. We shall assume first (the case *BIa*) that there are two separatrices of equal stability on the boundary, for example, two unstable ones. Since each of these belongs to the boundary of the region and cannot go to a saddle point, then its asymptotic behaviour is the same as that of the other paths. Both unstable separatrices of the saddle point tend to a stable node or focus, and there is a closed curve C consisting of the saddle point, two unstable separatrices and a stable focus (or node). The cell must lie either entirely outside C or entirely inside it. If it lies inside, then it is evident that the stable separatrix of the saddle point inside C is also on the boundary. It comes from an unstable node (or focus) which is bound to lie inside C. Thus, on the boundary of the cell there are three separatrices of the saddle point and three equilibrium states suitably placed. These six singular elements exhaust the elements on the boundary of this connected cell. We prove this by a *reductio ad absurdum*.

Suppose that somewhere inside the curve C there is a saddle point that is on the boundary of the cell. However, as a saddle point is on the boundary, then separatrices must be on the boundary. But, if one of the separatrices is on the boundary, then another contiguous separatrix must also be on the boundary. Thus there must be one stable and one unstable separatrix occurring on the boundary. Since these separatrices are bound to tend to the same stable and unstable elements, the cell is divided into two parts, and the curve C is no longer on the boundary of the cell. We have arrived at a contradiction, and there are no other singular elements on the boundary, other than the six enumerated.

The other alternative when the cell lies entirely outside the curve C can be discussed using arguments perfectly analogous to those above, leading to the same conclusion. Thus the case *BIa* has only one topological type of elementary cell (see Fig. 306, the case *BIa*).

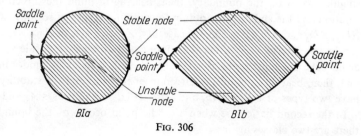

Fig. 306

We assume now (the case *BIb*) that there are two contiguous separatrices on the boundary one stable and one unstable and that the remaining two separatrices are not on the boundary of this cell. Since separatrices cannot join saddle point to saddle point, then necessarily the stable separatrix comes from an unstable node (or focus), and the unstable separatrix goes to a stable node (or focus). Since by hypothesis the remaining separatrices of the saddle point are not on the boundary, then there must be one more saddle point on the boundary. Here, clearly, there are two possible types of behaviour for the separatrices of the second saddle point (Fig. 307).

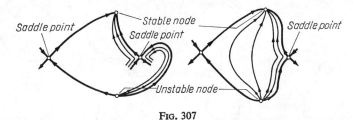

FIG. 307

It cannot be the first type when there are two separatrices of the same stability on the boundary, since we have already shown that there is not a second saddle point on the boundary. The second case remains. Now we can make two assumptions: either our cell lies entirely inside the closed curve *C* formed by four separatrices and four states of equilibrium or lies entirely outside it. Let us consider the first hypothesis, and the only singular paths that might be on the boundary are separatrices and saddle points (limit cycles cannot occur on the boundary by hypothesis; and there is already a sink and a source). However, if a saddle point is on the boundary then it necessarily has two contiguous separatrices which go to a stable and an unstable node (or focus) dividing the cell into two parts. The curve *C* can no longer be entirely on the boundary of the cell. We have arrived at a contradiction, and the cell is not entirely inside *C*. The assumption that our cell lies entirely outside the curve *C* is also easily refuted. Thus the case *BIb* again yields only one topological type of elementary cell (see Fig. 306, the case *BIb*).

In the more complicated case *BII*, there are both limit cycles and saddle points on the boundary of the cell, the cases possible are shown in Figs. 308 and 309. Observe that the cases *BII* are in a certain sense the opposite of the cases *BI*. Thus the cases *BII* are obtained from the cases *BI* by replacing one or two nodes by other elements of attraction or repulsion–limit cycles. The number of different types has increased

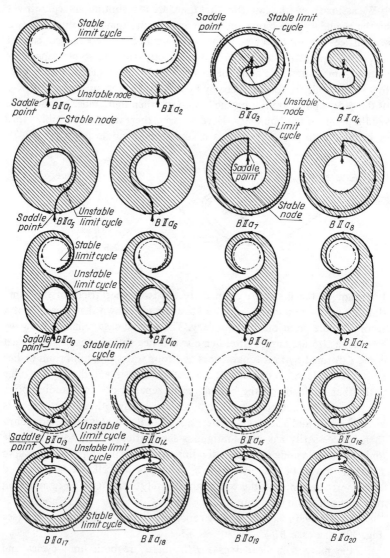

Fig. 308

markedly because one cycle can be situated inside or outside another one and because of the necessity of distinguishing the direction of rotation of the cycles. The detailed analysis of these cells and of the cells adjoining the cycle without contact will be omitted. The cases that can be realized are shown in Fig. 310.

Having considered the various types of elementary cells possible in coarse systems, the next step might be to examine the "laws of the simul-

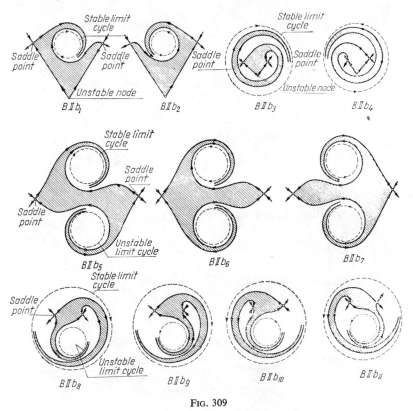

Fig. 309

taneous existence" of elementary cells of different types. These problems have not yet been solved completely and we only mention a concept that has some relation to these problems. It is sometimes convenient to use the concept of the *region of stability at large* of a given element of attraction; by such a region of stability at large we mean the aggregate of all elementary cells that have the singular element as their element of attraction. Finally, note one more simple but very important property of coarse systems: the qualitative structure of the phase portrait of every coarse system can be established by the approximate construction of all singular

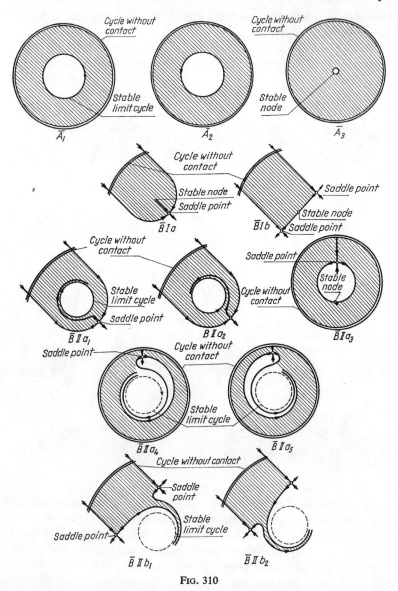

Fig. 310

paths (equilibrium states, limit cycles, and separatrices). The accuracy with which the singular paths must be constructed, is determined by a certain quantity—the "measure of coarseness" [31].

§ 5. Effect of a Parameter Variation on the Phase Portrait
[10–13]

We have repeatedly considered the case when the right-hand sides of the system equations contain a certain parameter and we have studied the variation of the qualitative structure of the phase portrait when this parameter varies (see Chapter II). Now we examine this question under more general assumptions than in Chapter II.

Every system of equations that corresponds to a physical system contains a certain number of parameters whose limits of variation are determined by the conditions of the problem. Such parameters can be, for example, the coefficient of self-induction, the resistance of an oscillating circuit, etc. We must assume that the qualitative phase portrait for given particular values of the parameters, represents real features of a physical system only in the case when this qualitative picture does not vary for "small" variations of the parameters.

However, for greater variations of the parameters the character of the motion of a physical system can vary markedly, and the qualitative appearance of the phase portrait will vary also.

We will assume for simplicity that the right-hand sides of the differential equations depend on one parameter and thus the system equations are

$$\left. \begin{aligned} \frac{dx}{dt} &= P(x, y, \lambda); \\ \frac{dy}{dt} &= Q(x, y, \lambda). \end{aligned} \right\} \quad (6.22)$$

The right-hand sides $P(x, y, \lambda)$ and $Q(x, y, \lambda)$ are analytic functions of x and y in the region G (independently of λ) and analytic functions of λ for values of λ in the region $\lambda_1 < \lambda < \lambda_2$, where λ_1 and λ_2 are certain constants.

The basic theorems necessary for investigating the variations of the paths for variations of a parameter, are enunciated in Appendix I, Theorems IV, V and VI. However, as already observed in Section 4 of this chapter, these theorems only settle the question of how a portion of a path corresponding to a finite interval of time varies for a change of the parameter and do not, by themselves, say anything about how an entire path or the qualitative appearance of the paths will vary.

1. Branch value of a parameter

We recall the definition introduced in Section 5, Chapter II of the branch value of a parameter. Let us assign a certain value $\lambda=\lambda_0$ ($\lambda_1<\lambda_0<\lambda_2$). If an $\varepsilon>0$ exists such that, for all values of λ satisfying the condition $|\lambda-\lambda_0|<\varepsilon$ the topological structure of the phase portrait is the same, then we say that $\lambda=\lambda_0$ is an ordinary value of the parameter; the value $\lambda=\lambda_0$ is called a branch value of the parameter if values of λ arbitrarily close to λ_0 are found for which the qualitative picture of the phase portrait is different from the picture that corresponds to $\lambda=\lambda_0$. By the definition of branch value of a parameter, the system cannot be coarse for such a value.

The qualitative structure of the phase portrait is determined by the so-called singular elements or singular paths (see Section 3 of this chapter). Therefore, we must study the dependence of the singular elements upon a parameter of the system.

It is evident that in a coarse system the value $\lambda=\lambda_0$ cannot be a branch value and in fact, we can always find an interval about λ_0 such that for values of λ within this interval the system will be coarse, the qualitative picture of the paths being the same as for $\lambda=\lambda_0$. It is clear therefore that there cannot be a last coarse value of λ, nor a first non-coarse value of λ.

We further assume, to simplify the argument, that for all variations of the parameter ($\lambda_1<\lambda<\lambda_2$) a cycle without contact will remain a cycle without contact. It is also clear that λ can have branch values when singular elements appear which are non-coarse. The simplest cases of such non-coarse elements are:

(1) multiple equilibrium states (such states can either appear or arise from the merging of simple points such as a node and a saddle point);

(2) a degenerated focus or centre;

(3) a double limit cycle (such a cycle can either appear to arise from the merging of a stable and an unstable cycle);

(4) a separatrix going from saddle point to saddle point.

2. The simplest branchings at equilibrium states

It is clear (and we have already discussed this in connexion with the σ, $\varDelta$ diagram) that for a variation of a parameter the character of an equilibrium state may only vary when either $\varDelta$ or σ reduce to zero. It is also easily seen that with the assumptions on $P(x, y, \lambda)$ and $Q(x, y, \lambda)$, the index of a closed curve

$$j = \frac{1}{2\pi} \int_N \frac{P\,dQ - Q\,dP}{Q^2 + P^2}$$

is a continuous, analytic function of the parameter λ provided that no equilibrium states appear on the curve N. Hence, since it is an integer, the index does not vary in a certain interval around λ_0. Thus, an equilibrium state with a non-zero index cannot either appear or disappear for a variation of λ. A simple singular point—a node—can disappear only after having previously merged with a saddle point, thus forming a multiple singular point with an index equal to zero. Conversely, a saddle point or a node can

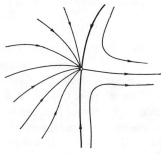

Fig. 311

appear from the splitting up of a multiple singular point with an index equal to zero. The simplest multiple singular point is obtained from merging a saddle point and a node. This singular point is called a "saddle node" and it can be shown that the neighbourhood of such a singular point has the structure shown in Fig. 311. If the saddle node appears for $\lambda = \lambda_0$ it can either disappear or split up when λ is close but not equal to λ_0.

3. Limit cycles emerging from multiple limit cycles

In § 4 we have already discussed the fact that, in a variation of the right-hand sides of a system of differential equations, limit cycles can be split into a number of cycles or even disappear, and we will pursue this topic further.

We will assume that for $\lambda = \lambda_0$ there is a segment without contact l and on it a sequence function. By Theorem VI of Appendix I we can state the following proposition: *It is always possible to find $\eta > 0$ such that, for all values of λ inside the interval $\lambda_0 - \eta < \lambda < \lambda_0 + \eta$ the segment without con-*

tact remains a segment without contact†, and on it there will be a sequence function $\bar{s}=f(s,\lambda)$ for $s_1<s<s_2$, where s_1 and s_2 can be taken independently of λ. The function $f(s,\lambda)$ is an analytic function of s and λ in the indicated intervals (see Section 4 of this chapter, Subsection 3).

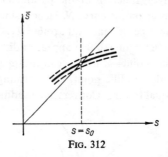

Fig. 312

We will examine how limit cycles can appear and disappear as λ varies. Let us construct the curve $\bar{s}=f(s,\lambda)$ and the straight line $\bar{s}=s$ on the $\bar{s}$, s plane. Closed paths (see Chapter V, Section 7) correspond to values of s for which $f(s,\lambda)-s=0$. We have seen (see Chapter V, Section 7) that from the behaviour of the function $\bar{s}=f(s,\lambda)$ near its common point with the straight line $\bar{s}=s$, we can draw conclusions about the paths close to the limit cycle.

Assume first that for $\lambda=\lambda_0$ the segment without contact l intersects a coarse limit cycle L_0 at a point corresponding to $s=s_0$. Then on Lamerey's diagram the curve $\bar{s}=f(s,\lambda_0)$ intersects $\bar{s}=s$ at a simple point R_0 when $s=s_0$ ($f'(s_0,\lambda_0)\neq 1$). Since the sequence function is analytic in λ, then the curve $\bar{s}=f(s,\lambda)$ will also intersect the straight line $\bar{s}=s$ at a point R close to R_0 and will not have other points of intersection with the straight line $\bar{s}=s$ close to R (Fig. 312).

Thus for all values of λ sufficiently close to λ_0 we have only one limit cycle (corresponding to values of s sufficiently close to s_0) that is stable or unstable according to whether the limit cycle L_0 is stable or unstable.

Now assume that $\lambda=\lambda_0$ is a branch value for which the system (6.22) has a double limit cycle (see Section 4, Subsection 3) intersecting the segment without contact l at a point corresponding to $s=s_0$, where $s_1<s_0<s_2$. Then the sequence function

$$\bar{s}=f(s,\lambda_0),$$

is, at this point, tangential to the straight line $\bar{s}=s$. For small variations

---

† It is easily seen that the same assumption will be valid when, instead of a fixed segment without contact, we consider a "moving" segment (or arc) without contact, the functions in the parametric equations of this segment (or arc) being analytic functions of λ.

of λ this intersection point may either vanish or split into two simple points of intersection (Fig. 313).

For example, for $\lambda < \lambda_0$ the common point may vanish while for $\lambda > \lambda_0$ it splits into two. Then, for a variation of λ from $\lambda_1 < \lambda_0 (|\lambda_1 - \lambda_0| < \eta)$ up to

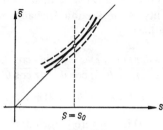

FIG. 313

$\lambda_2 > \lambda_0$ the system at first has no limit cycles intersecting the segment without contact l and then one double ("semi-stable") limit cycle that, for a further variation of the parameter is split into two simple limit cycles, one of which is stable and the other unstable.

A physical example with such branching will be considered in Section 10, Chapter IX (the hard excitation of oscillations in a valve generator).

We might indicate here that the above considerations can also be treated in a different geometric form and reduced to the ordinary branch theory, if we again introduce the function

$$\Psi(s, \lambda) = f(x, \lambda) - s.$$

It is clear that the roots of the equation (cfr. § 4)

$$\Psi(s, \lambda) = 0$$

correspond to limit cycles for every given λ.

4. Limit cycles emerging from a multiple focus

A multiple focus is an equilibrium state for which $\Delta > 0$, $\sigma = 0$. We have already met this in Section 4 but now we assume the system equations have been reduced to

$$\left.\begin{aligned} \frac{dx}{dt} &= a_1(\lambda)x - b_1(\lambda)y + g(x, y, \lambda), \\ \frac{dy}{dt} &= b_1(\lambda)x + a_1(\lambda)y + h(x, y, \lambda), \end{aligned}\right\} \quad (6.23)$$

with an equilibrium state at the origin. $g(x, y, \lambda)$ and $h(x, y, \lambda)$ are power series in x and y of at least the second degree, and $a_1(\lambda)$ and $b_1(\lambda)$ are the real and the imaginary parts of the characteristic roots (we can assume, without detracting from generality, that $b_1(\lambda) > 0$ for all λ). Then for $a_1(\lambda) < 0$ the singular point 0 is a stable focus, and for $a_1(\lambda) > 0$ it is an unstable focus; if, however, $a_1(\lambda) = 0$, then terms of the first degree cannot determine stability. As in Section 4, change to polar coordinates and rearrange the equations

$$\frac{dr}{d\theta} = r\frac{a_1(\lambda)r + g(r\cos\theta, r\sin\theta, \lambda)\cos\theta + h(r\cos\theta, r\sin\theta, \lambda)\sin\theta}{b_1(\lambda)r + h(r\cos\theta, r\sin\theta, \lambda)\cos\theta - g(r\cos\theta, r\sin\theta, \lambda)\sin\theta} =$$
$$= rR_1(\theta, \lambda) + r^2 R_2(\theta, \lambda) + \ldots \qquad (6.24)$$

Since, by hypothesis, $b_1(\lambda)$ is non-zero in the interval of λ, then the series converges for all θ, and all $\lambda_1 < \lambda < \lambda_2$, when $|r| < \varrho$, where ϱ is sufficiently small and does not depend on θ and $\lambda \cdot R_i(\theta, \lambda)$ are again periodic functions of θ and $R_1(\theta, \lambda) = a_1(\lambda) \cdot b_1(\lambda)$. As in Section 4, we look for a solution of (6.24) $r = f(\theta, r_0, \lambda)$ which reduces to r_0 for $\theta = 0$. This solution can be expanded, by virtue of Theorem VI, Appendix I and its corollary, into a power series in r_0, converging for all θ, $0 < \theta < 2\pi$ and for all λ in its interval when $|r_0| < \varrho_0$, where ϱ_0 can be taken independently of λ. Hence

$$r = r_0 u_1(\theta, \lambda) + r_0^2 u_2(\theta, \lambda) + \ldots,$$

The equations for $u_i(\theta, \lambda)$ have the form (6.12) except that $R_i(\theta)$ now depends on λ. The sequence function for the half line $\theta = 0$ is

$$r = f(2\pi, r_0, \lambda) = u_1(2\pi, \lambda)r_0 + \ldots,$$

where

$$u_1(2\pi, \lambda) = e^{\frac{a_1}{b_1} 2\pi}.$$

We could carry out a geometrical investigation analogous to that carried out for the case of an ordinary sequence function. However, we will use a different geometrical approach, and introduce the function

$$\Psi(r_0, \lambda) = f(2\pi, \lambda) - r_0 = \alpha_1(\lambda)r_0 + \alpha_2(\lambda)r_0^2 + \ldots$$

Let us consider what possibilities there are:

(1) Suppose that $a_1(\lambda) \neq 0$, and hence $\alpha_1(\lambda) \neq 0$. Then, the curve $\Psi(r_0, \lambda_0) = 0$ has no points at which simultaneously $\Psi'_r = 0$ and $\Psi'_\lambda = 0$. The sign of $a_1(\lambda)$ does not vary and the singular point, a focus, remains stable, no limit cycle emerging from it (or contracting to it).

(2) Consider now the above case, when $a_1(\lambda) = \sigma/2 = 0$ so that there is a value $\lambda = \lambda_0$ such that $a_1(\lambda_0) = 0$ and $\alpha_1(\lambda_0) = 0$ (the focus becomes degenerate). The curve $\Psi(r_0, \lambda) = 0$ a singular point at the point $r_0 = 0$ for (Ψ'_{r_0}) $r_0 = 0$, $\lambda = \lambda_0 = 0$ and $(\Psi'_\lambda)r_0 = 0$, $\lambda = \lambda_0 = 0$.

We recall that if $\alpha_1(\lambda_0) = , \ldots, = \alpha_{2n-1}(\lambda_0) = 0$, then $\alpha_{2n}(\lambda_0) = 0$.

Consider now the function $\Psi(r_0, \lambda) = 0$. Its second derivatives for the value $r_0 = 0$, $\lambda = \lambda_0$, are

$$(\Psi''_{r_0 r_0})_{\substack{r_0=0\\\lambda=\lambda_0}} = \alpha_2(\lambda_0) = 0, \quad (\Psi''_{\lambda\lambda})_{\substack{r_0=0\\\lambda=\lambda_0}} = 0;$$

$$(\Psi''_{r_0\lambda})_{\substack{r_0=0\\\lambda=\lambda_0}} = \left(\frac{d\alpha_1(\lambda)}{d\lambda}\right)_{\lambda=\lambda_0} = 2\pi \frac{a'_1(\lambda_0)}{b_1(\lambda_0)}.$$

But when

$$(\Psi''_{r_0\lambda})_{\substack{r_0=0\\\lambda=\lambda_0}} = \left(\frac{d\alpha_1(\lambda)}{d\lambda}\right)_{\lambda=\lambda_0} \neq 0;$$

then

$$(\Psi''_{r_0 r_0})_{\substack{r_0=0\\\lambda=\lambda_0}} (\Psi''_{\lambda\lambda})_{\substack{r_0=0\\\lambda=\lambda_0}} - (\Psi''_{r_0\lambda})^2_{\substack{r_0=0\\\lambda=\lambda_0}} < 0$$

and the point $r_0 = 0$, $\lambda = \lambda_0$ will be a simple double point (a node) for the curve $\Psi(r_0, \lambda) = 0$. In this case, for a variation of λ from below λ_0 to above, $\alpha_1(\lambda)$ and $a_1(\lambda)$ change their signs and this focus varies its stability.

The nature of the point $r_0 = 0$, $\lambda = \lambda_0$ is most simply investigated if we use the fact that $\Psi(r_0, \lambda) = 0$ is split into the straight line $r_0 = 0$ and the curve $\varphi(r_0, \lambda) = \alpha_1(\lambda) + \alpha_2(\lambda)r_0 + , \ldots, = 0$. To find how the curve $\varphi(r_0, \lambda) = 0$ behaves near the point $r_0 = 0$, $\lambda = \lambda_0$, we must calculate the values of $d\lambda/dr_0$ and $d^2\lambda/dr_0^2$ at this point. We find

$$\left(\frac{d\lambda}{dr_0}\right)_{\substack{\lambda=\lambda_0\\r_0=0}} = -\frac{\alpha_2(\lambda_0)}{\left(\dfrac{d\alpha_1(\lambda)}{d\lambda}\right)_{\lambda=\lambda_0}} = 0,$$

i.e. the curve $\varphi(r_0, \lambda) = 0$ has a vertical tangent at the point $r_0 = 0$, $\lambda = \lambda_0$. Also

$$\left(\frac{d^2\lambda}{dr_0^2}\right)_{\substack{r_0=0\\\lambda=\lambda_0}} = -\frac{2\alpha_3(\lambda_0)}{\left(\dfrac{d\alpha_1(\lambda)}{d\lambda}\right)_{\lambda=\lambda_0}}.$$

Suppose that $\alpha_3(\lambda_0) \neq 0$. Then near the point $r_0 = 0$, $\lambda = \lambda_0$, the curve $\varphi(r_0, \lambda) = 0$ lies entirely on one side of the tangent.

It is seen that, according to the signs of $a'_1(\lambda_0)$ and $\alpha_3(\lambda_0)$, there are four cases to be considered since $b_1(\lambda)$ does not reduce to zero; we

412 QUALITATIVE THEORY OF SECOND-ORDER DIFFERENTIAL EQUATION [VI

assume $b_1(\lambda) > 0$. We will show the branch diagrams corresponding to these cases, where, as is usual, the shaded regions correspond to $\Psi(r_0, \lambda) > 0$.

(a) $\quad a_1'(\lambda_0) > 0, \quad \alpha_3(\lambda_0) < 0;$
then
$$\left(\frac{d\alpha_1}{d\lambda}\right)_{\lambda=\lambda_0} > 0 \quad \text{and} \quad \left(\frac{d^2\lambda}{dr_0^2}\right)_{\substack{r_0=0\\\lambda=\lambda_0}} > 0.$$

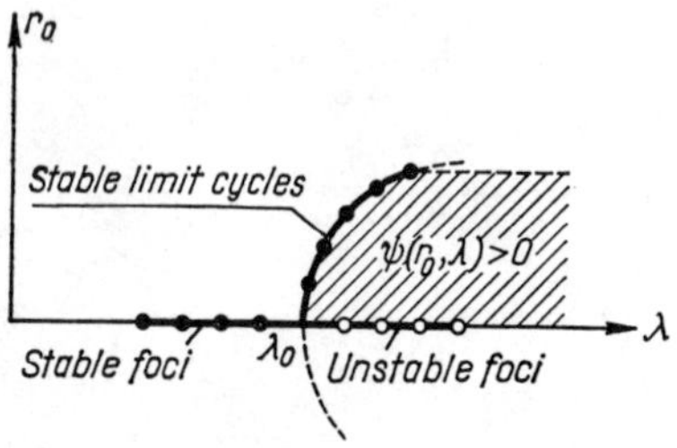

Fig. 314 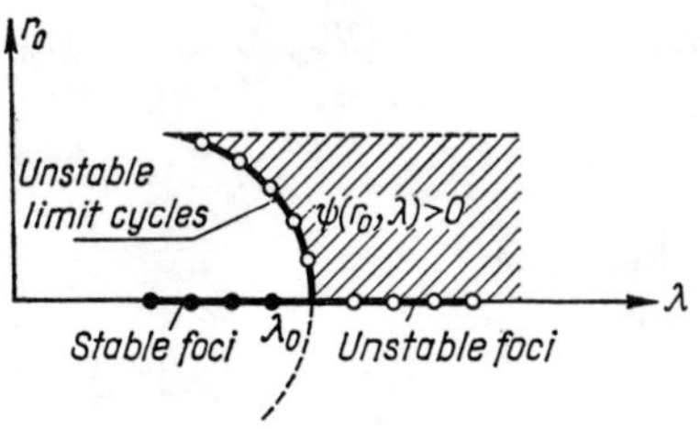   Fig. 315

The branch diagram is shown in Fig. 314. As λ increases, the focus changes from stable to unstable, and at the same time a stable limit cycle appears.

(b) $\quad a_1'(\lambda_0) > 0, \quad \alpha_3(\lambda_0) > 0;$
then
$$\left(\frac{d\alpha_1(\lambda)}{d\lambda}\right)_{\lambda=\lambda_0} > 0 \quad \text{and} \quad \left(\frac{d^2\lambda}{dr_0^2}\right)_{\substack{r_0=0\\\lambda=\lambda_0}} < 0.$$

The branch diagram is shown in Fig. 315. As λ increases, the focus changes from stable to unstable and single unstable limit cycle contracts into a focus.

(c) $\quad a_1'(\lambda_0) < 0, \quad \alpha_3(\lambda_0) > 0;$
$$\left(\frac{d\alpha_1(\lambda)}{d\lambda}\right)_{\lambda=\lambda_0} < 0 \quad \text{and} \quad \left(\frac{d^2\lambda}{dr_0^2}\right)_{\substack{r_0=0\\\lambda=\lambda_0}} > 0.$$

In this case an unstable focus becomes stable, and an unstable limit cycle appears as λ increases (Fig. 316).

(d) $\quad a_1'(\lambda_0) < 0, \quad \alpha_3(\lambda_0) < 0;$
$$\left[\frac{d\alpha_1(\lambda)}{d\lambda}\right]_{\lambda=\lambda_0} < 0 \quad \text{and} \quad \left[\frac{d_2\lambda}{dr_0^2}\right]_{\substack{r=r_0\\\lambda=\lambda_0}} < 0.$$

As λ increases an unstable focus becomes stable. A stable limit cycle contracts into a focus (Fig. 317).

We would have obtained the same results, if $\alpha_3(\lambda_0)$ had been equal to zero and the first non-zero coefficient had been $\alpha_{2k+1}(\lambda_0)$ where $k>1$. In this case one limit cycle appears (or disappears). The case of $a_1'(\lambda_0)=0$ will not be considered but, generally speaking, various numbers of limit cycles can appear as λ varies.

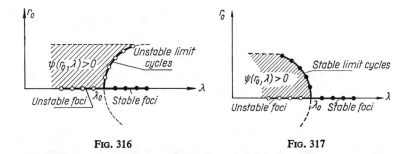

FIG. 316   FIG. 317

Finally, note that in the simplest and most interesting practical case it is sufficient to know the quantities $\alpha_k(\lambda)$ for $\lambda=\lambda_0$ only. Therefore, to simplify the calculations, we shall write the equation (6.24) for $\lambda=\lambda_0$ only:

$$\frac{dr}{d\theta} = \frac{1}{b_1(\lambda_0)}[(P_2)_0 \cos\theta + (Q_2)_0 \sin\theta]\left\{1 + \frac{(P_2)_0 \sin\theta - (Q_2)_0 \cos\theta}{b_1(\lambda_0)r} + \right.$$
$$\left. + \left(\frac{(P_2)_0 \sin\theta - (Q_2)_0 \cos\theta}{b_1(\lambda_0)r}\right)^2 + \ldots\right\}, \qquad (6.25)$$

where by $(P_2)_0$ and $(Q_2)_0$ denote $P_2(r\cos\theta, r\sin\theta, \lambda)$ and $Q_2(r\cos\theta, r\sin\theta, \lambda)$ for $\lambda=\lambda_0$. This also gives

$$\frac{dr}{d\theta} = R_2(\theta, \lambda_0)r^2 + R_3(\theta, \lambda_0)r^3 + \ldots,$$

since $R_1(\lambda_0, \theta) = 0$. Again we look for a solution of this simplified equation in the form of a series

$$r = r_0 u_1 + r_0^2 u_2 + r_0^3 u + \ldots,$$

where $u_1=1$ and for $u_k(\theta, \lambda_0)$ we have

$$\left.\begin{aligned}
\frac{du_2}{d\theta} &= R_2(\theta, \lambda_0); \\
\frac{du_3}{d\theta} &= 2u_2 R_2(\theta, \lambda_0) + R_3(\theta, \lambda_0); \\
\frac{du_4}{d\theta} &= (u_2^2 + 2u_3) R_2(\theta, \lambda_0) + 3u_2 R(\theta, \lambda_0) + R_4(\theta, \lambda_0); \\
\frac{du_5}{d\theta} &= (2u_4 + 2u_2 u_3) R_2(\theta, \lambda_0) + (3u_2^2 + 3u_3) R_3(\theta, \lambda_0) + \\
&\qquad + 4u_2 R_4(\theta, \lambda_0) + R_5(\theta, \lambda_0) \\
&\cdots\cdots\cdots\cdots\cdots\cdots\cdots\cdots\cdots\cdots\cdots\cdots
\end{aligned}\right\} \quad (6.26)$$

with the initial conditions $u_k(0, \lambda_0) = 0$ ($k = 2, 3, \ldots$).

We can find, from this, $\alpha_3(\lambda_0)$, and if $\alpha_3(\lambda_0) = 0$, $\alpha_5(\lambda_0)$ etc. $a_1'(\lambda_0)$, and then $a_1(\lambda)$ and $b_1(\lambda)$ can be found from the usual characteristic equation.

5. Physical example

We will consider the so-called soft excitation of self-oscillations in a valve generator. The equation of a valve generator with a tuned grid circuit, with the usual idealizations, and approximating to the valve characteristic by a polynomial of the third degree

$$i_a = i_{a0} + S_0 u + S_1 u^2 - S_2 u^3,$$

where i_a is the anode current, u is the variable component of the grid voltage, and the co-efficients i_{a0}, s_0, s_2 are positive, has the following form (see, for example, Chapter IX)

$$LC \frac{d^2 u}{dt^2} + u = (MS_0 - RC + 2MS_1 u - 3MS_2 u^2) \frac{du}{dt}.$$

If we introduce the dimensionless variables $\tau = \omega_0 t$, where $\omega_0 = (LC)^{-\frac{1}{2}}$, $y = u/u_0$, and $x = dy/d\tau$, then the equation of the oscillations is[†]

$$\frac{dx}{d\tau} = -y + (\alpha + \beta y - \gamma y^2) x, \qquad \frac{dy}{d\tau} = x, \qquad (6.27)$$

where

$$\alpha = \omega_0 (MS_0 - RC), \qquad \beta = 2\omega_0 MS_1 u_0,$$
$$\gamma = 3\omega_0 MS_2 u_0^2 \quad (\gamma > 0).$$

[†] The somewhat unusual notation for the coordinates on the phase plane has been chosen in order that the system (6.27) shall yield directly, for $\alpha = 0$, a system of the form (6.23) with $a_1 = 0$ and $b_1 = 1 > 0$.

We will consider the equilibrium state $x=0$, $y=0$ and investigate the possible emergence of a cycle from this point when the mutual inductance M varies. The characteristic equation for this equilibrium state is

$$v^2 + \alpha v + 1 = 0,$$

whence

$$a_1(M) = \frac{\alpha}{2} = \frac{\omega_0}{2}(MS_0 - RC),$$

$$b_1(M) = +\sqrt{1 - \frac{\alpha^2}{4}} = \sqrt{1 - \frac{\omega_0^2}{4}(MS_0 - RC)^2}.$$

The bifurcation (or branch) value of the parameter M is equal to

$$M_0 = \frac{RC}{S_0},$$

whence

$$a_1'(M_0) = \frac{\omega_0 S_0}{2}.$$

To evaluate $\alpha_3(M_0)$, we have from (6.25) and (6.27), where $\beta_0 = (\beta)_{M=M_0}$ and $\gamma_0 = (\gamma)_{M=M_0}$

$$R_2(\theta, M_0) = \beta_0 \cos^2 \theta \sin \theta,$$
$$R_3(\theta, M_0) = -\gamma_0 \sin^2 \theta \cos^2 \theta + 2\beta_0 \cos^3 \theta \sin^3 \theta,$$

whence, by the formulae (6.26), we find

$$u_2(\theta, M_0) = \frac{1}{3}\beta_0(1 - \cos^3 \theta),$$

$$u_3(2\pi, M_0) = \alpha_3(M_0) = -\frac{1}{8}\gamma_0 2\pi = -\frac{3\pi\omega_0}{4}RC\frac{S_2}{S_0}.$$

Since, by hypothesis, $S_0 > 0$, $S_2 > 0$, then $a_1'(M_0) > 0$, $\alpha_3(M_0) < 0$. We have, from our classification, the case (a): As M increases, a focus changes from stable to unstable and at the same time a stable limit cycle appears. If follows that for values of M larger than M_0 but sufficiently close to M_0, a stable self-oscillating process is certainly possible in the system. We observe that, if S_2 had been negative, then we would have had the case (b), when, for an increase of M, a focus changes from stable to unstable and, at the same time, an unstable cycle contracts into a point.

416 QUALITATIVE THEORY OF SECOND-ORDER DIFFERENTIAL EQUATION [VI

We make two more remarks: (1) if we had taken into account further terms in the expansion of the characteristic proportional to u^4, u^5, u^6, etc then, as is seen from the equations (6.26) if $S_2 \neq 0$, these terms would not have appeared in the solution of the problem of the emergence of a cycle; (2) all our results have been arrived at without any assumptions whatsoever on the smallness of the quantities α, β and γ. A similar but more thorough investigation of the onset of self-oscillations in a valve generator when M varies will be carried out in Chapter IX. There, however, we have to restrict the values of various physical parameters within certain small limits.

6. Limit cycles emerging from a separatrix joining two saddle points, and from a separatrix of a saddle-node type when this disappears

These cases are of great interest in the theory of differential equations and from the point of view of physics.

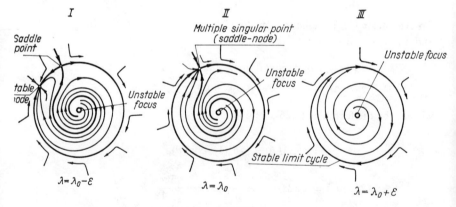

Fig. 318

We assume that for a certain value of the parameter λ in a given interval, the system of differential equations has a saddle point with two separatrices joining it to a node (Fig. 318, *I*). Suppose, in addition, that as λ increases, the saddle point and the node get nearer, finally merging into one multiple singular point of the saddle-node type at $\lambda = \lambda_0$ (Fig. 318, *II*). Then one of the separatrices L_0 from this saddle node will return to it (for $t \to +\infty$) (Fig. 318, *II*). If, for a further increase of λ, the multiple saddle node singular point disappears, then a limit cycle is bound to appear lying, for values of λ sufficiently close to λ_0, in an arbitrarily small neighbourhood of the separatrix L_0 (Fig. 318, *III*)†. It is also clear that a limit cycle can disappear.

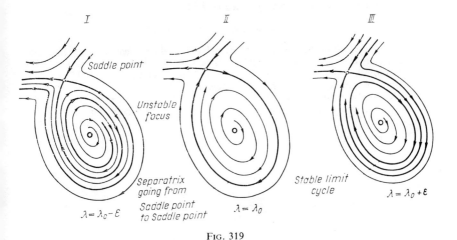

FIG. 319

Consider now the emergence of a cycle from a separatrix of a saddle point. We will assume that for a certain value of λ the separatrices of the saddle point have the disposition shown in Fig. 319, *I*, and that as λ increases (or decreases) they get nearer to each other and finally merge together at $\lambda = \lambda_0$ so forming a "separatrix loop" (Fig. 319, *II*). If, for a further increase (or decrease) of λ the separatrices of the saddle point split again, as is shown in Fig. 319, *III*, then at the same time, at least one limit cycle

---

† Let (x, y) be the saddle node singular point, existing by hypothesis for $\lambda = \lambda_0$. If
$$\sigma_0 = P'_x(x_0, y_0, \lambda_0) + Q'_y(x_0, y_0, \lambda_0)$$
then it is easily shown that, for a saddle node singular point, $\sigma \neq 0$. It can be shown that, in the case when $\sigma_0 \neq 0$, the limit cycle, appearing when the saddle node vanishes, is stable and unstable when $\sigma_0 > 0$.

A multiple saddle node singular point appears on the limit cycle at $\lambda = \lambda_0$ and subsequently is split into two.

separates (is "generated") from the separatrix loop†. It is clear that, conversely, a limit cycle can disappear by first merging with a separatrix loop. In both cases Fig. 319, *II* corresponds to the branch value of the parameter.

† Let the saddle point be x_0, y_0 when $\lambda = \lambda_0$. If

$$\sigma_0 = P'_x(x_0, y_0, \lambda_0) + Q'_y(x_0, y_0, \lambda_0)$$

then when $\sigma_0 < 0$ the "separatix loop" is stable (i.e. all paths passing through points sufficiently close to the loop and lying inside the loop, tend to it as $t \to +\infty$) and when $\sigma_0 > 0$, the loop is unstable. When $\sigma_0 = 0$, the problem of the stability of the loop cannot be solved by means of the quantity σ_0.

It can also be shown that when $\sigma_0 \neq 0$, a single limit cycle is generated from the separatrix loop, this limit cycle being stable if $\sigma_0 < 0$ and unstable if $\sigma_0 > 0$.

CHAPTER VII

SYSTEMS WITH A CYLINDRICAL PHASE SURFACE†

§ 1. Cylindrical phase surface

In representing the behaviour of a dynamic system on a phase surface, a *one-to-one continuous correspondence* between the states of the system and the points of the phase surface is required, with near-by points of the phase surface corresponding necessarily to near-by states of the system. This requirement establishes a certain connection between the behaviour of a given physical system and the basic features of the geometrical figure which is the phase surface for the system. So far we have considered systems (with 1 degree of freedom) for which a plane can serve as the phase surface. However, as we have seen in Chapters II and III, there exist systems for which a plane cannot serve as the phase surface, since a plane would not satisfy the requirement for a one-to-one correspondence. The ordinary physical pendulum serves as an example of such a system. In fact, the state of a pendulum is determined by the angle of its deviation from the state of equilibrium and by its velocity; however, if the angle of deviation varies by 2π, we obtain a state of the pendulum that physically does not differ from the initial one. We obtain therefore on a phase plane an infinite number of points that correspond to one and the same physical state of the system (all the points whose abscissae differ by $2k\pi$ from each other). A plane is not suitable, strictly speaking, as a phase surface for an ordinary physical pendulum, since the conditions for a one-to-one continuous correspondence cannot be satisfied. It is true that the use of a plane as the phase surface can hardly be the cause of misunderstandings, since we limit ourselves to motions that do not exceed the limits of one complete revolution. Still, if we do consider motions that exceed these limits, for one-to-one continuous correspondence to be observed, we must represent the motions of the pendulum on a *phase cylinder*‡. This circum-

† The chapter has been revised and Sections 1 and 4 completely rewritten by N. A. Zheleznov.

‡ We shall plot, around the axis of the cylinder, the angle ϑ which determines the position of the pendulum, and along the axis the angular velocity $z = \dot{\vartheta}$. It often proves convenient to take instead of the cylinder its development on the ϑ, z plane, as we have already done in Sections 4 and 5 of Chapter II.

stance is clearly connected with the existence of two qualitatively different types of periodic motion of the pendulum (the oscillations about the state of equilibrium and the motions of the pendulum involving a revolution around the axis).

We shall consider in this chapter several physical systems which need a phase cylinder to chart their behaviour. We shall take to describe the system of two first-order equations

$$\frac{d\vartheta}{dt} = \Phi(\vartheta, z), \quad \frac{dz}{dt} = F(\vartheta, z),$$

where ϑ and z are the coordinates of the cylindrical phase surface, and we shall elucidate, just as in the case of a phase plane, the basic elements of the phase portrait: its singular points, separatrices, and the limit cycles corresponding to periodic motions. On the phase cylinder, however, in addition to "ordinary" limit cycles that lie on the surface of the cylinder and encircle a state of equilibrium without encircling the cylinder itself (these curves are completely analogous to the closed paths on a phase plane), a completely new type of limit cycle can be met which encircles the cylinder itself. It is evident that these closed paths also correspond to periodic motions, and are of special interest for the reason that they correspond to periodic solutions of the equation of the integral curves

$$\frac{dz}{d\vartheta} = \frac{F(\vartheta, z)}{\Phi(\vartheta, z)}.$$

As these solutions are periodic with period 2π, they satisfy the condition $z(\vartheta+2\pi) = z(\vartheta)$ for any ϑ[†]. To detect the presence of such periodic solutions, we can use the following method. If there exist two particular solutions of the equation (7.2) $z_1(\vartheta)$ and $z_2(\vartheta)$ such that for a certain ϑ_0 we have

$$z_1(\vartheta_0+2\pi) \geqslant z_1(\vartheta_0), \quad z_2(\vartheta_0+2\pi) \leqslant z_2(\vartheta_0),$$

and if in the region between the integral curves corresponding to these

---

[†] We assume that ϑ and z are continuous functions of the time t; then, for a revolution of the representative point around the cylinder the angular coordinate ϑ will vary by $\pm 2\pi$. We attribute, therefore, to each point of the phase cylinder not one but an enumerable set of values of the angular coordinates that differ from each other by 2π. Thus, by retaining the continuity of the dependence of ϑ upon the time t, we are forced to renounce the one-to-one character of the correspondence between the points of the phase cylinder and their coordinates.

It is evident that the functions $\Phi(\vartheta, z)$ and $F(\vartheta, z)$, the right-hand sides of the equations of motion of the system (7.1), are bound to be periodic functions of the angle ϑ with period 2π.

solutions there are no singular points, then, owing to the continuity of the dependence of the solutions upon the initial conditions, we can affirm that between $z_1(\vartheta)$ and $z_2(\vartheta)$ there exists a periodic solution for which

$$z(\vartheta_0+2\pi) = z(\vartheta_0)$$

and, hence,

$$z(\vartheta+2\pi) \equiv z(\vartheta)^2\dagger$$

(in the general case, of course, we cannot affirm that this periodic solution is *unique*).

The search for limit cycles that encircle the cylinder, and the determination of their number and stability can be carried out by constructing the point transformation of a generator of the cylinder $\vartheta=\vartheta_0$ into itself. If phase paths that encircle the cylinder pass through a point of a certain segment (L) of the generator $\vartheta=\vartheta_0$ (Fig. 320), these have consecutive points on the same segment and we can construct the sequence function‡

$$z' = f(z)$$

The fixed points z^* of this point transformation, i.e. the points determined by the equation

$$z = f(z),$$

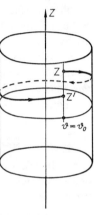

Fig. 320

are the points of intersection of closed phase paths (limit cycles), that encircle the cylinder, with the generator $\vartheta=\vartheta_0$ of the cylinder. According to Koenigs's theorem a limit cycle is stable if

$$|f'(z^*)| < 1$$

and is unstable if

$$|f'(z^*)| > 0.$$

If the solution itself is known, corresponding to a limit cycle encircling the cylinder, $\vartheta=\vartheta(t)$, $z=z(t)$, then the stability of this limit cycle can be

---

† We are assuming, of course, that the conditions of Cauchy's theorem on the existence and uniqueness of the solution of a differential equation are satisfied for the system of equations (7.1).

‡ Just as in the case of a phase plane, the evaluation of the sequence function is most simply carried out for piece-wise linear systems. An example of such system is given in Chapter VIII, Section 10.

determined by evaluating its characteristic exponent

$$h = \frac{1}{T}\int_0^T \{\Phi'_\vartheta[\vartheta(t), z(t)] + F'_z[\vartheta(t), z(t)]\}\,dt,$$

where T is the period of the periodic motion[†]. More precisely, the limit cycle is stable for $h<0$ and unstable for $h>0$ (the proof of this statement is analogous to that given in Chapter V, § 8).

In investigating the phase portrait of dynamic systems with a cylindrical phase surface, Bendixon's and Dulac's criteria, which have been stated earlier (Chapter V, §§ 9 and 11) for the case of a phase plane, can prove of some help. It is easily seen that, *if the conditions of Bendixon's criterion or Dulac's criterion are satisfied in a certain region, between two closed curves that encircle the phase cylinder, then no closed phase paths exist in this region that do not encircle the cylinder, and there cannot be more than one closed phase path that encircles the cylinder.*

§ 2. Pendulum with constant torque

A cylindrical phase space is convenient for representing the behaviour of a series of electromechanical systems; for example, a synchronous electric motor, or an a.c. generator working on a common bus-bar with other machines in parallel, etc. All these systems lead under certain simplifying conditions to the consideration of one and the same mathematical problem and we shall therefore consider only one, namely, an ordinary pendulum with "linear friction" which rotates under the action of a constant moment M_0. We shall obtain the following equation for the motion of the pendulum

$$I\frac{d^2\vartheta}{dt^2} + b\frac{d\vartheta}{dt} + mga\sin\vartheta = M_0,$$

where I is the moment of inertia of the pendulum, and b is the moment of the forces of friction acting on the pendulum at unit angular velocity. On introducing the new independent variable $\tau = (mga/I)^{\frac{1}{2}}\,t$ the equation reduces to

where
$$\left.\begin{array}{c}\dfrac{d^2\vartheta}{d\tau^2} + \alpha\dfrac{d\vartheta}{d\tau} + \sin\vartheta - \beta = 0,\\[2mm] \alpha = \dfrac{b}{\sqrt{Imga}} > 0 \quad\text{and}\quad \beta = \dfrac{M_0}{mga} > 0.\end{array}\right\} \quad (7.3)$$

[†] The function $z(t)$ is periodic, i.e. $z(t+T) \equiv z(t)$ whereas for the function $\vartheta(t)$, owing to its continuity, $\vartheta(t+T) \equiv \vartheta(t) \pm 2\pi$.

Before investigating this equation it is worth indicating how the problems of the synchronous motors and generators in parallel can be reduced to the same equation.

In the case of a synchronous motor we denote by ϑ the angle between the directions of the magnetic fields of the stator and of the rotor. M_0 is the constant load torque acting on the rotor. The load torque tends to slow down the rotation of the rotor, so the angle ϑ is taken to be positive when the field of the rotor lags behind the field of the stator. There is also a moment due to the forces of friction and to electromagnetic damping. The moment of these forces is assumed proportional to the angular velocity, and can be expressed in the form of a term $-b\,d\vartheta/dt$ on the right-hand side, $b > 0$. However, besides the load and frictional torques there also acts on the rotor the electromagnetic torque arising from the interaction of the fields of the stator and the rotor. This moment is a function of the angle and is directed so as to tend to reduce the angle (to accelerate the lagging rotor). Therefore, we can denote this moment by $-f(\vartheta)$, f being of the same sign as ϑ and goes to zero as $\vartheta \to 0$. The equation of motion is thus[†]

$$I\frac{d^2\vartheta}{dt^2} = M_0 - b\frac{d\vartheta}{dt} - f(\vartheta).$$

The form of the function $f(\vartheta)$ that characterizes the interaction of the fields of the stator and the rotor is approximately sinusoidal. On replacing $f(\vartheta)$ by $\sin \vartheta$, we obtain an equation completely analogous to the one obtained for the pendulum (7.3).

In the case of a generator working on a common bus-bar in parallel with other machines, denote by ϑ the angle of lead of the rotor of the generator with respect to the rotors of the other machines, and by M_0 the constant moment applied by the prime mover that drives the given generator (for such a choice, M_0 tends to increase ϑ). The damping moment is, as before, equal to $-b\,d\vartheta/dt$. But, in the case of a generator connected to a common bus-bar there arises an electromechanical moment, because of the phase shift ϑ between its generated e.m.f. and that of the other generators[‡].

[†] Note that, for a reference system at rest, the equation of the motor of the rotor has the form

$$I\frac{d^2\psi}{dt^2} = f(\omega t - \psi) - b\frac{d\psi}{dt} - M_0,$$

where ω is the angular velocity of the axis of the magnetic field of the stator and ψ is the angle of rotation of the rotor. On putting $\vartheta = \omega t - \psi$, we obtain the equation given in the text.

[‡] Strictly speaking, we should consider the *interaction* of the generators and not only the action of all remaining generators on the particular one.

This moment is a function of ϑ and tends to decrease $|\vartheta|$. Therefore, it must be equal to $-f(\vartheta)$, where $f(\vartheta)$ is of the same sign as ϑ. Under certain assumptions, it can be put equal to $\sin \vartheta$.

To investigate the equation (7.3) we shall introduce the new variable $z = d\vartheta/dt$. We obtain two equations of the first order

$$\frac{dz}{d\tau} = -\alpha z - \sin \vartheta + \beta; \quad \frac{d\vartheta}{d\tau} = z. \quad (7.4)$$

By eliminating τ, we obtain one equation of the first order

$$z \frac{dz}{d\theta} = -\alpha z - \sin \theta + \beta. \quad (7.5)$$

We shall investigate (7.5) for the particular case $\alpha = 0$, when the system is conservative. The equation takes the form

$$z \frac{dz}{d\vartheta} = \beta - \sin \vartheta. \quad (7.6)$$

After separating the variables and integrating we obtain

$$\frac{1}{2} z^2 = \cos \vartheta + \beta \vartheta + C_1 \quad \text{or} \quad z = \pm \sqrt{2(\cos \vartheta + \beta \vartheta) + C}. \quad (7.7)$$

First of all, according to (7.6), the singular points will be the points $(\bar{\vartheta}, 0)$ where $\bar{\vartheta}$ are the roots of the equation

$$f(\vartheta, \beta) \equiv \beta - \sin \vartheta = 0. \quad (7.8)$$

We have for $\beta < 1$, two positions of equilibrium: $\bar{\vartheta} = \bar{\vartheta}_1$ and $\bar{\vartheta} = \bar{\vartheta}_2 = \pi - \bar{\vartheta}_1$, where $\bar{\vartheta}_1 = \sin^{-1} \beta \, (0 \leq \vartheta_1 \leq \pi/2)$. For $\beta = 1$, these positions of equilibrium

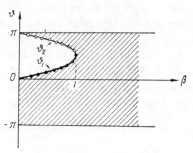

Fig. 321

merge with one another ($\vartheta_1 = \vartheta_2 = \pi/2$) and for $\beta > 1$ they do not exist. The corresponding branch diagram is shown in Fig. 321. The region in which $f(\vartheta, \beta) > 0$ is shaded; stable states of equilibrium (centres) correspond to the points of the thick line with black dots and for which $\bar{\vartheta}=\vartheta_1$, while the unstable states of equilibrium (the saddle points $\bar{\vartheta}=\vartheta_2$) are the points of the thin line with small circles. The value $\beta=1$ is clearly a branch value.

To construct the integral curves, we shall use the method of Section 3, Chapter II. By constructing on the auxiliary ϑ, y plane the curve

$$y = 2(\cos\vartheta + \beta\vartheta)$$

and putting below it the development of the phase cylinder, we construct, on the development for each assigned value of β, the family of the integral curves (7.7). The construction of the integral curves on the ϑ, z plane reduces to this: after choosing values of C we take the square root of the sum $C+y$ and plot it above and below the ϑ axis. For a given C, each value of y for which $y+C>0$ will correspond to two points on the ϑ, z plane, but for values of y for which $y+C<0$, this is not so since the values of z are imaginary.

These constructions give different results for different values of β.

The case of $\beta=0$, when there is no constant moment, has already been considered in Chapter II, Section 4. We have seen that all phase paths, except the two singular points (the centre and the saddle point) and the separatrices of the saddle point, are closed paths and correspond to periodic motions of the pendulum. Inside the separatrices there lie closed paths encircling the centre and not encircling the cylinder, while outside the separatrices there lie closed paths encircling the phase cylinder. The latter paths correspond to the new type of periodic motion already discussed.

For $\beta\neq 0$ we shall obtain different pictures depending on whether $\beta<1$ or $\beta>1$. Construct on the auxiliary plane the curve

$$y = 2\cos\vartheta + 2\beta\vartheta.$$

For $\beta<1$ this curve has a maximum for $\vartheta=\vartheta_1 = \sin^{-1}\beta\ (0\leq\vartheta_1\leq\pi/2)$ and a minimum for $\vartheta=\vartheta_2=\pi-\vartheta_1$ (in constructing the curve (7.9) we can restrict ourselves to the values $-\pi<\vartheta\leq+\pi$). For $\beta=1$, the curve $y=2(\cos\vartheta+\beta\vartheta)$ has no maximum or minimum but has, at $\vartheta=\pi/2$, a turning point with horizontal tangent. For $\beta>1$ the curve $y=2(\cos\vartheta+\beta\vartheta)$ increases monotonically and has neither extrema nor an inflexion point.

In the case $\beta<1$, there is one singular point of the centre type, one singular point of the saddle type and one separatrix, Fig. 322. We obtain on the cylinder the picture shown in Fig. 323. The curves that are found inside the separatrix are closed and correspond to periodic motions. The curves that lie outside the separatrix do not form a closed contour around the cylinder, since, as ϑ increases by 2π, z does not recover its previous value but increases in absolute value for each revolution. Hence, periodic motions of the "second type" are impossible in this case. For $\beta=1$ (Fig. 324)

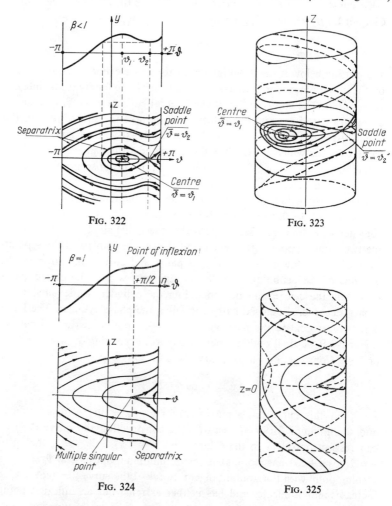

Fig. 322

Fig. 323

Fig. 324

Fig. 325

we shall obtain one singular point of higher order. In this case there are no closed curves on the cylinder (Fig. 325). For $\beta > 1$ there are no singular points (Fig. 326) and no closed curves on the cylinder (Fig. 327). Therefore, for $\beta \geq 1$ periodic motions of either the first or second type are impossible. The physical meaning of these results is quite clear. If the constant moment is not too large, so that it displaces the lowest position of equilibrium by less than $\pi/2$, then, for sufficiently small initial deviations (and initial velocities), oscillations about this displaced position of equilibrium are possible. If the initial deviation is large, then owing to the action of the constant external moment the pendulum passes through the upper position of equilibrium and then will move in the direction of the constant

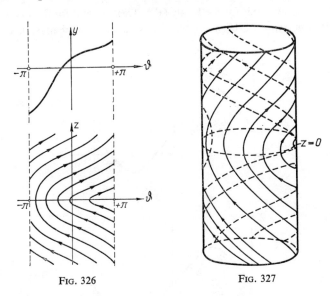

Fig. 326   Fig. 327

moment, the velocity of the pendulum increasing after each revolution. If, however, $\beta > 1$, then the external moment exceeds the maximum moment of the force of gravity. In such a case oscillations are in general impossible, and for all initial conditions the pendulum will in the end rotate in the direction of the constant moment with a velocity increasing monotonically, since now $dz/d\tau = d^2\vartheta/d\tau^2 > 0$ for all ϑ.

§ 3. Pendulum with constant torque. The non-conservative case [198]

Let us pass to consider the non-conservative system (7.4) for $\alpha > 0$. In this case the equation of the integral curves on the cylinder

$$z\frac{dz}{d\vartheta} = -\alpha z - \sin \vartheta + \beta \quad \text{or} \quad \frac{dz}{d\vartheta} = \frac{-\alpha z - \sin \vartheta + \beta}{z}$$

no longer lends itself to direct integration. We shall employ therefore the methods of qualitative integration.

First of all the isocline $dz/d\vartheta = 0$ is a displaced sinusoid. Its equation is

$$z = \frac{\beta - \sin \vartheta}{\alpha}.$$

It intersects the ϑ axis for $\beta < 1$ only (Fig. 328), and not for $\beta > 1$ (Fig. 329). In addition $dz/d\vartheta > 0$ between the sinusoid and the ϑ axis, i.e. in the

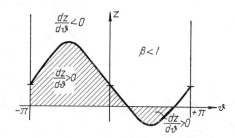

Fig. 328

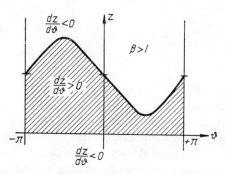

Fig. 329

regions shaded in Figs. 328 and 329. In all the remaining region $dz/d\vartheta < 0$.

The coordinates of the singular points are determined, as before, by the equations

$$\beta - \sin \vartheta = 0, \quad z = 0.$$

Therefore for $\beta > 1$ there are no singular points. For $\beta < 1$ there exist two singular points (two states of equilibrium): $\vartheta = \vartheta_1$, $z = 0$ and $\vartheta = \vartheta_2 = \pi - \vartheta_1$, $z = 0$, where, as before $\vartheta_1 = \sin^{-1} \beta$ $(0 \leqslant \vartheta_1 < \pi/2)$.

To establish the character of these states of equilibrium, put, in equations (7.4), $\vartheta = \vartheta_i + \xi (i = 1, 2)$ and expand $\sin \vartheta$ in a power series with respect to ξ. By restricting ourselves to the first power of ξ, we obtain a system of linearized equations that describe the behaviour of the system about the state of equilibrium $(\vartheta_i, 0)$

$$\frac{dz}{d\tau} = -\alpha z - \xi \cos \vartheta_i, \qquad (7.10)$$

$$\frac{d\xi}{d\tau} = z$$

with the characteristic equation

$$\lambda^2 + \alpha \lambda + \cos \vartheta_i = 0. \qquad (7.11)$$

Since $\cos \vartheta_1 > 0$ and $\cos \vartheta_2 = -\cos \vartheta_1 < 0$, the state of equilibrium $(\vartheta_1, 0)$ is a stable focus for $\alpha^2 < 4 \cos \vartheta_1$ and a stable node for $\alpha^2 > 4 \cos \vartheta_1$ while the state of equilibrium $(\vartheta_2, 0)$ is always a saddle point. For $\alpha = 0$ the first state of equilibrium reduces to a centre.

Next we shall denote the right-hand sides of the equations (7.11) by F and $\Phi (F = -\alpha z - \sin \vartheta + \beta, \Phi = z)$. Then

$$\Phi'_\vartheta + F'_z = -\alpha < 0 \qquad (7.12)$$

over the whole phase cylinder. Therefore, according to Bendixon's criterion for the paths on the phase cylinder (see Section 1 of this chapter), the dynamic system (7.4) does not have on the phase cylinder any closed paths *not* encircling the cylinder and can have at most *one* limit cycle encircling the cylinder. This limit cycle, if it exists, is bound to be stable, for, according to (7.12) its characteristic exponent $h = -\alpha < 0$ and lies entirely on the upper half of the cylinder $(z > 0)$.

It is clear, first of all, that the system (7.4) cannot have closed paths encircling the cylinder and intersecting the line $z = 0$. In fact, assuming the existence of closed paths that intersect the line $z = 0$ (intersection must take place at two points at least, see Fig. 330) we shall arrive at the conclusion that it cannot encircle the cylinder, since in passing through the $z = 0$ axis the sign of $d\vartheta/d\tau$ changes $(d\vartheta/d\tau = z)$ and there exists therefore an interval $\vartheta' < \vartheta < \vartheta''$ in which the closed path does not pass. Hence, closed phase paths encircling the cylinder can only lie either entirely in the region $z > 0$ or entirely in the region $z < 0$.

Then, integrating the equation (7.5), written in the form

$$\frac{1}{2} \frac{d(z^2)}{d\tau} = -\alpha z - \sin \vartheta + \beta,$$

along a closed path $z = z_0(\vartheta)$ that encircles the cylinder within the limits ϑ_0 and $\vartheta_0 + 2\pi$, we obtain

$$-\alpha \int_{\vartheta_0}^{\vartheta_0+2\pi} z_0(\vartheta)\, d\vartheta + 2\pi\beta = 0,$$

since $z_0(\vartheta_0 + 2\pi) = z_0(\vartheta_0)$. For a limit cycle encircling the cylinder and lying entirely in the $z < 0$ region, this equality cannot be satisfied owing

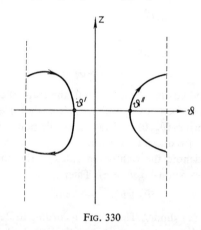

Fig. 330

to the condition $\beta > 0$. Thus if a limit cycle encircling the cylinder exists, it must lie entirely in the $z > 0$ region.

Let us consider now the question of the existence of this limit cycle, the cases $\beta > 1$ and $\beta < 1$ being considered separately.

I. $\beta > 1$

To verify the existence of periodic solutions $z(\vartheta)$ it suffices, as has already been shown, to find two particular solutions $z_1(\vartheta)$ and $z_2(\vartheta)$ such that the following conditions are satisfied

$$z_1(\vartheta + 2\pi) \leqslant z_1(\vartheta), \tag{A}$$

$$z_2(\vartheta + 2\pi) \geqslant z_2(\vartheta) \tag{B}$$

for any values of ϑ. The first of these solutions can be found at once. In fact, every solution $z_1(\vartheta)$ for which, for a certain $\vartheta_0, z_1(\vartheta_0) \geqslant (1+\beta)/\alpha$, will be just the required solution, since above the sinusoid $z = (\beta - \sin \vartheta)/\alpha$ and $dz/d\vartheta < 0$ always (Fig. 331). Hence, $z_1(\vartheta_0) \leqslant z_1(\vartheta_0 - 2\pi)$ which satisfies the condition (A).

3] THE NON-CONSERVATIVE CASE 431

To find the second solution that satisfies the condition (B) we shall consider the integral curve passing through the point A (Fig. 332) with coordinates $\vartheta = \pi/2$ and $z = (\beta-1)/\alpha$, i.e. through the point at which the

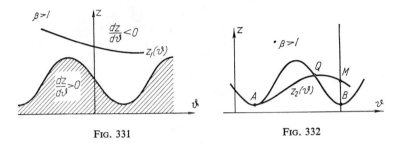

FIG. 331   FIG. 332

sinusoid $z = (\beta - \sin \vartheta)/\alpha$ has a minimum. Let us examine the behaviour of the integral curve on the right of the point A. Since between the sinusoid and the ϑ axis $dz/d\vartheta > 0$, then, as ϑ increases, the curve must go upwards and at a certain point Q must intersect the sinusoid. At this point the integral curve has a horizontal tangent since the sinusoid is the $dz/d\vartheta = 0$ isocline. Then the integral curve goes downwards and intersects the $\vartheta = 5\pi/2$ straight line at a point M which is not lower than the point B (as the sinusoid is the $dz/d\vartheta = 0$ isocline, the integral curve must intersect it with a horizontal tangent and this is only possible at the point B or after the point B). Therefore, the integral curve considered corresponds to the solution for which

$$z_2\left(\frac{\pi}{2} + 2\pi\right) \geqslant z_2\left(\frac{\pi}{2}\right),$$

i.e. satisfies the condition (B).

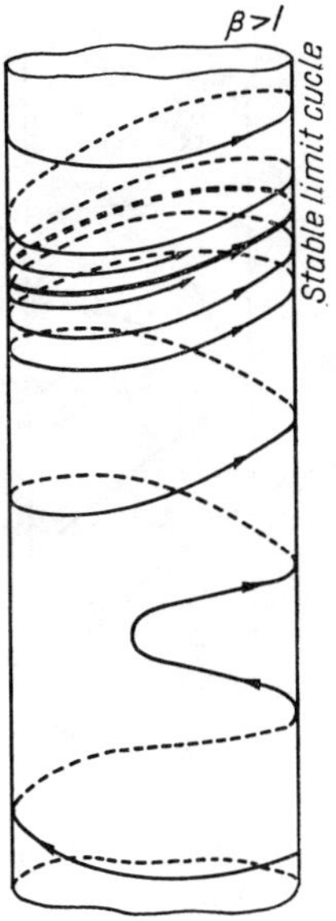

Since in this case ($\beta > 1$) there are no singular points between the solutions z_1 and z_2 there must exist, for reasons of continuity, a periodic solution for which $z_0(\vartheta + 2\pi) = z_0(\vartheta)$. We have already shown that this periodic motion is unique

FIG. 333

and stable. The limit cycle encircling the cylinder that corresponds to this solution is shown in Fig. 333.

II. $0 < \beta < 1$

To clarify the conditions for the existence of a limit cycle encircling the cylinder for $0 < \beta < 1$, we shall construct qualitatively the sequence function that transforms into itself the upper half of the generator $\vartheta = \vartheta_2$ passing

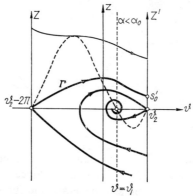

Fig. 334

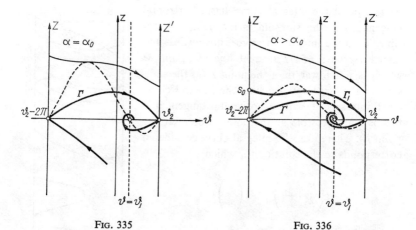

Fig. 335   Fig. 336

through the saddle point $(\vartheta_2, 0)$. On the development of the cylinder (Figs. 334–6) this transformation will be the transformation of the half straight line Z: $\vartheta = \vartheta_2 - 2\pi$, $z \geq 0$ into the half straight line Z': $\vartheta = \vartheta_2$, $z \geq 0$. We shall denote by s and s' the ordinates of the points of this generator and of their consecutive points, if the latter exist, $(s, s' > 0)$. The fixed point s^* of this transformation, if such a point exists, will be the point of intersection of the limit cycle encircling the cylinder with the generator $\vartheta=\vartheta_2$. As we have seen, this limit cycle can only lie entirely in the $z>0$ region. Therefore, a necessary and sufficient condition for its existence is the existence of a fixed point $s^*>0$ in the point transformation, of the half straight line $\vartheta=\vartheta_2$, $z>0$ into itself, generated by the paths of the system (7.4).

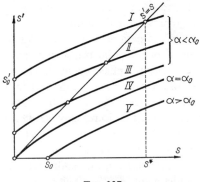

Fig. 337

First of all, just as in the previous case we can verify at once the existence of paths $z=z_1(\vartheta)$ that encircle the cylinder and satisfy the condition $z_1(\vartheta_2-2\pi)>z_1(\vartheta_2)$ (such paths will be those for which $z(\vartheta_2)>(1+\beta)/\alpha$). For them $s'<s$, and for sufficiently large s the graph of the sequence function $s'=f(s)$ always lies (for $\alpha>0$) below the bisector $s'=s$ (Fig. 337).

Let us consider now the behaviour of the separatrix Γ that leaves the saddle point with a positive slope (the three possible cases are shown in Fig. 334–6). In the conservative case, $\alpha=0$, the separatrix Γ is entirely on the upper half of the cylinder (see Figs. 322 and 323). Therefore, for a sufficiently small α when the system is close to a conservative one, the phase portrait is like that shown in Fig. 334: the separatrix Γ now intersecting the half straight line Z' at a certain point $s'=s_0'>0$†. Since, according to (7.5), $dz/d\vartheta$ decreases monotonically as α increases (and, moreover, decreases without limits for $\alpha \to +\infty$), then, as the parameter α increases, the separatrix Γ goes down monotonically (within the region $z>0$). And so s_0' decreases and, for a certain branch value $\alpha=\alpha_0$, tends to zero ($\alpha_0 = \alpha_0(\beta)$ and for $\alpha=\alpha_0$ the separatrix Γ returns to the saddle point; see Fig. 335). Therefore for all $\alpha<\alpha_0$, the separatrix Γ has the form shown in Fig. 334 and its initial point $s=0$ has a consecutive point $s_0'>0$. Moreover, since

† The solutions of the equation (7.5) depend continuously on the parameter α (this is ensured by a general theorem given in Appendix I).

above the separatrix $\varGamma$ there are no singular points of the equation (7.5), all paths that intersect the half straight line Z will encircle the phase cylinder, and correspondingly all the points $s>0$ of this half straight line will have consecutive points $s'(s'>s_0'>0)$. In other words, for $\alpha<\alpha_0$ the sequence function $s'=f(s)$ exists (is defined) for all $s\geqslant 0$, $f(0)=s_0'$ being greater than zero. By virtue of the continuity of the sequence function[†], its graph for $\alpha<\alpha_0$ is bound to intersect the bisector $s'=s$ (see the curves I, II and III in Fig. 337 representing the graphs of the sequence function for three different values of the parameter $\alpha>\alpha_0$.). This point of intersection is just the fixed point s^* of the point transformation and corresponds to the limit cycle that encircles the phase cylinder. The fixed point is unique since, as has been shown, there cannot be more than one limit cycle that encircles the cylinder. It is evident that $s^* \to +0$ for $\alpha \to \alpha_0$.

It follows, from examining the decrease of $dz/d\vartheta$ for an increase in α, that if, $\alpha>\alpha_0$ then the separatrix $\varGamma$ no longer meets the generator $\vartheta=\vartheta_2$ (Fig. 336) and the point $s=0$ has no consecutive sequence point. Then, however, the other separatrix $\varGamma_1$ of the saddle point, the one with a negative slope near the saddle point $(dz/d\vartheta<0)$, is bound to meet the generator $\vartheta=\vartheta_2$ (the half straight line Z of Fig. 336) at a certain point s_0, since it cannot leave the stable state of equilibrium $(\vartheta_1, 0)$. Therefore the point $s=s_0>0$ will have the consecutive point $s'=0$ and the graph of the sequence function (this will be a continuous curve for $s>s_0>0$) will pass through the point $(s_0, 0)$ situated below the bisector $s'=s$ (curve V in Fig. 337) and either will not intersect this bisector (then there are no fixed points nor limit cycles), or will intersect it at an even number of (fixed) points. The latter is impossible since the point transformation $s'=f(s)$ cannot have more than one fixed point since the system (7.4) cannot have more than one limit cycle encircling the cylinder.

Thus for $0<\beta<1$ we have two different cases: $\alpha<\alpha_0$ and $\alpha>\alpha_0$.

For $\alpha<\alpha_0$ there is a unique periodic motion of the system—a unique limit cycle that encircles the phase cylinder (Fig. 338). As α increases (remaining smaller than α_0) this limit cycle moves down (since s^* decreases) and for $\alpha=\alpha_0$ merges with the separatrix loop that now encircles the cylinder.

For $\alpha>\alpha_0$ no periodic motions of the system exist (Fig. 339).

The periodic solutions that we have found are both of the second type, i.e. encircle the cylinder (the first for an arbitrary α and $\beta>1$, the second for $\alpha<\alpha_0$ and $\beta<1$) and are stable, since all adjacent motions tend to these

---

[†] The continuity of the sequence function follows from the theorem on the continuity of the dependence of the solutions of differential equations upon the initial conditions (see Appendix I).

periodic motions. However, whereas in the first case ($\beta > 1$) the periodic solution is established for arbitrary initial conditions, in the second case there exists a region of initial values from which the system tends to a state of rest (to a stable focus or a node). This region is shown shaded in Fig. 338.

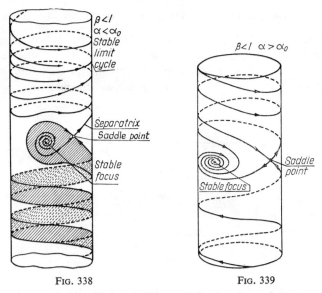

Fig. 338   Fig. 339

The physical meaning of these results is quite clear. If in the system there exists friction proportional to velocity, and a constant torque, then the work done to overcome friction clearly increases with velocity whereas the work of the external forces remains constant. Therefore, if $\beta > 1$ (the constant moment in the pendulum example is so large that it exceeds the maximum moment of the force of gravity), it will make the pendulum rotate on its axis for any initial conditions until a balance is established between the energy dissipated by friction and the work of the external forces. However, if $\beta < 1$ (the moment of the external forces is less than the maximum moment of the force of gravity), then the external moment by itself cannot make the pendulum rotate completely. But the pendulum will be able to complete a revolution if its starts from certain initial conditions, and even become periodic provided that the energy dissipated by friction in one revolution will be equal to the work done by the constant external moment. In addition, for a suitable initial velocity, in order that the loss by friction during one revolution shall not exceed the work of the external torque it is necessary that α be less than a certain critical value α_0.

Thus all our results admit a simple interpretation. In the cases of a synchronous motor, and of a.c. machines working in parallel this interpretation will be more complicated.

§4. ZHUKOVSKII'S PROBLEM OF GLIDING FLIGHT

In concluding the chapter, we shall consider Zhukovskii's problem [64] on gliding flight taking place in a vertical plane (Fig. 340). We shall introduce the notation: ϑ—the angle of slope of the trajectory, v—the velocity of the centre of gravity of the glider, m—the mass of the glider,

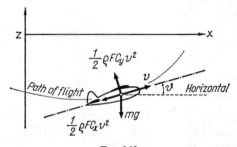

FIG. 340

F—the area of its wings, g—the acceleration of gravity, ϱ—the density of the air and C_x and C_y—the aerodynamical drag and lift coefficients of the bird or glider. Then the equations of motion of the centre of gravity to the glider for tangential and centripetal components of the acceleration are

$$m\frac{dv}{dt} = -mg\sin\vartheta - \frac{1}{2}\varrho F C_x v^2, \\ mv\frac{d\vartheta}{dt} = -mg\cos\vartheta + \frac{1}{2}\varrho F C_y v^2. \tag{7.13}$$

Let the moment of inertia of the glider (with respect to the centre of gravity) be so small, and the stabilizing moment of the forces developed by the tail unit be so large, that we can neglect the variations of the angle of attack of the glider (the angle between its longitudinal axis and the trajectory of its centre of gravity) and assume it to be constant. Then the coefficients C_x and C_y in the equation (7.13) will also be constant. On introducing the new variables

$$v = v_0 y,$$

where $v_0 = (2mg/\varrho FC_y)^{\frac{1}{2}}$ is the velocity of horizontal flight for which the weight of the glider is equalled by the lift force, and

$$t = \frac{v_0}{g} t_{\text{new}}.$$

we shall reduce the equations (7.13) to the following non-dimensional form:

$$\left. \begin{array}{l} \dot{y} = -\sin \vartheta - ay^2 = F(\vartheta, y), \\ \dot{\vartheta} = \dfrac{-\cos \vartheta + y^2}{y} = \varPhi(\vartheta, y), \end{array} \right\} \quad (7.14)$$

where a dot denotes differentiation with respect to the new time and

$$a = \frac{C_x}{C_y}.$$

Since the states $(\vartheta + 2\pi, y)$ and (ϑ, y) are physically coincident (the right-hand sides of the equations (7.14) are periodic functions of the angle ϑ with period 2π), we take a phase cylinder and plot the quantity y, proportional to the velocity v, along its axis, and the angle around the axis. For forward flight, we can consider phase paths only on the upper half of the cylinder ($y \geqslant 0$). The equation of the integral curves on the cylinder is clearly

$$\frac{dy}{d\vartheta} = \frac{y(\sin \vartheta + ay^2)}{\cos \vartheta - y^2}. \quad (7.15)$$

Note that this equation has the integral curve $y=0$, which is a *singular* phase path of the system (7.14) and corresponds to an *instantaneous tip-over* of the glider into the position $\vartheta = -\pi/2$ as soon as the velocity v (or y) reduces to zero. For according to (7.14) when $y=0$, $\dot{\vartheta}=+\infty$, if $-3\pi/2 < \vartheta < -\pi/2$ and $\dot{\vartheta} = -\infty$ if $-\pi/2 < \vartheta < +\pi/2$.

The appearance of this singular phase path, corresponding to physically impossible somersaulting of the glider at the instant of rest ($v=0$) is caused by assuming a constant angle of attack. This assumption, as is wellknown, is not satisfied for small velocities of motion of the glider for then the stabilizing moment developed by the tail unit is also small. For large velocities this moment ensures a nearly constant angle of attack.

I. Just as in the previous problem, we shall begin our analysis with the conservative case $a=0$ (no forces of resistance). This has been analysed in detail by N. E. Zhukovskii [64, 171]. The differential equation of the integral curves (7.15) has the integral

$$\frac{y^3}{3} - y\cos\vartheta = C\,(=\text{const}) \tag{7.16}$$

and the three singular points: (1) $\vartheta = 0$, $y = +1$; (2) $\vartheta = +\pi/2$, $y = 0$ and (3) $\vartheta = -\pi/2$, $y = 0$. Only the first of these is a state of equilibrium of equations (7.14) (for $a = 0$),

$$\dot{y} = -\sin\vartheta, \quad \dot{\vartheta} = \frac{-\cos\vartheta + y^2}{y}, \tag{7.14a}$$

and corresponds to horizontal flight of the glider with constant velocity $v = v_0$. The other two singular points lie on the singular integral curve $y = 0$

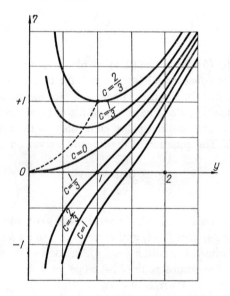

Fig. 341

which we have discussed above and are not states of equilibrium of the system (7.14a) since at these points $\dot{y} \neq 0$.

To construct the integral curves note that equation (7.16) can be solved with respect to θ

$$\vartheta = \pm\arccos\eta, \tag{7.16a}$$

where

$$\eta = \frac{y^2}{3} - \frac{C}{y} \quad (|\eta| \leq 1).$$

Fig. 341 shows a family of auxiliary curves $\eta=\eta(y, C)$ for $y<0$†, and in Figs. 342 and 343 the phase paths on the development of the cylinder and on the phase cylinder itself. The value $C=-\dfrac{2}{3}$ corresponds to the singular point of the centre type $\vartheta=0$, $y=+1$, and a state of equilibrium of the

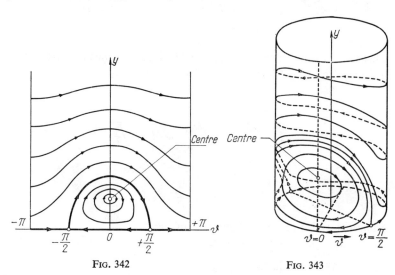

Fig. 342   Fig. 343

system of equations (7.14a). The remaining phase paths are closed: the phase paths for which $-\dfrac{2}{3}<C<0$ encircle the centre but do not encircle the cylinder, while the paths with $C>0$ encircle it‡. The first of them corresponds to flight of the glider along "wave-shaped" lines and the second ones to a flight for which the glider performs "dead" loops. The flight paths of the glider are shown in Fig. 344††.

---

† The curves $\eta = \eta(y, C)$ are monotonic for $C > 0$, and have minima that lie on the parabola $y^2 = \eta$ for $C < 0$, for, as is easily seen

$$\frac{d\eta}{dy} = \frac{2}{3}y + \frac{C}{y^2} = \frac{1}{y}(y^2-\eta);$$

For $C < -\frac{2}{3}$ these curves lie entirely above the straight line $\eta = +1$.

‡ These two types of closed phase paths are separated by the integral curve $C=0$, consisting of the circle $y=0$ and of the separatrix of the saddle points (the equation of the latter has the form: $\vartheta = \cos^{-1} y^{\frac{2}{3}}$).

†† The equations of these symmetric flight paths of the glider, in the absence of air resistance, were studied by N. E. Zhukovskii. These paths were later called "phugoids",

II. Let us pass now to the qualitative analysis of the flight taking into account the air resistance ($a>0$) [166]. As before there is a unique state of equilibrium of the system of equations (7.14); its coordinates will be

$$\left. \begin{array}{l} \vartheta_0 = -\arctan a \quad \left(-\dfrac{\pi}{2} < \vartheta_0 < 0\right), \\[1em] y_0 = \dfrac{1}{\sqrt[4]{1+a^2}} \qquad (0 < y_0 < 1). \end{array} \right\}$$

This state of equilibrium of the system (7.14) corresponds to a flight along a *descending* straight line with constant velocity $v<v_0$. On linearizing the

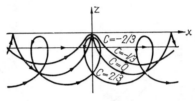

Fig. 344

equations (7.14) in a neighbourhood of the state of equilibrium (ϑ_0, y_0) it is easily verified that the latter is always *stable* and for sufficiently small a $\left(\text{for } a < 8^{\frac{1}{2}}\right)$ it is a focus.

We shall use Dulac's criterion to prove the absence of closed integral curves (except the circle $y=0$). On taking y as the multiplier $B(\vartheta, y)$ we obtain from the equations (7.14)

$$\frac{\partial}{\partial y}[yF] + \frac{\partial}{\partial \vartheta}[y\Phi] = -3ay^2 \leq 0, \qquad (7.18)$$

and may be obtained in the following manner. First of all, since $v \sin \vartheta = dz/dt$, we obtain from the equations (7.13), for the case $C_x = 0$,

$$\frac{1}{2}d(v^2) = -g\,dz, \quad \frac{1}{2}v^2 = -gz \quad \text{or} \quad y^2 = -\frac{2gz}{v_0^2} \qquad (\alpha)$$

(the constant of integration is equal to zero, as the height z is measured from the level that corresponds to the velocity $v=0$). Further,

$$\frac{dz}{dx} = \tan \vartheta \quad \text{or} \quad dx = \frac{dz}{\tan \vartheta} = \frac{dz}{R(z, C)} \qquad (\beta)$$

where $R(z, C)$ is a function of z that is obtained if we express $\tan \vartheta$ in terms of z by means of the relations (7.16a) and (α). On integrating the equation (β) by any approximate method (the integral of the right-hand side is not expressible in terms of elementary functions) we have x in terms of z, the graphs of which are shown in Fig. 344.

the equality to zero being true only on the circle $y=0$. Hence there are no closed integral curves (or closed contours consisting of integral curves) that do not encircle the phase cylinder, and there is not more than one closed integral curve encircling the cylinder. Since such a closed integral curve encircling the phase cylinder is the circle $y=0$ (it corresponds, as in the conservative case, to an instantaneous turn-over of the glider at $v=0$), we can assert that the system of equations (7.14) (for $a>0$) does not have any closed phase paths encircling the cylinder on the upper half of the phase cylinder ($y>0$). In other words, this system does not have any periodic oscillations.

The equation of the integral curves (7.15), as in the case $a=0$, has in addition to the state of equilibrium (ϑ_0, y_0) two more singular points of the saddle type ($-\pi/2, 0$) and ($\pi/2, 0$) that are not states of equilibrium of the system (7.14). However, in contrast to the case $a=0$, the separatrix of the saddle point ($-\pi/2, 0$), in the upper half of the phase cylinder, can no longer reach the saddle point ($\pi/2, 0$)[†]. Also note that all the circles $y = \text{const} \geqslant a^{-\frac{1}{2}}$ are such that on them $\dot{y} \leqslant 0$. Hence, all the phase paths go from distant regions of the upper half of the cylinder into the region comprised between the circle $y=0$ and $y=a^{-\frac{1}{2}}$ and containing the state of equilibrium (ϑ_0, y_0). We can assert, since there are no closed integral curves (except the circle $y=0$), that *all phase paths approach asymptotically the stable state of equilibrium*, the point (ϑ_0, y_0).

Enough has been discovered to construct the phase portrait on the phase cylinder for the system (7.14) for any $a>0$. Such a picture is shown in Fig. 345.

In the presence of air resistance the glider can have a unique stable flight with constant velocity $v=v_0 y_0$ along a descending straight line at an angle ϑ_0 to the horizontal. This flight path can arise for any initial conditions. If the initial velocity of the glider is sufficiently large, then the glider first performs a number of "dead" loops (this number being determined by the initial conditions) and approaches along a "wave-shaped" curve the final rectilinear flight path. Such a flight path is shown in Fig. 345[‡].

[†] Should this separatrix arrive at the saddle point ($\pi/2$, 0), then on the phase cylinder there would be two closed contours consisting of integral curves (of the separatrix of the saddle point and of one or other semicircle $y=0$) and not encircling the cylinder, which is impossible since the conditions of Dulac's criterion are verified.

[‡] In contrast to the conservative case $a=0$ the equations of the flight paths in the x, z plane are no longer obtainable by quadratures, since when $a>0$, the integral (7.16) is not true, nor is the equation (α) in the footnote on p. 440.

It is necessary in certain problems to introduce other types of phase surface, differing from the plane and the cylinder, for example a torus or

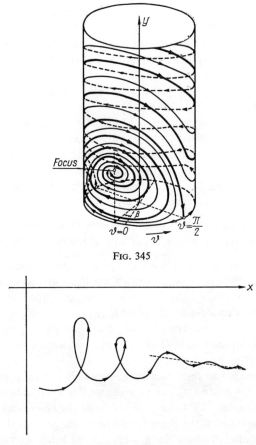

Fig. 345

Fig. 346

surfaces with many sheets. The systems with a phase surface in the form of a torus exceed the scope of this book. Certain systems with a phase surface of many sheets will be considered in the following chapter[†].

[†] One more dynamic system with a cylindrical phase surface (a simplified model of a steam engine) will be considered in the following chapter (in Section 10).

CHAPTER VIII

THE METHOD OF THE POINT TRANSFORMATIONS IN PIECE-WISE LINEAR SYSTEMS[†]

§ 1. INTRODUCTION

We shall proceed now to a quantitative investigation of non-linear dynamic systems, restricting ourselves as before to autonomous systems of the second order (1 degree of freedom). At the present state of the theory, a quantitative investigation by analytical methods can be satisfactorily carried out only for three classes of systems, which have, however, a considerable practical interest. One class includes systems that are approximately conservative, in particular, resembling harmonic oscillators (these are the most important in practice). A second class includes systems that sustain discontinuous oscillations. These two classes will be considered respectively in Chapters IX and X. A third class includes those for which a quantitative investigation can be carried out by the method of the point transformations[‡]. This method is relatively simple for the so-called *piecewise linear systems*, i.e. for systems where phase space consists of regions each of which has linear dynamic equations of motion. This chapter is devoted to just such piece-wise linear systems.

A few problems on the self-oscillations of piece-wise linear systems have already been investigated in Sections 4–6, Chapter III. In these problems the search for the limit cycles and the investigation of their stability was reduced to the construction of a certain point transformation of a straight line into itself (i.e. to the evaluation of the corresponding sequence function), and to the search for the fixed points of the point transformation and the investigation of their stability. In all the problems, the sequence function was obtained (or could have been obtained) in an explicit form.

In the great majority of problems, however, it is difficult to obtain the sequence function in an explicit form, while it is comparatively easy to

[†] Written by N. A. Zheleztsov.
[‡] The basic concepts of the method of the point transformations (the concepts of sequence function, of fixed point of a point transformation and of its stability) have been formulated in Section 7, Chapter V. Koenigs's theorem on the stability of a fixed point is also given there.

obtain it in a *parametric form*. Let, for example, the x, y phase plane of a certain dynamic system be divided by the straight lines $x=x_1$ and $x=x_2$ into three regions (I), (II) and (III) (Fig. 347) in each of which the equations of motion are linear. Let us denote by S, S_1, S_2 and S_3 the half straight lines which the representative point crosses in moving respectively from region (I) into region (II), from (II) into (III), from (III) into (II) and finally,

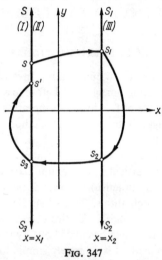

Fig. 347

from (II) back into (I); and by s, s_1, s_2 and s_3 the ordinates of the points of these half straight lines. The phase paths in the "regions of linearity" (I), (II) and (III) generate transformations of the half straight line S into S_1, of S_1 into S_2, of S_2 into S_3 and of S_3 into S, for the points of these half straight lines have a one-to-one continuous relationship. We shall denote these point transformations by Π_1, Π_2, Π_3 and Π_4 respectively. By integrating the linear differential equations of motion of the system in the corresponding region, we are able to find for each of these point transformations of a half straight line into a half straight line the *sequence function*† in a parametric form

$$\left.\begin{array}{lll} \Pi_1: & s_1 = \varphi_1(\tau_1), & s = \psi_1(\tau_1); \\ \Pi_2: & s_2 = \varphi_2(\tau_2), & s_1 = \psi_2(\tau_2); \\ \Pi_3: & s_3 = \varphi_3(\tau_3), & s_2 = \psi_3(\tau_3); \\ \Pi_4: & s' = \varphi_4(\tau_4), & s_3 = \psi_4(\tau_4), \end{array}\right\} \quad (8.1)$$

† The sequence function for the point transformation of a line into another line is sometimes called the *correspondence function*.

where τ_1, τ_2, τ_3 and τ_4 are the times of transit of the representative point through the corresponding regions[†].

If the phase paths leaving a certain segment of S return to it after passing through the regions (I), (II), (III) and (II); (see Fig. 347), then the point transformation Π of this segment of S into the half straight line S (having a sequence function $s' = f(s)$) is obtained by applying in succession the transformations Π_1, Π_2, Π_3 and Π_4. Thus,

$$\Pi = \Pi_1 \cdot \Pi_2 \cdot \Pi_3 \cdot \Pi_4.$$

Therefore the problem of obtaining the limit cycles passing through all the regions (I), (II) and (III) reduces to finding the fixed points of this "overall" transformation Π, i.e. to solving the system of (usually transcendental) equations

$$\left. \begin{array}{ll} \varphi_1(\tau_1) = \psi_2(\tau_2), & \varphi_3(\tau_3) = \psi_4(\tau_4), \\ \varphi_2(\tau_2) = \psi_3(\tau_3), & \varphi_4(\tau_4) = \psi_1(\tau_1). \end{array} \right\} \quad (8.2)$$

The stability of the fixed point and the corresponding limit cycle is easily determined using Koenigs's theorem and noticing that, at the fixed point

$$\frac{ds'}{ds} = \frac{\varphi_1'(\bar{\tau}_1)}{\psi_1'(\bar{\tau}_1)} \cdot \frac{\varphi_2'(\bar{\tau}_2)}{\psi_2'(\bar{\tau}_2)} \cdot \frac{\varphi_3'(\bar{\tau}_3)}{\psi_3'(\bar{\tau}_3)} \cdot \frac{\varphi_4'(\bar{\tau}_4)}{\psi_4'(\bar{\tau}_4)}$$

($\bar{\tau}_1$, $\bar{\tau}_2$, $\bar{\tau}_3$ and $\bar{\tau}_4$ denote the values of τ_1, τ_2, τ_3 and τ_4 for the fixed point.[‡]

[†] As the differential equations of the motion of the system in the region (II) are linear, the equations of the phase path that reaches this region at a point of the half straight line S with coordinate s at the instant $t=0$ will depend linearly on s

$$x = sf_1(t) + f_2(t), \qquad y = sf_3(t) + f_4(t).$$

Let the representative point, moving along this path, arrive on the half straight line S_1 at a point s_1 at $t = \tau_1$; then, clearly,

$$x_2 = sf_1(\tau_1) + f_2(\tau_1), \qquad s_1 = sf_3(\tau_1) + f_4(\tau_1).$$

On solving these relations we obtain the correspondence function for the transformation Π_1

$$s = \frac{x_2 - f_2(\tau_1)}{f_1(\tau_1)} \equiv \psi_1(\tau_1), \qquad s_1 = \psi_1(\tau_1) \cdot f_3(\tau_1) + f_4(\tau_1) \equiv \varphi_1(\tau_1).$$

In exactly the same manner we can find the correspondence functions for the other point transformations Π_1, Π_2 and Π_3.

[‡] Generally speaking, there are possible limit cycles that pass through only two regions, for example, through the regions (I) and (II). These limit cycles can be clearly found by constructing the point transformation $\Pi' = \Pi_5 \cdot \Pi_4$, where Π_5 is the transformation of the half straight line (s) into (s_3) generated by the phase paths lying entirely in the region (II).

In principle, we can obtain in this way the point transformations for any piece-wise linear dynamic system of the second order and carry out a quantitative investigation. In practice, of course, the difficulties in solving the system of equations for the fixed points and in assessing their stability, increase rapidly with increase of the number of the regions of linearity. Therefore, to avoid complicating our exposition, we shall restrict ourselves in this chapter to considering comparatively simple problems of self-oscillating systems, for which the "overall" point transformation is the product of not more than two point transformations of a straight line into a straight line, the point transformations being expressed in parametric form. In these problems the fixed points corresponding to limit cycles will be determined by two transcendental equations, conveniently examined by means of Lamerey's diagrams (see Chapter III).

§ 2. A Valve Generator

1. Equation of the oscillations

As a first problem we shall consider the self-oscillations of a valve generator with the resonant network in the grid circuit or in the anode circuit (Fig. 348). If we neglect the anode conductance, the grid currents

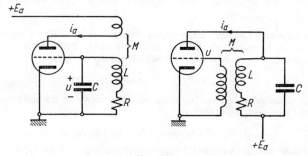

Fig. 348

and the inter-electrode capacitances, then, as we have seen in Chapter I, § 6, the equation of the oscillations of such a valve generator can be written as follows:

$$LC\frac{d^2u}{dt'^2}+[RC-MS(u)]\frac{du}{dt'}+u = 0 \dagger.$$

† We denote the time by t' since, later, t will denote "dimensionless" time.

A VALVE GENERATOR

We shall use in this section a piece-wise linear approximation to the characteristic of the valve $i_a = i_a(u)$, shown in Fig. 349,

$$i_a = \begin{cases} 0 & \text{for } u \leq -u_0, \\ S(u+u_0) & \text{for } u > -u_0, \end{cases} \tag{8.4}$$

where S is the positive slope of the valve characteristic and $-u_0$ is the cut-off voltage of the valve ($u_0 > 0$). We shall introduce the dimensionless

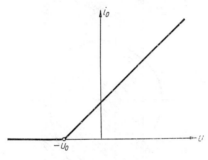

Fig. 349

variables $x = u/u_0$ and $t = \omega_0 t'$, where $\omega_0 = (LC)^{-\frac{1}{2}}$ is the undamped natural frequency of the resonant circuit. The equation (8.3) can now be written as

$$\begin{aligned} x < -1 & \quad \ddot{x} + 2h_1 \dot{x} + x = 0, \\ x > -1 & \quad \ddot{x} - 2h_2 \dot{x} + x = 0, \end{aligned} \tag{8.5}$$

where

$$h_1 = \frac{\omega_0}{2} RC \quad \text{and} \quad h_2 = \frac{\omega_0}{2} [MS - RC].$$

Thus, for such a piece-wise linear approximation to the characteristic of the valve, the phase plane $x, y (y = \dot{x})$ of the valve generator is divided by the straight line $x = -1$ into two regions (*I*) and (*II*) (Fig. 350) in each of which the phase paths are determined by the corresponding linear differential equations (8.5)[†]. We shall assume, of course, the phase paths to be

---

† It is clear that the assumptions used in constructing a mathematical model of a valve generator, cannot reproduce the properties of a real valve generator for sufficiently large positive values of the voltage *u*, when grid currents flow in the valve. Therefore, certain properties of the mathematical model are not properties of real valve generators.

continuous everywhere and in particular on the boundary between the regions of linearity, the straight line $x=-1$.

The only state of equilibrium $x=0$, $y=0$ lies in the region (II); it is stable for $h_2<0$ (i. e. for $MS<RC$) and unstable for $h_2>0$ (for $MS>RC$). We shall only consider the latter case, the case of a "self-exciting" generator[†]. As the state of equilibrium $x=0$, $y=0$ is an unstable focus for

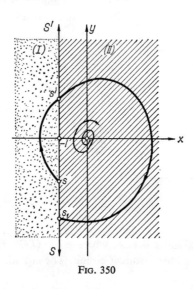

Fig. 350

$0<h_2<1$ and an unstable node for $h_2>1$ and is never a saddle point, the singular points that determine the qualitative character of the phase portrait are the known state of equilibrium and the limit cycles, if any. Our object, therefore, is the search for the limit cycles and the investigation of their stability.

Since the equations (8.5) are linear in each of the regions (I) and (II), then on the phase plane there cannot be limit cycles lying entirely in only one region. A limit cycle, if it exists, must pass through both regions and encircle the state of equilibrium. Hence it will intersect the boundary line $x=-1$.

Let us split this boundary line into two half lines, the straight line S: $x=-1$, $y=-s$ $(s>0)$, and the straight line S': $x=-1$, $y-s'>0$. These lines are clearly *lines without contact*: the line S is intersected by the phase

[†] If $h_2<0$, i.e. if $MS<RC$, then all phase paths will approach the origin asymptotically and the generator will not oscillate.

paths that go (as t increases) from (II) into (I) and the line S' by the paths that go from (I) into (II).

Consider a phase path that leaves point s of line S. This path, after traversing (I) will intersect the line S' at a point s' and then, if $h_2 < 1$, i. e. if the phase paths in (II) are spirals, will again reach the line S at a point s_1 (Fig. 350). Thus, for $0 < h_2 < 1$, the phase paths generate a point trans-

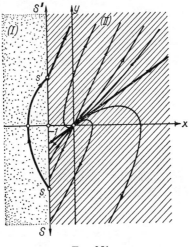

FIG. 351

formation of the line S into itself, by establishing a one-to-one continuous correspondence between the points s and s_1. The fixed point of this transformation is the point of intersection of a limit cycle with S.

If, however, $h_2 > 1$, then the state of equilibrium $(0, 0)$ will be an unstable node; in the region (II) there will be two rectilinear phase paths that recede to infinity (Fig. 351) and, hence, the paths that leave S' can no longer arrive on S but will recede to infinity. It is clear that in this case there are no limit cycles and all phase paths go to infinity where, incidentally, the mathematical model is not valid.

2. Point transformation

Consider the case $0 < h_2 < 1$. The point transformation Π of line S into itself is the product of two transformations: the transformation Π_1 of the points s of S into the points s' of S' generated by the paths in (I), and the transformation Π_2 of the points s' into points s_1 of line S generated by the paths in the region (II).

In the region (I) $(x < -1)$ the phase paths are determined by the first of the differential equation (8.5). Its solution, when $x = x_0$, $y = y_0$ at $t = 0$ is†

$$x = e^{-h_1 t}\left\{x_0 \cos \omega_1 t + \frac{y_0 + h_1 x_0}{\omega_1} \sin \omega_1 t\right\},$$
$$y = \dot{x} = e^{-h_1 t}\left\{y_0 \cos \omega_1 t - \frac{x_0 + h_1 y_0}{\omega_1} \sin \omega_1 t\right\},$$
(8.6)

where

$$\omega_1 = +\sqrt{1 - h_1^2}.$$

Therefore the equation of the path leaving S at $t = 0$ ($x_0 = -1$, $y_0 = -s$, where $s > 0$), will be

$$x = -e^{-h_1 t}\left[\cos \omega_1 t + \frac{s + h_1}{\omega_1} \sin \omega_1 t\right],$$
$$y = \dot{x} = e^{-h_1 t}\left[-s \cos \omega_1 t + \frac{1 + h_1 s}{\omega_1} \sin \omega_1 t\right].$$
(8.7)

The representative point moving along the path (8.7), will reach at time $t_1 = \tau_1/\omega_1$ the half line S' at a point s' ($x = -1$, $y = s' > 0$) (Fig. 350). Then

$$s' = e^{-\frac{h_1 \tau_1}{\omega_1}}\left[-s \cos \tau_1 + \frac{1 + sh_1}{\omega_1} \sin \tau_1\right],$$
$$-1 = -e^{-\frac{h_1 \tau_1}{\omega_1}}\left[\cos \tau_1 + \frac{s + h_1}{\omega_1} \sin \tau_1\right].$$

Solving these equations for s and s', we obtain the correspondence or sequence function for the transformation Π_1,

$$s = \frac{e^{\gamma_1 \tau_1} - \cos \tau_1 - \gamma_1 \sin \tau_1}{\sqrt{1 + \gamma_1^2} \cdot \sin \tau_1},$$
$$s' = \frac{e^{-\gamma_1 \tau_1} - \cos \tau_1 + \gamma_1 \sin \tau_1}{\sqrt{1 + \gamma_1^2} \cdot \sin \tau_1},$$
(8.8)

where

$$\gamma_1 = \frac{h_1}{\omega_1} = \frac{h_1}{\sqrt{1 - h_1^2}}$$

(as h_1 varies from 0 to $+1$, γ_1 increases monotonically from 0 to $+\infty$). On differentiating (8.8), we find

† See, for example, Chapter I, Section 4.

$$\frac{ds}{d\tau_1} = \frac{1-e^{\gamma_1\tau_1}(\cos\tau_1-\gamma_1\sin\tau_1)}{\sqrt{1+\gamma_1^2}\cdot\sin^2\tau_1} \quad \text{and}$$

$$\frac{ds'}{d\tau_1} = \frac{1-e^{-\gamma_1\tau_1}(\cos\tau_1+\gamma_1\sin\tau_1)}{\sqrt{1+\gamma_1^2}\cdot\sin^2\tau_1}.$$

Now introduce the auxiliary function

$$\varphi(\tau,\gamma) = 1-e^{\gamma\tau}(\cos\tau-\gamma\sin\tau),$$

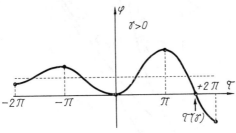

Fig. 352

the graph of which (for $\gamma > 0$) is shown qualitatively in Fig. 352. We note the following three properties of this function:

(1) $\varphi(-\tau, -\gamma) \equiv \varphi(\tau, \gamma)$;
(2) $\partial\varphi/\partial\tau = (1+\gamma^2)e^{\gamma\tau}\sin\tau$:
(3) for $\gamma > 0$, $\varphi(\tau, \gamma)$ reduces to zero for a certain $\tau = \tau^0(\gamma)$ ($\pi < \tau^0 < 2\pi$) and is greater than zero for $\tau < \tau^0$

Then

$$\left.\begin{array}{ll} s = \dfrac{e^{\gamma_1\tau_1}\varphi(\tau_1,-\gamma_1)}{\sqrt{1+\gamma_1^2}\cdot\sin\tau_1}, & \dfrac{ds}{d\tau_1} = \dfrac{\varphi(\tau_1,\gamma_1)}{\sqrt{1+\gamma_1^2}\sin^2\tau_1}; \\[2mm] s' = \dfrac{e^{-\gamma_1\tau_1}\varphi(\tau_1,\gamma_1)}{\sqrt{1+\gamma_1^2}\sin\tau_1}, & \dfrac{ds'}{d\tau_1} = \dfrac{\varphi(\tau_1,-\gamma_1)}{\sqrt{1+\gamma_1^2}\sin^2\tau_1}. \end{array}\right\} \quad (8.8a)$$

From these (8.8a) and from the properties of the function $\varphi(\tau,\gamma)$ it follows that, for s to have all values in the interval $0 < s < +\infty$, the parameter τ_1 must vary in the interval $0 < \tau_1 < \pi$. Also as τ_1 varies from 0 to π, s and s' increase monotonically from 0 to $+\infty$, $s, s', ds/d\tau_1$ and $ds'/d\tau_1$ remaining positive and continuous[†].

[†] The parameter of the transformation, τ_1, is the normalized time of transit across the region (I). Therefore, we must take, from all possible values of τ_1 corresponding to a given value of s (according to the first relation (8.8)), the *minimum positive* one.

The limit values of s and s' for $\tau_1 \to 0$ are found from (8.8) using L'Hôpital's rule.

To construct the graph (Fig. 353) connecting the values of s and s' it suffices to note the following:

(1) for $0 < \tau < \pi$

$$\frac{ds}{ds'} = \frac{\varphi(\tau_1, \gamma_1)}{\varphi(\tau_1, -\gamma_1)} > 0$$

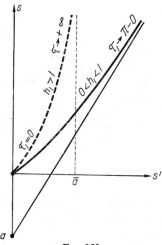

Fig. 353

and increases monotonically from 1 at $\tau_1 \to +0$ to $e^{\gamma_1 \pi}$ at $\tau_1 \to \pi - 0$, since

$$\frac{d^2s}{ds'^2} = \frac{\partial}{\partial \tau_1} \left\{ \frac{\varphi(\tau_1, \gamma_1)}{\varphi(\tau_1, -\gamma_1)} \right\} \frac{1}{\frac{ds'}{d\tau_1}} =$$

$$= \frac{2(1+\gamma_1^2)^{\frac{3}{2}} \sin^3 \tau_1}{[\varphi(\tau_1, -\gamma_1)]^3} [\sinh \gamma_1 \tau_1 - \gamma_1 \sin \tau_1] > 0 \quad (8.10)$$

for $0 < \tau < \pi$;

(2) for $\tau \to \pi - 0$ the graph of the correspondence function (8.8) has a rectilinear asymptote

$$s = e^{\gamma_1 \pi} s' + a, \quad (8.11)$$

where

$$a = \lim_{\tau \to \pi - 0} [s - e^{\gamma_1 \pi} s'] = -\frac{2\gamma_1(1+e^{\gamma_1 \pi})}{\sqrt{1+\gamma^2}} < 0;$$

(3) owing to the fact that $d^2s/ds'^2 > 0$ and $a < 0$ the curve (8.8) is situated above the asymptote (8.11).

For $h_1 > 1$ the solution of the equation (8.5) in region (I) is easily seen to be

$$\left. \begin{aligned} s &= \frac{e^{\bar{\gamma}_1 \tau_1} - \cosh \tau_1 - \bar{\gamma}_1 \sinh \tau_1}{\sqrt{\bar{\gamma}_1^2 - 1} \sinh \tau_1}, \\ s' &= \frac{e^{-\bar{\gamma}_1 \tau_1} - \cosh \tau_1 + \bar{\gamma}_1 \cdot \sinh \tau_1}{\sqrt{\bar{\gamma}_1^2 - 1} \sinh \tau_1}, \end{aligned} \right\} \qquad (8.12)$$

where

$$\bar{\gamma}_1 = \frac{h_1}{\sqrt{h_1^2 - 1}}$$

(as h_1 varies from 1 to $+\infty$, $\bar{\gamma}_1$ decreases monotonically from $+\infty$ to 1). It is also easily verified that, as τ_1 varies from 0 to $+\infty$, s increases monotonically from 0 to $+\infty$, while s' increases from 0 to

$$\bar{a} = \lim_{\tau_1 \to +\infty} s' = \sqrt{\frac{\bar{\gamma}_1 - 1}{\bar{\gamma}_1 + 1}}.$$

The graph of the correspondence function (8.11) in this case is shown with a dotted line in Fig. 353.

We shall pass now to the point transformation Π_2, i.e. the transformation of the points of the half straight line S' into points s_1 of the half straight line S as generated by paths in the region (II), restricting ourselves to the case $0 < h_2 < 1$[†]. Let the representative point, moving along a path in the region (II), arrive at the point s_1 at $t = 0$, having previously left at point s' on the line S' at a time $t = -\tau_2/\omega_2 < 0$ (Fig. 350). τ_2 is the normalized time of transit of the representative point in the region (II). By the same procedure as before the correspondence function for the transformation Π_2 is shown to be

$$\left. \begin{aligned} s_1 &= -\frac{e^{\gamma_2 \tau_2} - \cos \tau_2 - \gamma^2 \sin \tau_2}{\sqrt{1 + \gamma_2^2} \sin \tau_2}, \\ s' &= -\frac{e^{\gamma_{-2} \tau_2} - \cos \tau_2 + \gamma_2 \sin \tau_2}{\sqrt{1 + \gamma_2^2} \sin \tau_2}, \end{aligned} \right\} \qquad (8.13)$$

where

$$\gamma_2 = \frac{h_2}{\omega_2} = \frac{h_2}{\sqrt{1 - h_2^2}}$$

and $\omega_2 = +\sqrt{1 - h_2^2}$.

[†] It was shown above that for $h_2 > 1$ the transformation Π_2 does not exist and all phase paths go to infinity.

Clearly

$$s' = -\frac{e^{-\gamma_2\tau_2}\varphi(\tau_2, -\gamma_2)}{\sqrt{1+\gamma_2^2}\sin\tau_2}, \quad s' = -\frac{e^{\gamma_2\tau_2}\varphi(\tau_2, \gamma_2)}{\sqrt{1+\gamma_2^2}\sin\tau_2} \Bigg\} \quad (8.13a)$$

and

$$\frac{ds_1}{ds'} = \frac{\varphi(\tau_2, \gamma_2)}{\varphi(\tau_2, -\gamma_2)}.$$

The parameter τ_2 varies over such an interval of minimum positive values that $0 < s' < +\infty$. It follows from the properties of the function $\varphi(\tau, \gamma)$

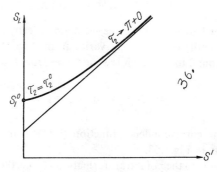

Fig. 354

(Fig. 352) and from the expressions (8.13a) that such an interval will be $\pi < \tau_2 < \tau_2^0$ where τ_2^0 is the smallest positive root of the equation $s'=0$, i.e. of

$$\varphi(\tau_2, \gamma_2) = 0$$

(clearly, $\pi < \tau_2^0 < 2\pi$). The following properties of the correspondence function (8.13) are easily verified: (1) as τ_2 decreases from τ_2^0 to π, s' increases monotonically from 0 to $+\infty$, and s_1 from a certain value s_1^0 to $+\infty$; (2) $d^2s_1/ds'^2 > 0$, therefore, as s' increases, ds_1/ds' increases monotonically from 0 at $s'=0$ (when $\tau_2=\tau_2^0$) to $e^{\gamma_2\pi}$ at $s' \to +\infty$ (when $\tau_2 \to \pi+0$); (3) for $\tau_2 \to \pi+0$ the curve (8.15) has a rectilinear asymptote

$$s_1 = e^{\gamma_2\pi}s' + \frac{2\gamma_2(1+e^{\gamma_2\pi})}{\sqrt{1+\gamma_2^2}}. \quad (8.14)$$

The graph of the sequence function (8.13) for the point transformation Π_2 is shown in Fig. 354.

3. *The fixed point and its stability*

To determine the fixed points of the transformation π of the half straight line S into itself, we shall superimpose on one diagram (Lamerey's diagram) the graphs of the correspondence functions (8.8) and (8.13) (Figs. 355, 356

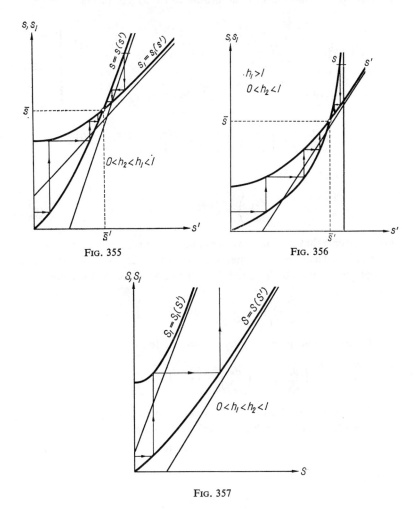

Fig. 355

Fig. 356

Fig. 357

and 357). For $0 < h_2 < h_1 < 1$ (Fig. 355) the curves $s = s(s')$ and $s_1 = s_1(s')$ have one point of intersection, since these curves are continuous and $s_1 > s$ for $s' = 0$, but $s_1 < s$ for $s' \to +\infty$ (as $\gamma_2 < \gamma_1$, the asymptote (8.14) has a smaller slope than the asymptote (8.11)). Therefore there is only one fixed point of the transformation Π and *only one limit cycle*, as shown formally below.

Quantities corresponding to this fixed point will be distinguished by a superscript bar and we have, according to (8.8a) and (8.13a)

$$\left. \begin{array}{l} \dfrac{e^{\gamma_1 \bar{\tau}_1}}{\sqrt{1+\gamma_1^2} \sin \bar{\tau}_1} \varphi(\bar{\tau}_1, -\gamma_1) = -\dfrac{e^{-\gamma_2 \bar{\tau}_2}}{\sqrt{1+\gamma_2^2} \sin \bar{\tau}_2} \varphi(\bar{\tau}_2, -\gamma_2), \\[2ex] \dfrac{e^{-\gamma_1 \bar{\tau}_1}}{\sqrt{1+\gamma_1^2} \sin \bar{\tau}_1} \varphi(\bar{\tau}_1, \gamma_1) = -\dfrac{e^{\gamma_2 \bar{\tau}_2}}{\sqrt{1+\gamma_2^2} \sin \bar{\tau}_2} \varphi(\bar{\tau}_2, \gamma_2) \end{array} \right\} \quad (8.15)$$

and

$$\left(\overline{\dfrac{ds_1}{ds}}\right) = \left(\overline{\dfrac{ds_1}{ds'}}\right) : \left(\overline{\dfrac{ds}{ds'}}\right) = \dfrac{\varphi(\bar{\tau}_2, \gamma_2)\varphi(\bar{\tau}_1, -\gamma_1)}{\varphi(\bar{\tau}_2, -\gamma_1)\varphi(\bar{\tau}_1, \gamma_1)}$$

or, according to (8.15),

$$\left(\overline{\dfrac{ds_1}{ds}}\right) = e^{2(\gamma_2 \bar{\tau}_2 - \gamma_1 \bar{\tau}_1)} > 0. \qquad (8.16)$$

Let us assume that the curves $s = s(s')$ and $s_1 = s_1(s')$ have several points of intersection. Then, since for small s', $s_1 > s$, for the first point of intersection (the one with smallest $\bar{s}'$) there is bound to take place the inequality

$$\left(\overline{\dfrac{ds_1}{ds'}}\right)_1 < \left(\overline{\dfrac{ds}{ds'}}\right)_1 \quad \text{or} \quad \left(\overline{\dfrac{ds_1}{ds}}\right)_1 < 1,$$

and for the second point of intersection, the following one

$$\left(\overline{\dfrac{ds_1}{ds'}}\right)_2 > \left(\overline{\dfrac{ds}{ds'}}\right)_2 \quad \text{or} \quad \left(\overline{\dfrac{ds_1}{ds}}\right)_2 > 1.$$

The latter is impossible, since to a larger $\bar{s}'$ there corresponds a larger $\bar{\tau}_1$ and a smaller $\bar{\tau}_2$ and, therefore, if there were a second point of intersection, it would, according to (8.16), satisfy the inequality $\overline{(ds_1/ds)}_2 < \overline{(ds_1/ds)}_1 < 1$. Thus there exists only one point of intersection of the curves $s = s(s')$ and $s_1 = s_1(s')$. At this point

$$0 < \left(\overline{\dfrac{ds_1}{ds}}\right) < 1.$$

Therefore, for $0 < h_2 < h_1 < 0$ there is a single stable limit cycle. All *phase paths* (Fig. 358) *approach* it asymptotically as $t \to +\infty$. The same result is obtained for the case $0 < h_2 < 1$, $h_1 > 1$ (Fig. 356). There is again a single stable limit cycle.

If, however, $0 < h_1 < h_2 < 1$, then the curves $s = s(s')$ and $s_1 = s_1(s')$ do not intersect each other (Fig. 357). In fact, if points of intersection did

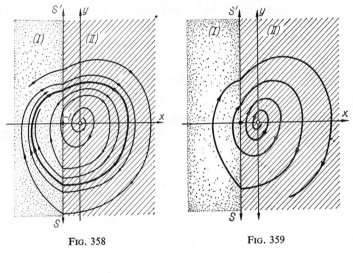

Fig. 358　　　　Fig. 359

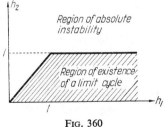

Fig. 360

exist in this case (there would be an even number of them), then, for the first of them (the one with smallest $\bar{s}'$) we would necessarily have

$$\left(\overline{\frac{ds_1}{ds}}\right)_1 = \left(\overline{\frac{ds_1}{ds'}}\right)_1 : \left(\overline{\frac{ds}{ds'}}\right)_1 < 1,$$

which is impossible owing to (8.16), since $\bar{\tau}_2 > \pi > \bar{\tau}_1$ and, for $h_2 > h_1$, $\gamma_2 > \gamma_1$. Thus, for $0 < h_1 < h_2 < 1$ the transformation Π has no fixed points, and there are no limit cycles. All phase paths go off to infinity (Fig. 359).

In this example the valve oscillator has two basic parameters h_1 and h_2 so we can construct a stability diagram on the (h_1, h_2) plane. Fig. 360 shows the first quadrant of this diagram divided into a region where unique and stable limit cycles exist, and a region of "absolute instability", for which all phase paths go to infinity. Obviously, the theory developed here does not reproduce correctly the properties of valve generators, since this theory neglects grid currents, etc. which play an important role for large amplitudes of oscillations.

4. Limit cycle

When the conditions $0 < h_2 < h_1$ and $h_2 < 1$ are satisfied there is a limit cycle, which is approached asymptotically by all other phase paths as $t \to +\infty$, and so the self-excitation is soft, occurring for any initial conditions. For $0 < h_2 < h_1 < 1$, the limit cycle is determined uniquely by the system of equations (8.15)

$$\left. \begin{aligned} \frac{e^{\gamma_1 \bar{\tau}_1} - \cos \bar{\tau}_1 - \gamma_1 \sin \bar{\tau}_1}{\sqrt{1+\gamma_1^2} \sin \bar{\tau}_1} &= -\frac{e^{\gamma_2 \bar{\tau}_2} - \cos \bar{\tau}_2 - \gamma_2 \sin \bar{\tau}_2}{\sqrt{1+\gamma_2^2} \sin \bar{\tau}_2}, \\ \frac{e^{\gamma_1 \bar{\tau}_1} - \cos \bar{\tau}_1 + \gamma_1 \sin \bar{\tau}_1}{\sqrt{1+\gamma_1^2} \sin \bar{\tau}_1} &= -\frac{e^{-\gamma_2 \bar{\tau}_2} - \cos \bar{\tau}_2 + \gamma_2 \sin \bar{\tau}_2}{\sqrt{1+\gamma_2^2} \sin \bar{\tau}_2}, \end{aligned} \right\} \quad (8.15a)$$

where
$$0 < \bar{\tau}_1 < \pi < \bar{\tau}_2 < \tau_2^0 < 2\pi.$$

If these two transcendental equations are solved then we can find the quantities that characterize the self-oscillatory mode of operation. For example, the period of the oscillations will be equal to

$$\tau = \frac{\bar{\tau}_1}{\omega_1} + \frac{\bar{\tau}_2}{\omega_2} \qquad (8.17)$$

in units of dimensionless time and

$$T = \sqrt{LC}\left(\frac{\bar{\tau}_1}{\omega_1} + \frac{\bar{\tau}_2}{\omega_2}\right) \qquad (8.17a)$$

in ordinary units.

The solution of (8.15a) can be carried out by numerical analysis, but here we will make an approximate evaluation of the period and amplitude of the self-oscillations for a practical case when h_1 and h_2 are small (i.e. for a generator with a high$-Q$ oscillating circuit and weak feedback coupling).

A VALVE GENERATOR

We shall denote by a and b the limiting values of the quantities $\bar{\tau}_1$ and τ_2 for $h_1, h_2 \to 0$. To calculate these limit values reduce the equations (8.5a) to the form

$$\left.\begin{array}{c}\dfrac{\cosh \gamma_1 \bar{\tau}_1 - \cos \bar{\tau}_1}{\sqrt{1+\gamma_1^2}\sin \bar{\tau}_1} = -\dfrac{\cosh \gamma_2 \bar{\tau}_2 - \cos \bar{\tau}_2}{\sqrt{1+\gamma_1^2}\sin \bar{\tau}_2}, \\ \dfrac{\sinh \gamma_1 \bar{\tau}_1 - \gamma_1 \sin \bar{\tau}_1}{\sqrt{1+\gamma_1^2}\sin \bar{\tau}_1} = -\dfrac{\sinh \gamma_2 \bar{\tau}_2 - \gamma_2 \sin \bar{\tau}_2}{\sqrt{1+\gamma_1^2}\sin \bar{\tau}_2}, \end{array}\right\} \quad (8.15b)$$

and substitute in them approximate relations, valid for $h_1, h_2 \ll 1$,

$$\gamma_1 = h_1, \quad \gamma_2 = h_2,$$
$$\cosh \gamma_1 \bar{\tau}_1 = 1, \quad \cosh \gamma_2 \bar{\tau}_2 = 1,$$
$$\sinh \gamma_1 \bar{\tau}_1 = h_1 a \quad \text{and} \quad \sinh \gamma_2 \bar{\tau}_2 = h_2 b^{2\dagger}.$$

Then the equations (8.15b) give

$$\frac{1-\cos a}{\sin a} = -\frac{1-\cos b}{\sin b},$$

$$h_1 \frac{a - \sin a}{\sin a} = -h_2 \frac{b - \sin b}{\sin b},$$

whence we obtain: $\tan a/2 = -\tan b/2$, or $a+b = 2\pi$, and

$$a - \sin a = 2\pi \frac{h_2}{h_1+h_2} = 2\pi \frac{MS-RC}{MS}. \quad (8.18)$$

The equation (8.18) has, for $h_2 < h_1$ the unique solution $0 < a < \pi$ (the graphical solution of this equation is shown in Fig. 361). Since for $h_1, h_2 \ll 1$,

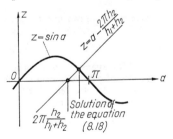

Fig. 361

† As is easily verified

$$\omega_1 = 1 + O(\gamma_1^2), \quad \omega_2 = 1 + O(\gamma_2^2), \quad h_1 = \gamma_1 + O(\gamma_1^3),$$
$$h_2 = \gamma_2 + O(\gamma_2^3), \quad \cosh \gamma_1 \bar{\tau}_1 = 1 + O(\gamma_1^2), \quad \cosh \gamma_2 \bar{\tau}_2 = 1 + O(\gamma_2^2),$$
$$\sinh \gamma_1 \bar{\tau}_1 = \gamma_1 a + O(\gamma_1^2) \quad \text{and} \quad \sinh \gamma_2 \bar{\tau}_2 = \gamma_2 b + O(\gamma_2^2).$$

$\omega_1 = 1$, $\omega_2 = 1$, the period of the self-oscillations is clearly equal to $\tau = a+b = 2\pi$ and is equal to the period of the undamped natural oscillations of the resonant circuit of the generator†. As to the amplitude of the self-oscillations, since for small values of h_1 and h_2 the limit cycle is nearly a circle, we have (Fig. 362):

$$A = \frac{1}{\cos\dfrac{a}{2}}. \qquad (8.19)$$

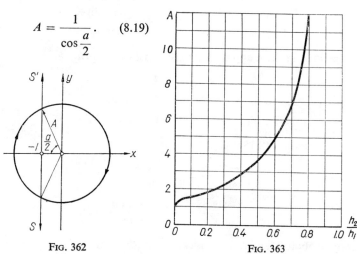

Fig. 362   Fig. 363

The dependence of the amplitude A upon the parameters of the generator is expressed in parametric form by the relations (8.18) and (8.19). Fig. 363 shows how the amplitude A depends upon the ratio h_2/h_1. For $h_2/h_1 \to 1$, $a \to \pi$ and, hence, $A \to \infty$.

† By putting $\gamma_2 = \beta\gamma_1$, where $\beta < 1$, we can find the solution of the system (8.15b) as power series in γ_1

$$\bar{\tau}_1 = a + a_1\gamma_1 + a_2\gamma_1^2 + \ldots, \quad \bar{\tau}_2 = b + b_1\gamma_1 + b_2\gamma_1^2 + \ldots$$

Substituting in the equations (8.15b) expansions in power series of all quantities depending on γ_1, we obtain a sequence of equations determining the coefficients a, b, a_1, b_1, a_2, b_2, ... The coefficients a and b are determined by the equations above, and $a_1 = 0$, $b_1 = 0$. Thus the period of the self-oscillations is

$$\tau = 2\pi + O(\gamma_1^2).$$

§ 3. VALVE GENERATOR (THE SYMMETRICAL CASE)

1. The equations of the oscillations and phase plane

Consider now a valve oscillator (Fig. 348), assuming that the valve characteristic saturates and is symmetrical about the static working point. We shall replace the real characteristic of the valve by a symmetrical piece-wise linear function

$$i_a = \begin{cases} 0 & \text{for } u < -u_0, \\ S(u+u_0) & \text{for } |u| < u_0, \\ 2Su_0 & \text{for } u > u_0, \end{cases} \quad (8.20)$$

shown in Fig. 364.

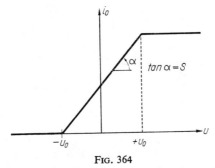

FIG. 364

As before we neglect the anode conductances, the grid currents and the parasitic capacitances (including the interelectrode ones). Introduce the new variables

$$x = \frac{u}{u_0} \quad \text{and} \quad t = \omega_0 t' \left(\omega_0 = \frac{1}{\sqrt{LC}} \right)$$

and noting that now

$$S(u) = \begin{cases} S & \text{for } |x| < 1, \\ 0 & \text{for } |x| > 1, \end{cases}$$

the equation of the valve generator (equation (8.3)) reduces to

$$\left. \begin{array}{l} \ddot{x} + 2h_1 \dot{x} + x = 0 \quad \text{for} \quad |x| > 1, \\ \ddot{x} - 2h_2 \dot{x} + x = 0 \quad \text{for} \quad |x| < 1, \end{array} \right\} \quad (8.21)$$

where, just as before

$$h_1 = \frac{\omega_0}{2} RC \quad \text{and} \quad h_2 = \frac{\omega_0}{2} [MS - RC].$$

Thus the phase plane (x, y) (where $y = \dot{x}$), is divided by the lines $x = -1$ and $x = +1$ into three "regions of linearity": (I) $x < -1$, (II) $|x| < 1$ and (III) $x > +1$, in each of which the appropriate linear equation (8.21) is valid (Fig. 365). Proceeding from physical premises (they have been re-

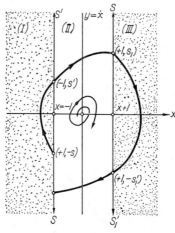

Fig. 365

peatedly mentioned before) the phase paths must be continuous on the phase plane and on the boundaries $x = -1$ and $x = +1$[†]. Also the equations (8.21) are invariant with respect to a change of the variables x, y into $-x, -y$. The same symmetry also holds for the paths in the upper and lower half of the region (II).

The dynamic system (8.21) has a single state of equilibrium at the origin (0, 0) which is a node or a focus, stable for $h_2 < 0$ (i.e. for $MS < RC$) and unstable for $h_2 > 0$ (for $MS > RC$). Below we shall mainly consider the self-excited generator in which $h_1 > 0$ and $h_2 > 0$[‡].

2. Point transformation

The x, y phase plane of the system considered is filled with sections of paths, corresponding to the linear equations (8.21); these sections of paths are joined together at their ends on the straight lines $x = -1$ and $x = +1$ thus forming entire phase paths.

[†] In Fig. 365 the phase paths are shown as spirals. This, of course, only takes place for $|h_1| < 1$, $|h_2| < 1$.

[‡] If $h_1 > 0$ while $h_2 < 0$, i.e. $MS < RC > 0$, then all phase paths approach asymptotically the stable state of equilibrium (0, 0); hence the system will not oscillate (whatever the initial conditions).

Now, for the case $h_1 > 0$, $h_2 > 0$, the point at infinity, as is easily seen, is unstable. The single state of equilibrium $(0, 0)$ is also unstable (it is an unstable focus for $0 < h_2 < 1$ and an unstable node for $h_2 > 1$). Therefore, there is at least one stable limit cycle (see Theorem V on p. 361). It is clear that the limit cycles must encircle the origin of the coordinates, the only state of equilibrium (see Section 8, Chapter V) and cannot lie entirely inside the region (II) ($|x| < 1$), since inside this region the equation (8.21) is linear. Moreover, since the system considered cannot have asymmetrical limit cycles†, the limit cycles will be symmetrical (with respect to the origin of the coordinates) and will traverse all three regions and intersect the half lines which are "lines without contact". To find all limit cycles we construct the point transformation of the half line into themselves and determine its fixed points.

Π is evidently the product of four transformations Π_1, Π_2, Π_3 and Π_4, between respectively, the half lines S and S', S' and S_1, S_1 and S'_1 and S'_1 and S (see Fig. 365). However,

$$\Pi_3 \equiv \Pi_1 \quad \text{and} \quad \Pi_4 \equiv \Pi_2$$

by virtue of the symmetry already mentioned. Therefore, the transformation Π is obtained by applying the transformation Π' twice, where

$$\Pi' = \Pi_1 \cdot \Pi_2, \quad \text{then} \quad \Pi = (\Pi')^2$$

relates S to S_1.

The transformation Π_1 is evidently identical to the transformation Π_1 of the preceding section (see (8.8) and (8.2)). Thus, for the case $0 < h_1 < 1$, the correspondence function for Π_1 has the form

$$\left. \begin{aligned} s &= \frac{e^{\gamma_1 \tau_1} - \cos \tau_1 - \gamma_1 \sin \tau_1}{\sqrt{1 + \gamma_1^2} \sin \tau_1}, \\ s' &= \frac{e^{-\gamma_1 \tau_1} - \cos \tau_1 + \gamma_1 \sin \tau_1}{\sqrt{1 + \gamma_1^2} \sin \tau_1}, \end{aligned} \right\} \quad (8.22)$$

where, just as before, τ_1 is the normalized time of transit of the representative point in the region (I) ($0 < \tau_1 < \pi$), and where

---

† Let us assume that the system (8.21) has an asymmetrical limit cycle Γ_1 (this must necessarily encircle the state of equilibrium). Then, owing to the symmetry of the phase paths (with respect to the origin of the coordinates), the system (8.21) will have another limit cycle Γ_2, symmetrical with Γ_1 and, hence, intersecting it. The latter is impossible. Thus, the system can only have symmetrical limit cycles.

$$\gamma_1 = \frac{h_1}{\omega_1} = \frac{h_1}{\sqrt{1-h_1^2}}.$$

The graph of (8.22) is shown in Fig. 353†.

For the phase path leaving the point s' of the half line S' ($x = -1$, $s' > 0$) at $t = 0$ and passing through the region (II), we have, according to (8.21) (see Section 4, Chapter I) for the case $0 < h_2 < 1$:

$$\left. \begin{array}{l} x = e^{h_2 t}\left[-\cos \omega_2 t + \dfrac{s'+h_2}{\omega_2} \sin \omega_2 t \right], \\[6pt] \dot{x} = y = e^{h_2 t}\left[s' \cos \omega_2 t + \dfrac{1+s'h_2}{\omega_2} \sin \omega_2 t \right], \end{array} \right\} \quad (8.23)$$

where

$$\omega_2 = \sqrt{1-h_2^2}.$$

The parametric expressions for the transformation Π_2 will be obtained by assuming that s_1 is reached at the point $x = +1$, $y = s_1 > 0$, at $t_2 = \tau_2/\omega_2 > 0$, and solving for s' and s_1

$$\left. \begin{array}{l} s_1 = \dfrac{e^{\gamma_2 \tau_2} + \cos \tau_2 + \gamma_2 \sin \tau_2}{\sqrt{1+\gamma_2^2}\,\sin \tau_2}, \\[10pt] s' = \dfrac{e^{-\gamma_2 \tau_2} + \cos \tau_2 - \gamma_2 \sin \tau_2}{\sqrt{1+\gamma_2^2}\,\sin \tau_2}, \end{array} \right\} \quad (8.24)$$

where

$$\gamma_2 = \frac{h_2}{\omega_2} = \frac{h_2}{\sqrt{1-h_2^2}}. \quad \ddagger$$

From (8.24) it is obvious that

(1) For $\tau_2 \to +0$, s' and $s_1 \to +\infty$.

(2) $s' = 0$ for a certain $\tau_2 = \tau_2'$ ($0 < \tau_2' < \pi$) determined by the equation $s'(\tau_2') = 0$ or $1 + e^{\gamma_2 \tau_2}(\cos \tau_2 - \sin \tau_2) = 0$, where $s_1(\tau_2') > 0$.

---

† If $h_1 > 1$, the expression for the correspondence function is obtained by the change in (8.22) of the trigonometric functions into corresponding hyperbolic functions. The graph of the correspondence function for this case is shown by a dotted line in Fig. 353.

‡ If $h_2 > 1$, the correspondence function for the transformation Π_2 is obtained from (8.24) by replacing the trigonometric functions by the corresponding hyperbolic functions, γ_2 by $\bar{\gamma}_2 = h_2(h_2^2-1)^{-\frac{1}{2}}$ and $(1+\gamma_2^2)^{\frac{1}{2}}$ by $(\bar{\gamma}_2^2-1)^{\frac{1}{2}}$.

(3) Differentiating (8.24) we have

$$\frac{ds_1}{d\tau_1} = -\frac{1+e^{\gamma_2\tau_2}(\cos\tau_2 - \gamma_2\sin\tau_2)}{\sqrt{1+\gamma_2^2\sin^2\tau_2}}$$

$$\frac{ds'}{d\tau_2} = -\frac{1+e^{-\gamma_2\tau_2}(\cos\tau_2 + \gamma_2\sin\tau_2)}{\sqrt{1+\gamma_2^2\sin^2\tau_2}}$$

and

$$\frac{ds_1}{ds'} = \frac{1+e^{\gamma_2\tau_2}(\cos\tau_2 - \gamma_2\sin\tau_2)}{1+e^{-\gamma_2\tau_2}(\cos\tau_2 + \gamma_2\sin\tau_2)}.$$

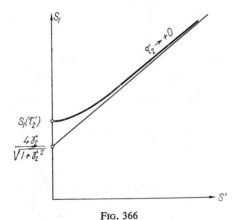

Fig. 366

Since $1+e^{\gamma_2\tau_2}(\cos\tau_2 - \sin\tau_2) > 0$ and

$$1+e^{-\gamma_2\tau_2}(\cos\tau_2 + \gamma_2\sin\tau_2) > 0$$

for $0 < \tau_2 < \tau_2'$, then for these values of τ_2, $ds'/d\tau_2$ and $ds_1/d\tau_2 > 0$. Also $ds_1/ds > 0$, so that as τ_2 varies from 0 to τ_2', *s' decreases monotonically* from $+\infty$ to 0, and s_1 decreases from $+\infty$ to $s_1(\tau_2') > 0$. Hence, the interval of smallest positive values of τ_2 needed to reach all points of the half line S', will be $0 < \tau_2 < \tau_2'$.

(4) Since

$$\frac{d^2s_1}{ds'^2} = \frac{2(1+\gamma_2^2)^{\frac{3}{2}}\sin^3\tau_2\,(\sinh\gamma_2\tau_2 - \gamma_2\sin\tau_2)}{[1+e^{-\gamma_2\tau_2}(\cos\tau_2+\gamma_2\sin\tau_2)]^3} > 0$$

for all values of τ_2 in the interval $0 < \tau_2 < \tau_2'$, then as s' increases from 0 to $+\infty$, ds_1/ds increases monotonically from 0 (at $s'=0$) to $+1$ (at $s' \to +\infty$). The curve (8.24) has the asymptote $s_1 = s' + 4\gamma_2/(1+\gamma_2^2)^{-\frac{1}{2}}$, and, owing to the fact that $d^2s_1/ds'^2 > 0$, this curve is situated above the asymptote. These properties are sufficient to construct a graph of the correspondence function (8.24); this is shown in Fig. 366†.

3. *Fixed point and limit cycle*

Construct the curves (8.22) and (8.24) on one plane, i.e. on Lamerey's diagram (Fig. 367). The fixed points are determined analytically by the following system of equations

$$\left.\begin{aligned}\frac{e^{\gamma_1\tau_1}-\cos\tau_1-\gamma_1\sin\tau_1}{\sqrt{1+\gamma_1^2}\sin\tau_1} &= \frac{e^{\gamma_2\tau_2}+\cos\tau_2+\gamma_2\sin\tau_2}{\sqrt{1+\gamma_2^2}\sin\tau_2},\\ \frac{e^{-\gamma_1\tau_1}-\cos\tau_1+\gamma_1\sin\tau_1}{\sqrt{1+\gamma_1^2}\sin\tau_1} &= \frac{e^{-\gamma_2\tau_2}+\cos\tau_2-\gamma_2\sin\tau_2}{\sqrt{1+\gamma_2^2}\sin\tau_2},\end{aligned}\right\} \quad (8.25)$$

which is obtained from the equations (8.22) and (8.24) by eliminating s' and putting $s_1 = s$.

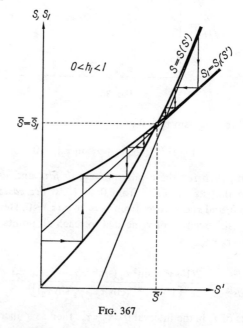

Fig. 367

† For $h_2 > 1$ the graph of the correspondence function of Π_2 has qualitatively the same form: as s' varies from 0 to $+\infty$, s_1, increases monotonically from a certain positive value to $+\infty$, and the derivative ds_1/ds' increases from 0 to $+1$.

It is easy to show that there exists only one point of intersection of the curves (8.22) and (8.24) (Fig. 367). In fact, the existence of at least one point of intersection follows from the continuity of these curves and from the inequalities

$s_1 - s > 0$ for $s' = 0$
$s_1 - s < 0$ for sufficiently large s'[†].

Further, if several points of intersection did exist, then we should have for the first of them (the one with smallest s') $ds_1/ds' < ds/ds'$, and for the following one $ds_1/ds' > ds/ds'$. The latter is impossible since $0 < ds_1/ds' < 1$ and $ds/ds' > 1$ (for any values of s'). Thus, there is only one point of intersection and therefore only one fixed point if $0 < h_1 < 1$ and $0 < h_2 < 1$. The fixed point is stable since at it $0 < ds_1/ds < 1$.

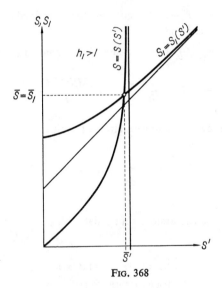

Fig. 368

The same situation occurs when $0 < h_1 < 1$, but $h_2 > 1$. Lamerey's diagram for the case $h_1 > 1$ and arbitrary $h_2 > 0$ is shown in Fig. 368. It appears that for these values also there is a unique and stable fixed point of the point transformation Π'.

† The slopes of the asymptotes of the curves (8.22) and (8.24) are equal respectively to $e^{\gamma_1 \pi}$ and to 1, i.e. the asymptote of the curve (8.22) is steeper than the asymptote of the curve (8.24).

Therefore, for arbitrary positive values of the parameters h_1 and h_2 there is a unique stable limit cycle, to which all phase paths tend (for $t \to +\infty$). Thus the generator has a *soft* mode of self-excitation.

The period of the self-oscillations is clearly equal to

$$\tau = 2\left[\frac{\bar{\tau}_1}{\omega_1} + \frac{\bar{\tau}_2}{\omega_2}\right]$$

(in units of the dimensionless time), where $\bar{\tau}_1$ and $\bar{\tau}_2$ are values of τ_1 and τ_2 in a limit cycle.

We now consider three limiting cases:

(1) $h_1 \to 0$, then $\bar{\tau}_1 \to \pi$, $\bar{\tau}_2 \to 0$ (the fixed point, and with it the limit cycle, go to infinity).

(2) $h_2 \to 0$, then $\bar{\tau}_1 \to 0$, $\bar{\tau}_2 \to \pi$; the coordinate of the fixed point $\bar{s} \to 0$ and the limit cycle is a circle $x^2 + y^2 = 1$.

(3) $h_1, h_2 \ll 1$ ($h_1, h_2 \to 0$). In this case $\bar{\tau}_1$ is determined by the equation

$$\bar{\tau}_1 - \sin \bar{\tau}_1 = \frac{\pi h_2}{h_1 + h_2} = \pi \frac{MS - RC}{MS},$$

$\bar{\tau}_2 = \pi - \bar{\tau}_1$, the limit cycle is almost a circle of radius

$$A = \frac{1}{\cos \dfrac{\bar{\tau}_1}{2}},$$

and the self-oscillations are almost sinusoidal with period 2π.

§4. VALVE GENERATOR WITH A BIASSED ∫ CHARACTERISTIC

In the preceding two sections we have considered examples of valve generators with a *soft* mode of excitation. We shall consider now a valve generator with a so-called biassed ∫ characteristic and a *hard* mode of excitation. The valve characteristic (as in Section 4, Chapter III) is supposed to be discontinuous:

$$i_a = \begin{cases} I_s & \text{for} \quad u_g > 0, \\ 0 & \text{for} \quad u_g < 0, \end{cases}$$

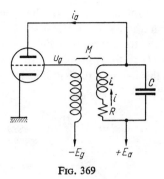

Fig. 369

and under static working conditions the valve is cut off by a negative bias—E_g (Fig. 369).

1. The equation of the oscillations

The phase plane. The equation of the circuit is

$$LC\frac{d^2i}{dt^2} + RC\frac{di}{dt} + i = \begin{cases} I_s & \text{for} \quad u_g > 0, \\ 0 & \text{for} \quad u_g < 0, \end{cases}$$

where

$$u_g = -E_g - M\frac{di}{dt}.$$

Below we shall assume that $M < 0$, for only in this case can the generator sustain self-oscillations.

By the change of variables

$$x = \frac{i}{I_s}, \quad t_{\text{new}} = \omega_0 t \,\dagger,$$

where $\omega_0 = (LC)^{-\frac{1}{2}}$, we reduce the equation of the generator to

$$\ddot{x} + 2h\dot{x} + x = \begin{cases} 1 & \text{for} \quad \dot{x} > b, \\ 0 & \text{for} \quad \dot{x} < b, \end{cases} \quad (8.26)$$

where $2h = \omega_0 RC$ is the damping constant of the resonant circuit and

$$b = \frac{E_g}{\omega_0 |M| I_s}. \quad (8.26\text{a})$$

† Differentiation with respect to the new dimensionless time will be denoted below by a dot, and the dimensionless time itself simply by t.

The x, y phase ($y = \dot{x}$) is divided by the line $y = b$ into two regions of linearity: (*I*) the region where $y > b$ and (*II*) the region where $y < b$ (Fig. 370). In each of these regions the appropriate linear equation holds. The phase paths of the regions (*I*) and (*II*) are joined (because of continuity) along the line $y = b$. Divide this line into the half lines S ($x = -s$, where $s > 2hb - 1$)

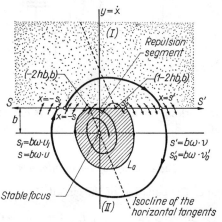

Fig. 370

and S' ($x = s' > -2hb$). The phase paths cross S into region (*I*) and cross S' into region (*II*), but on the segment $y = b$, $-2hb > x > 1 - 2hb$, which is common to S and S', the phase paths (as t increases) can enter into the region (*I*) (for $y = b + 0$) and into the region (*II*) for ($y = b - 0$). We shall refer to this segment as the "repulsion segment"†. The dynamic system (8.26), has a unique stable state of equilibrium at the origin (0, 0), which is a focus for $h < 1$ and a node for $h > 1$. For a node the system cannot have limit cycles‡, and the generator does not self-oscillate. Therefore, we shall restrict ourselves to considering only the case $0 < h < 1$.

† The isocline of the horizontal tangents ($dy/dx = 0$) is the straight line $y = -2hx$ in the region (*II*) and the straight line $y = 1 - 2hx$ in the region (*I*). On the left of the isocline $\dot{y} > 0$ and on the right $\dot{y} < 0$.

Also note that at the points of the "repulsion segment", as over the whole straight line $y = b$, the motion of the representative point is not determined by the equation (8.26) and must be defined in a suitable manner. At points of $y = b$ outside this segment the completion of the definition is trivial: the representative point leaves the straight line $y = b$ along a path going into the region (*I*) for $x < -2hb$ or in the region (*II*) for $x > 1 - 2hb$. The completion of the definition of the motion of the representative point at the points of the "repulsion segment" is less obvious and will be carried out later.

‡ For $h > 1$ there are two integral straight lines, in the the region (*II*), coming from infinity and passing through the node. The limit cycle, if it did exist, would intersect these straight lines, which is impossible.

2. Point transformation

The limit cycles, if they exist, must encircle the origin (the only state of equilibrium) and, on the other hand, cannot lie entirely in the region (I) (or in the region (II)). Hence, they will necessarily intersect the straight line $y=b$ and in particular the half line S. Therefore, to find the limit cycles of the equation (8.26) it will suffice to consider the point transformation Π of the half line S into itself (with the sequence function $s_1 = f(s)$). The transformation Π_1 refers to a path in (I) from a point $(-s, b)$ of S to a point (s', b) of S': the transformation Π_2 refers to a path in (II) from a point (s', b) on S' back on to the line S at a point $(-s_1, b)$ (Fig. 370). Then the overall transformation is

$$\Pi = \Pi_1 \cdot \Pi_2.$$

Parametric expressions for the correspondence functions of Π_1 and Π_2 are easily obtained. A path leaves a point $(-s, b)$ on S at $t=0$, moving into the region (I). According to (8.26) its equation is

$$x = 1 + e^{-ht}\left[-(1+s)\cos\omega t + \frac{b-h(1+s)}{\omega}\sin\omega t\right],$$

$$y = e^{-ht}\left[b\cos\omega t + \frac{1+s-hb}{\omega}\sin\omega t\right],$$

where

$$\omega = \sqrt{1-h^2}.$$

After a certain time t_1 the representative point reaches S' at a point (s', b)[†] where

$$s' = 1 + e^{-ht_1}\left[-(1+s)\cos\omega t_1 + \frac{b-h(1+s)}{\omega}\sin\omega t_1\right],$$

$$b = e^{-ht_1}\left[b\cos\omega t_1 + \frac{1+s-hb}{\omega}\sin\omega t_1\right].$$

Solving these relations for s and s', we obtain the correspondence function Π_1

$$s = b\omega\frac{e^{\gamma\tau_1} - \cos\tau_1 + \gamma\sin\tau_1}{\sin\tau_1} - 1,$$

$$s' = b\omega\frac{e^{-\gamma\tau_1} - \cos\tau_1 - \gamma\sin\tau_1}{\sin\tau_1} + 1.$$

[†] It is evident that $s' > 1 - 2hb$, since the phase paths in (I) leave the straight line $y=b$ where $x < 1 - 2hb$.

where

$$\gamma = \frac{h}{\omega} = \frac{h}{\sqrt{1-h^2}}$$

and $\tau_1 = \omega t_1$ is the normalized transit time of the representative point in the region (*I*). If we introduce

$$u = \frac{s}{b\omega} \quad \text{and} \quad v = \frac{s'}{b\omega},$$

then the correspondence function for Π_1 will be written more simply as

$$\left.\begin{aligned} u &= \frac{e^{\gamma\tau_1} - \cos\tau_1 + \gamma\sin\tau_1}{\sin\tau_1} - a, \\ v &= \frac{e^{-\gamma\tau_1} - \cos\tau_1 - \gamma\sin\tau_1}{\sin\tau_1} + a, \end{aligned}\right\} \quad (8.27)$$

where

$$a = \frac{1}{b\omega} = \frac{\omega_0 |M| I_s}{E_g \sqrt{1-h^2}}.$$

In a similar way the correspondence function for the transformation Π_2 of the points $(b\omega v, b)$ of S' into the points $(-b\omega u_1, b)$ of S is found to be

$$\left.\begin{aligned} u_1 &= -\frac{e^{-\gamma\tau_2} - \cos\tau_2 - \gamma\sin\tau_2}{\sin\tau_2}, \\ v &= -\frac{e^{\gamma\tau_2} - \cos\tau_2 + \gamma\sin\tau_2}{\sin\tau_2}, \end{aligned}\right\} \quad (8.28)$$

where $\tau_2 = \omega t_2$ is the normalized transit time of the representative point along a path in the region (*II*).

The investigation of the correspondence function (8.27) is perfectly analogous to the investigation of the function (8.8) (see §2). The parameter τ_1 need vary only in the interval $0 < \tau_1 < \pi$, during which variation u increases monotonically from $u_0 = 2\gamma - a$ to $+\infty$ and v from $v_0 = a - 2\gamma$ to $+\infty$. Also the initial points of the curves (8.27) lie, clearly, on the straight line $u + v = 0$. Further,

$$\frac{du}{dv} = \frac{1 - e^{\gamma\tau_1}(\cos\tau_1 - \gamma\sin\tau_1)}{1 - e^{-\gamma\tau_1}(\cos\tau_1 + \gamma\sin\tau_1)} > 0$$

$$\frac{d^2u}{dv^2} = \frac{2(1+\gamma^2)\sin^3\tau_1(\sinh\gamma\tau_1 - \gamma\sin\tau_1)}{[1 - e^{-\gamma\tau_1}(\cos\tau_1 + \gamma\sin\tau_1)]^3} > 0$$

for $0 < \tau_1 < \pi$. Finally, the curve (8.27) has the asymptote $u = e^{\gamma\pi}v - a(1 + e^{\gamma\pi})$ as $\tau_1 \to \pi$. The family of curves (8.27) is shown in Fig. 371 for a certain fixed value of γ and for various values if the parameter $a \geqslant 0$. For the correspondence function (8.28) we note at once that the representative point moving along the spiral in the region (II), to which the transformation Π_2 refers, sweeps out by an angle larger than π but smaller than 2π, this angle being the smaller the larger s' and s_1. Therefore

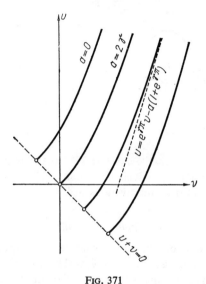

Fig. 371

(see Section 4, Chapter I) the parameter τ_2 must lie in the interval $\pi < \tau_2 < 2\pi$. The smaller τ_2 the larger v and u_1 (as $\tau_2 \to +0$, v and $u_1 \to +\infty$). Not all points, however, of S' are transformed by the paths in (II) into points on S. Draw in (II) the phase path L_0 that passes through the point $(-2hb, b)$; this path will isolate a region (shaded in Fig. 370) in which the paths arriving do not reach the straight line $y = b$, but spiral towards the stable focus. We shall denote by s_0' the abscissa of the point of intersection of the path L_0 with S' (Fig. 370); then, clearly, the points of the half line S' for which $-2hb < x < s_0'$ will no longer be transformed by paths of (II) into points of the half straight line S. The normalized time of transit τ_2^0 for the point $s' = s_0'$ (or $v = v^0 = s_0'/b\omega$) is given by

$$-s_1(\tau_2^0) = -2hb \quad \text{or} \quad u_1(\tau_2^0) = 2\gamma,$$

or,
$$1 - e^{\gamma \tau_2^0}(\cos \tau_2^0 - \gamma \sin \tau_2^0) = 0.$$

It is evident that $\pi < \tau_2^0 < 2\pi$ (a graphical solution is shown in Fig. 372).

Thus, by varying the parameter τ_2 from τ_2^0 to π, we determine the set of points on S' connected by the transformation Π_2 with points of the half

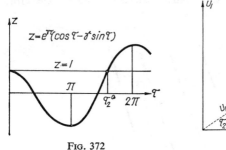

Fig. 372

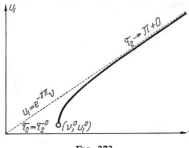

Fig. 373

straight line S. For $\tau_2 \to \pi$, the curve (8.28) has the asymptote $u_1 = e^{-\gamma \pi} v$. The graph of the correspondence function (8.28) of the transformation Π_2 can be determined qualitatively as before and is shown in Fig. 373.

3. Fixed points and limit cycles

As before we plot on the same plane the correspondence functions (8.27) and (8.28) (one axis will serve as the v axis and the other as the u and u_1 axis) and then examine Lamerey's diagrams (Fig. 372) for a certain fixed γ and various values of the parameter $a \geqslant 0$.

For $a \leqslant 2\gamma$ the curve (8.27) will not intersect the curve (8.28)[†]. Moreover, only the curve (8.27) depends on a, so as the parameter a increases, the cases (a), (b), (c), (d) and (e) shown in Fig. 374 will appear in that order.

The points of intersection $\bar{u}$, $\bar{v}$ of the curves (8.27) and (8.28) (at $\bar{\tau}_1$ and $\bar{\tau}_2$) determine the fixed points of Π and, thus, the limit cycles[‡]. The fixed

---

[†] For $a \leqslant 2\gamma$ the curve (8.27) is situated above the bisector $u = v$ whereas the curve (8.28) always lies below its asymptote $u_1 = e^{-\gamma \pi} v$ and, hence, below the bisector $u_1 = v$.

[‡] It is easily seen that the curves (8.27) and (8.28) cannot have more than two points of intersection. In fact, if the number of the points of intersection of these curves were more than two, then, for the second and third point of intersection (numbering them in the direction of increasing $\bar{v}$) the following inequality would take place:

$$\left(\frac{du_1}{dv}\right)_{v=\bar{v}_2} < \left(\frac{du}{dv}\right)_{v=\bar{v}_2} \quad \text{and} \quad \left(\frac{du_1}{dv}\right)_{v=\bar{v}_3} > \left(\frac{du}{dv}\right)_{v=\bar{v}_3},$$

which is impossible since, as v increases, du_1/dv decreases and du/dv increases.

points are determined analytically by

$$\left. \begin{array}{l} \dfrac{e^{\gamma \tau_1} - \cos \tau_1}{\sin \tau_1} - a = \dfrac{\cos \tau_2 - e^{-\gamma \tau_2}}{\sin \tau_2}, \\[2mm] \dfrac{e^{-\gamma \tau_1} - \cos \tau_1}{\sin \tau_1} + a = \dfrac{\cos \tau_2 - e^{\gamma \tau_2}}{\sin \tau_2} \end{array} \right\} \quad (8.29)$$

$$(0 < \bar{\tau}_1, < \pi, \pi < \bar{\tau}_2 < \tau_2^0 < 2\pi),$$

which is obtained from (8.27) and (8.28) with $u_1 = u$.

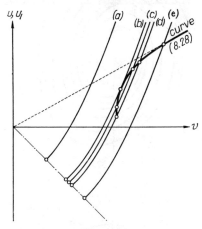

Fig. 374

If there are two fixed points (case (c) in Fig. 374), then for the one nearest the origin $\bar{u} = \bar{u}^{(1)}$, $\bar{v} = \bar{v}_1$ (the largest value of τ_2 and smallest value of τ_1)

$$\left(\frac{du_1}{du} \right)_{v=\bar{v}_1} = \left(\frac{du_1}{dv} \right)_{v=\bar{v}_1} : \left(\frac{du}{dv} \right)_{v=\bar{v}_1} > 1$$

and for the second one ($\bar{u} = \bar{u}^{(2)} > \bar{u}^{(1)}$, $\bar{v} = \bar{v}_2 > \bar{v}_1$)

$$0 < \left(\frac{du_1}{dv} \right)_{v=\bar{v}_2} = \left(\frac{du_1}{dv} \right)_{v=\bar{v}_2} : \left(\frac{du}{dv} \right)_{v=\bar{v}_2} < 1,$$

so the first is unstable and the second is stable. If, on the other hand, there is only one fixed point of the transformation Π (the case (e) in Figs. 374), then this is always stable, since the condition for stability is satisfied:

$$0 < \left(\frac{du_1}{du} \right)_{v=\bar{v}} < 1.$$

The various possible types of phase portrait corresponding to the cases (a), (b), (c), (d) and (e) of Lamerey's diagram (Fig. 374) are shown in Figs. 375-9. Fig. 380 is a stability diagram for the parameters γ and a,

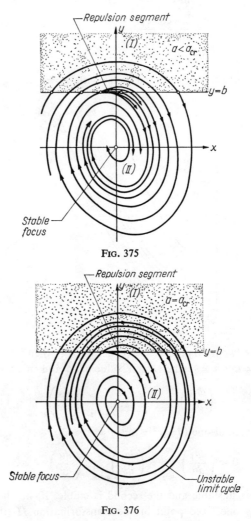

Fig. 375

Fig. 376

divided into regions of existence of the modes of operation of the generator. If large feedback coupling occurs such that the point (γ, a) lies in the non-shaded region (Fig. 380), corresponding to the case (a) of Lamerey's diagram, then the phase paths go towards the stable state of equilibrium (0, 0) for any initial conditions (Fig. 375). For a certain critical coupling (for $a = a_{\text{crit}} = f(\gamma)$) there appears on the phase plane a semi-stable limit

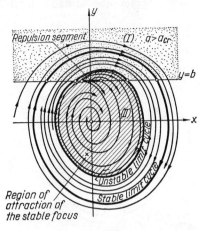

Fig. 377

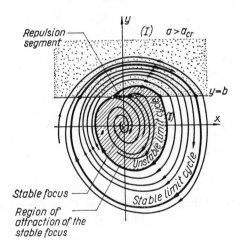

Fig. 378

cycle (Fig. 376), which corresponds to the tangential point of the curves on Lamerey's diagram in the case (b) of Fig. 374. For an arbitrarily small increase of the feedback coupling parameter a this limit cycle is split into two limit cycles, one of which is stable and the other is unstable (Fig. 377).

For a further increase of the parameter a, the dimensions of the unstable limit cycle decrease, and for a second branch value of this parameter $\left(a=a_{\text{crit}}^{(2)}=f_1(\gamma)\right)$, corresponding to the curve (d) in Fig. 380 and to the case (d) of Lamerey's diagram, the unstable limit cycle touches the

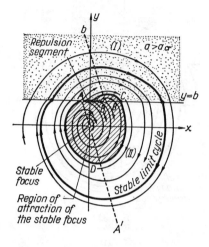

Fig. 379

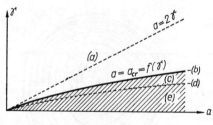

Fig. 380

repulsion segment (Fig. 378). Finally, Fig. 379 shows the phase portrait for $a > f_1(\gamma)$, when the point (γ, a) lies in the region (e) of Fig. 380 and the point transformation has a unique fixed point (the case (e) of Lamerey's diagram).

Thus, for $a > a_{\text{crit}} = f(\gamma)$ the generator cannot self-oscillate, and for $a > a_{\text{crit}} = f(\gamma)$, there is a *hard mode* of excitation. A periodic (self-oscillatory) mode of operation is possible only for initial conditions such that the representative point is outside the unstable limit cycle (Fig. 377) or outside the shaded region in Fig. 379.

The period of the stable self-oscillations is clearly equal to

$$\tau = \frac{1}{\omega}(\bar{\tau}_1 + \bar{\tau}_2),$$

where $\bar{\tau}_1$ and $\bar{\tau}_2$ are determined by the solution of the system (8.29)†.

The boundary curve $a_{\text{crit}} = f(\gamma)$ on the plane of the parameters of the generator separates the region of a non-excited generator (the region (a)) from the region of a hard mode of excitation (the rest of the first quadrant); it is clearly determined by the equations (8.29) and by the condition that for $a = a_{\text{crit}}$ the curves (8.27) and (8.28) are tangential to each other, i.e.

$$\left(\frac{du_1}{dv}\right)_{v=\bar{v}} = \left(\frac{du}{dv}\right)_{v=\bar{v}}$$

or

$$\frac{1 - e^{\gamma \bar{\tau}_1}(\cos \bar{\tau}_1 - \gamma \sin \bar{\tau}_1)}{1 - e^{-\gamma \bar{\tau}_1}(\cos \bar{\tau}_1 + \gamma \sin \bar{\tau}_1)} = \frac{1 - e^{-\gamma \bar{\tau}_2}(\cos \bar{\tau}_2 + \gamma \sin \bar{\tau}_2)}{1 - e^{\gamma \bar{\tau}_2}(\cos \bar{\tau}_2 - \gamma \sin \bar{\tau}_2)}.$$

It can be shown that this boundary curve (the curve (b) in Fig. 380) passes through the origin of the plane a, γ and that a_{crit} increases monotonically as γ increases.

4. *The case of small values of a and* γ

Let us find approximate expressions for the period and amplitude of the self-oscillations in the case of sufficiently small values of a and γ ($a, \gamma \ll 1$). We shall write the equations (8.29) in the form

$$\frac{\cosh \gamma \tau_1 - \cos \tau_1}{\sin \tau_1} = -\frac{\cosh \gamma \tau_2 - \cos \tau_2}{\sin \tau_2}, \quad \frac{\sinh \gamma \tau_1}{\sin \tau_1} - a = -\frac{\sinh \gamma \tau_2}{\sin \tau_2},$$

Then, for $a, \gamma \ll 1$, we have

$$\frac{1 - \cos \tau_1}{\sin \tau_1} = -\frac{1 - \cos \tau_2}{\sin \tau_2}, \quad \frac{\gamma \tau_1}{\sin \tau_1} - a = \frac{\gamma \tau_2}{\sin \tau_2},$$

whence $\bar{\tau}_1 + \bar{\tau}^2 = 2\pi$ and $\bar{\tau}_1$ is determined by the equation

$$\sin \bar{\tau}_1 = \frac{2\pi \gamma}{a},$$

---

† If the system has two solutions for $\bar{\tau}_1$ and $\bar{\tau}_2$ that satisfy the inequality $0 < \bar{\tau}_1 < \pi < \bar{\tau}_2 < \tau_2^0 < 2\pi$ (the case (c) in Fig. 374), then, evidently, to calculate the period of the oscillations we must take the larger of the two values of $\bar{\tau}_1$ (and correspondingly the smaller for $\bar{\tau}_2$).

which has two real roots $0<(\bar{\tau}_1)_1<\pi/2$ and $\pi/2<(\bar{\tau}_1)_2<\pi$, only when $2\pi\gamma/a<1$; hence

$$a_{\text{crit}} = 2\pi\gamma.$$

The radii of the limit cycles (very nearly circles) are equal to

$$R = \frac{b}{\cos\dfrac{\bar{\tau}_1}{2}} = \frac{\sqrt{2}b}{\sqrt{1+\cos\bar{\tau}_1}} = \frac{\sqrt{2}b}{\sqrt{1\mp\sqrt{1-\left(\dfrac{2\pi\gamma}{a}\right)^2}}},$$

a stable limit cycle corresponding to $\pi/2<\tau_1<\pi$, and to the larger value for the radius of the limit cycle.

§ 5. Valve generator with a two-mesh RC circuit

Two circuit diagrams of a generator with a two-mesh RC circuit (a cathode-coupled twin triode or a pentode under transitron conditions of operation) are shown in Fig. 381†. An investigation of the self-oscillations

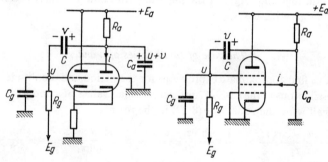

Fig. 381

in these circuits was carried out in Section 12, Chapter V, using the isocline method. These same circuits, if the capacitances C_a and C_g are assumed to be small parasitic capacitances, represent a multivibrator with one RC circuit (see Section 7, Chapter IV).

To oscillate, both circuits require a characteristic relating current i to the control voltage u which has a section with negative slope. We assume that i depends on u only $(i=i(u))$, but in contrast to Section 12, Chapter V, we shall represent this function approximately by a piece-wise linear

---

† One of the capacitances C_a and C_g can be absent.

function, shown in Fig. 382. We shall assume also that the grid bias E_g has been chosen so that the working point of the oscillator lies at the centre of the section with negative slope, [59].

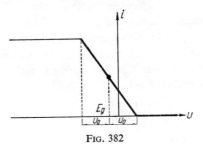

Fig. 382

The equations of the generator (Section 12, Chapter V) are

$$\left.\begin{aligned} \mu\dot{x} &= -x-y-K\varphi(x), \\ \dot{y} &= x, \end{aligned}\right\} \quad (8.30)$$

where x and y are related to u and v by

$$u = E_g + u_0 x,$$
$$v = E_a - E_g - R_a i(E_g) + u_0 \frac{C_g}{C} x + u_0 \left[1 + \frac{C_g}{C} + \frac{R_a}{R_g}\left(1 + \frac{C_a}{C}\right)\right] y$$

(u_0 has been taken equal to half the "length" of the sloping section of the characteristic; see Fig. 382); and

$$\varphi(x) = \frac{1}{u_0 S}\{i(E_g + u_0 x) - i(E_g)\} = \begin{cases} +1 & \text{for } x < -1, \\ -x & \text{for } |x| \leq 1, \\ -1 & \text{for } x > +1 \end{cases}$$

is the normalized (dimensionless) characteristic of the valve (or of the group of valves)[†].

$$t = \frac{t_{\text{ord}}}{R_a(C+C_a) + R_g(C+C_g)}$$

and

$$K = \frac{SR_a}{1 + \frac{C_g}{C} + \frac{R_a}{R_g}\left(1 + \frac{C_a}{C}\right)}, \quad \mu = \frac{R_a}{R_g} \frac{\frac{C_a}{C} + \frac{C_g}{C} + \frac{C_a}{C}\frac{C_g}{C}}{\left[1 + \frac{C_g}{C} + \frac{R_a}{R_g}\left(1 + \frac{C_a}{C}\right)\right]^2}$$

are dimensionless parameters ($K > 0$ and $0 < \mu < 1/4$)[‡].

[†] We denote by S the absolute value of the negative slope $i = i(u)$.
[‡] The latter inequality follows from the fact that in two-mesh RC circuits with the valve disconnected (so that) $K = 0$ all processes are aperiodically damped.

We note that the system of equations (8.30) is equivalent to the equation

$$\mu\ddot{x}+[1+K\varphi'(x)]\dot{x}+x = 0,$$

which we have already considered in Section 3 of this chapter. However, with a view to obtaining a detailed analysis of the oscillations of the generator that are close to discontinuous oscillations (they occur for C_a, $C_g \ll C$, i.e. for $0 < \mu \ll 1$), we shall carry out once more a brief investigation of the equations (8.30) restricting ourselves to the case of a self-excited generator when $K > 1$.

1. The phase plane

The point transformation. Just as in Section 3 of this chapter, the x, y phase plane of (8.30) is divided by the straight lines $x = +1$ and $x = -1$ into three regions: (*I*), (*II*) and (*III*) in each of which the equations (8.30) are linear (Fig. 383); and the paths are continuous curves over these boundaries as well as over the whole phase plane. Note also that the phase portrait is symmetrical with respect to the state of equilibrium (0, 0).

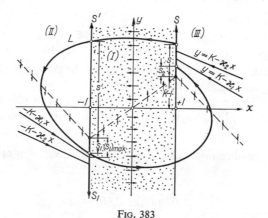

Fig. 383

The qualitative investigation of this system (8.30) is perfectly analogous to the investigation of the system (5.89) carried out in Section 12 of Chapter V. Thus the y axis ($x=0$) is the isocline of horizontal tangents (where $\dot{y} = 0$), and the broken line

$$y = -x - K\varphi(x)$$

is the isocline of vertical tangents (on it $\dot{x}=0$). In each of the regions (*II*) and (*III*) there are two rectilinear paths $y = \mp K - \varkappa_{1,2} x$, where $\varkappa_{1,2}$ are quantities reciprocal (and of opposite sign) to the roots of the equation

$$\mu\lambda^2+\lambda+1 = 0, \tag{8.31}$$

the characteristic equation of (8.30) in the regions (II) and (III); $\varkappa_1$ and $\varkappa_2$ are therefore positive.

The only state of equilibrium lies in (I), at the origin $(0, 0)$. As in the region (I) the characteristic equation is

$$\mu\lambda^2-(K-1)\lambda+1 = 0, \tag{8.32}$$

this state of equilibrium is unstable for $K>1$, being a focus for $4\mu>(K-1)^2$ and a node for $4\mu<(K-1)^2$ [†].

In addition, since the point at infinity is always unstable, we have for $K>1$, at least one stable limit cycle, symmetrical with respect to the origin of the coordinates. The results of Section 3 of this chapter enable us to affirm that this limit cycle is unique.

Since the limit cycle is symmetrical[‡], must encircle the state of equilibrium $(0, 0)$ and at the same time cannot lie entirely in the region (I), then it must pass through all three regions of linearity, intersecting, in particular, the straight lines $x=+1$ and $x=-1$. Proceeding from this, we shall take as a "segment without contact" the half line $S: x=+1$, $y=K-1+s$ (where $s>0$) where paths go from (III) to (I). Just as in Section 3, the transformation

$$\Pi = (\Pi')^2,$$

where Π' is the point transformation of the half line S into the half line S_1 ($x=-1, y=-(K-1)-s_1; s_1>0$), generated by the paths leaving the line S. In its turn the transformation Π' can be represented as the product of two transformations Π_1 and Π_2, i.e. the transformation of S into S': $x=-1, y=-(K-1)+s'$ ($s'>0$), and the transformation of S' into S_1, as generated by paths in the regions (I) and (II), respectively:

$$\Pi' = \Pi_1 \cdot \Pi_2.$$

To evaluate the correspondence function of the first transformation we shall return to the differential equations (8.30) in (I), which are conveniently rewritten as

$$\left.\begin{array}{r}\mu\ddot{y}-(K-1)\dot{y}+y = 0, \\ x = \dot{y}.\end{array}\right\} \tag{8.30a}$$

[†] For $K<1$ the state of equilibrium $(0,0)$ is stable and all paths of the system approach it asymptotically (as $t \to +\infty$).
[‡] The proof of the symmetry of the limit cycle exactly coincides with the proof carried out in Section 3 of this chapter.

The characteristic equation of this system (the equation (8.32)) has, for $4\mu > (K-1)^2$, complex roots $\lambda = h_1 \pm ja$ and, for $4\mu < (K-1)^2$, real (positive) roots $\lambda = h_1 \pm \omega_1$ $(h_1 > \omega_1)$, where

$$h_1 = \frac{K-1}{2\mu} \quad \text{and} \quad \omega_1 = +\frac{1}{2\mu}\sqrt{|(K-1)^2 - 4\mu|} = +\sqrt{\left|h_1^2 - \frac{1}{\mu}\right|}.$$

Let $4\mu > (K-1)^2$. Then the general solution of the equations (8.30a) is

$$\left.\begin{aligned} y &= e^{h_1 t}\left[\frac{x_0}{\omega_1}\sin \omega_1 t + y_0\left(\cos \omega_1 t - \frac{h_1}{\omega_1}\sin \omega_1 t\right)\right], \\ x &= e^{h_1 t}\left[x_0\left(\cos \omega_1 t + \frac{h_1}{\omega_1}\sin \omega_1 t\right) - \frac{y_0}{\mu\omega_1}\sin \omega_1 t\right] \end{aligned}\right\} \quad (8.33)$$

(x_0, y_0 being the initial values at $t=0$; see Section 4 of Chapter I). For the path L that leaves the point s of line S at $t=0$, we put in (8.33): $x_0 = +1$, $y_0 = K-1+s$. Let t_1 be the time of transit of the representative point along the path L in (I)
where

$$\tau_1 = \omega_1 t_1 \quad \text{and} \quad \gamma_1 = \frac{h_1}{\omega_1} = \frac{K-1}{+\sqrt{|K-1|^2 - 4\mu|}}.$$

Then for $t = t_1 > 0$, $x = -1$ and $y = -(K-1) + s'$ so that

$$-(K-1) + s' = e^{\gamma_1 \tau_1}\left[\frac{1}{\omega_1}\sin \tau_1 + (K-1+s)(\cos \tau_1 - \gamma_1 \sin \tau_1)\right],$$

$$-1 = e^{\gamma_1 \tau_1}\left[\cos \tau_1 + \gamma_1 \sin \tau_1 - \frac{1}{\mu\omega_1}(K-1+s)\sin \tau_1\right].$$

Solving the second of these relations for s, and then the first for s', we obtain the correspondence function of the transformation Π_1 (where $(4\mu > (K-1)^2)$, connecting s and s', in the parametric form

$$\begin{aligned} s &= \frac{K-1}{2}\frac{e^{-\gamma_1\tau_1} + \cos \tau_1 - \gamma_1 \sin \tau_1}{\gamma_1 \sin \tau_1} \\ s' &= \frac{K-1}{2}\frac{e^{\gamma_1\tau_1} + \cos \tau_1 + \gamma_1 \sin \tau_1}{\gamma_1 \sin \tau_1} \end{aligned} \quad (8.35)$$

since

$$\mu\omega_1 = \mu h_1 \frac{\omega_1}{h_1} = \frac{K-1}{2\gamma_1}.$$

Similarly, for $4\mu < (K-1)^2$, when the roots of the characteristic equation (8.32) are real we obtain, for the correspondence function of the transformation Π_1

$$\left.\begin{aligned} s &= \frac{K-1}{2} \frac{e^{-\gamma_1 \tau_1} + \cosh \tau_1 - \gamma_1 \sinh \tau_1}{\gamma_1 \sinh \tau_1}, \\ s' &= \frac{K-1}{2} \frac{e^{\gamma_1 \tau_1} + \cosh \tau_1 + \gamma_1 \sinh \tau_1}{\gamma_1 \sinh \tau_1} \end{aligned}\right\} \quad (8.36)$$

(the values of τ_1 and γ_1 are defined as before but now $\gamma_1 > 1$).

In the region (II) the equations (8.30) can be written in the form

$$\left.\begin{aligned} \mu \ddot{y} + \dot{y} + y &= -K, \\ x &= \dot{y}. \end{aligned}\right\} \quad (8.30b)$$

The characteristic equation (8.31) for this system always has real negative roots $\lambda = -h_2 \pm \omega_2$ (since $0 < \mu < \frac{1}{4}$) where

$$h_2 = \frac{1}{2\mu} \quad \text{and} \quad \omega_2 = \frac{\sqrt{1-4\mu}}{2\mu}$$

and $h_2 > \omega_2$. Hence, the general solution of the equations (8.30b) can be written in the form

$$\left.\begin{aligned} y &= -K + e^{-h_2 t}\left[\frac{x_0}{\omega_2}\sinh \omega_2 t + (y_0 + K)\left(\cosh \omega_2 t + \frac{h_2}{\omega_2}\sinh \omega_2 t\right)\right], \\ x &= e^{-h_2 t}\left[x_0\left(\cosh \omega_2 t - \frac{h_2}{\omega_2}\sinh \omega_2 t\right) - \frac{y_0 + K}{\mu \omega_2}\sinh \omega_2 t\right]. \end{aligned}\right\} \quad (8.37)$$

Let, $x = x_0 = -1$, $y = y_0 = -(K-1) + s'$ at $t = 0$, and for $x = -1$, $y = -(K-1) + s_1$ at $t = t_2$ where t_2 is the transit time along the path L in the region (II). Then the second relation (8.37) gives

$$-1 = e^{-\gamma_2 \tau_2}\left[-(\cosh \tau_2 - \gamma_2 \sinh \tau_2) - \frac{1}{\mu \omega_2}(1+s')\sinh \tau_2\right]$$

where

$$\tau_2 = \omega_2 t_2 \quad \text{and} \quad \gamma_2 = \frac{h^2}{\omega_2} = \frac{1}{\sqrt{1-4\mu}} > 1. \quad (8.38)$$

Solving for s', using the relation $\mu\omega_2 = \mu h_2 \omega_2 / h_2 = (2\gamma_2)^{-1}$ and then changing s' into $-s_1$ and τ_2 into $-\tau_2$, in the expression obtained for s', we shall obtain the correspondence function for the transformation Π_2.

$$\left. \begin{aligned} s_1 &= \frac{e^{-\gamma_2 \tau_2} - \cosh \tau_2 + \gamma_2 \sinh \tau_2}{2\gamma_2 \sinh \tau_2}, \\ s' &= \frac{e^{\gamma_2 \tau_2} - \cosh \tau_2 - \gamma_2 \sinh \tau_2}{2\gamma_2 \sinh \tau_2}. \end{aligned} \right\} \quad (8.39)$$

2. The correspondence functions

We shall begin with the transformation Π_1 for the case $4\mu > (K-1)^2$ ($K > 1$), when the state of equilibrium $(0, 0)$ is an unstable focus, the paths in (I) are spirals and the correspondence function is given by (8.35). The

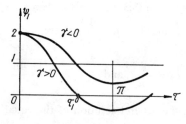

Fig. 384

representative point, moving from the point s to the point s' along an arc of a spiral path in the region (I) makes less than half a revolution round the focus $(0, 0)$. Therefore the parameter τ_1, the normalized time of transit of the representative point in the region (I), will satisfy the inequality $0 < \tau_1 < \pi$; a smaller τ_1 corresponds to larger s and s'.[†] τ_1^0 is the value of τ_1 corresponding to $s = 0$; and this boundary value of the parameter τ_1 is determined, clearly, by

$$\psi_1(\tau_1, \gamma_1) = 0,$$

where

$$\psi_1(\tau, \gamma) = 1 + e^{\gamma \tau}(\cos \tau - \gamma \sin \tau)$$

The graph of this function and the graphical solution for τ_1^0 are shown in Fig. 384 (it is clear that $0 < \tau_1^0 < \pi$). Then, as τ_1 varies from τ_1^0 to 0,

---

[†] This is quite understandable since the representative point moves more rapidly along paths with larger y; the larger y the larger $|\dot{x}|$ (by virtue of the first equation (8.30) and the smaller t_1 or τ_1; for $y \to +\infty$, $\dot{x} \to -\infty$ as $t_1 \to 0$.

s takes all values from 0 to $+\infty$; at the same time s' will also increase monotonically from a certain positive value

$$s'_0 = (K-1)\left[\frac{\sinh \gamma_1 \tau_1^0}{\gamma_1 \sin \tau_1^0}+1\right] > 2(K-1)$$

to $+\infty$.

On differentiating (8.35) we obtain

$$\frac{ds}{d\tau_1} = -\frac{K-1}{2\gamma_1} \cdot \frac{1+e^{-\gamma_1\tau_1}(\cos \tau_1 + \gamma_1 \sin \tau_1)}{\sin^2 \tau_1} = -\frac{K-1}{2\gamma_1} \frac{\psi_1(\tau_1, -\gamma_1)}{\sin^2 \tau_1},$$

$$\frac{ds'}{d\tau_1} = -\frac{K-1}{2\gamma_1} \cdot \frac{1+e^{\gamma_1\tau_1}(\cos \tau_1 - \gamma_1 \sin \tau_1)}{\sin^2 \tau_1} = -\frac{K-1}{2\gamma_1} \frac{\psi_1(\tau_1, \gamma_1)}{\sin^2 \tau_1},$$

$$\frac{ds}{ds'} = \frac{\psi_1(\tau_1, -\gamma_1)}{\psi_1(\tau_1, \gamma_1)}$$

and

$$\frac{d^2s}{ds'^2} = -\frac{4\gamma_1(\gamma_1^2+1)\sin^3 \tau_1}{(K-1)[\psi_1(\tau_1, \gamma_1)]^3}\{\sinh \gamma_1\tau_1 - \gamma_1 \sin \tau_1\}.$$

Since, for $0<\tau_1<\tau_1^0$, $\psi_1(\tau_1, \gamma_1)$ and $\psi_1(\tau_1, -\gamma_1) > 0$, then, for any value of τ_1 in this interval, $ds/d\tau_1 < 0$, $ds'/d\tau_1 < 0$ and $ds/ds' > 0$; moreover, ds/ds' varies monotonically from $+\infty$ to $+1$ as τ_1 varies from τ_1^0 to 0 (an increase of s from 0 to $+\infty$), since $d^2s/ds'^2 < 0$.

Note also that

$$s'-s = (K-1)\left\{\frac{\sinh \gamma_1 \tau_1}{\gamma_1 \sin \tau_1}+1\right\} > 2(K-1)$$

and so as $\tau_1 \to +0$, $s'-s \to 2(K-1)$. Hence the curve (8.35) has the asymptote

$$s = s' - 2(K-1).$$

These results are sufficient to construct the graph of the correspondence function (8.35), as shown in Fig. 385.

Similarly, for $4\mu < (K-1)^2$, when the correspondence function of the transformation Π_1 is written in the form (8.36), the parameter of the transformation, τ_1, is also within the interval $0 < \tau_1 < \tau_1^0$, where τ_1^0 is the value of the parameter τ_1 for the point $s=0$ and is determined now by the equation

$$\psi_2(\tau_1, \gamma_1) \equiv 1 + e^{\gamma_1\tau_1}(\cosh \tau_1 - \gamma_1 \sinh \tau_1) = 0.$$

Graphs of this function $\psi_2(\tau, \gamma)$ for $\gamma = \gamma_1 > 1$ and $\gamma = -\gamma_1 < -1$ are shown in Fig. 386. Again, as τ_1 decreases from τ_1^0 to 0 s takes all values from 0 to $+\infty$, and s' increases from a certain positive value

$$s_0' = (K-1)\left\{\frac{\sinh \gamma_1 \tau_1^0}{\gamma_1 \sinh \tau_1^0} + 1\right\} > 2(K-1)$$

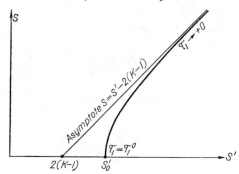

Fig. 385

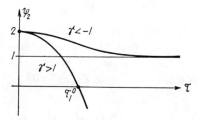

Fig. 386

to $+\infty$.† The derivative $ds/ds' = \psi_2(\tau_1, -\gamma_1)/\psi_2(\tau_1, \gamma_1)$ decreases monotonically from $+\infty$ to $+1$, since

$$\frac{d^2 s}{ds'^2} = -\frac{4\gamma_1(\gamma^2-1)\sinh^3 \tau_1}{(K-1)[\psi_2(\tau_1, \gamma_1)]^3}\{\sinh \gamma_1\tau_1 - \gamma_1 \sinh \tau_1\} < 0$$

for $0 < \tau_1 < \tau_1^0$. Thus, the graph of the correspondence function (8.36) has the same form as the graph of the correspondence function (8.35) (Fig. 385).

Now consider the correspondence function (8.39) of the transformation Π_2. Here, as τ_2 varies from 0 to $+\infty$, s' increases monotonically from 0 to $+\infty$, and s_1' from 0 to $(s_1)_{\max} = (\gamma_2-1)(2\gamma_2 > 0)$‡

---

† The curve (8.36) has the asymptote $s = s' - 2(K-1)$ for $\tau_1 \to 0$.

‡ What has been said can be inferred from the following elementary consideration of the paths in the region (II). In the first place, since the paths cannot intersect each

To prove that s' and s_1 increase monotonically as τ_2 increases from 0 to $+\infty$ it suffices to consider the derivatives $ds/d\tau_2$ and $ds_1/d\tau_2$. It will be seen that

$$\frac{ds_1}{d\tau_2} = \frac{\psi_3(\tau_2, -\gamma_2)}{2\gamma_2 \sinh^2 \tau_2}, \quad \frac{ds'}{d\tau_2} = \frac{\psi_3(\tau_2, \gamma_2)}{2\gamma_2 \sinh^2 \tau_2}, \quad \frac{ds_1}{ds'} = \frac{\psi_3(\tau_2, -\gamma_2)}{\psi_3(\tau_2, \gamma_2)}$$

and

$$\frac{d^2 s_1}{ds'^2} = -\frac{4\gamma_2(\gamma_2^2-1)\sinh^3 \tau_2}{[\psi_3(\tau_2, \gamma_2)]^3}\{\sinh \gamma_2 \tau_2 - \gamma_2 \sinh \tau_2\},$$

where

$$\psi_3(\tau, \gamma) = 1 - e^{\gamma \tau}(\cosh \tau - \gamma \sinh \tau) = 2 - \psi_2(\tau, \gamma).$$

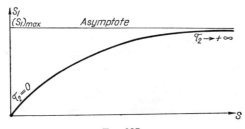

Fig. 387

Since $\gamma_2 > 1$ and for $|\gamma| > 1$ and $\tau > 0$, $\psi_3(\tau, \gamma) > 0$†, then, for $0 < \tau_2 < +\infty$

$$\frac{ds_1}{d\tau_2} > 0, \quad \frac{ds'}{d\tau_2} > 0, \quad 0 < \frac{ds_1}{ds'} < 1, \quad \frac{d^2 s_1}{ds'^2} > 0$$

Fig. 387 shows the graph of the correspondence function of the transformation Π_2.

3. Lamerey's diagram

Fig. 388 shows Lamerey's diagram for the correspondence functions of the transformation Π_1 and Π_2. These curves have a unique point of intersection, the fixed point of the transformation Π. It exists because $s_1 - s > 0$ for $s' = s'_0$ and $s_1 - s < 0$ for sufficiently large s'.

Since $0 < ds_1/ds' < 1 < ds/ds'$, for all $s' \geq s'_0$ then the fixed point is stable.

---

other then as s' increases, s also increases; larger s' will correspond to larger arcs of paths between the points s' and s_1 and, of course, larger times of transit t_2 (or τ_2). In the second place, all the paths that enter the region (II) from the s', go above the rectilinear path $y = -K - x_1 x$: therefore $s_1 < (s_1)_{max}$, where $(s_1)_{max}$ is the value of s_1 at the point of intersection of this rectilinear path with S_1.

† To prove the inequality it suffices to observe that $\psi_3(0, \gamma) = 0$ and, for $|\gamma| > |$ and $\tau > 0$, $\psi_2(\tau, \gamma) < 2$.

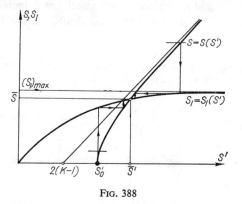

Fig. 388

The fixed point of Π' is determined analytically by the system of transcendental equations

$$\left. \begin{array}{l} \dfrac{K-1}{2}\dfrac{e^{-\gamma_1\tau_1}+\cos\tau_1-\gamma_1\sin\tau_1}{\gamma_1\sin\tau_1}=\dfrac{e^{-\gamma_2\tau_2}-\cosh\tau_2+\gamma_2\sinh\tau_2}{2\gamma_2\sinh\tau_2}, \\ \dfrac{K-1}{2}\dfrac{e^{\gamma_1\tau_1}+\cos\tau_1+\gamma_1\sin\tau_1}{\gamma_1\sin\tau_1}=\dfrac{e^{\gamma_2\tau_2}-\cosh\tau_2-\gamma_2\sinh\tau_2}{2\gamma_2\sinh\tau_2}. \end{array} \right\}$$

(8.40)

for $4\mu > (K-1)^2$ and by the system

$$\left. \begin{array}{l} \dfrac{K-1}{2}\dfrac{e^{-\gamma_1\tau_1}+\cosh\tau_1-\gamma_1\sinh\tau_1}{\gamma_1\sinh\tau_1}=\dfrac{e^{-\gamma_2\tau_2}-\cosh\tau_2+\gamma_2\sinh\tau_2}{2\gamma_2\sinh\tau_2}, \\ \dfrac{K-1}{2}\dfrac{e^{\gamma_1\tau_1}+\cosh\tau_1+\gamma_1\sinh\tau_1}{\gamma_1\sinh\tau_1}=\dfrac{e^{\gamma_2\tau_2}-\cosh\tau_2-\gamma_2\sinh\tau_2}{2\gamma_2\sinh\tau_2} \end{array} \right\}$$

(8.40a)

for $4\mu < (K-1)^2$.

Thus the point transformation Π' of the half line S into the half line S_1 has a unique stable fixed point $(s=s_1=\bar{s},\ s'=\bar{s}')$, and on the phase plane there is a unique, symmetrical, and stable limit cycle, to which all phase paths tend as $t \to +\infty$ (Fig. 389). Therefore for $K > 1$ and arbitrary initial conditions, the circuit self-oscillates[†].

---

† Fig. 389 is the phase portrait for the case $K > 1+2\mu^{\frac{1}{2}}$, when there is an unstable node at the origin.

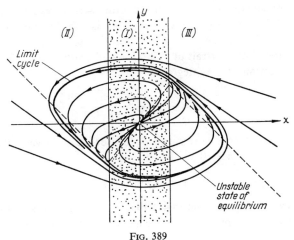

Fig. 389

The period of the self-oscillations is, clearly,

$$\tau = 2\left(\frac{\bar{\tau}_1}{\omega_1} + \frac{\bar{\tau}_2}{\omega_2}\right)$$

or

$$T = 2[R_a(C+C_1) + R_g C]\left(\frac{\bar{\tau}_1}{\omega_1} + \frac{\bar{\tau}_2}{\omega_2}\right)$$

(in ordinary time units), where $\bar{\tau}_1$, $\bar{\tau}_2$ ($0 < \bar{\tau}_1 < \tau_1^0$, $0 < \bar{\tau}_2 < +\infty$) are the values of τ_1 and τ_2 corresponding to the fixed point, and determined by (8.40) if $4\mu > (K-1)^2$ and by (8.40a) if $4\mu < (K-1)^2$.

There is a limiting case. If $K \to 1^+$, then the limit cycle tends to the circle $x^2 + y^2 = 1$, since $\bar{\tau}_1 \to \pi$ and $\tau_2 \to 0$, and the self-oscillations are nearly sinusoidal with period $2\pi[R_a(C+C_1) + R_g(C+C_g)]$.

4. Discontinuous oscillations

We shall consider now another very interesting limiting case:

$$\mu \ll 1, \quad \mu \ll \frac{1}{4}(K-1)^2.$$

With both $C_a \ll C$ and $C_g \ll C$, the circuit acts as a multivibrator with one RC circuit, generating self-oscillations of a discontinuous type. Discontinuous oscillations, as we shall see in Chapter X, are associated with differential equations whose higher order derivatives have small coefficients and the system (8.30) with a small μ is a fairly simple but typical example of such a system.

To find the phase portrait of (8.30) for *sufficiently small values of μ* we need the equation of the integral curves

$$\frac{dy}{dx} = -\frac{\mu x}{x+y+-K\varphi(x)} \tag{8.41}$$

and we shall construct on the x, y plane the isocline of the vertical tangents, the curve F on which $\dot{x}=0$ and $dy/dx=\infty$.

$$y = -x - K\varphi(x) \tag{8.42}$$

It follows from (8.30) and (8.41) that $\mu \to 0$, $|\dot{x}|$ increases rapidly, and $|dy/dx|$ decreases rapidly as we move away from the isocline F. In fact, at a distance from it of an order of magnitude μ, $\dot{x}$ and $dy/dx = 0$ (1), and at a distance of $\mu^{\frac{1}{2}}, \dot{x} \to 0\left(\mu^{-\frac{1}{2}}\right)$ and $dy/dx \to 0\left(\mu^{\frac{1}{2}}\right)^{\dagger}$. Therefore, for sufficiently small values of μ, the phase paths outside a strip of half-width $\mu^{\frac{1}{2}}$ along the contour F are arbitrarily close to the horizontal straight lines $y=$const, and the representative point moves along them *arbitrarily rapidly* $\Big(\dot{x} \to \infty$ at least as fast as $\mu^{-\frac{1}{2}} \to \infty$ as $\mu \to 0\Big)^{\ddagger}$. The representative point moves towards the right at the points lying below the isocline F (there, $-x-y-K(x) > 0$ and $\dot{x} = [-x-y-K\varphi(x)]/\mu \to \infty$), and towards the left at points above the isocline F (Fig. 390). These paths of arbitrarily rapid motions of the system (in the limit, of instantaneous jumps) go from the point at infinity and from the segment CA of the isocline F, to the half lines F_1^+ and F_2^+, which are the parts of the isocline F that lie in the regions (II) and (III). In the μ-neighbourhoods of the half straight lines F_1^+ and F_2^+, $\dot{x}$ remains finite as $\mu \to +0$, so that in these neighbourhoods there are paths of "slow" motions of the system where the phase velocities remain finite as $u \to +0^{\dagger\dagger}$. The slow motion of the representative point becomes

---

† Here and below we denote by $0[\bar{f}(\mu)]$ functions that behave, for small values of μ, like $f(\mu)$; the notation $g(x, y, \mu) = 0[f(\mu)]$ indicates that, for $\mu \to +0$, the ratio $g(x, y, \mu)/f(\mu)$ tends to a finite limit (depending, generally speaking, on x, y).

‡ By the ε-neighbourhood of a certain curve we mean, just as before, the set of all points the distance of which from the given curve does not exceed ε. Obviously, the $\mu^{\frac{1}{2}}$ - neighbourhood of the curve F reduces to F for $\mu \to 0$.

†† We shall recall that $\dot{y}=x$ and, hence, remains a finite quantity for $\mu \to +0$ both inside and outside the neighbourhood of the curve F.

arbitrarily rapid, almost instantaneously in the $\mu^{\frac{1}{2}}$-neighbourhoods of the points A and C.

Thus the motion of the representative point of the system (8.30) will consist, for sufficiently small values of μ, of alternate rapid jump-wise motions along paths arbitrarily close to horizontal straight lines $y=$const. and of "slow" motions along paths lying in the μ-neighbourhoods of the half straight lines F_1^+ and F_2^+. In Fig. 390 is shown the limiting phase

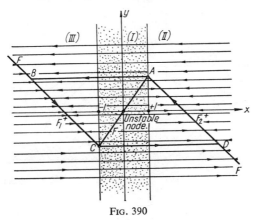

FIG. 390

portrait (for $\mu \to +0$) : the paths of the instantaneous jumps are represented by the straight lines $y=$const, and the paths of the "slow" motions by half straight lines F_1^+ and F_2^+. A limit cycle is the closed curve $ABCDA$[†].

† Approximate (asymptotic) equations of motion of the system for sufficiently small values of μ can be written in the form

$$\mu\dot{x} = -x - y_0 K\varphi(x) \tag{A}$$

during a jump-wise motion along the path $y = y_0 =$const. (but outside a certain neighbourhood of the curve F) and

$$\left.\begin{array}{l} \dot{y} = x \\ y = -x - K\varphi(x) \end{array}\right\} \quad \text{or} \quad \dot{x}[1 + K\varphi'(x)] + x = 0 \tag{B}$$

during the "slow" motion (in the vicinity of F_1^+ and F_2^+). We have already used the equation (B) in the analysis of a multivibrator in Section 7, Chapter IV, by substituting, for the dynamic analysis of a rapid process (when $C_a, C_g \ll C$), the jump postulate. This postulate (the system "jumps" instantaneously from the states $|x| \leqslant 1$ into states $|x| > 1$ while y, and so the voltage across the capacitor C, remains constant during the jump) follows now *as a consequence of the dynamic equations (8.30)* when $\mu \to +0$.

Integrating the equation of the "slow" motion $\dot{x} + x = 0$ (see Section 7, Chapter IV) over the segments BC and DA, we shall obtain a limiting expression for the period of the self-oscillations: $\lim_{\mu \to 0} \tau = 2\ln(2K-1)$, since for $\mu \to 0$ the time of transit of the representative point in the region (I) $\bar{t}_1 \to +0$ and in the region (II) $\bar{t}_2 \to \ln(2K-1)$.

494   POINT TRANSFORMATIONS AND PIECE-WISE LINEAR SYSTEMS   [VIII

To prove rigorously now that the curve $ABCDA$ is actually the limit cycle of the system (8.30) as $\mu \to 0$, we shall construct on the phase plane a region (ε) from which the phase paths *cannot leave* (as t increases) and such as to contain inside itself the curve $ABCDA$. The region reduces to $ABCDA$ as $\mu \to 0$. To this end draw (Fig. 391) the isoclines $\varkappa = 0$ (the y axis),

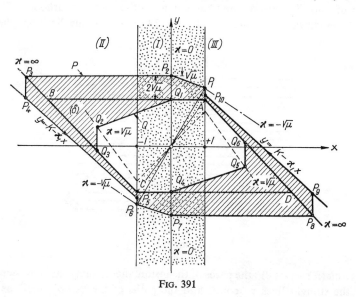

Fig. 391

$\varkappa = \infty$ (see curve F, Fig. 390), the isocline $\varkappa = -\mu^{\frac{1}{2}}$ and $\varkappa = +\mu^{\frac{1}{2}}$ †, and also the closed curves P and Q, symmetrical with respect to the origin and formed by straight segments in the following way.

The first contour P will begin from the point $P_1\!\left(1, K-1+\mu^{\frac{1}{2}}\right)$ at the intersection of the isocline $\varkappa = -\mu^{\frac{1}{2}}$ and the line $x = +1$. The segment $P_1 P_2$ has a slope $-\mu^{\frac{1}{2}}$ and connects P_1 to $P_2\!\left(0, K-1+2\mu^{\frac{1}{2}}\right)$. Segment $P_2 P_3$ is drawn horizontally from P_2 until it meets the isocline $\varkappa = \infty$, and

† According to (8.41) the equation of the isocline $dy/dx = \varkappa$ will be

$$\varkappa = -\frac{\mu x}{x+y+K\varphi(x)} \quad \text{or} \quad y = -\left(1+\frac{\mu}{\varkappa}\right)x - K\varphi(x),$$

hence the isoclines $\varkappa = \pm \mu^{\frac{1}{2}}$ will be the broken lines

$$y = -(1 \pm \sqrt{\mu})x - K\varphi(x).$$

the segment P_3P_4 is vertical and intersects the rectilinear path of the system (8.30) in the region (II)

$$y = -K - \varkappa_1 x, \quad \text{where} \quad \varkappa_1 = \frac{1+\sqrt{1-4\mu}}{2} = 1 + O(\mu);$$

P_4P_5 is a segment of this path and finally P_5P_6 is a segment of the straight line $x = -1$, the point $P_6\left(-1, -K+1-\mu^{\frac{1}{2}}\right)$ being symmetrical to the point P_1 and situated below the point $P_5(-1, -K-\varkappa_1)$. The second half of the contour P is symmetrical with the broken line just constructed.

One half of the contour Q consists of the segment AQ_1 of the horizontal straight line $y = K-1$ (the point Q_1 lies on the y axis), the segment Q_1Q_2 with slope $+\mu^{\frac{1}{2}}$ (the point Q_2 lies on the isocline $\varkappa = +\mu^{\frac{1}{2}}$), the vertical segment Q_2Q_3 intersecting the isocline $\varkappa = \infty$, and the segment Q_3C of the isocline $\varkappa = \infty$; the second half of the contour Q (the broken line $CQ_4Q_5Q_6A$) is symmetrical to the first one.

The region (ε) lies between the contours P and Q (shown shaded in Fig. 391). In the first place, (ε) contains in itself or on its boundary the curve $ABCDA$ and reduces to it as $\mu \to +0$, since the greatest distances of the curves P and Q from the curve $ABCDA$ do not exceed $2\mu^{\frac{1}{2}}$ and $(2K-1)\mu^{\frac{1}{2}}$ respectively and tend to zero as $\mu \to +0$. In the second place, the phase paths cannot leave the region (ε) as t increases since on its boundaries the paths are either tangent to the boundaries or cross them and enter the region (ε).

To prove the latter statement it suffices to consider the behaviour of the paths of the system (8.30) on the broken lines $P_1P_2P_3P_4P_5P_6$ and $AQ_1Q_2Q_3C$. On the segment P_1P_2 lying between the isoclines $\varkappa = -\mu^{\frac{1}{2}}$ and $\varkappa = 0$ and above the isocline F, $-\mu^{\frac{1}{2}} \leq dy/dx \leq 0$ and $\dot{x} < 0$; therefore the paths have a smaller slope than the segment itself and the representative points move to the left entering (ε) (an exception is the point P_1 at which the path is tangent to the segment P_1P_2).

Exactly similar argument for each segment of curves P and Q will verify that on the boundaries P and Q of the region (ε) the phase paths of the system (8.39) are either tangent or cross the boundaries into the region (ε). Since this region does not contain states of equilibrium, then as we have seen (see Theorem V, Section 2, Chapter VI), there exists in it a stable limit cycle.

We have thus shown that the unique and stable limit cycle of the system (8.30) is found in the region (ε) and, hence, tends to $ABCDA$ as $\mu \to +0$[†].

5. Period of self-oscillations for small values of μ

The asymptotic formula for the period of discontinuous oscillations
$$\tau = 2 \ln (2K-1), \tag{8.43}$$
which was found earlier (see, for example, Section 7, Chapter IV) is found to produce fairly large errors for the period of a multivibrator, if the parameter μ is not very small. For example, for $\mu=0{\cdot}05$ and $K=2$, when the self-oscillations are very close to discontinuous ones[‡], the error of the formula (8.43) amounts to about 20%. It is better to evaluate an asymptotic expression for the period of the self-oscillations of a multivibrator for small values of μ (for C_a, $C_g \ll C$) proceeding not from the case when $\mu \to +0$ but from the correspondence functions (8.36) and (8.39) and the equations (8.40a) which determine exactly the fixed point of the transformation Π' and the limit cycle[††]. The form in which the correspondence functions (8.36) and (8.39) and the equations (8.40a) are written is inconvenient for this purpose, so, we change from $\tau_1 = \omega_1 t_1$, $\tau_2 = \omega_2 t_2$ directly to t_1 and t_2, the times of transit in the regions (I) and (II). We shall denote the roots of the characteristic equation (8.32) for the region (I) by λ_1 and λ_1'

$$\left.\begin{aligned}
\lambda_1 &= h_1 - \omega_1 = \frac{K-1-\sqrt{(K-1)^2-4\mu}}{2\mu} = \\
&= \frac{1}{K-1} + \frac{\mu}{(K-1)^3} + 2\frac{\mu^2}{(K-1)^5} + \cdots, \\
\lambda_1' &= h_1 + \omega_1 = \frac{K-1+\sqrt{(K-1)^2-4\mu}}{2\mu} = \\
&= \frac{K-1}{\mu} - \frac{1}{K-1} - \frac{\mu}{(K-1)^3} - 2\frac{\mu^2}{(K-1)^5} - \cdots,
\end{aligned}\right\} \tag{8.44}$$

[†] This can also be formulated somewhat differently: the limit cycle of the system (8.30) is found in the δ-neighbourhood of the curve $ABCDA$, where $\delta = 0\!\left(\mu^{\frac{1}{2}}\right)$. For example, the δ-neighbourhood of the curve $ABCDA$ where δ is the largest among $2\mu^{\frac{1}{2}}$ and $(2K-1)\mu^{\frac{1}{2}}$ contains inside itself the region (ε) and, hence, the limit cycle.

[‡] For $\mu=0{\cdot}05$ and $K=2$ the self-oscillations of the multivibrator consist of alternate "slow" variations of x (with a velocity $\dot x$ of the order of unity) and "rapid" ones (with a velocity of the order of $\mu^{-1} = 20$).

[††] See also [114, 52, 93, 158, 159] where asymptotic expansions for the period of certain solutions of differential equations of the second order are presented. In [114], for example, an equation is considered that is equivalent to the system (8.30) with $K=2$.

and the absolute values of the roots of the characteristic equation (8.31) for the region (II) by λ_2 and λ_2'

$$\begin{aligned} \lambda_2 &= h_2 - \omega_2 = \frac{1-\sqrt{1-4\mu}}{2\mu} = 1+\mu+2\mu^2+\ldots, \\ \lambda_2' &= h_2 + \omega_2 = \frac{1+\sqrt{1-4\mu}}{2\mu} = \frac{1}{\mu} - 1 - \mu - 2\mu^2 - \ldots \end{aligned} \quad (8.44a)$$

Substituting $\tau_1 = \omega_1 t_1$ in (8.36) and multiplying the numerator and denominator of this expression by $e^{h_1 t_1}$ we have

$$s = \frac{K-1}{2}\frac{\omega_1}{h_1}\frac{1+\frac{1}{2}\left[e^{\lambda_1' t_1}+e^{\lambda_1 t_1}\right]-\frac{1}{2}\frac{h_1}{\omega_1}\left[e^{\lambda_1' t_1}-e^{\lambda_1 t_1}\right]}{\frac{1}{2}\left[e^{\lambda_1' t_1}-e^{\lambda_1 t_1}\right]}.$$

Noting that $(K-1)/2h_1 = \mu$, $h_1 = (\lambda_1' + \lambda_1)/2$, $\omega_1 = (\lambda_1' - \lambda_1)/2$, we obtain the following parametric expressions for the correspondence function of the transformation Π_1

$$\begin{aligned} s &= \mu\frac{\lambda_1'\left[1+e^{\lambda_1 t_1}\right]-\lambda_1\left[1+e^{\lambda_1' t_1}\right]}{e^{\lambda_1' t_1}-e^{\lambda_1 t_1}}, \\ s' &= \mu\frac{\lambda_1'\left[1+e^{-\lambda_1 t_1}\right]-\lambda_1\left[1+e^{-\lambda_1' t_1}\right]}{e^{-\lambda_1 t_1}-e^{-\lambda_1' t_1}} \end{aligned} \quad (8.45)$$

(s' is obtained from the equation for s by changing the sign and changing t_1 into $-t_1$). Similarly by changing τ_2 into $\omega_2 t_2$ in (8.38) and using the relations

$$\gamma_2\omega_2 = h_2, \quad \frac{1}{2h_2} = \mu, \quad h_2+\omega_2 = \lambda_2', \quad h_2-\omega_2 = \lambda_2,$$

$$h_2 = \frac{1}{2}(\lambda_2'+\lambda_2) \quad \text{and} \quad \omega_2 = \frac{1}{2}(\lambda_2'-\lambda_2),$$

we obtain for the transformation Π_2

$$\begin{aligned} s_1 &= \mu\frac{\lambda_2\left[e^{\lambda_2' t_2}-1\right]-\lambda_2'\left[e^{\lambda_2 t_2}-1\right]}{e^{\lambda_2' t_2}-e^{\lambda_2 t_2}}, \\ s' &= \mu\frac{\lambda_2'\left[1-e^{-\lambda_2 t_2}\right]-\lambda_2\left[1-e^{-\lambda_2' t_2}\right]}{e^{-\lambda_2 t_2}-e^{-\lambda_2' t_2}}. \end{aligned} \quad (8.46)$$

From the correspondence function of Π_1 and Π_2 it is now easy to obtain asymptotic expansions of the correspondence functions and so the period of the self-oscillations when μ is small by noting the fact that $\lambda_1, \lambda_2 = O(1)$ and $\lambda_1', \lambda_2' = O(\mu^{-1})$.

For motion along a phase path that intersects the line S (for example, along a limit cycle), its time of transit in the region (I) $t_1 \to +0$ as $\mu \to +0$, and in the region (II) its transit time t_2 tends to a finite limit $(t_2 = O(1))$†. Then, however, $e^{-\lambda_2 t_2} = O(1)$, and $e^{-\lambda_2' t_2} = O(e^{-1/\mu})$ so tending to zero as $\mu \to +0$ *more rapidly than any* power of μ. Therefore (to an accuracy of terms of the order of $e^{-1/\mu}$) the correspondence function of the transformation Π_2 can be written in the form‡

$$\left.\begin{aligned} s_1 &= \mu\lambda_2 + O\left(e^{-\frac{1}{\mu}}\right), \\ s' &= \mu[(\lambda_2' - \lambda_2)e^{\lambda_2 t_2} - \lambda_2'] + O\left(e^{-\frac{1}{\mu}}\right). \end{aligned}\right\} \quad (8.46a)$$

We have for the limit cycle, to the same order of accuracy,

$$\bar{s} = \bar{s}_1 = \mu\lambda_2 + O\left(e^{-\frac{1}{\mu}}\right). \quad (8.46b)$$

Substituting (8.46b) in the first relation (8.45), we obtain an equation determining the time of transit $\bar{t}_1$ of the representative point along the limit cycle in the region (I),

$$\left[\lambda_2 + \lambda_1 + O\left(e^{-\frac{1}{\mu}}\right)\right] e^{\lambda_1' \bar{t}_1} = (\lambda_1' + \lambda_2)e^{\lambda_1 \bar{t}_1} + \lambda_1' - \lambda_1. \quad (8.47)$$

This equation can be solved by successive approximations using the difference in the order of magnitude of the roots λ_1 and λ_1'. Since $e^{\lambda_1 \bar{t}_1} \to 1$ as $\mu \to 0$, then, to satisfy the equation (8.47) $e^{\lambda_1' \bar{t}_1}$ must be of the order of

---

† In the region (I) the abscissa x of the representative point varies from $+1$ to -1 with a velocity $\dot{x} \to -\infty$ for $\mu \to +0$; therefore for $\mu \to +0$, $t_1 \to +0$; it can be shown that for values of s in the interval $0 < s < M$, $0(\mu) < t_1 < 0$ ($\mu \ln 1(\mu)$). In the region (II) the ordinate y of the representative point varies by a finite quantity; from $y' = -(K-1)+s' > K-1$, since $s' > 2(K-1)$, to $y_1 = -(K-1)-s_1 < -(K-1)$, with a finite velocity $\dot{y} = x$; it hence follows that $t_2 = O(1)$.

‡ The asymptotic expression (8.46a) is valid for all values of s' larger than an arbitrarily small but fixed quantity $(s' > a)$. The first relation (8.46) shows that all paths that intersect the half straight line S' outside a certain fixed neighbourhood of the point $s' = 0$, and, in particular, all paths that come from the region (III), including the limit cycle, enter the region (II) in a very small (of the order of $e^{-1/\mu}$) neighbourhood of the rectilinear path $y = -K - \varkappa_1 x$.

magnitude of λ_1', i.e. $e^{\lambda_1 \bar{t}_1}=0\,(1/\mu)$, $\lambda_1'\bar{t}_1=0\,(\ln 1/\mu)$ and $\bar{t}_1=0(\mu \ln 1/\mu)$. Substituting $e^{\lambda_1'\bar{t}_1}=1+0(\lambda_1\bar{t}_1)=1+0(\ln 1/\mu)$ in (8.47) we have

$$e^{\lambda_1'\bar{t}_1} = \frac{2\lambda_1'+\lambda_2-\lambda_1+\lambda_1'O\left(\mu\ln\frac{1}{\mu}\right)}{\lambda_2+\lambda_1} = \frac{2(K-1)^2}{\mu K}\left[1+O\left(\mu\ln\frac{1}{\mu}\right)\right],$$

whence

$$\lambda_1'\bar{t}_1 = \ln\frac{1}{\mu}+\ln\frac{2(K-1)^2}{K}+\ln\left[1+O\left(\mu\ln\frac{1}{\mu}\right)\right],$$

or, since

$$\ln\left[1+O\left(\mu\ln\frac{1}{\mu}\right)\right] = O\left(\mu\ln\frac{1}{\mu}\right) \quad \text{and} \quad \frac{1}{\lambda_1'} = \mu\lambda_1 = \frac{\mu}{K-1}+O(\mu^2),$$

$$\bar{t}_1 = \frac{\mu}{K-1}\ln\frac{1}{\mu}+\frac{\mu}{K-1}\ln\frac{2(K-1)^2}{K}+O\left(\mu^2\ln\frac{1}{\mu}\right). \tag{8.48}$$

We shall find now, an asymptotic expression for $\bar{s}'$ (the coordinate of the point of intersection of the limit cycle with the half line S') in order to find $\bar{t}_2$ and so the period of the self-oscillations. According to (8.48) we have

$$e^{-\lambda_1'\bar{t}_1} = \frac{\mu K}{(K-1)^2}+O\left(\mu^2\ln\frac{1}{\mu}\right),$$

$$\lambda_1\bar{t}_1 = \frac{\mu}{(K-1)^2}\ln\frac{1}{\mu}+\frac{\mu}{(K-1)^2}\ln\frac{2(K-1)^2}{K}+O\left(\mu^2\ln\frac{1}{\mu}\right)$$

and

$$e^{\lambda_1\bar{t}_1} = 1+\lambda_1\bar{t}_1+O(\lambda_1^2\bar{t}_1^2) = 1+\frac{\mu}{(K-1)^2}\ln\frac{1}{\mu}+$$

$$+\frac{\mu}{(K-1)^2}\ln\frac{2(K-1)^2}{K}+O\left(\mu^2\ln^2\frac{1}{\mu}\right).$$

Therefore, by virtue of the second equation (8.45), which we shall rewrite in the form

$$s' = \frac{(\mu\lambda_1'-\mu\lambda_1)e^{\lambda_1 t_1}+\mu\lambda_1'-\mu\lambda_1 e^{\lambda_1 t_1}e^{-\lambda_1' t_1}}{1-e^{\lambda_1 t_1}\cdot e^{-\lambda_1' t_1}},$$

we obtain

500   POINT TRANSFORMATIONS AND PIECE-WISE LINEAR SYSTEMS   [VIII

$$\bar{s}' = \left[1 + \frac{\mu K}{2(K-1)^2}\right]\left\{\left(K-1-\frac{2\mu}{K-1}\right)e^{\lambda_1 \bar{t}_1} + K - 1 - \frac{\mu}{K-1}\right\} +$$
$$+ O(\mu^2) = 2(K-1) + \frac{\mu}{K-1}\ln\frac{1}{\mu} + \frac{\mu}{K-1}\left[K-3+\ln\frac{2(K-1)^2}{K}\right] +$$
$$+ O\left(\mu^2 \ln^2 \frac{1}{\mu}\right)^\dagger.$$

Then, using the second equation (8.46a), we shall obtain

$$e^{\lambda_2 \bar{t}_2} = \frac{\bar{s}' + \mu \lambda_2'}{\mu \lambda_2' - \mu \lambda_2} = (1+2\mu)(\bar{s}'+1-\mu) + O(\mu^2) = (K-1) \times$$

$$\times \left\{1 + \frac{\mu \ln \frac{1}{\mu}}{(K-1)(2K-1)} + \mu\left[\frac{2K(2K-3)}{(K-1)(2K-1)} + \frac{\ln \frac{2(K-1)^2}{K}}{(K-1)(2K-1)}\right] + \right.$$
$$\left. + O\left(\mu^2 \ln^2 \frac{1}{\mu}\right)\right\}^\ddagger$$

$$\lambda_2 \bar{t}_2 = \ln(2K-1) + \frac{\mu \ln \frac{1}{\mu}}{(K-1)(2K-1)} + \mu \frac{2K(2K-3) + \ln \frac{2(K-1)^2}{K}}{(K-1)(2K-1)} +$$
$$+ O\left(\mu^2 \ln^2 \frac{1}{\mu}\right)$$

and finally

$$\bar{t}_2 = \frac{1}{\lambda_2}(\lambda_2 \bar{t}_2) = (1-\mu)\lambda_2 \bar{t}_2 + O(\mu^2) \quad \text{or}$$

$$\bar{t}_2 = \ln(2K-1) + \frac{\mu \ln \frac{1}{\mu}}{(K-1)(2K-1)} +$$
$$+ \mu\left[\frac{2K(2K-3)}{(K-1)(2K-1)} + \frac{\ln \frac{2(K-1)^2}{K}}{(K-1)(2K-1)} - \ln(2K-1)\right] + O\left(\mu^2 \ln^2 \frac{1}{\mu}\right). \tag{8.49}$$

† The error in the numerator is $O(\mu^2 \ln 1/\mu)$; we therefore expand the expressions for $\mu \lambda_1'$ and $\mu \lambda_1$ to an accuracy $O(\mu^2)$.

It follows from the asymptotic expression obtained for $\bar{s}'$ that the point of the limit cycle with abscissa $x=1$ lies at a distance of the order of $\mu \ln 1/\mu$ from the curve $ABCDA$.

‡ We write all terms to an accuracy up to $O(\mu^2 \ln 1/\mu)$. In particular

$$\mu \lambda_2' = 1 - \mu + O(\mu^2) \quad \text{and} \quad \frac{1}{\mu \lambda_2' - \mu \lambda_2} = 1 + 2\mu + O(\mu^2).$$

TWO-POSITION AUTOMATIC PILOT FOR SHIPS

From these we obtain the following asymptotic expression for the period of the self-oscillations of a multivibrator (for the period of the solution of the equation (8.39) for small values of μ)

$$\tau = 2(\bar{t}_1+\bar{t}_2) = 2\ln(2K-1) + \frac{4K}{(K-1)(2K-1)}\mu\ln\frac{1}{\mu} +$$

$$+\left[\frac{4K(2K-3)}{(K-1)(2K-1)} + \frac{4K\ln\frac{2(K-1)^2}{K}}{(K-1)(2K-1)} - 2\ln(2K-1)\right]\mu + O\left(\mu^2\ln\frac{1}{\mu}\right). \quad (8.50)$$

§ 6. TWO-POSITION AUTOMATIC PILOT FOR SHIPS' CONTROLLER

1. Formulation of the problem

Let φ be the deviation of a ship from its assigned course (Fig. 392). Neglecting the lateral drift of the ship during its rotations, and taking into account both the moment $M = M(\psi)$ generated by the rudder and the

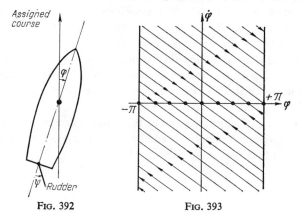

FIG. 392   FIG. 393

moment of the forces of resistance $-H\,d\varphi/dt$ then if I is the moment of inertia of the ship with respect to its main vertical axis, the equation of the rotation of the ship has the form

$$I\frac{d^2\varphi}{dt^2} + H\frac{d\varphi}{dt} = M. \quad (8.51)$$

The ship by itself has no course stability. To illustrate this Fig. 393 shows the phase plane (the development of the phase cylinder) of the ship with the rudder along the axis ($\psi=0$ and $M=0$) the phase paths form a family of straight lines along which the representative point approaches states of equilibrium that cover the *entire* φ axis, as $t \to +\infty$.

Stability on a given course can be achieved only by a steering apparatus making suitable rudder movements. In the language of the mapping of the phase plane by the paths, the problem of the steering apparatus (either a man or an automatic course controller) is to generate, instead of the continuum of states of equilibrium, a stable *state of equilibrium* corresponding to the given course ($\varphi=0$).

One of the simplest automatic course controllers is the so-called *two-position automatic pilot* for which the rudder can assume *two positions* $\psi=\pm\psi_0$, generating moments $M=\pm M_0$. We shall assume the position of the rudder to depend according to a certain law upon the state of ship, φ and $d\varphi/dt$. We have the simplest two-position course controller when the movement of the rudder from one side to the other occurs for a passage of the ship through the given course $\varphi=0$. Also we assume that the rudder movement takes place instantaneously. As we shall see later, such automatic equipment does in fact stabilize the course of the ship when certain requirements are met[†]. It is natural to think, however, if only on the basis of experience in steering an ordinary boat, that the stabilizing action of the equipment would be more effective if the rudder movement occurred somewhat *before* the ship swung through the assigned course. Such anticipatory control is usually achieved in practice by two methods: either by means of the so-called *velocity correction* or by the introduction of the so-called *parallel or hard feedback*[‡].

In the case of velocity correction or derivative action the rudder movement occurs when a certain linear combination of course deviation φ and angular velocity $\dot{\varphi}$ reduces to zero.

$$\sigma = \varphi + b\frac{d\varphi}{dt}$$

It is easily seen that for $b>0$ the switch of the rudder position will occur before the passage by the ship through $\varphi=0$.

[†] It is evident, for example, that for the correct operation of the automatic pilot it is necessary that, for a deviation of the ship on the starboard side of the assigned course (for $\varphi>0$) the rudder be put over on the port side of the ship ($\psi=-\psi_0$ and $M=-M_0$) and, vice versa, for $\varphi<0$ $\psi=+\psi_0$ and $M=+M_0$.

[‡] An automatic pilot with parallel feedback will be briefly considered in Sub-section 4 of this section.

A schematic diagram of such a two-position automatic pilot with velocity correction is shown in Fig. 394; also shown is the block diagram of the system "ship + automatic steering apparatus". This self-steering apparatus has two data transmitters: the data transmitter of the course deviation φ is a gyrocompass, the data transmitter of the angular velocity $d\varphi/dt$ is a

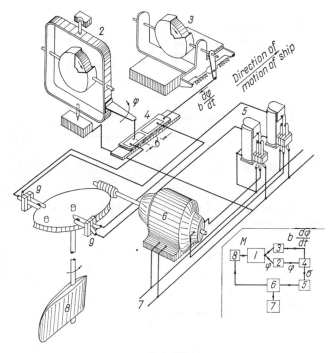

Fig. 394

so-called "rate gyroscope". The rate gyroscope has its vertical axis fixed to the ship only. About the horizontal axis are moments due to a spring and a special viscous damper. The angle of rotation of the horizontal axis of the gyroscope (after its free oscillations have been damped) is found to be proportional to the angular velocity of the ship.

These two data transmitters move the brush and the contacts of a switch in such a manner that for a change in sign of the quantity $\sigma = \varphi + b \, d\varphi/dt$ the switch causes (via auxiliary relays) the steering engine to operate in the required direction and put the rudder into one of the extreme positions, $\psi = \pm \psi_0$. It is clear that the rudder is on the port side ($\psi = -\psi_0$ and $M = -M_0$) for $\sigma > 0$ and on the starboard side ($\psi = +\psi_0$ and $M = +M_0$) for $\sigma < 0$. If, however, $\sigma = 0$, the switch is in the central off-position and the steering engine is disconnected so that the rudder can assume an arbitrary position between the two extremes: $-\psi_0 \leq \psi \leq +\psi_0$ and $-M_0 \leq M \leq +M_0$ (Fig. 395).

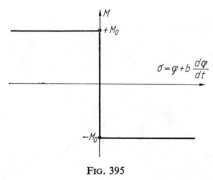

Fig. 395

We can write the equations of the two-position controller and steering engine with velocity correction in the form

$$M = M_0 Z\left(\varphi + b \frac{d\varphi}{dt}\right), \tag{8.52}$$

where

$$Z(\sigma) = \begin{cases} -1 & \text{for} \quad \sigma > 0 \\ +1 & \text{for} \quad \sigma < 0 \end{cases} \quad \text{and} \quad |Z(0)| \leq 1.$$

2. The phase plane

"Slip-motion". We shall first of all simplify the equations (8.51) and (8.52) of the system by introducing the dimensionless variables x, t_{new} and z defined by the relations

$$\varphi = Ax, \quad t_{\text{crit}} = T t_{\text{new}}, \quad M = M_0 z,$$

where

$$A = \frac{M_0 I}{H^2} \quad \text{and} \quad T = \frac{I}{H}.$$

These equations will take the form

$$\ddot{x} + x = z \quad \text{and} \quad z = Z(x + \beta \dot{x})$$

where

$$\beta = \frac{b}{T} = b\frac{H}{I}$$

and where a dot superscript denotes differentiation with respect to the new dimensionless time (below this will be denoted by t). In addition, since we are only interested in the case of small deviations of the ship from its true course so that $|\varphi|$ will always be smaller than π, we can take an ordinary plane for the phase surface x, $\dot{x}$.

We write $\dot{x}=y$. This phase plane x, y is divided by the "switching line"

$$x = \beta y = 0, \qquad (8.54)$$

into two regions (I) and (II) in each of which the appropriate linear equation of motion is valid in the region (I) ($x+\beta y>0$)

$$\left.\begin{array}{l} \dot{x} = y, \\ \dot{y} = -y-1 \end{array}\right\} \qquad (8.55)$$

and in the region (II) ($x+\beta y<0$).

$$\left.\begin{array}{l} \dot{x} = y, \\ \dot{y} = -y+1 \end{array}\right\} \qquad (8.55a)$$

Therefore the paths in the regions (I) and (II) are symmetrical to each other with respect to the origin.

On the "switching line" itself the motion of the representative point remains as yet *undetermined*, since there the equation of the controller and rudder action (the second equation (8.53)) does not determine uniquely the equation of motion of the ship. Therefore, to obtain a *complete* picture of the phase portrait the definition of the equations of motion, when $x+\beta y=0$, must be completed in a suitable way.

Let us introduce the normalized coordinate of the discriminating switch

$$\xi = x+\beta y$$

and let us calculate $\dot{\xi}$. For the region (I)

$$\dot{\xi} = \dot{x}+\beta\dot{y} = y-\beta(y+1) = (1-\beta)y-\beta.$$

The isocline $\dot{\xi}=0$ (the locus where the paths are parallel to the switching line) is, clearly, the horizontal straight line

$$y = \frac{\beta}{1-\beta}.$$

Let $0<\beta<1$. In this case, above this isocline $\dot{\xi}>0$ the phase paths move away from the switching line (8.54) and, below it, approach this line. There is a symmetrical situation in the region (*II*). Thus, on the switching line (8.54) there is a segment

$$|y| \leq \left|\frac{\beta}{1-\beta}\right|, \tag{8.56}$$

which is approached on both sides by phase paths. Outside this segment the phase paths approach the switching line on one side and move away

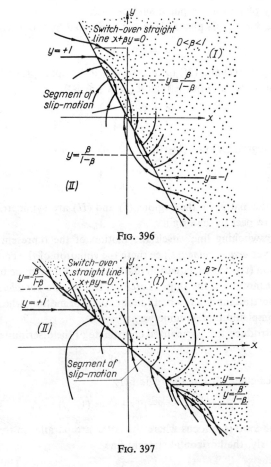

Fig. 396

Fig. 397

from it on the other (Fig. 396). The same picture is obtained for $\beta < 1$ (Fig. 397). Now, in region (*I*) the paths approach the switching line ($\dot{\xi} > 0$) if $y > \beta/(1-\beta)$ and move away from it ($\dot{\xi} < 0$) if $y < \beta(1-\beta)$.

This behaviour of phase paths near the switching line, the boundary of the regions (*I*) and (*II*), enables us to propound a definition for motion of the system with the discriminating switch in the off position

(1) if the representative point reaches the switching line outside the segment (8.56), then it intersects it, passing from the region (*I*) into the region (*II*) or vice versa;

(2) if the representative point of the system reaches the switching line within the segment (8.56), then *it will continue its motion along this segment*.

The law of motion of the system in the latter case is obtained from equation (8.54) with $y = \dot{x}$,

$$x + \beta \dot{x} = 0,$$

which gives

$$x = x_0 e^{-\frac{t}{\beta}}. \tag{8.57}$$

This is the so-called *slip-motion* of the two-position automatic pilot [98]. For these conditions the switch is found in the off-position and the position of the rudder varies from one extreme to neutral. The normalized coordinate z of the rudder varies, clearly, as

$$z = \ddot{x} + \dot{x} = x_0 \left(\frac{1}{\beta^2} - \frac{1}{\beta} \right) e^{-\frac{t}{\beta}}. \tag{8.57a}$$

In order to understand the mechanism of the slip-motion it is necessary to take into account certain (generally speaking, second-order) factors: for example, lags in the automatic pilot and the inertia of the steering engine which are always present in real apparatus. These lead to a delay in moving the rudder after the instant at which $\xi = 0$ (i. e. $\tau = 0$). These factors cause the slip-motion to be vibratory, the higher the frequency the smaller the switching time lags with oscillations of the rudder about an average position given by (8.57)[†]. The idealized slip-motion obtained in the simplified analysis is only the limiting case when lags and time delays are taken to be zero.

In the slip-motion, the deviation φ is aperiodically damped according to (8.57): the smaller β the greater the damping. However, as β decreases, the region of slip-motion is reduced and outside this region the oscillations of φ are now under-damped. Therefore, both too small and too large values of the parameter β, characterizing the velocity correction, lead to a slow approach to the true course.

† See the following section.

3. The point transformation

We shall consider now the behaviour of the phase paths outside the segment of slip-motion by reducing the problem to a point transformation of a straight line into a straight line. Let $0 < \beta < 1$[†], and consider a phase path intersecting the switching line (8.54) at a point $S_0(-\beta s_0, s_0)$ and entering (at $t=0$) the region (I). It is evident that $s_0 > \beta/(1-\beta)$. Integrating (8.55) we obtain the general solution in the region (I)

$$\left.\begin{array}{l} y = -1+(y_0+1)e^{-t}, \\ x = x_0-t+(y_0+1)(1-e^{-t}), \\ \xi = x+\beta y = \xi_0-t+(1-\beta)(y_0+1)(1-e^{-t}) \end{array}\right\} \quad (8.58)$$

(x_0, y_0, ξ_0) are the values of x, y, ξ for $t=0$). For the path considered $x_0 = -\beta s_0$ and $\xi_0 = 0$; then for a certain $t = \tau > 0$, uniquely determined by the equation

$$(1-\beta)(y_0+1)(1-e^{-\tau}) - \tau = 0, \quad (8.59)$$

the switch coordinate ξ reduces to zero and the representative point returns to the switching line at the point $S_1(\beta s_1, -s_1)$, where

$$s_1 = 1 - (s_0+1)e^{-\tau}. \quad (8.59\text{a})$$

Two cases can occur here. If $s_1 \leqslant \beta/(1-\beta)$ then the phase path arrives on the segment of slip-motion and the subsequent motion obeys (8.57). If, however, $s_1 > \beta/(1-\beta)$, then the path passes into the region (II) and again reaches the switching line at a point $S_2(-\beta s_2, s_2)$[‡] etc.

We shall thus obtain a sequence of points of intersection with the switching line.

It is easily seen that, owing to the symmetry of the phase paths in the regions (I) and (II) with respect to the origin, each consecutive point S_{k+1} is obtained from the preceding one S_k by the same transformation as transforms the point S_0 into the point S_1, so the transformation has the sequence function

$$\left.\begin{array}{l} s_0 = -1 + \dfrac{\tau}{(1-\beta)(1-e^{-\tau})}, \\ s_1 = +1 - \dfrac{\tau}{(1-\beta)(e^\tau-1)}. \end{array}\right\} \quad (8.60)$$

[†] The case $\beta > 1$ is of no particular interest owing to the low damping of Φ during the slip-motion. In this case, as is easily seen from Fig. 397, the slip-motion occurs at the second switching point; if not at the first.

[‡] It is easily seen that the case $s_1 < -\beta/(1-\beta)$ is impossible.

However, the point S_k has the consecutive one S_{k+1} only on condition that $s_k > \beta(1-\beta)$; otherwise the point S_k has no consecutive point on the half line (S), since the corresponding $|y_k|$ is smaller than $\beta/(1-\beta)$ and the phase path arrives on the segment of slip-motion.

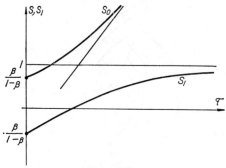

Fig. 398

Graphs of the functions (8.60) are shown in Fig. 398. For $\tau=0$, $s_0 = \beta/(1-\beta)$ and $s_1 = -\beta/(1-\beta)$, and s_0 and s_1 increase monotonically towards $+\infty$ and $+1$ respectively as $\tau \to \infty$†. These curves do not intersect each other and the transformation (8.60) has no fixed points, since

$$s_1 - s_0 = -2\left[\frac{\dfrac{\tau}{2}}{(1-\beta)\operatorname{th}\dfrac{\tau}{2}} - 1\right] < 0,$$

for $\beta < 1$. Thus each consecutive point of intersection of the given phase path with the switching line lies nearer the origin than the preceding one, and after a finite number of oscillations conditions of slip-motion are reached.

The number of oscillations of the ship, up to the onset of slip-motion, clearly depends on the initial conditions and on the value of the parameter β. If $+\dfrac{1}{2} \leqslant \beta < 1$, then for any τ (i.e. for any s_0) $s_1 < 1 < \beta/(1-\beta)$ and after

† In fact, for $\tau > 0$.
$$\frac{ds_0}{d\tau} = \frac{e^{-\tau}[e^\tau - (\tau+1)]}{(1-\beta)(1-e^{-\tau})^2} > 0 \quad \text{and} \quad \frac{ds_1}{d\tau} = \frac{e^\tau]\tau - (1-e^{-\tau})]}{(1-\beta)(e^\tau-1)^2} > 0,$$

not more than *one* intersection of the switching line the phase path reaches the segment of slip-motion. The same action also takes place for $\beta > 1$. Now an increase of the time constant β decreases the rate of approach to

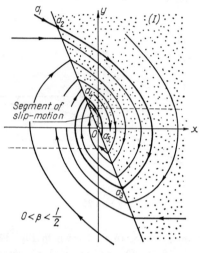

Fig. 399

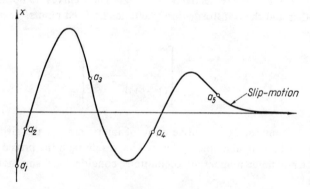

Fig. 400

the true course in the slip-motion so that normally it is inexpedient to use velocity correction with $\beta > \dfrac{1}{2}$. Fig. 399 is the phase portrait for the case $\beta < \dfrac{1}{2}$ and Fig. 400 is the oscillogram of the deviation φ of the course of a ship corresponding to the phase path $a_1 a_2 a_3 a_4 a_5 0$ in Fig. 399. Whatever the initial conditions slip-motion is reached ultimately after which the true course is approached aperiodically. For $\beta=0$ there is no slip-motion and the variation of the course of the ship is oscillatory and comparatively

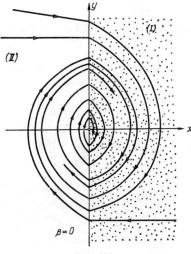

Fig. 401

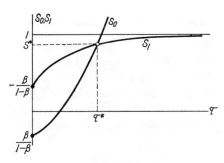

Fig. 402

lightly damped as shown in Fig. 401. For $\beta<0$, i.e. for a reversed connexion of the rate gyroscope so that switching occurs *after* the passage of the ship through $\varphi=0$ a self-oscillatory motion arises.

Lamerey's diagram for the case $\beta<0$ is shown in Fig. 402. The point transformation (8.60) has for $\beta<0$ a unique and stable fixed point which corresponds on the phase plane (Fig. 403) to a stable and symmetrical limit cycle. The half-period of the oscillations is determined from the equations (8.60), if we put in them $s_1 = s_0$ or from the equation

$$1 - \beta = \frac{\tau^*}{2} \operatorname{cth} \frac{\tau^*}{2}.$$

It is easily shown that the amplitude and the period of the self-oscillations tend to zero as $\beta \to -0$.

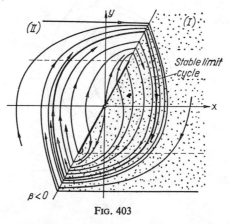

Fig. 403

4. Automatic pilot with parallel feedback

We shall now briefly consider a second method of achieving anticipatory switching of the rudder which is also applied to automatic regulators and consists in the introduction of a *parallel*† *feedback* path. The schematic diagram of such a modified steering controller and the block diagram of the system are shown in Fig. 404. In the presence of parallel feedback the steering engine is controlled by an electric switch, the coordinate of which is

$$\sigma^* = \varphi + B\psi$$

(ψ is the angle of rotation of the rudder about a vertical axis of the ship); for $\sigma^* > 0$ the rudder is hard over on port side ($\psi = -\psi_0$), for $\sigma^* < 0$ on the starboard side ($\psi = +\psi_0$). For $\sigma^* = 0$ the switch is in the neutral position (the steering engine is disconnected and the rudder can assume any position $-\psi_0 \leqslant \psi \leqslant +\psi_0$‡).

† The Russian authors call it "hard" feedback.
‡ It is easily seen that an anticipatory action is obtained for $B > 0$ only.

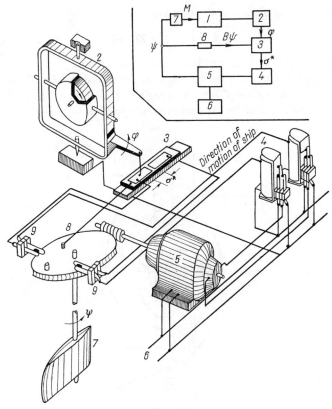

Fig. 404

Neglecting, as before, the time of action of the rudder and assuming the moment M to be related to the rudder angle ψ by

$$M = \frac{M_0}{\psi_0} \psi,$$

the equation of automatic pilot is now†

$$M = +M_0 Z\left(\varphi + B\psi_0 \frac{M}{M_0}\right).$$

Then, in terms of the new variables

$$x = \frac{H^2}{M_0 I}\varphi, \quad t_{\text{new}} = \frac{H}{I} t_{\text{old}} \quad \text{and} \quad z = \frac{M}{M_0} = \frac{\psi}{\psi_0}$$

† For the definition of the function $Z(\sigma^*)$ see Sub-section 1 of this section.

the equations of motion of the ship are

$$\ddot{x}+\dot{x} = z, \quad z = Z(x+\beta^*z), \tag{8.61}$$

where $\beta^* = B\psi_0$ is the normalized feedback coefficient.

The x, y phase plane of the system, where $y=\dot{x}$, is divided by the straight lines $x=+\beta^*$ and $x=-\beta^*$ into three regions: (I), (II) and (III) (Fig. 405). In (I) the rudder is in the extreme port-side position so $z=-1$ and hence

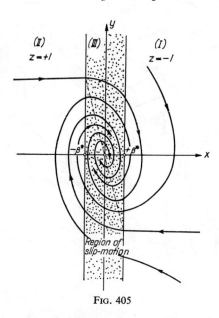

Fig. 405

$\xi^* = x-\beta^* \geqslant 0$. In this region the equations of motion will be the equations (8.55). The region (II) is where $z=+1$ (the rudder is put over in the extreme starboard-side position) and $\xi^* = x +\beta^* \leqslant 0$. In the region (III) where $-\beta^*<x<+\beta^*$, z cannot be equal to $+1$ or to -1 (the rudder cannot assume either extreme position); and therefore, $\xi^* = x+\beta^*z \equiv 0$, i.e. the switch is found in the neutral position, but the rudder may assume positions anywhere between its extremes.

$$z = -\frac{x}{\beta^*}. \tag{8.62}$$

Again there is a slip-motion, which in the presence of a parallel feedback occurs in the whole region $-\beta^*<x<+\beta^*$ of the phase plane. In contrast to the case with velocity correction, all rudder movements cause the ship to enter or emerge from the region of slip-motion.

In a real two-position automatic pilot with parallel feedback the slip motion consists of frequent on-and-off switchings of the steering engine causing small rudder movements determined by (8.62). The frequency of the on-off switchings of the steering engine is determined by the time lags and delays in the apparatus and by the inertia of the steering engine itself, i.e. by the factors we have neglected in this section. This frequency is

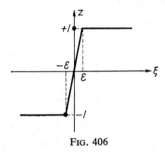

FIG. 406

larger the smaller the time lags. Our idealization of the slip-motion, as expressed in the additional definition (8.62) is a limiting case. Note that the additional definition (8.62) for the slip-motion can be obtained from the continuous characteristic shown in Fig. 406. For such a characteristic, $|\xi^*| < \varepsilon$, where $\xi^* = x + \beta^* z$

$$z = \frac{\xi^*}{\varepsilon} = \frac{1}{\varepsilon}(x + \beta^* z), \quad \text{i.e.} \quad z = -\frac{x}{\beta^* - \varepsilon},$$

whence, in the limit as $\varepsilon \to 0$, we obtain (8.62).

The variations of φ during the slip-motion in the region (*III*) are described from the first equation (8.61) and the equation (8.62), by a linear equation

$$\ddot{x} + \dot{x} + \frac{x}{\beta^*} = 0. \tag{8.62a}$$

These variations of the course will always be damped, the damping being oscillatory for $\beta^* < 4$ and aperiodic for $\beta^* > 4$.

Fig. 405 shows a typical phase portrait of the system: ship + two-position controller with parallel feedback. It can be shown by finding the usual transformations and correspondence functions that all paths tend to the stable state of equilibrium $x = 0$ as $t \to \infty$. Whatever the initial conditions, the ship will reach the true course, the last stage of the process being accomplished under conditions of slip-motion.

5. Other automatic controlling systems

In concluding the section we observe that the dynamic systems considered here are types which represent many other automatic regulating and controlling systems.

As an example, consider the system for the speed regulation of a prime mover using a "floating" or "constant-velocity" servo-motor without "dead zone", and with an ideal detecting element. The schematic diagram

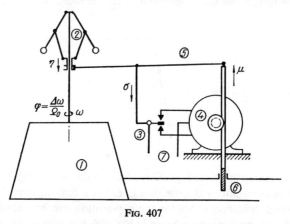

FIG. 407

of such an automatic speed regulator is shown in Fig. 407. Its equations of motion for variations in angular velocity about the state of equilibrium $\omega = \Omega$ can be written in the following form[†]

$$T_a \frac{d\varphi}{dt} + k\varphi = \mu;$$

The linearized equation of the speed-serving element or governor is

$$\delta\eta + \varphi = 0.$$

The feedback equation is

$$\sigma = \eta - \gamma\mu$$

and the equation of the servo-motor which moves either one way or the other is

$$\frac{d\mu}{dt} = \begin{cases} +\dfrac{1}{T_s} & \text{for} \quad \sigma > 0, \\ -\dfrac{1}{T_s} & \text{for} \quad \sigma < 0. \end{cases}$$

[†] See, for example, [120, 99, 1].

The meaning of φ, η, τ and μ should be obvious from the diagram. μ is the relative displacement of the valve controlling the flow of fuel or electricity into the motor. T_a and k are the so-called time constant and sensitivity of the motor, δ is the sensitivity of the detecting element, γ is the feedback coupling coefficient and T_s is the time constant of the servo-motor. From these equations we obtain

$$T_a T_s \frac{d^2\varphi}{dt^2} + kT_s \frac{d\varphi}{dt} = \begin{cases} +1 & \text{for } \sigma > 0, \\ -1 & \text{for } \sigma < 0 \end{cases}$$

$$\sigma = -\left(\frac{1}{\delta} + \gamma k\right)\varphi - \gamma T_a \frac{d\varphi}{dt}.$$

If now we introduce the new variables x and ξ and the new dimensionless time t_{new} connected with the old variables by the relations

$$t = \frac{T_a}{k} t_{\text{new}} \quad \varphi = \frac{T_a}{k^2 T_s} x \quad \text{and} \quad \sigma = -\frac{T_a\left(\frac{1}{\delta} + \gamma k\right)}{k^2 T_s} \xi,$$

then the equations of the system are reduced to the form

$$\ddot{x} + \dot{x} = \begin{cases} -1 & \text{for } \xi > 0, \\ +1 & \text{for } \xi < 0, \end{cases}$$

where

$$\xi = x + \beta \dot{x} \quad \text{and} \quad \beta = \frac{\gamma k}{\frac{1}{\delta} + \gamma k} \; †$$

i.e. to the system of equations (8.53).

§ 7. Two-position automatic pilot with delay

As already said, to clarify the mechanism of slip-motion of a two-position feedback controller it is necessary to take into account some factors that act in a real apparatus and lead to a time delay between the final positioning of the rudder and the previous reduction to zero of the control variable (switching signal) σ. We shall consider in this section the dynamics of a ship with automatic steering when two factors are approximately (and separately) taken into account. We restrict ourselves to the case of a two-position controller with velocity correction‡.

† Clearly, for this system, $0 < \beta < 1$.
‡ The analysis of the mechanism of slip-motion for a two-position controller with parallel feedback requires a study of dynamic systems with $\frac{1}{2}$ a degree of freedom; the consideration of such systems is not the object of this book.

As a first cause of delay in positioning the rudder, we shall consider the so-called "spatial delay" of the contactor switch. In this case the rudder moves instantaneously not when $\sigma=0$ but for $\sigma=\pm\sigma_0$. If σ increases to σ_0 the rudder is put into the position $\psi=-\psi_0$, and if σ decreases to $\sigma=-\sigma_0$ the rudder is put over into the position $\psi=+\psi_0$. This hysteresis-type characteristic is shown in Fig. 408. Such a "spatial" delay can be caused, for example, by the presence of backlash in the levers connecting the rate

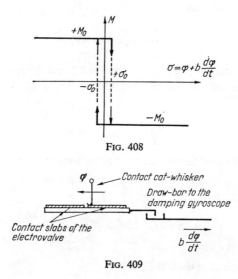

Fig. 408

Fig. 409

gyroscope with the contacts of the switch (see Fig. 409). Such a formulation of the problem leads to consideration of a dynamic system with a two-sheet phase surface.

Another cause of delay in rudder action is the finite velocity of the position of the rudder when being moved by the steering engine. Immediately after the coordinate σ reduces to zero, rudder motion starts with a certain constant velocity $\dot\psi = -A\,\text{sgn}\,\sigma$ ($A=$const) until the rudder reaches the extreme position $\psi=\pm\psi_0$ or until σ reduces to zero again. Such a dynamic system will have $1\frac{1}{2}$ degrees of freedom (the phase space is three-dimensional) and its investigation does not reduce to that of a point transformation of a straight line into a straight line.

We shall replace, therefore, this fairly accurate assumption by the simplifying assumption that rudder positioning is instantaneous but happens after an interval of time Δ following the instant at which $\sigma=0$. We obtain a piece-wise linear system with *time delay*† the motion of which is described by a difference–differential equation of the second order; the investigation of the dynamics of such a system (for a certain basic class of motions) can be reduced to the investigation of a point transformation of a straight line into itself.

1. *Ship's Automatic Pilot with "spatial" delay*

The switching characteristic of such a controller has been shown in Fig. 408; the controller putting over the rudder instantaneously from one extreme position into another at $\sigma = +\sigma_0$ when σ is increasing and at $\sigma = -\sigma_0$

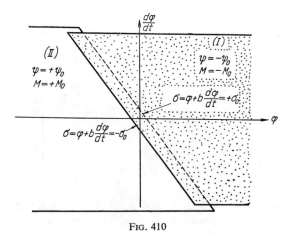

Fig. 410

when σ is decreasing. For $-\sigma_0 < \sigma < +\sigma_0$ the position of the rudder is no longer determined uniquely by σ for it can be either $\psi = +\psi_0$ or $\psi = -\psi_0$, but is determined by the *preceding states* of the system. The rudder remains in the same position that it occupied at the preceding instant of time. Obviously, we cannot take the φ, $d\varphi/dt$ plane as the phase surface of the system. The phase surface will be a *two-sheet* surface consisting of the sheets (*I*) and (*II*), representing respectively the set of the states of the system with the rudder in the port and starboard-side positions, and overlapping each other over the "non-unique zone": $|\sigma| < \sigma_0$ (Fig. 410). We must also assume that the passage of the representative point from sheet (*I*) to the sheet (*II*)

---

† Such a delay by a certain constant interval of time is usually called *time* delay, in contrast with *space* delay when the delay in the working of the relay system is determined, not by a fixed interval of time, but by a displacement of the representative point in the phase space by a certain constant distance (a variation of $|\sigma|$ by σ_0 in our problem).

occurs only on the boundary of the sheet (*I*), and the converse passage only on the boundary of the sheet (*II*)[†].

We shall introduce, as in the preceding section, the new variables x, ξ and z and the new dimensionless time t_{new}, connected with the old variables by the relations

$$\varphi = \frac{M_0 I}{H^2} x, \quad \sigma = \frac{M_0 I}{H^2} \xi, \quad M = +M_0 z \quad \text{and} \quad t = \frac{I}{H} t_{\text{new}}.$$

Then the equations of motion of the dynamic system are reduced to a form, similar to (8.53),

$$\left.\begin{aligned}
&\dot{x} = y, \\
&\dot{y} = -y + z, \\
&z = Z^*(\xi) = \begin{cases} -1 & \text{for } \xi > -\dfrac{\alpha}{2}, \\ +1 & \text{for } \xi < +\dfrac{\alpha}{2} \end{cases} \\
&\text{and} \\
&\xi = x + \beta y, \quad \alpha = \frac{2H^2}{M_0 I}\sigma_0, \quad \beta = \frac{bH}{I}\ddagger.
\end{aligned}\right\} \quad (8.63)$$

On the sheet (*I*) corresponding to the states of the system with the rudder in the port-side extreme position, $\psi = -\psi_0$, and represented by the half-plane

$$\xi = x + \beta y > -\frac{\alpha}{2},$$

then $M = M_0$, i.e. $z = -1$ and the equations of motion will be in the form (8.55) and their solutions will be the relations (8.58). The phase portrait on this sheet is shown in Fig. 411, for the case $0 < \beta < \dfrac{1}{2}$. As is easily seen, there are no states of equilibrium on the sheet (*I*) and all paths arrive at its boundary

$$\xi = x + \beta y = -\frac{\alpha}{2}$$

---

† See also Section 6, Chapter III, where two dynamic systems with two-sheet phase surfaces have been considered.

‡ For $|\xi| < \alpha/2$, $z = +1$ or -1 according to the value it had at the preceding instants of time.

and then proceed on to sheet (*II*). Note that the paths on (*II*) are symmetrical with the paths on (*I*) with respect to the origin. This symmetry is a direct consequence of the invariance of the equations (8.63) with respect to a change of the variables x, y into $-x, -y$.

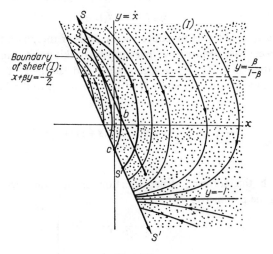

Fig. 411

Let us draw on sheet (*I*) (Fig. 411) the half line S'

$$x+\beta y = -\frac{\alpha}{2}, \quad \xi = (1-\beta)y-\beta \leqslant 0,$$

from which the representative point passes on to the sheet (*II*), and the half line S

$$x+\beta y = +\frac{\alpha}{2}, \quad (1-\beta)y+\beta \geqslant 0$$

from which paths leave the sheet (*II*) for the sheet (*I*). Choose the coordinates: $s=y$ on the half line S and $s=-y$ on the half line S' (symmetrical points on these half lines correspond to the same values of the coordinate s) and consider the point transformation Π of the half line S into the half line S'

$$s' = \Pi(s),$$

generated by the paths on the sheet (I)†. The points of the half line S' are transformed in their turn into points of the half line S by paths on the sheet (II) by virtue of the symmetry paths on the sheets (I) and (II), this transformation will be identical with Π. Thus this transformation Π determines a consecutive value s_{k+1} from the preceding one s_k

$$s_{k+1} = \Pi(s_k),$$

which, clearly, enables us to reduce the analysis of the phase portraits on the two-sheet phase surface to an investigation of this point transformation.

Substituting in (8.58) $\xi_0 = \alpha/2$ and $y_0 = s$ and letting τ be the time of transit along a path on the sheet (I) from the line S to the line S' where $\xi = -\alpha/2$ and $y = -s'$, we obtain the relations

$$-\frac{\alpha}{2} = +\frac{\alpha}{2} - \tau + (1-\beta)(s+1)(1-e^{-\tau}),$$

$$-s' = -1 + (s+1)e^{-\tau},$$

from which the correspondence function of the transformation Π is derived in the parametric form

$$\left.\begin{aligned} s &= -1 + \frac{\tau - \alpha}{(1-\beta)(1-e^{-\tau})}, \\ s' &= +1 - \frac{\tau - \alpha}{(1-\beta)(e^\tau - 1)}. \end{aligned}\right\} \quad (8.64)$$

To construct Lamerey's diagram we shall introduce the auxiliary functions

$$\left.\begin{aligned} \Psi_1(\tau) &= \frac{\tau - \alpha}{1 - e^{-\tau}}, \\ \Psi_2(\tau) &= \frac{\tau - \alpha}{e^\tau - 1} = \Psi_1(\tau)e^{-\tau}. \end{aligned}\right\} \quad (8.65)$$

The graphs of these functions (for $\tau > 0$) are shown in Fig. 412. The first of them is a monotonically increasing function, while the second has a maximum for the value $\tau = \tau_1$ determined by the condition

$$\Psi_1(\tau_1) = 1;$$

† It is evident that every point s of the half line S has a consecutive point s' on the half line S'. This transformation will be single-valued and continuous; the transformation Π, however, is not such that its reciprocal is single-valued, since the half line S is not a half line without contact (at the point $s = y = \beta/(1-\beta)$, $\dot{\xi} = 0$) and so phase paths are tangents. For example, the points a and b of the half line S in Fig. 411, belonging to one and the same paths, are transformed by the transformation Π into the same point c of the half line S'.

Evidently

$$s = -1 + \frac{\Psi_1(\tau)}{1-\beta}, \quad s' = +1 - \frac{\Psi_2(\tau)}{1-\beta},$$
$$\frac{ds}{d\tau} = \frac{\Psi_1'(\tau)}{1-\beta}, \quad \frac{ds'}{d\tau} = -\frac{\Psi_2'(\tau)}{1-\beta} \quad \text{and} \quad \frac{ds'}{ds} = -\frac{\Psi_2'(\tau)}{\Psi_1'(\tau)}, \qquad (8.64a)$$

the initial point $s = s_0 = -\beta/(1-\beta)$ of the half line S corresponding to the value $\tau = \tau_0$ determined by the equation†

$$\Psi_1(\tau_0) = 1 - 2\beta.$$

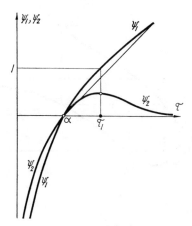

FIG. 412

Let $0 < \beta < 1$. In this case, on the half line S, $s = y \geq s_0 = -\beta/(1-\beta)$, therefore the points of this half line correspond to values of the parameter of the transformation $\tau \geq \tau_0$ (for $\beta < 1$, $ds/d\tau > 0$). Using the graphs of the functions Ψ_1 and Ψ_2 and the relations (8.64a) it is easy to construct Lamerey's diagram; this is shown in Fig. 413 for the case $0 < \beta < 1$. It will be seen that the curves (8.64) have a unique point of intersection, and hence the transformation Π has a *unique fixed point* ($s = s' = s^*$, $\tau = \tau^*$). This follows immediately from the fact that the difference

$$s - s' = -2 + \frac{1}{1-\beta}[\Psi_1(\tau) + \Psi_2(\tau)] = -2 + \frac{\tau - \alpha}{1-\beta}\operatorname{cth}\frac{\tau}{2}$$

† τ_0 and τ_1 are determined unequivocally by the equations above owing to the fact that the function $\Psi_1(\tau)$ is monotonic. It is evident that $\tau_0 < \tau_1$.

is a continuous and monotonically increasing function of τ, since $\beta < 1$ and for α and τ are both positive. Also

$$\frac{d}{d\tau}\left[\Psi_1(\tau)+\Psi_2(\tau)\right] = \frac{d}{d\tau}\left[(\tau-\alpha)\operatorname{cth}\frac{\tau}{2}\right] = \frac{\alpha+\operatorname{sh}\tau-\tau}{2\operatorname{sh}\frac{\tau}{2}} > 0, \quad (8.65a)$$

so this difference tends to $+\infty$ when $\tau \to +\infty$, while, for $\tau = \tau_0$, it is equal to

$$s_0 - s_0' = -2 + \frac{1}{1-\beta}(1-2\beta)(1+e^{-\tau_0}) = -\frac{1-e^{-\tau_0}+2\beta e^{-\tau_0}}{1-\beta} < 0.$$

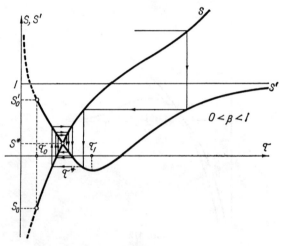

Fig. 413

The value of the parameter $\tau = \tau^*$ at the fixed point of the transformation Π is uniquely determined by the equation

$$-2 + \frac{\tau^*-\alpha}{1-\beta}\operatorname{cth}\frac{\tau^*}{2} = 0$$

or

$$\tau^* - 2(1-\beta)\operatorname{th}\frac{\tau^*}{2} = \alpha, \qquad (8.66)$$

and the coordinate s^* of the fixed point is determined by the relation

$$s^* = -1 + \frac{\tau^*-\alpha}{(1-\beta)(1-e^{-\tau^*})}$$

or, since $(\tau^* - \alpha)/(1-\beta) = 2\tanh\tau^*/2 = 2(1-e^{-\tau^*})/(1+e^{-\tau^*})$

$$s^* = -1 + \frac{2}{1+e^{-\tau^*}} = \frac{1-e^{-\tau^*}}{1+e^{-\tau^*}} = \tanh\frac{\tau^*}{2}. \qquad (8.66a)$$

Note that for small values of α (for a small width of the "non-unique zone" of the characteristic of the controller) τ^* and s^* are also small; and neglecting terms of the order of α^3,

$$\tau^* = \frac{\alpha}{\beta} \quad \text{and} \quad s^* = \frac{\alpha}{2\beta}. \qquad (8.66b)$$

This fixed point $(s=s'=s^*, \tau=\tau^*)$ is *stable* since the condition of its stability

$$\left|\frac{ds'}{d\tau}\right| < \left|\frac{ds}{d\tau}\right| \quad \text{for} \quad \tau = \tau^*$$

reduces, by virtue of (8.64a), to the inequality

$$\frac{d\Psi_1}{d\tau} > \left|\frac{d\Psi_2}{d\tau}\right| \quad \text{for} \quad \tau = \tau^*,$$

which is always satisfied. If, however, $\tau^* < \tau_1$, then for $\tau = \tau^*$, $d\Psi_2/d\tau > 0$, and the condition of stability assumes the form

$$\frac{d}{d\tau}(\Psi_1 - \Psi_2) > 0 \quad \text{for} \quad \tau = \tau^*$$

and is also always satisfied, since

$$\Psi_1 - \Psi_2 = \tau - \alpha \quad \text{and} \quad \frac{d}{d\tau}(\Psi_1 - \Psi_2) = 1.$$

Thus, for $0 < \beta < 1$, the point transformation Π has a *unique*, and moreover, *stable fixed point* which, is the limiting point of the sequences

$$s, s_1, s_2, \ldots$$

with arbitrary initial point s.

The same situation exists for $\beta \geq 1$. For $\beta = 1$ the half lines S and S' reduce to the straight lines $x + \beta y = \alpha/2$ and $x + \beta y = -\alpha/2$, and (see (8.58)) the time of transit of the representative point from S to S' does not depend on s and is equal to $\tau = \alpha$. In this case, the sequence function of Π is

$$s' = 1 - (s+1)e^{-\alpha},$$

and the transformation Π itself has a unique and stable fixed point $s^* = \tanh \alpha/2$.

For $\beta > 1$, at the points of the half line S: $s = y \leqslant -\beta/(1-\beta) = s_0$, but $ds/d\tau < 0$ and the points of this half line correspond to values $\tau \geqslant \tau_0$. Lamerey's diagram for this case is shown in Fig. 414. The existence of a fixed

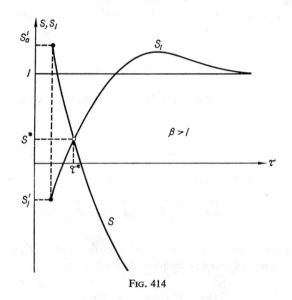

Fig. 414

point follows from the fact that the functions (8.64) are continuous, and from the inequalities $s-s' > 0$ for $\tau = \tau_0$ and $s-s' < 0$ $(s-s' \to -\infty)$ for $\tau \to +\infty$. The fixed point is also unique since $s-s'$ decreases monotonically as τ varies from τ_0 to $+\infty$. It is easily proved, by the method above, that it is a stable fixed point.

Thus, for positive values of the parameter β there is a unique, stable, and symmetric limit cycle, to which all phase paths tend (for $t \to +\infty$). In other words, the ship ultimately oscillates about the true course ($\varphi = 0$ or $x = 0$). The amplitude and the period of these oscillations are the smaller the smaller the delay in the controller and the larger the velocity correction[†].

[†] The equation of the part of the limit cycle that is situated on the sheet (I) follows from (8.58) $(y_0 = s, x_0 = (\alpha)2 - \beta^* s)$,

$$y = -1 + (1+s^*)e^{-t}, \quad x = \frac{\alpha}{2} - \beta s^* - t + (1+s^*)(1-e^{-t}) \quad \text{[continued on next page]}$$

The period of the self-oscillations (in units of the dimensionless time) is equal to $2\tau^*$; and so is equal to $2\alpha/\beta$ for $\alpha \ll \beta$. An oscillogram of the deviation from course φ (or $\varkappa$) corresponding to the path A in Fig. 415, is shown in Fig. 416 (the circles on the curve indicate the instants of switching the rudder from one extreme position to the other).

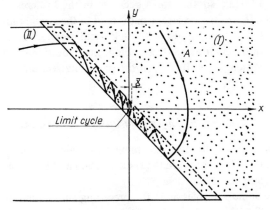

Fig. 415

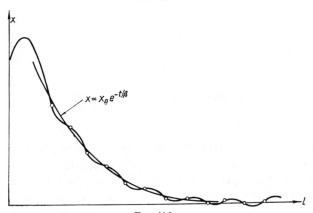

Fig. 416

Denote by $\bar{t}$ the interval of time after which the representative point intersects the x axis. It is evident that

$$e^{\bar{t}} = 1+s^*, \quad \text{i.e.} \quad \bar{t} = \ln(1+s^*),$$

and, hence, the amplitude of the self-oscillations is

$$\bar{x} = \frac{\alpha}{2} + (1-\beta)s^* - \ln(1+s^*).$$

For small values of s^* (for $\alpha \ll \beta$) we have, according to (8.65b): $\bar{x} = s^{*2}/2 = \alpha^2/8\beta^2$.

The paths that arrive at the switching-lines $x+\beta y = \pm \alpha/2$ with $y <$ $(1 = \beta)$ move afterwards between these straight lines. These zigzag paths correspond, to *slip-motion* when the controller, without leaving the "non-unique zone", positions the rudder to one or other extreme (separated by intervals of time Δt of the order of α/β). During the slip-motion $y=\dot{x}=-x/\beta+0\,(\alpha)$, so that for a decrease in the hysteresis zone of the characteristic (for $\alpha \to 0$) the zigzag paths tend to the straight line $x+\beta y=0$; $x \to x_0 e^{-t/\beta}$ and the amplitude of the self-oscillations or chatter tends to zero. Thus, in the limit as $\alpha \to 0$ we obtain the "additional definition" for slip-motion which we postulated in the preceding section.

2. Automatic Ship's Pilot with pure time delay

Results that are qualitatively the same are obtained when the controller has a pure time delay Δ between the rudder action and the reduction to zero of the switching signal σ.

Using the same variables as in the preceding sub-section, the equations of motion of a ship with a two-position controller having a time delay will be

where
$$\left.\begin{aligned} \dot{x} &= y, \\ \dot{y} &= -y+z, \\ z(t) = Z[\xi(t-\theta)] &= \begin{cases} -1 & \text{for} \quad \xi(t-\theta) > 0, \\ +1 & \text{for} \quad \xi(t-\theta) < 0, \end{cases} \end{aligned}\right\} \quad (8.67)$$

$\xi = x+\beta$, y and $\theta = (H/I)\Delta$ is the normalized delay. Now, in contrast to the dynamic systems considered earlier, the equations of motion are no longer differential equations but *finite-difference-differential* equations. The velocity $\dot{x}$ and the acceleration $\dot{y}$ at time t are determined not only by y at this instant, but also by the value of ξ at an earlier instant of time $t'=t-\theta$.

As a consequence of this the motion for $t>t^*$ is not uniquely determined by the values of x and y at time t^*, but requires for its complete determination that the function $x(t)$ be given over the interval of time $t^*-\theta \leqslant t \leqslant t^*$. In other words, the state of the system at an arbitrary time t^* is determined by choosing x and y at the instant of time t^* and the piece-wise constant function $Z[\xi(t)]$ for $t^*-\theta \leqslant t \leqslant t^*$; correspondingly the phase space of the system will be not an ordinary surface but a *functional space*[†].

In order to simplify the problem and reduce it to the investigation of a point transformation of a straight line into a straight line, we shall restrict ourselves below to considering only a certain particular class of motions of the system, which can be represented by paths on a certain two-sheet surface which is part of the functional phase space. Denote by K_0 the set of the states at arbitrary instants of time t^* that satisfy the condition that for $t^*-\theta < t \leq t^*$ the coordinate ξ will not reduce to zero. We only consider, below, the motions of the system that begin from these states. States of the type K_0 are determined uniquely by the values of x and y at some instant of time, and, therefore, we shall represent them by the (x, y) points on the x, y plane from which we exclude the straight line $\xi = x + \beta y = 0$[‡]. As a particular case, consider the points (x_0, y_0) representing initial states of the type K_0 at $t=0$ which lie on the half-plane K_0': $x+\beta y > 0$, so that the initial values of the coordinate ξ are $\xi_0 = \xi(0) = x_0 + \beta y_0 > 0$. Then, by definition of the set of states K_0, $\xi > 0$ for $-\theta < t \leq 0$ and hence, at least for $0 < t \leq \theta$, $z = -1$ and the motions of the system will be described by the differential equations (8.55)

$$\dot{x} = y,$$
$$\dot{y} = -y-1.$$

These equations remain valid until the rudder moves across and z changes sign from -1 into $+1$: if $t_1 = t_1(x_0, y_0)$ is the instant of time at which $\xi = 0$

[†] This functional space can be considered as a space "*with an infinite number of dimensions*", since the function $x(t)$ (or $Z[\xi(t)]$) can be assigned on the segment $t^*-\theta \leq t \leq t^*$ by the infinite enumerable set of coefficients of the expansion of this function as a Fourier series.

[‡] By assigning x and y (outside the straight line $x+\beta y = 0$) for $t = t^*$, we also determine the coordinate $\xi \neq 0$. Now ξ had the same sign for $t^*-\theta < t \leq t^*$ (since the state belongs to the set K_0), thus determining $Z(t+\theta) = Z[\xi(t)]$ for $t^*-\theta < t \leq t^*$ as equal to $+1$ or -1 depending on the sign of ξ. Therefore, by assigning x and y at $t = t^*$, we completely determine the motion of the system for $t > t^*$, and thus determine completely the state of the system of the type K_0 at the instant of time t^*.

If, in the general case, the states of the system are assigned by the values of x and y at the instant t^* and by the Fourier coefficients $\{a_n, b_n\}$ for the function $Z[\xi(t)]$ on the interval $t^*-\theta \leq t \leq t^*$ $Z[\xi(t)] = a_0/2 + \sum_{n=1}^{\infty} a_n \cos(2\pi nt/\theta) + b_n \sin(2\pi nt/\theta)$ for $t^*-\theta < t \leq t^*$, then the set of the states K_0 is the sub-space: $a_n = b_n = 0$ for $n = 1, 2, \ldots$, and $a_0/2 = -\text{sgn}(x+\beta y)$.

To isolate from the total (functional) space this two-dimensional set of states K_0, through which the system passes for motions of a certain type, is only possible because this system is a *relay* system ($\xi(t-\theta)$ is the argument of the piece-wise constant function Z, which is completely determined by the sign of ξ).

for a motion beginning at the state (x_0, y_0), then the equations (8.55) will, clearly, describe the motions of the system for $0 < t < t_1 + \theta$[†]. It is clear too, that in the interval $0 < t < t_1$, the system passes through states belonging to set K_0 as indicated by the motion of the representative point (x, y) along the phase paths (8.58) of the equation (8.55) on the half-plane K_0'. However, the states of the system for $t_1 < t < t_1 + \theta$ no longer belong to this set, since for $t = t_1$, $\xi = 0$. Therefore we shall represent these states by the

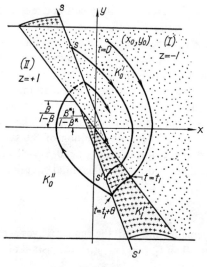

Fig. 417

points of another region K_1' connected to the half-plane K_0' but superimposed on the half-plane $K_0''(x + \beta y < 0)$ as shown in (Fig. 417)[‡]. This additional region K_1' together with the half-plane K_0' form the sheet (I) of the phase surface corresponding to the set of the states through which the system passes for motions that begin from states of the type K_0 (with $\xi_0 > 0$) and are determined by the equations (8.55).

For $t = t_1 + \theta$, a time θ after $\xi = 0$, the rudder moves and the rudder coordinate γ becomes $+1$, but the states of the system again belong to the set K_0 (the corresponding representative points lie on the line S' on the

---

[†] It follows from the general solution (8.58) of the equation (8.55) that, for arbitrary values (x_0, y_0), there exists an unique instant of time $t_1 = t_1(x_0, y_0)$ such that, for $t = t_1$, $\xi = 0$ and for $t > t_1$, $\xi < 0$.

[‡] Fig. 417 corresponds to the case $0 < \beta < 1$ and fairly small values of θ.

half-plane K_0''). The differential equations (8.55a) are now valid

$$\dot{x} = y,$$
$$\dot{y} = -y+1,$$

and the system passes through points of the sheet (*II*) of the phase surface. (*II*) is symmetrical to the sheet (*I*) and is formed by the half-plane K_0'' ($x+\beta y<0$) and the additional region K_1''. In due course ξ becomes zero, again its sign changing again from negative to positive, and then after an interval θ the rudder coordinate z changes from $+1$ to -1 on the line S, and the representative point passes on to sheet (*I*). This whole process is now repeated and a spiral type path obtained.

Thus, if the initial state of the system belong to the set K_0, then subsequently the system will only pass through states that belong to the set $K=K_0+K_1'+K_1''$ and form the two-sheet phase surface K shown in Fig. 417. The phase paths on sheets (*I*) and (*II*) will be symmetrical to each other with respect to the origin. Therefore, if we restrict ourselves to only considering motions beginning from states of the type K_0, the investigation reduces to finding a point transformation of the line S into the line S' generated by the paths (8.56) on the sheet (*I*)†.

To determine the boundary line S' of sheet (*I*), we note that ξ reduces to zero, changing its sign from positive to negative, only at points of the half line

$$\xi = x+\beta y = 0, \quad \dot{\xi} = \dot{x}+\beta\dot{y} = (1-\beta)y+\beta < 0,$$

i.e. at points

$$x = -\beta u, \quad y = u,$$

where

$$u < \frac{\beta}{1-\beta} \quad \text{for} \quad \beta < 1 \quad \text{and} \quad u > \frac{\beta}{1-\beta} \quad \text{for} \quad \beta > 1.$$

Since the boundary S' of the sheet (*I*) corresponds to the states at which the system arrives at a time θ after $\xi=0$ then we obtain the equations of this line from the general solution (8.58) if we use as initial points the points $x_0 = -\beta u$, $y_0 = u$ of the half line: $\xi=0$, $\dot{\xi}<0$ and put $t=\theta$

$$y = -1+(u+1)e^{-\theta},$$
$$x = -\beta u-\theta+(u+1)(1-e^{-\theta}),$$

† The coordinates s and s' on the lines S and S' must be chosen in such a manner that symmetrical points of these lines correspond to equal values of s and s'.

The line S' is the half line

$$x + \beta^* y = -\frac{\alpha^*}{2},$$

where

$$\beta^* = \beta e^\theta - (e^\theta - 1), \quad \frac{\alpha^*}{2} = (\beta - 1)(e^\theta - 1) + \theta \qquad (8.68)$$

with the initial point (this corresponds to $u = \beta/(1-\beta)$)

$$y_{\text{init}} = -1 + \frac{1}{1-\beta} e^{-\theta} = \frac{\beta^*}{1-\beta^*}.$$

The half line S is symmetrical to the half line S': its equation will be

$$x + \beta^* y = +\frac{\alpha^*}{2}$$

and its initial point will be the point with $y = -\beta^*/(1-\beta^*)$ (on the half line $S, y > -\beta^*/(1-\beta^*)$ if $\beta^* < 1$, and $y < \beta^*/(\beta^* - 1)$ if $\beta^* > 1$). The form of the two-sheet phase surface is shown in Figs. 418–21 for differing signs of α^* and β^*.†

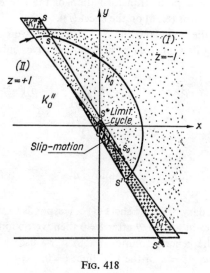

Fig. 418

If $\beta < 1 - e^{-\theta}$ (θ is not too large) or if $\beta > 1$, then $\alpha^* > 0$, $\beta^* > 0$ and the example of automatic steering being considered here has the same dynamics as a ship with a two-position controller possessing a certain equivalent spatial delay α^* and velocity correction β^*.

This case reduces to the one considered in the first sub-section of this section: all motions of the system (beginning from states of the type K_0) lead

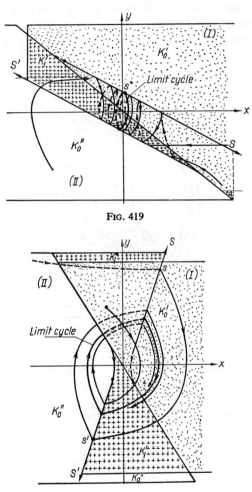

Fig. 419

Fig. 420

to rudder chatter, the amplitude and period being smaller, the smaller the delay θ and the larger the velocity correction β.† For small θ, there are

† For small values of θ, $\beta^* \approx \beta - (\beta-1)\theta \approx \beta$ and $\alpha^*/2 = \beta\theta$, the period of the self-oscillatory chatter is $2\tau^* = 2\alpha^*/\beta^* \approx 4\theta$. The amplitude of these self-oscillations is equal, for small values of θ, to $\bar{x} \approx \alpha^{*2}/8\beta^{*2} = \theta^2/2$ (see footnote at page 526).

zigzag phase paths between the half lines S and S' corresponding to high frequency switching of the steering engine and chatter of the rudder.

For larger values of θ so that $1-\theta/(e^\theta-1) \leqslant \beta < 1-e^{-\theta}$, $\alpha^* \geqslant 0$ but $\beta^* < 0$. There is thus an equivalent spatial delay α^*, but negative equivalent velocity correction. This case does not differ from that considered in the

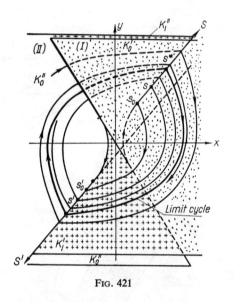

Fig. 421

preceding sub-section having the same Lamerey's diagram (Fig. 413) but with $\tau_0 > \tau_1$, which excludes the existence of slip-motion. All paths lead to a unique and stable limit cycle (Fig. 420).

To conclude consider the last case: $\beta < 1-\theta/(e^\theta-1)$ when both α^* and β^* are negative quantities. The phase surface is shown in Fig. 421. Choose $s = y$ and $s' = -y$ as the coordinates on the half lines S and S' (s and $s' > s_0 = -\beta^*/(1-\beta^*)$). Then, the correspondence function of the point transformation of the half line S into the half line S' generated by paths on the sheet (I), will, as before, have the parametric equations (8.64) with the coefficients α and β replaced by the negative quantities α^* and β^*:

$$\left. \begin{array}{l} s = -1 + \dfrac{\tau - \alpha^*}{(1-\beta^*)(1-e^{-\tau})}, \\[2mm] s' = +1 - \dfrac{\tau - \alpha^*}{(1-\beta^*)(e^\tau - 1)}. \end{array} \right\} \quad (8.69)$$

TWO-POSITION AUTOMATIC PILOT WITH DELAY

In contrast to the case $\alpha > 0$ considered in sub-section 1 of this section, when $\alpha = \alpha^* < 0$ the function $\Psi_1(\tau)$ (see (8.65)) has a minimum value of $e^{\tau_1'}$ at $\tau = \tau_1'$, and so $\Psi_2(\tau)$ is a monotonically decreasing function of τ (Fig. 422).

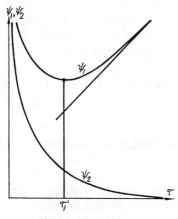

Fig. 422

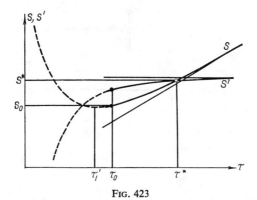

Fig. 423

Now to every value $s > s_0 = -\beta^*/(1-\beta^*)$, according to (8.69), there are two values of the parameter τ, one of which is larger and the other smaller than τ_1'†. In fact, all phase paths leaving points of the half line S intersect the continuation of S' before reaching the half line S' itself. Therefore the time of transit of the representative point along a path from a point s of S to a point s' of S' will be the larger of the two values of τ calculated from (8.69)

† It is easily shown from (8.68), that $s_0 = -\beta^*/(1-\beta^*) > s_{\min} = (s)_\tau = \tau_1'$.

for a given value of s. If τ_0 is the value of τ corresponding to the initial point s_0 on S†, then, since for $\tau > \tau_1'$, $d\psi/d\tau > 0$ and $ds/d\tau > 0$, the points $s > s_0$ on S correspond to values of $\tau_0 > \tau > \tau_1'$. As τ varies from τ_0 to $+\infty$, s increases monotonically from $s_0 = -\beta^*/(1-\beta^*)$ to $+\infty$.

Lamerey's diagram for $\beta < 1 - \theta/(e^\theta - 1)$ is shown in Fig. 413. It is easily shown that the point transformation of the half line S into the half line S' has a unique and stable fixed point‡. There is thus a unique and stable limit cycle, to which all phase paths tend as $t \to +\infty$ (Fig. 421).

§ 8. RELAY OPERATED CONTROL SYSTEMS
(with dead zone backlash and delay)

In Sections 6 and 7 we have considered, using the example of a self-steering ship, an automatic regulating system of the two-position type, in which, for arbitrary initial conditions, self-oscillations are caused by delays in the controller. In this section we shall study the dynamics of a somewhat more complicated relay type regulator or controller of the second order. The controller or relay characteristic is discontinuous and symmetrical, as shown in Fig. 424, [122, 102]††.

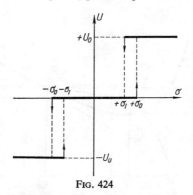

Fig. 424

The output of the relay element U can assume one of three possible values: $+U_0$, 0 or $-U_0$ depending on the value of the input σ and on the state

---

† τ_0 is determined as before by the equation $\psi_1(t_0) = 1 - 2\beta^*$, but now we shall mean by τ_0 the larger root of this equation so that $\tau_0 > \tau_1'$.

‡ The proof of the existence, uniqueness and stability of the fixed point of the point transformation of the half line S into the half line S' is analogous to the proof given in the sub-section 1 of this section.

†† An element of this kind is usually called a relay when its output coordinate can only assume certain discrete values. A system is called a relay system if it consists of relay elements and linear elements.

of the element at the preceding instant of time. If the output U had been zero at the preceding instant of time (only possible for $|\sigma|<\sigma_0$), it would remain zero until the input σ exceeds the limits of the "dead zone": $|\sigma| = \sigma_0$. When $|\sigma|>\sigma_0$, U varies instantaneously jump-wise from 0 to $+U_0$ for $\sigma = +\sigma_0$. The relay element has backlash or hysteresis and does not switch off ($u=0$) unless $|\sigma|$ decreases to σ_1 where $\sigma_1<\sigma_0$. In particular for $-\sigma_0<\sigma<-\sigma_1$ and for $\sigma_1<\sigma<\sigma_0$ the output U is determined exclusively by its preceding state, i.e. by whether the element had been "on" or "off" at the preceding instant of time. Such an element is usually characterized by three parameters: the maximum absolute value of the output U_0, by the width of the dead zone σ_0, and by the so-called backlash coefficient of the relay $\lambda=\sigma_1/\sigma_0$ ($-1\leqslant \lambda \leqslant 1$).

Thus the relay element has both spatial delay (i.e. a "hysteresis" characteristic with two non-unique zones) and a *dead zone*. This is a good model of ordinary electromagnetic relays (or relay groups), of various electrical contact devices, and also "constant velocity" servo-motors when the dead zone and the backlash are taken into account.

Below we shall see that self-oscillations are possible in the system, and how they can be eliminated by the introduction of sufficiently large parallel feedback or velocity correction.

1. *The equations of motion of certain relay systems*

Consider a position servo the block diagram of which is shown in Fig. 425. The servo-motor is controlled by a relay or contactor with the characteristic considered earlier (Fig. 424). The armature voltage U is the

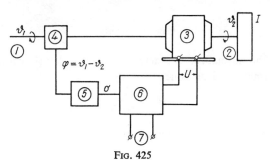

Fig. 425

output of this relay controlled in its turn by the position error $\varphi = \vartheta_1 - \vartheta_2$ and its derivative, so that the input to the relay element is $\sigma = \varphi + B \, d\varphi/dt$, I is the moment of inertia at the output shaft, $k d\vartheta_2/dt$ the counter-electromotive force in the armature of the motor, i the armature current, Ai the motor torque and R the resistance of the armature circuit[†]. Then the equations of motion of the output shaft, and of the current in the armature circuit, will be

$$I\frac{d^2\vartheta_2}{dt^2} = Ai, \quad Ri = U - k\frac{d\vartheta_2}{dt}$$

or

$$I\frac{d^2\vartheta_2}{dt^2} + \frac{Ak}{R}\frac{d\vartheta_2}{dt} = \frac{A}{R}U$$

where friction is neglected.

Let us consider a particular case of motion when the input shaft is stationary (so $\theta_1 = $ const). The equations of motion are then

$$I\frac{d^2\varphi}{dt^2} + \frac{Ak}{R}\frac{d\varphi}{dt} = -\frac{A}{R}U,$$

$$\sigma = \varphi + B\frac{d\varphi}{dt},$$

or in dimensionless variables

$$\left.\begin{array}{r}\ddot{x} + \dot{x} = -u(\xi), \\ \xi = x + \beta\dot{x}.\end{array}\right\} \qquad (8.70)$$

where

$$x = \frac{Ak^2}{IRU_0}\varphi, \quad \xi = \frac{Ak^2}{IRU_0}\sigma, \quad u = \frac{U}{U_0}.$$

A dot denotes differentiation with respect to the dimensionless time

$$t_{\text{new}} = \frac{Ak}{IR}t_{\text{crit}},$$

$u = u(k)$ is the normalized characteristic of the relay shown in Fig. 426 with the normalized dead zone

$$\varepsilon = \frac{Ak^2\sigma_0}{IRU_0}$$

and $\beta = (Ak/IR)B$ is the normalized coefficient of the velocity-error control signal.

[†] We assume that, when the relay group is in the neutral position the voltage across the motor armature $U=0$ but the armature of the motor is connected in series with a "dynamic braking" resistance chosen so that the total resistance in the armature circuit is the same for all three positions of the relay.

The equations (8.70) follow from the equations of a regulating system with a "constant-velocity" servo-motor, with parallel feedback (Fig. 427; see also Fig. 407). If one takes into account the backlash and dead zone

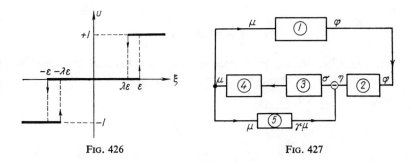

FIG. 426   FIG. 427

in the contactor of the "constant-velocity" servo-motor, then the equations of the latter will be

$$\frac{d\mu}{dt} = \begin{cases} +\dfrac{1}{T_s} & \text{for} \quad \sigma > \lambda\sigma_0, \\ 0 & \text{for} \quad |\sigma| < \sigma_0, \\ -\dfrac{1}{T_s} & \text{for} \quad \sigma < -\lambda\sigma_0, \end{cases}$$

where σ_0 and λ are the width of the dead zone and the backlash coefficient of the relay element, i.e. of the contactor of the servo-motor (the characteristic of such a servo-motor coincides with the one shown in Fig. 424, if we replace U by $d\mu/dt$ and U_0 by T_s^{-1})†. The equations of the remaining linear elements will be written in the same manner as in Sub-section 5 of Section 6 of this chapter:

$$T_a \frac{d\varphi}{dt} + k\varphi = \mu,$$

$$\delta\eta + \varphi = 0,$$

$$\sigma = \eta - \gamma\mu.$$

† If we denote by l and Δ the gap between the contacts of the contactor (Fig. 407) and the backlash in the linkage, measured in the same relative units as σ, then

$$\sigma_0 = \frac{l+\Delta}{2} \quad \text{and} \quad \lambda = \frac{l-\Delta}{l+\Delta}.$$

These equations are reduced by the change of variables

$$\varphi = \frac{T_a}{k^2 T_s} x, \quad \sigma = -\frac{T_a\left(\frac{1}{\delta}+\gamma k\right)}{k^2 T_s} \xi, \quad T_s \frac{d\mu}{dt} = -u$$

and

$$t_{\text{old}} = \frac{T_a}{k} t_{\text{new}}$$

to the equations (8.70), the dimensionless width of the dead zone and the feedback coefficient being equal respectively to

$$\varepsilon = \frac{k^2 T_s \sigma_0}{T_a\left(\frac{1}{\delta}+\gamma k\right)} \quad \text{and} \quad \beta = \frac{\gamma k}{\frac{1}{\delta}+\gamma k} \quad (0 < \beta < 1).$$

2. *The phase surface*

As usual we shall use the variables $y = \dot{x}$ and ξ. In terms of these variables the equations of motion are

$$\left.\begin{aligned} \dot{y} &= -y - u(\xi), \\ \dot{\xi} &= (1-\beta)y - \beta u(\xi). \end{aligned}\right\} \tag{8.71}$$

Owing to the presence of two "non-unique" zones in the relay characteristic $u = u(\xi)$, the phase surface (ξ, y) will be a *three-sheet* surface shown in Fig. 428, consisting of the regions $(I): |\xi| < \varepsilon$ at the points of which

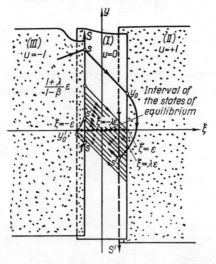

Fig. 428

$u=0$ (the relay is in the neutral position), and of the half-planes, partially superimposed on it, (II): $\xi > \lambda\varepsilon$ and (III): $\xi < -\lambda\varepsilon$, corresponding to "on" states of the relay (on (II) $u = +1$ and on (III) $u = -1$). The representative point moves from the region (I) on the half-planes (II) and (III)

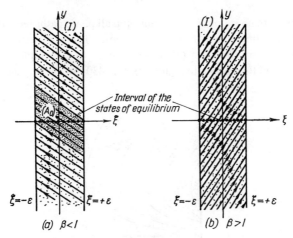

FIG. 429

only at its boundaries where $\xi = \pm \varepsilon$, and the converse passage only occurs at the boundaries of the half-planes where $\xi = +\lambda\varepsilon$ and $\xi = -\lambda\varepsilon$ respectively. The coordinates ξ, y vary continuously during these passages. Again, the phase paths are symmetrical with respect to the origin.

In the region (I) $u=0$ and the equations of motion (8.71) have the form

$$\left.\begin{aligned} \dot{y} &= -y, \\ \dot{\xi} &= (1-\beta)y \\ \frac{dy}{d\xi} &= -\frac{1}{1-\beta}. \end{aligned}\right\} \quad (8.71a)$$

Therefore all the points of the ξ axis within this region $(|\xi|<\varepsilon, y=0, u=0)$ are *states of equilibrium* and, moreover, *stable*, since the representative points move in the region (I) along the straight lines

$$\xi + (1-\beta)y = \text{const.}$$

towards the ξ axis (Fig. 429). These states of equilibrium are approached asymptotically by all paths the points of which satisfy the inequality

$$|\xi + (1-\beta)y| < \varepsilon.$$

On the half-plane (*II*), corresponding to a switched-on state of the relay, $u = +1$, the equations of motion of the system are

$$\left. \begin{array}{l} \dot{y}+y = -1, \\ \dot{\xi} = (1-\beta)y-\beta. \end{array} \right\} \quad (8.71\text{b})$$

On (*II*) there are no states of equilibrium and all phase paths tend asymptotically to the line $y = -1$, $\xi = -t$ +const. The representative point, having arrived on the sheet (*II*) must proceed to the boundary $\xi = +\lambda\varepsilon$ of the sheet and then pass into region (*I*). Fig. 430 shows the phase portrait

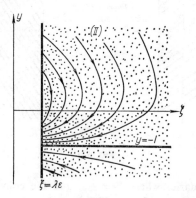

Fig. 430

on (*II*) for the case $\beta<1$. The phase paths on the sheet (*III*) ($u = -1$) are symmetric with the paths on sheet (*II*) just considered.

The character of the motions depends substantially on the sign of $\beta-1$†.

3. *The point transformation for $\beta<1$*

The phase portraits of the region (*I*) and of (*II*) are shown separately in Figs. 429a and 430. It is evident that there are two types of phase paths. The paths that start at points of (*I*) satisfying the inequality $|\xi+(1-\beta)y| < \varepsilon$, approach asymptotically the equilibrium states *without leaving the region (I)*. They correspond to motions with the relay always switched off.

---

† In the border-line case $\beta=1$ the phase paths in (*I*) are vertical straight lines, along which the representative points approach the states of equilibrium: $|\xi|<\varepsilon$, $y=0$, $u=0$. Paths that start on (*II*) and (*III*) must reach the boundaries of these sheets and then approach (as $t \to +\infty$) the states of equilibrium $\xi = \pm\lambda\varepsilon$, $y=0$, $u=0$. Thus, for $\beta=1$ all the motions are damped and end at states of equilibrium.

All remaining paths pass from sheet to sheet (but necessarily traverse the region (I)) and correspond to motions during which the relay is switched to one or both extremes.

The study of the paths of the second type, just as in earlier problems, produces a point transformation of a straight line into a straight line. We construct, on the phase surface (Fig. 428) two half lines without contact S and S': $\xi = -\lambda\varepsilon$, $y > -\beta/(1-\beta)$ and $\xi = +\lambda\varepsilon$, $y < \beta/(1-\beta)$. Through these lines the paths pass, respectively, from the sheets (III) and (II) on to region (I)[†]. With coordinate $s = y$ on S and $s = -y$ on S', symmetrical points of the half lines correspond to the same value of the coordinate s (evidently $s > -\beta/(1-\beta)$). The lines S and S' intersect all the paths except the ones that remain within the region (I). Since the paths that pass through symmetrical points s of the half lines S and S' are also symmetrical, their consecutive points of intersection with these half lines will have the same coordinate s'. Clearly, we need not distinguish between the half lines S and S' and we need to find a single point transformation $s' = \Pi(s)$, of use irrespective of which of the lines, S or S' intersects the path being studied.

To evaluate the sequence function of this point transformation we consider an arbitrary phase path, passing on to (I) from (III) at a point s on the half line S (Fig. 428). Within the boundaries of (I) the path will be the straight line

$$\xi + (1-\beta)y = -\lambda\varepsilon + (1-\beta)s. \tag{8.72}$$

Therefore, for $|-\lambda\varepsilon + (1-\beta)s| < \varepsilon$, i.e. for

$$-\frac{1-\lambda}{1-\beta}\varepsilon < s < \frac{1+\lambda}{1-\beta}\varepsilon, \tag{8.72a}$$

the representative point never leaves the region (I), and approaches asymptotically a state of equilibrium. The path has no consecutive points of intersection with the half lines S' and S.

For

$$s \geq \frac{1+\lambda}{1-\beta}\varepsilon \tag{8.72b}$$

the representative point reaches the right-hand boundary of the region

---

† By virtue of the second equation (8.71) we have on the half line S

$$\dot{\xi} = (1-\beta)y + \beta > 0$$

and on the half line S'

$$\dot{\xi} = (1-\beta)y - \beta < 0.$$

(*I*) at the point

$$\xi = +\varepsilon, \quad y = y_0 = s - \frac{1+\lambda}{1-\beta}\varepsilon \geqslant 0,$$

and then moves within (*II*) along a path determined by the differential equation (8.71b) and by the initial conditions: $t=0$, $\xi = +\varepsilon$, $y=y_0$. This path

$$y = -1+(1+y_0)e^{-t},$$
$$\xi = \varepsilon - t + (1-\beta)(1+y_0)(1-e^{-t}),$$

will necessarily reach the boundary S' of this sheet. τ is the time of transit across the sheet (*II*). Then, for $t=\tau>0$ $\xi = +\lambda\varepsilon$, $y = -s'$, which gives, after using the relation $s=y_0+\varepsilon(1+\lambda)/(1-\beta)$, the following parametric expressions for the sequence function

$$\left. \begin{array}{c} s \geqslant \dfrac{1+\lambda}{1-\beta}\varepsilon \quad s = -1 + \dfrac{1+\lambda}{1-\beta}\varepsilon + \dfrac{\tau-(1-\lambda)\varepsilon}{(1-\beta)(1-e^{-\tau})}, \\[2mm] s' = +1 - \dfrac{\tau-(1-\lambda)\varepsilon}{(1-\beta)(e^{\tau}-1)}. \end{array} \right\} \quad (8.73a)$$

Finally, for

$$s \leqslant -\frac{1-\lambda}{1-\beta}\varepsilon, \tag{8.72c}$$

which is only possible for $\beta > (1-\lambda)\varepsilon$, the representative point, moving along the path (8.72) can reach the left-hand boundary of the region (*I*) at a point

$$y_0' = s + \frac{1-\lambda}{1-\beta}\varepsilon \leqslant 0$$

Then, after moving on the half-plane (*III*), returns after an interval τ to the line S (at a point s'). It is easily seen that in this case the sequence function is expressed by

$$\left. \begin{array}{c} s \leqslant -\dfrac{1-\lambda}{1-\beta}\varepsilon \quad s = +1 - \dfrac{1-\lambda}{1-\beta}\varepsilon - \dfrac{\tau-(1-\lambda)\varepsilon}{(1-\beta)(1-e^{-\tau})}, \\[2mm] s' = +1 - \dfrac{\tau-(1-\lambda)\varepsilon}{(1-\beta)(e^{\tau}-1)}. \end{array} \right\} \quad (8.73b)$$

The relations (8.73a) and (8.73b) completely determine the point transformation $s' = \Pi(s)$ of the lines S and S' into each other or into themselves, as generated by the phase paths of the system. The coordinate s of a point, outside the interval (8.72a), completely determines the parameter τ

and the coordinate s' of the consecutive point. The dependence of τ on s and of s' in τ are expressed by single-valued, continuous functions $\tau = f(s)$ and $s' = g(\tau)$.

4. Lamerey's diagram

We shall use the auxiliary functions $\Psi_1(\tau)$ and $\Psi_2(\tau)$ introduced in the preceding section (see (8.65) and Fig. 412). If we put $\alpha = (1-\lambda)\varepsilon \geqslant 0$ in (8.65), then the sequence function $s' = \Pi(s)$ can be written in the form

$$s = \begin{cases} -1 + \dfrac{1+\lambda}{1-\beta}\varepsilon + \dfrac{\Psi_1(\tau)}{1-\beta} & \text{for} \quad s \geqslant \dfrac{1+\lambda}{1-\beta}\varepsilon, \\ +1 - \dfrac{1-\lambda}{1-\beta}\varepsilon - \dfrac{\Psi_1(\tau)}{1-\beta} & \text{for} \quad s \leqslant -\dfrac{1-\lambda}{1-\beta}\varepsilon, \end{cases} \quad (8.73)$$

$$s' = 1 - \frac{\Psi_2(\tau)}{1-\beta}.$$

Let τ_0 be the value of τ that corresponds to values $s = \varepsilon(1+\lambda)/(1-\beta)$ and $s = -\varepsilon(1-\lambda)/(1-\beta)$; this clearly, is uniquely determined by

$$\Psi_1(\tau_0) = 1 - \beta \quad (8.74)$$

or

$$(1-\beta)(1 - e^{-\tau_0}) = \tau_0 - (1-\lambda)\varepsilon$$

The coordinate of the consecutive point is then

$$(s')_{\tau = \tau_0} = s'_0 = 1 - e^{-\tau_0} = \frac{\tau_0 - (1-\lambda)\varepsilon}{1-\beta}. \quad (8.74a)$$

Note that $(1-\lambda)\varepsilon < \tau_0 < \tau_1$, where τ_1 is the value of the parameter τ for which the function Ψ_2 reaches a maximum[†]. Then, since Ψ_1 is a monotonically increasing function of τ, to obtain the values $s \geqslant \varepsilon(1+\lambda)/(1-\beta)$ and $s \leqslant -\varepsilon(1-\lambda)/(1-\beta)$ the parameter τ of the transformation must lie in the interval $\tau_0 \leqslant \tau < +\infty$.

We shall construct Lamerey's diagram (Figs. 431-3), plotting against τ the coordinates s and s' of a point and its consecutive point Observe,

---

[†] As in the preceding section, τ_1 is determined by the equation $\Psi_1(\tau_1) = 1$ and is a monotonically increasing function of the quantity $\alpha = (1-\lambda)\varepsilon$. For small values of α, $\tau_1 = (2\alpha)^{\overline{2}}$. The corresponding minimum value is

$$(s')_{\min} = 1 - \frac{e^{-\tau_1}}{1-\beta} > -\frac{\beta}{1-\beta}.$$

first, that since $d\Psi_1/d\tau > |d\Psi_2/d\tau|$† and $\beta < 1$

$$\left|\frac{ds'}{d\tau}\right| < \left|\frac{ds}{d\tau}\right|$$

for every $\tau \geqslant \tau_0$. Further, it follows from the relations (8.73) that

$$\left.\begin{aligned}s &\leqslant -\frac{1-\lambda}{1-\beta}\varepsilon \quad s'-s = \frac{\tau}{1-\beta} \geqslant \frac{\tau_0}{1-\beta} > 0; \\ s &\geqslant \frac{1+\lambda}{1-\beta}\varepsilon \quad s'-s = 2 - \frac{1+\lambda}{1-\beta}\varepsilon - \frac{1}{1-\beta}[\Psi_1 + \Psi_2]\end{aligned}\right\}$$

and $s'-s$ decreases monotonically from $s_0' - \varepsilon(1+\lambda)/(1-\beta)$ to $-\infty$ as τ varies from τ_0 to $+\infty$, since $d(\Psi_1+\Psi_2)/d\tau > 0$ (as $\tau \to +\infty$ then $s' \to +1$, while $s \to +\infty$). Therefore the point transformation $s' = \Pi(s)$ cannot have fixed points on the half line S where $s \leqslant -\varepsilon(1-\lambda)/(1-\beta)$. A unique fixed point exists only on the half line S where $s \geqslant \varepsilon(1+\lambda)/(1-\beta)$ and subject to the condition that $s_0' \geqslant \varepsilon(1+\lambda)/(1-\beta)$. Thus, depending on the sign of the expression $s_0' - \varepsilon(1+\lambda)/(1-\beta)$ two types of Lamerey's diagrams are possible.

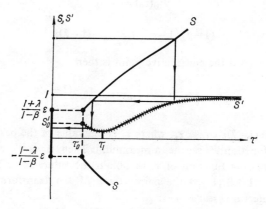

Fig. 431

For $s_0' < \varepsilon(1+\lambda)/(1-\beta)$ the curves $\tau = f(s)$ and $s' = g(\tau)$, defined by the relations (8.73), do not intersect each other (Fig. 431) and the point transformation $s' = \Pi(s)$ has no fixed points. It is easily seen that every

† This follows from the inequalities $d\Psi_1/d\tau > 0$, $d(\Psi_1+\Psi_2)/d\tau > 0$ and $d(\Psi_1-\Psi_2)/d\tau > 0$, which have been proved in the preceding section for all values of $\tau > 0$.

consecutive point s' lies closer to the interval (8.72a) than the preceding point s. Therefore, every path, after a finite number of intersections with the lines S and S' will arrive at a coordinate s such that

$$-\frac{1-\lambda}{1-\beta}\varepsilon < s < \frac{1+\lambda}{1-\beta}\varepsilon \qquad (8.72a)$$

and, afterwards never leave the region (I) as it approaches asymptotically a state of equilibrium. In this case there is *absolute stability*.

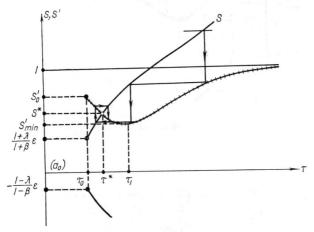

Fig. 432

Another Lamerey's diagram is obtained for $s'_0 > \varepsilon(1+\lambda)/(1-\beta)$ (Figs. 432 and 433). Now the curves $\tau = f(s)$ and $s' = g(\tau)$ have a unique point of intersection and the point transformation $s' = \Pi(s)$ has *a unique and stable fixed point* $s^* > \varepsilon(1+\lambda)/(1-\beta)$, which corresponds to *a symmetric limit cycle* on the phase surface (Fig. 434)[†]. The value τ^* corresponding to the fixed point s^* is determined by the equation $s' = s$, or

$$[\tau^* - (1-\lambda)\varepsilon]\,\text{cth}\,\frac{\tau^*}{2} = 2(1-\beta) - (1+\lambda)\varepsilon^{\ddagger}. \qquad (8.76)$$

[†] The existence of a fixed point $s^* > \varepsilon(1+\lambda)/(1-\beta)$ follows from the fact that the curves $\tau = f(s)$ and $s' = g(\tau)$ are continuous and that the difference $s' - s$ has different signs for $\tau = \tau_0$, and for $\tau \to +\infty$; the uniqueness from the fact that $s' - s$ is a monotonic function of τ; the stability from the inequality (8.75). A symmetric limit cycle has a point of intersection with each of the half lines S and S'.

[‡] Since $\tau^* < \tau_0 = (1-\lambda)\varepsilon$, the fixed point s^* cannot exist if $(1+\lambda)\varepsilon > 2(1-\beta)$.

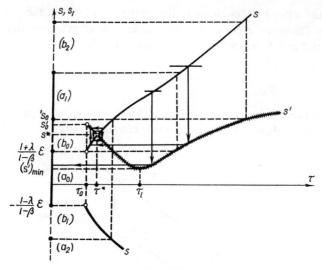

Fig. 433

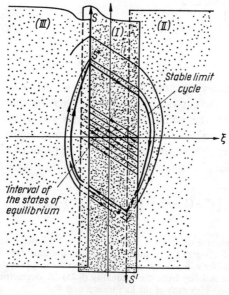

Fig. 434

The condition for the existence of a limit cycle is

$$s_0' = \frac{\tau_0 - (1-\lambda)\varepsilon}{1-\beta} > \frac{1+\lambda}{1-\beta}\varepsilon$$

or

$$\tau_0 > 2\varepsilon \qquad (8.77)$$

and is satisfied by certain allowed combinations of β, λ and ε. It is evident that the limiting relation is the surface in the stability space of the para-

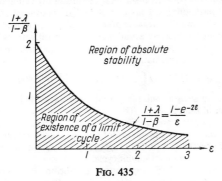

Fig. 435

meters β, ε, λ, that divides the region of existence of a symmetric limit cycle from the region of "absolute stability". This is determined by

$$\tau_0 = 2\varepsilon.$$

Whence, using (8.74), the equation of this dividing surface is

$$(1-\beta)(1-e^{-2\varepsilon}) = 2\varepsilon - (1-\lambda)\varepsilon = (1+\lambda)\varepsilon$$

or

$$\frac{1+\lambda}{1-\beta} = \frac{1-e^{-2\varepsilon}}{\varepsilon}. \qquad 8.77a)$$

Fig. 435 shows the stability diagram of the parameters ε and $(1+\lambda)/(1-\beta)$ with the curve (8.77a). From (8.74) τ_0 increases as β decreases (for fixed values of the parameters ε and λ) so below the curve (8.77a) in Fig. 435 the condition (8.77) is satisfied. Hence below this curve (8.77a) there lies the region instability (limit cycle) and above it a region of "absolute stability".

5. Structure of the phase portrait

Let us consider in greater detail the case

$$\frac{1+\lambda}{1-\beta} < \frac{1-e^{-2\varepsilon}}{\varepsilon}, \qquad (8.77b)$$

when in addition to the segment of stable states of equilibrium in the region (I) ($|\xi| < \varepsilon$, $y=0$, $u=0$), there is also a symmetric stable limit cycle (Fig. 434). Both the segment of equilibrium states and the limit cycle have their *regions of attraction*, wherein paths approach them asymptotically.

It has to be shown that between them these regions of attraction do cover the whole phase surface. Until we do this it will not be certain that, for example, the two-valuedness of the function $s = s(\tau)$ and the presence of a section of the curve $s' = g(\tau)$ with negative slope does not make possible a multiple and, generally speaking, asymmetrical periodic motion determined by the fixed point not of the transformation $s' = \Pi(s)$ but by a multiple of this transformation.

$$s' = \Pi\{\Pi\{\Pi\{\ldots\{\Pi(s)\}\ldots\}\}\} = \Pi^n(s).$$

On Lamerey's diagram such a multiple periodic motion would correspond not to a point of intersection of the curves $\tau = f(s)$ and $s' = g(\tau)$ but to a certain closed contour consisting of alternate vertical and horizontal segments with their ends on these curves. Therefore, to ensure that multiple periodic motion cannot exist we must carry out a more detailed analysis of the structure of the regions of attraction of the segment of the equilibrium states and the symmetric limit cycle.

The region of attraction of the segment of equilibrium states contains, under all conditions, the area A_0 of (I) where $|\xi + (1-\beta)y| < \varepsilon$ (Fig. 429a), which corresponds, on the half straight lines S and S', to the interval (a_0) where $-\varepsilon(1-\lambda)/(1-\beta) < s < \varepsilon(1+\lambda)/(1-\beta)$. Using the inverse transformation of s' into s† we can find other intervals (a_1), (a_2), ..., the points of which are connected by phase paths related by the transformation $s' = \Pi(s)$ to points of the interval (a_0). Depending on the relation between $(s')_{\min} = (s')_{\tau=\tau_1}$ and $\varepsilon(1+\lambda)/(1-\beta)$, two cases are possible.

If $(s')_{\min} \geqslant \varepsilon(1+\lambda)/(1-\beta)$ then Lamerey's diagram is that of Fig. 432 (for the sake of definiteness, the case $\tau^* < \tau_1$ is illustrated). Consider the sequence of points of intersection of a path with the half straight lines S and S'

$$s_0, \ s_1, \ s_2, \ldots, \ s_k, \ s_{k+1}, \ldots,$$

where $s_{k+1} = \Pi(s_k)$ and the initial point s_0 is outside the interval (a_0). It is clear that this sequence will be unending with all s_k greater than or equal to $\varepsilon(1+\lambda)/(1-\beta)$ (for $k \geqslant 1$)‡ and, as follows from (8.75), the following inequality is valid:

$$|s_{k+1} - s^*| < |s_k - s^*|. \qquad (8.78)$$

Therefore, any such sequences $s_0, s_1, s_2, \ldots$, have as their limit point the fixed point s^*, and the corresponding phase paths approach asymptotically the symmetrical limit cycle. Thus, for this case, the region of attraction of the segment of the equilibrium states consists of the area (A_0) and of small segmental regions on the sheets (II) and (III). All the remaining part of the phase surface is the region of attraction of the simple symmetric cycle (Fig. 436).

---

† Note that this inverse transformation is a multi-valued one, since to each value $s' \geqslant s'_{\min}$ there correspond one or two values of τ and to each value of τ two values of s, one of which is larger than $\varepsilon(1+\lambda)/(1-\beta)$ and the other is smaller than $-\varepsilon(1-\lambda)/(1-\beta)$.

‡ s_0 can be smaller than $-\varepsilon(1-\lambda)/(1-\beta)$.

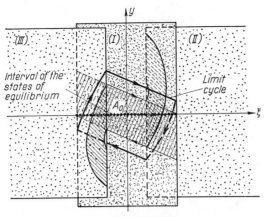

Fig. 436

A different and more complicated structure of the regions of attraction is obtained when

$$(s')_{\min} < \frac{1+\lambda}{1-\beta}\varepsilon\,\dagger.$$

For brevity, we shall consider in detail the case when $-\varepsilon(1-\lambda)/(1-\beta) < (s')_{\min} < \varepsilon(1+\lambda)/(1-\beta)$ and then Lamerey's diagram has the form shown in Fig. 433 (the same analysis can be carried out with the same conclusions for $(s')_{\min} \le -\varepsilon(1-\lambda)/(1-\beta)$). Now $\tau^* < \tau_1$ always and the consecutive points of intersection of the phase paths with the half lines S and S' can be found not only outside the interval (a_0) but also on it.

Let us denote by $'s_0$ the coordinate of the point on the half line $s \ge \varepsilon(1+\lambda)/(1-\beta)$ that is transformed by $s' = \Pi(s)$, with the smallest value of τ, into the point $\varepsilon(1+\lambda)/(1-\beta)$, and by (b_0) the segment $\varepsilon(1+\lambda)/(1-\beta) \le s \le {}'s_0 \ddagger$. It is evident that the sequence of the points of intersection with the half lines S and S'

$$s_0,\ s_1,\ s_2,\ \ldots,\ s_k,\ \ldots$$

with initial point s_0 on the segment (b_0), will be unending and that all the points s_k will belong to this segment.

The sequence will have as its limit point the fixed point s^*, since the inequality (8.78) is valid for this sequence. Thus the segment (b_0) is the "segment of attraction" of the fixed point s^* and all paths that intersect the half lines S and S' at points on the segment (b_0) tend to the symmetric limit cycle as $t \to \infty$. Thus the points of these paths form the region of attraction of the limit cycle, the non-shaded region in Fig. 437.

† It is easily verified that this inequality occurs for $\dfrac{1+\lambda}{1-\beta} < \dfrac{1 - \dfrac{e^{-\beta}}{1-\beta}e^{-2\varepsilon}}{\varepsilon}$.

‡ On the half straight line $s \ge \varepsilon(1+\lambda)/(1-\beta)$ there are two points whose consecutive point is the point $\varepsilon(1+\lambda)/(1-\beta)$. By $'s_0$ we denote the one of these two points that corresponds to the smallest value of the parameter τ. It is evident that $'s_0 > s'_0$ owing to the inequality (8.75), and that the segment (b_0) contains the fixed point s^*.

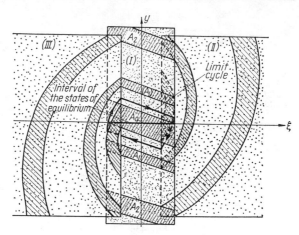

Fig. 437

Then, carry out the inverse transformation of the transformation $s' = \Pi(s)$, so determining the points s that have consecutive points s'. Proceeding from the interval (a_0) and from (b_0), we divide the half straight line $s \geqslant -\beta/(1-\beta)$ into a finite number of intervals $(a_1), (a_2), \ldots, (a_m)$ and segments $(b_0), (b_1), (b_2), \ldots, (b_m)$ the points of which are transformed by the phase paths into points of interval (a_0) and of segment (b_0) respectively†. Note that since the interval (a_0) and the segment (b_0) have a common boundary point (at $s = \varepsilon(1+\lambda)/(1-\beta)$) the intervals and segments indicated above will alternate, being contiguous to each other‡. We assert therefore, that every point s of the half line $s \geqslant -\varepsilon(1-\lambda)/(1-\beta)$ is transformed, after a finite number of transformations $s' = \Pi(s)$ into a point that belongs either to the interval (a_0) or to the segment (b_0) and the corresponding phase path will approach asymptotically either an equilibrium state or the symmetric limit cycle, depending on whether the first point of intersection of this path with the S (or S') was found in an interval (a_i) or on a segment (b_j).

The same construction of intervals (a_i) and segments (b_j) can also be carried out, with the same results, for

$$(s')_{\min} \leqslant -\frac{1-\lambda}{1-\beta}\varepsilon.$$

We have therefore proved that, when the conditions (8.77b) are satisfied for the existence of a simple symmetric limit cycle the phase surface consists only of "regions of attraction" of the segment of the states of equilibrium and of the limit cycle. Therefore, no other stable stationary types of motions exist in the system. Furthermore, when the conditions (8.77b) are satisfied, the self-oscillations have a *hard mode* of excitation.

† The construction of the intervals (a_i) and of the segments (b_j) must be carried out on the half line $s \geqslant -\beta/(1-\beta)$ only, since only the points of this line are transformed by the transformation $s' = \Pi(s)$ into points of the half line S'. Then, since $(s')_{\min} \leqslant s < 1$, the number of the intervals (a_i) and of the segments (b_j) will be finite, the latter containing points with arbitrarily large values of s.

‡ More precisely, the boundary points of each interval (a_i) are the end points of the adjacent segments (b_j), and the end points of every segment (b_j) are in their turn the boundary points of the adjacent intervals (a_i).

It is interesting to note that the boundary separating the regions of attraction is not an unstable limit cycle, as was the case in the dynamic systems with a plane phase surface considered earlier. This boundary consists of the phase paths that pass through the end points of the segment of the equilibrium states. This is a comparatively unusual structure.

6. *The dynamics of the system with large velocity correction*

We have still to consider the motion of the system when $\beta > 1$†. The phase portrait in region (I) (the relay off) for this case is shown in Fig. 429. As before, the representative points in this region move along rectilinear paths

$$\xi + (1-\beta)y = \text{const}$$

towards the ξ axis but now, in contrast to the situation when $\beta < 1$,

$$\frac{dy}{d\xi} = \frac{1}{\beta - 1} > 0.$$

It is easy to obtain, by the same method as for $\beta < 1$, the following equations of the point transformation of the half lines S and S' into each other:

$$s = \begin{cases} -1 - \dfrac{1+\lambda}{\beta-1}\varepsilon - \dfrac{\tau-(1-\lambda)\varepsilon}{(\beta-1)(1-e^{-\tau})} & \text{for} \quad s \leq -\dfrac{1+\lambda}{\beta-1}\varepsilon, \\ 1 + \dfrac{1-\lambda}{\beta-1}\varepsilon + \dfrac{\tau-(1-\lambda)\varepsilon}{(\beta-1)(1-e^{-\tau})} & \text{for} \quad s \geq \dfrac{1-\lambda}{\beta-1}\varepsilon, \end{cases}$$

$$s' = 1 + \frac{\tau-(1-\lambda)\varepsilon}{(\beta-1)(e^{\tau}-1)}$$

(8.79)

s is the coordinate of the initial point on the half line S or S', and s' is the coordinate of the consecutive point. The points of S and S' with coordinates $-\varepsilon(1+\lambda)/(\beta-1) < s < \varepsilon(1-\lambda)/(\beta-1)$ are transformed by phase paths that do not leave the region (I) into the segment containing the equilibrium states. Also note that the parameter τ occurring in (8.79), being the time of transit of the representative point on the sheet (II) or on the sheet (III), can assume values $\tau_0 < \tau < +\infty$ where τ_0, just as before, is the root of the equation (8.74).

Lamerey's diagram for $\beta > 1$ is shown in Fig. 438. The s curve cannot intersect the s' curve, and there are no fixed points and no limit cycles.

† This can only be realized in a system with derivative action in the forward path (Fig. 425), since in the two-position regulating system with a "constant-velocity" servomotor and parallel feedback (Fig. 427) the coefficient β is less than one.

Since

$$\frac{ds'}{d\tau} < \frac{ds}{d\tau} \quad \text{for} \quad s \geq \frac{1-\lambda}{\beta-1}\varepsilon,$$

then, for any initial conditions, the system reaches a state of equilibrium. All "Lamerey's ladders" (see Fig. 438) arrive at the interval

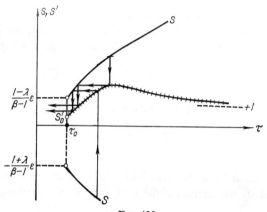

Fig. 438

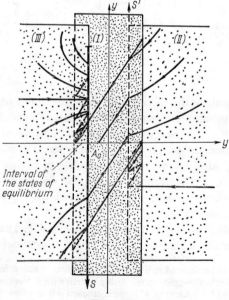

Fig. 439

$$-\frac{1+\lambda}{\beta-1}\varepsilon < s < \frac{1-\lambda}{\beta-1}\varepsilon,$$

the points of which, as has been shown, are transformed into points of the segment of the equilibrium states. The phase portrait is shown in Fig. 439.

To sum up, we can assert that in this automatic control system the hysteretic features of the relay characteristic are the cause of self-oscillations. However, the latter can be suppressed by introducing a sufficiently strong feedback or a sufficiently strong velocity correction. The value of the feedback (or of the velocity correction) necessary to suppress self-oscillations becomes smaller as the dead zone becomes larger.

§9. Oscillator with Square-Law Friction

We shall consider once more an oscillator with square-law friction the oscillations of which are described by the equation

$$2\ddot{x} + \dot{x}^2 \operatorname{sgn} \dot{x} + x = 0$$

or by the system of equations

$$\dot{x} = y, \quad 2\dot{y} = -x - y^2 \operatorname{sgn} y \tag{8.80}$$

(see Section 1, Chapter III), and shall show, by constructing a suitable point transformation of a straight line into a straight line that all its phase paths on the x, y plane are spirals which approach the origin $(0, 0)$ as $t \to +\infty$. The origin is thus a stable focus[†]. The system considered is *piece-wise conservative*, but not piece-wise linear. It is interesting from the point of view of the theory of the point transformations in having a peculiar form of correspondence function. Furthermore, the stability of the fixed point corresponding to the state of equilibrium cannot be assessed by means of Koenigs' Theorem.

Let us recall the basic results of the analysis of such an oscillator carried out in Section 1, Chapter III. On the x, y phase plane where $y > 0$ the equations of the oscillations have the form

$$\dot{x} = y, \quad 2\dot{y} = -x - y^2,$$

---

† It follows from the equations (8.80) that the only state of equilibrium of the system is the point $(0, 0)$, but the problem of the stability of this state of equilibrium cannot be solved by a linearization of these equations at the point $(0, 0)$. In fact, by neglecting the second order term $y^2 \operatorname{sgn} y$, in the process of linearizing the equations we obtain the equation of the harmonic oscillator with a singular point $(0, 0)$ of the *centre* type.

and the phase paths will be the curves

$$(y^2+x-1)e^x = C \qquad (8.81)$$

(C is an integration constant such that $C \geqslant -1$); the parabola $y^2=1-x$ corresponding to the value $C=0$ is a *separatrix* separating the paths that proceed from infinity (for them $C>0$) from the paths that start at points of the x axis on the left of the origin (for these $-1<C<0$).

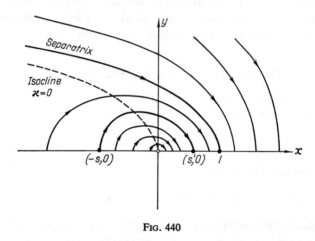

Fig. 440

The value $C=-1$ corresponds to the isolated singular point $(0, 0)$, the state of equilibrium of the system. All paths on the upper half-plane reach the x axis on the right of the origin (Fig. 440).

Since the equations (8.80) do not vary their form for a change of the variables x, y into $-x, -y$, the paths on the lower part of the phase plane ($y<0$) are symmetrical with respect to the origin with the paths on the upper half-plane. To study the behaviour of the paths we need only investigate the point transformation of the positive and negative parts of the x axis (the half straight lines U and V in Fig. 441) into each other, as generated by the paths of equations (8.80). As usual, introduce as coordinates on the positive and negative x half-axes U and V the distance s from the origin ($s>0$). Then symmetrical points on the x axis correspond to the equal values of s, and so in the sequence of the points of intersection of a certain arbitrary phase path with the x axis

$$s_1, s_2, \ldots, s_k, s_{k+1}, \ldots$$

each consecutive point is determined from the preceding one by a *single point transformation*, or *single* correspondence function

$$s_{k+1} = f(s_k),$$

irrespective of whether the first point lies on U or V. Consider a phase path that starts at $(-s, 0)$ on the half line U and meets the half line V at

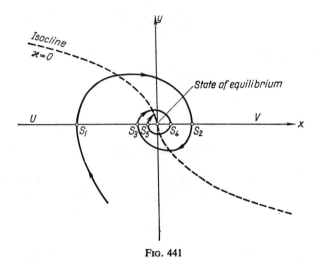

FIG. 441

$(s', 0)$ (Fig. 441). According to (8.81) the coordinates of these points are connected by the equation

$$-(s+1)e^{-s} = (s'-1)e^{s'} \quad (=C). \tag{8.82}$$

with $-1 < C < 0$. Since $d[(s'-1)e^{s'}]/ds' = s'e^{s'} > 0$, this equation determines a *single-valued continuous* correspondence function $s' = f(s)$. In order to find the fixed points of this transformation, we shall introduce the parameter $\xi = -C \, (0 < \xi < 1)$. Then the equation (8.82) can be written in a parametric form as

$$\left. \begin{array}{l} (1+s)e^{-s} = \xi, \\ (1-s')e^{s'} = \xi, \end{array} \right\} \tag{8.82a}$$

expressing implicity s and s' in terms of the parameter ξ. We can write

$$s = \varphi(\xi), \quad s' = \psi(\xi),$$

where as is easily seen φ and ψ are single-valued smooth functions. Lamerey's diagram is shown in Fig. 422. It is evident that these curves intersect each other at the point $s=s'=0$, $\xi=1$, i.e. the point $\bar{s}=0$ is a fixed point of the point transformation. This fixed point corresponds to a state of equilibrium of the system. Since

$$\frac{d\xi}{ds} = -se^{-s} < 0,$$

$$\frac{d\xi}{ds'} = -s'e^{s'} < 0$$

then, for equal s and s'

$$\frac{d\xi}{ds} > \frac{d\xi}{ds'},$$

and so the curve $s=\varphi(\xi)$ lies further on the right than the curve $s'=\psi(\xi)$.

$$s' = \psi(\xi) < s = \varphi(\xi)$$

for every $0<\xi<1$. Thus the fixed point $\bar{s}=0$ of the point transformation is *unique*; moreover, the point $\bar{s}=0$ is *stable*, since any sequence of points

$$s_1, s_2, \ldots, s_k, s_{k+1}, \ldots,$$

where $s_{k+1}=f(s_k)$, converges to it (see Lamerey's "ladder" in Fig. 442)[†].

[†] Since $f'(0) = \lim_{s \to 0} f'(s) = 1$, then we cannot use Koenigs's Theorem to prove analytically the stability of the fixed point $\bar{s}=0$. We give the following simple proof, basing our arguments on its uniqueness. Consider a sequence of numbers

$$s_1, s_2, \ldots, s_k, s_{k+1}, \ldots,$$

where $s_{k+1} = f(s_k)$ is the sequence of the coordinates of the points of intersection of a certain phase path with the x axis. Any such sequence is a monotonically decreasing sequence of positive numbers; therefore the sequence considered has a limit. Let us assume that this is equal to $a>0$. Then $\lim_{k \to +\infty} s_k = a$ and $\lim_{k \to +\infty} s_{k+1} = a$, but $s_{k+1} = f(s_k)$ where $f(s)$ is a continuous function for $a>0$, therefore, according to the theorem on the limit of a continuous function

$$a = \lim_{k \to +\infty} s_{k+1} = \lim_{k \to +\infty} f(s_k) = f(a),$$

i.e. the point $s=a$ is also a fixed point of the transformation which is impossible since the fixed point $\bar{s}=0$ is unique. Thus every sequence

$$s_1, s_2, \ldots, s_k, \ldots \to 0,$$

and the unique fixed point $\bar{s}=0$ is stable.

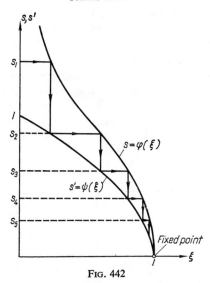

Fig. 442

Correspondingly any phase path approaches asymptotically the state of equilibrium $(0, 0)$, the number of its intersections with the x axis increasing without limit. Hence all paths are spirals winding in towards the state of equilibrium $(0, 0)$, and so the origin is a *stable focus*.

§ 10. Steam-engine

In concluding the chapter we shall consider an example of *self-oscillating* system using a very simple dynamic model of a steam-engine. A schematic diagram is shown in Fig. 443.

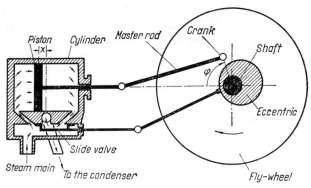

Fig. 443

As is well known, a steam-engine is a system with "feed-back": the slide-valve of the engine, connected with its main shaft by means of the eccentric, controls the admission of steam into the cylinder (into its left-hand or right-hand cavity, depending on the position of the shaft) and produces a variable force on the piston from a constant source of energy (from the steam main with *constant* steam pressure P_0). Such a *variable*

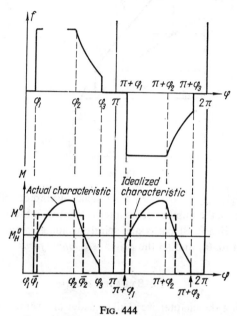

Fig. 444

action on the piston causes it to oscillate and this is used to produce rotary motion of the shaft, even in the presence of forces of resistance due to friction and to its load†. The force f that the steam pressure exerts on the piston depends on the position of the slide-valve and on the position of the piston. The slide-valve connects one side or other of the piston with the steam main or with steam condenser, and also isolates the cylinder. The steam pressure at one end of the cylinder depends on the volume of this cavity if its opening is closed by the slide-valve. Since the positions of the piston, of the shaft and of the slide-valve of the engine are uniquely determined by the angle of rotation of the shaft, φ, then, firstly, the

† In order to obtain a variable force on the piston and maintain the rotation of the shaft the eccentric of the slide-valve must be rotated about 90° ahead of the crank of the connecting rod.

force f will be a certain *single-valued* and periodic function of the angle φ (a qualitative graph of this function is shown in Fig. 444), and secondly, the state of the dynamic model of the steam engine will be determined uniquely by assigning the angle of rotation φ and the angular velocity $d\varphi/dt$ of the shaft. Accordingly the phase surface of this model will be a cylindrical phase surface†.

With simplifiying assumptions the equation of motion of the shaft is

$$I\frac{d^2\varphi}{dt^2} = M - M_H, \qquad (8.83)$$

where I is the moment of inertia of the engine and fly-wheel and load referred to the shaft, M is the torque on the shaft due to the piston and M_H is the torque due to resistance and the load. The torque M is related to f by

$$M = f(\varphi) A \sin \varphi$$

provided the "throw" A of the crank is small in comparison with the length of the connecting rod. Therefore the driving torque M is also a *single-valued* periodic function of the angle φ, but now the period is equal to π (Fig. 444). We assume that the driving torque $M \geqslant 0$, for all φ, reaches a maximum M_{max} for angles φ close to $\pi/2$ and $3\pi/2$, and reduces to zero in the vicinity of the "dead points" $\varphi = 0$, π, 2π. The load torque M_H can be assumed in the majority of cases to depend only on the angular velocity $d\varphi/dt$.

---

† This graph of the dependence of the force f upon the angle φ (Fig. 444) needs explanation. For $\varphi_1 < \varphi < \varphi_2$, when the steam main is connected with the left-hand end of the cylinder, the steam pressure in the cavity is constant ($=P_0$), and the force on the piston is constant. The admission of steam into the left-hand cavity is interrupted at $\varphi = \varphi_2$, and therefore, for a further increase of travel φ the force f decreases, since the volume of the left-hand cavity of the cylinder increases while the mass of steam in it remains constant. Finally for $\varphi = \varphi_3$, the left-hand cavity is connected with the condenser and the force f reduces to zero (the right-hand cavity of the cylinder is connected with the condenser for all values of φ, $0 < \varphi < \pi$). For $\pi < \varphi < 2\pi$ the same processes occur with the only difference that now the left-hand cavity of the cylinder is connected to the condenser while steam from the steam main is admitted (for $\pi + \varphi_1 < \varphi < \pi + \varphi_2$) into the right-hand cavity. It is clear that, owing to the symmetry of the steam-distributing apparatus of the engine, the function $f(\varphi)$ satisfies the condition $f(\varphi + \pi) = -f(\varphi)$.

The function $f(\varphi)$ is obtained by experiment. f is not a single-valued function of the coordinate x of the piston, so the x, dx/dt plane cannot be taken as the phase surface.

1. Engine working with a "constant" load and without a regulator

Let us assume that the load on the engine is generated by friction forces obeying Coulomb's law (see Section 2, Chapter III), so that

$$\text{for} \quad \frac{d\varphi}{dt} \neq 0 \quad M_H = M_H^0 \operatorname{sgn} \frac{d\varphi}{dt},$$

$$\text{for} \quad \frac{d\varphi}{dt} = 0 \begin{cases} M_H = M, & \text{if } M \leqslant M_H^0, \\ M_H = M_H^0, & \text{if } M \geqslant M_H^0 \end{cases}$$

where M_H^0 is the maximum torque of the forces of friction of rest, where $M_H^0 < M_{\max}$†. For such a load, the engine will have stable states of equilibrium in the vicinity of the "dead points". In fact, if $M = M_H$ at angles $\bar{\varphi}_1$ and $\bar{\varphi}_2$ (see Fig. 444), then for $0 < \varphi < \bar{\varphi}_1$, for $\bar{\varphi}_2 < \varphi < \pi + \bar{\varphi}_1$ and for $\pi + \bar{\varphi}_2 < \varphi < 2\pi$, $M(\varphi) < M_H^0$; therefore all states $(\varphi^*, 0)$, where φ^* is any angle in one of these three intervals, are states of equilibrium, since in them, according to (8.84), $M_H = M(\varphi^*)$ and, hence, $d^2\varphi/dt^2 = 0$.

This circumstance suggests a simple idealization of the torque characteristic $M = M(\varphi)$, which considerably simplifies the analysis yet retains the states of equilibrium indicated above. We replace the actual torque characteristic by the discontinuous piece-wise constant function

$$M = \begin{cases} 0 & \text{for} \quad 0 \leqslant \varphi < \bar{\varphi}_1, \quad \bar{\varphi}_2 \leqslant \varphi < \pi + \bar{\varphi}_1, \\ & \qquad \pi + \bar{\varphi}_2 \leqslant \varphi < 2\pi, \\ M_0 & \text{for} \quad \bar{\varphi}_1 \leqslant \varphi < \bar{\varphi}_2, \quad \pi + \bar{\varphi}_1 \leqslant \varphi < \pi + \bar{\varphi}_2 \end{cases} \qquad (8.85)$$

whose graph is shown in the lower half of Fig. 444 by a dashed line. The constant "amplitude" M_0 of the idealized driving torque will be chosen so that, during every half-turn of the shaft, the work done (8.85) is equal to the work done by the actual torque, i.e.

$$M_0 = \frac{1}{\theta} \int_0^\pi M(\varphi) \, d\varphi > M_H^0,$$

where $\theta = \bar{\varphi}_2 - \bar{\varphi}_1$ is the so-called "cut-off angle".

It is easily seen that motion will occur with increasing angular velocity if $M_0 > \pi M_H^0 / \theta$ since after each half-turn the work done by the driving torque $(= M_0 \theta)$ is larger than the work absorbed by the load $(+ M_H^0 \pi)$. If

---

† For $M_H^0 > M_{\max}$, either the shaft of the engine does not rotate or is stopped for a finite interval of time that depends on the initial conditions. This interesting little case is not considered.

$M_0 = (\pi/\theta)M_H^0$ the model will be "quasi-conservative": it will have a continuum of periodic motions with $d\varphi/dt > 0$ (these motions will correspond to closed phase paths encircling the phase cylinder in the region $d\varphi/dt > 0$)). Finally, for $M_0 < (\pi/\theta)M_H^0$ the engine will stop for any initial conditions. The quasi-stationary state is evidently not a *coarse* one.

These results reproduce, to a certain extent, properties of real steam engines which possess very little self-regulation, so that the shaft velocity varies considerably for comparatively small variations of the load or of the steam pressure.

Even though a steam engine without a regulator under constant load conditions is seldom used, we shall carry out a brief analysis of the phase portrait on the phase cylinder, as it will be used in later analysis of other models.

Introduce the variables

$$\vartheta = \varphi - \bar{\varphi}_1 \quad \text{and} \quad t_{\text{new}} = \sqrt{\frac{M_H^0}{I}}\, t;$$

then the equation (8.83) can be written as

$$\dot{\vartheta} = z, \quad \dot{z} = \lambda\Phi(\vartheta) - \Psi(z, \vartheta)\dagger, \qquad (8.83a)$$

where

$$\lambda = \frac{M_0}{M_H^0} > 1$$

and

$$\Phi(\vartheta) = \begin{cases} 1 & \text{for } 0 \leq \vartheta < \theta \quad \text{and for} \quad \pi \leq \vartheta < \pi + \theta, \\ 0 & \text{for } \theta \leq \vartheta < \pi \quad \text{and for} \quad \pi + \theta \leq \vartheta < 2\pi \end{cases} \qquad (8.85a)$$

$$\Psi(z, \vartheta) = \begin{cases} \operatorname{sgn} z & \text{for } z \neq 0 \quad \text{and} \quad \text{arbitrary } \vartheta, \\ 1 & \text{for } z = 0 \quad \text{and} \quad 0 \leq \vartheta < \theta \quad \text{or} \quad \pi \leq \vartheta < \pi + \theta, \\ 0 & \text{for } z = 0 \quad \text{and} \quad \theta \leq \vartheta < \pi \quad \text{or} \quad \pi + \theta \leq \vartheta < 2\pi \end{cases}$$
$$(8.84a)$$

are the normalized idealized characteristic of the driving torque, and of the torque of the "constant" load (Fig. 445).

Let us underline certain features of the portrait on the phase cylinder ϑ, z associated with the equations (8.83a):

(1) near $z = 0$ there are two *"segments of rest"* $\theta \leq \vartheta < \pi$ and $\pi + \theta \leq \vartheta < 2\pi$, consisting of *stable states of equilibrium* which are approached by phase paths from both halves of the phase cylinder, as t increases, since for these values of ϑ, $\dot{z} = -1$ for $z > 0$, and $\dot{z} = +1$ for $z < 0$;

† As usual, a dot denotes differentiation with respect to the new dimensionless time.

(2) on the lower half of the phase cylinder (for $z<0$) $\dot{z}>0$, and therefore all paths go towards the circle $z=0$, either approaching the "segments of rest" or passing to the upper half of the cylinder;

(3) there are no phase paths that pass (as t increases) from the upper half of the phase cylinder to the lower half.

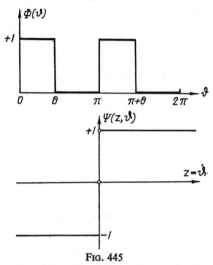

Fig. 445

The paths on the upper half of the phase cylinder ($z>0$) determine two point transformations: the transformation Π relating a point of the half line $U(\vartheta=0, z=u>0)$ to a point of the half line $U'(\vartheta=\pi, z=u'>0)$, and the point transformation Π' between the half line U' and the half line $U(\vartheta=2, z>0)$. Since the driving torque and hence the right-hand sides of the equations (8.83) are periodic functions in ϑ with a *period equal to* π then the phase portraits on the halves of the phase cylinder $0 \leqslant \vartheta < \pi$ and $\pi \leqslant \vartheta < 2\pi$ will be *identical* to each other. Hence the point transformations Π and Π' will *coincide* ($\Pi \equiv \Pi'$). Therefore, in the sequence of points of intersection by any path with the half lines U and U'

$$u_1, u_2, \ldots, u_k, u_{k+1}, \ldots$$

each consecutive point is determined from the preceding by the transformation $\Pi^\dagger$.

In the region (I): $0 \leqslant \vartheta < \theta$, $z>0$, the equations of motion (8.83a) have the form

$$\dot{\vartheta} = z, \quad \dot{z} = \lambda - 1,$$

† The sequence of the points $u_1, u_2, \ldots,$ can be *finite*, as when part of the half line U there is transformed by the phase paths not into U', but into a "segment of rest".

whence
$$\frac{dz}{d\vartheta} = \frac{\lambda-1}{z}.$$

On integrating we obtain arcs of the parabolae for the paths in the region (I)

$$z^2 - 2(\lambda-1)\vartheta = \text{const.} \tag{8.86a}$$

Similarly, in the region (II): $\theta \leqslant \vartheta < \pi$, $z > 0$ where

$$\dot{\vartheta} = z, \quad \dot{z} = -1 \quad \text{and} \quad \frac{dz}{d\vartheta} = -\frac{1}{z}$$

the paths will be arcs of other parabolae (Fig. 446),

$$z^2 + 2\vartheta = \text{const.} \tag{8.86b}$$

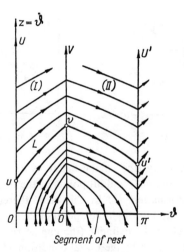

Fig. 446

Let L be a path of the equations (8.83a) beginning at a point $z = u > 0$ on the line U. Its equation for $0 \leqslant \vartheta < \theta$ will be

$$z^2 = u^2 + 2(\lambda-1)\vartheta$$

and it will reach the boundary V of region (I), which is the half line ($\vartheta = \theta$, $z = v > 0$), at $z = v$ where

$$v^2 = u^2 + 2(\lambda-1)\theta.$$

In the region (II) the equations of the path L will be

$$z^2 = v^2 - 2(\vartheta - \theta),$$

and it intersects the half line U' at $z = u'$ where

$$u'^2 = v^2 - 2(\pi - \theta)\dagger$$
$$u'^2 = u^2 + 2(\lambda\theta - \pi).$$

This relation also determines (in explicit form) the correspondence function for the transformation.

Lamerey's diagrams in the three possible cases: (a) $\lambda\theta > \pi$, (b) $\lambda\theta = \pi$ and (c) $\lambda\theta < \pi$ are shown in Figs. 447–9 where u^2 and u'^2 are plotted so that the graphs of the correspondence functions will be straight lines. For $\lambda\theta > \pi$, $u'^2 > u^2$ (Fig. 447), and therefore the phase paths encircling the

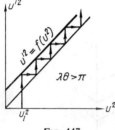
Fig. 447

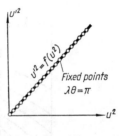
Fig. 448

cylinder recede to infinity (the engine increases its velocity indefinitely). For $\lambda\theta = \pi$, $u'^2 \equiv u^2$ (Fig. 448) and all points of the line U (or U') are fixed points of the transformation Π. In this case, therefore, the system is "quasi-conservative" and through each point of the half line U there passes a closed path encircling the cylinder (the engine runs with an arbitrary average velocity determined by the initial conditions).‡ Note that both for $\lambda\theta > \pi$ and for $\lambda\theta = \pi$ there are "segments of rest" on the phase cylinder with a certain region of attraction, so that initial conditions can be chosen

---

† The path that leaves the point (θ, v) reaches the line U' only if $v \geqslant [2(\pi-\theta)]^{\frac{1}{2}}$. If, however, $v < [2(\pi-\theta)]^{\frac{1}{2}}$ then the path that leaves the point (θ, v) arrives at the "segment of rest" $\theta \leqslant \vartheta < \pi$, $z = 0$ and the point $\dot{v}$ of the line V will not have a consecutive point on the line U'.

‡ For $\lambda\theta \geqslant \pi$ all points of the half line U have consecutive points on the half line U'; and if $\lambda\theta < \pi$, consecutive points on U' only occur for the points on U for which

$$u^2 \geqslant 2(\pi - \lambda\theta).$$

such that the engine will stop. Finally, for $\lambda\theta < \pi$ (Fig. 449) when $u'^2 < u^2$ all sequences of points of intersection by a path with the lines U and U' are monotonically decreasing and *finite*. If u_1 is the first point of intersection with the half line U or U' then, for the nth point of intersection,

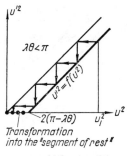

Fig. 449

where $n = E[u_1^2/2(\pi - \lambda\theta)] + 1^\dagger$, the ordinate $u_n < [2(\pi - \lambda\theta)]^{\frac{1}{2}}$. Hence this point u_n will not have a consecutive point in the transformation Π, and the corresponding path arrives at one of the "segments of rest" without intersecting U or U' again. Thus for $\lambda\theta < \pi$ all paths reach equilibrium, i.e. the engine stops. The phase portraits for these three cases are shown in Figs. 450–2. Since $\lambda = M_0/M_H^0$, these results coincide, evidently, with the results of the simple energy analysis given above.

Thus the steam engine cannot work stably on a "constant" load: it either stops or increases its velocity without limit. Therefore in order to operate with a stable shaft speed the steam engine must be equipped with a speed regulator.

2. Steam-engine working on a "constant" load but with a speed regulator

A widely used regulating scheme incorporates a speed measuring device (tachometer or governor) on the engine shaft, which controls (either directly or via a suitable servo-system) the slide-valve of the engine. By decreasing the cut-off angle θ as speed increases, the driving torque is reduced and the engine slows down. Assuming this regulator to be capable

† $E(x)$ denotes the integral part of x, i.e. $E(x) = l$, $(l = 0, 1, 2, \ldots)$, for $l \leq x < l+1$.

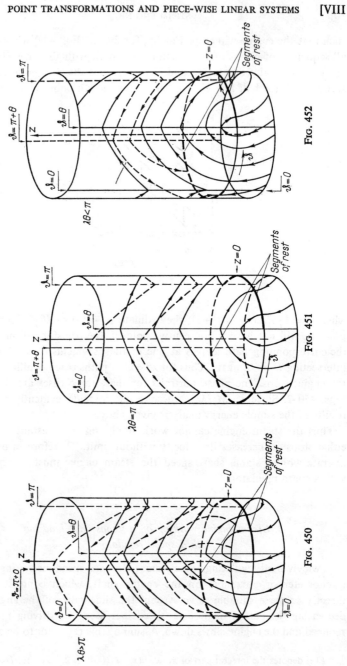

Fig. 450

Fig. 451

Fig. 452

of instantaneous operation, we shall suppose now that the cut-off angle θ is a simple *linear* function of the shaft velocity $\dot{\vartheta} = z$

$$\theta = \theta_0 - kz,$$

where $k > 0$ is a coefficient of the regulator and θ_0 is the cut-off angle for $\dot{\vartheta} = 0$.

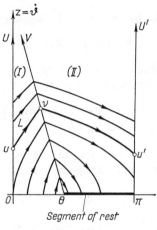

Fig. 453

The phase portrait on the phase cylinder ϑ, z of this model is similar in many respects to the one just described. On the circle $z = 0$ there are two "segments of rest" $\theta_0 \leq \vartheta < \pi$ and $\pi + \theta_0 \leq \vartheta < 2\pi$, and all paths on the lower half of the cylinder move towards these "segments of rest" or else pass to the upper half of the cylinder. The study of the paths on the upper half of the phase cylinder ($z > 0$) reduces as before to finding a transformation of the half line $U(\vartheta = 0, z = u > 0)$ into the half line $U'(\vartheta = \pi, z = u' > 0)$, generated by these paths. Again, the paths on the upper half of the phase cylinder are the parabolae (8.86a) in region (I), where $M = M_0$ and $\Phi(\vartheta) = 1$, and the parabolae (8.86b) in region (II) where torque $M = 0$. Now, however, the boundary V between the regions (I) and (II) is not the generator $\vartheta = \theta = $ const. but the line $\vartheta = \theta - kz$, $0 < z \leq \theta_0/k$ (Fig. 453). This circumstance substantially alters the correspondence function of the transformation.

To evaluate the correspondence function, consider a path L leaving any point $u = z$ on the half line U. If $u < u_0/k$ the path L is in the region (I) ($0 < \vartheta < \theta_0 - kz$, $0 < z < \theta_0)k$, and its equation is

$$z^2 = u^2 + 2(\lambda - 1)\vartheta,$$

It will reach the boundary V of the region (I) at a point whose ordinate v is determined by
$$v^2 = u^2 + 2(\lambda - 1)(\theta_0 - kv). \tag{8.87a}$$

The path L enters the region (II) and its equation is
$$z^2 = v^2 - 2(\vartheta - \theta_0 + kv),$$

Hence it will reach the line U' at a point whose ordinate u' is given by
$$u'^2 = v^2 - 2(\pi - \theta_0 + kv). \tag{8.87b}$$

The relations (8.87a) and (8.87b) determine the correspondence function of the transformation between U and U', for $u < \theta_0/k$, in parametric form
$$\left. \begin{array}{l} u^2 = v^2 - 2(\lambda - 1)(\theta_0 - kv), \\ u'^2 = v^2 - 2(\pi - \theta_0 + kv). \end{array} \right\} \tag{8.87}$$

Finally, a path reaches U' and transforms points on U into U' only if $u'^2 = v^2 - 2(\pi - \theta_0 + kv) \geq 0$, i.e. the paths must have
$$v \geq v_0 = k + \sqrt{k^2 + 2(\pi - \theta_0)} > k;$$

The points of U on paths for which $v < v_0$ are transformed into points of a "segment of rest".

If, however, $u \geq \theta_0/k$ then the path L enters the region (II) at once. Its equation will be
$$z^2 = u^2 - 2\vartheta,$$
and the correspondence function for $u \geq \theta_0/k$ is determined by the equation
$$u'^2 = u^2 - 2\pi.$$

It is clear that for $u \geq \theta_0/k$ there are no fixed points of the transformation, and each point of the half lines U and U' with ordinates $u \geq \theta_0/k$ is transformed after a finite number of transformations into a point with ordinate $u < \theta_0/k$. Below, therefore, in constructing Lamerey's diagram we can restrict u to the interval $0 \leq u \leq \theta_0/k$.

The curves (8.87) are hyperbolae and in the quadrant of Lamerey's diagram of interest to us $(u, u' > 0, v > 0)$ these hyperbolae have either no points of intersection (Fig. 454), or else intersect each other at one point, corresponding to a fixed point $\bar{u}$ (Fig. 455). For the fixed point $(u = u' = \bar{u}, v = \bar{v})$ we have
$$\bar{v}^2 - 2(\lambda - 1)(\theta_0 - k\bar{v}) = \bar{v}^2 - 2(\pi - \theta_0 + k\bar{v}),$$

i.e.
$$\theta_0 - k\bar{v} = \frac{\pi}{\lambda} \quad \text{or} \quad \bar{v} = \frac{1}{k}\left(\theta_0 - \frac{\pi}{\lambda}\right);$$

Hence
$$\bar{u}^2 = \bar{v}^2 - 2(\lambda-1)(\theta_0 - k\bar{v}) = \frac{1}{k^2}\left(\theta_0 - \frac{\pi}{\lambda}\right)^2 - 2(\lambda-1)\frac{\pi}{\lambda}.$$

If the fixed point exists, then $\bar{v} > 0$ and $\bar{u}^2 > 0$, so the condition for its

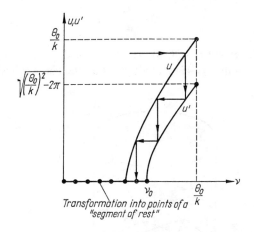

Fig. 454

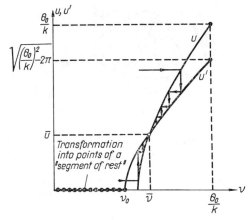

Fig. 455

existence is that the following inequalities are satisfied:

$$\lambda\theta_0 > \pi \quad \text{and} \quad k^2 < \frac{(\lambda\theta_0-\pi)^2}{2\pi\lambda(\lambda-1)}. \tag{8.88}$$

This fixed point is stable, since $\bar{v} > v_0 > k > 0$ and so

$$\frac{du'}{dv} = 2\bar{v}-2k > 0, \quad \frac{du}{dv} = 2\bar{v}+2k(\lambda-1) > \frac{du'}{dv}$$

and hence

$$0 < \frac{du'}{du} < 1†$$

Thus, if the conditions (8.88) are not satisfied, Lamerey's diagram has the form shown in Fig. 454 and all paths on the phase cylinder end even-

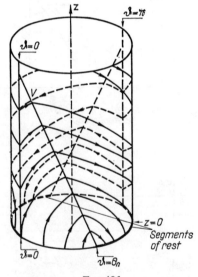

Fig. 456

tually at points on the "segments of rest" (Fig. 456), i.e. the engine stops. If, however, the conditions (8.88) are satisfied, then all sequences of points

† In the case $k < 0$, when the regulator is incorrectly connected, a fixed point will also exist if the conditions (8.88) are satisfied but will necessarily be unstable, since now

$$0 < \frac{du}{dv} < \frac{du'}{dv} \quad \text{or} \quad \frac{du'}{du} > 1.$$

of intersection by the paths with the lines U and U' converge to a single and stable limit point $\bar{u}$ (Fig. 455). There exists then a unique and stable limit cycle encircling the cylinder and corresponding to an operating model which is a shaft rotation. This limit cycle is approached asymptotically by all paths that intersect the lines U or U' at least once (Fig. 457)†.

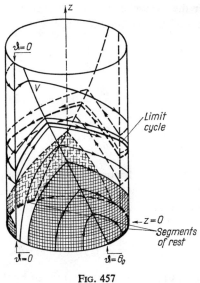

FIG. 457

In addition there are also paths that reach the "segments of rest" without intersecting the half lines U and U'. Therefore, when the conditions (8.88) are satisfied, the limit cycle motion has a *hard mode* of excitation. Self-rotating motion will be established in the engine if the initial conditions are so chosen that the representative point at $t=0$ is outside the regions of attraction of the "segments of rest". The boundaries of these regions are the paths that arrive at the points $(\theta, 0)$ and $(\theta + \pi, 0)$; these regions are shaded in Fig. 457.

† The period of this self-rotating mode can be evaluated in the following manner. From the second equation (8.63) $z = \lambda - 1$ in the region (*I*) and $z = -1$ in the region (*II*). Therefore the time of transit of the representative point moving along the limit cycle in the region (*I*) is equal to $\tau_1 = (\bar{v} - \bar{u})/(\lambda - 1)$ and in the region (*II*) to $\tau_2 = \bar{v} - \bar{u}$. Thus the period of rotation of the shaft is (in units of the dimensionless time)

$$\tau = 2(\tau_1 + \tau_2) = \frac{2\lambda}{\lambda - 1}(\bar{v} - \bar{u}) = \frac{2\lambda}{\lambda - 1}\left[\frac{1}{k}\left(\theta_0 - \frac{\pi}{\lambda}\right) - \sqrt{\frac{1}{k^2}\left(\theta_0 - \frac{\pi}{\lambda}\right)^2 - 2(\lambda - 1)\frac{\pi}{\lambda}}\right].$$

Thus with a suitably chosen speed regulator, stable working of the steam-engine on a "constant" load becomes possible. A similar stabilizing action occurs when load torque increases with shaft speed.

3. Engine with a speed-dependent load torque

Let us consider the dynamics of a steam engine whose load torque is generated by forces of dry and fluid friction:

$$M_H = \begin{cases} M_H^0 \operatorname{sgn} \dfrac{d\varphi}{dt} + H \dfrac{d\varphi}{dt} & \text{for} \quad \dfrac{d\varphi}{dt} \neq 0, \\ M, & \text{if} \quad \dfrac{d\varphi}{dt} = 0 \quad \text{and} \quad M \leqslant M_H^0, \\ M_H^0, & \text{if} \quad \dfrac{d\varphi}{dt} = 0, \quad \text{but} \quad M > M_H^0, \end{cases}$$

where M_H^0 is the maximum friction torque at rest and H is the viscous friction coefficient ($H>0$). The steam engine of a steamship has a load of such a type. The equations of motion of the shaft (in terms of the variables introduced in Sub-section 1 of this section) will be

$$\dot{\vartheta} = z, \quad \dot{z} = \lambda \Phi(\vartheta) - \Psi(z, \vartheta) - hz, \tag{8.89}$$

where $\Phi(\vartheta)$ and $\Psi(z, \vartheta)$ are functions determined by the relations (8.85a) and (8.84a), and

$$\lambda = \frac{M}{M_H^0} > 1 \quad \text{and} \quad h = \frac{H}{\sqrt{M_H^0 I}} > 0,$$

The cut-off angle θ is again considered constant.

The phase portrait in the region $0 \leqslant \vartheta < \pi$ for paths of the equations (8.89) is shown in Fig. 458. The portrait in the region $\pi \leqslant \vartheta < 2\pi$ is identical with that of the region $0 \leqslant \vartheta < \pi$, since the right-hand side of the second equation (8.89) has a period equal to π. On the circle $z = 0$ there are, as before, two "segments of rest" $\theta \leqslant \vartheta < \pi$ and $\pi + \theta \leqslant \vartheta < 2\pi$ consisting of stable states of equilibrium. The paths on the lower half of the phase cylinder will either arrive at the "segments of rest" or pass to the upper half of the cylinder. In addition, there are no paths that pass from its upper half on to the lower. Therefore, just as in Sub-section 1 of this section, the study of the dynamics of this steam engine reduces to analysing the point transformation of the half line $U(\vartheta = 0, z = u > 0)$ into the half line $U'(\vartheta = \pi, z = u' > 0)$ generated by the paths on the upper half of the phase cylinder. Again, consider the path L passing through an arbitrary point u on the line U (Fig. 458). Integrating the equation (8.89) in

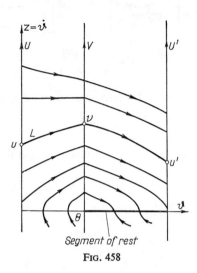

Fig. 458

the region (*I*): $0 \leq \vartheta < \theta$, $z > 0$, we obtain the equations of this path in the region (*I*)

$$z = a + (u-a)e^{-ht},$$

$$\vartheta = at + \frac{1}{h}(u-a)(1-e^{-ht}),$$

where

$$a = \frac{\lambda - 1}{h} > 0.$$

This path is bound to intersect the right-hand boundary V of the region (*I*) which is the half line ($\vartheta = \theta$, $z = v > 0$), since in the region (*I*) $\dot{\vartheta} = z > 0$ and $\dot{z} > 0$ on the arc $0 \leq \vartheta < \theta$ of the circle $z = 0$. If $t_1 = \tau_1/h$ is the time of transit of the representative point along the path L in region (*I*) then τ_1, and the ordinate v at which L intersects the half line V, are determined by the following equations:

$$\theta = \frac{a}{h}\tau_1 + \frac{1}{h}(u-a)(1-e^{-\tau_1}),$$

$$v = a + (u-a)e^{-\tau_1}.$$

On solving for u and v, we obtain in parametric form the correspondence function for the point transformation Π_1 relating U to V, as generated by paths in (I).

$$u = a\left[1 + \frac{\alpha - \tau_1}{1 - e^{-\tau_1}}\right], \quad v = a\left[1 + \frac{\alpha - \tau_1}{e^{\tau_1} - 1}\right], \tag{8.90}$$

where

$$\alpha = \frac{h\theta}{a} = \frac{h^2\theta}{\lambda - 1}.$$

The point $u=0$ on the line U corresponds to $\tau = \tau_1^0$ where

$$\tau_1^0 - 1 + e^{-\tau_1^0} = \alpha \tag{8.90a}$$

It is evident that $\tau_1^0 > a$, and has a consecutive point on the half line V with ordinate

$$v = v_0' = a(1 - e^{-\tau_1^0}) = a(\tau_1^0 - \alpha) \tag{8.90b}$$

and that $0 < v_0' < a$. In addition, since

$$\frac{du}{d\tau_1} = a\frac{-(1-e^{-\tau_1}) - (\alpha-\tau_1)e^{-\tau_1}}{(1-e^{-\tau_1})^2} = -\frac{v}{1-e^{-\tau_1}} < 0,$$

$$\frac{dv}{d\tau_1} = a\frac{-(e^{\tau_1}-1) - (\alpha-\tau_1)e^{\tau_1}}{(e^{\tau_1}-1)^2} = -\frac{u}{e^{\tau_1}-1} < 0$$

and

$$\frac{du}{dv} = \frac{du}{d\tau_1}\bigg/\frac{dv}{d\tau_1} = \frac{v}{u}e^{\tau_1} > 0, \tag{8.90c}$$

u and v are monotonically decreasing functions of τ_1; therefore to the set of values of u from 0 to $+\infty$ there corresponds a set of values of τ_1 from τ_1^0 to 0, and of the coordinates v of the consecutive points from v_0' to $+\infty$. The graph of the correspondence function (8.90) of the transformation π_1 is shown in Fig. 459.

The phase path L, after intersecting the line V, passes to the region (II): $\theta \leq \vartheta < \pi$, $z > 0$, where its equations will be

$$z = -b + (v+b)e^{-ht},$$

$$\vartheta = \theta - bt + \frac{1}{h}(v+b)(1-e^{-ht})$$

($b = h^{-1}$; in addition, we have chosen a new origin of time so that $\vartheta = \theta$ and $z = v$ for $t = 0$). If the path L intersects U', then the ordinate u' will be determined by

$$\pi = \theta - \frac{b}{h}\tau_2 + \frac{1}{h}(v+b)\left(1-e^{-\tau_2}\right)$$

$$u' = -b + (v+b)e^{-\tau_2}.$$

where $\tau_2 = ht_2$, and t_2 is the time of transit of the representative point along the path L in the region (II) from V to U'. Solving these equations

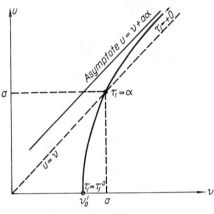

Fig. 459

for u' and v, we obtain in parametric form the correspondence function of the transformation Π_2 as generated by the paths in the region (II) which connect V and U,

$$u' = b\left[-1 + \frac{\beta + \tau_2}{e^{\tau_2}-1}\right], \quad v = b\left[-1 + \frac{\beta + \tau_2}{1-e^{-\tau_2}}\right],$$

where

$$\beta = \frac{h(\pi-\theta)}{b} = h^2(\pi-\theta).$$

Let τ_2^0 be the value of τ_2 corresponding to $u'=0$ and determined therefore by

$$e^{\tau_2^0} - 1 - \tau_2^0 = \beta; \tag{8.91a}$$

τ_2^0 corresponds to the point of the line V whose ordinate is

$$v = v_0'' = b(e^{\tau_2^0}-1) = b(\beta + \tau_2^0). \tag{8.91b}$$

It is evident that only points $v \geq v_0''$ on the line V are transformed by paths in region (II) into points on the line U' ($u' \geq 0$). The points on the line V such that $0 < v < v_0''$ are transformed into points on a "segment of rest".

Differentiating (8.91) we shall obtain

$$\frac{du'}{d\tau_2} = -\frac{v}{e^{\tau_2}-1} < 0, \quad \frac{dv}{d\tau_2} = -\frac{u}{1-e^{-\tau_2}} < 0$$

and

$$\frac{du'}{dv} = \frac{v}{u'} e^{-\tau_2} > 0, \tag{8.91c}$$

i.e. u and v are monotonically decreasing functions of the parameter τ_2, and, hence, to the set of points $v \gg v_0'$ on the line V, which are transformed by paths in region (II) into points on the line U', there corresponds a set of values of the parameter τ_2: $0 < \tau_2 \leq \tau_2^0$. The graph of the correspondence function (8.91) of the point transformation Π_2 is shown in Fig. 460.

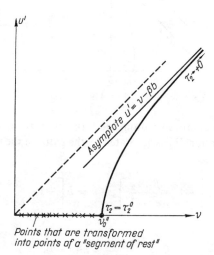

Points that are transformed
into points of a "segment of rest"

Fig. 460

The "overall" point transformation Π is the product of the transformations Π_1 and Π_2:

$$\Pi = \Pi_1 \cdot \Pi_2.$$

The fixed point of this transformation ($u' = u = \bar{u}$, $v = \bar{v}$, $\tau_1 = \bar{\tau}_1$, $\tau_2 = \bar{\tau}_2$) corresponds to a limit cycle which encircles the phase cylinder (equivalent to a self-rotating mode of operation of the steam engine). It is determined by the following system of transcendental equations:

$$a\left[1+\frac{\alpha-\bar{\tau}_1}{1-e^{-\bar{\tau}_1}}\right] = b\left[-1+\frac{\beta+\bar{\tau}_2}{e^{\bar{\tau}_2}-1}\right],$$
$$a\left[1+\frac{\alpha-\bar{\tau}_1}{e^{\bar{\tau}_1}-1}\right] = b\left[-1+\frac{\beta+\bar{\tau}_2}{1-e^{-\bar{\tau}_2}}\right] \quad (8.92)$$

(it is clear that $\alpha < \bar{\tau}_1 \leq \tau_1^0$ and $0 < \bar{\tau}_2 \leq \tau_2^0$). According to (8.90c) and (8.91c), we have at the fixed point,

$$0 < \frac{du'}{du} = \frac{dv'}{dv}\bigg/\frac{du}{dv} = e^{-(\bar{\tau}_1+\bar{\tau}_2)} < 1; \quad (8.92a)$$

Hence, the fixed point, if it exists, is stable and unique[†].

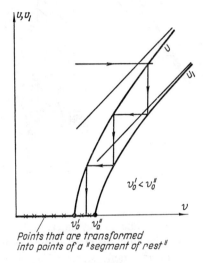

Points that are transformed into points of a "segment of rest"

Fig. 461

Depending on the values of h, λ, θ two qualitatively different cases are possible. If the parameters of the system are such that $v_0' < v_0''$, then there

---

† The stability of the fixed point follows directly from Koenigs's theorem, and its uniqueness follows from the fact that if the transformation Π had several fixed points, then, at least one of them (owing to the continuity of u and u' and their derivatives (du/dv and du'/dv as functions of v) would satisfy the inequality

$$\frac{du'}{dv} > \frac{du}{dv} > 0,$$

which is impossible according to (8.92a).

exist no fixed points of the transformation Π (Fig. 461)†, and all paths of the system arrive at the "segments of rest" (i.e. the steam engine stops for any initial conditions).

If, however, $v_0' > v_0''$, the point transformation Π has a unique and stable fixed point, to which all the sequences of points of intersection by the paths with the lines U and U' converge (Fig. 462)‡. Therefore there is a unique

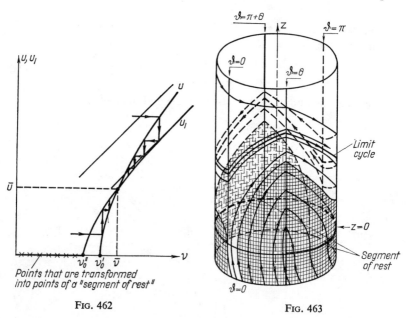

Fig. 462

Fig. 463

and stable limit cycle encircling the cylinder, and this limit cycle is approached asymptotically by all paths intersecting the half lines U and U'. Since, in addition, there are stable states of equilibrium forming the "segments of rest" the excitation of the self-rotating mode of operation is hard. The phase portrait for this case is shown in Fig. 463. The period of

---

† If, then $v_0' < v_0''$, $u' < v$ both for $v = v_0^*$ and for $v \to +\infty$, and the graphs of the correspondence functions (8.90) and (8.91) if plotted on Lamerey's diagram, either do not intersect or else have an even number of points of intersection (i.e. there is an even number of fixed points of the transformation Π). The latter case, according to what has been proved above, is impossible.

‡ Now $u' > u$ for $v = v_0'$, but as before $u' < u$ for $v \to +\infty$, then, owing to the continuity of the correspondence functions (8.90) and (8.91), there is at least one point of intersection on Lamerey's diagram (see Fig. 462). We have proved above, on the basis of the inequality (8.92a) that this fixed point is stable and unique.

the self-rotating motion of the shaft is equal, clearly (in units of the dimensionless time) to

$$\tau = 2(\bar{\tau}_1 + \bar{\tau}_2),$$

where $\bar{\tau}_1$ and $\bar{\tau}_2$ are the roots of the system of equations (8.92) that determines the fixed point of the transformation Π.

Thus the condition for the existence of a steady state rotating mode of operation of the engine reduces to the inequality

$$v_0' > v_0''$$

or according to (8.90b) and (8.91b)

$$(\lambda - 1)\left(1 - e^{-\tau_1^0}\right) > e^{\tau_2^0} - 1. \tag{8.93}$$

The dynamic model of the steam engine considered here has three independent parameters: λ, θ and h in terms of which the parameters a, α, b and β are expressed

$$a = \frac{\lambda - 1}{h}, \quad \alpha = \frac{h^2 \theta}{\lambda - 1}, \quad b = \frac{1}{h}, \quad \beta = h^2(\pi - \theta).$$

Thus we can draw a stability diagram in the three-dimensional space λ, θ, h, in the portion of this space for which $\lambda > 1$, $0 < \theta < \pi$ and $h > 0$. It divides into two regions: the region where a self-rotating mode of operation (with hard excitation) can exist and in which the condition (8.93) is satisfied, and the region where the engine stops for any initial conditions and where the condition (8.93) is no longer satisfied. The equation of the boundary surface that separates these regions can be written in the form

$$(\lambda - 1)\left(1 - e^{-\tau_1^0}\right) = e^{\tau_2^0} - 1$$

together with the equations (8.90a) and (8.91a)

$$\frac{h^2 \theta}{\lambda - 1} = \tau_1^0 - 1 + e^{-\tau_1^0}, \quad h^2(\pi - \theta) = e^{\tau_2^0} - 1 - \tau_2^0,$$

or in the parametric form

$$\left.\begin{aligned}
\lambda - 1 &= \frac{e^{\tau_2^0} - 1}{1 - e^{-\tau_1^0}}, \\
\frac{\pi}{\theta} &= 1 + \frac{1}{\lambda - 1} \frac{e^{\tau_2^0} - 1 - \tau_2^0}{\tau_1^0 - 1 + e^{-\tau_1^0}}, \\
h^2 &= \frac{1}{\pi}\left[e^{\tau_2^0} - 1 - \tau_2^0 + (\lambda - 1)\left(\tau_1^0 - 1 + e^{-\tau_1^0}\right)\right].
\end{aligned}\right\} \tag{8.94}$$

The equations (8.94) enable us to construct cross-sections of the boundary surface in the planes $\lambda=$const†. This boundary surface is shown in Fig. 464. Since, as θ increases (λ and h being fixed) α and τ_1^0 increase while β

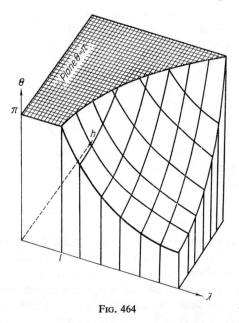

Fig. 464

and τ_2^0 decrease, then in the region above the boundary surface the condition (8.93) will be satisfied. So, above the boundary surface there is the region of self-rotating modes of operation.

† If we assign a value $\lambda>1$, and a series of values of τ_1^0, the first equation (8.94) will determine a series of values of τ_2^0, and the two remaining equations will determine the corresponding values of θ and h.

CHAPTER IX

NON-LINEAR SYSTEMS WITH APPROXIMATELY SINUSOIDAL OSCILLATIONS[†]

§1. Introduction

We shall proceed now to the quantitative analysis of autonomous dynamic systems (with one degree of freedom) which are approximately conservative. We shall restrict ourselves in this chapter to the simplest case, when the system is approximately *a linear conservative* system producing almost sinusoidal oscillations. For such systems the equations of motion can be written in the form[‡]:

$$\ddot{x}+x = \mu f(x, \dot{x}) \tag{9.1}$$

or, in the form

$$\dot{x} = y, \quad \dot{y} = -x+\mu f(x, y). \tag{9.2}$$

[†] Sections 1 and 3 have been revised and Sub-section 2 of Sections 3 and 4, and Sub-section 1 of Section 7 have been rewritten by N. A. Zheleztsov.

[‡] The equation of a system similar to a harmonic oscillator, in the usual variables, has the form

$$\frac{d^2v}{d\tau^2}+\omega_0^2 v = \mu F\left(v, \frac{dv}{d\tau}, \mu\right) \tag{α}$$

where τ is the time, ω_0 is the angular frequency, v is a dependent variable, a voltage or a current, for example, and μ is the so-called *small parameter* which we shall assume to be dimensionless and which determines the closeness of the system to a linear conservative system. By introducing the dimensionless independent variable $t=\omega_0\tau$ and the dimensionless variable $x=v/v_0$ where v_0 is a fixed quantity having the same dimensions as v, such as the saturation voltage or the saturation current, then equation (α) becomes

$$\ddot{x}+x = \mu \frac{1}{v_0\omega_0^2} F(v_0 x, v_0\omega_0\dot{x}; \mu)$$

or, using the notation

$$\frac{1}{v_0\omega_0^2} F(v_0 x, v_0\omega_0\dot{x}; \mu) = f(x, \dot{x}; \mu),$$

we have

$$\ddot{x}+x = \mu f(x, \dot{x}; \mu) \tag{β}$$

To simplify the discussion the theory is expounded in the text for the particular case when $f(x, \dot{x};\mu)$ does not depend on μ. If $f(x, \dot{x}; \mu)$ is a polynomial in μx and $\dot{x}$ then the formulae developed for the first approximation to the solutions of (9.2), the formulae (9.13a) and (9.14a) for example, remain valid for the equation (β) provided that $f(\xi, \eta)$ is replaced by $f(\xi, \eta; 0)$.

Here μ is a dimensionless positive parameter, which we assume to be sufficiently small. The magnitude of μ determines the closeness of the system to the simple harmonic oscillator.

A typical example of systems which are approximately sinusoidal oscillators is an electronic oscillator with the resonant circuit in the grid circuit (Fig. 465(a)) or in the anode circuit (Fig. 465(b)). With various assumptions,

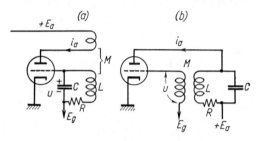

Fig. 465

the equation of motion of this oscillator was derived in Chapter 1 (Section 6, equation 1.36), and for the circuit of Fig. 465(a) is:

$$LC\frac{d^2u}{dt^2} + [RC - MS(E_g + u)]\frac{du}{dt} + u = 0$$

$$\frac{di}{dt} = -\frac{u}{M}.$$

Putting $t_{\text{new}} = \omega_0 t_{\text{old}}$, $x = u/u_0$, $\omega_0 = (LC)^{-\frac{1}{2}}$, and u_0 is a certain constant voltage, the above equation is transformed into

$$\ddot{x} + x = \mu[-1 + \alpha s(x)]\dot{x}, \qquad (9.3)$$

where $\mu = \omega_0 RC$ is the damping coefficient of the resonant circuit. $\alpha = MS_0/RC$ where $S_0 = S(E_g)$ is the slope of the tube characteristic at the working point, and $s(x) = S(E_g + u_0 x)/S_0$ is the normalized transconductance of the electronic tube and has a magnitude close to unity. Since

$$\mu \ll 1, \quad \mu\alpha = \omega_0 MS_0 \ll 1,$$

this valve generator will produce nearly sinusoidal oscillations†.

† If, $\mu = \omega_0 RC$ is not small, then we must reduce the equation to the form (9.1) by a different method. Introduce $S_1(x) = S(E_g + u_0 x) - S_0 = S_1^0 \varphi(x)$ where S_1^0 is the value of $S_1(x_0)$ for a fixed x_0, in the required interval of values of x. Then the equation of the valve generator is

$$\ddot{x} + x = \omega_0[-RC + MS_0 + MS_1^0 \varphi(x)]\dot{x}.$$

[*continued on next page*]

To solve (9.1) when μ is small we will use two approximate methods in this chapter: Van der Pol's method of slowly varying amplitudes [186], and Poincaré's method [184, 185]. The first method enables asymptotic solutions of (9.1) to be found, whose accuracy depends on the size of μ. The second method enables periodic solutions of (9.1) to be found in the form of a power series in μ, and to any degree of accuracy, provided that these series converge[†].

§ 2. VAN DER POL'S METHOD

To investigate (9.2) we shall use the "method of the slowly varying amplitudes" or Van der Pol's method [186, 187, 190, 35, 36], in which (9.2) is replaced by Van der Pol's auxiliary on *truncated equations*, whose solutions are the approximate solutions of the original equations. In particular, the problem of finding the limit cycles reduces to the much simpler problem of finding the states of equilibrium of the truncated equations. Van der Pol's method takes into account the specific non-linearity since the auxiliary equations are also non-linear.

For $\mu = 0$, the system (9.2) reduces to the equations of the ordinary simple harmonic oscillator whose solutions (see, for example, Sections 1 and 2 of Chapter I), have the form

$$\left. \begin{array}{l} x = a \cos t + b \sin t, \\ y = -a \sin t + b \cos t \end{array} \right\} \quad (9.4)$$

or

$$\left. \begin{array}{l} x = K \cos(t+\vartheta), \\ y = -K \sin(t+\vartheta) \end{array} \right\} \quad (9.5)$$

This equation will be approximately the equation of the harmonic oscillator for

$$\omega_0 |MS_0 - RC| \ll 1 \quad \text{and} \quad \omega_0 M S_1^0 \ll 1.$$

By now introducing the notation $\mu = \omega_0(MS - RC)$ and $\beta = MS_1^0 (MS_0 - RC)$, we reduce this equation to the form (9.1).

† Van der Pol's and Poincaré's methods are also suitable for the solution of non-autonomous equations such as

$$\ddot{x} + x = \mu f(x, \dot{x}, t)$$

where μ is a sufficiently small positive number. They can also be extended to autonomous and non-autonomous systems which are nearly conservative and have an arbitrary number of degrees of freedom.

Other methods, developed for the investigation of almost conservative systems (for example, the method of the mean slope [18, 136, 178, 73, 74] the method of the harmonic balance [78, 79, 46, 47, 2] and others [118] assume the oscillations to be almost sinusoidal, and are essentially modified versions of the methods of this chapter).

with phase paths which are circles about the origin. The representative points move with unit angular velocity $\omega = 1$, along these circles.

We shall seek a similar solution for (9.2), ($0 < \mu \ll 1$), in the same form as (9.4) or (9.5) but, now assuming a and b (or K and ϑ) no longer constants but *slowly varying* functions of time. We can interpret a and b as co-

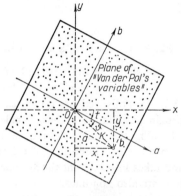

Fig. 466

ordinates on a phase plane which rotates with unit angular velocity relative to the x, y phase plane (in a clockwise direction). a and b are rectangular coordinates, and K and ϑ polar coordinates (Fig. 466) such that according to (9.4) and (9.5),

$$a = K \cos \vartheta, \quad b = -K \sin \vartheta. \tag{9.6}$$

For $\mu = 0$, the representative point moves with the a, b axes and so each point of the a, b plane is a state of equilibrium.

In terms of a and b the equations (9.2) become

$$\frac{da}{dt} \cos t + \frac{db}{dt} \sin t = 0,$$

$$-\frac{da}{dt} \sin t + \frac{db}{dt} \cos t = \mu f(a \cos t + b \sin t, -a \sin t + b \cos t)$$

or

$$\left. \begin{array}{l} \dfrac{da}{dt} = -\mu f(a \cos t + b \sin t, -a \sin t + b \cos t) \sin t, \\[2mm] \dfrac{db}{dt} = \mu f(a \cos t + b \sin t, -a \sin t + b \cos t) \cos t. \end{array} \right\} \tag{9.7}$$

The right-hand sides are periodic in t with period 2π and can be expanded into a Fourier series

$$\left.\begin{aligned}\frac{da}{dt} &= \mu\left\{\frac{\varphi_0(a,b)}{2} + \varphi_1(a,b)\cos t + \bar{\varphi}_1(a,b)\sin t +\right.\\ &\qquad \left.+ \varphi_2(a,b)\cos 2t + \bar{\varphi}_2(a,b)\sin 2t + \ldots\right\},\\ \frac{db}{dt} &= \mu\left\{\frac{\psi_0(a,b)}{2} + \psi_1(a,b)\cos t + \tilde{\psi}_1(a,b)\sin t +\right.\\ &\qquad \left.+ \psi_2(a,b)\cos 2t + \tilde{\psi}_2(a,b)\sin 2t + \ldots\right\},\end{aligned}\right\} \quad (9.7a)$$

where $\varphi_i(a,b)$, $\bar{\varphi}_i(a,b)$, $\psi_i(a,b)$ and $\tilde{\psi}_i(a,b)$ are the respective Fourier coefficients of the functions

$$-\mu f(a\cos t + b\sin t, -a\sin t + b\cos t)\sin t$$

and

$$+\mu f(a\cos t + b\sin t, -a\sin t + b\cos t)\cos t$$

with fixed a and b.

The equations (9.7) (or (9.7a)) represent (9.2) in terms of a and b, where da/dt and db/dt are of the same order of magnitude as μ. Since the formulae (9.4) involved t explicitly, the new system of equations may not be autonomous, though the original system was. Now consider the approximate *truncated equations of Van der Pol*

$$\frac{da}{dt} = \mu\frac{\varphi_0(a,b)}{2}, \quad \frac{db}{dt} = \mu\frac{\psi_0(a,b)}{2}, \quad (9.8)$$

obtained by neglecting in the right-hand sides of (9.7a) all "oscillatory" terms.

The solution of (9.8) approximates for sufficiently small values of μ to the solutions of (9.7), and (9.2). If we find, solutions $a = a_0(t)$, $b = b_0(t)$ of (9.8), then from (9.4), an approximate solution of (9.2) is

$$\left.\begin{aligned}x_0(t) &= a_0(t)\cos t + b_0(t)\sin t,\\ y_0(t) &= -a_0(t)\sin t + b_0(t)\cos t.\end{aligned}\right\} \quad (9.9)$$

In particular, the states of equilibrium $a_0(t) \equiv$ const. $b_0(t) \equiv$ const. correspond to approximate sinusoidal solutions of system (9.2). This will be proved in the next section.

The truncated equation (9.7) is autonomous and can be investigated particularly simply in polar coordinates K, ϑ, for then the variables can be separated.

In polar coordinates (9.2) becomes

$$\frac{dK}{dt} \cos(t+\vartheta) - K\frac{d\vartheta}{dt} \sin(t+\vartheta) = 0,$$

$$-\frac{dK}{dt} \sin(t+\vartheta) - K\frac{d\vartheta}{dt} \cos(t+\vartheta) = \mu f[K\cos(t+\vartheta), -K\sin(t+\vartheta)]$$

or,

$$\left. \begin{aligned} \frac{dK}{dt} &= -\mu f[K\cos(t+\vartheta), -K\sin(t+\vartheta)] \sin(t+\vartheta), \\ \frac{d\vartheta}{dt} &= -\frac{\mu}{K} f[K\cos(t+\vartheta), -K\sin(t+\vartheta)] \cos(t+\vartheta). \end{aligned} \right\} \quad (9.10)$$

Averaging the right-hand sides with respect to t, as it occurs explicitly (or with respect to $u = t+\vartheta$) we obtain the truncated equations

$$\frac{dK}{dt} = \mu \Phi(K), \quad \frac{d\vartheta}{dt} = \mu \Psi(K), \quad (9.11)$$

where

$$\left. \begin{aligned} \Phi(K) &= -\frac{1}{2\pi} \int_0^{2\pi} f[K\cos u, -K\sin u] \sin u \, du, \\ \Psi(K) &= -\frac{1}{2\pi K} \int_0^{2\pi} f[K\cos u, -K\sin u] \cos u \, du \end{aligned} \right\} \quad (9.12)$$

are the mean values with respect to u of the periodic functions[†]

$$-f[K\cos u, -K\sin u] \sin u \quad \text{and} \quad -\frac{1}{K} f[K\cos u, -K\sin u] \cos u,$$

[†] In fact, the right-hand sides of the truncated equations for K and ϑ are the zero-order terms in the Fourier expansions of the right-hand sides of the equations (9.9), and are equal respectively to

$$\Phi = -\frac{1}{2\pi} \int_0^{2\pi} f[K\cos(\xi+\vartheta), -K\sin(\xi+\vartheta)] \sin(\xi+\vartheta) \, d\xi$$

and

$$\Psi = -\frac{1}{2\pi K} \int_0^{2\pi} f[K\cos(\xi+\vartheta), -K\sin(\xi+\vartheta)] \cos(\xi+\vartheta) \, d\xi$$

The integration is carried out for fixed values of K and ϑ s and we obtain (9.12)

[*continued on next page*]

We shall investigate the system of truncated equations and shall construct their phase paths on the plane of Van der Pol's variables.

Consider the first of the equations (9.11),

$$\frac{dK}{dt} = \mu \Phi(K); \tag{9.11a}$$

the qualitative nature of its solutions is completely determined, as we have seen, by the distribution of the states of equilibrium on the phase line (unidimensional phase space). These states of equilibrium are the roots of

$$\Phi(K) = 0 \tag{9.13}$$

or

$$-\frac{1}{2\pi}\int_0^{2\pi} f(K\cos u, -K\sin u)\sin u\, du = 0. \tag{9.13a}$$

A state of equilibrium $K=K_i$ will be stable, if

$$\Phi'(K_i) < 0. \tag{9.14}$$

or if

$$\frac{1}{2\pi}\int_0^{2\pi} f_y'(K_i\cos u, -K_i\sin u)\, du < 0, \tag{9.14a}$$

and unstable if

$$\Phi'(K_i) > 0.$$

The remaining motions are either asymptotic to an equilibrium state both for $t \to \pm \infty$ or asymptotic to an equilibrium state for $t \to +\infty$, and receding to infinity for $t \to -\infty$.

Analytical expressions can always be found for such motions. In fact, from (9.11a)

$$\mu(t-t_0) = \int_{K_0}^{K} \frac{dK}{\Phi(K)},$$

where $K=K_0$ at $t=t_0$. Hence, solving for K,

$$K = K\{\mu(t-t_0)\}.$$

In the case of non-autonomous systems of the form $\ddot{x}+x = \mu f(x, \dot{x}; t)$, the truncated equations obtained are also autonomous but with non-separable variables K and ϑ.

Now consider the second equation (9.11)

$$\frac{d\vartheta}{dt} = \mu \, \Psi(K). \tag{9.11b}$$

Two cases must be distinguished. In the first case, which is often met in practice,

$$\Psi(K) \equiv 0$$

or

$$\frac{1}{2\pi K} \int_0^{2\pi} f(K \cos u, \, -K \sin u) \cos u \, du \equiv 0.$$

This equation is integrable so

$$\frac{d\vartheta}{dt} = 0 \quad \text{and} \quad \vartheta = \text{const} = \vartheta_0,$$

and integral curves are straight line through the origin with slope ϑ =const. The motion is the same along each line and is determined by (9.11a). The roots of (9.13) $K=K_i$ give the radii of circles which are curves of states of equilibrium of the truncated system. Fig. 467 shows, on the plane

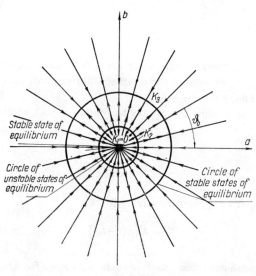

Fig. 467

of Van der Pol's variables (the a, b plane), a particular case of three states of equilibrium of the truncated equation (9.11a).

If we pass from the rotating a, b plane to the fixed x, y phase plane then, the circles of equilibrium on the a, b plane become circular limit cycles on

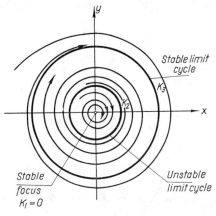

Fig. 468

the x, y plane having the same radii K_i (Fig. 468). The motion of the representative point along a cycle of radius K_i is given by

$$x = a\cos t + b\sin t = K_i \cos(t+\vartheta_0),$$
$$y = -a\sin t + b\cos t = -K_i \sin(t+\vartheta_0),$$
(9.15)

where ϑ_0 is arbitrary (because the states of equilibrium of the truncated equations form whole circles on the a, b plane).

Clearly a limit cycle will be *orbitally* stable if the corresponding states of equilibrium on the a, b plane are stable, and conversely. The remaining paths, represented by segments of straight lines on the a, b plane are transformed on the x, y plane into spirals that wind on to limit cycles either for $t \to +\infty$ or for $t \to -\infty$.

Now consider the second case, when $\Psi(K) \not\equiv 0$. Let the equation $\Psi(K) = 0$ have a number of roots $K_1, K_2, \ldots, K_m$, all different from $K_1, K_2 \ldots, K_n$.

Then, from (9.11) it is easily seen that the states of equilibrium of the equation (9.11a) correspond to circular limit cycles on the a, b phase plane, again with radii $K_1, K_2, \ldots, K_n$. Motion of the representative point along a limit cycle of radius K_j on the a, b plane is given by

$$K = K_j = \text{const}, \quad \vartheta = \mu\, \Psi(K_j)t + \vartheta_0$$

or
$$a = K_j \cos\{\mu \, \Psi(K_j)t + \vartheta_0\}, \quad b = -K_j \sin\{\mu \, \Psi(K_j)t + \vartheta_0\}$$

The stability or instability of this limit cycle is determined by the stability or instability of the corresponding equilibrium state given by equation (9.11a), and the direction of rotation by the sign of $\Psi(k_j)$.

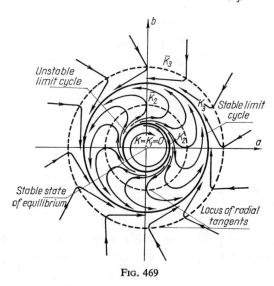

Fig. 469

The remaining curves are spirals, winding on to limit cycles (or about a state of equilibrium) either for $t \to +\infty$ or for $t \to -\infty$ (Fig. 469). If in this second case we use the fixed system of coordinates x, y we shall obtain a picture similar to that in the first case. Again there is a series of limit cycles with radii $K_1, K_2, \ldots, K_n$, and the motion along any of these is given by the equations

$$\left. \begin{array}{l} x = K_j \cos\{[1 + \mu \, \Psi(K_j)]t + \vartheta_0\}, \\ y = -K_j \sin\{[1 + \mu \, \Psi(K_j)]t + \vartheta_0\}. \end{array} \right\} \quad (9.15a)$$

This case differs from the first one only in having a certain frequency correction $\Delta\omega = \mu\Psi(K_j)$, which corresponds, in the first approximation with respect to μ, to a correction for the period of amount $\tau = -2\pi\mu\Psi(K_j)$. The remaining paths are again spirals that wind about the limit cycles or about the singular point 0,0 if it exists and is stable.

We must now use information about the truncated equations (9.8) or (9.11) to illuminate the properties of (9.2).

It can be shown (and this is the justification of Van der Pol's method) that the phase portrait of equation (9.8) is, for small μ, a good approximation to the phase portrait of (9.2). More precisely, if $\Phi(K) = 0$ has simple roots, then the equation (9.2) actually has limit cycles close to circles with radii K_i, and no others. These limit cycles correspond to almost sinusoidal motions, stable if $\Phi' \cdot (K_i) < 0$, i.e. stable in the sense of Liapunov.

The transient solutions corresponding to motion being initiated has been discussed by Mandel'shtam and Papaleksi [90]. We shall prove that the solutions of the truncated equations differ arbitrarily little from the solutions of equations (9.2) for similar initial conditions.

§ 3. Justification of Van der Pol's method

1. The justification of Van der Pol's method for transient processes [90, 149]

To prove the statement at the end of the preceding section it suffices to prove the following proposition:

Let $a = a(t)$, $b = b(t)$ be the solution of the "complete system" (9.7a) and $a = a_0(t)$, $b = b_0(t)$ the solution of the system of truncated equations (9.8) such that, initially, for $t = t_0$, $a(t_0) = a_0(t_0)$, $b(t_0) = b_0(t_0)$: *then for assigned positive ε and D (ε can be arbitrarily small, D arbitrarily large) it is always possible to find a sufficiently small value of μ such that*

$$|a(t) - a_0(t)| < \varepsilon, \quad |b(t) - b_0(t)| < \varepsilon$$

for

$$t_0 \leqslant t \leqslant t_0 + \frac{D}{\mu}.$$

To simplify the calculations, we shall prove the proposition above for the case of one equation of the first order, since its extension to higher orders is easy:

$$\frac{da}{dt} = \mu F(a, t), \tag{9.16}$$

whose auxiliary equation is

$$\frac{da}{dt} = \mu f(a), \tag{9.17}$$

$F(a, t)$ is periodic in t (with period 2π), and

$$f(a) = \frac{1}{2\pi} \int_0^{2\pi} F(a, \xi) \, d\xi.$$

We can write $F(a, t)$ as

$$F(a, t) = f(a) + \varphi(a, t), \tag{9.18}$$

where, clearly, $\varphi(a, t)$ is periodic in and has a time-average over 2π equal to zero.

The solution $a=a(t)$ of the "complete" equation (9.16) and the solution $a=a_0(t)$ of the truncated equation (9.17) satisfy the same initial condition for $t=t_0$

$$a(t_0) = a_0(t_0) = \eta.$$

$$\int_t^{t+2\pi} \varphi(a,\xi)\,d\xi \equiv 0. \tag{9.19}$$

We shall assume that on a certain interval of variation of a

$$|a-\eta| < A \tag{9.20}$$

that, for any t, the functions $f(a)$ and $F(a,t)$ are continuous, bounded, and satisfy Lipshitz's conditions. These are, that positive numbers M, P, Q and B exist such that for arbitrary a, a', a'' in the interval (9.20) the following inequality is satisfied:

$$\left.\begin{array}{l} |f(a)| < M, \quad |\varphi(a,t)| < P, \\ |\varphi(a'',t)-\varphi(a',t)| < Q|a''-a'|, \\ |F(a'',t)-F(a',t)| < B|a''-a'|. \end{array}\right\} \tag{9.21}$$

We need to prove that, for *arbitrary assigned positive numbers ε and D (ε arbitrarily small) it is always possible to choose a sufficiently small μ such that for all values of t satisfying the condition*

$$0 \leq \mu(t-t_0) \leq D,$$

the following inequality is satisfied:

$$|a(t)-a_0(t)| < \varepsilon.$$

D must be such that the solution $a_0(t)$, for any μ and all values of t satisfying the inequality $0 \leq \mu(t-t_0) \leq D$, shall not exceed the chosen limits of the interval (9.20) so that D must be such that

$$|a_0(t)-\eta| < A \tag{9.22}$$

for

$$0 \leq \mu(t-t_0) \leq D.$$

Such a D can always be chosen, since the solution $a_0(t)$ is a function of $\mu(t-t_0)$ only. Note that by choosing the initial value η and the interval D we select, on the path $a=a_0(t)$ of the truncated equation, a certain segment of *finite* length. In essence, we want to prove that the solution $a_0(t)$ approximates (for sufficiently small values of μ) to the solution $a(t)$ over this segment, i.e. for *finite* variations of the variable a. We find the first approximation to the solution of (9.16) by substituting $a_0(t)$ into the right-hand side and integrating

$$a_1(t) = \eta + \mu \int_{t_0}^t F[a_0(t),t]\,dt. \tag{9.23}$$

Similarly the second approximation is

$$a_2(t) = \eta + \mu \int_{t_0}^t F[a_1(t),t]\,dt, \tag{9.24}$$

and, in general, the nth approximation will be

$$a_n(t) = \eta + \mu \int_{t_0}^t F[a_{n-1}(t),t]\,dt. \tag{9.25}$$

Now when the conditions (9.21) are satisfied, $\lim_{n\to\infty} a_n(t)$ exists and is the only solution of

(9.16) that satisfies the initial condition: $a(t_0) = \eta$†. For the first approximation we have

$$a_1(t) = \eta + \mu \int_{t_0}^{t} F[a_0(t), t]\, dt = \eta + \mu \int_{t_0}^{t} f[a_0(t)]\, dt + \mu \int_{t_0}^{t} \varphi[a_0(t), t]\, dt.$$

However

$$\eta + \mu \int_{t_0}^{t} f[a_0(t)]\, dt \equiv a_0(t),$$

therefore

$$a_1(t) - a_0(t) = \mu \int_{t_0}^{t} \varphi[a_0(t), t]\, dt.$$

It is easy to show that this integral is finite. Let N be the integral part of the ratio $(t - t_0)2\pi$, or the number of whole periods of the integrand in the interval of integration $[t_0, t]$. Then

$$\int_{t_0}^{t} \varphi[a_0(t), t]\, dt = \sum_{k=0}^{N-1} \int_{t_0 + 2\pi k}^{t_0 + 2\pi(k+1)} \varphi[a_0(t), t]\, dt + \int_{t_0 + 2\pi N}^{t} \varphi[a_0(t), t]\, dt =$$

$$= \sum_{k=0}^{N-1} \int_{t_0 + 2\pi k}^{t_0 + 2\pi(k+1)} \{\varphi[a_0(t), t] - \varphi[a_0(t_0 + 2\pi k), t]\}\, dt + \int_{t_0 + 2\pi N}^{t} \varphi[a_0(t), t]\, dt.$$

Using the inequality (9.21) and Lagrange's theorem on the finite increments of functions, we have

$$|\varphi[a_0(t), t] - \varphi[a_0(t_0 + 2\pi k), t]| < Q\, |a_0(t) - a_0(t_0 + 2\pi k)| <$$
$$< \mu M Q\, |t - (t_0 + 2\pi k)|,$$

hence

$$\left| \int_{t_0 + 2\pi k}^{t_0 + 2\pi(k+1)} \{\varphi[a_0(t), t] - \varphi[a_0(t_0 + 2\pi k), t]\}\, dt \right| <$$

$$\mu M Q \int_{t_0 + 2\pi k}^{t_0 + 2\pi(k+1)} [t - (t_0 + 2\pi k)]\, dt = 2\pi^2 k \mu M Q,$$

while

$$\left| \int_{t_0}^{t} \varphi[a_0(t), t]\, dt \right| < 2\pi^2 M Q D + 2\pi P,$$

since $\mu N \leq D$ and

$$\left| \int_{t_0 + 2\pi N}^{t} \varphi[a_0(t), t]\, dt \right| < 2\pi P.$$

Therefore

$$|a_1(t) - a_0(t)| < \mu S, \tag{9.26}$$

† See advanced texts on differential equations.

where
$$S = 2\pi^2 MQD + 2\pi P,$$
and the difference is a quantity of the order of μ.

In order to estimate $a_2(t) - a_0(t)$, we shall observe that
$$|a_2(t) - a_0(t)| \leq |a_2 - a_1| + |a_1 - a_0|.$$
However,
$$a_2 - a_1 = \mu \int_{t_0}^{t} [F(a_1, t) - F(a_0, t)] \, dt;$$
so, using the last of the equations (9.21), we have
$$|a_2(t) - a_1(t)| < \mu B \int_{t_0}^{t} |a_1 - a_0| \, dt < \mu^2 BS |t - t_0| \leq \mu BSD.$$

Therefore we have
$$|a_2(t) - a_0(t)| < \mu S(1 + BD). \tag{9.27}$$

Next, by a similar process we find
$$|a_3 - a_0| < \mu S \left[1 + BD + \frac{(BD)^2}{1 \cdot 2} \right]. \tag{9.28}$$

Continuing further in the same manner, we obtain finally
$$|a_n - a_{n-1}| < \mu S \frac{(BD)^{n-1}}{(n-1)!},$$
$$|a_n - a_0| < \mu S \left[1 + BD + \frac{(BD)^2}{1 \cdot 2} + \ldots \frac{(BD)^{n-1}}{(n-1)!} \right] < \mu S e^{BD}. \tag{9.29}$$

Since $\lim_{n \to \infty} a_n(t)$ is a solution of the equation (9.16), we have now that
$$|a - a_0| \leq \mu S e^{BD}. \tag{9.30}$$

The estimates that have been carried out using the inequality (9.21) are only valid when the functions $a_j(t)$ satisfy
$$|a_j(t) - \eta| < A \tag{9.31}$$
for
$$0 \leq \mu(t - t_0) \leq D$$

This can be verified, for owing to the inequality (9.22), there is a positive number α such that[†]
$$|a_0(t) - \eta| \leq A - \alpha$$
for all t that satisfy
$$0 \leq \mu(t - t_0) \leq D.$$

For $a_1(t)$ (for the same values of t) we deduce from (9.26) that
$$|a_1(t) - \eta| \leq |a_1 - a_0| + |a_0 - \eta| < |\mu S + A - \alpha|,$$
whence it follows that if $a_1(t)$ is to satisfy the inequality (9.31) then $\mu S < \alpha$.

---

† The quantity α is defined as soon as D is assigned and A is chosen. The choice of μ does not affect α.

Next, for $a_2(t)$ we have
$$|a_2(t)-\eta| \leq |a_2-a_0|+|a_0-\eta| < |\mu S(1+BD)+A-\alpha|,$$
and in order that $|a_2-\eta|$ be smaller than A, it suffices to take
$$\mu S(1+BD) < \alpha.$$

By continuing this argument it is easy to show that all the estimates made are valid, if $\mu S e^{BD} < \alpha$, and however small ε may be, we can always find μ such that
$$\mu S e^{BD} < \alpha \tag{9.33}$$
and
$$|a(t)-a_0(t)| < \mu S e^{BD} < \varepsilon$$
for all t that satisfy the inequality $\mu(t-t_0) \leq D$. To do this we choose μ smaller than the smallest of the quantities $\alpha e^{-BD}/S$ and $\varepsilon e^{-BD}/S$.

Thus the proposition is proved for a single order equation but the theorem enunciated at the beginning of the section for the system of the second order (9.7a) can be proved in exactly the same manner and with analogous assumptions for the properties of the right-hand sides of the equations.

2. *Justification of Van der Pol's method for steady-state oscillations*

We shall prove now that, *if the equation $\Phi(K)=0$ has a simple root K_i and $\Phi_i'(K_i) \neq 0$ then for any given positive small number ε it is always possible to find a sufficiently small μ such that the system (9.2) has a limit cycle lying in the ε-neighbourhood of the circle $x^2+y^2 = K_i^2$. This limit cycle is stable if $\Phi'(K_i) < 0$ and unstable if $\Phi'(K_i) > 0$.* The function $\Phi(K)$ has a continuous derivative (at least in the neighbourhood of the root K_i) if the function $f(x, y)$ in (9.2) has continuous derivatives. Suppose that $\Phi(K) = 0$, $\Phi'(K_i) < 0$†, then $K = K_i$ is a stable state of equilibrium of the first truncated equation
$$\frac{dK}{dt} = \mu \Phi(K), \tag{9.11a}$$
which has a stable limit cycle on the x, y phase plane: — the circle of radius K_i. Choose a sufficiently small ε-neighbourhood of this circle (Fig. 470) such that for $K_i-\varepsilon \leq K \leq K_i+\varepsilon$
$$\Phi'(K) \leq -\beta, \tag{9.34}$$
where β is a positive number; this can always be done since $\Phi'(K)$ is a continuous function and $\Phi'(K_i)<0$.

Consider on the x, y phase plane the path Γ:
$$x = x(t), \quad y = y(t)$$
of the equations (9.2) and the path Γ_0
$$\left.\begin{array}{l} x_0(t) = K_0(t)\cos[t+\vartheta_0(t)], \\ y_0(t) = -K_0(t)\sin[t+\vartheta_0(t)] \end{array}\right\} \tag{9.9}$$
of the truncated equations. Both paths pass at $t=0$ through the point $A(0, K_i+\varepsilon)$. As before, $K_0(t), \vartheta_0(t)$ is a solution of the truncated equations (9.11). The path Γ_0 is a spiral winding on the circle $x^2+y^2 = K_i^2$ as $t \to +\infty$, since, for $K_i \leq K \leq K_i+\varepsilon$
$$\Phi(K) \leq -\beta(K-K_i) < 0 \tag{9.34a}$$

† The proof for the case $\Phi'(K_i)>0$ amounts to the change of t into $-t$. The case $\Phi'(K_i)=0$ is impossible since K_i is a simple root of the equation $\Phi(K)=0$.

Hence, $K_0(t)$ decreases monotonically towards K_i as $t \to +\infty$. Let us choose D such that $K_0(t) - K_i \leq \varepsilon/2$ for $t = D/\mu$ and also that the path Γ_0 completes more than one revolution about the origin in an interval D/μ†.

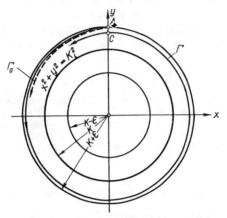

Fig. 470

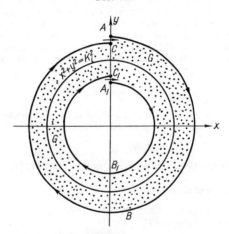

Fig. 471

† According to (9.34a), for $K_i \leq K \leq K_i + \varepsilon$

$$\frac{dK}{dt} = \mu \Phi(K) \leq -\mu\beta(K - K_i),$$

i.e. for the path Γ_0 we have

$$0 < K_0(t) - K_i \leq \varepsilon e^{-\mu\beta t}.$$

We can take, therefore, as the required interval $D = \beta^{-1} \ln 2$. The number of revolutions of the spiral Γ_0 during this interval of time can be made arbitrarily large, provided that we choose a sufficiently small μ.

According to the theorem enunciated in Sub-section 1 of this section, there exists a $\mu = \mu(\varepsilon, D)$ such that the representative point $[x(t), y(t)]$ does not leave the $\varepsilon/2$-neighbourhood of the point $[x_0(t), y_0(t)]$ in the interval of time $0 \leq t \leq D/\mu$. We shall take this value of μ in the system (9.2). For this value of μ, the point $[x(D/\mu), y(D/\mu)]$ of the path Γ will evidently be found *inside* the ε-neighbourhood of the circle $x^2 + y^2 = K_i^2$, and the Γ path makes more than one revolution about the origin during the interval $0 \leq t \leq D/\mu$. Since Γ is a phase path of the autonomous system (9.2) and cannot therefore intersect itself, then the first point of its intersection C with the y axis will have the ordinate

$$y_C < K_i + \varepsilon.$$

Therefore the closed curve $ABCA$ (Fig. 471) consisting of the arc ABC of the path Γ and of the segment CA of the y axis can only be crossed by phase paths of the system (9.2) from outside the area bounded by ABCA, since on CA $y = \dot{x} > 0$.

In exactly the same way we can construct another closed curve $A_1B_1C_1A_1$ consisting of the arc $A_1B_1C_1$ of the path of the system (9.2) that passes through the point $A_1(0, -K_i - \varepsilon)$ and of the segment C_1A_1 of the y axis; the phase paths of the system (9.2) can only cross this curve from the area inside it. Thus, there is an annular region G bounded by the curves $ABCA$ and $A_1B_1C_1A_1$ (Fig. 471) such that the *paths* of the system (9.2) *cannot leave* it (as $t \to \infty$). In G equation (9.2) has no equilibrium states†. Then, according to a theorem in the qualitative theory of differential equations of the second order (see Chapter VI, Section 2), there is in the ε-neighbourhood of the circle $x^2 + y^2 = K_i^2$, a stable limit cycle.

The proof of the existence of an *unstable* limit cycle of the system (9.2) (for a sufficiently small μ) lying in a neighbourhood of the circle $x^2 + y^2 = K_i^2$, where $\Phi(K_i) = 0$ but now $\Phi'(K_i) > 0$, reduces to replacing t by $-t$ in the proof just given. Thus, the proposition enunciated at the beginning of subsection 2 of this section is proved‡.

To conclude, we now prove that for sufficiently small values of μ the system (9.2) has no limit cycles that lie *outside* near neighbourhoods of the circles $x^2 + y^2 = K_i^2$. More precisely, we prove:

If $\Phi(K) \neq 0$ for $0 < R_1 \leq K \leq R_2$: then there exist sufficiently small values of the parameter μ

$$0 < \mu \leq \mu_0,$$

such that the system of equations (9.2) *has no limit cycles in the annular region* R

$$R_1^2 \leq x^2 + y^2 \leq R_2^2.$$

Let $\Phi(K) > 0$ for $R_1 \leq K \leq R_2$. Then, since $\Phi(K)$ is a continuous function there are positive numbers ε and Φ_0 such that for $R_1 \leq K \leq R_2 + \varepsilon$

$$\Phi(K) > \Phi_0 > 0. \tag{9.35}$$

---

† The unique equilibrium state of the system (9.2) with $\mu \ll 1$ lies on the x axis near the origin, its abscissa is determined by the equation

$$-x + \mu f(x, 0) = 0.$$

‡ In the proof we have made use of the theorem on the existence of a limit cycle, which is only valid for autonomous systems of the second order. The proof of the analogous proposition for systems with an arbitrary number of degrees of freedom is contained in the work by N.N. Bogoliubov [35, 36].

Let γ be a path

$$x = x(t), \qquad y = y(t)$$

of the system (9.2) and let γ_0 be a path

$$\left.\begin{array}{l} x_0(t) = K_0(t) \cos [t+\vartheta_0(t)], \\ y_0(t) = -K_0(t) \sin [t+\vartheta_0(t)] \end{array}\right\}$$

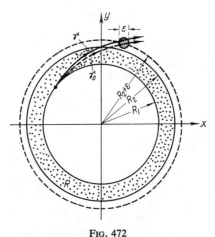

Fig. 472

of the truncated equation, both paths starting at $t = 0$ from a point on the circle $x^2+y^2 = R_i^2$ (Fig. 472). Then, clearly, on the segment $R_1 \leqslant K \leqslant R_2+\varepsilon$

$$\frac{dK}{dt} > \mu \Phi_0 > 0,$$

for the solution $K = K_0(c)$ of the first auxiliary equation, i.e. for the path γ_0

$$K_0(t) > R_1 + \mu \Phi_0 t.$$

Hence

and $\qquad t = \dfrac{R_2+\varepsilon-R_1}{\mu\Phi_0} = \dfrac{D}{\mu} \qquad K_0\left(\dfrac{D}{\mu}\right) > R_2+\varepsilon,$

i.e., during the interval $0 \leqslant t \leqslant D/\mu$, the path γ_0 will intersect the annular region R and will move away beyond the circle $x^2+y^2 = (R_2+\varepsilon)^2$.

However, according to the theorem proved in the first sub-section of this section, a $\mu_0 = \mu_0(\varepsilon, D)$ exists such that, for any given $0 < \mu \leqslant \mu_0$ and for any $0 \leqslant t \leqslant D/\mu$, the representative point $[x(t), y(t)]$ of the system (9.2), moving along the path γ, will not leave the ε-neighbourhood of the point $[x_0(t), y_0(t)]$. Therefore, during the interval of time $0 \leqslant t \leqslant D/\mu$, not only the curve γ_0 but also the path γ of the system (9.2) will intersect the region R and move beyond its boundary.

Since the annular region R does not contain states of equilibrium of the system (9.2) (for sufficiently small values of μ), it can only contain closed phase paths (limit cycles) of the system (9.2) which surround the circle $x^2+y^2 = R_1^2$. The system (9.2), however, cannot have such limit cycles, since, if such a cycle existed, it would intersect the path γ of the same system of equations (9.2) which is impossible†.

Thus we have shown that for sufficiently small values of μ the system of equations (9.2) has limit cycles *close* to the circles $x^2+y^2 = K_i^2$, where K_i are the roots of the equation $\Phi(K)=0$, and has *no other* limit cycles.

§ 4. Application of Van der Pol's Method

We shall use Van der Pol's method to analyse a valve oscillator with a tuned grid or tuned anode circuit (Fig. 465), neglecting, as is usual, the anode conductance and the grid currents. The damping ratio of the tuned circuit is

$$\omega_0 RC \ll 1.$$

The equation of the generator reduces (see Section 1 of this chapter) to the following equation

$$\ddot{x}+x = \mu[-1+\alpha s(x)]\dot{x}, \qquad (9.3)$$

where $x = u/u_0$ (u_0 is a fixed voltage), $\mu = \omega_0 RC \ll 1$, $\alpha = MS_0/RC$, and $s(x) = S(E_g+u_0 x)/S_0$ is the normalized dimensionless slope of the valve characteristic.

The truncated or auxiliary equations (8.11) for this equation are clearly,

$$\frac{dK}{dt} = \mu \Phi(K), \quad \frac{d\vartheta}{dt} = \mu \Psi(K),$$

where

$$\Phi(K) = \frac{1}{2\pi} \int_{-\pi}^{+\pi} [-1+\alpha s(K\cos\xi)]K\sin^2\xi \, d\xi =$$

$$= \frac{K}{\pi} \int_0^{\pi} [-1+\alpha s(K\cos\xi)]\sin^2\xi \, d\xi$$

and

$$\Psi(K) = \frac{1}{2\pi K} \int_{-\pi}^{+\pi} [-1+\alpha s(K\cos\xi)]K\sin\xi\cos\xi \, d\xi \equiv 0 \qquad (9.36)$$

---

† The proof for the case $\Phi(K)<0$ for $R_1 \leq K \leq R_2$ is analogous to the one outlined above, except for the fact that in this case the initial point of the paths γ and γ_0 must be taken on the circle $x^2+y^2 = R_2^2$.

Thus $\Psi(K) \equiv 0$ for any valve characteristic and the period of the self-oscillations (neglecting terms of the order of μ^2) coincides with the period of the undamped oscillations of the tuned circuit.

1. The valve generator with soft operating conditions

The valve characteristic is represented by a third degree polynomial

$$i_a = f(E_g+u) = i_{a0}+S_0u+S_1u^2-S_2u^3. \tag{9.37}$$

The transconductance will be

$$S(E_g+u) = \frac{di_a}{du_g} = S_0+2S_1u-3S_2u^2 \; \dagger.$$

Now $u = u_0x$, so to arrange that the coefficient of x^2 is unity we choose $u_0 = (S_0/3S_2)^{\frac{1}{2}}$. The dimensionless slope or transconductance $s(x) = S/S_0$ is thus

$$s(x) = 1+\beta_1x-x^2,$$

where

$$\beta_1 = \frac{2S_1}{S_0}u_0.$$

The voltage u_0 has some physical meaning. It is a "saturation voltage" for which $S = 0$. It is evident that a cubic only approximately represents a real characteristic for $|u| \leqslant u_0$ i.e. for $|x| \leqslant 1$ (Fig. 473).

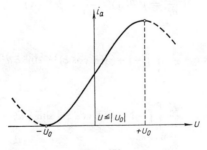

Fig. 473

The equation of the generator is now

$$\ddot{x}+x = \mu[-1+\alpha(1+\beta_1x-x^2)]\dot{x}. \tag{9.38}$$

---

† The slope S_0 at the state of equilibrium is positive: we shall assume also that S_2 is positive. This will ensure a decrease of the slope $S(E_g+U)$ for a large increase of u.

The truncated equations in Van der Pol's polar variables K and ϑ will be

$$\frac{dK}{dt} = \mu \frac{K}{2}\left[\alpha - 1 - \alpha \frac{K^2}{4}\right], \quad \frac{d\vartheta}{dt} = 0. \tag{9.39}$$

The radii of the limit cycles on the x, y plane are given by

$$\Phi(K) \equiv \frac{K}{2}\left[\alpha - 1 - \alpha \frac{K^2}{4}\right] = 0. \tag{9.40}$$

Two cases are possible.

If $\alpha < 1$, (i.e. $MS_0 < RC$ and the generator is not self-excited) then the only real root of Φ is $K = 0$ corresponding to an equilibrium state at $(0,0)$. This singular point is stable since

$$\Phi'(0) = \frac{\alpha - 1}{2} < 0.$$

All remaining paths are spirals that approach the origin asymptotically for $t \to +\infty$. The phase portrait is typical of damped oscillations (Fig. 474(a)).

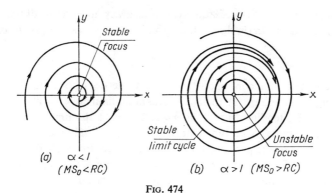

Fig. 474

For $\alpha > 1$ (i.e. for $MS_0 > RC$ when the generator is self-excited) the equation (9.40) has two roots of interest:

$$K = 0 \quad \text{and} \quad K = 2\sqrt{\frac{\alpha - 1}{\alpha}} = K_0.$$

The first corresponds to an unstable singular point since

$$\Phi'(0) = \frac{\alpha-1}{2} > 0.$$

The second corresponds to a stable limit cycle of radius

$$K_0 = 2\sqrt{\frac{\alpha-1}{\alpha}} = 2\sqrt{\frac{MS_0-RC}{MS_0}}, \qquad (9.41)$$

since

$$\Phi'(K_0) = -\frac{\alpha K_0^2}{4} < 0.$$

The remaining paths are divided into two classes: paths that wind externally on to the limit cycle for $t \to +\infty$ and recede to infinity for $t \to -\infty$, and paths that wind internally on to the limit cycle for $t \to +\infty$ and tend to the origin for $t \to -\infty$. The phase plane picture is typical of the simplest self-oscillating system with *soft* excitation (Fig. 474(*b*)).

The amplitude of the self-oscillations is given by the radius of the limit cycle K_0 and in dimensional units is evidently equal to

$$U_0 = u_0 K_0 = 2\sqrt{\frac{MS_0-RC}{3MS_2}}.$$

The period of the self-oscillations (to an accuracy up to terms of the order of μ^2) is equal to 2π (in dimensionless units), since $\Psi(K) \equiv 0$, or in ordinary units

$$T = 2\pi\sqrt{LC}.$$

If α is decreased continuously from a value $\alpha > 1$ (for example, by decreasing the coupling M), then the radius of the limit cycle will also decrease continuously, tending to zero as $\alpha \to 1$. For $\alpha = 1$ the limit cycle merges with the unstable focus and the origin becomes a stable focus. $\alpha = 1$ is a branch value of the parameter α[†]. If we vary α continuously from $\alpha < 1$ to $\alpha > 1$, then self-oscillations begin at $\alpha = 1$, their amplitude increasing continuously. As α varies in the opposite direction, the amplitude of the oscillations decreases continuously to zero. The generator then behaves as a damped oscillator (Fig. 475). This behaviour is called a *soft build-up of oscillation* in contrast to the hard build-up of self-oscillations, when oscillations of a finite amplitude are established suddenly even though some parameter varies continuously.

[†] The bifurcation theory for the case considered is given in a general form at Section 10 of this chapter.

The auxiliary truncated equations enable approximate analytical expressions for the oscillations to be found when $\alpha > 1$. Integrating the equations (9.39) we find

$$K = \frac{K_0}{\sqrt{1+Ce^{-\mu(\alpha-1)t}}}, \quad \vartheta = \vartheta_0 = \text{const}$$

where C is a constant determined by the initial value of K.

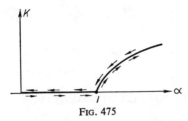

Fig. 475

It is easily seen that $-1 < C \leqslant +\infty$.† Hence,

$$\left. \begin{array}{l} x(t) = \dfrac{K_0 \cos(t+\vartheta_0)}{\sqrt{1+Ce^{-\mu(\alpha-1)t}}}, \\[2mm] y(t) = -\dfrac{K_0 \sin(t+\vartheta_0)}{\sqrt{1+Ce^{-\mu(\alpha-1)t}}}. \end{array} \right\} \qquad (9.42)$$

This is an approximate expression for the general integral of the equations (9.38), since there occur two arbitrary constants: C and ϑ_0 ($C=0$ corresponds to the limit cycle and $C=\infty$ to the equilibrium state). The expression (9.37) contains a quadratic term which does not enter at all into the zero-order approximation for the general solution (its presence affects the subsequent approximations only.) This is a general property of all even terms of the characteristic. This is due to the fact that the expansion of even powers of sines and cosines will only contain sines and cosines of even multiples of the angles and, therefore, their expansion will not contain the fundamental (resonance) frequency.

2. The valve generator whose characteristic is represented by a polynomial of the fifth degree

Let us consider again the same valve generator, but with a more accurate expression for the valve characteristic

$$i_a = i_{a0} + S_0 u + S_1 u^2 + S_2 u^3 + S_3 u^4 - S_4 u^5. \qquad (9.43)$$

† It is evident that the smaller $\mu(\alpha-1) = \omega_0(MS_0 - RC)$ (i.e. the closer the generator to the threshold of self-excitation) the slower the build-up of self-oscillations.

The slope or transconductance is

$$S(E_g+u) = \frac{di_a}{du} = S_0+2S_1u+3S_2u^2+4S_3u^3-5S_4u^4 \dagger.$$

Putting $u=u_0x$, where $u_0=(S_0/5S_4)^{\frac{1}{4}}$, we shall obtain for the normalized slope $s=S/S_0$

$$s(x) = \frac{S(E_g+u_0x)}{S_0} = 1+\beta_1x+\beta x^2+\beta_3x^3-x^4,$$

and the equation of the valve generator is now

$$\ddot{x}+x = \mu[-1+\alpha(1+\beta_1x+\beta x^2+\beta_3x^3-x^4)]\dot{x}. \qquad (9.44)$$

According to (9.11) and (9.12) the auxiliary equations are

$$\frac{dK}{dt} = \mu\Phi(K), \quad \frac{d\vartheta}{dt} = 0, \qquad (9.45)$$

where

$$\Phi(K) = \frac{\alpha K}{2}\left[\frac{\alpha-1}{\alpha}+\frac{\beta K^2}{4}-\frac{K^4}{8}\right].$$

$K = 0$ is a root of $\Phi = 0$, so there is a singular point at the origin. Since

$$\Phi'(0) = \frac{\alpha-1}{2},$$

then this equilibrium state is stable for $\alpha<1$, and unstable for $\alpha>1$. The remaining roots of $\Phi(K)=0$ are different from zero and are the radii of limit cycles. They are clearly the roots of the biquadratic equation

$$\frac{K^4}{8}-\frac{\beta K^2}{4}-\frac{\alpha-1}{\alpha} = 0, \qquad (9.46)$$

which cannot have more than two positive roots. Construct a diagram (Fig. 476) with $\gamma=(\alpha-1)/\alpha$ as the horizontal axis and $\varrho=K_i^2$ as the vertical axis. If $\beta<0$ (Fig. 476(a)), then γ is a monotonic increasing function of ϱ (for $\varrho>0$) and the equation (9.46) has no positive roots for $\gamma<0$ (i.e. for $\alpha<1$), but has a single positive root K_1 for $\gamma>0$ (for $\alpha>1$). If, however,

---

† S_0 and S_4 are positive. The condition $S_4>0$ ensures a decrease of the slope S for a large increase of $|u|$ which, as we shall see, is a necessary condition for the existence of stable self-oscillations.

$$\beta_1 = \frac{2S_1u_0}{S_0}, \quad \beta = \frac{3S_2u_0^2}{S_0}, \quad \beta_3 = \frac{4S_3u_0^3}{S_0}.$$

$\beta > 0$ (Fig. 476(b), then the parabola (9.46a) intersects the ϱ axis at two points, where $\varrho = 0$ and $\varrho = 2\beta$. The vertex is at the point $\varrho = \beta$, $\gamma = -\beta^2/8$, and the parabola is open to the right. Therefore, for $\beta > 0$ the equation

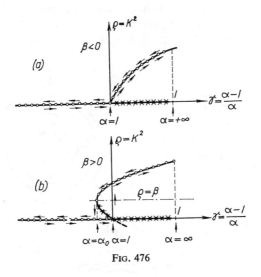

Fig. 476

(9.46) has no positive roots for $\gamma < -\beta^2/8$ (i.e. for $\alpha < \alpha_0 = [1+(\beta^2/8)]^{-1}$), has two positive roots K_1 and K_2 for $-\beta^2/8 < \gamma < 0$ (for $\alpha_0 < \alpha < 1$), and finally one positive root for $\gamma > 0$ (for $\alpha > 1$).

Since
$$\Phi'(K_i) = -\frac{\alpha K_i^2}{2}(K_i^2 - \beta), \tag{9.47}$$

then, for $\beta < 0$, and $\alpha > 1$ there is a unique stable limit cycle. The phase portrait is qualitatively the same as in the last example (Fig. 474) and is typical of a *soft build-up of oscillations*.

For $\beta > 0$ (i.e. for $S_2 > 0$), the only stable limit cycle is one whose radius is

$$K_i > \sqrt{\beta},$$

and lies on the upper half of the parabola (9.46a) (indicated in Fig. 476(b) by hollow circles). The arc of the parabola between the axis of the parabola and the horizontal axis corresponds to unstable limit cycles. Thus, for $\beta > 0$ we have three qualitatively different phase portraits (Fig. 477). For $\alpha < \alpha_0$ (Fig. 477(a)) there is a stable focus at the origin. The generator is not self-excited and any oscillatiosn in it are damped.

For $\alpha > 1$ (Fig. 477(b)) the origin is an unstable singular point and all paths tend (for $t \to +\infty$) to a single stable limit cycle, with *soft operating conditions*.

Finally, for $\alpha_0 < \alpha < 1$ (Fig. 477(c) a stable equilibrium state (0,0) and a stable limit cycle of radius K_2 are separated by an unstable limit cycle of radius K_1. Therefore the paths that start inside the unstable limit cycle

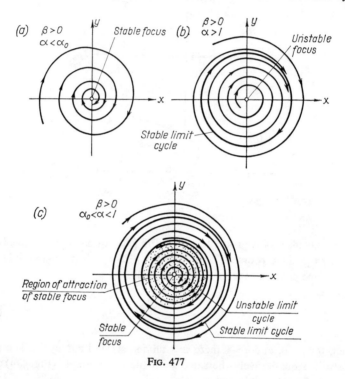

Fig. 477

move towards the state of equilibrium and only the paths that start *outside* the unstable limit cycle will wind on the stable limit cycle[†]. The system has *hard operating conditions* and requires an "impulse" to initiate self-oscillations.

Fig. 478 is a stability diagram on the α, β plane. If at first the generator is not excited, $\beta > 0$ but $\alpha < \alpha_0$, then, for a slow and continuous increase of α, the representative point remains at or near the equilibrium state at the

---

† An unstable limit cycle does not correspond, of course, to self-oscillating processes in the generator. It is the boundary that separates "the region of attraction" of stable self-oscillations from the "region of attraction" of a stable state of equilibrium.

origin until $\alpha = 1$, when the equilibrium state becomes unstable, and there is a stable limit cycle of radius $K = (2\beta)^{\frac{1}{2}}$. As α increases further self-oscillations of a finite amplitude are established. For a further increase of α, the amplitude increases monotonically.

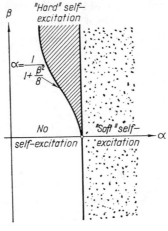

Fig. 478

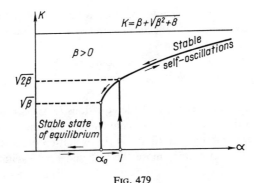

Fig. 479

If α now decreases oscillations persist, even at $\alpha = 1$, until α becomes equal to α_0. As α passes through this branch value the stable limit cycle disappears and the self-oscillations vanish (the final amplitude was $\beta^{\frac{1}{2}}$) and the system passes on to stable equilibrium state.

For $\beta > 0$, the onset and quenching of self-oscillations occur for different values of the excitation coefficient $\alpha = MS_0/RC$.

The self-oscillations arise or are quenched with different (but finite) amplitudes. This is typical of a *hard build-up* of self-oscillations (Fig. 479)[†].

Since

$$S_2 = \frac{1}{2}\left[\frac{d^2 S}{du_g^2}\right]_{u_g = E_g},$$

then the intervals of values of grid bias E_g in which there occur a soft or a hard excitation can be determined in the following manner. Construct, according to the given approximate characteristic $i_a = f(u_g)$ (Fig. 480),

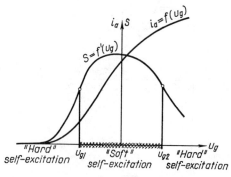

Fig. 480

the graph of the transconductance or slope $S = f'(u_g)$ against u_g and mark on this curve the points of inflexion u_{g1} and u_{g2}. Then, for $u_{g1} < E_g < u_{g2}$, $(d^2 S/du_g^2)_{u_g = E_g} < 0$ and for these values of grid bias we shall have a soft excitation of self-oscillations. On the contrary outside this interval (for $E_g < u_{g1}$ or $E_g > U_{g2}$) $(d^2 S/du_g^2)_{u_g = E_g} > 0$ and there is a hard excitation of oscillations.

We observe, in conclusion, that the equations (9.45) can be integrated as was done in the case of the cubic characteristic for the valve. The solutions so obtained describe quantitatively the excitation processes (the transients). We will discuss once more the soft and hard excitation of self-oscillations in a valve generator in Section 10 of this chapter in connexion with bifurcation or branch theory.

[†] We use the words "soft" and "hard" in two senses. In the first place, we speak of a soft or hard *mode of operation* of a self-oscillatory system for *assigned values of its parameters* depending on whether a self-oscillating process is established for all or not for all *initial* conditions. In the second place we speak of a soft or a hard *excitation* (onset) of self-oscillations depending on the way the amplitude of the self-oscillations varies for a slow and continuous *variation* of a *parameter*. It is clear that for a hard onset of self-oscillations the system will be under hard operating conditions for certain values of its parameter.

3. Self-oscillations in a valve generator with a two-mesh RC circuit

The equations of the generator with a two-mesh RC circuit (Fig. 481(a)) are

$$C\frac{dv}{dt} = \frac{u-E_g}{R}, \quad \frac{E_a-(u+v)}{R_a} = i(u)+C\frac{dv}{dt}+C_a\frac{d(u+v)}{dt}$$

(see also Section 12, Chapter V and Section 5, Chapter VIII), or

$$R_a R_g C C_a \frac{d^2u}{dt^2} + R_g C\left[1+\frac{R_a}{R_g}\left(1+\frac{C_a}{C}\right)+R_a S(u)\right]\frac{du}{dt}+u = E_g$$

or, after introducing the new dimensionless variables

$$t_{\text{new}} = \frac{t}{\sqrt{R_a R_g C C_a}} \quad \text{and} \quad x = \frac{u-E_g}{u_0}$$

(u_0 is a certain voltage) to the form

$$\ddot{x}+x = -\sqrt{\frac{R_g C}{R_a C_a}}\left[1+\frac{R_a}{R_g}\left(1+\frac{C_a}{C}\right)+R_a S(E_g+u_0 x)\right]\dot{x}, \quad (9.48)$$

where $S(u)=di/du$ is the slope of the characteristic of the valve-pair. Since this characteristic $i=i(u)$ is a *descending* characteristic (Fig. 481(b)), then

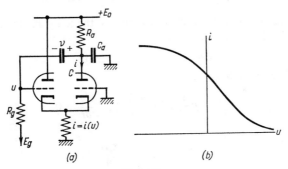

Fig. 481

$S(u) < 0$. For self-excitation of the generator, so that $x = 0$, $\dot{x} = 0$ is an unstable singular point, it is necessary that

$$R_a S_0 > 1+\frac{R_a}{R_g}\left(1+\frac{C_a}{C}\right), \quad (9.49)$$

where $S_0 = -S(E_g)$ is the numerical value of the slope of the characteristic of the valve pair in the state of equilibrium (working point).

Let this condition be satisfied and the generator be self-excited. Again we represent the characteristic $i=i(u)$ by means of a polynomial of the third degree. Then

$$S(E_g+u_0x) = -S_0+S_1x+S_2x^2 \;\dagger$$

and the equation of the generator (9.48) will be

$$\ddot{x}+x = \sqrt{\frac{R_gC}{R_aC_a}}\left\{R_aS-\left[1+\frac{R_a}{R_g}\left(1+\frac{C_a}{C}\right)\right]-R_aS_1x-R_aS_2x^2\right\}\dot{x}.$$

The oscillations of the generator are nearly sinusoidal only when

$$R_aS_0-\left[1+\frac{R_a}{R_g}\left(1+\frac{C_a}{C}\right)\right] \ll 1, \quad R_a|S_1| \ll 1, \quad R_aS_2 \ll 1,$$

i.e. when the generator is close to the threshold of self-excitation and the departure from linearity of the characteristic is small.

Now introduce the small parameter

$$\mu = \sqrt{\frac{R_gC}{R_aC_a}}\left\{R_aS_0-\left[1+\frac{R_a}{R_g}\left(1+\frac{C_a}{C}\right)\right]\right\}$$

($0 < \mu \ll 1$) and the notation

$$\sqrt{\frac{R_gC}{R_aC_a}}R_aS_1 = \mu\alpha_1, \quad \sqrt{\frac{R_gC}{R_aC_a}}R_aS_2 = \mu\alpha_2.$$

Then the equations of the generator reduce to the following form, suitable for Van der Pol's method:

$$\ddot{x}+x = \mu[1-\alpha_1x-\alpha_2x^2]\dot{x}. \qquad (9.50)$$

The auxiliary equations are

$$\frac{dK}{dt} = \mu\frac{K}{2}\left(1-\frac{\alpha_2K^2}{4}\right), \quad \frac{d\vartheta}{dt} = 0.$$

From the function

$$\Phi(K) = \frac{K}{2}\left(1-\frac{\alpha_2K^2}{4}\right),$$

we see that the system (9.50) has an equilibrium state ($x=0$, $\dot{x}=0$) corresponding to the root $K=0$, and a limit cycle of radius

$$K_0 = 2\sqrt{\frac{1}{\alpha_2}} = 2\sqrt{\frac{R_aS_0-\left[1+\frac{R_a}{R_g}\left(1+\frac{C_a}{C}\right)\right]}{R_aS_2}}. \qquad (9.51)$$

† The coefficients S_0, S_1, S_2, which have the dimensions of a conductance, depend, for a given value of u_0, upon E_g. Now $S_0 > 0$ and in order that self-oscillations exist we also assume that $S_2 > 0$.

The equilibrium state is unstable, since

$$\Phi'(0) = \frac{1}{2} > 0^\dagger,$$

and the limit cycle is stable, since

$$\Phi'(K_0) = -\frac{\alpha_2}{4} K_0^2 = -1 < 0.$$

Thus a soft mode of operation exists and nearly sinusoidal self-oscillations with an amplitude K_0 are established for any initial conditions[‡]. Their period is clearly very nearly equal to

$$T = 2\pi \sqrt{R_a R_g C C_a}.$$

§ 5. Poincaré's method of perturbations

We shall consider here the integration of non-linear equations, given by Poincaré in his works on celestial mechanics [184, 185]. Notwithstanding the restrictions imposed on the choice of the equation this method is most useful in practice for finding *periodic* solutions of a non-linear equation depending on a parameter μ. We assume that for some $\mu = \mu_0$ (for example, for $\mu = 0$) the equation or system of equations has a well-known periodic solution.

The method is useful only for values of μ that differ little from μ_0, and there is no loss of generality if we assume that for $\mu = \mu_0 = 0$ the equations have linear constant coefficients. The line of reasoning, however, is quite general and is applicable under different assumptions. The system of non-linear equations is

$$\frac{dx}{dt} = ax + by + \mu f_1(x, y, \mu); \quad \frac{dy}{dt} = cx + dy + \mu f_2(x, y, \mu), \quad (9.52)$$

where a, b, c, d and μ are constants, and μ is small. We shall assume also that it is possible to expand f_1 and f_2 into convergent power series in x and y (at least for small values) or that they are polynomials.

Consider first the reduced equations when $\mu = 0$:

$$\frac{dx}{dt} = ax + by, \quad \frac{dy}{dt} = cx + dy, \quad (9.53)$$

---

† Remember that we are considering the case of a self-exciting generator, when $R_a S_0 > 1 + (1 + C_a/C) R_a/R_g$.

‡ If we approximate the characteristic to a polynomial of the fifth degree, then we obtain both soft and hard modes of excitation.

or
$$\ddot{x} - (a+d)\dot{x} + (ad-bc)x = 0. \qquad (9.53a)$$

(9.53a) must have periodic solutions, so that we must have

$$(a+d) = 0, \qquad \begin{vmatrix} a & b \\ c & d \end{vmatrix} > 0. \qquad (9.54)$$

Then the roots of the characteristic equation are

$$\lambda_1 = +j\sqrt{ad-bc}, \quad \lambda_2 = -j\sqrt{ad-bc},$$

and there is a periodic solution, whose amplitude and phase angle depend on initial conditions, but with a well-defined angular frequency $\omega_1 = |\lambda_1| = |\lambda_2| = (ad-bc)^{\frac{1}{2}}$. These solutions have the form

$$x = K\cos(\omega_1 t + \chi); \qquad y = kK\sin(\omega_1 t + \chi + \chi'),$$

where k and χ' depend on a, b, c, and d but K and χ are arbitrary. Otherwise

$$x = \varphi_0(t, \chi, K); \qquad y = \psi_0(t, \chi, K),$$

where φ_0 and ψ_0 are periodic in t with period $2\pi/\omega_1$. Such a general type of solution will also be obtained if the reduced system is *non-linear but conservative*, and has a *centre*. By choosing the origin of t we make $\chi = 0$ and then

$$x = x_0(t, K) = K\cos\omega_1 t; \qquad y = y_0(t, K) = kK\sin(\omega_1 t + \chi'). \qquad (9.55)$$

1. The procedure in Poincaré's method

Suppose that the solution of equations (9.53) is $x = x_0(t, K)$: $y = y_0(t, K)$. Now suppose the solutions of equations (9.52) (for $\mu \neq 0$.) to be $x = x(t, \mu, K), y = y(t, \mu, K)$, taking the values $x = x_0(0, K) + \beta_1, y = y_0(0, K) + \beta_2$ at $t = 0$. Poincaré searches for solutions in the form of power series in β_1, β_2 and μ and proves that they converge uniformly for sufficiently small values of β_1, β_2 and μ, within any finite interval $0 < t < t_1$. The coefficients of these power series are functions of t and can be evaluated by equating the coefficients of equal powers of β_1, β_2 and μ after substitution of the power series in the equation (9.52). Thus we obtain for x and y certain expressions,

$$x = x(t, \mu, \beta_1, \beta_2, K), \qquad y = y(t, \mu, \beta_1, \beta_2, K), \qquad (9.56)$$

If (9.56) is periodic we can write its new period as $T+\tau$ where τ is small and T is the period of the solution of the reduced equation. It is easily seen that the conditions for periodicity are

$$x(T+\tau, \mu, \beta_1, \beta_2, K) - x(0, \mu, \beta_1, \beta_2, K) = 0,$$
$$y(T+\tau, \mu, \beta_1, \beta_2, K) - y(0, \mu, \beta_1, \beta_2, K) = 0,$$

or, since T is known can be rewritten as

$$\Phi(\tau, \mu, \beta_1, \beta_2, K) = 0, \quad \Psi(\tau, \mu, \beta_1, \beta_2, K) = 0. \tag{9.57}$$

Thus we have two equations with three unknowns τ, β_1 and β_2, but since the original equations are autonomous and the phase arbitrary, we can choose $\beta_2 = 0$.

For $\mu = 0$ the periodic solutions must have a period T, and clearly, for $\mu = 0$ and $\tau = 0$ the condition of periodicity is satisfied. Therefore μ is a common factor and the conditions for periodicity (9.57) can be rewritten as

$$\mu\Phi_1\left(\frac{\tau}{\mu}, \mu, \beta_1, \beta_2, K\right) = 0, \quad \mu\Psi_1\left(\frac{\tau}{\mu}, \mu, \beta_1, \beta_2, K\right) = 0$$

or

$$\Phi_1\left(\frac{\tau}{\mu}, \mu, \beta_1, \beta_2, K\right) = 0, \quad \Psi_1\left(\frac{\tau}{\mu}, \mu, \beta_1, \beta_2, K\right) = 0.$$

So that $\beta_1 = \tau = 0$ when $\mu = 0$, these equations must not contain zero-order terms. By equating to zero these zero-order terms we determine the amplitude K and the first order correction $(\tau/\mu)_{\mu \to 0}$ to the period.

Periodic solutions can exist, but not with arbitrary values of K.

2. Poincaré's method for almost linear systems

We consider a differential equation of a nearly linear system

$$\ddot{y} + y = \mu f(y, \dot{y}), \tag{9.2}$$

$f(y, \dot{y})$ can be expanded as a power series in y and $\dot{y}$. For $\mu = 0$ the solution is

$$y = K \cos t, \tag{9.58}$$

when the time origin is suitably chosen.

Since equation (9.2) does not contain t explicitly then the phase paths are non-intersecting curves on the $y, \dot{y}$ plane. For $\mu = 0$ let

$$y = K \cos t = \varphi_0(t), \tag{9.59}$$

and the paths are concentrical circles.

From the point of view of the $y, \dot{y}$ phase plane, the first part of our problem is: the integral curves for $\mu = 0$ are a family of circles but for $\mu \neq 0$ circles turn into spirals and only certain integral curves remain closed as limit cycles. It is required to determine the value of K for those circles near to which limit cycles form. With $\beta_1 = y(0) - \varphi_0(0)$ and $\beta_2 = \dot{y}(0) - \dot{\varphi}_0(0)$, we write the solution of (9.2)

$$y = \varphi_0(t) + A\beta_1 + B\beta_2 + C\mu + D\beta_1\mu + E\beta_2\mu + F\mu^2 + \ldots, \qquad (9.60)$$

where $A, B, C, D, E, F, \ldots$, are so-far unknown functions of time. It follows that

$$\dot{y} = \dot{\varphi}_0(t) + \dot{A}\beta_1 + \dot{B}\beta_2 + \dot{C}\mu + \dot{D}\beta_1\mu + \dot{E}\beta_2\mu + \dot{F}\mu^2 + \ldots, \qquad (9.61)$$
$$\ddot{y} = \ddot{\varphi}_0(t) + \ddot{A}\beta_1 + \ddot{B}\beta_2 + \ddot{C}\mu + \ddot{D}\beta_1\mu + \ddot{E}\beta_2\mu + \ddot{F}\mu^2 + \ldots, \qquad (9.62)$$

Since y and $\dot{y}$ are nearly equal to $\varphi_0(t)$ and $\dot{\varphi}_0(t)$, then we can expand the function $f(y, \dot{y})$ as a Taylor series about $\varphi_0(t)$ and $\dot{\varphi}_0(t)$, replacing y and $\dot{y}$ by the series (9.60) and (9.61).

$$f(y, \dot{y}) = f[\varphi_0(t), \dot{\varphi}_0(t)] +$$
$$+ f'_y[\varphi_0(t), \dot{\varphi}_0(t)] [A\beta_1 + B\beta_2 + C\mu + D\beta_1\mu + E\beta_2\mu + F\mu^2 + \ldots] +$$
$$+ f'_{\dot{y}}[\varphi_0, \dot{\varphi}_0] [\dot{A}\beta_1 + \dot{B}\beta_2 + \dot{C}\mu + \ldots] + \frac{1}{2} f''_{yy}[\varphi_0, \dot{\varphi}_0] [A\beta_1 + B\beta_2 +$$
$$+ C\mu + \ldots]^2 + f''_{y\dot{y}}[\varphi_0, \dot{\varphi}_0] [A\beta_1 + B\beta_2 + C\mu + \ldots] [\dot{A}\beta_1 + \dot{B}\beta_2 +$$
$$\dot{C}\mu + \ldots] + \frac{1}{2} f''_{\dot{y}\dot{y}}[\varphi_0, \dot{\varphi}_0] [\dot{A}\beta_1 + \dot{B}\beta_2 + \dot{C}\mu + \ldots]^2 + \ldots \qquad (9.63)$$

Substituting the expressions for $y, \ddot{y}$ and $f(y, \dot{y})$ in the original equation (9.2) and equating coefficients of similar terms in β_1, β_2 and μ, we shall obtain a system of non-homogeneous linear differential equations of the second order with constant coefficients and with periodic right-hand sides. The first six equations that determine A, B, C, D, E and F are

$$\left.\begin{array}{ll} \ddot{A} + A = 0, & \ddot{D} + D = f'_y(\varphi_0, \dot{\varphi}_0)A + f'_{\dot{y}}(\varphi_0, \dot{\varphi}_0)\dot{A}, \\ \ddot{B} + B = 0, & \ddot{E} + E = f'_y(\varphi_0, \dot{\varphi}_0)B + f'_{\dot{y}}(\varphi_0, \dot{\varphi}_0)\dot{B}, \\ \ddot{C} + C - (\varphi_0, \dot{\varphi}_0), & \ddot{F} + F = f'_y(\varphi_0, \dot{\varphi}_0)C + f'_{\dot{y}}(\varphi_0, \dot{\varphi}_0)\dot{C}. \end{array}\right\} \qquad (9.64)$$

Remembering

$$\beta_1 = y(0) - \varphi_0(0) \quad \text{and} \quad \beta_2 = \dot{y}(0) - \dot{\varphi}_0(0) \qquad (9.65)$$

and using (9.60) and (9.61) we obtain

$$\left.\begin{array}{l} A(0) = 1, \quad \dot{A}(0) = 0, \quad B(0) = 0, \quad \dot{B}(0) = 1, \\ C(0) = \dot{C}(0) = D(0) = \dot{D}(0) = E(0) = \\ \dot{E}(0) = F(0) = \dot{F}(0) = 0 \end{array}\right\} \qquad (9.66)$$

POINCARÉ'S METHOD

The first two equations of (9.64) have with initial conditions (9.66), the solutions

$$A = \cos t, \quad B = \sin t.$$

Since the solution of $\ddot{x} + x = \Phi(t)$ with the initial conditions $x = \dot{x} = 0$ at $t = 0$, is

$$x = \int_0^t \Phi(u) \sin(t-u)\, du, \quad \dot{x} = \int_0^t \Phi(u) \cos(t-u)\, du$$

Therefore for C, D, etc. we find

$$\left.\begin{aligned}
A(t) &= \cos t, & \dot{A}(t) &= -\sin t, \\
B(t) &= \sin t, & \dot{B}(t) &= \cos t, \\
C(t) &= \int_0^t [f] \sin(t-u)\, du, & \dot{C}(t) &= \int_0^t [f] \cos(t-u)\, du, \\
D(t) &= \int_0^t \left\{\left[\frac{\partial f}{\partial y}\right] \cos u - \left[\frac{\partial f}{\partial \dot{y}}\right] \sin u\right\} \sin(t-u)\, du, \\
\dot{D}(t) &= \int_0^t \left\{\left[\frac{\partial f}{\partial y}\right] \cos u - \left[\frac{\partial f}{\partial \dot{y}}\right] \sin u\right\} \cos(t-u)\, du, \\
E(t) &= \int_0^t \left\{\left[\frac{\partial f}{\partial y}\right] \sin u + \left[\frac{\partial f}{\partial \dot{y}}\right] \cos u\right\} \sin(t-u)\, du, \\
\dot{E}(t) &= \int_0^t \left\{\left[\frac{\partial f}{\partial y}\right] \sin u + \left[\frac{\partial f}{\partial \dot{y}}\right] \cos u\right\} \cos(t-u)\, du, \\
F(t) &= \int_0^t \left\{\left[\frac{\partial f}{\partial y}\right] C(u) + \left[\frac{\partial f}{\partial \dot{y}}\right] \dot{C}(u)\right\} \sin(t-u)\, du, \\
\dot{F}(t) &= \int_0^t \left\{\left[\frac{\partial f}{\partial y}\right] C(u) + \left[\frac{\partial f}{\partial \dot{y}}\right] \dot{C}(u)\right\} \cos(t-u)\, du.
\end{aligned}\right\} \quad (9.67)$$

Here, and below, the square brackets around f, $\partial f/\partial y$ and $\partial f/\partial \dot{y}$ indicate that we evaluate these expressions with y and $\dot{y}$ replaced by

$$\varphi_0(u) = K \cos u \quad \text{and} \quad \dot{\varphi}_0(u) = -K \sin u.$$

We need the values of these functions for $t=2\pi$†,

$$\begin{aligned}
&A(2\pi) = 1, & &\dot{A}(2\pi) = 0, \\
&B(2\pi) = 0, & &\dot{B}(2\pi) = 1, \\
&C(2\pi) = -\int_0^{2\pi} f(K\cos u, -K\sin u)\sin u\, du, \\
&\dot{C}(2\pi) = \int_0^{2\pi} f(K\cos u, -K\sin u)\cos u\, du, \\
&D(2\pi) = \int_0^{2\pi} \left\{-\frac{1}{2}\left[\frac{\partial f}{\partial y}\right]\sin 2u + \left[\frac{\partial f}{\partial y}\right]\sin^2 u\right\} du, \\
&\dot{D}(2\pi) = \int_0^{2\pi} \left\{\left[\frac{\partial f}{\partial y}\right]\cos^2 u - \frac{1}{2}\left[\frac{\partial f}{\partial y}\right]\sin 2u\right\} du, \\
&E(2\pi) = \int_0^{2\pi} \left\{-\left[\frac{\partial f}{\partial y}\right]\sin^2 u - \frac{1}{2}\left[\frac{\partial f}{\partial y}\right]\sin 2u\right\} du, \\
&\dot{E}(2\pi) = \int_0^{2\pi} \left\{\frac{1}{2}\left[\frac{\partial f}{\partial y}\right]\sin 2u + \left[\frac{\partial f}{\partial y}\right]\cos^2 u\right\} du, \\
&F(2\pi) = -\int_0^{2\pi} \left\{\left[\frac{\partial f}{\partial y}\right]C(u) + \left[\frac{\partial f}{\partial \dot{y}}\right]\dot{C}(u)\right\} \sin u\, du, \\
&\dot{F}(2\pi) = \int_0^{2\pi} \left\{\left[\frac{\partial f}{\partial y}\right]C(u) + \left[\frac{\partial f}{\partial \dot{y}}\right]\dot{C}(u)\right\} \cos u\, du.
\end{aligned} \qquad (9.68)$$

The expressions for $D(2\pi)$, $\dot{D}(2\pi)$, $E(2\pi)$ and $\dot{E}(2\pi)$ can be simplified by using the identities

$$\frac{1}{K}\frac{d}{du}\{[f]\cos u\} = -\frac{[f]}{K}\sin u - \frac{1}{2}\left[\frac{\partial f}{\partial y}\right]\sin 2u - \left[\frac{\partial f}{\partial y}\right]\cos^2 u,$$

$$\frac{1}{K}\frac{d}{du}\{[f]\sin u\} = \frac{[f]}{K}\cos u - \left[\frac{\partial f}{\partial y}\right]\sin^2 u - \frac{1}{2}\left[\frac{\partial f}{\partial y}\right]\sin 2u.$$

They become

† Note that the time scale is such that a period of oscillations is 2π.

POINCARÉ'S METHOD

$$D(2\pi) = \int_0^{2\pi} \left[\frac{\partial f}{\partial y}\right] du - \frac{C(2\pi)}{K},$$

$$\dot{D}(2\pi) = \int_0^{2\pi} \left[\frac{\partial f}{\partial \dot{y}}\right] du - \frac{\dot{C}(2\pi)}{K},$$

$$E(2\pi) = -\frac{1}{K}\dot{C}(2\pi), \quad \dot{E}(2\pi) = \frac{1}{K}C(2\pi).$$

(9.68a)

Now if $C(2\pi) = 0$, then

$$D(2\pi) = \int_0^{2\pi} \left[\frac{\partial f}{\partial \dot{y}}\right] du, \quad \dot{D}(2\pi) = \int_0^{2\pi} \left[\frac{\partial f}{\partial \dot{y}}\right] du - \frac{\dot{C}(2\pi)}{K},$$

$$E(2\pi) = -\frac{1}{K}\dot{C}(2\pi), \quad \dot{E}(2\pi) = 0.$$

(9.68b)

and if $C(2\pi) = \dot{C}(2\pi) = 0$, then

$$D(2\pi) = \int_0^{2\pi} \left[\frac{\partial f}{\partial y}\right] du, \quad \dot{D}(2\pi) = \int_0^{2\pi} \left[\frac{\partial f}{\partial \dot{y}}\right] du,$$

$$E(2\pi) = 0, \quad \dot{E}(2\pi) = 0.$$

(9.68c)

$D(2\pi)$ and $\dot{D}(2\pi)$ are obviously the constant terms in the Fourier expansions of $[\partial f/\partial y]$ and of $[\partial f/\partial \dot{y}]$ multiplied by 2π. $C(2\pi)$ and $\dot{C}(2\pi)$ are the coefficients of $\sin t$ and $\cos t$ in the expansions as Fourier series of $f(\varphi_0, \dot{\varphi}_0)$, multiplied by 2π. If $f(\varphi_0, \dot{\varphi}_0)$ is a polynomial, these quantities can be readily evaluated.

We are trying to find periodic solutions among the solutions (9.60) of equation (9.2) when $\mu \neq 0$. Let the period of a certain periodic solution be equal to $2\pi + \tau$, where τ is small (for $\mu \to 0$ $\tau \to 0$). Then, equating $y(2\pi + \tau)$ and $\dot{y}(2\pi + \tau)$ respectively to $y(0) = \varphi_0(0) + \beta_1$ and $\dot{y}(0) = \dot{\varphi}_0(0) + \beta_2$, we form

$$\left. \begin{array}{l} y(2\pi + \tau) - y(0) \equiv \psi_1(\beta_1, \beta_2, \tau, \mu) = 0, \\ \dot{y}(2\pi + \tau) - \dot{y}(0) \equiv \psi_2(\beta_1, \beta_2, \tau, \mu) = 0, \end{array} \right\}$$

(9.69)

which determine this periodic solution. These are two equations with three unknowns β_1, β_2 and τ. Now if we determine any one periodic solution, then an infinite number exist, differing from each other by an arbitrary phase. Therefore one of the β is arbitrary, and we can put it equal to zero. In our problem $\beta_2=0$ leads to definite results. If, however, we had proceeded from the solution $\varphi_0=K\sin t$, i.e. if we had put in the generating solution $\delta=-\pi/2$ (and not to zero), then we would need to use the second alternative $\beta_1=0$.

Since τ is small in comparison with 2π we can expand y and $\dot{y}$ in series about 2π, and keeping terms of the first and second order of smallness, we have

$$y(2\pi+\tau) = y(2\pi)+\tau\dot{y}(2\pi)+\frac{\tau^2}{2}\ddot{y}(2\pi)+\ldots,$$

$$\dot{y}(2\pi+\tau) = \dot{y}(2\pi)+\tau\ddot{y}(2\pi)+\frac{\tau^2}{2}\dddot{y}(2\pi)+\ldots$$

The values $y(2\pi)$, $\dot{y}(2\pi)$, $\ddot{y}(2\pi)$, etc. can be determined from the series (9.60)–(9.62) by replacing the functions A, B, C, etc. by their values at $t=2\pi$, i.e. $A(2\pi)$, $B(2\pi)$ etc. Neglecting the terms in μ and τ of an order higher than the second we obtain

$$y(2\pi+\tau) = \varphi_0(2\pi)+A(2\pi)\beta_1+B(2\pi)\beta_2+C(2\pi)\mu+D(2\pi)\beta_1\mu+$$
$$+E(2\pi)\beta_2\mu+F(2\pi)\mu^2+\tau\dot{\varphi}_0(2\pi)+\tau\dot{A}(2\pi)\beta_1+\tau\dot{B}(2\pi)\beta_2+$$
$$+\tau\dot{C}(2\pi)\mu+\frac{\tau^2}{2}\ddot{\varphi}_0(2\pi)+\ldots,$$

$$\dot{y}(2\pi+\tau) = \dot{\varphi}_0(2\pi)+\dot{A}(2\pi)\beta_1+\dot{B}(2\pi)\beta_2+\dot{C}(2\pi)\mu+\dot{D}(2\pi)\beta_1\mu+$$
$$+\dot{E}(2\pi)\beta_2\mu+\dot{F}(2\pi)\mu^2+\tau\ddot{\varphi}_0(2\pi)+\tau\ddot{A}(2\pi)\beta_1+\tau\ddot{B}(2\pi)\beta_2+$$
$$+\tau\ddot{C}(2\pi)\mu+\frac{\tau^2}{2}\dddot{\varphi}_0(2\pi)+\ldots$$

Substituting in these expressions the values of φ_0, A and B and their derivatives at $t=2\pi$ and substituting in (9.69) we have, finally

$$\left.\begin{aligned}y(2\pi+\tau)-y(0) &= -K\frac{\tau^2}{2}+\tau\beta_2+C(2\pi)\mu+\dot{C}(2\pi)\tau\mu+\\ &\quad+D(2\pi)\beta_1\mu+E(2\pi)\beta_2\mu+F(2\pi)\mu^2=0,\\ \dot{y}(2\pi+\tau)-\dot{y}(0) &= -K\tau-\tau\beta_1+\dot{C}(2\pi)\mu+\ddot{C}(2\pi)\tau\mu+\\ &\quad+\dot{D}(2\pi)\beta_1\mu+\dot{E}(2\pi)\beta_2\mu+\dot{F}(2\pi)\mu^2=0.\end{aligned}\right\} \quad (9.70)$$

These two equations determine τ and a β (in our case β_1) in terms of μ, provided we give the other β (β_2) a fixed value such as zero. We write τ and β_1 as power series in μ†

---

† In the expansions of τ and β_1 as power series in μ the zero-order terms must be absent, since $\tau\to 0$ and $\beta_1\to 0$ for $\mu\to 0$. If $\beta_2\neq 0$ is chosen this term must be of the order of μ.

$$\left.\begin{array}{l}\tau = \mu\tau_1+\mu^2\tau_2+\ldots,\\ \beta_1 = \mu\beta_{11}+\ldots\end{array}\right\} \quad (9.71)$$

and substitute in (9.70). Then equate to zero the terms in μ obtaining

$$C(2\pi) = 0, \quad -K\tau_1+\dot{C}(2\pi) = 0.$$

The first of these expressions

$$C(2\pi) = -\int_0^{2\pi} f(K\cos u, -K\sin u)\sin u\, du = 0, \quad (9.72)$$

or, according to (9.12),

$$\Phi(K) = 0,$$

determines the radii of those circles near which there are limit cycles. The second equation determines the correction τ, and so the first approximation to the period

$$\tau_1 = \frac{\dot{C}(2\pi)}{K_i} = \frac{1}{K_i}\int_0^{2\pi} f(K\cos u, -K\sin u)\cos u\, du, \quad (9.73)$$

or, according to (9.12)

$$\tau_1 = -2\pi\Psi(K_i).$$

By equating to zero terms of the order of μ^2 in (9.70) we obtain[†]

$$\left.\begin{array}{l}K\dfrac{\tau_1^2}{2}+D(2\pi)\beta_{11}+F(2\pi) = 0,\\ -K\tau_2+\ddot{C}(2\pi)\tau_1+[\dot{D}(2\pi)-\tau_1]\beta_{11}+\dot{F}(2\pi) = 0.\end{array}\right\} \quad (9.74)$$

which determine β_{11} and the second order correction τ_2, provided that $D(2\pi) \neq 0$.

An interesting case of practical importance is when

$$\dot{C}(2\pi) = 0,$$

and so $\tau_1 = 0$. Thus τ is, generally speaking, a quantity of the order of μ^2. We now write the equations (9.74) as

$$D(2\pi)\beta_{11}+F(2\pi) = 0,$$
$$-K\tau_2+\dot{D}(2\pi)\beta_{11}+\dot{F}(2\pi) = 0$$

[†] It is easily seen that β_2 vanishes from these equations, since the coefficients of β_2
$$\tau_1+E(2\pi) = 0 \quad \text{and} \quad \dot{E}(2\pi) = 0.$$
K being a root of the equation (9.72).

which give (for $D(2\pi) \neq 0$)

$$\left.\begin{aligned}\beta_{11} &= -\frac{F(2\pi)}{D(2\pi)}, \\ \tau_2 &= \frac{D(2\pi)\dot{F}(2\pi) - \dot{D}(2\pi)F(2\pi)}{D(2\pi)K}.\end{aligned}\right\} \quad (9.75)$$

Introduce $A(t)$, $B(t)$, $C(t)$ and $\beta_1 = \beta_{11}\mu +, \ldots$, into (9.60) and return to an arbitrary time origin by replacing t by $t+\delta$ we find that an approximate solution of (9.2) in the form

$$y = K\cos(t+\delta) + \mu\left\{\int_0^{t+\delta} f[\varphi_0(u), \dot{\varphi}_0(u)]\sin(t+\delta-u)\,du - \frac{F(2\pi)}{D(2\pi)}\cos(t+\delta)\right\} + O(\mu^2), \quad (9.76)$$

where K is a root of the equation (9.72). This first approximation (9.76), as with the zero-order approximation (9.59), has a period 2π, whereas the solution (9.59) must have a period somewhat different from $2\pi\,(2\pi + \mu^2\tau_2 +, \ldots,)$. The latter is ensured by the fact that the expression (9.60) is a power series expansion in μ of a Fourier series such that not only its "amplitude" but also its period depend on μ.

Finally note that, from the general theory (see Chapter V, Section 8) we can write the condition of stability for the periodic solution $y(t)$ in the form

$$\int_0^{2\pi} f'_y[y(t), \dot{y}(t)]\,dt < 0 \quad (9.77)$$

or, restricting ourselves to the first term of the expansion of $y(t)$ in μ

$$2\pi\Phi'(K_i) = \int_0^{2\pi} f'_y[\varphi_0(t), \dot{\varphi}_0(t)]\,dt < 0. \quad (9.78)$$

The expression appearing in the right-hand side of this inequality is the constant term (multiplied by 2π) in the expansion as a Fourier series of the function

$$f'_y(K_i\cos t, -K_i\sin t),$$

where K_i is the corresponding root of the equation (9.72).

§6. Application of Poincaré's Method

1. *A valve generator with soft self-excitation*

To illustrate Poincaré's method we investigate a familiar valve oscillator (Fig. 465). As we have seen, we can restrict the discussion to a cubic valve characteristic (9.37). We shall not assume here that $\omega_0 RC$ is small (as before $\omega_0 = 1/\sqrt{LC}$). Then the equation of the generator is

$$\ddot{x} + x = \{\omega_0(MS_0 - RC) + 2\omega_0 MS_1 u_0 x - 3\omega_0 MS_2 u_0^2 x^2\}\dot{x},$$

where $x = u/u_0$ (u_0 is a certain voltage) and a dot denotes differentiation with respect to $t_{\text{new}} = \omega_0 t_{\text{old}}$.

By introducing the notation

$$\omega_0(MS_0 - RC) = \mu\alpha', \quad 2\omega_0 MS_1 u_0 = \mu\beta'$$

and

$$3\omega_0 MS_2 u_0^2 = \mu\gamma', \tag{9.79}$$

where $0 < \mu \ll 1$ and α', β' and γ' are quantities of the order of unity, the reduced equation is

$$\ddot{x} + x = \mu(\alpha' + \beta' x - \gamma' x^2)\dot{x}. \tag{9.80}$$

We now proceed along known lines. The periodic solutions of the equation (9.80) are very approximately

$$\varphi_0(t) = K \cos t, \quad \dot{\varphi}_0(t) = -K \sin t, \tag{9.81}$$

K being determined by

$$C(2\pi) = -\int_0^{2\pi} (\alpha' + \beta' K \cos u - \gamma' K^2 \cos^2 u)(-K \sin u) \sin u \, du =$$

$$= \pi K \left(\alpha' - \frac{\gamma'}{4} K^2\right) = 0,$$

or

$$K^2 = 4\frac{\alpha'}{\gamma'}. \tag{9.81a}$$

It is easily verified that $\dot{C}(2\pi) = 0$. Thus, in the first approximation there is no "correction" to the period.

Next

$$\frac{\partial f}{\partial x} = (\beta' - 2\gamma' x)\dot{x}, \quad \frac{\partial f}{\partial \dot{x}} = \alpha' + \beta' x - \gamma' x^2$$

and therefore

$$\left[\frac{\partial f}{\partial x}\right] = (\beta' - 2\gamma' K \cos u)(-K \sin u),$$

$$\left[\frac{\partial f}{\partial \dot{x}}\right] = \alpha' + \beta' K \cos u - \gamma' K^2 \cos^2 u.$$

Integrating these expressions from 0 to 2π, we have (see (9.68c))

$$D(2\pi) = 2\pi\left(\alpha' - \frac{\gamma'}{2} K^2\right) = -2\pi\alpha', \quad \dot{D}(2\pi) = 0,$$

therefore the correction for the period (see (9.75)) is

$$\tau = \mu^2 \tau_2 = \frac{\dot{F}(2\pi)}{K}.$$

We need to evaluate $C(t)$ and the expressions $\dot{F}(2\pi)$ and $F(2\pi)$, and we find

$$C(t) = \int_0^t (\alpha' + \beta' K \cos u - \gamma' K^2 \cos^2 u)(-K \sin u) \sin(t-u)\, du =$$

$$= -\frac{\beta' K^2}{6}(2 \sin t - \sin 2t) + \frac{\gamma' K^3}{32}(3 \sin t - \sin 3t),$$

$$\dot{F}(2\pi) = \int_0^{2\pi} \left\{\left[\frac{\partial f}{\partial x}\right] C(u) + \left[\frac{\partial f}{\partial \dot{x}}\right] \dot{C}(u)\right\} \cos u\, du =$$

$$= \frac{\pi}{12} \beta'^2 K^3 + \frac{\pi}{128} \gamma'^2 K^5 = \pi \alpha' K \left(\frac{\beta'^2}{3\gamma'} + \frac{\alpha'}{8}\right),$$

whilst

$$F(2\pi) = -\int_0^{2\pi} \left\{\left[\frac{\partial f}{\partial x}\right] C(u) + \left[\frac{\partial f}{\partial \dot{x}}\right] \dot{C}(u)\right\} \sin u\, du = 0,$$

since the integrand is an odd periodic function.

Thus the correction for the period is

$$\tau = \mu^2 \pi \alpha' \left(\frac{\beta'^2}{3\gamma'} + \frac{\alpha'}{8}\right)^\dagger, \tag{9.82}$$

and a periodic solution in the form (9.76), (i.e. without secular terms), can

† In particular, with a symmetric cubic characteristic for which $\beta=0$, the correction for the period is

$$\tau = \frac{\pi\mu^2 \alpha'^2}{8} = \frac{\pi}{8}[\omega_0(MS_0 - RC)]^2.$$

be written, neglecting terms of the order of μ^2, as[†]

$$x(t) = 2\sqrt{\frac{\alpha'}{\gamma'}}\cos\left[\left(1-\frac{\tau}{2\pi}\right)t+\delta\right]+$$
$$+\mu\left\{\left(-\frac{4\alpha'\beta'}{3\gamma'}+\frac{3\alpha'}{4}\sqrt{\frac{\alpha'}{\gamma'}}\right)\sin\left[\left(1-\frac{\tau}{2\pi}\right)t+\delta\right]+\right.$$
$$\left.+\frac{2\alpha'\beta'}{3\gamma'}\sin 2\left[\left(1-\frac{\tau}{2\pi}\right)t+\delta\right]-\frac{\alpha'}{4}\sqrt{\frac{\alpha'}{\gamma'}}\sin 3\left[\left(1-\frac{\tau}{2\pi}\right)t+\delta\right]\right\}+O(\mu^2)$$
(9.82a)

In most practical cases, only the expression (9.18a) for the amplitude is of major interest. We have evaluated the second approximation in order to show how to carry out the calculations, and also to emphasize that the solution contains higher harmonics, which we neglect when using the linear approach.

For stable motion it is necessary that the constant term in the Fourier expansion of the coefficient of $\dot{x}$ on the right-hand side of (9.80) (after putting $x = K\cos t$ in the coefficient) be negative, i.e. that $\alpha' - 3\gamma' K^2/2 < 0$ or $K^2 > 2\alpha'/3\gamma'$. However, as we have found, $K^2 = 4\alpha'/3\gamma'$. Therefore the periodic solution is always stable.

2. The significance of the small parameter μ

Poincaré has proved that the series which represent a periodic solution in his theory possess a non-zero radius of convergence μ_0, so that for all $\mu \leq \mu_0$ these series converge absolutely and uniformly. Thus for all $\mu < \mu_0$ a periodic solution exists represented by the sums of the corresponding series (such a solution may exist for $\mu > \mu_0$). The fact that these series converge does not answer the question of how close the periodic solution is to a sinusoidal oscillation. We can only assert from Poincaré's theory, that we can always choose μ so that the solution is arbitrarily close to a sinusoid.

We usually use in the analysis a zero-order approximation ($x = K\cos t$, $\Phi(K) = 0$). Therefore we are interested in how the amplitude of the zero order approximation differs (for a given μ) from the amplitude of the fundamental component of the exact solution; and how much the first fre-

---

[†] The even term of the characteristic plays no role, if we restrict ourselves to the zeroorder approximation, but occurs both in the correction for the period and in the first approximation (9.82a) for the periodic solution $x(t)$.

quency correction differs from the true frequency correction, and perhaps, even in the true departure from sinusoidality (defined by the harmonic factor). If permissible errors are given, then in principle we can determine an upper bound for μ. Now μ has a well-determined value in a real system and we cannot necessarily assume it as small as we please, without losing physical meaning. If the system parameters determine $\mu = \mu_1$, then two questions arise: firstly is $\mu_1 > \mu_0$ or is $\mu_1 < \mu_0$, and, secondly is μ_1 such that the zero-order or the first approximation gives the required accuracy? At the present state of the theory, these questions are very difficult to answer and the problems remain, although a rough estimate of μ_0 can be found, as shown by Poincaré, but this estimate is very rough and often has no practical meaning. A useful engineering approach is to evaluate the numerical value of the expression $\mu_{\max}\{f(K_i \cos u, -K_i \sin u)\}/K_i$ for practical values of the parameters and the amplitude K_i of the zero order approximation. If this quantity is equal, say to $\frac{1}{10}$, it is assumed that the amplitude of the zero-order approximation is within 10% of the amplitude of the fundamental frequency. One can have no real confidence in this approach.

§ 7. A VALVE GENERATOR WITH A SEGMENTED CHARACTERISTIC

In the analysis of a valve generator we represented its characteristic by a polynomial, but we can use other analytical expressions. It sometimes proves very useful to approximate to a real characteristic by one consisting of segments of straight-lines, which, of course, are non-holomorphic functions. We have assumed that $f(x, \dot{x})$ is holomorphic but we can consider the non-holomorphic function as the limit of some holomorphic function. We then evaluate all the integrals needed in terms of the limit (which usually simplifies the calculations) and discuss the results not for the broken-line characteristic (which in general, would not be correct) but for a holomorphic one close to it.

1. A valve generator with a discontinuous ∫ characteristic

A moderately good representation of a valve characteristic for large amplitudes of oscillation is the discontinuous (step) characteristic (see Section 3, Chapter III). The equation of oscillations in such a generator

A VALVE GENERATOR WITH A SEGMENTED CHARACTERISTIC

(3.15) reduce, after introducing the dimensionless variables

$$x = \frac{i}{i_0} \quad \text{and} \quad t_{\text{new}} = \omega_0 t_{cr}$$

(i_0 is a certain current, $\omega_0 = (LC)^{-\frac{1}{2}}$) to the form

$$\ddot{x} + x = -\omega_0 RC\dot{x} + \begin{cases} \dfrac{I_s}{i_0} & \text{for} \quad \dot{x} > 0, \\ 0 & \text{for} \quad \dot{x} < 0. \end{cases}$$

This equation approximates to that of the harmonic oscillator if the following two conditions are satisfied:

$$\omega_0 RC \ll 1 \quad \text{and} \quad \frac{I_s}{i_0} \ll 1,$$

i.e. when the damping of the tuned circuit is small and the saturation current I_s is small. Let us introduce $\mu = \omega_0 RC \ll 1$ and $\beta = I_s/\omega_0 RC i_0$ — a quantity of the order of magnitude of unity. Then the equation of oscillations reduces to

$$\ddot{x} + x = \mu[-\dot{x} + \beta \cdot 1(\dot{x})]^\dagger,$$

suitable for Van der Pol's method. Since

$$\Phi(K) = -\frac{1}{2\pi} \int_0^{2\pi} [+K \sin u + \beta \cdot 1(-K \sin u)] \sin u \, du =$$

$$= -\frac{K}{2} - \beta \int_\pi^{2\pi} \sin u \, du = -\frac{K}{2} + \frac{\beta}{\pi}$$

while

$$\Psi(K) = -\frac{1}{2\pi K} \int_0^{2\pi} [K \sin u + \beta \cdot 1(-K \sin u)] \cos u \, du \equiv 0$$

then, the zero-order approximation to the amplitude of the self-oscillations is

$$K = \frac{2\beta}{\pi} = \frac{2}{\pi} \frac{\dfrac{I_s}{i_0}}{\omega_0 RC}, \tag{9.84}$$

† As before

$$1(z) = \begin{cases} 1 & \text{for} \quad z > 0, \\ 0 & \text{for} \quad z < 0. \end{cases}$$

and the period of the self-oscillations is equal to 2π. These self-oscillations are stable since

$$\Phi'(K) = -\frac{1}{2} < 0.$$

2. A valve oscillator with a segmented characteristic without saturation

The vacuum tube characteristic does not saturate and is represented by two rectilinear segments; one horizontal and one inclined (Fig. 482).

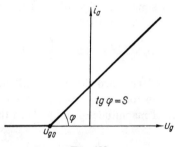

Fig. 482

As we have seen in Chapter VIII, Section 2, when such a characteristic is bounded on one side, stable self-oscillations are possible under certain conditions.

In the case when the tuned circuit is in the grid circuit (Fig. 465(a) the equation for the voltage across the capacitor (the "dimensionless" voltage) will be (see Section 1 of this chapter)

$$\ddot{x}+x = \mu[-1+\alpha s(x)]\dot{x}, \qquad (9.3)$$

where $x = u/u_0 = (u_g - E_g)/u_0$, $\mu = \omega_0 RC (0 < \mu \ll 1)$, $\alpha = MS/RC$,

$$s(x) = \frac{S(E_g + u_0 x)}{S} = 1(x-b) = \begin{cases} 1 & \text{for} \quad x > b, \\ 0 & \text{for} \quad x < b, \end{cases}$$

$b = (u_{g0} - E_g)/u_0$ is the normalized cut-off voltage, and differentiation is with respect to the "dimensionless time" $t' = \omega_0 t \left(\omega_0 = (LC)^{-\frac{1}{2}} \right)$.

Evidently for $b > 0$ there are no self-oscillations, and for $b < 0$ self-oscillations occur or not according to whether α is greater or smaller than unity.

We can take as an approximate periodic solution

$$x = \varphi_0(t) = K \cos t.$$

We shall assume K positive; since the phase is arbitrary this does not detract from generality. The amplitude K is determined by the condition that $C(2\pi) \neq 0$

$$C(2\pi) = K \int_0^{2\pi} [-1 + \alpha \cdot 1(K\cos u - b)] \sin^2 u \, du =$$
$$= K\left[-\pi + 2\alpha \int_0^{\xi} \sin^2 u \, du \right] = 0, \qquad (9.85)$$

where ξ is the value of u for which $K\cos u - b = 0$, i.e.

$$\xi = \arccos \frac{b}{K} \quad (0 < \xi < \pi).$$

ξ exists for $K \geq |b|$ only. Since K is positive, then $0 < \xi < \pi/2$ for $b > 0$ and $\pi/2 < \xi < \pi$ for $b < 0$. The amplitude K is determined by the relation

$$K = \frac{b}{\cos \xi}, \qquad (9.85a)$$

where ξ is determined in its turn by the equation (9.85). After integration (9.85) gives
$$-2\pi + \alpha(2\xi - \sin 2\xi) = 0 \qquad (9.85b)$$
or
$$\alpha = \frac{2\pi}{2\xi - \sin 2\xi}. \qquad (9.85c)$$

The relations (9.85a) and (9.85c) determine K for a given α_0.

Since the denominator of (9.85c) increases monotonically with ξ, lying between 0 and π for $0 < \xi < \pi/2$ and between π and 2π for $\pi/2 < \xi < \pi$ then,

$$\left. \begin{array}{ll} \alpha > 2 & \text{for} \quad b > 0, \\ 1 < \alpha < 2 & \text{for} \quad b < 0 \end{array} \right\} \qquad (9.86)$$

and for every α, ξ is uniquely determined by equation (9.85b), and amplitude K by (9.85a). If, however, the inequalities (9.86) are not satisfied, then the equation (9.85b) has no solution and the original equation (9.3) has no periodic solutions. Thus, only when the conditions (9.86) are satisfied does a limit cycle exist. The condition for stability is that the constant term of the Fourier expansion of the function

$$f'_x(K\cos t, -K\sin t) = -1 + \alpha \cdot 1(K\cos t - b)$$

is negative, i.e.
$$-2\pi + 2\alpha \int_0^{\xi} dt < 0,$$
or, using (9.85b)
$$2\pi\xi - 2\pi = \sin 2\xi < 0. \tag{9.87}$$

This condition is satisfied for $\pi/2 < \xi < \pi$ and so for $b < 0$, and is not satisfied for $b > 0$ when $0 < \xi < \pi/2$.

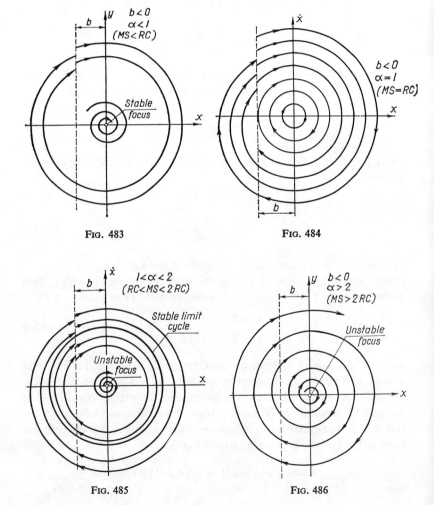

Fig. 483

Fig. 484

Fig. 485

Fig. 486

Thus for $b<0$ (i.e. for $E_g>u_{g0}$) according to the value of $\alpha=MS/RC$, there are three qualitatively different phase portraits (Figs. 483–6).

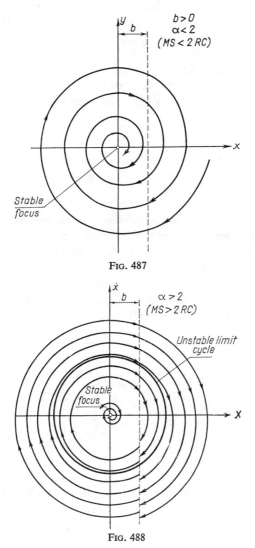

Fig. 487

Fig. 488

For $\alpha<1$ all paths tend to the stable equilibrium state as $t\to\infty$ (Fig. 483). For $1<\alpha<2$ there is a stable limit cycle (Fig. 485) where radius $K\to+\infty$ for $\alpha\to 2$. For $\alpha>2$ all paths go off to infinity (Fig. 486). The minimum value of K is equal to $|b|$ and corresponds to $\xi=\pi$ and $\alpha=1$. Therefore, as α passes through the value $\alpha=1$, a limit cycle of finite size appears immediately and periodic oscillations are possible with any amplitude $K<|b|$ so that in this case the stable equilibrium state at the origin is a *centre* (Fig. 484). For $b>0$ (i.e. $E_g<u_{g0}$) the equilibrium state is always a stable focus and all paths approach this focus if $\alpha<2$ (Fig. 487), but if $\alpha>2$, an unstable limit cycle exists outside of which the paths go off to infinity (Fig 488)[†]. These receding phase paths when $\alpha>2$ (for $MS>2RC$) clearly indicates the inadequacy of the idealized valve characteristic as a model of reality.

§ 8. The effect of grid currents on the performance of a valve oscillator

In the analysis of various valve circuits, we have always neglected the grid currents. This assumption, which simplifies the problem substantially, is very often but not always confirmed experimentally. Generally speaking, taking grid currents into account complicates the problem to a great extent and increases the order of the differential equation. However, it proves possible in certain particular cases to introduce grid currents without raising the order of the equations as, for example, in the valve generator with the tuned grid circuit (Fig. 489).

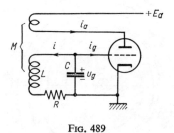

Fig. 489

We still neglect the anode conductance in the analysis and we make very simple choices for the anode and grid current characteristics, assuming that both anode and grid currents are third-degree polynomials of the grid voltage u_g.

---

† As is easily verified, all these results are found in full agreement with the results of Section 2, Chapter VIII for the case $h_1, h_2 \ll 1$, when the oscillations of the generator are nearly sinusoids.

8] EFFECT OF GRID CURRENTS ON VALVE OSCILLATOR

Using the notation shown in Fig. 489 and employing Kirchhoff's laws, we can eliminate i and obtain the equation

$$LC\frac{d^2u_g}{dt^2} + RC\frac{du_g}{dt} - M\frac{di_a}{dt} + L\frac{di_g}{dt} + u_g + Ri_g = 0. \qquad (9.88)$$

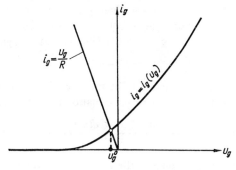

Fig. 490

Evidently, the generator has a single equilibrium state $u_g = u_g^0$ determined by the equation

$$u_g^0 + Ri_g(u_g^0) = 0$$

(the graphical solution of this equation is shown in Fig. 490). Let us introduce the variable component of the grid voltage

$$u = u_g - u_g^0$$

and let

$$\begin{aligned} i_a &= i_{a0} + S_1 u + S_2 u^2 - S_3 u^3, \\ i_g &= i_{g0} + p_1 u + p_2 u^2 + p_3 u^3. \end{aligned} \qquad (9.89)$$

Then the equation of the generator becomes

$$\frac{d^2u}{dt^2} + \left[\frac{R}{L} - \frac{MS_1}{LC} + \frac{p_1}{C}\right]\frac{du}{dt} + \frac{1+Rp_1}{LC}u - \left[\frac{MS_2}{LC} - \frac{p^2}{C}\right]\frac{d(u^2)}{dt} +$$

$$+ \left[\frac{MS_3}{LC} + \frac{p_3}{C}\right]\frac{d(u^3)}{dt} + \frac{Rp_2}{LC}u^2 + \frac{Rp_3}{LC}u^3 = 0.$$

Also let us set

$$\left[\frac{R}{L} - \frac{MS_1}{LC} + \frac{p_1}{C}\right] = -\alpha_1; \quad \frac{MS_2}{LC} - \frac{p_2}{C} = \beta_1; \quad \frac{MS_3}{LC} + \frac{p_3}{C} = \gamma_1;$$

$$\frac{1}{LC} = \omega_0^2; \quad \omega_0^2(1+Rp_1) = \omega_1^2; \quad \frac{Rp_2}{LC} = n; \quad \frac{Rp_3}{LC} = m.$$

With suitable assumptions about the size of the coefficients, this equation can be easily reduced to the form: $\ddot{x}+x = \mu f(x, \dot{x})$ (x is a dimensionless variable and μ is a small parameter) for which Van der Pol's and Poincaré's theories were developed. There are general formulae for the amplitude of periodic solutions, for the correction to the frequency, etc.

Here, however, we shall not use these general formulae but will show that it is possible to obtain similar results with a minimum of calculation. We introduce the "mistuning" a^2, where Ω is the actual angular frequency,

$$a^2 = \Omega^2 - \omega_1^2,$$

and assume $\alpha_1, \beta_1, \gamma_1, m, n$ and a^2 to be small (of the order of μ) in comparison with the frequency ω_0. Now the equation of motion is

$$\frac{d^2u}{dt^2} + \Omega^2 u = a_1 \frac{du}{dt} + \beta_1 \frac{d(u^2)}{dt} - \gamma_1 \frac{d(u)^3}{dt} - nu^2 - mu^3 + a^2 u. \quad (9.88a)$$

To determine the amplitude and the correction to the frequency we put $u = K \sin \Omega t$ and equate to zero the coefficients of $\sin \Omega t$ and $\cos \Omega t$ on the right-hand side. We obtain two equations giving K^2 and a^2

$$K\alpha_1 \Omega - \frac{\gamma_1 K^3 \Omega^3}{4} = 0,$$

$$-m \frac{3}{4} K^3 + a^2 K = 0,$$

whence

$$K^2 = \frac{\alpha_1}{\frac{3}{4}\gamma'}, \quad a^2 = \frac{3}{4} Km^2 = \frac{\alpha_1 m}{\gamma_1}. \quad (9.90)$$

Therefore, the frequency is given by[†]

$$\Omega^2 = \frac{1}{LC} + \frac{Rp_1}{LC} + \frac{m\alpha_1}{\gamma_1}.$$

The oscillation $u = K \sin \Omega t$ is stable if the constant term of the Fourier series of the derivative of the right-hand side of (9.88a) with respect to u is negative, i.e. if the constant term of the expansion of $\alpha_1 + 2\beta_1 u - 3\gamma_1 u^2$ is negative. This means

$$\alpha_1 - \frac{3}{2}\gamma_1 K^2 < 0$$

[†] If we consider the resistance R of the tuned circuit to be also small (if we assume $\omega_0 RC$ to be a quantity of the order of μ) then the correction for the period will be a quantity of the order of μ^2.

or

$$K^2 > \frac{1}{2} \frac{\alpha_1}{\frac{3}{4}\gamma_1},$$

which, by virtue of (9.90) is always true. Therefore the periodic motion is always stable.

Finally the condition of self-excitation is $\alpha_1 > 0$ or

$$\frac{R}{L} + \frac{p_1}{C} - \frac{MS_1}{LC} < 0.$$

As regards self-excitation the grid current acts as an additional load or resistance across the tuned circuit.

§ 9. THE BIFURCATION OR BRANCH THEORY FOR A SELF-OSCILLATING SYSTEM CLOSE TO A LINEAR CONSERVATIVE SYSTEM [89]

We consider as before a self-oscillating system with one degree of freedom, close to a linear conservative system and assume that the behaviour of this self-oscillating system depends substantially on a parameter to which we can attribute various fixed values. The equation of motion in such a case can be written

$$\ddot{x} + x = \mu f(x, \dot{x}; \lambda), \qquad (9.91)$$

where x is the coordinate of the system (displacement, voltage, etc.), μ is a small parameter which characterizes the degree of closeness to a linear conservative system, λ is the parameter (an inductance, etc.) whose influence on the system is to be determined and $f(x, \dot{x}, \lambda)$ is a non-linear function determined by the physical nature of the system.

Using small-parameter methods (Van der Pol's method and Poincaré's method) we have shown that for small $\mu \neq 0$ there remain only isolated closed curves, almost circles on the phase plane. The radii K are given by

$$\Phi(K; \lambda) = 0, \qquad (9.92)$$

where

$$\Phi(K; \lambda) = -\frac{1}{2\pi} \int_0^{2\pi} f(K \cos u, -K \sin u; \lambda) \sin u \, du.$$

The remaining integral curves are spirals, differing little from circles if μ is sufficiently small. The periodic solutions, corresponding to the isolated closed paths — Poincaré's limit cycles — will be stable (both orbitally and in the sense of Liapunov) if

$$\Phi'_K(K; \lambda) < 0. \tag{9.93}$$

The conditions (9.92) and (9.93) are completely analogous with the conditions that we found for an equilibrium state of a conservative system (Chapter II, Section 5), except that instead of the coordinates of the singular points $\bar{x}_1, \bar{x}_2, \ldots, \bar{x}_s$, we must consider $K_1, K_2, \ldots K_s$. These are the amplitudes of the stationary motions, which include limit cycles (in this case almost circles) and the singular point $K=0$.

The dependence of the *stationary motions* upon a parameter is similar to that discovered in Chapter II, Section 5 for the dependence of the equilibrium states upon a parameter. We obtain again not sets or "linear series" of equilibrium states, but sets of stationary motions which retain their stability or instability up to the branch points. The sets of stationary motions are determined by the equation (9.92), and their stability can be determined in the same manner as in Chapter II, Section 5: we mark out on the λ, K plane the region where $\Phi(K; \lambda) > 0$; then the sets situated *above* this region correspond to stable stationary motions, and the sets situated *below* the region $\Phi(K; \lambda) > 0$ correspond to unstable stationary motions. As we shall see below, branch points have an important physical meaning; they are the values of the parameter for which qualitative changes occur in the processes taking place in the system, for example, the build-up or quenching of oscillations, etc. Thes tationary motions that we have discussed here, are similar to the states of equilibrium of conservative systems, in that they form a closed system of elements, among which there occurs an "exchange of stability".

Before considering a concrete example from the point of view of branch theory, note that in a number of problems the study of how the motion depends upon a parameter λ is conveniently carried out on the λ, ϱ plane and not on the λ, K plane, where

$$\varrho = K^2$$

is the square of the amplitude of a stationary motion. If we consider instead of the function $\Phi(K; \lambda)$ the function

$$\bar{\Phi}(\varrho; \lambda) = 2\sqrt{\varrho}\, \Phi(\sqrt{\varrho}; \lambda) =$$
$$= -\frac{1}{\pi} \int_0^{2\pi} f(\sqrt{\varrho} \cos u, -\sqrt{\varrho} \sin u; \lambda) \sqrt{\varrho} \sin u\, du; \tag{9.94}$$

the linear series of stationary motions are determined by the equation

$$\Phi(\varrho; \lambda) = 0, \tag{9.92}$$

and their stability by the condition

$$\Phi'_\varrho(\varrho; \lambda) < 0.\text{†} \tag{9.93a}$$

§ 10. Application of Branch Theory in the Investigation of the Modes of Operation of a Valve Generator [14]

Let us consider the case of soft and hard excitation in a valve generator, and to avoid repetition, a valve generator with a tuned anode circuit (Fig. 465(b) page 584). The equation of the current in the oscillating circuit can be written (neglecting grid current and anode conductance) in the form

$$LC\frac{d^2i}{dt^2} + RC\frac{di}{dt} + i = i_a \tag{9.95}$$

where the anode current $i_a = \varphi(u_g)$ depends only on the grid voltage $u_g = E_g + u$, and where $u = M di/dt$.

We will use a fifth degree polynomial to represent the valve characteristic (see Section 4 of this Chapter)

$$i_a = \varphi(E_g + u) = i_{a0} + S_0 u + S_1 u^2 + S_2 u^3 + S_3 u^4 - S_4 u^5 \text{ ‡}.$$

Let us introduce the new, dimensionless variables

$$t_{\text{new}} = \frac{t}{\sqrt{LC}} \quad \text{and} \quad x = \frac{M}{\Phi_0}(i - i_{a0}),$$

where Φ_0 is a certain constant quantity having the dimensions of a magnetic flux. The small parameter is $\mu = \omega_0 M S_0$ the equation of the oscillations (9.95) reduces to

$$\ddot{x} + x = \mu[\alpha\dot{x} + \beta(\dot{x})^2 + \gamma(\dot{x})^3 + \delta(\dot{x})^4 - \varepsilon(\dot{x})^5], \tag{9.95a}$$

---

† An investigation using "bifurcation or branch diagrams" on the λ, ϱ plane is convenient when $f(x, \dot{x}, \lambda) = F(x, \lambda)\dot{x}$ or when $f(x, \dot{x}, \lambda) = F_1(\dot{x}, \lambda)$, where $F(x, \lambda)$ and $F_1(\dot{x}, \lambda)$ are polynomials.

‡ As in Section 4 of this chapter, S_0 is positive and the coefficient of the highest-order odd term is negative, since stable self-oscillations only exist under these conditions.

where

$$\alpha = \frac{MS_0 - RC}{MS_0}, \quad \beta = \frac{\omega_0 \Phi_0 S_1}{S_0}, \quad \gamma = \frac{\omega_0^2 \Phi_0^2 S_2}{S_0}, \quad \delta = \frac{\omega_0^3 \Phi_0^3 S_3}{S_0},$$

$$\varepsilon = \frac{\omega_0^4 \Phi_0^4 S_4}{S_0}$$

are dimensionless parameters.

According to (9.92a) and (9.93a) we can write the conditions determining the amplitudes and stability (except for a positive factor) as follows:

$$\left. \begin{aligned} \Phi(\varrho, \lambda) &\equiv (MS_0 - RC)\varrho + \frac{3}{4} MS_0 \gamma \varrho^2 - \frac{5}{8} MS_0 \varepsilon \varrho^3 = 0, \\ \Phi'_\varrho(\varrho, \lambda) &\equiv (MS_0 - RC) + \frac{3}{2} MS_0 \gamma \varrho - \frac{15}{8} MS_0 \varepsilon \varrho^2 < 0. \end{aligned} \right\} \quad (9.96)$$

Let the mutual inductance M be the parameter whose effect we wish to study. Therefore we construct the M, ϱ bifurcation or branch diagram for soft and hard excitation. We shall restrict our analysis to $M > 0$ which is essential for a valve oscillator. Note also that only $\varrho \geqslant 0$ has a physical meaning.

To simplify the calculations the simplest possible mathematical model will be chosen[†].

1. Soft-excitation of oscillations

This is obtained for $S_2 < 0$ (see Section 4 of this chapter). Therefore take $S_2 < 0$, $S_4 = 0$ (i.e. $\gamma < 0$, $\varepsilon = 0$) as the simplest assumptions which will reproduce the basic features of a soft excitation. With $3S_0\gamma/4 = 3\omega_0^2\Phi_0^2 S_2/4 = -a(a > 0)$, we write $\Phi(\varrho, M)$ as

$$\Phi(\varrho, M) = \{MS_0 - RC - aM\varrho\}\varrho. \quad (9.97)$$

Thus on the M, ϱ plane the curve $\Phi(\varrho, M)$ splits into the straight line $\varrho = 0$ and the hyperbola

$$MS_0 - RC - aM\varrho = 0.$$

We can isolate on the M, ϱ plane the region $\Phi(\varrho, M) > 0$ and by the use of general rules expounded in Chapter II, Section 5, mark out the stable parts (white circles) and the unstable parts (black circles) of the sets, (Fig. 491.)

---

† In the expressions (9.96) only coefficients of odd powers in the series representing the characteristic occur. Thus the remaining coefficients have no effect in the first approximation on the amplitudes or the stability of the stationary motions, but can play an important role when external forces are present.

A branch point of M will be $M_1 = RC/S_0$, where the sets or linear series on the straight line and the hyperbola intersect. The straight line is stable up to the value $M = M_1$ at which branch point it is the turn of the hyperbola to become stable. Now let us investigate the phase plane for

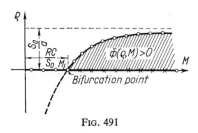

Fig. 491

various values of M. For $M < M_1$ there is one stable stationary state only — a stable focus (Fig. 492). Wherever the representative point may be found, by moving along a spiral it will reach the vicinity of the stable singular point.

As M passes M_1, a stable limit cycle separates from the singular point (Fig. 493). A representative point which was at the singular point will pass on to a limit cycle, since for $M > M_1$ the equilibrium state is unstable. This means that oscillations occur and are self-excited. As M increases the

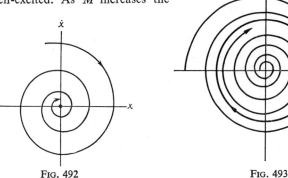

Fig. 492 Fig. 493

radius of the limit cycle increases and tends to an asymptotic value corresponding to $\varrho = S_0/a$. If now M decreases the limit cycle shrinks and the representative point "clings" to the limit cycle. For $M = M_1$ the limit cycle reduces to a point and the representative point will be at the origin of the coordinates, which at this instant has become a stable focus.

An instrument measuring the amplitude K of the oscillations as M is varied, will show a smooth ("soft") transition from zero to some large value and conversely (Fig. 494).

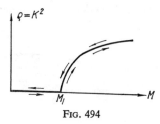

Fig. 494

2. Hard excitation of oscillations

The assumptions $S_2 > 0$ and $S_4 < 0$ reproduce as we have seen, the essential features of a hard excitation. By introducing

$$\frac{3}{4} S_0 \gamma = a \quad (a > 0), \qquad \frac{5}{8} S_0 \varepsilon = b \quad (b > 0),$$

we obtain

$$\Phi(\varrho, M) = \{MS_0 - RC + aM\varrho - bM\varrho^2\} \varrho. \tag{9.98}$$

In the M, ϱ plane the bifurcation or branch diagram splits into a straight line $\varrho = 0$ and a curve of the third order

$$MS_0 - RC + aM\varrho - bM\varrho^2 = 0.$$

The approximate situation of these curves, the regions where $\Phi(\varrho, M) > 0$, and the stable (black circles) and unstable (white circles), parts of the linear series, are shown in Fig. 495.

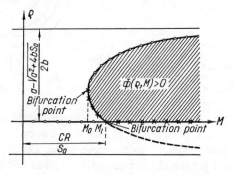

Fig. 495

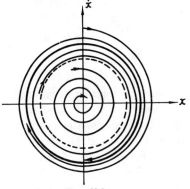

Fig. 496

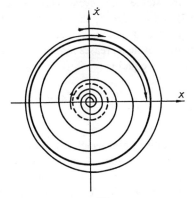

Fig. 497

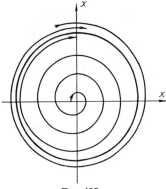

Fig. 498

For $M>0$ there are two branch values of M:

$$M_0 = \frac{RC}{S_0 + \frac{a^2}{4b}} \quad \text{and} \quad M_1 = \frac{RC}{S_0}.$$

M_0 corresponds to a merging together of two sets of states[†] and M_1 to the intersection of two sets.

For $0 < M < M_1$ there is on the phase plane, just as in the previous case, only one stationary motion — a stable focus at the origin (Fig. 492). As M passes through the branch value M_0 two limit cycles appear having finite radii (Fig. 496). The larger of these cycles is stable, and the smaller is unstable, while the singular point at the origin remains stable. As M is further increased, the stable limit cycle increases and the unstable one decreases (Fig. 497) and at $M = M_1$ the unstable cycle vanishes, merging with the singular point and making it become unstable (Fig. 498). For larger M the stable limit cycle increases monotonically and its radius tends asymptotically to a value equal to

$$\varrho = \frac{a + \sqrt{a^2 + 4bS_0}}{2b}.$$

Consider a representative point which is found for small values of M in the vicinity of the equilibrium state. It is evident that it will remain there until this state of equilibrium becomes unstable, when $M = M_1$. The fact that for $M = M_0$ a pair of limit cycles appears (one of them stable) does not affect this representative point since the stable nature of the equilibrium state is not changed.

For $M > M_1$ the singular point is no longer stable; the representative point "is launched" and passes through $M = M_1$ and moves on an integral curve until it arrives at the stable limit cycle to which it now "clings" for any further increase of M. As the parameter M is decreased, a different picture results. The representative point remains on the limit cycle right up to $M = M_0$, when the stable limit cycle merges with the unstable one and disappears. The representative point remains on the stable limit cycle at $M = M_1$ not being affected by the now stable singular point. At $M = M_0$, however, the representative point follows an integral curve to the equilibrium state and remains there for a further decrease of M.

[†] This branch point corresponds to the so-called "limiting" stationary motion.

The instrument, measuring the amplitude of the current in the oscillating circuit (or of the grid voltage) will register jumps at $M=M_1$ during the increase of M, and at $M=M_0$ during the decrease of M. We are dealing with hard excitation, and a phenomenon having an irreversible "hysteretic" nature (Fig. 499).

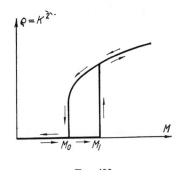

Fig. 499

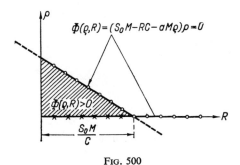

Fig. 500

We have obtained branch diagrams for soft and hard excitation in terms of a parameter M, the mutual inductance coefficient. We could have obtained analogous diagrams for other parameters that characterize this system.

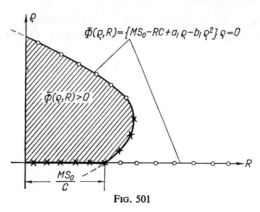

Fig. 501

Figs. 500 and 501 show R, ϱ branch diagrams where R is the resistance for soft and hard self-excitation. The corresponding relations can be derived easily from the expressions (9.97) and (9.98).

CHAPTER X

DISCONTINUOUS OSCILLATIONS†

§ 1. INTRODUCTION

As has already been repeatedly said, in the analysis of any real physical system we must choose from all its properties the ones essential to our purposes and then construct a simplified dynamic (mathematical) model whose equations reproduce adequately the behaviour of the real system. But in employing such an idealization, we run the risk that we may neglect the very essential properties and that our assumptions may not enable us to answer correctly all the questions raised. In constructing a simpli-

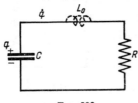

Fig. 502

fied dynamic model of a real physical system, the so-called "small" or "parasitic" parameters of the system are often neglected. Thus, for example, when we analyse the oscillations in the RC circuit of Fig. 502 by means of the equation

$$R\dot{y} + \frac{q}{C} = 0, \qquad (10.1)$$

we neglect, in particular, the small parasitic inductance L_0. As we have seen in Chapter I, Section 5, this parameter is not important provided that it is sufficiently small ($L_0 \ll CR^2$). If we take into account L_0, and obtain the "more exact" equation

$$L_0\ddot{y} + R\dot{y} + \frac{q}{C} = 0, \qquad (10.1a)$$

† Section 1, Sub-section 2 of Section 2, Sections 3–5, Section 7, Sub-section 4 of Section 8, Sections 9–11, Sub-section 2 of Section 12 and Section 13 have been written and Sections 6, 8 and 12 substantially revised by N. A. Zheleztsov.

we do not introduce anything new and only find small corrections to the solution of the equation (10.1)[†].

Similarly, neglecting small parasitic parameters, we can analyse accurately the processes in a series $L-R$ circuit

$$L\frac{di}{dt}+Ri = 0, \qquad (10.2)$$

provided that these parasitic parameters are small. Taking the small coil stray capacitance C_0 into account (Fig. 503) leads to the differential equation

$$C_0RL\frac{d^2i}{dt^2}+L\frac{di}{dt}+Ri = 0, \qquad (10.2a)$$

but does not alter substantially the results of our analysis, provided that $C_0 \ll L/R^2$.

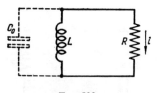

Fig. 503

In the examples already investigated[‡] the neglect of small parameters, and other simplifying assumptions, gave dynamic models adequately representing the behaviour of oscillating systems for $0 < t < +\infty$. Often, however, it is not permissible to discard every small parameter since one or more may be very important in elucidating the processes in an oscillating system. For example, in the analysis of the generator shown in

---

[†] Here, as everywhere in the book, we shall only consider motions of the system *that start from states compatible with the equations of the dynamic model*. In the equation (10.1) we can arbitrarily assign at $t=0$ only one of the quantities that characterize the state of the RC circuit such as q_0. But if we are interested in processes in the RC circuit that correspond to initial conditions not satisfying the equation (10.1) (for example, for $q_0 \neq 0$ and $\dot{q}_0=0$), then the analysis requires the use of the equation (10.1a), which includes the parasitic inductance L_0. As we have seen in Chapter I, Section 5, during an initial stage of duration L_0/R or so, $R\dot{q}+q/C$ and therefore, $L_0\ddot{y}$ is not small. For small values of L_0, there is thus a rapid initial variation of $\dot{q}$ until it reaches a value close to $-q_0/RC$. Subsequently, even in this case, the phenomena are satisfactorily described by the equation (10.1). The equation (10.1a), or a suitably formulated postulate on the current jump, are necessary for the analysis of motion during the initial stage, when the states of the system are "in conflict" with the equation (10.1).

[‡] Except the multivibrator with one RC circuit (Chapter IV, Section 7).

Fig. 504, we cannot neglect the parasitic capacitance C_{ag}. It is just this capacitance that provides the feedback coupling of the tuned anode circuit to the grid necessary for the excitation of self-oscillations.

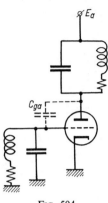

As a second example consider the processes in the simple circuit shown in Fig. 505. If the charge on the capacitor at $t=0$ is $q_0=0$, then current i is given by

$$L\frac{di}{dt}+Ri+\frac{1}{C}\int_0^t i\,dt = E. \qquad (10.3)$$

FIG. 504   FIG. 505

The capacitor voltage is

$$\frac{1}{C}\int_0^t i\,dt,$$

and if $1/C$ is "small" (since C is "large"), we discard this term and obtain the equation

$$L\frac{di}{dt}+Ri = E, \qquad (10.3a)$$

This, however, only reproduces correctly the current variation during the initial stage $t \gg CR$. In fact, according to the simplified equation (10.3a) the current tends to E/R for $t \to +\infty$, while in reality and according to the equation (10.3) the current almost reaches E/R (after time of the order of L/R) and subsequently, as the capacitor voltage increases, tends slowly (as $e^{-t/RC}$) to 0 for $t \to +\infty$ (Fig. 506). Thus, even though

$$\frac{1}{C}\int_0^t i\,dt$$

has a "small" coefficient $1/C$ and is in fact small at first compared with other terms in the equation (10.3), it cannot be neglected if we want to describe the entire current transient.

Finally, there exist systems where the solutions cannot be formulated without allowing for certain small parameters. Examples are the multivibrator with one RC circuit and other oscillating systems which produce *discontinuous oscillations*, i.e. oscillations in which slow variations of the state alternate with very rapid "jump-wise" variations.

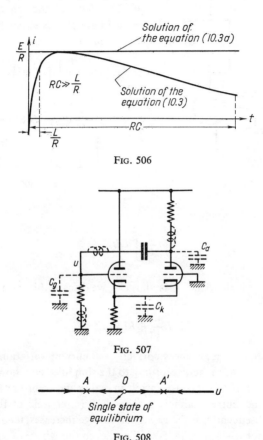

Fig. 506

Fig. 507

Fig. 508

In the analysis of such a multivibrator in Chapter IV, Section 8 (Fig. 507) we neglected all parasitic parameters. The first order dynamic model (its phase line is shown in Fig. 508) proved defective or, "degenerate" in the sense that it did not enable us to describe the behaviour of the system at all instants. Whatever the initial conditions, the equation of the first order dynamic model "leads" the system to one of the states at the "junction points of phase paths" A and A' (Fig. 508) which are not states of equilibrium nor have apparently any phase paths leading away from them.

Since all "large" parameters were taken into account, we seek the cause of such a "defective" model in some neglected small parasitic parameter. They turn out to be the small parasitic capacitances C_a and C_g or C_k (Fig. 507). These play a determining role for the rapid "jump-wise" variations of the grid voltage u which are typical of a multivibrator and when taken into account produce a quite "satisfactory" model of the second order (see Chapter VIII, Section 5 and Chapter V, Section 12)[†]. The important fact to note is that in the course of its oscillations the multivibrator periodically reaches states in which the terms of the differential equations having the small parasitic capacitances as coefficients are *not* small in comparison with other terms in these equations.[‡]

Before discussing approximate methods of analysis of discontinuous oscillations (in Section 3) we shall try to show the influence of those terms of differential equations with small coefficients on the stability of equilibrium states.

§ 2. Small parameters and stability of states of equilibrium [127]

Suppose that the behaviour in the vicinity of a state of equilibrium can be described (when small parameters are neglected) by the linearized equation of the nth order

$$a_0 \frac{d^n x}{dt^n} + a_1 \frac{d^{n-1} x}{dt^{n-1}} + \ldots + a_n x = 0. \tag{10.4}$$

The characteristic equation is

$$a_0 \lambda^n + a_1 \lambda^{n-1} + \ldots + a_n = 0. \tag{10.5}$$

[†] Not all parasitic parameters are important in a multivibrator. If, for example, we take into account one of the parasitic inductances indicated by a dotted line in Fig. 507 and omit the parasitic capacitances, we obtain a dynamic model of the second order which however, as before, is "degenerate" and inadequate even for a qualitative explanation of the operation of a multivibrator (see Section 8 of this chapter).

[‡] In Chapter IV Section 8 we analysed the self-oscillations of a multivibrator using a "defective" model of the first order completed by a postulate on the jumps of the grid voltage u. This jump postulate is an indirect manner of allowing for the essential parasitic parameters and is obtained as a consequence of the dynamics of a "satisfactory" model of the second order (see Section 4 of this chapter and also Section 5 of Chapter VIII).

As is well known, the stability of the equilibrium state is stable if all n roots are negative or have negative real parts.

Suppose that, by taking into consideration a small parameter, the order of the differential equation increases by one (when the order increases by more than one the analysis is similar)[†]. This increase of the order in the linearized equation (10.4) can occur in two ways — either the addition of a small term $\mu\, d^{n+1}x/dt^{n+1}$ or of a small term $\mu \int_0^t x\,dt$ where μ is small. In the first case the characteristic equation assumes the form

$$\mu\lambda^{n+1} + a_0'\lambda^n + a_1'\lambda^{n-1} + \ldots + a_n' = 0, \tag{10.6}$$

and in the second case

$$a_0'\lambda^{n+1} + a_1'\lambda^n + \ldots + a_n'\lambda + \mu = 0. \tag{10.7}$$

Both these equations have $n+1$ roots, of which n roots $\lambda_1, \lambda_2, \ldots, \lambda_n$, since μ is small, must have values close to the roots of the original characteristic equation and in particular have the same signs for the real parts[‡]. The stability of the equilibrium state can only be affected by the new root λ_{n+1}.

Let us begin with the first case. We know that for $\mu \to 0$, $\lambda_{n+1} \to \infty$. Therefore if μ is sufficiently small, we evaluate λ_{n+1} by neglecting all terms in (10.6) of a degree less than n. We shall obtain the following asymptotic expression for λ_{n+1} valid for small values of μ:

$$\lambda_{n+1} \approx -\frac{a_0}{\mu}. \tag{10.8}$$

[†] The introduction of a small parameter that does not increase the order of the equation cannot vary the stability of the equilibrium state if the original system is coarse.

The concept of "coarseness" of an autonomous system defined, for example, by two differential equations of the first order can be generalized to the case when the small additional terms contain the first derivatives, i.e. when the new system is

$$\frac{dx}{dt} = P(x, y) + p_1\left(x, y; \frac{dx}{dt}, \frac{dy}{dt}\right);$$

$$\frac{dy}{dt} = Q(x, y) - q_1\left(x, y; \frac{dx}{dt}, \frac{dy}{dt}\right).$$

If, however, the small additional terms contain derivatives of higher orders, the idea of "coarseness" of the system fails, since phase space with more than two dimensions is needed. In this latter case, as we shall see later, we cannot be sure (without special restrictions) that the smallness of the additional terms will have no effect on the stability of the equilibrium state.

[‡] We are assuming that the initial system is "coarse" and that, therefore, the real parts of all roots of the initial characteristic equation are different from zero.

In the second case we can rewrite the characteristic equation (10.7):

$$\lambda[a_0\lambda^n + a_1\lambda^{n-1} + \ldots + a_n] + \mu = 0.$$

For $\mu \to 0$, $\lambda_{n+1} \to 0$. Therefore, by neglecting the higher powers of λ_{n+1} we shall obtain the following asymptotic expression for λ_{n+1} in this second case:

$$\lambda_{n+1} \approx -\frac{\mu}{a_n}. \tag{10.9}$$

Near the equilibrium state the behaviour of the system is determined by the equation

$$x = be^{\lambda_{n+1}t} + c_1 e^{\lambda_1 t} + c_2 e^{\lambda_2 t} + \ldots + c_n e^{\lambda_n t}. \tag{10.10}$$

If among the "old" roots $\lambda_1, \lambda_2, \ldots, \lambda_n$ there is at least one with positive real part, λ_{n+1} cannot affect this state which will be unstable. If the real parts of all n roots are negative, the stability of the state is decided by the real part of λ_{n+1}. If this is negative it alters nothing, but if it is positive, the equilibrium state is unstable and thus completing the original system has produced instability. The sign of λ_{n+1}, however, depends on the sign of μ and on the sign of the coefficient a_0 or a_n of the original equation. If we choose $a_0 > 0$, then a_n must be positive in order that the original state be stable (Routh-Hurwitz criterion). Therefore instability can occur when the introduced coefficient μ is negative. As we shall see there are such cases in real systems.

Although there can be no general methods, we shall indicate one method with which it is sometimes possible to "unmask" equilibrium states which appear stable but in reality are unstable. We introduce successively into the equation various small physical parameters determining the signs which they have in the equation. If we discover one with a negative sign, then it is possible that the state of equilibrium is in reality unstable. We only say "it is possible" since another positive parasitic parameter might occur in this same term of the equation and the ultimate sign of the coefficient will depend on some unknown relation between parameters. Strictly speaking, therefore, we can never be certain whether an equilibrium state, which from the point of view of model theory appears to be stable, is in reality stable.

In order to demonstrate this danger we consider a few concrete examples, restricting ourselves to simple examples with "complete" non-linear equations of the second order.

1. Circuit with a voltaic arc

As a first example consider the now well-known circuit shown in Fig. 509. Let the dependence of the arc voltage v on current c be $v = \Psi(i)$ (Fig. 510).

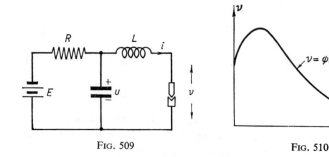

FIG. 509   FIG. 510

We shall repeat and supplement the results obtained earlier (see Chapter V, Section 5) for this circuit. The basic equations are

$$\left. \begin{array}{l} L\dfrac{di}{dt} = u - \psi(i), \\[6pt] C\dfrac{du}{dt} = \dfrac{E-u}{R} - i. \end{array} \right\} \qquad (10.11)$$

The equilibrium states (I, U) are determined by

$$\psi(i) = E - Ri$$

and are intersections of $u = \psi(i)$ and the "load" line $u = E - Ri$; and there can be either one or three states of equilibrium (Fig. 511). Three states

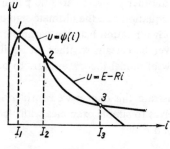

FIG. 511

of equilibrium is the more interesting case. We linearize the equations (10.11) near the point (I, U) by writing the variable component of current $\xi = i - I$ and that of the capacitor voltage $\eta = u - U$:

$$L\frac{d\xi}{dt} = \eta - \psi'(I)\xi, \quad CR\frac{d\eta}{dt} = -\eta - R\xi.$$

The characteristic equation is thus

$$\begin{vmatrix} L\lambda + \psi'(I) & -1 \\ R & CR\lambda + 1 \end{vmatrix} = 0$$

or

$$LCR \cdot \lambda^2 + [L + CR\psi'(I)]\lambda + R + \psi'(I) = 0. \tag{10.12}$$

It is evident that the character of the singular point (I, U) depends on the sign and value of $\psi'(I)$, i.e. on the "differential resistance" of the arc. Consider the equilibrium states *1, 2* and *3* in Fig. 511. Point *1* is stable since $\psi'(I_1) > 0$ and both roots of (10.12) have negative real parts. It is either a stable focus or a stable node depending on the relative magnitudes of L, C, R and $\psi'(I_1)$. At point 2, $\psi'(I)$ is negative and $R + \psi'(I_2) > 0$ so the singular point *2* is a saddle point; and the corresponding equilibrium state is unstable. Finally, at point *3*, $\psi'(I_3)$ although negative is such that $R + \psi'(I_s) > 0$, and the singular point *3* is also either a focus or a node. This singular point is unstable if $|\psi'(I_s)| > L/CR$ and is stable otherwise. Since $\psi'(I_3)$ is relatively small (the characteristic is descending but with a slight slope), then the equilibrium state *3* is always unstable for small values of L and stable for small values of C. In general the transition from a stable to an unstable state takes place at this point only for a "critical" value of the varying parameter (R or L or C). Thus with three equilibrium states their stability is represented by one of the two combinations shown in Figs. 512 and 513[†].

Let us examine now whether the character of these states varies if we neglect C or L (the circuits are shown in Fig. 514 and Fig. 515). We have already considered these circuits in Chapter IV, Section 6). The values of i and u at the equilibrium states in the three cases: the general case $L \neq 0$, $C \neq 0$; and $C = 0$, $L \neq 0$; and $C \neq 0$ $L = 0$; remain unaltered. As C reduces to the case $C = 0$ there is no change in the stability of these equilibrium states, *1* and *3* remain stable and *2* unstable, so that a small C in the circuit is not important for the stability of any equilibrium state.

[†] The points *1* and *3* are shown in the Figures as nodes. They can also be foci but as to their stability the picture remains the same.

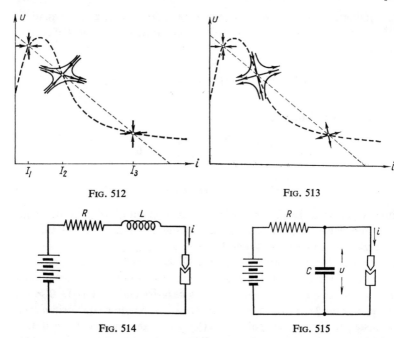

Fig. 512 Fig. 513

Fig. 514 Fig. 515

A different picture is obtained when we reduce the inductance L to zero. The characteristic equation for $L=0$ is

$$CR\psi'\lambda + R + \psi' = 0 \qquad (10.12\text{a})$$

and has the single root

$$\lambda = -\frac{R+\psi'}{CR\psi'},$$

which is negative for *1* and *2* and is positive for *3*. For the state of equilibrium *2* both ψ' and $R+\psi'$ are negative and therefore $\lambda<0$. Thus when we neglect the inductance the equilibrium states *1* and *2* are found to be stable and the state *3* unstable.

But the characteristic equation (10.12) when L is small (more precisely for $L \to +0$) has roots

$$\lambda_1 = -\frac{R+\psi'}{CR\psi'}$$

and

$$\lambda_2 = -\frac{\psi'}{L} \to \infty,$$

the first of which coincides with the root of the characteristic equation (10.12a). For the first state of equilibrium $\psi'(I_1)>0$, $\lambda_2 \to -\infty$ and $\lambda_1<0$;

this state of equilibrium is also stable for small values of L, and the small inductance L is not an important parameter. But for the second and third state of equilibrium $\psi' < 0$ and $\lambda_2 \to +\infty$, and thus these equilibrium states are unstable for arbitrarily small values of L. Therefore the analysis of the behaviour of a circuit with a Voltaic arc in the vicinity of the states 2 and 3 (on the descending section of the characteristic) necessarily requires that we take into account the inductance L, however small this may be. In particular the state of equilibrium 2 which was "stable" for $L=0$ is in reality unstable since there is always a small parasitic inductance. This change in stability will occur when $a > 0$, $b < 0$ and $c < 0$ in the characteristic equation $a\lambda^2 + b\lambda + c = 0$ and a vanishes when some parameter is put equal to zero. This case can be referred to briefly as that of a "negative saddle", in contrast to that of a "positive saddle" for which only $c < 0$. A "positive saddle" retains its instability if $a = 0$.

Thus, without suitable verification the state of equilibrium being considered is just such a one as could be mistaken for a stable state. That is what happened to Friedländer (151, 152) who gave a Voltaic arc in a circuit without self-inductance as an example of the system with two stable states of equilibrium. He took as a stable state the saddle point which in fact only "seems" to be stable.

2. Self-excitation of a multivibrator

As a second example consider the self-excitation of an ordinary multivibrator with one RC circuit but take into account the two small parasitic inductances L_a and L (Fig. 516). Neglecting grid currents and assuming

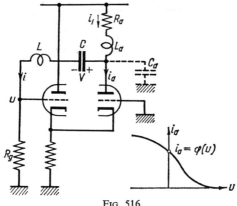

Fig. 516

the anode current i_a of the right-hand triode to depend only on grid voltage u of the left-hand triode: $i_a = \varphi(u)$, we obtain the following equations:

$$i_1 = i + i_a; \qquad i = C\frac{dV}{dt} = \frac{u}{R_g};$$

$$L_a \frac{di_1}{dt} + R_a i_1 = E_a - \left[u + L\frac{di}{dt} + V \right]$$

or, after simple transformations

$$\left. \begin{aligned} \left[\frac{L_a}{R_g} + L_a \varphi'(u) + \frac{L}{R_g}\right] \frac{du}{dt} &= \\ = E_a - R_a \varphi(u) - \left(1 + \frac{R_a}{R_g}\right) u - V, & \\ CR_g \frac{dV}{dt} &= u. \end{aligned} \right\} \qquad (10.13)$$

The only state of equilibrium in this circuit is determined by

$$u_0 = 0, \quad V_0 = E_a - R_a \varphi(0).$$

Now expand $\varphi(u)$ in a power series

$$i_a = \varphi(u) = \varphi(0) - Su + \ldots,$$

where $S = -\varphi'(0) > 0$, since the characteristic has a negative slope at the origin. For the first approximation then, we have

$$\left. \begin{aligned} \mu \frac{du}{dt} &= -\varrho u - R_g v, \\ CR_g \frac{dv}{dt} &= u, \end{aligned} \right\} \qquad (10.13\text{a})$$

where

$$\varrho = R_g + R_a(1 - SR_g), \quad \mu = L + L_a(1 - SR_g) \quad \text{and} \quad v = V - V_0.$$

The characteristic equation of (10.13a) is

$$\begin{vmatrix} \mu\lambda + \varrho & R_g \\ -1 & CR_g \lambda \end{vmatrix} = 0$$

or

$$\mu \lambda^2 + \varrho \lambda + \frac{1}{C} = 0, \qquad (10.14)$$

which gives the following stability conditions for the equilibrium state

$$\mu > 0, \quad \varrho > 0.$$

For $\mu<0$ the singular point $(0, V_0)$ is a saddle point, and for $\mu>0$ this same singular point can be either a node or a focus and is unstable for $\varrho<0$. The complete stability diagram in the plane μ, ϱ is shown in Fig. 517.

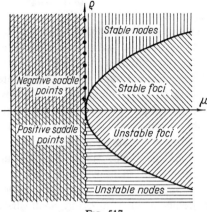

Fig. 517

If we had neglected the parasitic inductances L_a and L, and so put $\mu=0$, we would have obtained (after linearization) an equation of the first order

$$\varrho C \frac{du}{dt} + u = 0$$

and the stability of the state of equilibrium $u=0$ would only depend on the sign of the parameter ϱ; being stable for $\varrho>0$. The "stability" of the state (for $\varrho>0$) is caused once more by our "naiveté". However, as we shall see now, "not to be naive" in this case is even more difficult than in the preceding one.

Suppose first there is a parasitic inductance L_a and that $L=0$ in the circuit with the capacitance. In addition we assume that $1<SR_g<(R_g/R_a)+1$, i.e. that $0<\varrho<R_g$. Then the equilibrium state, stable for $L_a=0$ from the point of view of our ordinary criteria, loses its stability for an arbitrary small L_a and becomes a saddle-point. Therefore, in a real system such a state of equilibrium is unstable.

In fact, self-inductance, however small, must be present in the circuit with the capacitance, i.e. in a real system $L\neq 0$. If we take L into account we do not destroy the conditions of stability that are satisfied by the equilibrium state of the degenerate system when $L_a=0$ and $\varrho>0$, but we can even impart "stability" to the equilibrium state which proves unstable when L_a is present. In fact, if L is sufficiently large, we have $\mu = L+L_a$ $(1-SR_h) > 0$ even though $SR_g > 1$, and the equilibrium state which is unstable for $L_a \neq 0$ and $L=0$ (a saddle point) will be stable. Thus, stability is affected not only by the presence of parasitic parameters but also by their relative magnitudes.

But, while we can be certain that parasitic parameters do exist, their magnitude is usually not known, and even less do we know their ratios. In general, there may exist a region in which nothing can be said about the stability of the equilibrium state in the degenerate system. In our example, this is the region for ϱ contained in the limits $r > \varrho > 0$. In this region $SR_g > 1$, and the stability of the equilibrium states depends on the values of the parasitic parameters. Therefore, in considering a degenerate system and its condition of instability $\varrho < 0$, we must recognise the existence of a region of "undetermined" states of equilibrium extending from $\varrho = r$ to $\varrho = 0$. It is quite possible that the so-called "parasitic self-excitation" which appears and disappears without any noticeable variation of the parameters of the circuits, is often caused by small variations of the parameters in these regions of "undetermined" equilibrium states.

An analogous result is obtained when we take into account the small parasitic inductance L_a, and the small parasitic capacitance C_a in the anode circuit. In this case the characteristic equation is

$$\lambda^3 C_a L_a R_g + \lambda^2 [L_a(1-SR_g) + C_a R_a R_g] + $$
$$+ \lambda [R_a(1-SR_g) + R_g] + \frac{1}{C} = 0$$

(here $C_a \ll C$). The condition of stability of the equilibrium state is

$$R_a(1-SR_g) + R_g > 0 \quad \text{and} \quad L_a(1-SR_g) + C_a R_a R_g > 0.$$

Hence, self-excitation of the circuit is possible for $1 < SR_g < 1+(R_g/R_a)$ if the parasitic inductance L_a is sufficiently large:

$$L_a > \frac{C_a R_a R_g}{SR_g - 1} > C_a R_a^2.$$

For the usual values of the parameters of a multivibrator ($C_a \sim 10\,pF$, $R_a \sim 10^3 - 10^4\,\Omega$) the quantity $C_a R_a^2 \sim 10 - 10^3$ microhenrys is considerably larger than the usual parasitic inductance of the anode circuit. Therefore such a multivibrator will only be excited if $SR_g > 1 + (R_g/R_a)$.

In conclusion it should be noted that when the equilibrium state is unstable in the presence of a small parameter the velocity with which the system moves away from this seemingly stable state is completely different in the two possible cases. Thus, in the first case (the more interesting one in practice) the small parameter occurs in the coefficient of the highest order derivative and the new root of the characteristic equation $\lambda_{n+1} = -a_0/\mu$. Thus the root and the velocity with which the system moves away from the equilibrium state increases as μ decreases. In the limit for $\mu \to 0$ the departure from the equilibrium state occurs with an instantaneous jump. In the second case when the small parameter occurs in the coefficient of the integral, the departure from the equilibrium state occurs with a slow "creeping" motion where the velocity decreases as μ decreases.

§ 3. SMALL PARASITIC PARAMETERS AND DISCONTINUOUS OSCILLATIONS

That which has been said above about the role of small parameters in an oscillatory system can be translated into the language of phase space. In a real system, as we consider more small parameters, we introduce a greater number of degrees of freedom, the number of the dimensions of the associated phase space increases. But the complexity of the mathematical analysis also increases, and we must restrict the number of independent variables characterizing the state of the system, choosing them so as to reproduce qualitatively, and possibly quantitatively, the oscillating processes. Thus, we assume that the system can be represented in a phase space whose dimensions seldom exceed two or three.

Suppose the dynamic model, obtained from the real system when a few small (parasitic) parameters are included is represented by a system of differential equations of the nth order

$$\left. \begin{array}{l} \mu \dot{x}_i = F_i(x_1, \ldots, x_s; y_1, \ldots, y_{n'}), \\ \dot{y}_j = G_j(x_1, \ldots, x_s; y_1, \ldots, y_{n'}) \end{array} \right\}$$

$(i = 1, 2, \ldots, s; j = 1, 2, \ldots, n'; n = n'+s)$ or in a vectorial notation

$$\mu \dot{x} = F(x; y), \quad \dot{y} = G(x; y), \tag{10.15}$$

where μ is a small positive parameter which reduces to zero when certain small physical quantities of the system are put equal to zero. We will only consider the most interesting case, when the new small parameters appear in the equations of motion as small coefficients of the higher-order derivatives. Further we assume the functions $F(x, y)$ and $G(x, y)$ to be bounded and differentiable over the region of interest. The functions F and G can also depend upon μ: we then assume that they tend to finite limits as

$\mu \to +0$. Putting $\mu = 0$, and so neglecting these small parameters we obtain a more simplified dynamic model

$$F(x; y) = 0, \quad \dot{y} = G(x; y) \qquad (10.16)$$

whose set of states have a smaller number of dimensions $n'(n' < n)$ and only corresponds in the "complete" phase space to a certain subspace F, defined by the equation

$$F(x; y) = 0.$$

The question of whether the processes occurring in the system can be investigated by neglecting these parameters evidently amounts to asking whether the motion of the representative point of the complete system (10.15) in a small neighbourhood of the sub-space F can be satisfactorily replaced by the motion of a representative point within the limits of the sub-space F itself—a space with a smaller number (n') of dimensions. This will be so for sufficiently small values of the positive parameter μ and for some time interval of interest such as $(0 < t < +\infty)$.

1. *The mapping of the "complete" phase space by the paths*

To answer the above question we analyse the general features of the phase portrait in the x, y phase space of the complete system (10.15) when μ is sufficiently small [61]. Consider first the region of the phase space that lies *outside* the small $0 \, (\mu^a)$-neighbourhood of the sub-space F $(0 < a < 1)$ and which reduces to F as $\mu \to +0$†. In this region

$$|F(x; y)| \geqslant O(\mu^a) \quad \text{and} \quad |\dot{x}| \geqslant O(\mu^{a-1}).$$

Therefore there are "rapid" motions of the representative point, which are the more rapid, the smaller the value of μ, for $\mu \to +0$, $\dot{x} \to +\infty$. We will call this region the *region of "rapid" motions*. Since in this region $\dot{y} = G(x; y)$ remain bounded for $\mu \to +0$ and

$$\left|\frac{dy_j}{dx_i}\right| = \mu \left|\frac{G_j(x; y)}{F_j(x; y)}\right| \leqslant O(\mu^{1-a}) \to 0 \quad \text{for} \quad \mu \to +0,$$

then for finite increments of the x variables in small intervals of time $\Delta t \leqslant 0(\mu^{1-a})$ the y variables only change by quantities of the order of not more than μ^{1-a}. Thus the phase paths in the region of "rapid" motions lie near the s-dimensional space

$$y = \text{const.}$$

† By $0[g(\mu)]$-neighbourhood of the sub-space F we mean the set of all points whose distance from F does not exceed a quantity of the order of $g(\mu)$; here $g(\mu) = \mu^a$ $(0 < a < 1)$.

Hence, for sufficiently small values of μ the motions of the representative point in this region in small intervals of time $\Delta t \leqslant O(\mu^{1-a})$ are almost *instantaneous jumps*, in which the x variables vary rapidly (jump-wise) and the y variables remain unchanged. Therefore, the approximate differential equations of motion in the region of "rapid" motions can be written as

$$y \equiv y^0 = \text{const}, \quad \dot{x} = \frac{1}{\mu} F(x; y^0). \tag{10.17}$$

Evidently the equations (10.16) of the simplified or incomplete model are inadequate to describe the behaviour of the system and can represent the system only within the limits of a small $O(\mu)$-neighbourhood of the n'-dimensional sub-space F, where $|F(x; y)| \leqslant O(\mu)$ and where, therefore the rate of change of the state of the system (both $\dot{x}$ and $\dot{y}$) remain finite for arbitrary small values of μ.

2. Condition for small (parasitic) parameters to be unimportant

According to the distribution of phase paths of "rapid" motions in the vicinity of the n'-dimensional sub-space F, two basic cases are possible.

It is possible that all paths of "rapid" motions remain inside a small neighbourhood of the sub-space F (as t increases), and the representative point, if initially inside this neighbourhood, will remain there. In this case the representative point will move comparatively slowly ($\dot{x}$ and $\dot{y}$ are bounded for $\mu \to +0$) as described by the equations (10.16) [119, 42]. These motions of the representative point for which $\dot{x}$ and $\dot{y}$ remain bounded during finite intervals of time (for small $\mu \to +0$) will be called "*slow*", and the small $O(\mu)$-neighbourhood of the sub-space F will be called the region of "slow" motions. Thus the parasitic parameters in the "complete" equations (10.15) are unimportant for these slow processes which begin from states compatible with the equations (10.16).

If, however, the initial state of the representative point is not near the sub-space F, then it moves along a path of "rapid" motion to the region of "slow" motions and continues therein. The duration Δt of the rapid motion or jump will clearly be the smaller, the smaller the initial distance of the representative point from the sub-space F and the smaller the value of the parameter μ (it can be shown [42] that $\Delta t \leqslant O(\mu \ln \mu^{-1})$). It is not really necessary to make a detailed analysis of the "rapid" motions of the systems during the initial stage using equations (10.15) or the approximate equations (10.17) but merely postulate that the representative point jumps instantaneously on to the corresponding point of the n'-dimensional sub-space F. This is especially convenient if the condition that $y=$const. during a jump of x enables us to determine the end-point of the jump in sub-space F. If $y=$const. is not satisfied it is necessary to make, at least a qualitative analysis of the "rapid" motions (even if only by using the approximate equations (10.17)) or by recourse to additional considerations. The points of the n'-dimensional sub-space $F(x, y) = 0$ are equilibrium states for the approximate equations of "rapid" motions (10.17) and therefore the behaviour of the paths of "rapid" motions near F is completely determined by the stability of these states. Let us introduce the new "rapid" time

$$t' = \frac{t}{\mu}$$

then the approximate differential equations of "rapid" motions (10.17) can be written as

$$\frac{dx}{dt'} = F(x; y), \qquad y \equiv \text{const.}$$

Linearizing these equations in a neighbourhood of the point $(\bar{x}; y)$ of the sub-space F, we obtain the equations of the first approximations

$$\frac{dx_i}{dt'} = \sum_{k=1}^{s} \frac{\partial F_i}{\partial x_k} \xi_k \quad (i = 1, 2, \ldots, s),$$

where $\xi_i = x_i - \bar{x}_i$. The characteristic equations are

$$\begin{vmatrix} \dfrac{\partial F_1}{\partial x_1} - \lambda & \dfrac{\partial F_1}{\partial x_2} & \cdots & \dfrac{\partial F_1}{\partial x_s} \\ \dfrac{\partial F_2}{\partial x_1} & \dfrac{\partial F_2}{\partial x_2} - \lambda & \cdots & \dfrac{\partial F_2}{\partial x_s} \\ \vdots & \vdots & \cdots & \vdots \\ \dfrac{\partial F_s}{\partial x_1} & \dfrac{\partial F_s}{\partial x_2} & \cdots & \dfrac{\partial F_s}{\partial x_s} - \lambda \end{vmatrix} = 0. \qquad (10.18)$$

If *all s roots of the characteristic equation* (10.18) *have negative real parts for arbitrary values of x and y satisfying the equations* $F(x; y) = 0$, then the points of the sub-space F are stable equilibrium states for the approximate equations of "rapid" motions (10.17) and all paths of "rapid" motions in the vicinity of the sub-space F enter a small neighbourhood of the latter.

Therefore, in this case *the small parasitic parameters taken into account in the equations* (10.15) *are unimportant at least for processes that start from states compatible with the approximate equations of "slow" motions* (10.16). This theorem can be proved rigorously [49, 50, 119]. An equivalent statement of this condition can be formulated using the Routh-Hurwitz stability conditions for equations (10.18).

There are two particular cases which will be needed:

(1) if in the "complete" equations (10.15) there is only one equation with a derivative having a small coefficient (i.e. if $s=1$) then the equation (10.18) will be of the first degree

$$F'_x(x; y) - \lambda = 0.$$

and the condition for the small parameter to be unimportant is

$$F'_x(x; y) < 0, \qquad (10.19)$$

to be satisfied at all points of the sub-space $F(x:y) = 0$;

(2) if in the system (10.15) there are two equations with derivatives with small coefficients (i.e. if $s=2$) then the characteristic equation takes the form

$$\begin{vmatrix} \dfrac{\partial F_1}{\partial x_1} - \lambda & \dfrac{\partial F_1}{Fx_2} \\ \dfrac{\partial F_2}{\partial x_1} & \dfrac{\partial F_2}{\partial x_2} - \lambda \end{vmatrix} = 0$$

or

$$\lambda^2 - \left(\dfrac{\partial F_1}{\partial x_1} + \dfrac{\partial F_2}{\partial x_2}\right)\lambda + \dfrac{\partial(F_1, F_2)}{\partial(x_1, x_2)} = 0,$$

and the condition for the small parameters to be unimportant will be that at all points of the sub-space F

$$\dfrac{\partial F_1}{\partial x_1} + \dfrac{\partial F_2}{\partial x_2} < 0 \text{ and } D = \dfrac{\partial(F_1, F_2)}{\partial(x_1, x_2)} = \begin{vmatrix} \dfrac{\partial F_1}{\partial x_1} & \dfrac{\partial F_1}{\partial x_2} \\ \dfrac{\partial F_2}{\partial x_1} & \dfrac{\partial F_2}{\partial x_2} \end{vmatrix} > 0. \qquad (10.19a)$$

It is easily seen that (10.19) is satisfied for the RC and RL circuits in Figs. 502 and 503 with parasitic inductance L_0 (in the RC circuit) and the parasitic capacitance C_0 (in the RL circuit) as the small parameter. For example, for the RC circuit after introducing the dimensionless time

$t'=t/RC$ we reduce the equation (10.1a) to the form (10.15)

$$\mu \frac{di}{dt'} = -q-i = F(q, i), \quad \frac{dq}{dt'} = i,$$

where $\mu = L_0/CR^2$ is a small positive parameter since $L_0 \ll CR^2$. Therefore, $F' = -1 < 0$ is satisfied at all points of the phase line $q+i = 0$ of the incomplete system where $\mu = 0$.

Outside the line $q+i = 0$ there are "jumps" in the intensity of the current i with the capacitor charge q almost unchanged. Under these circumstances all paths of "rapid" motions on the q, i phase plane move into a small neighbourhood of the straight line $q+i = 0$, which is the region F on the q, i plane (Fig. 518).

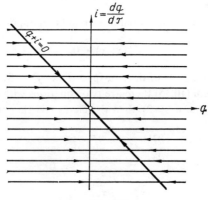

Fig. 518

There is a second case possible, when the condition for the small parasitic parameters to be unimportant is *not* satisfied at least on part of the n'-dimensional subspace F. This portion of F we will call F^-. Then the characteristic equation (10.18) determined at points of the sub-space F^-, must have roots with positive real parts and the points of the sub-space F^- are unstable equilibrium states for the approximate equations of "rapid" motions (10.17). In the complete x, y phase space there are paths of "rapid" motions that *leave* a small neighbourhood of the sub-space F^- (for example the $0(\mu^a)$-neighbourhood, where $0 < a < 1$). The representative point cannot remain near this part F^- of the n'-dimensional sub-space F and must enter the region of "rapid" motions. Therefore there exist motions of the system in this case which start from states compatible with the equations (10.16), i.e. from states belonging to the n'-dimensional sub-space F' and which cannot be analysed by means of the equations (10.16). Thus, in the case considered, the arbitrary small parasitic parameters taken into account in setting out the "complete" equations (10.15) are *essential* for the processes occurring in the physical system.

For example, let us consider once more the circuit of a Voltaic arc with small value of C and L (Fig. 509), whose oscillations are described by the "complete" equations (10.11) (see Section 2, Sub-section 1 of this chapter). The small capacitance C proves in general to be unimportant since when $C=0$ the phase line F is

$$F = \frac{E-u}{R} - i = 0,$$

and on it

$$\frac{\partial F}{\partial u} = -\frac{1}{R} < 0.$$

The portrait on the i, u phase plane for the limit case $C \to +0$ is shown in Fig. 519(a); all paths of "rapid" motions ("jumps" of the voltage u for i-const) reach the phase line $u = E - Ri$ of the system without capacitance.

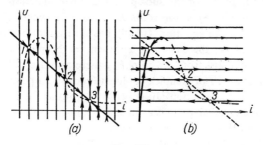

Fig. 519

A small inductance, however, is an important parameter. In fact, the phase line of the system without inductance is (on the i, u plane) the arc characteristic $u = \psi(i)$—and the condition for an arbitrarily small inductance to be unimportant is not satisfied over the section of this characteristic with negative slope, since

$$\frac{\partial}{\partial i}[u - \psi(i)] = -\psi'(i) > 0.$$

The states of the circuit on this section are unstable with respect to "rapid" motions, and its paths leave the descending section of the arc characteristic (Fig. 519(b)). Thus the inductance L (even if very small) must be taken into account if the analysis is to give results in some agreement with experimental data.

3. Discontinuous oscillations [61, 94, 105, 114, 158, 159]

A very interesting case, useful in the theory of discontinuous oscillations, occurs when $F(x; y) = 0$, the n'-dimensional phase space of the incomplete or "degenerate" model splits into two parts: a part F^+ here the condition for some small parameters to be unimportant is satisfied (all roots of the characteristic equation (10.18) have negative real parts) and a part F^- where this condition is not satisfied. Then, *only a small 0 (μ)-neighbourhood of the sub-space F^+* in the complete n-dimensional x, y phase space is a *region of "slow" motions of the representative point*. Therefore if the parasitic parameters considered are sufficiently small (i.e. if $\mu \ll 1$) we can use the approximate equations of "slow" motions (10.16) to describe a "slow" motion of the representative point in the vicinity of F^+. The motion itself may be considered as occurring approximately within this part F^+ of the sub-space $F(x, y) = 0$.

Before proceeding to the analysis of discontinuous oscillations let us consider the boundary between the sub-spaces F^+ and F^-. The x, y coordinates of the points of the sub-space F are parameters in the characteristic equation (10.18) whose roots λ, therefore, depend upon the x, y coordinates. Since the equation (10.18) has no root with a positive real part at the points of the sub-space F^+ then, as the point $(x; y)$ passes through the boundary from F^+ into F^-, there appears either one real positive root or a pair of complex conjugate roots with positive real parts in the characteristic equation (10.18) (the appearance of more than two roots with positive real parts is a singular "non-coarse" case and is only met for a special choice of functions $F(x; y)$).

Here we shall only consider the first case. Then, for points on the boundary surface γ, the characteristic equation (10.18) will have one zero root (the remaining $s-1$ roots have negative real parts) from which it follows that the constant term of this equation

$$D(x; y) = \frac{\partial(F_1, F_2, \ldots, F_s)}{\partial(x_1, x_2, \ldots, x_s)}$$

is equal to zero. So the boundary γ of the sub-spaces F^+ and F^- belongs to the $(n'-1)$-dimensional surface

$$F_i(x; y) = 0 \quad (i = 1, 2, \ldots, s), \qquad D(x; y) = 0. \quad (10.20)$$

Differentiating the equations $F_i(x; y) = 0$ with respect to t and using equations (10.16) we obtain (for the "slow" motions)

$$\sum_{j=1}^{s} \frac{\partial F_i}{\partial x_j} \dot{x}_j + \sum_{k=1}^{n'} \frac{\partial F_i}{\partial y_k} G_k = 0$$

or, solving with respect to $\dot{x}_j$

$$\dot{x}_j = \frac{D_j(x; y)}{D(x; y)},$$

where $D_j(x; y)$ is the determinant formed from the Jacobian $D(x; y)$ by replacing the jth column by the column $-\sum_{k=1}^{n'}(\partial F_i/\partial y_k)G_k$. Thus the points of the boundary surface γ on which $D(x; y)$ reduces to zero are points of infinite velocities $\dot{x}$ for the equations (10.16) of the "degenerate" model and also junction points of the paths of these equations since, in moving through γ, $D(x; y)$ and hence all $\dot{x}$ change their signs[†].

Now consider a representative point moving in sub-space F^+, its motion described by the equations (10.16). On reaching the surface γ the representative point cannot move further in the vicinity of F and will "escape" into the region where the x variables vary arbitrarily rapidly for $\mu \to +0$ according to a law that is represented approximately by equations (10.17) but not by the equations (10.16). Suppose that in the s-dimensional sub-space, $y=$const, there are no other "elements of attraction" except the stable equilibrium states of sub-space F^+. Then the paths of "rapid" motions go away from γ and back to the region of "slow" motions, i.e. into a small $0(\mu)$-neighbourhood of the sub-space F^+. Thus, in the "total" phase space there are paths passing through regions of "rapid" and "slow" motions. In the limit, as $\mu \to +0$, each such path will consist of alternate sections of two types: paths of "slow" motions lying in the n'-dimensional sub-space F^+ and on which the representative point moves according to the equations (10.16) during finite intervals of time, and paths of "rapid" motions each of which lies in the s-dimensional sub-space $y=$const. and moves instantaneously according to the equations (10.17). A "slow" motion of the system turns into a "rapid" one on the boundary surface γ

$$F(x; y) = 0, \quad D(x; y) = 0,$$

[†] If, as the point $(x : y)$ passes from F^+ into F^-, there appears among the roots of the characteristic equation a pair of complex conjugate roots with positive real part (this is only possible for $s \geqslant 2$), then, for the points of the boundary between F^+ and F^-, the equation (10.18) has a pair of purely imaginary roots differing from each other by their signs. In this case, as is well known, the last but one Hurwitz's determinant Δ_{s-1} formed by the coefficients of the equation (10.18) reduces to zero. Thus the boundary between the sub-space F^+ and F^- will belong to another $(n'-1)$-dimensional surface

$$F(x; y) = 0, \quad \Delta_{s-1} = 0.$$

The points of this boundary, for the equations of the "degenerate" model (10.16), are evidently neither points of infinitely large velocities of the x variables nor junction points of phase paths.

and we assume that for each point on γ there is only a single path of "rapid" motion.† In fact, it can be proved [105] that *the paths of the system of equations (10.15) for sufficiently small values of the positive parameter μ do move in fact in the vicinity of the paths constructed by means of the approximate equations* (10.16) *and* (10.17) *by the method indicated above.* A proof is given in Section 4 of this chapter for periodic motions in second order systems.

The paths considered above are mathematical models of *discontinuous oscillations* which may include closed paths or *discontinuous limit cycles* which evidently correspond to *discontinuous self-oscillations*.

Thus the investigation of oscillations which are approximately discontinuous for sufficiently small values of the certain parameters ($0 < \mu \ll 1$) can be carried out using the approximate equations of the "slow" motions

$$F(x; y) = 0, \quad \dot{y} = G(x; y) \qquad (10.16)$$

in the portion F^+ of the sub-space $F(x; y) = 0$, and by means of the approximate equations of the short-duration "rapid" motions (instantaneous jumps of the x variables)

$$y = \text{const}, \quad \mu\dot{x} = F(x; y) \qquad (10.17)$$

in the remaining part of the "complete" x, y phase space. Note further, that in an instantaneous jump of the variables x (for $\mu \to +0$) the y variables do not vary, and the initial point of the jump (x^-, y^-) (a point on surface γ) and the end point of the jump (x^+, y^+) lie in the same sub-space F, so that their coordinates are clearly connected by the following equations:

$$\left. \begin{array}{l} F(x^-, y^-) = 0, \quad D(x^-, y^-) = 0, \\ y^+ = y^-, \quad F(x^+, y^+) = 0, \end{array} \right\} \qquad (10.21)$$

which can be called the jump equations. In many problems it is known from experiment that "rapid" motions of the system suddenly become "slow" motions and the equations (10.21) suffice to determine the point (x^+, y^+). In such problems a detailed analysis can be replaced by the introduction of a jump postulate, indicating those points of the n'-dimensional phase space of the "degenerate" system where "slow" motions are im-

---

† For the approximate equations (10.17) the points of the boundary surface γ are multiple singular points: for them one root of the characteristic equation (10.18) is zero and the remaining roots have negative real parts. In the basic case these points are analogous to the singular points of the saddle-node type on a phase plane and only a single path of the equations (10.17) leaves each of them.

possible (region F^- and its boundary γ), and from which jumps start, together with the "jump law"

$$y^+ = y^-, \quad F(x^-, y^-) = 0, \quad F(x^+, y^+) = 0,$$

that determines the end point of a jump. This method will be used later in the analysis of discontinuous oscillations.

The conditions that variables $y=$const. in an instantaneous jump of x have usually a clear physical meaning. For example, in electrical systems they usually mean that during instantaneous jumps of the state of the system, voltages across capacitors or currents in inductances remain constant. However, the theoretical determination of the set of points in phase space from which the jumps start, of whether the "rapid" motions are of short duration and whether they turn again into "slow" motions, require an investigation of the differential equations (10.17). This investigation is particularly necessary when the jump conditions (10.21) admit several end points to a jump, e.g. oscillations of coupled multivibrators [37]. In such problems the analysis of the paths of the jumps with the approximate equations (10.17) removes this ambiguity without introducing any additional hypotheses.

To conclude this section we will briefly examine the case where the approximate equations (10.17) with $y=$const. have ω-limit paths that differ from stable states of equilibrium, e.g. when the equations (10.17) have a stable periodic or quasi-periodic solution

$$x = x^*\left(\frac{t}{\mu}, y\right) \tag{10.22}$$

(the y variables in this solution, as in the equations (10.17) are considered as constant parameters). It is evident that this can only occur for $s \geqslant 2$, when at least two equations of the system (10.15) have a small parameter multiplying a derivative. Now, in contrast to the case considered above, "rapid" motions of the system exist for the intervals of finite duration which do not tend to zero for $\mu \to +0$.

Therefore the assertion that the y variables vary little during a "rapid" motion is no longer correct. To see how they do vary, substitute (10.22) in the second equation (10.15); then we have

$$\dot{y} = G\left[x^*\left(\frac{t}{\mu}, y\right); y\right] \tag{10.23}$$

or after introducing the "rapid" time $t' = t/\mu$

$$\frac{dy}{dt'} = \mu G[x^*(t', y); y]. \quad (10.23a)$$

Since the y variables are slowly varying functions of the "rapid" time $t'[dy/dt' = 0\,(\mu)]$, then to find an approximate solution of equations (10.23a) we employ the "averaging method", used in Van der Pol's method (see Sections 2 and 3 of Chapter IX). In fact, the solution of (10.23a) for small values of μ is close to the solution of the auxiliary equations obtained by averaging the right-hand sides with respect to the time (as it occurs explicitly),

$$\frac{dy}{dt'} = \mu \bar{G}(y) \quad (10.24)$$

or

$$\dot{y} = \bar{G}(y), \quad (10.24a)$$

$\bar{G}(y)$ are the functions $G[x^*(t', y); y]$ averaged with respect to t'. Integrating these auxiliary equations, we obtain the approximate law of variation

$$y = y^*(t),$$

and a more accurate law of variation of x

$$x = x^*\left[\frac{t}{\mu},\ y^*(t)\right] \quad (10.22a)$$

during the "rapid" motion of the system. A "slow" variation of y during a "rapid" change of x can cause the motion to stop. Examples of rapid motions that last finite (or infinitely long) intervals of time t can be found in the literature [48, 53, 57, 109].

§ 4. Discontinuous Oscillations in Systems of the Second Order

To illustrate what has been stated in Section 3, we shall consider in greater detail discontinuous oscillations in a dynamic system described by two differential equations of the first order

$$\left.\begin{array}{l} \mu\dot{x} = F(x, y), \\ \dot{y} = G(x, y), \end{array}\right\} \quad (10.15a)$$

where $F(x, y)$ and $G(x, y)$ are single-valued continuous functions, having continuous partial derivatives, and μ is a small positive parameter. In such a system "rapid" motions that last finite or infinitely long intervals of time are impossible. We shall assume that the phase surface is an ordinary x, y plane.

The space of the "degenerate" system (when $\mu=0$) is a continuous line F defined on the x, y plane by

$$F(x, y) = 0.$$

In a small neighbourhood of this line (with dimensions of the order of μ) the phase velocity of the representative point will be finite, and outside a small neighbourhood of the line F, $\dot{x} \to \infty$ (for $\mu \to +0$) while $\dot{y}$ remains bounded and $dy/dx = \mu G(x, y)/F(x, y) \to 0^{\dagger}$. Hence, outside F, the phase paths of the system are close to the straight lines $y = $const, along which the representative point moves with large velocities. The approximate (but more accurate, the smaller μ), equations of these "rapid" motions along a path close to the straight line $y \equiv y^0 = $const. will be

$$y \equiv y^0 = \text{const}, \quad \mu \dot{x} = F(x, y^0). \qquad (10.17a)$$

For these approximate equations the points of intersection of the straight line $y \equiv y^0 = $const. and the line F are singular points (stable if $F'_x < 0$ and unstable if $F'_x > 0$) and determine the motion along the straight line $y \equiv y^0$. If for sufficiently large values of $|x|$ the sign of $F(x, y)$ is opposite to that of x, then jump paths move from infinity, and from the F^- sections of the line $F(x, y) = 0$ where $F'_x(x, y) > 0$ towards the F^+ sections of the same line where $F'_x(x, y) < 0$. Therefore, "slow" motions of the system, with bounded values of $\dot{x}$ and $\dot{y}$, will only occur in small neighbourhoods of the F^+ sections and will be represented approximately by

$$F(x, y) = 0, \quad \dot{y} = G(x, y), \qquad (10.16a)$$

which are the equations of the "degenerate or incomplete" system.

In the limiting case $\mu \to +0$, the entire plane (outside the line F) is filled with paths of "rapid" jumpwise motions, $y = $const, moving towards the F^+ line to the right ($\dot{x} \to +\infty$) in the region $F(x, y) > 0$ and to the left ($\dot{x} \to -\infty$) in the region $F(x, y) < 0$. The F^+ line itself contains the paths of "slow" motions where the phase velocity is finite.

† For example outside the $\mu^{\frac{1}{2}}$ neighbourhood of the line $F|\dot{x}| \geqslant 0\left(\mu^{-\frac{1}{2}}\right) \to \infty$ and $|dy/dx| \leqslant 0\left(\mu^{\frac{1}{2}}\right) \to 0$ for $\mu \to +0$.

Suppose that on the line F there are both F^+ sections $(F'_x(x, y) < 0)$ and F^- sections $(F'_x(x, y) > 0)$ separated by the boundary points γ. At these points, clearly,

$$F'_x(x, y) = 0$$

and the tangent to the line F is horizontal[†]. If the representative point moving "slowly" along a path F^+, reaches a point γ then subsequently it will move "rapidly" (with a jump) along the path $y =$ const. that leaves this point, until it again arrives on a F^+ line of "slow" motion. In

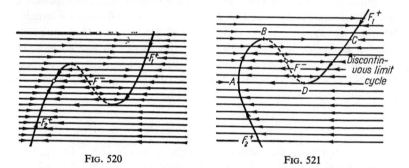

Fig. 520   Fig. 521

this case there can occur discontinuous oscillations consisting of alternate "slow" and "rapid" jump-wise motions.

A few examples of phase portraits of the equations (10.15a) are shown in Figs. 520–3 for the limiting case $\mu \to +0$. In Fig. 520 there is a stable equilibrium state which is attained after a few oscillations. The phase portraits shown in Figs. 521, 522 and 523 contain *discontinuous limit cycles*, corresponding to periodic discontinuous oscillations. The phase portrait in Fig 523 is the case of a hard mode of excitation when there is a stable state of equilibrium on the section F_2^+ in addition to a (stable) discontinuous limit cycle *ABCDA*. The closed line *abcda* is an unstable limit cycle and divides the phase plane into a region with attraction to the equilibrium state and a region with attraction to the limit cycle *ABCDA*.

[†] Here, and later, we assume that at the points of γ $F'_y(x, y) \neq 0$ so that these points are not singular points of the curve F.

By differentiating $F(x, y) = 0$ with respect to t and using the differential equation (10.16a) we find that during a "slow" motion $F'_x \dot{x} + F'_y G = 0$. Therefore, for the equations (10.16a) of the degenerate model the points γ are points at which $\dot{x}$ becomes infinite and are junction points of paths at which $\dot{x}$ changes its sign. The latter is also true when $F'_y(x, y)$ is not continuous at the points γ, which is usually the case in piece-wise linear systems.

Suppose that the system of equations

$$\mu\dot{x} = F(x, y), \quad \dot{y} = G(x, y), \qquad (10.15a)$$

has, in the limiting case when $\mu \to +0$, a discontinuous limit cycle C_0: $A_1 B_1 A_2 B_2, ..., A_m B_m A_1$, consisting of m sections of paths of "slow" motions on the F^+ line: $A_1 B_1, A_2 B_2, ..., A_m B_m$, alternating with m sections of paths of "rapid" motions (y=const): $B_1 A_2, B_2 A_3, ..., B_{m-1} A_m, B_m A_1$. From

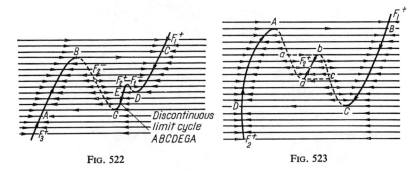

Fig. 522   Fig. 523

the definition of the discontinuous limit cycle, on each of its section of "slow" motion $A_k B_k$ (excluding the point B_k) we have

$$F(x, y) = 0, \quad F'_x(x, y) < 0, \quad G(x, y) \neq 0,$$

the sign of the function $G(x, y) = \dot{y}$ on the segment $A_k B_k$ being such that the representative point moves from the point A_k to the point B_k. On each of the sections of paths of "rapid" motions $B_{k-1} A_k$ (the points B_{k-1} and A_k being excluded) we have

$$F(x, y) \neq 0$$

the sign of $F(x, y) = \mu \dot{x}$ on the interval $B_{k-1} A_k$ is such that the representative point "jumps" from the point B_{k-1} to the point A_k. At the points of transition B_k

$$F(x, y) = 0, \quad F'_x(x, y) = 0, \quad G(x, y) \neq 0;$$

and we assume in addition that at these points $F'_y(x, y)$ and $F''_{xx}(x, y)$ are different from zero. Then, at each point B_k, the tangent to the line $F(x, y) = 0$ is horizontal and y has a maximum or minimum value.

y is a maximum at a point B_k if $G(x, y) > 0$ on $A_k B_k$ and a minimum if $G(x, y) < 0$. Therefore the sign of $F''_{xx}(x, y)/F'_y$ at the point B_k is the same as the sign of $G(x, y)$ on the segment $A_k B_k$. Also since the sign of $\dot{x}$ does not vary as the representative point goes through B_k from slow to rapid motion, and $\dot{x} = -F'_y G/F'_x$ and $F'_x < 0$ along the section $A_k B_k$ and $\dot{x} = F/\mu$ along the section $B_k A_{k+1}$, then the sign of $F'_y G$ at the point B_k is the same as the sign of the function $F(x, y)$ on the subsequent interval of "rapid" motion $B_k A_{k+1}$.

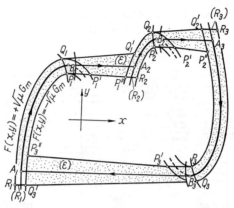

FIG. 524

Thus the signs of F''_{xx} and $F'_y G$ at the point B_k and the sign of $F(x, y)$ on $B_k A_{k+1}$ are the same. The sign of $G(x, y)$ at the point B_k is of course the same as its sign on the segment $A_k B_k$.

For example, Fig. 524 shows a discontinuous limit cycle $A_1 B_1 A_2 B_2 A_3 B_3 A_1$, for which $\dot{y} = G(x, y) > 0$ on the "slow" section $A_1 B_1$; $F(x, y) > 0$, and $\dot{x} \to +\infty$ since $\mu \to +0$, on the "rapid" section $B_1 A_2$; $G(x, y) > 0$ on $A_2 B_2$; $F(x, y) > 0$ on $B_2 A_3$; $G(x, y) < 0$ on $A_3 B_3$; $F(x, y) < 0$ on $B_3 A_1$; $F''_{xx} > 0$ and $F'_y > 0$ at the point B_1; $F''_{xx} > 0$ and $F'_y > 0$ at the point B_2; and $F''_{xx} < 0$ and $F'_y > 0$ at the point B_3.

We shall prove that for the equations (10.15a) *in a sufficiently small neighbourhood of the discontinuous limit cycle C_0 there lies a single stable limit cycle of the system* (10.15a) *only if the positive parameter μ is sufficiently small*. We shall prove, in other words, that *each discontinuous limit cycle is the limiting position for $\mu \to +0$ of just one and only one stable limit cycle of the system* (10.15a) [60][†].

We first choose, as in Section 5 of Chapter VIII, for a given small positive μ a closed doubly connected region (ε) with the following properties: (1) in (ε) there are no equilibrium states of (10.15a), (2) the region (ε) reduces to the *discontinuous* limit

---

† This theorem is demonstrated in [196] for a particular form of the equations (10.15a) when $G(x, y) = x$, and $F(x, y) = G(x) - y$.

cycle C_0 as $\mu \to +0$ and (3) the paths of (10.15a) for $\mu > 0$ enter the region (ε). Evidently, according to the Theorem V, Section 2, Chapter VI. this region contains at least one stable limit cycle of the system (10.15a).

We now isolate on the x, y plane a certain bounded simply connected closed region D containing C_0 and in which $F(x, y)$ and $G(x, y)$ and their derivatives (up to the order needed) are continuous and hence bounded. Also we assume that μ has been chosen so small that the region (ε) lies entirely inside D.

To construct the region (ε) draw in the region D the lines

$$F(x, y) = +\sqrt{\mu}\, G_m \quad \text{and} \quad F(x, y) = -\sqrt{\mu}\, G_m,$$

where G_m is the maximum absolute value of the function $G(x, y)$ in D. These lines are such that in the (open) region (α) between them, and which contains the line $F(x, y) = 0$

$$|F(x, y)| < \sqrt{\mu}\, G_m,$$

and in the remaining part of the region D

$$|F(x, y)| \geq \sqrt{\mu}\, G_m{}^\dagger$$

Also draw lines

$$\frac{F'_x(x, y)}{F'_y(x, y)} = \pm \sqrt{\mu},$$

on which the slope of the curve $F(x, y) = \text{const}$ is $\pm \mu^{\frac{1}{2}}$ and which select from (α) open regions (β_k) enclosed between them and containing the point B_k, such that in (β_k)

$$\left|\frac{F'_x(x, y)}{F'_y(x, y)}\right| < \sqrt{\mu}$$

and in the rest of the region (α)

$$\left|\frac{F'_x(x, y)}{F'_y(x, y)}\right| \geq \sqrt{\mu}.$$

Lines with such properties can be constructed within the region D by choosing a sufficiently small value for μ. Now the region (α) will lie inside a certain $0\left(\mu^{\frac{1}{2}}\right)$-neighbourhood of the line $F(x, y) = 0$ and the regions (β_k) inside certain $0\left(\mu^{\frac{1}{2}}\right)$-neighbourhoods of the points B_k.‡

† Note that the line $F(x, y) = +\mu^{\frac{1}{2}} G_m$ lies on the left of the sections of "slow" motions $A_k B_k$ of the discontinuous limit cycle C_0, while the line $F(x, y) = -\mu^{\frac{1}{2}} G_m$ lies on the right of them, since on the sections $A_k B_k$ $F(x, y) = 0$ and $F'(x, y) < 0$.

‡ This follows from the following simple lemma: let us suppose that in a certain bounded region A there is a line $\Phi(x, y)$ at the points of which there exist continuous derivatives Φ'_x and Φ'_y that do not reduce simultaneously to zero; then there is a positive number δ_0 such that for any δ in $0 < \delta \leq \delta_0$: (1) in the region A there exist lines $\Phi(x, y) = +\delta$ and $\Phi(x, y) = -\delta$ lying in a certain $0(\delta)$-neighbourhood of the line $\Phi(x, y) = 0$, and (2) in the open region comprised between these lines, and containing the line $\Phi(x, y) = 0$, $|\Phi(x, y)| < \delta$ and in remaining part of the region A $|\Phi(x, y)| \geq \delta$. It is evident that the functions $F(x, y)$ in the region D and $F'_x(x, y)/F'_y(x, y)$ in the region (α) satisfy the conditions of this lemma.

Finally, note that the points P_k and Q_k are the points of intersection (in $0\left(\mu^{\frac{1}{2}}\right)$-neighbourhoods of the points B_k) of the line $F'_x(x, y) = -\mu^{\frac{1}{2}} |F'_y(x, y)|$, which is one boundary of (β_k), with the lines $F(x, y) = \pm \mu^{\frac{1}{2}} G_m$. The point P_k is that point of intersection which has the smallest ordinate if $G(x, y) > 0$ on the segment $A_k B_k$, and the largest ordinate if $G(x, y) < 0$ on the same segment.

The construction of the boundaries of (ε) in the small $0\left(\mu^{\frac{1}{2}}\right)$-neighbourhood of the section $B_k A_{k+1} B_{k+1}$ of C_0 will be demonstrated first for the case shown in Fig. 525, when $G(x, y) > 0$ on the segments $A_k B_k$ and $A_{k+1} B_{k+1}$ and $F(x, y) > 0$ on the interval $B_k A_{k+1}$.

(1) Let us draw the horizontal rectilinear segment $P_k P'_k$ from P_k to the point P'_k that lies on the line $F(x, y) = +\mu^{\frac{1}{2}} G_m$ in the region $F'_x(x, y) > 0$ and in the $0\left(\mu^{\frac{1}{4}}\right)$ neighbourhood of the point B_k†. Since at B_k the function $G(x, y) > 0$ and continuous and the segment $P_k P'_k$ lies in a certain $0\left(\mu^{\frac{1}{4}}\right)$-neighbourhood of this point, then, we can choose a sufficiently small value of the parameter μ, so that $G(x, y) > 0$ on the segment $P_k P'_k$. Then some paths of the system (10.15a) will intersect this segment from below since $\dot{y} > 0$ on $P_k P'_k$.

(2) Now draw the rectilinear segment $P'_k P''_k$ with slope $-\mu^{\frac{1}{2}}$ and the rectilinear segment $Q_k Q'_k$ with slope $+\mu^{\frac{1}{2}}$. P''_k and Q'_k are on the left-hand boundary of the region (α) in a $0\left(\mu^{\frac{1}{2}}\right)$-neighbourhood of the point A_{k+1}. If μ is sufficiently small, both segments will lie outside the region (α): therefore we have on them

$$F(x, y) \geqslant \sqrt{\mu}\, G_m \geqslant \sqrt{\mu}\, |G(x, y)|$$

($F(x, y) > 0$ on the segments $P'_k P''_k$ and $Q_k Q'_k$ since they lie in a small $0\left(\mu^{\frac{1}{2}}\right)$-neighbourhood of the interval $B_k A_{k+1}$ on which the continuous function $F(x, y) > 0$. $P'_k P''_k$ and $Q_k Q'_k$, lie on the straight lines $\mu^{\frac{1}{2}} x \pm y = $ const. so it is easily seen that the paths of (10.15a) which intersect $P'_k P''_k$ and $Q_k Q'_k$ must enter the region between these segments and containing the section $B_k A_{k+1}$ of the discontinuous limit cycle C_0.

(3) Now draw in the $0\left(\mu^{\frac{1}{2}}\right)$-neighbourhood of the point A_{k+1} a horizontal segment (R_{k+1}) from the point P''_k to the point R_{k+1} of the right-hand boundary of the region (α). Since the continuous function $G(x, y) > 0$, near A_{k+1} then, for sufficiently small μ, $G(x, y) > 0$ on the segment (R_{k+1}). Again the paths of the system (10.15a) intersect this segment from below.

(4) Connect the points Q'_k and R_{k+1} to the points P_{k+1} and Q_{k+1} by arcs of the lines $F(x, y) = \pm \mu^{\frac{1}{2}} G_m$ lying outside the regions (β_i). On them

$$-F'_x(x, y) \geqslant \sqrt{\mu}\, |F'_y(x, y)|;$$

† Such a point P'_k exists in the $0\left(\mu^{\frac{1}{4}}\right)$-neighbourhood of the point B_k, since at the point B_k, $F''_{xx} \neq 0$.

and therefore for motion along paths of the system (10.15a)

$$\frac{1}{2}\frac{d}{dt}\{F(x,y)\}^2 = F\{F'_x\dot{x}+F'_y\dot{y}\} = \frac{F^2}{\mu}F'_x + F'_yFG = G_m^2\left\{F'_x + \sqrt{\mu}\,F'_y\,\frac{G}{G_m}\right\} \leq 0$$

at the points of these arcs, and so the phase paths of (10.15a) which intersect these arcs enter the region (α).

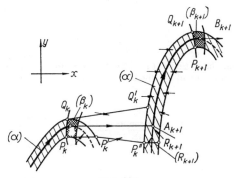

Fig. 525

The lines constructed now join the points P_k and Q_k with the points P_{k+1} and Q_{k+1} and delimit a portion of the region (ε) near the section $B_kA_{k+1}B_{k+1}$ of C_0. The phase paths of the system (10.15a) enter this region, as t increases. In a similar manner we can construct boundaries of the region (ε) in the $0\left(\mu^{\frac{1}{2}}\right)$-neighbourhoods of sections of C_0 when the signs of $F(x,y)$ and $G(x,y)$ differ from the ones assumed in Fig. 525. In fact, the construction of the sections $P_kP'_kP''_k$ and $Q_kQ'_k$ of the boundaries of the region (ε) reduces to the one described if x is replaced by $-x$ when $F(x,y)<0$ on the interval B_kA_{k+1}, and y is replaced by $-y$ when $G(x,y)<0$ on the segment $A\,B_k$. The horizontal segment (R_{k+1}) intersecting the region (α) near the point A_{k+1} is drawn through the lower of the points P''_k and Q'_k if $G(x,y)>0$ on $A_{k+1}B_{k+1}$, and through the upper ones if $G(x,y)<0$ on $A_{k+1}B_{k+1}$. The upper one of P''_k and Q'_k, if $G(x,y)>0$ on $A_{k+1}B_{k+1}$, or the lower, if $G(x,y)<0$ on $A_{k+1}B_{k+1}$ and the point R_{k+1} are connected by arcs of lines $F(x,y) = \pm\mu^{\frac{1}{2}}G_m$ to the points P_{k+1} and Q_{k+1}.

Having constructed the boundaries of the region (ε) about each of the sections of the discontinuous limit cycle C_0, we obtain a doubly connected region (ε) inside which is the discontinuous limit cycle, and into which phase paths of the system (10.15a) enter (see, for example, the region (ε) plotted in Fig. 524). For sufficiently small values of μ this region will not contain equilibrium states of (10.15a) since there are no such points on C_0 and $F(x,y)$ and $G(x,y)$ are continuous functions. Then the region (ε) will contain at least one stable limit cycle of the system (10.15a).

We shall prove now that any limit cycle C (10.15a) lying in the region (ε) so constructed is stable for sufficiently small values of μ, and thus that there is only one limit cycle of the system (10.15a), in the region (ε), since if several limit cycles existed there, some of them would of necessity be unstable. To this end consider the characteristic exponent of the limit cycle C, i.e. the integral

$$I = \int_C \left\{\frac{F'_x}{\mu} + G'_y\right\}dt,$$

taken along the limit cycle. The condition $I<0$ is (see Section 7, Chapter V) a sufficient condition for the stability of the limit cycle C. Let us split the limit cycle C by points C_j into sections $C_k C_{k+1}$ where C_k is a point of intersection with the boundary of the region (α) in the $0\left(\mu^{\frac{1}{2}}\right)$-neighbourhood of the points A_k. Then

$$I = \sum I_k,$$

where

$$I_k = \int\limits_{C_k C_{k+1}} \left\{\frac{F'_x}{\mu} + G'_y\right\} dt$$

is the integral along the section $C_k C_{k+1}$. To determine the sign of the integral I_k (for sufficiently small values of μ) we divide the section $C_k C_{k+1}$ into three by its points

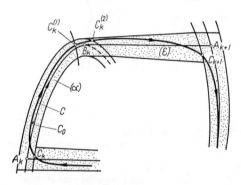

Fig. 526

of intersection with the line $F'_x(x, y) = -\mu(G'_y)_m$, where $(G'_y)_m$ is the maximum absolute value of the function $G'_y(x, y)$ in the region D, and with the boundary of the region (α) in a small neighbourhood of the point B_k (Fig. 526). The first of them, the point $C_k^{(1)}$, lies in the $0\left(\mu^{\frac{1}{2}}\right)$-neighbourhood of the point B_k, the second, $C_k^{(2)}$, in the $0\left(\mu^{\frac{1}{4}}\right)$-neighbourhood of the same point†. Correspondingly

$$I_k = I_k^{(1)} + I_k^{(2)} + I_k^{(3)},$$

† The line $F'_x(x, y) = -\mu(G'_y)_m$ lies in the $0(\mu)$-neighbourhood of the line $F'_x(x, y) = 0$, the segment of which within the region (α) is contained in the $0\left(\mu^{\frac{1}{2}}\right)$-neighbourhood of the point $B_k(x_k^*, y_k^*)$. Therefore the point $C_k^{(1)}$ also lies in the $0\left(\mu^{\frac{1}{2}}\right)$-neighbourhood of the point B_k, i.e. $|y_k^{(1)} - y_k^*| \leq 0\left(\mu^{\frac{1}{2}}\right)$, where $y_k^{(1)}$ is the ordinate of the point $C_k^{(1)}$. We have for the points $C_k^{(2)}(x_k^{(2)}, y_k^{(2)})$: $|y_k^{(2)} - y_k^*| \leq 0\left(\mu^{\frac{1}{2}}\right)$ and, therefore, $|x_k^{(2)} - x_k^*| \leq 0\left(\mu^{\frac{1}{4}}\right)$ as can easily be proved.

where

$$I_k^{(1)} = \int_{C_k C_k^{(1)}} \left\{ \frac{F_x'}{\mu} + G_y' \right\} dt = \int_{y_k}^{y_k^{(1)}} \left\{ \frac{F_x'}{\mu} + G_y' \right\} \frac{dy}{G}$$

$$I_k^{(2)} = \int_{C_k^{(1)} C_k^{(2)}} \left\{ \frac{F_x'}{\mu} + G_y' \right\} dt = \int_{y_k^{(1)}}^{y_k^{(2)}} \left\{ \frac{F_x'}{\mu} + G_y' \right\} \frac{dy}{G},$$

$$I_k^{(3)} = \int_{C_k^{(2)} C_{k+1}} \left\{ \frac{F_x'}{\mu} + G_y' \right\} dt = \int_{x_k^{(2)}}^{x_{k+1}} \left\{ F_x' + \mu G_y' \right\} \frac{dx}{F}$$

(here $x_k, y_k, x_k^{(1)}, y_k^{(1)}, x_k^{(2)}, y_k^{(2)}$ and x_{k+1}, y_{k+1} are the coordinates of the points $C_k, C_k^{(1)}, C_k^{(2)}$ and C_{k+1}).

Since, for sufficiently small values of μ, we have on the section $C_k C_k^{(1)}$; $F_x'(x, y) < -\mu(G_y)_m \leq -\mu |G_y'(x, y)|$ and $|G(x, y)| > a > 0$, then $I_k^{(1)} < 0$, and, since on this section we can isolate a segment on which $F_x'(x, y) < -b < 0$ (a and b are positive numbers), then

$$I_k^{(1)} = O\left(\frac{1}{\mu}\right).$$

On the section $C_k^{(1)} C_k^{(2)}$ lying inside a $O\left(\mu^{\frac{1}{4}}\right)$-neighbourhood of the point B_k, $|x - x_k^*| \leq$
$\leq O\left(\mu^{\frac{1}{4}}\right)$, $|y - y_k^*| \leq O\left(\mu^{\frac{1}{2}}\right)$, $|F_x'(x, y)| \leq O\left(\mu^{\frac{1}{4}}\right)$ since, at the point B_k, $F_x'' \neq 0$, and $|G(x, y)| > a$. Therefore, according to a mean-value theorem

$$|I_k^{(2)}| \leq O\left(\mu^{-\frac{3}{4}}\right) |y_h^{(2)} - y_k^{(1)}| \leq O\left(\mu^{-\frac{1}{4}}\right),$$

since

$$|y_k^{(2)} - y_k^{(1)}| \leq O(\sqrt{\mu}).$$

Finally, on the section $C_k^{(2)} C_{k+1}$ lying outside the region (α), $|F(x, y)| \geq \mu^{\frac{1}{2}} G_m$, so that

$$|I_k^{(3)}| \leq O\left(\mu^{-\frac{1}{2}}\right)^\dagger$$

Thus the integral $I_k = \int_{C_k C_{k+1}} \left\{ F_x'/\mu + G_y' \right\} dt$ is equal to the sum of a negative quantity $I_k^{(1)}$ of the order of μ^{-1} and of quantities $I_k^{(2)}$ and $I_k^{(3)}$ that although they tend to infinity for $\mu \to +0$, do so more slowly than μ^{-1}; therefore for sufficiently small values of μ, $I_k < 0$ on all sections $C_k C_{k+1}$. Hence the characteristic exponent of the limit cycle C

$$I = \int_C \left\{ \frac{F_x'}{\mu} + G_y' \right\} dt = \sum I_k < 0,$$

and a limit cycle C lying in the region (ε) is stable.

† The more accurate estimate $|I_k| \leq O(\ln \mu)$ is easily obtained.

As has already been indicated since it is stable it is also unique. The theorem enunciated above is thus proved.

This theorem enables us to use the discontinuous limit cycle of the system

$$\mu\dot{x} = F(x, y), \quad \dot{y} = G(x, y)$$

as an initial (zero-order) approximation for evaluating the characteristics of the self-oscillations occurring in the system (10.15a) for small values of the parameter μ. Thus, for example, the zero-order approximation to the period of self-oscillations is

$$T_0 = \int_{\overset{\frown}{C_0}} \frac{dy}{G} = \sum \int_{A_k B_k} \frac{dy}{G}.$$

For a more detailed investigation [93, 94, 158, 159] of the behaviour of the phase paths of (10.15a) near the discontinuous limit cycle, asymptotic expansions can be used.

In particular, the period of the self-oscillations can be expressed as

$$T = T_0 + A\mu^{\frac{2}{3}} + B\mu \ln \frac{1}{\mu} + C\mu + O\left(\mu^{\frac{4}{3}}\right),$$

where A, B and C are numbers determined by the values of the functions $F(x, y)$ and $G(x, y)$ on the discontinuous limit cycle[†].

Below we shall study discontinuous oscillations in physical systems, the "slow" and "rapid" motions of which are represented under suitable simplifying assumptions by equations (10.16) and (10.17) of an order not higher than the second.

§5. Multivibrator with one RC circuit

We shall consider once more a familiar oscillating system which under certain conditions generates discontinuous oscillations. This is the multivibrator with one RC circuit (Fig. 527) as discussed in Section 8, Chapter IV, Section 12 of Chapter V and Section 5 of Chapter VIII. As we have already seen in Chapter IV, Section 8, the model of a multivibrator constructed by neglecting all parasitic parameters is a "degenerate" or "defective" model in the sense that without the additional jump

---

[†] The coefficient A depends on the curvature of the line $F(x, y) = 0$ at the points B_k. In particular, when the radii of curvature of this line tend to zero at all points B_k, i.e. when the line F tends to a line with breaks at the points B_k, the coefficient $A \to 0$ and the correction for the period becomes a quantity of the order of $\mu \ln \mu^{-1}$ (see, for example, Section 5, Chapter VIII).

postulate it did not even explain the qualitative features of the oscillations. The small inter-electrode and wiring capacitances C_a and C_g, which are always present in a real system, are important parameters and to make a satisfactory model at least one must be taken in account.

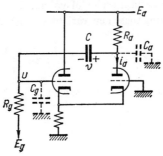

FIG. 527

1. Equations of the oscillations

The equations of the multivibrator will be developed taking into account these small parasitic capacitances, but assuming that $C_a \ll C$ and $C_g \ll C$. Neglecting grid currents and anode reaction and assuming the characteris-

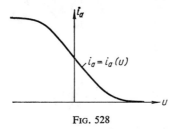

FIG. 528

tic of the valve group $i_a = i_a(u)$ given in Fig. 528, we obtain from Kirchhoff's laws

$$\left. \begin{array}{l} \dfrac{E_a-(u+v)}{R_a} = C\dfrac{dv}{dt}+C_a\dfrac{d(u+v)}{dt}+i_a(u), \\[2mm] C\dfrac{dv}{dt} = \dfrac{u-E_g}{R_g}+C_g\dfrac{du}{dt}. \end{array} \right\} \quad (10.25)$$

It is evident that the multivibrator has a single equilibrium state

$$u = E_g, \quad v = v_0 = E_a - R_a i_a(E_g) - E_g.$$

Now introduce, to simplify the calculations, new dimensionless variables x, y, t_{new} related to u, v, t by

$$u = E_g + u_0 x; \qquad v = v_0 + \alpha u_0 x + \beta u_0 y; \qquad t_{\text{new}} = T_0 t,$$

where u_0 and T_0 are certain scales of voltage and time and α and β are coefficients. Substituting these new variables in the second equation (10.25) we find

$$\alpha u_0 \frac{CR_g}{T_0} \dot{x} + \beta u_0 \frac{CR_g}{T_0} \dot{y} = u_0 x + \frac{C_g R_g}{T_0} u_0 \dot{x}.$$

A dot denotes differentiation with respect to t_{new}. Now choosing

$$\alpha = \frac{C_g}{C} \quad \text{and} \quad \beta C R_g = T_0,$$

we have

$$\dot{y} = x.$$

Similarly the first equation (10.25) can be written as

$$\frac{R_a(C+C_a)}{T_0}(\alpha u_0 \dot{x} + \beta u_0 \dot{y}) + \frac{R_a C_a}{T_0} u_0 \dot{x} =$$
$$= -u_0 x - \alpha u_0 x - \beta u_0 y - R_a[i_a(E_g + u_0 x) - i_a(E_g)]$$

or

$$\frac{\alpha R_a(C+C_a) + R_a C_a}{T_0} \dot{x} = -x \left[1 + \frac{C_g}{C_a} + \frac{R_a}{R_g}\left(1 + \frac{C_a}{C}\right) \right] -$$
$$- \beta y - \frac{R_a}{u_0}[i_a(E_g + u_0 x) - i_a(E_g)].$$

On choosing

$$\beta = 1 + \frac{C_g}{C} + \frac{R_a}{R_g}\left(1 + \frac{C_a}{C}\right)$$

we have

$$T_0 = (C+C_a)R_a + (C+C_g)R_g,$$

and (10.25) reduces to a form typical of systems with discontinuous oscillations

$$\left. \begin{aligned} \mu \dot{x} &= F(x, y) \equiv -x - y - K \cdot \varphi(x), \\ \dot{y} &= x, \end{aligned} \right\} \qquad (10.26)$$

where

$$\mu = \frac{R_a}{R_g} \frac{\dfrac{C_a + C_g}{C} + \dfrac{C_a C_g}{C^2}}{\left[1 + \dfrac{C_g}{C} + \dfrac{R_a}{R_g}\left(1 + \dfrac{C_a}{C}\right)\right]^2}$$

is a small positive parameter characterizing the smallness of the parasitic capacitances. For $C_a \ll C$ and $C_g \ll C$ the parameter $\mu \ll 1$,

$$K = \frac{SR_a}{1+\dfrac{C_g}{C}+\dfrac{R_a}{R_g}\left(1+\dfrac{C_a}{C}\right)}$$

is a transmission factor, S is the numerical value of the slope of the characteristic of the valve group at the equilibrium state

$$S = -\left(\frac{di_a}{du}\right)_{u=E_g}$$

and

$$\varphi(x) = \frac{1}{u_0 S}\left[i_a(E_g+u_0 x)-i_a(E_g)\right]$$

is the reduced, dimensionless, characteristic of the valve group, with $\varphi'(x) \leq 0$, $\varphi(0)=0$, $\varphi'(0)=-1$. For $C_a \ll C$ and $C_g \ll C$ the variables $x=(u-E_g)/u_0$ and $y \approx (v-v_0)/\beta u_0$ are proportional to the variable components of the grid voltage u of the left-hand triode and of the voltage v across the capacitor C respectively, and

$$\mu \approx \frac{R_a}{R_g}\frac{\dfrac{C_a}{C}+\dfrac{C_g}{C}}{\left[1+\dfrac{R_a}{R_g}\right]^2}, \quad K \approx \frac{SR_a}{1+\dfrac{R_a}{R_g}}.$$

The time scale is approximately

$$T_0 \approx C(R_a+R_g)$$

Note that we can choose u_0 to simplify the expression for $\varphi(x)$.

Now to simplify the analysis, we will consider the symmetrical case only, when the reduced characteristic $\varphi(x)$ is an even function of x ($\varphi(-x) \equiv -\varphi(x)$) with the numerical value of $\varphi'(x)$ decreasing monotonically as x increases (then $-1 \leq \varphi'(x) \leq 0$).

2. The x, y phase plane for $\mu \to +0$

The jumps of the voltage u. Let us consider the phase portrait for the limiting case $\mu \to +0$. First of all mark out on the phase plane the curve F,

$$F(x, y) \equiv -x-y-K\cdot\varphi(x) = 0$$

or

$$y = -x-K\cdot\varphi(x), \tag{10.27}$$

which is the phase line of the "degenerate" model of the multivibrator when $C_a = C_g = 0$ or $\mu = 0$. It follows from the equations (10.26) that for $\mu \to +0$ the phase velocity remains finite only in a small $0(\mu)$-neighbourhood of the line F. Outside a small neighbourhood of this curve (with dimensions, for example, of the order of $\mu^{\frac{1}{2}}$) there are "rapid" motions or "jumps" of x where $+0 < \dot{x} \to +\infty$ below the curve F and $\dot{x} \to -\infty$ above

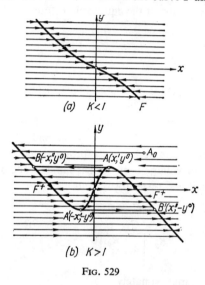

Fig. 529

it but $\dot{y} = x$ remains finite. Thus the phase paths of jumps are almost horizontal straight lines $y = \text{const.}$ (during the jumps of the grid voltage u the voltage v across the capacitor remains constant). The approximate equation of the "jump" of the representative point along a path close to the straight line $y \equiv y^0 = \text{const.}$ is obtained from the first equation (10.26) by substituting y^0 for y

$$\mu \dot{x} = -y^0 - x - K \cdot \varphi(x). \tag{10.28}$$

Two cases are possible.
For $K < 1$

$$\frac{\partial F}{\partial x} = -1 - K \cdot \varphi'(x) < 0$$

over the entire phase line of the "degenerate" system, and therefore all jumps proceed towards this line F (Fig. 529(a)).

Consequently if the initial state of the multivibrator is represented by a point close to the curve F, the representative point moves subsequently on or near the line F†. Correspondingly the oscillations of the multivibrator for $K<1$ and $\mu \to +0$ ($C_a \ll C$ and C and $C_g \ll C$) are described by the equation of "slow" motions in which $\mu=0$, or, in other words, by substituting the equation (10.27) into the second equation (10.26)

$$[1+K\varphi'(x)]\dot{x}+x = 0. \qquad (10.29)$$

so for $K<1$, very small stray capacitances do not have an important role in the oscillating processes of the multivibrator and can be neglected. Whatever the initial conditions, the state of equilibrium $x=0$, $y=0$ (or $u=E_g$, $v=v_0$) is reached ultimately, since $1+K\varphi'(x)>0$ for all x and $x \to 0$ for $t \to +\infty$.

A different picture is obtained when $K>1$. Now the state of equilibrium $(0,0)$ is unstable when $\mu \geqslant 0$. On the phase line F of the "degenerate" model there is a segment $-x \leqslant x \leqslant +x'$ where $x'>0$ is the only root of the equation $1+K\varphi'(x)=0$ on which the condition for the small parameter to be unimportant is not satisfied, for on this segment

$$\frac{\partial F}{\partial x} = -1-K\varphi'(x) \geqslant 0.$$

The phase paths of "rapid" motions move away from this segment of the phase line of the incomplete or degenerate system which contains the equilibrium state $(0,0)$ (Fig. 529(b)) Thus, for $\mu \to +0$, the multivibrator recedes with a jump from all states with $|x| \leqslant x'$. During the "jump", the x variable (the grid voltage of the left-hand triode) varies with a jump, while the value of the y variable (the voltage v across the capacitor C) remains constant. Thus, by including in the model the small parasitic capacitances C_a and C_g, we obtain the jump postulate (for $K>1$) used in Chapter IV, Section 8.

Obviously all paths of "rapid" motions move into the neighbourhood of the portions F^+ of the phase line F of the "degenerate" model. On this line the condition for the small parameter to be unimportant is satisfied:

$$\frac{\partial F}{\partial x} = -1-K\cdot\varphi'(x) < 0,$$

† If, however, the initial state of the multivibrator is at a point well outside a small neighbourhood of the curve F, then the representative point, having made a "jump" along a path of "rapid" motion into a small neighbourhood of the line F, will move subsequently in the vicinity of the phase line F. In the limit, for $\mu \to +0$, this "slow" motion will be along the line F itself.

and the paths go towards the sections of the curve $F: |x| > x'$ that have a negative slope[†]. Only in small $0(\mu)$ — neighbourhoods of these sections do "slow" motions occur which obey the approximate equation (10.29). In the limit for $\mu \to +0$, these paths lie on the sections of F with $|x| > x'$. The phase portrait is shown in Fig. 529(b) but the portrait is not substantially different when μ is small but finite.

Since on F where $|x| > x'$, $1 + K\varphi'(x) > 0$, the quantity $|x|$ decreases with time and (equation (10.20)) the representative point reaches either point A or A' from which it "jumps" along a path $y = $ const. to the point $B(-x'', y^0)$ or to the point $B'(x'', -y^0)$ respectively after which a "slow" motion begins again, etc. To find the end point of the jump there is no need to use (10.28) for it is determined from the initial point of the "jump" by the condition that y remains constant during the jump. Thus using (10.27), we have

$$x'' + K \cdot \varphi(x'') = -\{x' + K\varphi'(x')\}. \tag{10.30}$$

It is evident that the closed curve $ABA'B'A$ (Fig. 529(b)) is a limit cycle to which paths will tend whatever the initial conditions, and is the representation of "discontinuous" self-oscillations in which "slow" motions alternate periodically with instantaneous "jump-wise" motions. As shown above, there exists a limit cycle for small values of μ (Fig. 530) close to the cycle $ABA'B'A$. The waveforms of the oscillations of x and y for a phase path beginning at the point A_0 (Fig. 529(b)) are shown qualitatively in Fig. 531.

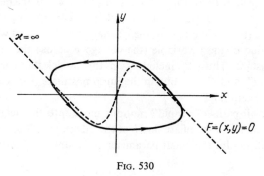

Fig. 530

To find the approximate period of the self-oscillations we need to evaluate the time of transit of the representative point along the sections

[†] On the line $F(x, y) = 0$, $F'_x + F'_y(dy/dx)_{F=0} = 0$, i.e. $(dy/dx)_{F=0} = -F'_x/F'_y = +F'_x$ since $F'_y = -1$.

5] MULTIVIBRATOR WITH ONE RC CIRCUIT 687

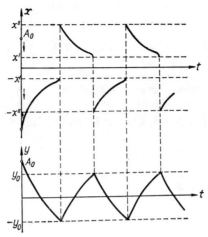

Fig. 531

$B'A$ and BA' of the limit cycle (we neglect the duration of the jumps). On the paths F^+ of "slow" motions we have

$$dt_{\text{new}} = -\frac{1+K\varphi'(x)}{x}\,dx,$$

therefore the period of the self-oscillations is

$$\tau = 2\int_{x'}^{x''} \frac{1+K\varphi'(x)}{x}\,dx$$

in units of dimensionless time, or

$$T = T_0 \cdot \tau = 2C(R_a+R_g)\int_{x'}^{x''} \frac{1+K\varphi'(x)}{x}\,dx$$

in ordinary units.

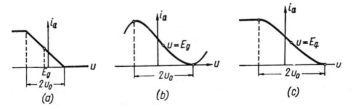

Fig. 532

For a piece-wise linear characteristic of the valve-pair (Fig. 532a)

$$\varphi(x) = \begin{cases} +1 & \text{for } x < -1, \\ -x & \text{for } |x| < 1, \\ -1 & \text{for } x > 1 \end{cases} \quad \text{and} \quad \varphi'(x) = \begin{cases} 0 & \text{for } |x| > 1, \\ -1 & \text{for } |x| < 1, \end{cases}$$

we choose as the voltage scale u_0 one-half the "width" of the descending section of the characteristic. Thus $x'=1^\dagger$, $x''=2K-1$ and, in agreement with the results of Section 8, Chapter IV and Section 5, Chapter VIII the period is

$$T = 2C(R_a + R_g) \ln (2K-1).$$

When we approximate to the characteristic by a polynomial of the third degree (Fig. 532(b)) we have

$$\varphi(x) = -x + \frac{x^3}{3} \quad \text{and} \quad \varphi'(x) = x^2 - 1,$$

and now we take for u_0 half the difference of the voltages for which the slope of the characteristic reduces to zero. The abscissae of the initial points of the jumps will be found from the equation

$$1 + K(x^2 - 1) = 0, \quad \text{i.e.} \quad x' = \sqrt{\frac{K-1}{K}}.$$

Then, according to (10.26) the abscissae $\pm x''$ of the end points of the jumps are determined by

$$x'' + K\left(-x'' + \frac{x''^3}{3}\right) = -\left\{x' + K\left(-x' + \frac{x'^3}{3}\right)\right\} =$$
$$= \frac{2}{3}(K-1)\sqrt{\frac{K-1}{K}},$$

which, as is easily verified, has a single real solution

$$x'' = 2\sqrt{\frac{K-1}{K}}.$$

Therefore

$$\tau = 2\int_{x'}^{x''} \left[Kx - \frac{K-1}{x}\right] dx = (3 - 2\ln 2)(K-1) \approx 1 \cdot 6(K-1)$$

† In the case of a piece-wise linear characteristic, the initial points of the jumps after the slow motions are the break-points $x = \pm 1$ of the characteristic, since the condition for the parasitic capacitances to be unimportant is not satisfied when $|x| < 1$ (for $F'_z = K - 1 > 0$) but only when $|x| > 1$, where $F'_z = -1 < 0$.

and
$$T \approx 1\cdot 6\, C(R_a+R_g)(K-1)^\dagger$$

Let us evaluate finally the period for a characteristic the descending section of which is represented by a polynomial of the third degree (Fig. 532(c)), as assumed in Chapter V, Section 12 in the analysis of the oscillations of a generator with a two-mesh RC circuit and of a multivibrator with one RC circuit. For this characteristic, if we take u_0 equal to half the width of the descending section of the characteristic we have

$$\varphi(x) = \begin{cases} +\dfrac{2}{3} & \text{for}\quad x < -1, \\ -x+\dfrac{x^3}{3} & \text{for}\quad |x| \leq 1, \\ -\dfrac{2}{3} & \text{for}\quad x \geq 1; \end{cases} \qquad \varphi'(x) = \begin{cases} 0 & \text{for}\quad |x| \geq 1, \\ x^2-1 & \text{for}\quad |x| \leq 1; \end{cases}$$

$x' = (K-1)^{\frac{1}{2}} K^{-\frac{1}{2}}$ (as in the previous example) and x'' is determined by the equation

$$x'' - \frac{2}{3} K = -\{x' + K\varphi(x)\} = \frac{2}{3}(K-1)\sqrt{\frac{K-1}{K}},$$

i.e.
$$x'' = \frac{2}{3}\left\{(K-1)\sqrt{\frac{K-1}{K}}+K\right\}$$

where $x'' > 1$, which is true for $K > 4/3$.

$$\tau = 2\int_{x'}^{1}\left\{Kx - \frac{K-1}{x}\right\}dx + 2\int_{1}^{x''}\frac{dx}{x} = 1-(K-1)\ln\frac{K}{K-1} +$$
$$+ 2\ln\frac{2}{3}\left[(K-1)\sqrt{\frac{K-1}{K}}+K\right]$$

† A. A. Dorodnitsyn [52] has obtained an asymptotic expansion for the period of the self-oscillations of a multivibrator for a cubic characteristic which in our notation is

$$\tau = 1\cdot 614(K-1) + 7\cdot 014(K-1)^{-\frac{1}{3}}\mu^{\frac{2}{3}} - \frac{11}{9}\frac{\mu}{K-1}\ln\frac{1}{\mu} -$$
$$- \frac{\mu}{K-1}\left\{\frac{22}{9}\ln(K-1) - 0\cdot 087\right\} + \ldots$$

and

$$T = C(R_a - R_g)\left\{1 - (K-1)\ln\frac{K}{K-1} + 2\ln\frac{2}{3}\left[(K-1)\sqrt{\frac{K-1}{K}} + K\right]\right\}.$$

If, however, $1 < K < 4/3$, then in the presence of self-oscillations the valve-pair will work only on the descending section of the characteristic, and the period will be expressed by the formula obtained in the previous example.

§ 6. MECHANICAL DISCONTINUOUS OSCILLATIONS

Mechanical oscillations can be generated under certain conditions by a body subject to large friction but having a small mass [69]. A physical example is the brake system, shown diagrammatically in Fig. 533[†]. This system has the following equation of motion

$$J\ddot{\varphi} = -k\varphi + M(\Omega - \omega)$$

or the equivalent system

$$\dot{\varphi} = \omega, \quad J\dot{\omega} = -k\varphi + M(\Omega - \omega), \qquad (10.31)$$

where φ is the angle of rotation of the brake-shoe with respect to the position in which the torque due to the spring-forces is zero, J is the moment of inertia of the brake-shoe, k is the coefficient of elasticity of the system,

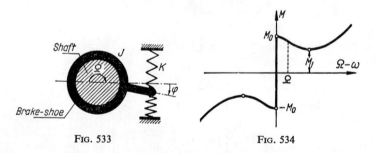

Fig. 533   Fig. 534

Ω is the angular velocity of the shaft which is assumed constant, and $M(\Omega - \omega)$ is a function expressing the dependence of the dry friction torque upon the relative velocity $\Omega - \omega$ (Fig. 534).

[†] This system represents an idealized model of ordinary brakes, and of Prony brakes which are used to measure power.

Strictly speaking the frictional torque M is a function not only of the relative velocity $\Omega - \omega$ but also of the angle φ; so that for $\Omega - \omega = 0$

$$M = \begin{cases} M_0 & \text{for} \quad k\varphi > M_0, \\ k\varphi & \text{for} \quad |k\varphi| \leq M_0, \\ -M_0 & \text{for} \quad k\varphi < -M_0, \end{cases}$$

where M_0 is the maximum value of the frictional torque at rest. Below we assume that the characteristic of friction has a descending section where the frictional torque M decreases with an increase of the relative velocity $\Omega - \omega$. It is only the presence of such descending sections that enables us to explain the occurrence of self-oscillations in this mechanical system.

The system has a single equilibrium state

$$\varphi = \varphi_0, \quad \omega = 0,$$

the angle φ_0 of equilibrium being evidently determined by

$$k\varphi_0 = M(\Omega).$$

This state of equilibrium is unstable (the system is self-excited and self-oscillations will build up) if

$$M'(\Omega) < 0$$

(see also Section 6 of Chapter I) and we now assume this condition to be satisfied. If $-M'(\Omega) \ll (kJ)^{\frac{1}{2}}$ and the non-linearity of the friction characteristic is small, then the self-oscillations of the system will be almost sinusoidal and can be investigated by using Van der Pol's and Poincaré's methods.

We consider here another limiting case, when the brake-shoe has a *small* moment of inertia.

$$J \ll \frac{[M'(\Omega)]^2}{k},$$

as a consequence of which, as we shall see, the self-oscillations will be of the "relaxation" type and markedly different from sinusoidal ones. The oscillatory process is sharply divided into two types of motions which alternate and differ substantially from each other.

(1) In states in which the spring torque nearly equals the friction torque the system has comparatively small accelerations $\dot{\omega}$ even though J is small. Hence, for motions through these states, ω varies *comparatively* slowly. In this region the term $J\dot{\omega}$ is small

$$J|\dot{\omega}| \ll k|\varphi| \approx |M(\Omega - \omega)|$$

and can be neglected. Therefore these states are represented on the phase plane φ, ω by points that lie in a small neighbourhood of the line F.

$$F(\varphi, \omega) \equiv -k\varphi + M(\Omega - \omega) = 0,$$

which is clearly the phase line of the "degenerate" system (with $J=0$); this neighbourhood reduces to F when $J \to +0$.

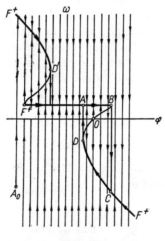

Fig. 535

(2) In states in which the friction and spring torques do not balance each other, the accelerations $\dot{\omega}$ are very large since the moment of inertia J is sufficiently small. These states, in which the *velocity ω varies very rapidly* correspond on the φ, ω phase plane to points lying outside a certain small neighbourhood of the same line F.

In this region of rapid variations of the velocity ω, when $J \to +0$, $\dot{\omega} \to \infty$ ($\dot{\omega} \to +\infty$ on the left of the line F and $\dot{\omega} \to -\infty$ on the right of it) whereas the velocities $\dot{\varphi} = \omega$ remain finite. In this region

$$\frac{d\varphi}{d\omega} = \frac{J\omega}{-k\varphi + M(\Omega - \omega)} \to 0$$

as $J \to +0$ and it will be filled with phase paths close to the vertical straight lines $\varphi = $ const. along which the representative point will move with a "jump".

The phase portrait for the limiting case $J \to +0$ is shown in Fig. 535. The entire plane outside the line F is filled with vertical rectilinear phase paths of "infinite" accelerations which come from infinity, or move away from the sections of the line F on which

$$F'_\omega = -M'(\Omega-\omega) > 0.$$

These sections therefore correspond to the descending sections of the friction characteristic and the phase point moves from there to the rest of the line F, which is denoted by F^+ and thick lines in Fig. 535. The sections are clearly the limiting position paths on which ω is constant or nearly so, and on them the small moment of inertia J is not important and the motion of the system can be described approximately by the "degenerate" equations

$$-k\varphi+M(\Omega-\omega) = 0, \qquad \dot{\varphi} = \omega. \qquad (10.32)$$

If we put $J=0$, and assume the equations (10.32) to be always applicable, then we obtain a "defective" model of the first order with the phase line F containing junction points of phase paths (the points B, D, D' in Fig. 535). These points are not equilibrium states yet on the line F there are no phase paths of the system (10.32) that leave these points. Thus we cannot, using this model, observe the motion of the system.

Note that the junction points F^+ of phase paths separate the arcs of the line F that correspond to descending sections of the friction characteristic and near which the representative point cannot move for arbitrarily small positive values of the moment of inertia J. In other words the phase line F of the "degenerate" system contains sections on which the moment of inertia of the brake-shoe, however small, is a parameter which is essential in determining the character of the motion of the system.

It is easily seen from the equations (10.32), that sooner or later the representative point moving along F^+ will arrive at one of the points B, D or D' and will "break away" into the region of "infinite" accelerations. The representative point moving along the proper jump path (BC, DA or $D'A'$) will return to one of the paths of "finite" accelerations, F^+ etc. (during the jump-wise variations of states the velocity ω of the system varies with a jump while the coordinate φ remains constant). As a result, after at most two motions with finite accelerations, such as one starting from state A in Fig. 535, discontinuous periodic self-oscillations will be generated. They correspond on the phase plane to the limit cycle $ABCDA$ (Fig. 535) and consist of alternate motions with finite and "infinite" accelerations.

The section AB of this limit cycle corresponds to the brake-shoe rotating together with the shaft with a uniform velocity Ω. Under these circumstances the springs are deformed and the spring torque increases as does the equal frictional torque. Finally, when the spring torque becomes equal to the maximum frictional torque (at the point B) a jump-wise variation of brake shoe velocity occurs both in magnitude and direction, the spring forces remaining constant. The representative point "jumps" from the point B to the point C that corresponds to the same value of the friction torque

$$M(\Omega - \omega_C) = M_0$$

and again lies on a phase path F^+ where the velocity ω is nearly constant[†].

Next there is a continuous variation of the velocity and of the coordinate φ determined by the equation (10.32) until the representative point moving along the path CD arrives at D ($\omega = \omega_D$) corresponding to a minimum in the friction characteristic. Starting from D there is another velocity jump from ω_D to Ω while the coordinate φ remains nearly constant at $\varphi = M_1/k$, where M_1 is the minimum frictional torque (see Fig. 554). These actions repeat and the brake-shoe performs self-oscillations. The oscillograms of the angle φ and the velocity ω are shown qualitatively in Fig. 556.

The "amplitude" (half a complete swing) of the self-oscillations of angle φ is equal to

$$\varphi_0 = \frac{1}{2} \frac{M_0 - M_1}{k}.$$

[†] In practice the tension of the springs during a "jump" varies a little, since in reality the jump does not occur instantaneously, but the smaller the moment of inertia J the shorter the duration of the jump and the smaller the variation of the coordinate φ and of the tension of the springs. The order of magnitude of the variations of the coordinate φ accompanying the jump of velocity (from Ω to ω_c) can be estimated approximately as follows. Since a real system always possesses a certain moment of inertia J, then during the jump of velocity the kinetic energy of the system varies by $J\Delta(\omega^2)/2$. This variation of kinetic energy must be equal to the work of the forces of tension of the springs and of the force of friction

$$\frac{1}{2} J\Delta(\omega^2) = \int_{\varphi}^{\varphi+\Delta\varphi} [-k\varphi + M(\Omega - \dot{\varphi})] \, d\varphi \approx \int_{\varphi}^{\varphi+\Delta\varphi} [-M_0 + M(\Omega - \omega)] \, d\varphi,$$

since during the jump

$$\varphi \approx \text{const} = \frac{M_0}{k}.$$

If we introduce the mean value of the frictional torque M_{AV} during the time of the jump, then $J\Delta(\omega^2)/2 \approx -(M_0 - M_{av}) \Delta\varphi$. Hence the variation of the coordinate φ during the jump in the angular velocity is

$$\Delta\varphi \approx -\frac{1}{2} \frac{J\Delta(\omega^2)}{M_0 - M_{av}}.$$

To evaluate the period we calculate the time of motion of the representative point along F^+ from A to B and from C to D (Fig. 535), ignoring the durations of the jumps. On the section $AB\dot\varphi=\Omega$, and its transit time is

$$T_1 = \frac{M_0-M_1}{k\Omega}.$$

The transit time T_2 for the section CD is determined by integrating over this section the equations (10.32)

$$T = \int_{M_0/k}^{M_1/k} \frac{d\varphi}{\omega} = -\frac{1}{k}\int_{\omega_C}^{\omega_D} \frac{M'(\Omega-\omega)}{\omega}\,d\omega,$$

where the equations (10.32) have been written in the form

$$k\omega = -M'(\Omega-\omega)\frac{d\omega}{dt}. \tag{10.32a}$$

The total period of the self-oscillations is then $T=T_1+T_2$.

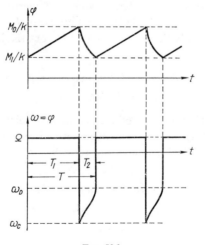

Fig. 536

For $J \neq 0$ the phase paths with "rapid" variations of velocity will not be rectilinear but must have a certain curvature, which, however, is inappreciable if J is small and k is large. But the character of the process changes markedly when J is large as is shown in the experimental curves of

Figs. 537 and 538. As J increases the form of the oscillations approximates more and more to the form of harmonic oscillations, and the "relaxation pendulum" turns into Froude's pendulum. The portrait on the

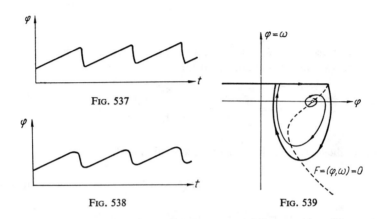

Fig. 537

Fig. 538

Fig. 539

phase plane (Fig. 539) plotted according to the oscillograms of the oscillations of a brake-shoe with large moment of inertia (Fig. 538) differs appreciably from the limit cycle in the limiting case $J \to +0$ (Fig. 535).

The inertia of the brake-shoe has smoothed out the jumps of velocity.

§ 7. Two electrical generators of discontinuous oscillations

Good electrical analogues of the mechanical relaxation system just considered are provided by two simple electrical generators of relaxation oscillations: a circuit with a neon tube (or a Voltaic arc), and a dynatron generator.

1. Circuit with a neon tube

The discontinuous oscillations in this circuit (Fig. 540) have already been discussed in Section 6 of Chapter IV (Sub-section 2), using a jump postulate for the intensity of current. This jump postulate is not a consequence of the first order model used (equation (4.30)) but followed from physical considerations.

Let us briefly consider the same circuit but taking into account essential parameters. We will obtain, in particular, the "jumps" in the current through the neon valve as a consequence of the dynamics of the system. The small stray inductance L of the circuit shown with a dotted line in Fig. 540†, is not neglected and the circuit equations are

$$C\frac{du}{dt} = \frac{E-u}{R} - i,$$
$$L\frac{di}{dt} = u - v, \qquad (10.33)$$
$$i = \varphi(v),$$

where $\varphi(v)$ is the dependence of the neon tube current i upon the tube voltage v. The inverse function is single-valued for a conducting neon tube and is $v = \psi(i)$. The characteristic of a neon tube (Fig. 541) has a

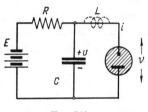

Fig. 540

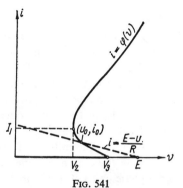

Fig. 541

descending section (for $0 < i < I_1$), and we assume that the equilibrium state (u_0, i_0), determined by the relations,

$$u = v, \quad i = \frac{E-v}{R} = \varphi(v),$$

lies on this section and is therefore unstable.

The characteristic equation at this equilibrium state is

$$LCR\lambda^2 + [L + CR\psi'(i_0)]\lambda + R + \psi'(i_0) = 0.$$

Therefore if the state (u_0, i_0) is on the descending section of the characteristic and is unique, i.e. if $-R < \psi'(i_0) < 0$, then the state is unstable for $L < -CR\psi'(i_0)$. This is true for small values of L. For the state

† Another small parasitic parameter of importance for the processes occurring in the circuit is the inertia of the gaseous discharge. We can approximately represent this inertia by introducing a certain small "equivalent" inductance in series with the neon tube. L may be understood to be the sum of this "equivalent" inductance and of the inductance of the connecting wires.

(u_0, i_0) to be stable and lie on the descending section it is necessary that $E > V_3$ and $R > R_{crit}$. For further details, see Section 7 of Chapter IV, Sub-section 2.

We shall plot the limiting case as $L \to +0$ of the phase portrait on the u, i plane for system (10.33). Draw first the line F

$$F(u, i) \equiv u - \psi(i) = 0$$

of the "degenerate" model (Fig. 542), and mark out the sections F_1^+ and F_2^+ on which the small parasitic inductance L is unimportant as given by

$$F_i' = -\psi'(i) < 0.$$

For $L \to +0$ we have on the right of the line F $di/dt \to +\infty$ and on its left $di/dt \to -\infty$, for bounded values of du/dt. Therefore the region outside the line F is a region of "jump-wise" variations of the states of the system and is filled with vertical rectilinear paths $u =$ const, coming from infinity and departing from the section of line F that corresponds to the descending section of the neon tube characteristic, towards the sections F_1^+ and F_2^+. In other words, in the region outside the lines F_1^+ and F_2^+, there are jumps of current intensity, the voltage u across C remaining constant. Otherwise, for sufficiently small values of L, there will be very rapid variations in the current i while the voltage u remains nearly constant, and the phase paths of these rapid variations of state are very close to the straight lines $u =$ const. "Slow" variations of states, with finite rates of change of i and u, only occur on or near the lines F_1^+ and F_2^+ for small L. On F_1^+ or F_2^+ the circuit is satisfactorily represented by the "degenerate" equation

$$i = \varphi(u),$$

$$RC \frac{du}{dt} = E - u - R\varphi(u).$$

Since on the paths F_1^+ and F_2^+ there are no equilibrium states the representative point moves along them towards B or D at which the current changes discontinuously. So, whatever the initial conditions, discontinuous (relaxation) self-oscillations begin, corresponding on the phase plane to the limit cycle $ABCDA$ (Fig. 542). The oscillations of the voltage u have a "sawtooth" form (Fig. 543). The amplitude and period of the self-oscillations are expressed by the formulae obtained in Section 6 of Chapter IV.

Thus in this circuit the inductance of the wiring and inertia of the gaseous discharge are small but essential factors in the oscillating process. Only by taking them into account does the dynamic model adequately represent the dynamics of the circuit, yielding results which are in qualitative and quantitative agreement with experimental data.

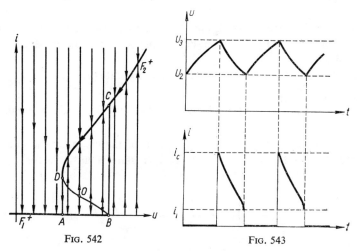

FIG. 542   FIG. 543

2. Dynatron generator of discontinuous oscillations

The circuit of a dynatron relaxation oscillator is shown in Fig. 544. The equations are

$$\left.\begin{aligned} L\frac{di}{dt} &= E_a - u - Ri, \\ C_a\frac{du}{dt} &= i - i_a, \\ i_a &= \varphi(u), \end{aligned}\right\} \quad (10.34)$$

where $i_a = \varphi(u)$ is the anode current characteristic represented in Fig. 545.

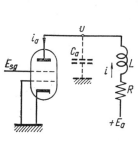

FIG. 544

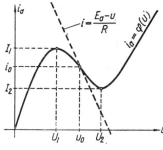

FIG. 545

The generator has equilibrium states evidently determined by

$$i = \frac{E_a - u}{R} = \varphi(u).$$

We assume the parameters E_a and R to be such that there is a unique state of equilibrium (u_0, i_0) on the descending section of the characteristic (Fig. 545).

The characteristic equation at the equilibrium state (u_0, i_0) is

$$\begin{vmatrix} L\lambda + R & 1 \\ -1 & C_a\lambda + \varphi'(u_0) \end{vmatrix} \equiv C_a L\lambda^2 + [C_a R + L\varphi'(u_0)]\lambda + 1 + R\varphi'(u_0) = 0.$$

Since (u_0, i_0) lies on the descending section of the characteristic and is unique, then $-1/R < \varphi'(u_0) < 0$. Hence this state is unstable for $C_a < -\varphi'(u_0)L/R$, which is satisfied for sufficiently small values of C_a.

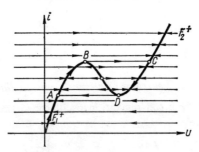

Fig. 546

As usual, we can investigate the oscillations by plotting the phase portrait on the u, i plane in the limiting case $C_a \to +0$. This portrait is shown in Fig. 546. Everywhere outside the line $F(u, i) = i - \varphi(u) = 0$, outside the phase line of the "degenerate" model, the rates of change $du/dt \to \infty$, and can be arbitrarily large for sufficiently small values of C_a, $du/dt \to +\infty$ above F and $\to -\infty$ below F, but di/dt is bounded everywhere. Therefore this region is filled with paths $i = \text{const}$, along which the representative point moves with an "infinitely large" phase velocity, so that these paths represent "jumps" of the voltage u. Some of these paths are shown in Fig. 546, where F_1^+ and F_2^+ are parts of F on which the value C_a is unimportant and the conditions

$$\frac{\partial F}{\partial u} = -\varphi'(u) < 0$$

are satisfied.

Thus, the representative point, having arrived as a result of a "jump" on F_1^+ or F_2^+ moves along these curves with a finite rate of change of voltage u. For C_a small but not zero the phase paths for "slow" variations in state lie near F_1^+ or F_2^+. On F_1^+ and F_2^+, therefore, a small capacitance C_a can be neglected and we write the equation of the circuit as a first order equation

$$i = \varphi(u), \quad L\frac{di}{dt}+Ri = E_a-u.$$

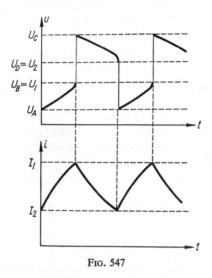

Fig. 547

On the paths F_1^+ and F_2^+, $du/dt \neq 0$ and $di/dt \neq 0$, du/dt being positive on F_1^+ and negative on F_2^+: therefore points B or D will be reached by the representative point. Then along a path i=const. it will "jump" to the point C or A. Thus, there exists a limit cycle $ABCDA$. Whatever the initial conditions, self-oscillations will be established in the circuit as, in contrast to a neon tube, the oscillations of the voltage u are discontinuous, while the oscillations of the current i have a "saw-tooth" form (Fig. 547). The current and voltage swings are clearly equal to $I_B-I_A=I_1-I_2$ and U_c-U_A respectively. The period T of the self-oscillations is the sum of the time spent on the paths AB and CD since we neglect the durations of the "jumps" and is

$$T = L\int_{U_A}^{U_B} \frac{\varphi'(u)\,du}{E_a-u-R\varphi(u)} + L\int_{U_c}^{U_D} \frac{\varphi'(u)\,du}{E_a-u-R\varphi(u)}.$$

§ 8. Frühhauf's circuit

We shall consider now the circuit of a relaxation oscillator suggested by Frühhauf [155, 142]. In this circuit (Fig. 548) it is important to note that the valves are connected in series and therefore we must pay attention to the division of supply voltage between the valves and the resistance R. Also, we cannot neglect anode reaction by assuming the anode conductances to be zero. Therefore we assume that the anode current is a function of the grid voltage and the anode voltage, but is single-valued and a monotonically increasing function of the controlling voltage $u_{\text{con}} = u_g + Du_a$

$$i_a = f(u_g + Du_a),$$

where D is the *durchgriff* of the valve, or the reciprocal of the amplification factor ($D<1$). We assume further that this function has an inverse

$$u_g + Du_a = U(i_a)$$

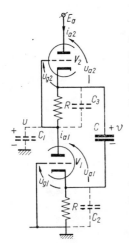

Fig. 548

which is also single-valued.

1. "Degenerate" model

Neglecting all parasitic parameters (in particular the stray capacitances) and the grid currents and assuming the two valves to be identical, we obtain the following circuit equations:

$$i_{a1} = f(u_{g1} + Du_{a1}), \quad i_{a2} = f(u_{g2} + Du_{a2}),$$

$$C \frac{dv}{dt} = i_{a2} - i_{a1},$$

$$u_{g1} = -Ri_{a2}, \quad u_{g2} = -Ri_{a1},$$

$$v = u_{a1} - u_{g2} = E_a - u_{a2} + u_{g1}$$

where u_{g1} and u_{a1}, u_{g2} and u_{a2} are respectively the grid and anode voltages of the valves V_1 and V_2 (measured with respect to the cathodes).

Let us now set

$$x = \frac{i_{a1}}{I_0} \quad \text{and} \quad y = \frac{i_{a2}}{I_0},$$

where I_0 is the saturation current or any other scale of intensity of current. Then

$$u_{g1} = -RI_0 y, \qquad u_{g2} = -RI_0 x,$$
$$u_{a1} = RI_0 \frac{\psi(x)+y}{D}, \qquad u_{a2} = RI_0 \frac{\psi(y)+x}{D},$$

where

$$\psi(z) = \frac{1}{RI_0} U(I_0 z)$$

The voltage across the capacitor C is

$$v = \frac{RI_0}{D}\{\psi(x)+Dx+y\} = \frac{RI_0}{D}\left\{\frac{DE_a}{RI_0}-\psi(y)-Dy-x\right\} \quad (10.35)$$

$$RC\frac{dv}{dt} = RI_0(y-x);$$

from which we find

$$\left.\begin{array}{l}\dfrac{dx}{dt} = -\dfrac{D}{CR}\dfrac{x-y}{\Delta(x,y)}\{1+D+\psi'(y)\}, \\[2mm] \dfrac{dy}{dt} = \dfrac{D}{CR}\dfrac{x-y}{\Delta(x,y)}\{1+D+\psi'(x)\},\end{array}\right\} \quad (10.36)$$

where

$$\Delta(x,y) = [D+\psi'(x)][D+\psi'(y)]-1. \quad (10.36a)$$

Thus the system is of the first order (a system with half a degree of freedom), since the x and y variables (see equ. (10.35)) are connected by the relation

$$\Phi(x,y) \equiv \psi(x)+\psi(y)+(1+D)(x+y) = \frac{DE_a}{RI_0} \quad (10.37)$$

while one of the equations is a consequence of the other and of (10.37).

It is difficult to eliminate x or y but equation (10.37) shows that the representative point must move along the line Φ

$$\Phi(x,y) = \frac{DE_a}{RI_0},$$

plotted on the x, y plane. This phase line, as well as its mapping by the paths of the equations (10.36), is symmetrical with respect to the bisector $y=x$. In addition we have on it

$$\frac{dy}{dx} = -\frac{1+D+\psi'(x)}{1+D+\psi'(y)} < 0, \quad (10.36b)$$

since $\varphi'(x) > 0$ and $\psi'(y) > 0$. Therefore, if ψ' is a continuous function, the phase line is everywhere smooth and cannot be closed. Hence, this system cannot generate continuous periodic oscillations, since the right-hand sides of (10.36) are single-valued functions of the point on the phase line.

Let us consider the motion of the representative point along the Φ phase line. The equations (10.36) have a unique equilibrium state (x_0, y_0) on the bisector $x = y$ determined according to (10.37) by $\psi(x_0) + (1+D)x_0 = DE_a/2RI_0$. In addition, according to the equations (10.36) the representative point moves along the Φ line in a direction towards the equilibrium state (x_0, y_0) at the points of Φ where $\Delta(x, y) > 0$ and in a direction away from this equilibrium state at the points where $\Delta(x, y) < 0$. Therefore the equilibrium state is stable if $\Delta(x_0, y_0) > 0$ and unstable if $\Delta(x_0, y_0) < 0$.

Now plot on the x, y plane the curve (symmetrical with respect to the bisector $y = x$),

$$\Delta(x, y) = 0, \qquad (10.38)$$

which we call the Γ curve. If this curve exists[†] two cases can occur:

(1) The parameter DE_a/RI_0 is such that the Φ phase line does not intersect the Γ curve. Then $\Delta(x, y) > 0$ everywhere on Φ and the system approaches, whatever the initial conditions, the equilibrium state (x_0, y_0) as t increases[‡].

(2) The parameter DE_a/RI_0 is such that the phase line Φ intersects the curve Γ. There must then be points $\gamma(x', y')$ on the line Φ disposed symmetrically with respect to the straight line $y = x$ such that $\Delta(x', y') = 0$ and which therefore are junction points of the phase paths of the equations (10.36). These points are not states of equilibrium but whatever the initial conditions representative points move towards them and cannot leave the line Φ. The mathematical model cannot describe a system with discontinuous oscillations and to investigate the latter we must either take into account some essential small parameters or else complete our "defective" model of the first order by a suitably defined jump hypothesis.

[†] It certainly exists for $\psi'_{\min} < 1 - D$, since then on the bisector $y = x$ there are points at which $\Delta(x, y) < 0$ and on the other hand, in the vicinity of the axes of the coordinates (for small values of x or y) $\Delta(x, y) > 0$, since there $\psi'(x)$ or $\varphi'(y)$ can be made as large as we please; hence owing to the continuity of the function $\Delta(x, y)$ there will exist a locus of points where $\Delta(x, y) = 0$, i.e. the curve Γ will exist. If the characteristic of the valve has the saturation current I_s, the $\psi'(z) \to +\infty$ for $z \to I_s/I_0$ and the curve Γ will be closed.

[‡] The same picture will be obtained for all values of E_a, if the curve Γ does not exist, for then $\Delta(x, y) > 0$ at all points of every phase line Φ (for every value of E_a).

2. The jump postulate

Let us at first complete the degenerate model of the first order (the equations (10.36)) by a jump postulate. Suppose, to be definite, that the phase line Φ intersects the curve Γ at two points $\gamma_1(x_1', y_1')$ and $\gamma_2(x_2', y_2')$ where $x_2' = y_1'$ and $y_2' = x_1'$ (Fig. 549). Since these junction points of phase paths are always boundary points of segments of a phase line on which some

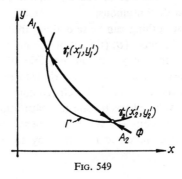

Fig. 549

small parameters are important in the oscillating processes, we must find which of the three sections $A_1\gamma_1$, $\gamma_1\gamma_2$ and γ_2A_2 of the phase line Φ have motions represented adequately by the equations (10.36).

We now make the following hypothesis about the oscillations:

(1) On the sections $A_1\gamma_1$ and $A_2\gamma_2$ of Φ; where $\Delta(x, y) > 0$, the small parasitic parameters, so far neglected, are unimportant for the processes in the system. Thus there are on them "slow" variations of state described by the equations (10.36), but on the section $\gamma_1\gamma_2$ there are only "rapid" motions away from $\gamma_1\gamma_2$; there the equations (10.36) cannot represent quantitatively or qualitatively the laws of motion[†].

(2) When the representative point, moving along the section $A_1\gamma_1$ (or $A_2\gamma_2$) arrives at the point γ_1 (or γ_2) it makes next an instantaneous jump to a point $B_1(x_1'', y_1'')$ or to $B_2(x_2'', y_2'')$ situated inside one of the intervals $A_1\gamma_1$ or $A_2\gamma_2$ and determined by the following jump conditions:

$$y'' + \psi(x'') + Dx'' = y' + \psi(x') + Dx',$$
$$x'' + \psi(y'') + Dy'' = x' + \psi(y') + Dy',$$

where
$$\Delta(x'', y'') > 0.$$

[†] In Sub-section 4 this hypothesis will be justified by constructing a "satisfactory" model of the circuit. However, note that the model of the first order is apparently satisfactory since on its phase line there are no junction points if the whole phase line lies where $\Delta(x, y) > 0$. The sections $A_1\gamma_1$ and $A_2\gamma_2$ lie just in this region.

These jump conditions follow, as usual, from the postulate that the voltage v across the capacitor C (see (10.35)) is constant during a jump in the anode currents x and y and are proportional respectively to i_{a1} and i_{a2}.

If the equations (10.39) determine uniquely the end-point B of a jump from the initial point γ then the hypothesis above will enable us to analyse the oscillations that start from states represented by points of the intervals $A_1\gamma_1$ and $A_2\gamma_2$ of the phase line Φ. These oscillations of the circuit will be clearly periodic and discontinuous.

In the general case nothing can be said about the existence and number of real branches of the curve (B) (the set of the points B that correspond according to (10.39) to points γ for all possible values of the parameter DE_a/RI_0). If (B) exists, then it is symmetric with respect to the bisector $y=x$ and is tangent to the curve Γ at their (common) points. It is, in addition, closed if the valve characteristic has saturation.

Obviously if the equations (10.39) determine several points B then the jump postulate must be modified in some way so that a B is uniquely determined by γ.

3. Discontinuous oscillations in the circuit

To analyse these oscillations further it is necessary to adopt an analytical expression for the valve characteristic. A suitable approximate form for the characteristic is (Fig. 550(a))

$$i_a = \frac{I_s}{2} + \frac{I_s}{\pi}\arctan\frac{\pi S}{I_s}(u_{\text{con}} - u_0),$$

where I_s is the saturation current, S is the maximum slope of the characteristic and u_0 is the value of the control voltage for which $i_a = I_s/2$ and $di_a/du_{\text{con}} = S$. Solving for $u_{\text{con}} = u_g + Du_a$ and putting $I_0 = I_s/\pi$, with

$$x = \pi\frac{i_{a1}}{I_s},$$

$$y = \pi\frac{i_{a2}}{I_s} \quad (0 \leqslant x \leqslant \pi,\ 0 \leqslant y \leqslant \pi),$$

we obtain for the reduced inverse characteristic

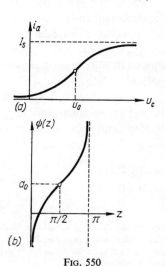

Fig. 550

$$\frac{u_{g1}+Du_{a1}}{RI_0} = \psi(x) = a_0 - a\cot x, \quad \frac{u_{g2}+Du_{a2}}{RI_0} = \psi(y) = a_0 - a\cot y,$$

where $a=1/RS$ and $a_0=\pi u_0/RI_s$. Under these circumstances $\psi' \geqslant a$ (Fig. 550(b)), and according to (10.36), (10.37), (10.38) and (10.39) we find the equation of the phase line Φ

$$(1+D)(x+y) - a(\cot x + \cot y) = b, \tag{10.37'}$$

with

$$b = \frac{\pi DE_a}{RI_s} - 2a_0;$$

and the equations of motion along this phase line are

$$\left.\begin{aligned}\frac{dx}{d\tau} &= -\frac{x-y}{\Delta(x,y)}\left\{1+D+\frac{a}{\sin^2 y}\right\}, \\ \frac{dy}{d\tau} &= \frac{x-y}{\Delta(x,y)}\left\{1+D+\frac{a}{\sin^2 x}\right\},\end{aligned}\right\} \tag{10.36'}$$

where $\tau = Dt/RC$ is a new dimensionless time.

The equation of the curve Γ, the locus of the initial points of the jumps, is now

$$\Delta(x,y) \equiv \left(D+\frac{a}{\sin^2 x}\right)\left(D+\frac{a}{\sin^2 y}\right) - 1 = 0 \tag{10.38'}$$

and the jump conditions are

$$\left.\begin{aligned}y''-a\cot x''+Dx'' &= y'-a\cot x'+Dx', \\ x''-a\cot y''+Dy'' &= x'-a\cot y'+Dy'.\end{aligned}\right\}$$

Since the anode currents can only vary within the limits $0 \leqslant i_a \leqslant I_s$, then only the points in the quadrant $0 \leqslant x \leqslant \pi$, $0 \leqslant y \leqslant \pi$ have a physical meaning. It is easily seen that the phase line Φ passes through the points $A_1(0,\pi)$ and $A_2(\pi,0)$ (Fig. 551), and that for $b=\pi(1+D)$ the phase line Φ is the straight line $x+y=\pi$. In addition, since $\Delta(x,y) \geqslant \Delta(\pi/2,\pi/2) = (D+a)^2-1$, the curve Γ determined by the equation (10.38) exists for $D+a<1$, i.e. for

$$RS(1-D) > 1;$$

Γ moreover is a closed curve symmetrical with respect to the straight lines

$$x=y; \quad x+y=\pi;$$
$$x=\frac{\pi}{2}; \quad y=\frac{\pi}{2}.$$

For $D+a<1$ the curve (B) determined by the equations (10.39') exists also and lies in the region $\Delta(x,y)>0$, outside the curve Γ. It is closed and symmetrical with respect to the straight lines $x=y$ and $x+y=\pi$, so that to each initial point γ of a jump there is an unique end point B, lying on the other side of the bisector $y=x$ (Fig. 551).

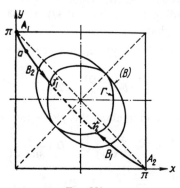

FIG. 551

A representative point starting from the point a moves along $A_1\gamma_1$ to the point γ_1, and then jumps to the point B_1 on the phase line $A_2\gamma_2$. Next, moving along $A_2\gamma_2$, it reaches the curve Γ again at γ_2, and then jumps to B_2 from where it moves along $A_1\gamma_1$ to the point γ_1, and so on. Thus the periodic discontinuous oscillations build up, corresponding to the "discontinuous" limit cycle $B_1\gamma_2 \dashrightarrow B_2\gamma_1 \dashrightarrow B_1$, having two paths of slow "motions" $B_1\gamma_2$ and $B_2\gamma_1$ and two jumps $\gamma_1 \dashrightarrow B_1$ and $\gamma_2 \dashrightarrow B_2$.

For the generation of discontinuous oscillations it is necessary that: (1) the curve Γ have real branches; (2) the constant b in the equation (10.37') of Φ is such that the line Φ intersects the curve Γ. But the curve Γ exists if
$$a < 1-D$$
and the second condition is satisfied at the equilibrium state (x_0, y_0) determined by
$$x_0 = y_0, \quad 2(1+D)x_0 - 2a \cot x_0 = b,$$
if $\Delta(x_0, y_0) = [D + (a/\sin^2 x_0)]^2 - 1 < 0$, or $\sin x_0 \leq [a/(1-D)]^{\frac{1}{2}}$. Hence the necessary condition for the line Φ to intersect the curve Γ is
$$2(1+D)\arcsin\sqrt{\frac{a}{1-D}} - 2a\sqrt{\frac{1-D}{a}-1} < b <$$
$$< 2(1+D)\left\{\pi - \arcsin\sqrt{\frac{a}{1-D}}\right\} - 2a\sqrt{\frac{1-D}{a}-1},$$
where the angle lies in the first quadrant, and the root is positive.

There are mathematical difficulties in the evaluation of the period along an arbitrary phase line Φ. For this reason, we will calculate the period for $b = \pi(1+D)$, when Φ is the diagonal line $x+y = \pi$. Then

$$\frac{dx}{d\tau} = -\frac{dy}{d\tau} = \frac{\pi - 2x}{\dfrac{a}{\sin^2 x} + D - 1},$$

whence the period (in ordinary time units) is

$$T = \frac{RC}{D}\left\{2a\int_{x_1}^{x_2}\frac{dx}{(\pi-2x)\sin^2 x} - (1-D)\ln\frac{\pi-2x_1}{\pi-2x_2}\right\},$$

where x_1 and x_2 are respectively the abscissae of the intersection of $x+y=\pi$ and the curves (B) and Γ.

4. Including the stray capacitances

To conclude this section we will show that the hypothesis made above about the oscillations in Frühhauf's circuit is inherent in a "satisfactory" model constructed by including one of the small parasitic capacitances (see Fig. 548). The circuit equations are

$$C\frac{dv}{dt} = i_{a2} + \frac{u_{g2}}{R} = -i_{a1} - \frac{u_{g1}}{R},$$

$$C_1\frac{du}{dt} = -\frac{u_{g2}}{R} - i_{a1},$$

$$v = u_{a1} - u_{g2} = E_a - u_{a2} + u_{g1},$$

$$u = u_{a1} - u_{g1} = E_a - u_{a2} + u_{g2},$$

$$i_{a1} = f(u_{g1} + Du_{a1}), \quad i_{a2} = f(u_{g2} + Du_{a2}).$$

Now set

$$x = \frac{i_{a1}}{I_0}, \quad y = \frac{i_{a2}}{I_0}$$

$$z = \frac{v}{RI_0}, \quad w = \frac{u}{RI_0}$$

$$t' = \frac{t}{RC}$$

$$\mu = \frac{C_1}{C},$$

and the circuit equations can be reduced to

$$\left.\begin{aligned}\mu\dot{w} &= \frac{Dx+y+\psi(x)-Dz}{1-D} = F(w, z), \\ \dot{z} &= \frac{Dy-x-\psi(x)+Dz}{1-D} = G(w, z),\end{aligned}\right\} \quad (10.40)$$

since

$$(1+D)(x+y)+\psi(x)+\psi(y) = \frac{DE_a}{RI_0} \quad (10.40a)$$

and

$$w = \frac{\psi(y)-\psi(x)+(1+D)z}{1-D}, \quad (10.40b)$$

$\psi=(u_g+Du_a)/RI_0$ is, as before, the dimensionless controlling voltage.

Let us take as the phase surface of the second-order system the cylindrical surface Φ^* defined in the x, y, z space by the equation (10.40a). This cylin-

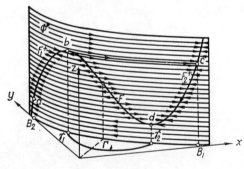

Fig. 552

der meets the x, y plane in the line Φ (see Fig. 552). Now draw on it the line F, the phase line when $\mu=0$.

$$F(w, z) = 0,$$

To evaluate $\partial F/\partial w$ on the line we have from (10.40)

$$(1-D)\frac{\partial F}{\partial w} = [D+\psi'(x)]\frac{\partial x}{\partial w}+\frac{\partial y}{\partial w},$$

$$[1+D+\psi'(x)]\frac{\partial x}{\partial w}+[1+D+\psi'(y)]\frac{\partial y}{\partial w} = 0,$$

$$-\psi'(x)\frac{\partial x}{\partial w}+\psi'(y)\frac{\partial y}{\partial w} = 1-D,$$

whence,

$$\frac{\partial F}{\partial w} = -\frac{\Delta(x,y)}{[1+D+\psi'(x)]\psi'(y)+[1+D+\psi'(y)]\psi'(x)}$$

$\psi' > 0$, so the denominator of this expression is always positive and therefore the condition for the small parasitic capacitance C_1 to be unimportant is

$$\Delta(x,y) > 0,$$

which is satisfied on the "ascending" sections F^+ of the line F (with $\partial z/\partial x > 0$.†) Therefore in small neighbourhoods of F^+ the motion of the representative point is described (for $C_1 \ll C$) by the degenerate equations

$$F = 0 \quad \text{or} \quad z = \frac{y+Dx+\psi'(x)}{D}$$

and

$$\dot{z} = y-x.$$

These are equivalent to (10.36) and so the first part of the hypothesis made in Sub-section 2 is justified.

In addition, outside the line F, $\dot{w} \to \pm \infty$ as $\mu \to +0$, whereas $\dot{z}$ remains finite. Hence the region of the surface Φ^* outside the line F is filled with paths of "rapid" motions $z=\text{const}$, along which the representative point "jumps" to the right above F, and to the left below it.

If the $\Delta < 0$ at the equilibrium state the circuit is self-excited and the line F has points of maxima and minima in z which are the boundary points of the sections F^+, since at them $\Delta(x,y) = \partial F/\partial w = 0$.

† On the line F

$$1+D+\psi'(x)+[1+D+\psi'(y)]\frac{dy}{dx} = 0$$

and

$$D\frac{dz}{dx} = D+\psi'(x)+\frac{dy}{dx} = D+\psi'(x)-\frac{1+D+\psi'(x)}{1+D+\psi'(y)} = \frac{\Delta(x,y)}{1+D+\psi'(y)}.$$

Since $1+D+\psi'(y) > 0$, then the sign of $(dz/dx)_{F=0}$ is the same as the sign of $\Delta(x,y)$.

At these points a "slow" motion passes into a "rapid" jump-wise one along a path $z=$const. that leads again to one of the sections F^+. During these jumps the voltage v across C remains constant or almost constant if μ is small but finite. This is the jump condition (10.39).

Now in contrast to the jump conditions (10.39), there is a one-to-one correspondence between the initial and end point of a jump even when the line F has more than one maximum or minimum in z. A jump is made from a point of maximum (minimum) z on the curve F along a path $z=$const. towards the right (towards the left) to the first point of intersection between the path $z=$const. and a section F^+.

The limiting case where $\mu \rightarrow +0$ is shown in Fig. 552 for a self-excited circuit. The excitation is "soft" and discontinuous oscillations build up as represented by the limit cycle *abcda* whose projection on the x, y plane is the discontinuous limit cycle $B_2\gamma_1 \dotdiv B_1\gamma_2 \dotdiv B_2$.

§9. A MULTIVIBRATOR WITH AN INDUCTANCE IN THE ANODE CIRCUIT

We have now seen that the investigation of a self-oscillating system is considerably simplified if one of the important oscillation parameters is small, so that the motions can be split into comparatively simple "rapid"

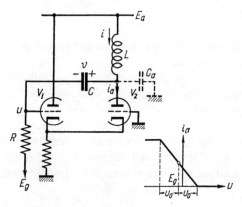

Fig. 553

and "slow" motions. The first being described by the equations (10.17) or a suitably defined jump postulate and the second by the equations (10.16) in which the selected small parameter is neglected.

Let us now consider a more complicated system with discontinuous oscillations in which the equations of the "slow" motions are now of the second order. An example is the multivibrator with one RC circuit, but with an inductive anode load (Fig. 553). To simplify the problem we neglect the ohmic resistance of this load.

1. *The equations of "slow" motions*

Neglecting all parasitic parameters, grid currents and anode conductance and using Kirchhoff's laws the circuit equations are (in the notation of Fig. 553):

$$L\frac{di}{dt} = E_a - (u+v),$$

$$i = i_a(u) + \frac{u-E_g}{R},$$

$$C\frac{dv}{dt} = \frac{u-E_g}{R}.$$

The circuit has evidently a single equilibrium state

$$u = E_g, \quad i_a = i_a^0 = i_a(E_g), \quad i = i_a^0, \quad v = v_0 = E_a - E_g.$$

To simplify the analysis we shall approximate the characteristic of the valve-pair $i_a = i_a(u)$ by a piece-wise linear function (Fig. 553), assuming that the grid bias E_g is so chosen that the working point (equilibrium state) lies at the centre of the section with the negative slope.

Introduce the new dimensionless variables

$$x = \frac{u-E_g}{u_0}, \quad y = \beta\frac{v-v_0}{u_0}, \quad z = \frac{R}{u_0}(i-i_a^0)$$

where $2u_0$ is the "width" of the descending section of the characteristic. The new dimensionless time is

$$t_{\text{new}} = \frac{t}{\sqrt{LC}},$$

and the dimensionless characteristic of the valve-pair is

$$\varphi(x) = \frac{1}{u_0 S}[i_a(E_g + u_0 x) - i_a(E_g)] = \begin{cases} +1 & \text{for} \quad x < -1, \\ -x & \text{for} \quad |x| \leq 1, \\ -1 & \text{for} \quad x > +1, \end{cases}$$

where S is the numerical value of the negative slope of the characteristic. Then putting $\beta = R\left(C/L^{\frac{1}{2}}\right)$, the reduced equations are

$$\left.\begin{aligned} \dot{z} &= -2hx - y, \\ z &= x + k\varphi(x), \\ \dot{y} &= x, \end{aligned}\right\} \quad (10.41)$$

where

$$k = SR \quad \text{and} \quad h = \frac{\beta}{2} = \frac{R}{2}\sqrt{\frac{C}{L}},$$

or, eliminating z

$$\left.\begin{aligned} \dot{x} &= -\frac{2hx + y}{1 + k\varphi'(x)}, \\ \dot{y} &= x. \end{aligned}\right\} \quad (10.41\text{a})$$

The x, y phase plane is evidently divided into three regions of linearity: (1) $x < -1$; (2) $x > +1$ and (3) $|x| < 1$. Since for $|x| < 1$, $\varphi'(x) = -1$, then in the region containing the unique equilibrium state $(x = 0, y = 0)$ the equations (10.41a) will be linear

$$(1-k)\dot{x} = -2hx - y,$$
$$\dot{y} = x$$

with the characteristic equation

$$(1-k)\lambda^2 + 2h\lambda + 1 = 0.$$

Therefore the singular point of the origin is stable for $k < 1$ and unstable for $k > 1$.

We need only consider the case $k > 1$, since for $k < 1$ there are no self-oscillations. As we pass through the lines $x = \pm 1$ the expression $1 + k\varphi'(x)$ changes its sign (since $k > 1$), and so the points of the half straight lines $x = +1, y > -1$ and $x = -1, y < +1$ are junction points of phase paths. These points are *not* states of equilibrium, although phase paths approach them on both sides. Thus, neglecting parasitic parameters, we have obtained a "defective" model of a multivibrator, since it does not enable us to investigate the oscillations of the multivibrator. To obtain a "satisfactory" model of a multivibrator, we must either complete the equations (10.41a) with a suitable jump postulate or take into account small essential parameters.

In the problem of the oscillations of a ship, controlled by a two-position automatic pilot (Chapter VIII, Section 6) we also found a line of junction points and subsequently completed the definition of the system of differential equations of motion so that a motion was possible along this line. This motion of the representative point corresponded to the so-called "slip-motion" operation of the automatic pilot which can be observed in practice. In this multivibrator, however, such a postulate about the motion of the representative point along the half straight lines $x=+1$, $y>-1$ and $x=-1$, $y<+1$ would only give motions receding to infinity which is not in agreement with experiment.

Just such a type of "defective" model (with a line of junction of phase paths) results from a multivibrator with one RC circuit (Section 4 of this Chapter) by taking into account the parasitic inductance of the anode circuit.

2. Equations of a multivibrator with stray capacitance C_a

The complete equations of a multivibrator (Fig. 553) will provide a "satisfactory" model of the multivibrator. The equations are

$$L\frac{di}{dt} = E_a - (u+v),$$

$$I = i_a(u) + C\frac{dv}{dt} + C_a\frac{d(u+v)}{dt},$$

$$C\frac{dv}{dt} = \frac{u-E_g}{R},$$

or, in the x, y, z and t_{new} variables but neglecting the small capacitance C_a in the expression $C+C_a$

$$\left.\begin{array}{l}\mu\dot{x} = z-x-k\varphi(x) \equiv F(x,z),\\ \dot{y} = x,\\ \dot{z} = -2hx-y,\end{array}\right\} \qquad (10.42)$$

$\mu = RC_a/(LC)^{\frac{1}{2}}$ is a small positive parameter characterizing the stray capacitance C_a otherwise the notation is the same as in (10.41). When $\mu=0$, (10.42) degenerates into (10.41) or (10.41a), the equations of "slow" motions which, however, are not valid over the whole phase surface

$$F(x, z) = 0 \quad \text{or} \quad z = x+k\varphi(x),$$

but only on the part F^+ where

$$F'_x = -1-k\varphi'(x) < 0,$$

or $|x|>1$, and where a small C_a is unimportant anyway.

(If, however, $k<1$, then the condition for the small capacitance C_a to be unimportant is satisfied over the whole surface $F=0$; correspondingly when $C_a=0$ there are no lines of junction points of phase paths.)

We shall observe that the phase surface F of the "degenerate" model and the x, y plane are homeomorphic with respect to each other (a one-to-one continuous correspondence exists between their points). Therefore we can represent "slow" motions of the system by a motion of the representative point, not on the surface F^+, but on the x, y plane ($|x| > 1$).

Outside a small neighbourhood of the surface F^+, $x \to \infty$ as $\mu \to +0$, and so x varies in a jump-wise fashion. The phase paths of these "rapid" motions are the straight lines $y=\text{const}$, $z=\text{const}$ and recede from F in the positive x direction (for $z > x + k\varphi(x)$) and in the negative x direction (for $z < x + k\varphi(x)$). The limiting phase portrait (for $\mu \to +0$) of the system

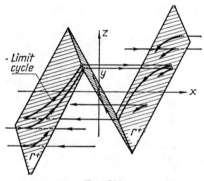

Fig. 554

(10.42) is shown qualitatively in Fig. 554; in particular the limit cycle is shown which as we shall see, actually exists for $\mu \to +0$. We draw the following conclusion on the character of the oscillations in the presence of C_a (for $C_a \to +0$):

(1) Discontinuous oscillations are present since paths of "rapid" motions go to the surface F^+ and become paths of "slow" motions, which in turn become paths of "rapid" motions on the boundaries $x = \pm 1$ of the surface F^+.

(2) "Slow" motions only occur on the surface F^+

$$z = x + k\varphi(x), \quad |x| > 1;$$

and the equations (10.41) or (10.41a) are valid only for $|x| > 1$.

(3) During "rapid" motions (jumps) of the representative point the x variable (the voltage u on the grid of the triode V_1) varies instantaneously, but the variables y and z (the capacitor voltage v and the anode current i) remain constant (voltages and currents must remain bounded).

3. Discontinuous oscillations of the circuit

We know that "slow" motions of the system only occur for $|x|>1$, and are described by the linear equations

$$\left.\begin{aligned}\dot{x} &= -2hx-y, \\ \dot{y} &= x,\end{aligned}\right\} \quad (10.41\text{b})$$

since $\varphi'(x) = 0$ for $|x|>1$. The characteristic equation is then

$$\lambda^2+2h\lambda+1 = 0, \quad (10.43)$$

and the behaviour of the circuit during "slow" variations of state depends on the parameter $h = R(C/L)^{\frac{1}{2}}/2$. If $h>1$, (i.e. if $L<CR^2/4$) then both roots of (10.43) will be real and negative and the system will behave aperiodically. Its phase paths outside the shaded region $|x|<1$ will be similar to those of a linear oscillator whose singular point is a stable node (Fig. 555)[†]. If, however, $h<1$ i.e. (if $L>CR^2/4$) then the system behaves

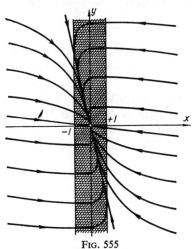

Fig. 555

[†] We shall represent the oscillations of the circuit not in the three-dimensional x, y, z phase space but on the x, y phase plane. It is evident that, in such a representation, the phase paths of "slow" and "rapid" motions can intersect each other since they are the projection of phase paths in the x, y, z, space.

during a "slow" motion as a linear system whose singular point is a stable focus at the origin. The phase paths outside the shaded region $|x|<1$ will resemble arcs of spirals (Fig. 556). In these cases the $\varkappa = \infty$ isocline is the

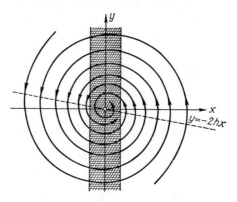

FIG. 556

straight line $y = -2hx$. Note that the representative point moves along the phase paths (10.41b) not in a clockwise direction, as is usual, but anti-clockwise, since y is not simply $\dot{x}$ but is related to x and $\dot{x}$ by

$$y = -\dot{x} - 2hx.$$

Whatever the value of the parameter h, phase paths of "slow" motions reach the straight lines $x = \pm 1$ and then the representative point recedes along a path of "rapid" motion: $y = $const, $z = $const. Since the end point of a "jump" again lies on the surface $z = x + k\varphi(x)$, its coordinates (x_2, y_2) are related to the coordinates (x_1, y_1) at the beginning of the jump $(x_1 = \pm 1)$ by the equations

$$y_2 = y_1,$$
$$x_2 + k\varphi(x_2) = x_1 + k\varphi(x_1)$$

and so

$$\left. \begin{array}{l} y_2 = y_1, \\ x_2 = -(2k-1)x_1. \end{array} \right\} \qquad (10.44)$$

Hence, the representative point jumps from the line $x = +1$ along a path $y = $const. to a point $x = -(2k-1)$, and conversely, from the point $x = -1$, to the line $x = 2k-1$. After the jump the representative point will move again along a phase path of "slow" motion until it reaches the straight line $x = \pm 1$ etc. (Figs. 557 and 558). We will show that these paths approach a stable limit cycle as $t \to +\infty$.

We begin with the case when $L > CR^2/4$, and $h < 1$. The phase portrait is shown in Fig. 557 and is symmetrical with respect to the origin. There-

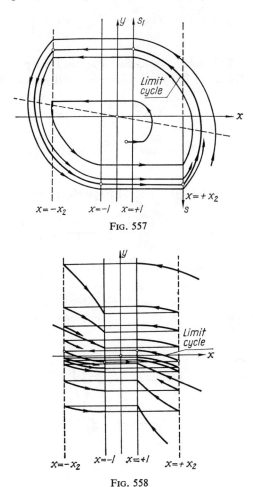

Fig. 557

Fig. 558

fore, we need only investigate the transformation $s_1 = f(s)$ of the points $y = -s$ of the line $x = x_2 = 2k-1$ into the points $y = s_1$ of the line $x = +1$, generated by the phase paths of "slow" motions on the half-plane $x > 1$. The fixed point s of this transformation corresponds to the limit cycle.

The phase path of the system (10.41b) leaving the point $(x_2, -s)$ at $t=0$ is

$$x = e^{-ht}\left[x_2 \cos \omega t - \frac{hx_2 - s}{\omega} \sin \omega t\right],$$

$$y = e^{-ht}\left[-s \cos \omega t + \frac{x_2 - hs}{\omega} \sin \omega t\right],$$

where $\omega = (1-R^2)^{\frac{1}{2}}$ and $h<1$ still. Suppose that the representative point arrives on the line $x=+1$ at $y=s_1$ when $t = \tau/\omega$ $(0<\tau<\pi)$. Then

$$1 = e^{-\gamma\tau}\left[x_2 \cos \tau + \left(\frac{s}{\omega} - \gamma x_2\right) \sin \tau\right],$$

$$s_1 = e^{-\gamma\tau}\left[-s \cos \tau + \left(\frac{x_2}{\omega} - \gamma s\right) \sin \tau\right],$$

where $\gamma = h/\omega = h/(1-h^2)^{\frac{1}{2}}$. The correspondence function $s_1 = f(s)$ is given by

$$\left.\begin{aligned} s &= \omega \frac{e^{\gamma\tau} - x_2(\cos \tau - \gamma \sin \tau)}{\sin \tau}, \\ s_1 &= \omega \frac{x_2 e^{-\gamma\tau} - (\cos \tau + \gamma \sin \tau)}{\sin \tau}. \end{aligned}\right\} \quad (10.45)$$

Graphs of these functions are shown in Fig. 559. Since for $\tau \to +0, s \to -\infty$ and $s_1 \to +\infty$, and for $\tau \to \pi-0, s \to +\infty$ and $s_1 \to +\infty$, s_1 being smaller

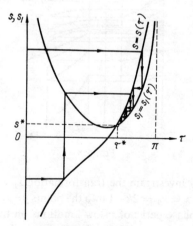

Fig. 559

than s, there is at least one point of intersection, which is the fixed point s^*, for which $\tau = \tau^*$ and is determined uniquely by

$$e^{\gamma\tau} - x_2(\cos\tau - \gamma\sin\tau) = x_2 e^{-\gamma\tau} - (\cos\tau + \gamma\sin\tau)$$

or

$$e^{\gamma\tau} + \cos\tau + \gamma\sin\tau = x_2[e^{-\gamma\tau} + \cos\tau - \gamma\sin\tau]. \quad (10.46)$$

On the x, y phase plane there is a unique limit cycle. This limit cycle is stable, since at $s = s^*$.

$$\left|\frac{ds_1}{ds}\right| < 1.$$

Thus we have verified the existence of a limit cycle and discontinuous self-oscillations differing appreciably from sinusoids.

The period of the self-oscillations is evidently equal to $2\tau^*/\omega$ in dimensionless units or

$$T = \frac{2\sqrt{LC}}{\sqrt{1-h^2}} \tau^* \quad (10.47)$$

in real time, provided the duration of the "rapid" motions is neglected. It is evident that T is less than $T' = 2\pi(LC)^{\frac{1}{2}}/(1-h^2)^{\frac{1}{2}}$, the "conditional period" of an oscillator described everywhere by the equations (10.41b), since the "instantaneous" jumps cause the representative point to make its revolution more rapidly than that of such an oscillator, and so $\tau^* < \pi$.

For the case of very large $L \gg CR^2/4$ or when $h \ll 1$ and $\gamma \ll 1$, we may put $\gamma = 0$ in (10.46) and obtain a zero-order approximation for τ^*

$$\tau^* = \pi,$$

and the period T is close to $T_0 = 2\pi(LC)^{\frac{1}{2}}$. To determine the corrections to the period for small values of γ we introduce

$$a = \pi - \tau^*.$$

Then, substituting $\tau^* = \pi - a$ in (10.46) we find for a

$$e^{\gamma(\pi-a)} - \cos a + \gamma \sin a = x_2[e^{-\gamma(\pi-a)} - \cos a - \gamma \sin a] \quad (10.46a)$$

or, expanding the functions in power series

$$\gamma\pi + \frac{\gamma^2(\pi-a)^2}{2} + \ldots + \frac{a^2}{2} + \frac{a^4}{4!} + \ldots - \gamma\frac{a^3}{3!} + \ldots =$$
$$= x_2\left[-\gamma\pi + \frac{\gamma^2(\pi-a)^2}{2} + \ldots + \frac{a^2}{2} - \frac{a^4}{4!} + \ldots + \gamma\frac{a^3}{3!} - \ldots\right] \quad (10.46b)$$

Hence it follows that a is of order of magnitude $\gamma^{\frac{1}{2}}$ and is determined by

$$\gamma\pi + \frac{a^2}{2} = x_2\left(-\gamma\pi + \frac{a^2}{2}\right) + O(\gamma^2),$$

or

$$a = \sqrt{\frac{2\pi k}{k-1}}\,\gamma^{\frac{1}{2}} + O\left(\gamma^{\frac{3}{2}}\right)^\dagger \qquad (10.47a)$$

Thus, for $L \gg CR^2/4$

$$T = 2\pi\sqrt{LC}\left\{1 - \sqrt{\frac{2k}{\pi(k-1)}}\,h^{\frac{1}{2}} + O\left(h^{\frac{3}{2}}\right)\right\}. \qquad (10.47b)$$

The correction to the period is of the order of $h^{\frac{1}{2}}$ and is therefore comparatively large (recall, for comparison, that in an ordinary valve generator the correction to the period is of the order of h^2).

Substituting (10.47a) in (10.45), we find $\bar{s}$, the amplitude of the oscillations of the y variable

$$\bar{s} = \frac{2k + O(\gamma)}{\sin \alpha} = \sqrt{\frac{2k(k-1)}{\pi h}} + O\left(h^{\frac{1}{2}}\right).$$

Since for small values of h the phase paths of "slow" motions are nearly the circles $x^2 + y^2 = $ const, then, approximately, the amplitude of the x variable will be the same. Thus the amplitude of the self-oscillations of the grid voltage u of V_1 and capacitor voltage v across C are

$$U_0 = u_0 \bar{s} = u_0 \sqrt{\frac{2k(k-1)}{\pi}}\,h^{-\frac{1}{2}} + O\left(h^{\frac{1}{2}}\right),$$

$$V_0 = u_0 \frac{\bar{s}}{2h} = u_0 \sqrt{\frac{k(k-1)}{2\pi}}\,h^{-\frac{3}{2}} + O\left(h^{-\frac{1}{2}}\right).$$

† On putting $a = a_0 \gamma^{\frac{1}{2}} + a_1$ in (10.46b) where $a_1 = O\left(\gamma^{\frac{3}{2}}\right)$ and $a_0 = (2\pi k/(k-1))^{\frac{1}{2}}$, we obtain the equation determining the following term of asymptotic expansion:

$$\frac{\gamma^2 \pi^2}{2} + a_0 \gamma^{\frac{1}{2}} a_1 - \frac{a_0^4 \gamma^2}{4!} = x_2\left[\frac{\gamma^2 \pi^2}{2} + a_0 \gamma^{\frac{1}{2}} a_1 - \frac{a_0^4 \gamma^2}{4!}\right] + O\left(\gamma^{\frac{5}{2}}\right),$$

or

$$a_1 = \left(\frac{a_0^3}{4!} - \frac{\pi^2}{2a_0}\right)\gamma^{\frac{3}{2}} + O(\gamma^2)$$

and, hence,

$$a = \sqrt{\frac{2\pi k}{k-1}}\left\{\gamma^{\frac{1}{2}} + \left[\frac{1}{12}\frac{\pi k}{k-1} - \frac{\pi}{4}\frac{k-1}{k}\right]\gamma^{\frac{3}{2}}\right\} + O(\gamma^2),$$

Since $V_0 \gg U_0$ when $h \ll 1$ the amplitude of the anode voltage $u+v$ of V_2 is approximately equal to V_0. All these amplitudes increase as L increases, so the theory here is invalid when L is sufficiently large.

Now let us consider the case of *small* values of L when $h \gg 1$. The roots of the characteristic equation (10.43) are real and negative and the system behaves during the "slow" motions as if it were linear with a node at the origin (Fig. 555). In this region where $|x| > 1$, there are two rectilinear phase paths with slopes $1/\lambda_1$ and $1/\lambda_2$. The phase portrait when L is small is shown in Fig. 558.

A detailed analysis shows that, as before, there is a stable periodic motion which consists of two motions with finite velocity and two jumps, and which is established for arbitrary initial conditions. The limit cycle represents the usual discontinuous self-oscillations in the multivibrator. The amplitude of these oscillations can be determined at once, equal to $x_2 = 2k-1$, whence the amplitude of the oscillations of the grid voltage u of V_1 is $U_0 = (2k-1)u_0$.

The period can be evaluated by direct integration, but is much simplified when $L \ll CR^2/4$, (but, as before, $L \gg R^2 C_a^2/C$ and $\mu \ll 1$). Then $h \gg 1$ and the phase paths of a "slow" motion (but outside a certain small neighbourhood of $y = -2hx$) are nearly horizontal straight lines[†]. Correspondingly the limit cycle will be close to the $y = 0$ axis; therefore, during a "slow" motion of the representative point along the limit cycle, $x \approx -2hx$ or $dx/dt = -(R/L)x$. Integrating within the limits from x_2 to 1, we obtain the half-period of the self-oscillations:

$$\frac{T}{2} = -\frac{L}{R} \int_{x_2}^{1} \frac{dx}{x} = \frac{L}{R} \ln x_2,$$

so the period is

$$T = \frac{L}{R} \ln (2k-1). \qquad (10.47c)$$

In this case C does not affect appreciably the period; because when L is small the discontinuous self-oscillations are of a relatively high frequency, and the alternating voltages across C are small, $(V_0 = u_0(k-1)L/RC^2 \ll u_0)$.

[†] $dy/dx = -x/(2hx+y) \simeq -1/2h \approx -0$ outside a small neighbourhood of the straight line $y = 2hx$. Therefore, in particular, the amplitude of the self-oscillations of the y variable will be equal approximately, when $h \gg 1$, to $(x_2-1)/4h = (k-1)/2h$ and the amplitude of the voltage v across C is

$$V_0 \approx \frac{u_0}{2h} \frac{k-1}{2h} = \frac{(k-1)L}{CR^2} u_0.$$

724   DISCONTINUOUS OSCILLATIONS   [X

We have restricted ourselves to the cases of large and small values of L, only in order to simplify the exposition; it is quite possible to investigate intermediate values of L. Fig. 560 shows photographs of the portrait on the u, $u+v$ plane obtained by means of a cathode-ray oscilloscope. The photographs are arranged in order of decreasing values of L and the character of the periodic process agrees with our theoretical investigation.

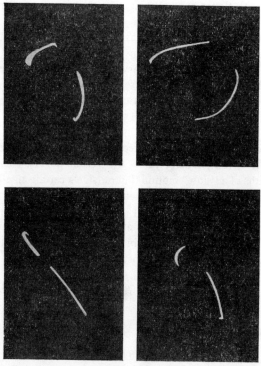

Fig. 560

§ 10. The "universal" circuit

There are systems in which continuous and discontinuous self-oscillations are possible, changing into one another as some parameter passes through a certain critical or branch value. An example is the so-called "universal" circuit [125] shown in Fig. 561 (see Section 4, Chapter V). This circuit is an "intermediate" one between a generator with a two-mesh RC circuit (see Section 12, Chapter V and Section 5, Chapter VIII) and a multivibrator with one RC circuit, reducing to the first for $\beta = r_1/r = 0$ and to the second for $\beta = r_1/r = 1$. We may expect that for a displacement of the contact of the potentiometer from the lower position to the upper one a transition will take place from continuous self-oscillations to discontinuous ones. The investigation confirms this.

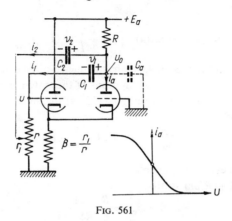

Fig. 561

The equations of the "universal" circuit, when grid currents, anode conductance, and all parasitic parameters except the small stray capacitance C_a (Fig. 561) are neglected, will be

$$\frac{E_a - u_a}{R} = i_a(u) + C_a \frac{du_a}{dt} + i_1 + i_2,$$

$$C_1 \frac{dv_1}{dt} = i_1, \quad C_2 \frac{dv_2}{dt} = i_2,$$

$$u = r(i_1 + \beta i_2), \quad u_a = u + v_1, \quad v_2 = v_1 + (1-\beta)ri_1,$$

or, in the variables, u, v_1 and $z = i_1 r$ (for $C_a \ll C_1, C_2$)

$$\left. \begin{array}{l} \mu \dot{u} = E_a - Ri_a(u) - \left(1 + \dfrac{R}{\beta r}\right) u + (1-\beta) \dfrac{R}{\beta r} z - v_1, \\[2mm] \dot{v}_1 = z, \quad (1-\beta)\dot{z} = \dfrac{C_1}{\beta C_2} u - \left(1 + \dfrac{C_1}{\beta C_2}\right) z, \end{array} \right\}$$

where $\mu = RC_a/rC$ is a small positive parameter, and the new dimensional time is $t' = t/rC_1$. The characteristic of the valve-pair $i_a = i_a(u)$ is shown in Fig. 561. We shall assume that the numerical magnitude of the slope

$$S(u) = -\frac{di_a}{du}$$

has a maximum S_0 at the equilibrium state $u=0$ and decreases monotonically to zero as $|u|$ increases.

If we neglect C_a, put $\mu=0$ in the equations (10.48) then the set of the equilibrium states form the surface F in u, z, v_1 space

where
$$F(u, z, v_1) = 0,$$
$$F(u, z, v_1) = E_a - Ri_a(u) - \left(1 + \frac{R}{\beta r}\right)u + (1-\beta)\frac{R}{\beta r}z - v_1. \quad (10.49)$$

The surface F in the u, z, v_1 space has a one-to-one correspondence with the u, z plane; therefore, "slow" motions of the system can be represented by paths on this plane. According to (10.19) the small capacitance C_a is unimportant if

$$\frac{\partial F}{\partial u} = RS(u) - 1 - \frac{R}{\beta r} < 0. \quad (10.49a)$$

Two cases are evidently possible.

For $RS_0 < 1 + (R/\beta r)$ so that

$$\beta < \beta_{cr} = \frac{R}{r}\frac{1}{RS_0 - 1}$$

the condition (10.49a) is satisfied over the whole phase surface F of the incomplete system. Hence the small parasitic capacitance C_a is not important to the oscillating processes and can be neglected. The oscillations of the "universal" circuit can be considered as that of a system with 1 degree of freedom described by (10.49) and the last two of the equations (10.48). This evidently results from the fact that for $\beta < \beta_{crit}$ all paths of "rapid" motions found for small values of C_a (for $\mu \to +0$) outside the surface F lead to this surface (Fig. 562). Eliminating V_1 from these equations we obtain two differential equations

$$\left.\begin{array}{l}\dot{u} = \dfrac{\dfrac{R}{\beta r}\dfrac{C_1}{\beta C_2}u - \left[1 + \dfrac{R}{\beta r}\left(1 + \dfrac{C_1}{\beta C_2}\right)\right]z}{1 + \dfrac{R}{\beta r} - RS(u)}, \\[2ex] \dot{z} = \dfrac{C_1}{\beta(1-\beta)C_2}n - \left(1 + \dfrac{C_1}{\beta C_2}\right)\dfrac{z}{1-\beta}\end{array}\right\} \quad (10.50)$$

with regular right-hand sides, which describe the behaviour of the circuit

If the condition of self-excitation

$$RS_0 > 1 + \frac{R}{r}\frac{C_1 + C_2}{C_1 + \beta C_2} \quad (10.51)$$

is not satisfied, then the unique equilibrium state ($u=0$, $z=0$) is a stable focus or a node, towards which all phase paths go, and whatever the initial conditions, the circuit reaches the singular point.

If, however, the condition of self-excitation (10.51) is satisfied but as before, $RS_0 < 1 + R/\beta r$, then the singular point (0, 0) will be an unstable

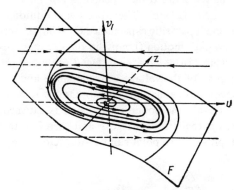

Fig. 562

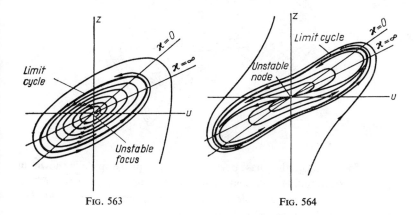

Fig. 563   Fig. 564

focus or node, and with the assumed characteristic $i_a = i_a(u)$, there will be one stable limit cycle on the u, v plane towards which all remaining phase paths tend (Figs. 563 and 564). The phase portraits shown in Figs. 563 and 564 are constructed by the method of isoclines from (10.50). Fig. 563 shows an unstable focus, and Fig. 564 an unstable node. Thus, whatever the initial conditions, *continuous* self-oscillations are built up. This is true when μ, and so C_a, is small but not zero.

A different picture is obtained for $RS_0 > 1 + (R/\beta r)$, when

$$\beta > \beta_{cr} = \frac{R}{r} \frac{1}{RS_0 - 1}. \tag{10.52}$$

Now, because of the form of $S(u)$ there evidently exists a grid voltage u^* on the grid of the left-hand triode such that $RS(+u^*) = 1 + (R/\beta r)$ and $RS(u) > 1 + (R/\beta r)$, for $|u| < u^*$. Therefore the condition (10.49a) is not satisfied and the small parasitic capacitance C_a is important. Therefore the phase paths of "rapid" motion (jumps) in the u, z, v_1 phase space recede away from the region $|u| \leqslant u^*$ of the surface F (obtained with $C_a = 0$). For $|u| \leqslant u^*$ only jumps of the voltage u are possible, and not described of course by the incomplete equations (10.50). On the remaining part of the surface F (for $|u| > u^*$) the condition (10.49a) is satisfied, the paths of "rapid" motion approach the surface F and therefore in its vicinity the motion of the representative point can be satisfactorily represented by the equations (10.50).

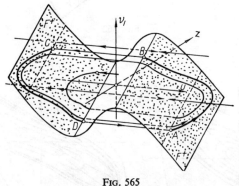

Fig. 565

Thus for $\beta > \beta_{crit}$, the phase portrait in u, z, v_1 phase space is shown in Fig. 565, corresponding to discontinuous oscillations. On the portion F^+ of F where $|u| > u^*$ there are "slow" motions along paths determined (for sufficiently small values of C_a) by the equations (10.50). Outside F^+ $u \to \infty$, for $\mu \to +0$ but $\dot{z}$ and $\dot{v}_1$ remain finite, therefore there are "rapid" motions along the paths $z = \text{const}$, $v_1 = \text{const}$,[†] which lead to the surface F^+

[†] The conditions for z and v_1 to remain constant during the jump of the voltage u can also be obtained from the additional (physical) assumption that the currents and voltages in the circuit remain bounded. If the currents are bounded then the voltages v_1 and v_2 across the capacitors C_1 and C_2 cannot vary with a jump ($\dot{v}_1$ and $\dot{v}_2$ must be bounded). Then, during an instantaneous jump of u, v_1 and $v_2 - v_1 = (1-\beta)ri_1 = (1-\beta)z$ remain constant.

where they pass into paths of "slow" motions. In due course all paths of "slow" motion pass into discontinuous jumps at $u = +u^*$ or at $u = -u^*$. It can easily be shown that all phase paths tend to a unique and stable limit cycle for $t \to +\infty$. Thus for $\beta > \beta_{\text{crit}}$, whatever the initial conditions, discontinuous oscillations build up in the system.

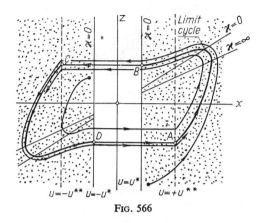

Fig. 566

Fig. 566 is the projection of the u, z plane of the phase portrait in the u, z, v_1 space. The paths of the jumps are the straight lines $z = \text{const.}$ and the end points ($\mp u^{**}, z_2$) of these paths which begin at the points ($+u^*, z_1$) are determined by the relations

$$z_2 = z_1, \quad Ri_a(\mp u^{**}) \mp \left(1 + \frac{R}{\beta r}\right) u^{**} = Ri_a(\pm u^*) \pm \left(1 + \frac{R}{\beta r}\right) u^*.$$

The limit cycle $ABCDA$, consists of two sections, the "slow" motions AB and CD, and the "rapid" motions BC and DA†.

The transition of continuous self-oscillations into discontinuous ones at $\beta = \beta_{\text{crit}}$ takes place continuously: for values of β approaching β_{crit} from below the rate of change of the voltage u on the line $u = 0$ increases without limit and for $\beta = \beta_{\text{crit}}$ becomes infinitely large: on the other hand for $\beta > \beta_{\text{crit}}$, the variation $u^{**} + u^*$ of u in a jump increases monotonically from zero as β increases from β_{crit}.

---

† A rigorous proof of the existence of a limit cycle and that all remaining paths tend to it can be carried out, for example, for a piece-wise characteristic of the valve pair, by constructing the point transformation of the straight line $u = +u^*$ into the straight line $u = -u^*$.

§ 11. THE BLOCKING OSCILLATOR

In radio engineering practice the so-called blocking oscillator [65, 71, 91], is sometimes used to generate short voltage pulses. A version of it is shown in Fig. 567.

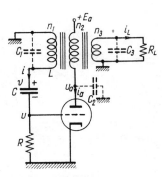

FIG. 567

The blocking oscillator has inductive feedback, and the anode current is cut-off during most of the period of an oscillation. Suppose that, after being cut-off, anode current begins to flow. Then a positive voltage is induced at the grid through the transformer, as a result of which the anode current increases further and the grid voltage u rapidly reaches a large positive value (up to several hundred volts), and considerable anode and grid currents flow. These currents, flowing through the windings of the transformer, induce a voltage pulse in its output winding. At the same time a grid-current pulse charges the capacitor C, thus decreasing the grid voltage u, and after a short interval of time the valve is again cut-off. As the anode current decreases, a negative voltage is induced at the grid causing a cumulative effect and rapid reduction of the anode current[†].

During the remaining part of the period the valve is cut off, grid currents are absent and the capacitor C discharges via the resistance R. The grid voltage u increases gradually and after a certain time (of a duration of the order of RC) reaches a value at which the valve is triggered (anode current flows) and the blocking oscillator generates another impulse.

---

† This is the so-called capacitive restoration of state in a blocking oscillator, which is realized in practice for sufficiently large values of L/C.

1. The equations of the oscillations

In the analysis we cannot neglect either grid current or anode conductance since they play an important role in the operation of the blocking oscillator: during the generation of an impulse considerable grid current flows in the valve charging the capacitor C and finally biassing off the valve at the end of the impulse. Meanwhile the anode voltage u_a decreases to a very small value, which limits maximum anode current in the impulse and so the grid voltage. We assume, therefore, the anode current to be a function,

$$i_a = i_a(u, u_a);$$

and to simplify the problem we assume the grid current to depend on the grid voltage only:

$$i_g = i_g(u).$$

In practice, to generate rectangular impulses it is necessary that the transformer has small magnetic leakage flux and small stray capacitance (across the windings, the transformer is usually wound on a toroidal ferromagnetic core). It is natural therefore to assume as a first approximation that the magnetic leakage flux is zero. With this assumption the magnetic flux through each turn of any of the windings is determined by the total number of ampere-turns in all windings and is given by

$$\Phi = \frac{L}{n_1^2}(n_1 i - n_2 i_a + n_3 i_L) = \frac{L}{n_1}(i - k i_a + k' i_L),$$

where L is the inductance of the grid circuit winding of the transformer, n_1, n_2 and n_3 are the number of turns in the grid, anode and output windings, and $k = n_2/n_1$ and $k' = n_3/n_1$ are the voltage ratios of the anode and output windings with respect to the grid winding, i is the current flowing in the grid winding and the capacitor C and i_L is the current in the load and the output-winding if C_1 and C_3 are neglected (Fig. 567). We call

$$I = i - k i_a + k' i_L \tag{10.53}$$

the magnetization current of the transformer†. Then the induced e.m.f.'s

---

† The grid and anode windings of the transformer are so connected that the partial magnetic fluxes generated in the transformer by the (positive) currents i and i_a have opposite signs. With this condition, positive feedback is obtained and the blocking oscillator is self-excited.

in the grid, anode and output windings of the transformer will be, respectively,

$$-\frac{d}{dt}(n_1\Phi) = -L\frac{dI}{dt}, \quad +\frac{d}{dt}(n_2\Phi) = kL\frac{dI}{dt}$$

and

$$-\frac{d}{dt}(n_3\Phi) = -k'L\frac{dI}{dt}.$$

Neglecting any other parasitic parameters of the circuit besides the stray capacitances of the windings and the inter-electrode capacitances of the valve, we obtain the equations of the blocking oscillator

$$\left. \begin{array}{l} -L\dfrac{dI}{dt} = u+v = \dfrac{E_a-u_a}{k} = \dfrac{R_L i_L}{k'}, \\[6pt] C\dfrac{dv}{dt} = i = \dfrac{u}{R}+i_g(u), \end{array} \right\} \quad (10.54)$$

where R_L is the resistance of the external load. Substituting (10.53) we find

$$I = \frac{u}{R} + i_g(u) - k i_a(u, u_a) + \frac{k'^2}{R_H}\frac{E_a-u_a}{k} = I(u, u_a) \quad (10.53\text{a})$$

and eliminating the voltage across C

$$v = \frac{E_a-u_a}{k} - u, \quad (10.54\text{a})$$

we have

$$L\left\{\frac{1}{R}+S_g(u)-kS(u, u_a)\right\}\frac{du}{dt} + L\left\{\frac{k^2}{R_i(u, u_a)}+\frac{k'^2}{R_H}\right\}\frac{1}{k}\frac{du_a}{dt} = \frac{E_a-u_a}{k},$$

$$-\frac{du}{dt} - \frac{1}{k}\frac{du_a}{dt} = \frac{1}{C}\left[\frac{u}{R}+i_g(u)\right],$$

where

$$S_g(u) = \frac{di_g}{du}, \quad S(u, u_a) = \frac{\partial i_a}{\partial u} \quad \text{and} \quad \frac{1}{R_i(u, u_a)} = \frac{\partial i_a}{\partial u_a}.$$

R_1 is the internal anode resistance of the valve. Solving these equations for the time derivatives, then

$$Lg(u, u_a)\frac{du}{dt} = \frac{u_a-E_a}{k} - \frac{L}{C}\left\{\frac{k^2}{R_i(u, u_a)}+\frac{k'^2}{R_H}\right\}\left[\frac{u}{R}+i_g(u)\right]$$

(10.55)

$$Lg(u, u_a)\frac{1}{k}\frac{du_a}{dt} = \frac{E_a-u_a}{k} - \frac{L}{C}\left\{\frac{1}{R}+S_g(u)-kS(u, u_a)\right\}\left[\frac{u}{R}+i_g(u)\right],$$

where

$$g(u, u_a) = \frac{1}{R} + S_g(u) + \frac{k^2}{R_i(u, u_a)} + \frac{k'^2}{R_H} - kS(u, u_a).$$

It is evident that the blocking oscillator has one equilibrium state determined by

$$u_a - E_a = 0 \quad \text{and} \quad u + Ri_g(u) = 0,$$

Therefore the equilibrium state, if we assume $i_g = 0$ for $u \leqslant 0$, is

$$u = 0, \quad u_a = E_a.$$

The characteristic equation for the equilibrium state is

$$\begin{vmatrix} Lg(0, E_a)\lambda + \dfrac{L}{C}\left[\dfrac{k^2}{R_i(0, E_a)} + \dfrac{k'^2}{R_H}\right]\left[\dfrac{1}{R} + S_g(0)\right] & -1 \\ \dfrac{L}{C}\left[\dfrac{1}{R} + S_g(0) - kS(0, E_a)\right]\left[\dfrac{1}{R} + S_g(0)\right] & Lg(0, E_a)\lambda + 1 \end{vmatrix} = 0$$

or

$$Lg(0, E_a)\lambda^2 + \left\{1 + \frac{L}{C}\left[\frac{k^2}{R_i(0, E_a)} + \frac{k'^2}{R_H}\right]\left[\frac{1}{R} + S_g(0)\right]\right\}\lambda +$$
$$+ \frac{1}{C}\left[\frac{1}{R} + S_g(0)\right] = 0.$$

Now we shall only consider the case when the equilibrium state is unstable and the blocking oscillator is self-excited, so that the condition of self-excitation is satisfied

$$g(0, E_a) = \frac{1}{R} + S_g(0) + \frac{k^2}{R_i(0, E_a)} + \frac{k'^2}{R_H} - kS(0, E_a) < 0. \quad (10.56)$$

2. Jumps of voltages and currents

However, when the condition (10.56) is satisfied there will be a certain curve on the u, u_a phase plane on which

$$g(u, u_a) = 0 \quad (10.57)$$

and the rates of change of u and u_a become infinite. In fact, for sufficiently large values of $|u|$ the anode current slope $S(u, u_a)$ is small and, $g(u, u_a)$ is positive[†]. On the other hand, $g(0, E_a) < 0$ by virtue of condition (10.56).

[†] For large positive grid voltages when anode current is large, the anode current nearly ceases to depend on the grid voltage and is mainly determined by the anode voltage. For large negative grid voltages the anode current is cut off.

It follows from this and the assumptions that $S_g(u)$, $S(u, u_a)$, $[R_i(u, u_a)]^{-1}$, and therefore $g(u, u_a)$, are continuous functions, that there exists on the u, u_a plane a continuous locus on which $g(u, u_a) = 0$; i.e. the curve Γ exists. This curve is shown in Fig. 568.

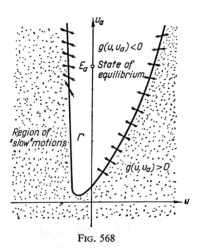

Fig. 568

The important point emerges that the function $g(u, u_a)$ changes its sign as we pass through the curve Γ; therefore part of this curve is formed by junction points of phase paths. Again, by neglecting all small parameters of the circuit, including the stray capacitances and the magnetic leakage fluxes, we have obtained a "defective" model of a blocking oscillator, on the phase plane of which there are junction points of phase paths. We are forced therefore to modify the dynamic model of a blocking generator either by completing it with postulates based necessarily on empirical knowledge about the oscillations or else by taking into account a few of the small parasitic parameters that are essential to the processes in a blocking oscillator.

We now postulate the following additional assumptions about the oscillations of the blocking oscillator.

(1) In the region

$$g(u, u_a) > 0$$

small parasitic parameters of the circuit have no major role to play and the oscillations ("slow" motions) are repesented by the equations (10.55) (since the region $g(u, u_a) > 0$ comprises values of u and u_a which cut the valve off, there is some justification for this postulate).

(2) If the representative point, on the u, u_a plane where $g(u, u_a) > 0$, moves along a path of (10.55) and meets the curve Γ, it will then make an instantaneous jump to another point in the region of "slow" motions $g(u, u_a) > 0$.

(3) All voltages and currents in the circuit are bounded, so that during an instantaneous jump the values of the voltage v across C and the magnetic flux Φ through the transformer windings remain constant. During the "slow" motion before and after a jump of the representative point, the voltage v is connected with u and u_a by the relation (10.54a) and the magnetic flux is determined by the magnetizing current $I = I(u, u_a)$, so the end point of the jump (u^*, u_a^*) is connected with the initial point (u, u_a) on the curve Γ, by the following relations

$$u^* + \frac{u_a^*}{k} = u + \frac{u_a}{k}, \\ I(u^*, u_a^*) = I(u, u_a). \tag{10.58}$$

These additional assumptions together with the equations (10.15) are sufficient for an analysis of the blocking oscillator.

We can prove the postulate about the discontinuous character of the oscillations from an analysis of the dynamics of a model of the third order obtained when the small self-capacitances of the windings (Fig. 567) are taken into account. The capacitance C_2 is the sum of the anode winding capacitance and the output capacitance of the valve, the capacitance C_3 is the sum of the output winding capacitances and of the load circuit. We still neglect other small parasitic parameters including the magnetic leakage flux in the transformer. For such a model, the equations are

$$-L\frac{dI^*}{dt} = u + v = \frac{E_a - u_a}{k} = \frac{u_{\text{out}}}{k'},$$

$$C\frac{dv}{dt} = \frac{u}{R} + i_g(u) = i(u),$$

where

$$I^* = i + C_1\frac{d(u+v)}{dt} - k\left[i_a + C_2\frac{du_a}{dt}\right] + k'\left[\frac{u_{\text{out}}}{R_H} + C_3\frac{du_{\text{out}}}{dt}\right] =$$
$$= i(u) - ki_a(u, u_a) + \frac{k'^2}{R_H}\frac{E_a - u_a}{k} - \frac{C'}{k}\frac{du_a}{dt} = I(u, u_a) - \frac{C'}{k}\frac{du_a}{dt}$$

is the magnetization current ($C' = C_1 + k^2 C_2 + k'^2 C_3$ and is the equivalent capacitance connected with the grid winding of the transformer). Eliminating $u = (E_a - u_a)/k - v$, we find

$$L\frac{dI^*}{dt} = \frac{u_a - E_a}{k}, \\ C\frac{dv}{dt} = i = i\left(\frac{E_a - u_a}{k} - v\right), \\ \frac{C'}{k}\frac{du_a}{dt} = I\left(\frac{E_a - u_a}{k} - v, u_a\right) - I^* = F(u_a, v, I^*). \tag{α}$$

From these equations (α), for $C' \to +0$, $du_a/dt \to \infty$, but dv/dt and dI^*/dt remain finite, in the u_a, v, I^* phase space outside the surface F

$$F(u_a, v, I^*) \equiv I - I^* = 0.$$

Therefore outside the surface F "rapid" motions of the representative point take place along the paths

$$v = \text{const}, \quad I^* = \text{const}, \tag{β}$$

(during which the voltages u and u_a vary step-wise but the voltage v across C and the magnetic flux remain constant.

These paths (β) of "rapid" motions lead towards that part of the surface F on which

$$\frac{\partial F}{\partial u_a} < 0 \quad \text{or} \quad g(u, u_a) > 0,$$

and since

$$k \left[\frac{\partial F}{\partial u_a} \right] = -\frac{\partial I}{\partial u} + k \frac{\partial I}{\partial u_a} = -\frac{I}{R} - S_\rho(u) + kS(u, u_a) - \frac{k^2}{R_i(u, u_a)} - \frac{k'^2}{R_H} = -g(u, u_a);$$

they lead away from that part of the surface F on which $g(u, u_a) < 0$. Therefore, "slow" motions of the representative point with finite rates of change of the variables occur only in a near neighbourhood of the surface F^+:

$$I^* = I(u, u_a), \quad u = \frac{E_a - u_a}{k} - v, \quad g(u, u_a) > 0 \tag{F^+}$$

in the limit, for $C' \to +0$, on the surface F^+ itself). Consequently the equations of "slow" motions can be written almost exactly in the form of equations (10.54) or (10.55).

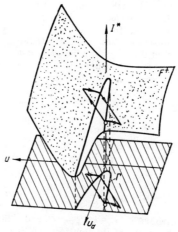

Fig. 569

These paths of "rapid" motions outside F^+ project on the u, u_a plane as lines $v = \text{const}$, or

$$u + \frac{u_a}{k} = \text{const.} \tag{β'}$$

The boundary surface between slow and rapid motions projects into the curve Γ on the u, u_a plane. In their turn the paths with "rapid" motions move towards the surface F^+, where they become paths with "slow" motions. The initial and end points of the paths of jumps are clearly connected by the conditions (10.58), since during a jump v and I^* do not vary, and $I^* = I(u, u_a)$ on the surface F^+. Fig. 569 shows the surface F^+ and certain phase paths in the u, u_a, I^* phase space and their projections on the u, u_a coordinate plane.

Thus taking into account small parasitic capacitances of the circuit, important during the jumps, produces a satisfactory model of the third order of a blocking oscillator.

It is interesting to note that the postulates above are justified from an analysis of a model of the third order obtained by taking into acccount the small transformer leakage flux but neglecting all stray capacitances, or of a fifth order model with both the small stray capacitances and the small magnetic leakage flux. The paths of "rapid" motions, or their projections oo the u, u_a plane are no longer the straight lines (β'); for example, with the magnetic leakage flux only, the projections of "rapid" motions on the u, u_a plane will be the lines

$$I(u, u_a) = \text{const.} \qquad (\beta'')$$

3. *Discontinuous oscillations*

To take further the analysis of the discontinuous self-oscillations of the blocking oscillator, we now assume a piece-wise linear approximation to the characteristics of the valve, as is shown in Fig. 570, and which is an idealization of the real characteristics of pentodes:

$$i_a = i_a(u, u_a) = \begin{cases} 0 & \text{for } u_a \leqslant 0 \text{ and for } u \leqslant -u_0, \\ S(u+u_0) & \text{for } 0 < S(u+u_0) < \dfrac{u_0}{R_0}, \\ \dfrac{u_a}{R_0} & \text{for } S(u+u_0) \geqslant \dfrac{u_0}{R_0} > 0, \end{cases}$$

$$i_g = i_g(u) = \begin{cases} 0 & \text{for } u < 0, \\ S_g u & \text{for } u \geqslant 0, \end{cases}$$

where $-u_0$ is the cut-off voltage of the valve and S and S_g are the slopes of anode and grid current characteristics. R_0 is the anode resistance of the valve in the small region where the anode current is dependent on u_a rather than u, i.e. where $u_a \leqslant SR_0(u+u_0)$. Otherwise, for $u_a > SR_0(u+u_0)$, the anode current depends only on the grid voltage.

The u, u_a plane is now divided into six regions of "linearity" shown in Fig. 571: the regon (*I*) and (*Ia*) correspond to a cut off valve ($i_a \equiv 0$), the regions (*II*) and (*IIa*) to where the anode current depends only on u: and the regions (*III*) and (*IIIa*) are regions of anode reaction where the anode current depends only on u_a. In the regions (*Ia*), (*IIa*) and (*IIIa*) $u > 0$ and anode current flows.

Let

$$kS > \frac{1}{R} + S_g + \frac{k'^2}{R_H} \quad \text{and} \quad E_a > SR_0 u_0. \quad (10.59)$$

Then the equilibrium state $(0, E_a)$ will lie on the boundary between the regions (II) and (IIa) in which $R_i = 1/(\partial i_a/\partial u_a) = \infty$ and therefore $g(u, u_a) < 0$.

Therefore this state is unstable, and in the regions (II) and (IIa) there are only "rapid" motions (jumps) of the representative point. However,

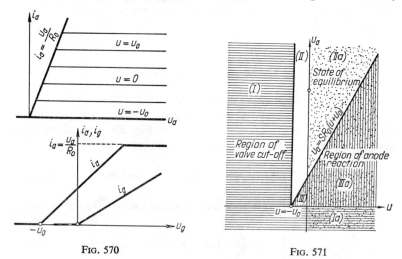

Fig. 570   Fig. 571

in the regions (I), (Ia), (III) and $(IIIa)$, where $S(u, u_a) = 0$ and $g(u, u_a) > 0$, "slow" motions are possible given by the equations (10.55). It is clear that the boundary Γ of "slow" motions, with a piece-wise linear characteristics, is the straight lines

$$u = -u_0, \quad u_a > 0 \quad \text{and} \quad u_a = SR_0(u+u_0) > 0.$$

Introduce new dimensionless variables x, y, t_{new},

$$u = u_0 x, \quad u_a = k u_0 y, \quad t_{\text{old}} = L \cdot g(u, u_a) t_{\text{new}}. \quad (10.60)$$

As usual, for brevity, we now denote t_{new} by t, and the ordinary time by t_{old}: and the time scale is clearly different in different regions of "linearity". Then the equations (10.55) are

$$\left. \begin{array}{l} \dot{x} = y - A - a(x, y)x, \\ \dot{y} = A - y - b(x)x, \end{array} \right\} \quad (10.61)$$

where

$$a(x, y) = \frac{L}{C} \frac{G(x)}{r(x, y)}, \quad b(x) = \frac{L}{C} [G(x)]^2,$$

$$G(x) = \begin{cases} \dfrac{1}{R} & \text{for} \quad x < 0, \\ G = \dfrac{1}{R} + S_g & \text{for} \quad x \geqslant 0, \end{cases}$$

$$\frac{1}{r(x, y)} = \frac{k^2}{R_i} + \frac{k'^2}{R_H} = \begin{cases} \dfrac{1}{r_1} = \dfrac{k'^2}{R_H} & \text{in the regions } (I) \text{ and } (Ia), \\ \dfrac{1}{r^2} = \dfrac{k^2}{R_0} + \dfrac{k'^2}{R} & \text{in the regions } (III) \text{ and } (IIIa) \end{cases}$$

$$A = \frac{E_a}{ku_0}.$$

The boundary Γ on the x, y phase plane will now be the lines

$$x = -1, \quad y > 0 \quad \text{and} \quad y = \frac{SR_0}{k} \quad (x+1) > 0, \quad (10.62)$$

and the jump conditions (10.58) state that the quantities

$$G(x)x - \frac{y}{r(x, y)} \quad \text{and} \quad x+y \quad (10.58\text{a})$$

have the same value before and after a jump from the half lines (10.62).

To prove this condition (10.58a) we observe that in the regions (I), (Ia), (III) and $(IIIa)$, the anode current is

$$i_a = \frac{u_a}{R_i} = \frac{ku_0 y}{R_i},$$

where the anode resistance $R_i = \infty$ in (I) and (Ia) and $R_i = R_0$ in (III) and $(IIIa)$. Therefore the magnetization current is

$$I = u_0 \left\{ G(x)x - \frac{y}{r(x, y)} + \frac{A}{r_1} \right\}. \tag{α}$$

The voltage across C is

$$v = u_0 \{A - (x+y)\}. \tag{β}$$

The jump conditions (10.58a) follow from the jump conditions (10.58) and from these expressions (α) and (β).

We now assume, for the sake of definiteness, that the parasitic capacitances are of most importance during the jumps of states in the blocking oscillator, and we neglect the small transformer leakage flux. Then the paths of jumps on the x, y plane will be the lines

$$x+y = \text{const}, \tag{γ}$$

since, in the absence of leakage flux the voltage v is given by the expression (β) both during the "slow" and "rapid" variations of the state and hardly varies during an instantaneous jump. The magnetization current also remains constant during the jump but owing to the presence of parasitic capacitances becomes equal to Eq. (α) only at the end of the jump.

The paths of jumps (γ) on the x, y plane are shown by thin lines.

The following equation

$$\frac{d}{dt}(x+y) = -\{a(x, y)+b(x)\}x, \qquad (10.61a)$$

derived from (10.61) will prove useful. It shows that $x+y$ increases (v decreases) for $x<0$ (i.e. for $u<0$) and, on the contrary, $x+y$ decreases (v increases) for $x>0$ (for $u>0$).

To construct a phase portrait of practical interest we may assume the following inequalities:

$$A \gg 1 \quad \text{and} \quad \frac{R_0}{k^2}, \ \frac{1}{S_g}, \ \frac{R_L}{k'^2}, \ \sqrt{\frac{L}{C}} \ll \sqrt{\frac{RR_L}{k'^2}} \ll R, \quad (10.63)$$

In the region (I) where $i_a \equiv 0$, $i_g \equiv 0$, $G(x) = 1/R$ and $r(x, y) = r_1 = R_L/k'^2$, the equations of "slow" oscillations are

$$\left. \begin{array}{l} \dot{x} = y-A-a_1 x, \\ \dot{y} = A-y-b_1 x, \end{array} \right\} \qquad (10.64)$$

where

$$a_1 = \frac{L}{CRr_1} \quad \text{and} \quad b_1 = \frac{L}{CR^2},$$

b_1 and a_1 would, in practice, satisfy the inequalities $0 < b_1 \ll a_1 \ll 1$. In the region (I) $t_{\text{old}} \approx (L/r_1)t_{\text{new}}$, since there $g(u, u_a) = (1/R)+(k'^2/R_L) \approx$ $\approx 1/r_1$.

The characteristic equation

$$\lambda^2+(1+a_1)\lambda+a_1+b_1 = 0 \qquad (10.64a)$$

has, for $0 < b_1 \ll a_1 \ll 1$, two real negative roots $-\gamma_1'$ and $-\gamma_1$ where

$$\gamma_1 = a_1+b_1[1+O(a_1)] \approx a_1 \quad \text{and} \quad \gamma_1' = 1-b[1+O(a_1)] \approx -1.$$

The general solution of (10.64) will be

$$\left. \begin{array}{l} x = B_1 e^{-\gamma_1 t}+B_1' e^{-\gamma_1' t} \approx B_1 e^{-a_1 t}+B_1' e^{-t}, \\ y = A+B_1(a_1-\gamma_1)e^{-\gamma_1 t}+B_1'(a_1-\gamma_1)e^{-\gamma_1' t} \approx \\ \qquad \approx A-B_1 b_1 e^{-a_1 t}-B_1' e^{-t}. \end{array} \right\} \qquad (10.64b)$$

The mapping of the region (I) by these paths of "slow" motions of the representative point is shown in Fig. 572. There are two rectilinear paths $y = A + \varkappa_1 x$ and $y = A + \varkappa_2 x$, where $\varkappa_1 = -b_1 \approx 0$ and $\varkappa_2 \approx -1$†. The remaining paths, outside a small neighbourhood of the path $y = A + \varkappa_1 x$,

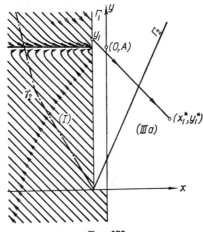

FIG. 572

are close to straight lines parallel to the second rectilinear path $y = A + \varkappa_2 x \approx A - x$ and the representative points move towards $y = A + \varkappa_1 x \approx A$. All paths moving where $y > 0$ reach the boundary line Γ_1 of the region (I), where Γ_1 is

$$x = -1, \quad y \geqslant A - a_1$$

(since $\dot{x} \geqslant 0$ only for $y \geqslant A - a_1$).

A motion of the representative point along or near $y = A + \varkappa_1 x \approx A$ corresponds to discharging C via R when the transformer emf.'s are very small. x (or grid voltage u) varies as $\exp(-a_1 t)$ (as $\exp(-t_{\text{old}}/RC)$), whereas, for a motion along the remaining paths the x and y variables vary as $\exp(-t)$ (as $\exp(-t_{\text{old}} r_1/L)$, i.e. appreciably more rapidly.

If the representative point has reached Γ_1, the half line $x = -1$, at y_1 $y_1 \geqslant A - a_1$, it will subsequently "jump" along a corresponding path of

---

† The slope of the phase path $y = A + \varkappa x$, according to (10.64), is given by

$$\varkappa^2 + (1 - a_1)\varkappa + b_1 = 0,$$

whence

$$\varkappa_1 = -b_1[1 + O(a_1)] \approx 0 \quad \text{and} \quad \varkappa_2 = -1 + a_1 + b_1[1 + O(a_1)] \approx -1.$$

"rapid" motion $x+y = y_1-1$ to the point (x_1^*, y_1^*) which is determined uniquely by (10.58a) (the conditions for the voltage across C and the magnetization current to remain constant). The end point of the jump (x_1^*, y_1^*) is determined by the initial point of the jump according to the equations

$$Gx_1^* - \frac{y_1^*}{r_2} = -\frac{1}{R} - \frac{y_1}{r_1} \approx -\frac{y_1}{r_1}, \quad \text{since usually} \quad r_1 \ll R,$$

and

$$x_1^* + y_1^* = -1 + y_1,$$

whence

$$x_1^* \approx \frac{y_1\left(\dfrac{1}{r_2} - \dfrac{1}{r_1}\right) - \dfrac{1}{r_2}}{G + \dfrac{1}{r_2}} \quad \text{and} \quad y_1^* \approx \frac{y_1\left(G + \dfrac{1}{r_1}\right) - G}{G + \dfrac{1}{r_2}} \qquad (10.65)$$

Since $y_1 \geqslant A - a_1 \gg 1$, then $x_1^* > 0$ and $0 < y_1^* < (SR_0/k)(x_1^* + 1)$, and the point (x_1^*, y_1^*) lies in (*IIIa*); also $y_1^* < y_1$.

In the region (*IIIa*) where $R_i = R_0$ and $G(x) = G \approx 1/S_g$, $(S_g \gg 1/R)$, the equations (10.61) will be

$$\begin{aligned} \dot{x} &= y - A - a_2 x, \\ \dot{y} &= A - y - b_2 x, \end{aligned} \Bigg\} \qquad (10.66)$$

where

$$a_2 = \frac{L}{C}\frac{G}{r_2} \quad \text{and} \quad b_2 = \frac{L}{C}G^2;$$

and now $(t_{\text{old}} = L[G+(1/r_2)]t_{\text{new}}$. The differential equation of the phase paths is

$$\frac{dy}{dx} = \frac{A - y - b_2 x}{y - A - a_2 x} \qquad (10.66a)$$

and the characteristic equation of (10.66)

$$\lambda^2 + (1 + a_2)\lambda + a_2 + b_2 = 0 \qquad (10.66b)$$

has either two real negative roots (for $(1-a_2)^2 > 4b_2$) or two complex conjugate roots with a negative real part (for $(1-a_2)^2 < 4b_2$).

a_2 and b_2, of course, depend on the resistive quantities $\varrho = (L/C)^{\frac{1}{2}}$ and

$$\frac{1}{G} \approx \frac{1}{S_g}\sqrt{\frac{r_2}{G}} \approx \left\{S_g\left(\frac{k^2}{R_0} + \frac{k'^2}{R_L}\right)\right\}^{-\frac{1}{2}}.$$

Hence, the paths in the region (*IIIa*) are similar to the parabolic paths leading to a stable node at (0, A) outside the region (*IIIa*), or of arcs of spirals leading to a stable focus at the same point.

For sufficiently large values of ϱ, when a_2, $b_2 \gg A$, then away from the y axis $dy/dx \approx +b_2/a_2$ (as follows from (10.66a)) and the phase paths are approximately the lines $y-(b_2/a_2)x = \text{const.}$ or $(y/r_2)-Gx = \text{const.}$ These are lines of constant magnetization current since in (*IIIa*) this current is

$$I = u_0 \left\{ Gx - \frac{y}{r_2} \right\}.$$

As the representative point moves along these paths x and y decrease (so u, u_a and i_a all decrease), accompanied by a relatively sharp decrease of $x+y$ (the voltage v increases). In other words, for a_2, $b_2 \gg A$, (or $\varrho \gg 1/G$, $r_2/G)^{1/2}$). The magnetization current I hardly varies during the generation of an impulse of anode current. The decrease of the grid voltage u, ultimately cutting off the anode current is due to the rapid increase of the voltage v across the capacitor C (since grid current is flowing). This decrease in grid voltage u occurs even though the grid winding voltage increases. Such a mechanism for suddenly cutting off the anode current is called *capacitive restoration* of the cut-off state.

On the other hand, for sufficiently small values of $\varrho = (L/C)^{\frac{1}{2}}$ ($a_2, b_2 \ll 1$), then $dy/dx \approx -1$ away from the lines $y = A+a_2x$ and $y = A-b_2x$, and so the phase paths are nearly straight lines $x+y = \text{const.}$ There are lines of constant voltage across C. For $y < A-b_2x$ x decreases and y increases, so there is a comparatively large decrease of the magnetization current I. Hence the decrease of grid voltage u, which cuts off the anode current impulse is mainly the result of the decreasing magnetization current I in the transformer producing a voltage equal to $-L\,dI/dt_{\text{old}}$ in the grid winding. This mechanism of stopping current flow at the end of the impulse of anode current is usually called *inductive restoration* to the cut-off state.

Phase paths from (*IIIa*) enter the region (*III*) and the equations of motion will be obtained from the equations (10.66) by replacing G by $1/R$. The phase paths in (*III*) will be those associated with a stable node at the point (0, A), and are, approximately, straight lines $x+y = \text{const.}$ Region (*III*) is associated with the absence of grid current but anode reaction.

The representative point, moving "slowly" along paths in the regions (*IIIa*) and (*III*) finally reaches the boundary Γ_2 of these regions, which is the line

$$y = \frac{SR_0}{k}(x+1) > 0.$$

From Γ_2 the representative point moves "rapidly" or jumps into the region (I). If the jump begins at the point (x_2, y_2) on Γ_2, the end point of the jump (x_2^*, y_2^*) is determined by the jump conditions

$$\frac{x_2^*}{R} - \frac{y_2^*}{r_1} = G(x_2)x_2 - \frac{y_2}{r_2} \quad \text{and} \quad x_2^* + y_2^* = x_2 + y_2,$$

so that

$$\left.\begin{aligned}
y_2^* &\approx \frac{r_1}{r_2}y_2 - r_1 G(x_2)x_2 = r_1\left\{\left[\frac{SR_0}{kr_2} - G(x_2)\right]x_2 + \frac{SR_0}{kr_2}\right\}, \\
x_2^* &= x_2 + y_2 - y_2^* \approx \left(1 - \frac{r_1}{r_2}\right)y_2 + [1 + r_1 G(x_2)]x_2 = \\
&= -r_1\left\{\left[kS - \frac{1}{r_1} - G(x_2)\right]x_2 + kS\right\}.
\end{aligned}\right\} \quad (10.67)$$

The locus of all such points (x_2^*, y_2^*) is shown in Fig. 572 by the dash-dot line (γ_2) and it is easy to show that y_2 is in (I), (x_2^*, y_2^*), the representative point follows a "slow" motion in (I) along a path of the equations (10.64) until it reaches Γ_1 again, where it jumps into $(IIIa)$ etc.

Thus, when the self-excitation conditions (10.59) are satisfied the blocking oscillator generates discontinuous oscillations corresponding to "slow" motion in the regions (I) and $(IIIa) + (III)$ alternating with (instantaneous) jumps from (I) into $(IIIa)$ and from the $(IIIa)$ (or (III)) into (I). The region (I) corresponds to an interval in the oscillating process during which anode current is cut off, and motions in $(IIIa)$ and (III) correspond to the generation of impulses of anode current (the valve conducts but saturation current is not reached).

4. Discontinuous self-oscillations of the blocking oscillator

To investigate the stability of the periodic discontinuous oscillations we need the point transformation Π of the half straight line Γ_1 into itself, as is generated by suitable phase paths (Fig. 573). Let s be the ordinate of an initial point on the half straight line Γ_1

$$x = -1, \quad y \geqslant A - a_1.$$

From this point there is a path of "rapid" motion to the point (x_1^*, y_1^*) determined by (10.65); next, a path of "slow" motion in the region $(IIIa)$

THE BLOCKING OSCILLATOR

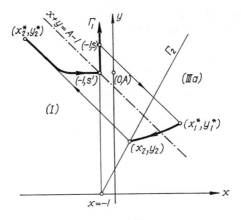

Fig. 573

(or in (IIIa) and (III)) to the point (x_2, y_2) on line Γ_2 from where there is jump motion to the point (x_2^*, y_2^*) in the region (I); and, finally, there is a path of "slow" motion in (I), ending on the half line Γ_1 at a certain point $(-1, s')$ which will be the consecutive point of

$$s' = \Pi(s).$$

There evidently exists an interval $A - a_1 \leq s \leq s_1$ on the half line Γ_1 for which the points (x_2, y_2) lie below the straight line $x + yA - 1$†. The next points (x_2^*, y_2^*) are also below this line so that for $b_1 \ll 1$ the phase paths in the region (I) leaving (x_2^*, y_2^*) arrive near the path $y \approx A$ and so will reach the line Γ_1 near to $(-1, A)$. Thus for $A - a_1 \leq s < s_1, s' \approx A$, and the graph of the correspondence function $s' = \Pi(s)$ over this interval of s is very nearly a horizontal straight line. It is easily shown, by a direct evaluation of the sequence function, that for $s \geq s_1$, $s' < s_1$. Therefore the graph of the sequence function (Lamerey's diagram) of Π has the form shown in Fig. 574. The sequence function has a unique point of intersection $s = s^* \approx A$ with the bisector $s' = s$, $\left(1 \, ds'/ds \right| \ll 1$ at this point), and so Π has a unique, and stable, fixed point $s = s^*$ corresponding to a unique,

† To prove this it suffices to observe that during a jump of the representative point $x + y$ does not vary (since the voltage v across the capacitor C does not vary), and in the regions (IIIa) and (III)

$$\frac{d}{dt}(x+y) = -[a+b]x < 0.$$

stable, limit cycle. This cycle intersects Γ_1 at a point close to the point $(-1, A)$ for $b_1 \ll 1$. All remaining paths tend to this limit cycle (as $t \to +\infty$), so there is only one mode of operation.

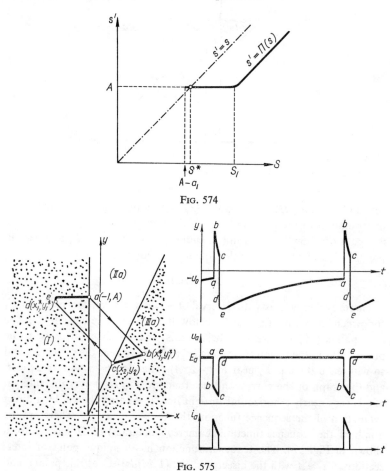

Fig. 574

Fig. 575

The form of the limit cycle, and, hence, of the discontinuous self-oscillations depend mainly on the shape of the phase paths in the region $(IIIa)$, which in its turn depends on the values of the parameters a_2 and b_2. Figures 575–7 show the limit cycles and the corresponding waveforms of the grid and anode voltages and the anode current, for various values of $\varrho = (L/C)^{\frac{1}{2}}$. In Fig. 575 $\varrho \gg 1/G$, $(r_2/G)^{\frac{1}{2}}$ (the case of capacitive restora-

11] THE BLOCKING OSCILLATOR 747

tion): in Fig. 576 ϱ is of the order of $1/G$, $(r_2/G)^{\frac{1}{2}}$, and in Fig. 577 $\varrho \ll 1/G$, $(r_2/G)^{\frac{1}{2}}$ (the case of inductive restoration). As can be seen the impulse of anode current i_a and also of anode voltage u_a have the flattest tops for $(L/C)^{\frac{1}{2}} \sim 1/G$, $(r_2/G)^{\frac{1}{2}}$, i.e. a case of "mixed" restoration.

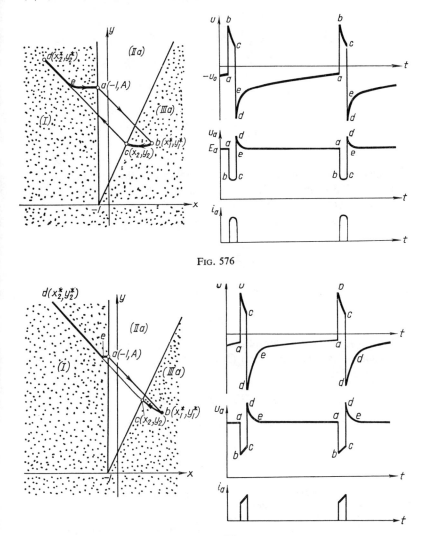

Fig. 576

Fig. 577

The evaluation of the period of the oscillations, the duration of the impulses and the voltage and current amplitudes is substantially simplified when $b_1 \ll 1$, $\left((L/C)^{\frac{1}{2}} \ll R\right)$, because the limit cycle intersects the line Γ_1 at a point close to $(-1, A)$. Thus the coordinates of the end point of a jump (x_1^*, y_1^*), determined by (10.65), are equal to

$$x_1^* \approx \frac{A\left(\frac{1}{r_2}-\frac{1}{r_2}\right)-\frac{1}{r_2}}{G+\frac{1}{r_2}} \approx A\frac{\frac{k^2}{R_0}}{S_g+\frac{k^2}{R_0}+\frac{k'^2}{R_L}},$$

$$y_1^* \approx \frac{A\left(G+\frac{1}{r_1}\right)-G}{G+\frac{1}{r_2}} = A\frac{S_g+\frac{k'^2}{R_L}}{S_g+\frac{k^2}{R_0}+\frac{k'^2}{R_L}} < A$$

and, in practice, do not depend on either C or L. Let $x=\varphi(t)$, $y=\psi(t)$ be the equation of the phase path of "slow" motion in the region $(IIIa)$ that starts (at $t=0$ from (x_1^*, y_1^*), and is an arc of the limit cycle. That is $x=\varphi(t)$, $y=\psi(t)$ is a solution of the equations (10.66) with $\varphi(0)=x_1^*$, $\psi(0)=y_1^*$. This path meets the line Γ_2 at the point (x_2, y_2). Let us form the function

$$\Phi(t) = \frac{SR_0}{k}[1+\varphi(t)] - \psi(t)$$

$(\Phi(t) > 0$ inside $(IIIa))$: then the root t' of the equation

$$\Phi(t) = 0,$$

will, clearly, be the duration of the impulse of anode current (in the units of dimensionless time for region $(IIIa)$). The point where the limit cycle reaches the line Γ_2 is given by

$$x_2 = \varphi(t'), \quad y_2 = \psi(t')\dagger$$

The duration of the impulses in units of ordinary time will be

$$\tau = L\left(G+\frac{1}{r_2}\right)t'\ddagger$$

---

† If the equation $\Phi(t)=0$ has several roots (which can only be for $(1-a_2)^2 < 4b_2$), then we shall mean by t' the smallest positive root of this equation.

‡ If, however, the limit cycle passes into the region (III) and then reaches the line Γ_2, then after integrating the equations of motion in $(IIIa)$ and (III) and using the evident condition of continuity at the boundary we find the equation of the arc of the limit cycle

11] THE BLOCKING OSCILLATOR

From the point (x_2, y_2), the representative point makes an instantaneous jump along a segment of $x+y=$const to the point (x_2^*, y_2^*) determined by (10.67) and lying in the region (I) and then moves in the region (I) along a path of "slow" motion (10.64b) starting (say for $t=0$) from (x_2^*, y_2^*)†

$$x \approx -[A-(x_2+y_2)]e^{-a_1 t}+(A-y_2^*)e^{-t},$$
$$y \approx A-(A-y_2^*)e^{-t}.$$

The representative point moving along this arc of the limit cycle reaches the line Γ_1 at t_1 which is clearly determined by the equation $x=-1$, or, since $a_1 \ll 1$ and $e^{-t_1} \ll e^{-t_1 a_1}$, by

$$-1 \approx -[A-(x_2+y_2)]e^{-a_1 t_1}.$$

Thus the time during which the valve is cut off is

$$t_1 = \frac{1}{a_1}\ln[A-(x_2+y_2)]$$

in the units of dimensionless time that apply to the region (I), or

$$T_1 = \frac{L}{r_1}t_1 = CR \cdot \ln[A-(x_2+y_2)]$$

in units of ordinary time.

Usually the duration of an impulse τ is less than T_1, and the period of the self-oscillations is

$$T = T_1 + \tau \approx T_1$$

---

lying in the regions $(IIIa)$ and (III)
$$x = \varphi_1(t_{\text{old}}), \quad y = \psi_1(t_{\text{old}}),$$
We form the function
$$\Phi_1(t_{\text{old}}) = \frac{SR_0}{k}[1+\varphi_1(t_{\text{old}})]-\psi_1(t_{\text{old}}).$$
Then the root of the equation
$$\Phi(t_{\text{old}}) = 0$$
will be equal to the duration τ of the impulse and the point (x_2, y_2) will be
$$x_2 = \varphi_1(\tau), \quad y_2 = \psi_1(\tau).$$
† Assuming that at $t=0$, $x=x_2^*$ and $y=y_2^*$ we obtain from (10.64b)
$$B_1 + B_1' = x_2^* \quad \text{and} \quad A-B_1' = y^*$$
since $b_1 \ll 1$, or
$$B_1 = x_2^* + y_2^* - A = -A[-(x_2+y_2)] \quad \text{and} \quad B_1' = A-y_2^*.$$

§ 12. Symmetrical Multivibrator

The symmetrical multivibrator (Fig. 578) suggested by Abraham and Bloch is a well-known generator of discontinuous voltage oscillations [131, 6, 61].

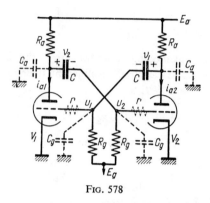

Fig. 578

1. The equations of the oscillations

By neglecting grid currents, anode reaction and all parasitic parameters including stray capacitances, and assuming the same characteristics for V_1 and V_2, then (in the notation of Fig. 578) we find

$$\frac{E_a - (u_1 + v_1)}{R_a} = i_{a2} + \frac{u_1 - E_g}{R_g},$$

$$\frac{E_a - (u_2 + v_2)}{R_a} = i_{a1} + \frac{u_2 - E_g}{R_g},$$

$$C\frac{dv_1}{dt} = \frac{u_1 - E_g}{R_g},$$

$$C\frac{dv_2}{dt} = \frac{u_2 - E_g}{R_g},$$

where the anode currents i_{a1} and i_{a2} are related to the grid voltages u_1 and u_2 by

$$i_{a1} = i_a(u_1), \quad i_{a2} = i_a(u_2).$$

The circuit has a single state of equilibrium at which $u_1 = u_2 = E_g$, $i_{a1} = i_{a2} = i_a(E_g)$ and $v_1 = v_2 = v_0 = E_a - R_a i_a(E_g) - E_g$.

To simplify the calculations, introduce dimensionless variables x_1, x_2, y_1, y_2 related to u_1, u_2, v_1, v_2 by

$$u_{1,2} = E_g + u_0 \cdot x_{1,2},$$

$$v_{1,2} = v_0 + \left(1 + \frac{R_a}{R_g}\right) u_0 \cdot y_{1,2},$$

$$t_{\text{new}} = \frac{t_{\text{old}}}{C(R_a + R_g)}$$

$$\varphi(x) = \frac{1}{u_0 S}[i_a(E_g + u_0 \cdot x) - i_a(E_g)],$$

where u_0 is a scale voltage and S is the slope of the valve characteristic at the working point for $u = E_g$. We assume the valve characteristics and the grid bias E_g to be such that the slope $\varphi'(x)$ is an even continuous function of x, which decreases monotonically to zero as $|x|$ increases (Fig. 579).

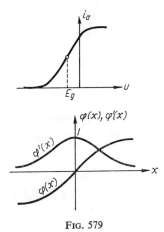

Fig. 579

The equations of the multivibrator are now

$$\left.\begin{array}{ll} -y_1 = x_1 + k\varphi(x_2), & -y_2 = x_2 + k\varphi(x_1), \\ \dot{y}_1 = x_1, & \dot{y}_2 = x_2, \end{array}\right\} \quad (10.68)$$

where

$$k = \frac{S R_a R_g}{R_a + R_g}.$$

Eliminating y_1 and y_2, we obtain two first order differential equations

$$\left.\begin{array}{l} \dot{x}_1 + k\varphi'(x_2)\dot{x}_2 + x_1 = 0, \\ \dot{x}_2 + k\varphi'(x_1)\dot{x}_1 + x_2 = 0 \end{array}\right\} \quad (10.68a)$$

from which

$$\begin{aligned}\dot{x}_1 &= \frac{x_1 - k\varphi'(x_2)x_2}{k^2\varphi'(x_1)\varphi'(x_2) - 1}, \\ \dot{x}_2 &= \frac{x_2 - k\varphi'(x_1)x_1}{k^2\varphi'(x_1)\varphi'(x_2) - 1}.\end{aligned} \right\} \quad (10.68b)$$

Thus the multivibrator is a second order system when all parasitic parameters are neglected, and can be represented by points on the x_1, x_2 plane. The equation of the integral curve is therefore

$$\frac{dx_1}{dx_2} = \frac{P(x_1, x_2)}{Q(x_1, x_2)} = \frac{x_1 - k\varphi'(x_2)x_2}{x_2 - k\varphi'(x_1)x_1},$$

and Bendixson's criterion shows that the system of equations (10.68b) has no continuous periodic solutions, since

$$\frac{\partial P}{\partial x_1} + \frac{\partial Q}{\partial x_2} = +2$$

is everywhere positive.

The characteristic equation for the single equilibrium state ($x=0, y=0$) is

$$(1-k^2)\lambda^2 + 2\lambda + 1 = 0,$$

for $\varphi'(0)=1$. For $k<1$ this state is a stable node, and whatever the initial conditions, there cannot be any self-oscillations.

We need only consider the case

$$k > 1,$$

when the single equilibrium state is an unstable saddle-point, and the multivibrator is self-excited. On the x_1, x_2 plane there is clearly a set of points at which

$$k^2\varphi'(x_1)\varphi'(x_2) - 1 = 0; \quad (10.69)$$

and which lie on a closed continuous curve Γ, symmetric with respect to the bisector and to the axes x_1 and x_2. On this curve Γ, x_1 and x_2 become infinite and a part of this curve contains junction points of phase paths of the equations (10.68b). The equation (10.68b), set out by neglecting parasitic parameters, cannot give further information about the motion, and again we have a "defective" dynamic model.

2. Jumps of the voltages u_1 and u_2

By taking into account the small stray capacitances C_a and C_g (Fig. 578) a dynamic model of the multivibrator is obtained in which discontinuous oscillations occur for $k > 1$ [61].

The equations of the oscillations of the multivibrator are now

where
$$\begin{rcases} \mu \dot{x}_1 = -y_1 - x_1 - k\varphi(x_2) \equiv F_1(x_1, x_2, y_1), \\ \mu \dot{x}_2 = -y_2 - x_2 - k\varphi(x_1) \equiv F_2(x_1, x_2, y_2), \\ \dot{y}_1 = x_1, \quad \dot{y}_2 = x_2, \\ \mu = \dfrac{\dfrac{R_a}{R_g}}{\left[1 + \dfrac{R_a}{R_g}\right]^2} \cdot \dfrac{C_a + C_g}{C} \end{rcases} \quad (10.70)$$

is a small positive parameter since $C_a, C_g \ll C$.

Let us consider the limiting case when $\mu \to +0$. In the four-dimensional phase space x_1, x_2, y_1, y_2 there is the surface F

$$-y_1 = x_1 + k\varphi(x_2), \quad -y_2 = x_2 + k\varphi(x_1)$$

which is the phase surface of the "degenerate" system (when $\mu = 0$). The points of the surface F and of the x_1, x_2 plane are in a one-to-one continuous correspondence. At every point (x_1, x_2, y_1, y_2) outside this surface F, $\dot{x}_1 \to \infty$ and $\dot{x}_2 \to \infty$, while $\dot{y}_1$ and $\dot{y}_2$ remain finite. Therefore, in the limit when $\mu \to +0$, all the phase space outside the surface F is filled with phase paths lying on the planes

$$y_1, y_2 = \text{const.}$$

The representative point jumps along these paths, and the state of the multivibrator changes rapidly, the grid voltages u_1 and u_2 varying jumpwise while the capacitor voltages v_1 and v_2 remain constant.

For $\mu \to +0$ the approximate equations of the "rapid" motions lying in or near the plane $y_1 = y_1^0$, $y_2 = y_2^0$ are obtained from the first two equations (10.70) by replacing y_1 and y_2 by the constants y_1^0 and y_2^0:

$$\begin{rcases} \mu \dot{x}_1 = -y_1^0 - x_1 - k\varphi(x_2) \equiv F_1(x_1, x_2, y_1^0), \\ \mu \dot{x}_2 = -y_2^0 - x_2 - k\varphi(x_1) \equiv F_2(x_1, x_2, y_2^0). \end{rcases} \quad (10.71)$$

Of course, these equations are only valid outside a small neighbourhood of the intersection points of the plane $y_1 = y_1^0$, $y_2 = y_2^0$ with F.

Since
$$\frac{\partial F_1}{\partial x_1}+\frac{\partial F_2}{\partial x_2} = -2,$$

then, according to Bendixson's criterion, the approximate equations (10.71) cannot have closed phase paths. Therefore the behaviour of all paths of "rapid" motions is determined by the singular points of the equations (10.71) and their separatrices. The singular points are clearly the intersection points of the plane $y_1=y_0^1$, $y_2=y_2^0$ with the surface F; the point (x_1, x_2, y_1^0, y_2^0) of the surface F being a stable node of (10.71), if

$$k^2\varphi'(x_1)\varphi'(x_2)-1 < 0, \qquad (10.72)$$

and a saddle point if

$$k^2\varphi'(x_1)\varphi'(x_2)-1 > 0 \dagger \qquad (10.72a)$$

Therefore, all phase paths of "rapid" motions, when $\mu \to +0$, lie on the planes $y_1, y_2 =$ const. and come from infinity and from points (x_1, x_2, y_1, y_2) of the surface F at which

$$k^2 \cdot \varphi'(x_1)\varphi'(x_2)-1 > 0,$$

into small neighbourhoods of that part F^+ or the surface F on which

$$k^2\varphi'(x_1)\varphi'(x_2)-1 < 0.$$

The phase paths of "slow" motions of the representative point lie only in or near the surface F^+:

$$\left.\begin{array}{c} -y_1 = x_1+k\varphi(x_2), \quad -y_2 = x_2+k\varphi(x_1), \\ k^2\varphi'(x_1)\varphi'(x_2)-1 < 0. \end{array}\right\} \qquad (10.73)$$

The equations of these paths on the surface F^+ are identical with the equations (10.68). The boundary of the region F^+ is a closed line γ on F determined by

$$k^2\varphi'(x_1)\varphi'(x_2)-1 = 0;$$

---

† The characteristic equation for the point (x_1, x_2, y_1^0, y_2^0) of the surface F, i.e. for the singular point of the approximate equations of rapid motions (10.71), has the form (see also (10.18) in Section 3 of this chapter)

$$\begin{vmatrix} \frac{\partial F_1}{\partial x_1}-\lambda & \frac{\partial F_1}{\partial x_2} \\ \frac{\partial F_2}{\partial x_1} & \frac{\partial F_2}{\partial x_2}-\lambda \end{vmatrix} = 0$$

or $\lambda^2-2\lambda+1-k^2\varphi'(x_1)\varphi'(x_2) = 0,$

and because of our assumptions about the valve characteristics, the region F^+ lies outside the curve γ. In F^+ there are no equilibrium states or closed phase paths nor do the paths recede to infinity, so the representative point moves on the surface F^+ to the boundary γ, after which it "jumps" along a path $y_1, y_2 = $ const. to another part of F^+. The coordinates of the end point of the jump $x_1 = x_1^+$ are related to those of the initial point $x_1 = x_1^-$, $x_2 = x_2^-$ by

$$\left. \begin{aligned} x_1^+ - k\varphi(x_2^+) &= x_1^- + k\varphi(x_2^-), \\ x_2^+ + k\varphi(x_1^+) &= x_2^- + k\varphi(x_1^-). \end{aligned} \right\} \quad (10.74)$$

During the jump, y_1 and y_2 (i.e. the voltages v_1 and v_2 across the capacitors C) do not vary.

The use of the differential equations of the jumps (10.71), is not necessary for the determination of the end points, but they do enable us to find the paths of rapid motions when μ is small but finite. The initial point of the jump on the curve γ is a singular point of the saddle-node type for the approximate equations (10.71) and there is only one path leaving it.

3. Discontinuous oscillations of the multivibrator.

Thus, for $k > 1$, the multivibrator has periodic discontinuous oscillations, and by making use of the homeomorphicity of the x_1, x_2 plane and the planes $y_1, y_2 = $ const. these discontinuous oscillations can be studied by considering the "phase portrait" on the x_1, x_2 plane, i.e. the plane of the grid voltages u_1, u_2. Note that the paths on the x_1, x_2 plane are the projections of phase paths in the four-dimensional x_1, x_2, y_1, y_2 phase space and can therefore intersect each other.

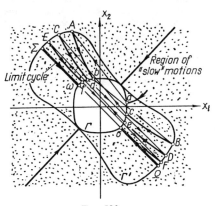

Fig. 580

This mapping of the x_1, x_2 plane is shown qualitatively in Fig. 580. Since the curve γ projects into the curve Γ on the x_1, x_2 plane (see (10.69)), then the region of "slow" motions (the projection of the surface F^+) will be that part of the x_1, x_2 plane which lies outside the closed curve Γ. In the region inside the curve Γ there can be no slow motions but only jump-wise motions of the representative point. The representative point, moving along a path of the equations (10.68) in the region of slow motions, is bound to arrive on the boundary curve Γ of this region, and then moving along a path of rapid motion (along the corresponding path of the equations (10.71)) it jumps again into the region of slow motions. The locus of the end points of the jumps (x_1^+, x_2^+) corresponding according to the initial points (x_1^-, x_2^-) on the curve Γ is shown in Fig. 580 as the curve Γ'' which also is closed and continuous, symmetrical with respect to the bisector of the x_1, x_2 plane and surrounding Γ.

If the representative point is a point a on Γ (Fig. 580), then its subsequent motion is along the path $aAbBcCd,\ldots$, consisting of segments of slow motions Ab, Bc, Cd, etc. and of segments of jumps aA, bB, cC, etc. It can be shown that the system approaches asymptotically (for $t \to +\infty$) the limit cycle $\Sigma\omega\Omega\sigma\Sigma$ consisting of two segments of paths of slow motions $\Sigma\omega$ and $\Omega\sigma$ and of two paths of jumps $\omega\Omega$ and $\sigma\Sigma$†.

For such a periodic motion the following equalities are always satisfied:
$$x_2 = -x_1, \quad y_2 = -y_1.$$
This symmetry of the oscillations is, of course, a consequence of the sym-

† The build-up of the periodic discontinuous oscillations can be demonstrated by a graphical integration [6] or by the method of the point transformation using a piecewise linear approximation to the characteristic of the valve. [58].

In this problem the limit cycle is a closed phase path in the four-dimensional x_1, x_2, y_1, y_2 phase space with a projection on the segment $\Sigma\Omega$ of the bisector $x_2 = x_1$ of the x_1, x_2 plane. Thus the representative point (x_1, x_2) moves along this segment sometimes in one direction and sometimes in the other. However, we can arrange things so that the discontinuous periodic processes are represented by a motion of the representative point along an ordinary limit cycle on a certain phase surface. We have seen that the representative point, found on the closed curve Γ (Fig. 580), jumps on the curve Γ'', after which the paths of "slow" motions are comprised in the region between these two curves. Suppose the point a to coincide with A, the point b with B, etc., so that the paths of jumps are compressed into points, then we can represent this region of slow motions on the surface of a sphere. Discontinuous oscillations will be represented now by a limit cycle (for example, the equator). In addition there are two unstable nodes on the sphere situated on either side of the cycle (for example, at the poles), corresponding to the points of contact of the curves Γ and Γ''. It is seen at once after such a representation, that in the multivibrator there cannot be either quasi-periodical oscillations (such oscillations could only exist if the phase surface were a torus), or periodic motions of the representative point along a closed path encircling the sphere twice. These results are not evident *a priori*.

metry of the circuit and of the valve characteristics. If we had assumed from the beginning that the steady-state self-oscillations were symmetrical, then in the equations (10.70) we would have put $x_2 \equiv -x_1$, $y_2 \equiv -y_1$ and $\varphi(x_2) \equiv -\varphi(x_1)$ and obtained the second order system

$$\mu \dot{x} = -x - y + k\varphi(x),$$
$$\dot{y} = x.$$

A system of this type has already been considered in Section 5 of this chapter in the study of the discontinuous oscillations of a multivibrator with one RC circuit. The results obtained in Section 5 and, in particular, the expressions for the period of the self-oscillations are therefore valid for the steadystate oscillations of a symmetrical multivibrator. Such an approach, however, would not have allowed a discussion of the build-up of these oscillations.

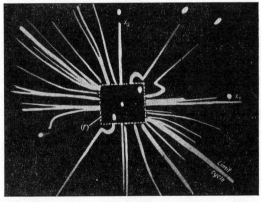

Fig. 581

Figure 581 is a photograph of the paths in the plane of the grid voltages u_1, u_2 (the x_1, x_2 plane) obtained by means of a cathode-ray oscilloscope[†]. This photograph fully confirms the results arrived at above on the discontinuous character of the oscillations of the multivibrator and on the onset in it of periodic discontinuous oscillations (self-oscillations). It is clearly seen that the jumps of the voltages u_1 and u_2 start not only from the points of the curve Γ but also from points of the region lying inside it. The representative points, brought at the initial instant of time into the region inside Γ, will move away from it with a jump.

† Grid currents in the valves have been limited by the resistances r in Fig. 578, shown with a dotted line.

§ 13. SYMMETRICAL MULTIVIBRATOR (WITH GRID CURRENTS)

To conclude the chapter we finally consider in greater detail the discontinuous self-oscillations of a symmetrical multivibrator (Fig. 582) neglecting anode reaction, as we may if we assume the valves are pentodes, or triodes with a large internal resistance R_i, and sufficiently large anode voltages. However, we do not omit the grid currents which usually have an important

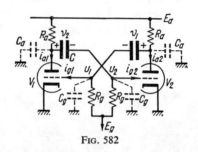

FIG. 582

role in the operation of a multivibrator [62]. To obtain quantitative results, we approximate to the valve characteristics by a piece-wise linear function, without saturation.

$$i_a = i_a(u) = \begin{cases} 0 & \text{for } u \leq -u_0; \\ S(u+u_0) & \text{for } u > -u_0; \end{cases}$$

$$i_g = i_g(u) = \begin{cases} 0 & \text{for } u \leq 0, \\ S_g u & \text{for } u > 0; \end{cases}$$

S and S_g are the slope s of the ascending sections of the anode and grid current characteristics respectively.

1. Equations of the oscillations

The jumps of the voltages u_1 and u_2. Neglecting the stray capacitances, we find from Kirchhoff's laws, and in the notation of Fig. 582, the following equations for the slow variations of the state of the multivibrator, where t' is real time.

$$C\frac{dv_1}{dt'} = \frac{u_1 - E_g}{R_g} + i_g(u_1),$$

$$\frac{E_a - (u_1 + v_1)}{R_a} = i_a(u_2) + C\frac{dv_1}{dt'},$$

SYMMETRICAL MULTIVIBRATOR (WITH GRID CURRENTS)

$$C\frac{dv_2}{dt'} = \frac{u_2-E_g}{R_g}+i_g(u_2),$$

$$\frac{E_a-(u_2+v_2)}{R_a} = i_a(u_1)+C\frac{dv_2}{dt'}.$$

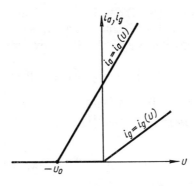

Fig. 583

Now introduce new dimensionless variables x_1, x_2, y_1, y_2, related to u_1, u_2, v_1, v_2 by

$$u_{1,2} = u_0 x_{1,2}, \quad v_{1,2} = E_a + \frac{R_a}{R_g}E_g + u_0\left(1+\frac{R_a}{R_g}\right)y_{1,2},$$

and a new time parameter

$$t = \frac{t'}{C(R_a+R_g)}.$$

The equations of slow variations are now

$$\left.\begin{array}{l} -y_1 = x_1+\alpha\cdot\psi(x_1)+k\varphi(x_2), \\ -y_2 = x_2+\alpha\cdot\psi(x_2)+k\varphi(x_1), \\ \dot{y}_1 = x_1+\beta\cdot\psi(x_1)-\sigma, \\ \dot{y}_2 = x_2+\beta\cdot\psi(x_2)-\sigma, \end{array}\right\} \quad (10.75)$$

where

$$k = \frac{SR_aR_g}{R_a+R_g},$$

$$\alpha = \frac{S_gR_aR_g}{R_a+R_g},$$

$$\beta = S_gR_g,$$

$$\sigma = \frac{E_g}{u_0}$$

$$\varphi(x) = \frac{1}{u_0 S} i_a(u_0 x) = \begin{cases} 0 & \text{for } x < -1, \\ x+1 & \text{for } x \geqslant -1, \end{cases}$$

$$\psi(x) = \frac{1}{u_0 S_g} i_g(u_0 x) = \begin{cases} 0 & \text{for } x < 0, \\ x & \text{for } x \geqslant 0 \end{cases}$$

The first two equations (10.75) establish a functional relationship between x_1, x_2 and y_1, y_2 (i.e. between the grid voltages u_1, u_2 and the voltages v_1, v_2 across the capacitors C) which is true only during slow variations of state.

Eliminating from (10.75) y_1 and y_2 we obtain the following system:

$$\left. \begin{array}{l} [1+\alpha\psi'(x_1)]\dot{x}_1 + k\varphi'(x_2)\dot{x}_2 + x_1 + \beta\psi(x_1) = \sigma, \\ k\varphi'(x_1)\dot{x}_1 + [1+\alpha\psi'(x_2)]\dot{x}_2 + x_2 + \beta\psi(x_2) = \sigma, \end{array} \right\} \quad (10.75\text{a})$$

or

$$\dot{x}_1 = \frac{P_1(x_1, x_2)}{\Delta(x_1, x_2)}, \quad \dot{x}_2 = \frac{P_2(x_1, x_2)}{\Delta(x_1, x_2)} \quad (10.75\text{b})$$

where

$$P_1(x_1, x_2) = \varphi k'(x_2)[x_2 + \beta\psi(x_2) - \sigma] - [1+\alpha\psi'(x_2)][x_1 + \beta\psi(x_1) - \sigma],$$
$$P_2(x_1, x_2) = k\varphi'(x_1)[x_1 + \beta\psi(x_1) - \sigma] - [1+\alpha\psi'(x_1)][x_2 + \beta\psi(x_2) - \sigma],$$
$$\Delta(x_1, x_2) = [1+\alpha\psi'(x_1)][1+\alpha\psi'(x_2)] - k^2\varphi'(x_1)\varphi'(x_2).$$

Thus we can represent the states of a multivibrator during their slow variations by points on the x_1, x_2 plane, and slow processes by phase paths of (10.76b) on this plane. However, *the region of slow motions* of the system where the equations (10.75) are applicable is only that part of the x_1, x_2 plane—*the region M*—in which

$$(x_1, x_2) = [1+\alpha\psi'(x_1)][1+\alpha\psi'(x_2)] - k^2\varphi'(x_1)\cdot\varphi'(x_2) > 0.$$

Outside M there are only possible rapid jump-wise variations of the state of the multivibrator, i.e. instantaneous jumps of the grid voltages u_1, u_2 (or of the x_1, x_2 variables).

A physical argument can serve to justify this jump postulate. If $x_1 > -1$ and $x_2 > -1$ so that both valves are conducting, and if the voltage amplification k is sufficiently large to make $\Delta(x_1, x_2) < 0$, then it is easily shown by calculation that the total loop gain or amplification, from one valve grid, through both valves and back to the grid, is larger than unity. There is thus positive feedback and any deviation of a grid voltage (for example, u_1) results in its immediate augmentation and rapid increase as long as both valves remain conducting.

On the contrary, for $x_1 < -1$ or for $x_2 < -1$, when at least one of the valves is cut off while $\Delta(x_1, x_2) > 0$, there is no positive feedback or loop amplification and a variation of a grid voltage will not be increased which makes "slow" variations of state quite possible.

The boundary of M is the line Γ where $\varDelta(x_1, x_2)$—the denominator of the right-hand sides of the equations (10.75b)—changes its sign: therefore a part of the line Γ contains junction points of phase paths of the equations (10.75b). If the representative point, moving slowly along a path of the equations (10.75b) in M reaches the line Γ at a certain point (x_1^-, x_2^-), it will make an instantaneous jump to the point (x_1^+, y_1^+) that is also in M. Since the capacitor voltages v_1 and v_2 and hence the values of y_1 and y_2 cannot vary during an instantaneous jump and since the first two equations (10.75) are valid in M, the initial and end-points of a jump are related by the *jump conditions*

$$\left.\begin{array}{l} x_1^+ + \alpha\psi(x_1^+) + k\varphi(x_2^+) = x_1^- + \alpha\psi(x_1^-) + k\varphi(x_2^-), \\ x_2^+ + \alpha\psi(x_2^+) + k\varphi(x_1^+) = x_2^- + \alpha\psi(x_2^-) + k\varphi(x_1^-). \end{array}\right\} \quad (10.76)$$

To justify that the equations (10.75) are applicable in a region M, and jumps do occur it is necessary to take into account at least some of the small parameters that are important during rapid motions of the system. Such parameters are the small stray capacitances C_a and C_g (Fig. 582). With these capacitances and with $C_a, C_g \ll C$ the equations of the multivibrator are

$$\left.\begin{array}{l} \mu\dot{x}_1 = -y_1 - x_1 - \alpha\psi(x_1) - k\varphi(x_2) \equiv F_1(x_1, x_2, y_1), \\ \mu\dot{x}_2 = -y_2 - x_2 - \alpha\psi(x_2) - k\varphi(x_1) \equiv F_2(x_1, x_2, y_2), \\ \dot{y}_1 = x_1 + \beta\psi(x_1) - \sigma \equiv G(x_1), \\ \dot{y}_2 = x_2 + \beta\psi(x_2) - \sigma \equiv G(x_2), \end{array}\right\} \quad (10.77)$$

where $\mu = [R_a R_g/(R_a + R_g)^2][(C_a + C_g)/C]$ is a small parasitic parameter. Now the phase space will be the four-dimensional x_1, x_2, y_1, y_2 space.

Since

$$\frac{\partial F_1}{\partial x_1} + \frac{\partial F_2}{\partial x_2} = -2 - \alpha[\psi'(x_1) + \psi'(x_2)] < 0$$

and

$$\frac{\partial(F_1, F_2)}{\partial(x_1, x_2)} = [1 + \alpha\psi'(x_1)][1 + \alpha\psi'(x_2)] - k^2\varphi'(x_1)\varphi'(x_2) \equiv \varDelta(x_1, x_2)$$

then, according to Section 3 of this chapter, slow motions with bounded values of $\dot{x}_1$ and $\dot{x}_2$ even when $C_a/C, C_g/C \to +0$ will only occur in a small $0(\mu)$-neighbourhood of the surface F^+, determined by

$$F_1(x_1, x_2, y_1) = 0, \quad F_2(x_1, x_2, y_2) = 0, \quad \varDelta(x_1, x_2) > 0,$$

or on the surface F^+ itself in the limit for $\mu \to +0$. Thus, the equations (10.75) are the proper approximate equations of slow motions. The surface F^+ has the region M, as a homeomorphic projection on the x_1, x_2 plane.

Outside the surface F^+ the limiting rapid motions are along paths $y_1, y_2 = \text{const.}$ and so there are jumps of the grid voltages u_1 and u_2 while the voltages v_1 and v_2 across the capacitors C remain constant. Approximate differential equations of the "rapid" motions along paths lying in or near the plane $y_1 \equiv y_1^0$, $y_2 \equiv y_2^0$ follow from the first two equations (10.77):

$$\left.\begin{array}{l} \mu\dot{x}_1 = -y_1^0 - x_1 - \alpha\psi(x_1) - k\varphi(x_2) \equiv F_1(x_1, x_2, y_1^0), \\ \mu\dot{x}_2 = -y_2^0 - x_2 - \alpha\psi(x_2) - k\varphi(x_1) \equiv F_2(x_1, x_2, y_2^0). \end{array}\right\} \quad (10.77a)$$

Let us note the following: (1) since $(\partial F_1/\partial x_1)+(\partial F_2/\partial x_2)<0$ for all values of x_1, x_2, y_1^0, y_2^0, Bendixson's criterion states that the equations (10.77a) have no closed phase paths: (2) the points of intersection of the plane $y_1=y_1^0$, $y_2=y_2^0$ with the surface F^+ are stable equilibrium states for the approximate equations (10.77a), and (3) there are no paths of rapid motions that recede into infinity. Therefore all paths of rapid motions lead (for $\mu \to +0$) into a small $0(\mu)$-neighbourhood of the surface F^+ where they become paths of slow motions, which in their turn, on reaching the boundary γ of F^+ (the projection of γ on the x_1, x_2 plane is the curve Γ) change into paths of jumps. The end points of such jumps lie again on the surface F^+ and satisfy the jump conditions (10.76)

Thus discontinuous oscillations are possible in the multivibrator that consist of alternate slow motions with finite rates of change of the grid voltages u_1 and u_2 and rapid ones with very large rates of change when μ is small.

2. Discontinuous oscillations

The most interesting case for practical applications is where

$$k > 1+\alpha, \quad \beta \gg 1+\alpha, \sigma. \tag{10.78}$$

The first condition (10.78) ensures that the multivibrator self-excites for $\sigma>-1$ (i.e. $E_g>-u_0$), whilst the second one simplifies the analysis of the slow motions in the presence of grid current[†]. Owing to the first condition (10.78), $\Delta(x_1, y_2)>0$ only for $\varphi'(x_1)\varphi'(x_2) = 0$, i.e. for either x_1 or x_2 less than -1. Thus the boundary Γ of the region M on the x_1, x_2 plane is the half lines $\Gamma_1: x_1 = -1, x_2>-1$ and $\Gamma_2: x_2 = -1, x_1>-1$. The region M lies on the left of and below these half lines and in it at least one of the valves is cut off. On the right of and above the half straight lines Γ_1 and Γ_2 both valves conduct, $\Delta(x_1, x_2)<0$ and only "rapid" jump-wise variations of state are possible (Fig. 584).

For sufficiently large values of u_1 or u_2, anode reaction increases and the anode current i_a becomes dependent on the anode voltage u_a and nearly independent of the grid voltage (see, for example, Section 10 of this chapter). The boundary of the region of anode reaction, assuming that in it $i_a = u_a/R_0$, is represented qualitatively in Fig. 584 by the shaded line Γ_a.

[†] The conditions (10.78) are equivalent to $SR_a > 1+S_gR_a$, $R_a \ll R_g$, $S_gR_g \gg 1$, and $S_gR_g \gg E_g/u_0$, which are usually satisfied in practical circuits where $R_a \sim 10^3$–$10^4\,\Omega$, $R_g \sim 10^4$–$10^6\,\Omega$, $S \sim 3$–10 mA/V, $S_g \sim 0.1$–1 mA/V, $E_g \sim 0$–300 V and $u_0 \sim$ 5–20 V, i.e. $k \sim 5$–20, $\alpha \sim 0.1$–1, $\beta \sim 10$–10^3 and $\sigma \sim 0$–60.

Note that for $k<1$ $\Delta(x_1, x_2)>0$ over the whole x_1, x_2 plane, and all paths go towards a stable equilibrium state.

The case $1<k<1+\alpha$, although discontinuous self-oscillations are possible, is of no practical interest.

The points (x_1, x_2) that lie to the right and above Γ_a are in the region of anode reaction and belong, as is easily seen, to the region of "slow" motions. We now assume that during oscillations the representative point (x_1, x_2) is not found in the region of anode reaction.

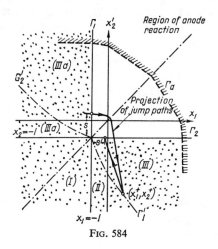

Fig. 584

The single state of equilibrium lies on $x_1 = x_2$ and is determined by

$$x + \beta \psi(x) = \sigma,$$

Hence the equilibrium state is at

$$x_1 = x_2 = x_0(\sigma) = \begin{cases} \sigma & \text{for } \sigma \leq 0, \\ \dfrac{\sigma}{1+\beta} \approx 0 & \text{for } \sigma > 0. \end{cases}$$

The characteristic equation at the state is

$$\begin{vmatrix} \mu\lambda + a^* & k^* & 1 & 0 \\ k^* & \mu\lambda + a^* & 0 & 1 \\ -\beta^* & 0 & \lambda & 0 \\ 0 & -\beta^* & 0 & \lambda \end{vmatrix} \equiv [\lambda(\mu\lambda + a^*) + \beta^*]^2 - k^{*2}\lambda^2 = 0,$$

where $\alpha^* = 1 + a\psi'(x_0)$, $\beta^* = 1 + \beta\psi'(x_0)$ and $k^* = k\varphi'(x_0)$. For $0 < \mu \ll 1$, the roots are

$$\lambda_{1,2} = -\frac{\beta^*}{\alpha^* \pm k^*} \quad \text{and} \quad \lambda_{3,4} = -\frac{1/\mu}{\alpha^* \pm k^*}$$

λ_1 and λ_2 are of the order of unity and the others of the order of $1/\mu$.

For $\sigma<-1$ (i.e. for $E_g<-u_0$), when $x_0=\sigma$ and $k^*=0$, all four roots are negative and the equilibrium state is a stable node for the paths of "slow" motions on the x_1, x_2 plane. If, however, $\sigma>-1$ ($E_g>-u_0$), the equilibrium state is in the region of "rapid" motions ($x_0>-1$) and is unstable since, for $x_0>-1$, $k^* = k>1+\alpha$ (from (10.78)). The characteristic equation has two positive roots of which one is of the order of $1/\mu$, and hence the representative point will move away from the equilibrium state with a velocity of the order of $1/\mu$ (with a jump for $\mu\to +0$). In the case $\sigma>-1$ the multivibrator is self-excited and, as will be seen, periodic discontinuous oscillations are set up. Stable discontinuous self-oscillations are also possible for $\sigma<-1$, when the equilibrium state is stable: but now there is a hard mode of self-excitation and only some initial states lead to self-oscillation.

Owing to the piece-wise-linear approximation used for the valve characteristics, the region M of "slow" motions is divided by the straight lines $x_1=-1, x_1=0, x_2=-1$ and $x_2=0$ into five regions: (*I*), (*II*), (*IIa*), (*III*) and (*IIIa*) (Fig. 584) in each of which the equations of "slow" motions are linear. In the region (*I*): $x_1<-1, x_2<-1$, and both valves are cut off. The equations are written, clearly, as

$$\left.\begin{array}{l}\dot{x}_1+x_1 = \sigma, \\ \dot{x}_2+x_2 = \sigma.\end{array}\right\} \qquad (10.79a)$$

In the region (*II*): $-1<x_1<0, x_2<-1$, and valve V_2 is cut off, while valve V_1 conducts but grid currents are absent. The equations are

$$\left.\begin{array}{l}\dot{x}_1+x_1 = \sigma, \\ \dot{x}_2+x_2 = \sigma-k\dot{x}_1.\end{array}\right\} \qquad (10.79b)$$

Finally, in the region (*III*): $x_1>0, x_2<-1$, the valve V_2 is cut off, but both anode and grid currents occur in the valve V_1. The equations are

$$\left.\begin{array}{l}(1+\alpha)\dot{x}_1+(1+\beta)x_1 = \sigma, \\ \dot{x}_2+x_2 = \sigma-k\dot{x}_1.\end{array}\right\} \qquad (10.79c)$$

Obviously the phase portrait in M of the paths of the equations (10.75) will be *symmetrical* about the line $x_1=x_2$, which is always an integral curve. In particular the equations of "slow" variations of state in the regions (*IIa*) and (*IIIa*), where valve V_1 is cut off while valve V_2 conducts, are obtained from the equations (10.79b) and (10.79c) by interchanging x_1 and x_2.

In the region (*I*) the integral curves are the straight lines $(x_2-\sigma)/(x_1-\sigma)$ = const, which pass through the point (σ, σ). This region contains the equilibrium state (σ, σ), when $\sigma < -1$.

In the region (*II*) the isocline $\varkappa = 0$ is the straight line $x_2-\sigma-k(x_1-\sigma) = 0$, passing through the point (σ, σ) and intersecting the half line Γ_2 (on the segment $-1 \leqslant x_1 \leqslant 0$) for $\sigma \leqslant 1/(k-1)$ only. On the left of this isocline $\dot{x}_2 < 0$ and paths move away from Γ_2, on the right $\dot{x}_2 > 0$ and the paths approach Γ_2. In particular for $\sigma \geqslant 1/(k-1)$ all paths in the region (*II*) pass into (*III*) without meeting Γ_2. In addition, in the region (*II*) there is a rectilinear phase path $x_1 \equiv \sigma$ on which $\dot{x}_2 > 0$.

In the region (*III*)

$$\frac{dx_2}{dx_1} = -k + \frac{1+\alpha}{1+\beta} \frac{x_2-\sigma}{x_1-\dfrac{\sigma}{1+\beta}}$$

and is approximately equal to $-k$ outside a small neighbourhood of the axis $x_1 = 0$ as follows from (10.78). Therefore all paths in (*III*) outside this neighbourhood are close to the straight lines

$$x_2 + kx_1 = \text{const},$$

and the velocities $\dot{x}_1$ and $\dot{x}_2$ on these paths are of the order of magnitude of $(1+\beta)/(1+\alpha) \gg 1$. We shall refer to these comparatively rapid motions of the representative point as "semi-rapid"[†]. In addition, for $\sigma \geqslant 0$, there is a rectilinear phase path of "slow" motion in (*III*)

$$x_1 \equiv \frac{\sigma}{1+\beta} = x_0 \approx 0$$

(on it x_2 is positive and of the order of unity).

The phase portrait in the region *M* of "slow" motions is shown in Fig. 585 for the three possible cases: $\sigma < -1$, $-1 < \sigma < 0$ and $\sigma \geqslant 0$. For $\sigma > -1$, it can be seen, or proved from Bendixson's Criterion, that there are no closed phase paths in the region of "slow" motions, and self-oscillations

---

[†] In order that "semi-rapid" motions of the representative point in the region (*III*) may belong to the class of "slow" motions, it is necessary, clearly, to assume that

$$\frac{1+\beta}{1+\alpha} \ll \frac{1}{\mu}.$$

"Semi-rapid" motions correspond to a comparatively rapid charging of the capacitor *C* (in a time of the order of $C/S_g \ll CR_g$) by the grid currents flowing in valve V_1. These are much larger than the currents through the resistor R_g.

766 DISCONTINUOUS OSCILLATIONS [X

in the multivibrator, if such exist are bound to be discontinuous, and consist of alternate slow and "jump-wise" variations of the state of the multivibrator.

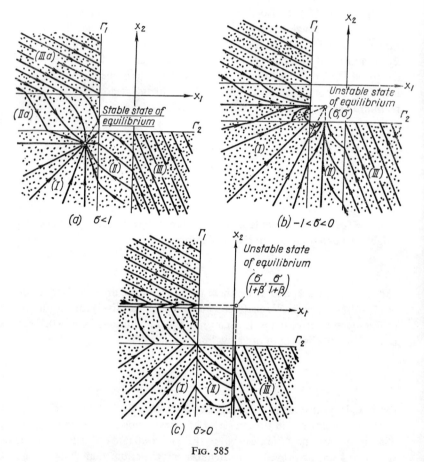

Fig. 585

The corresponding paths on the x_1, x_2 plane will intersect the half lines Γ_1 and Γ_2. Therefore the analysis of the oscillations reduces to the construction of the point transformations of the half straight lines Γ_1 and Γ_2 into themselves or into each other, as generated by the paths of the system, and to the investigation of these transformations.

Introduce on the half lines Γ_1 and Γ_2 the distance $s = 1 + x_2$ of a point of the half line Γ_1 from the point $(-1, -1)$: and the distance $s = 1 + x_1$

on the half line Γ_2, from $(-1, 1)$. Suppose the representative point leaves a point with coordinate s on the line Γ_1, i.e. the point $(-1, s-1)$. The representative point jumps from here along a path of rapid motion to a point (x_1^+, x_2^+) determined by the jump conditions (10.76). If we assume that $x_1^+ \le -1$, from (10.76) we find that $x_1^+ = -1$ and $x_2^+ = s-1$, so the end point of the jump coincides with the initial one, which is impossible since an end point of a jump can only lie inside the region of slow motions and not on its boundary. Therefore the end point of the jump can only lie in the regions (II) and (III) (i.e. $x_1^+ > -1$, $x_2^+ < -1$) and so it is determined by the following equations, obtained from (10.76):

$$x_1^+ + \alpha\psi(x_1^+) = ks-1, \quad x_2^+ + kx_1^+ = s-1+\alpha\psi(s-1)-k; \quad (10.80)$$

In particular, for $s < 1/k$ the point (x_2^+, y_2^+) lies in the region (II)

$$x_1^+ = ks-1 < 0, \quad x_2^+ = -(k^2-1)s-1,$$

and for $s > 1/k$ the point (x_1^+, x_2^+) lies in the region (III)[†].

From the point (x_1^+, x_2^+), the representative point moves "slowly" along a path and either meets the line Γ_2 at $s = s'$ or else approaches asymptotically the equilibrium state (σ, σ) lying in the region (I). In the first case the point s of the half straight line Γ_1 has a consecutive point s' on the half straight line Γ_2 and the correspondence function

$$s' = \Pi(s)$$

is single-valued and continuous. In the second case (for which $\sigma < -1$) the point s on Γ_1 has no consecutive point on either Γ_2 or Γ_1.

Owing to the symmetry of the circuit, the transformation of the points $(s-1, -1)$ on Γ_2 into points $(-1, s'-1)$ on Γ_1 will be just the same as the transformation above. Therefore, we need consider one point transformation Π of the half lines Γ_1 and Γ_2 into one another. By using this transformation repeatedly we can obtain a sequence of intersection points of the path with the lines Γ_1 and Γ_2:

where
$$s, s_1, s_2, \ldots, s_k, s_{k+1}, \ldots,$$

$$s_1 = \Pi(s),$$
$$s_2 = \Pi(s_1), \ldots,$$
$$s_{k+1} = \Pi(s_k), \ldots$$

irrespective of whether the preceding point of intersection lies on one or the other of the half lines Γ_1 or Γ_2.

[†] The locus of the end points (x_1^+, x_2^+) of the jumps of the representative point from points of line Γ_1 is shown in Fig. 584 by the dotted line Γ_1'. The broken line Γ_2', symmetrical with the line Γ_1', is the locus of the end points of jumps from the line Γ_2.

3. The point transformation Π

Let us proceed to evaluate and investigate the correspondence function $s' = \Pi(s)$ (shown graphically in Fig. 591) by considering in detail the behaviour of the paths of the system that begin at points of the line Γ_1. Let L be the positive half-path of "slow" motion that starts at the point (x_1^+, x_2^+) on Γ_1', reaching the half line Γ_2 at the coordinate s', provided it does reach Γ_2. Firstly we make the following statements:

(1) Since x_1^+ and x_2^+ are continuous piece-wise-linear functions of s, while the right-hand sides of the differential equations of "slow" motions (10.75b) are piece-wise-linear functions of x_1 and x_2, then *the correspondence function $s' = \Pi(s)$ will be a continuous piece-wise-differentiable function* whose derivative ds'/ds has a discontinuity at the points $s = 1/k$ and $s = 1$.

(2) Let s_1 and s_2 be the coordinates of two initial points of the transformation Π, s_2 being larger than s_1. Then the consecutive point s_2' will be larger than the consecutive points s_1'. Therefore the correspondence function $s' = \Pi(s)$ *is a monotonically increasing function* and

$$\frac{ds'}{ds} > 0$$

(at the points $s = 1/k$ and $s=1$ this inequality is satisfied by the left-hand and right-hand limits of ds'/ds).

To prove this we note that according to (10.80) we have, for the locus Γ_1' of the end points of the jumps,

$$\frac{dx_1^+}{ds} = \frac{k}{1+\alpha\psi'(x_1^+)} > 0,$$

and

$$\frac{dx_2^+}{dx_1^+} = -k + \frac{1}{k}[1+\alpha\psi'(x_1^+)][1+\alpha\psi'(s-1)] \geq -\left(k-\frac{1}{k}\right) > -k$$

(the equality sign applies when $s<1/k$ and the points (x_1^+, x_2^+) lie in the region (II). In addition the half-paths L reach the region situated *above* the line Γ_1' either in (III) or, when $s > -(1+\sigma)(k-1)/k^3$ is satisfied, in (II). The half-paths L in the region (II) corresponding to $s < +(1+\sigma)(1-k)/k^3$ (such half-paths exist for $\sigma < -1$ only) do not intersect Γ_1' again but remain below it and so do not reach the line Γ_2.†

---

† In (III) (outside a small neighbourhood of the x_2 axis) the representative point moves along paths close to the straight lines $x_2 + kx_1 = $ const. so that x_1 decreases; therefore the half-paths L in the region (III) reach the region above the line Γ_1' since,

To prove the statement (2) let $(x_1^+)_1$, $(x_2^+)_2$, $(x_2^+)_2$ be the end points of jumps starting from the points s_1 and s_2 on Γ_1, and let L_1 and L_2 be positive half-paths of "slow" imotions starting respectively from the points $((x_1^+)_1, (x_2^+)_1)$ and $((x_1^+)_2, (x_2^+)_2)$. The half-path L_1 reaches Γ_2 and consequently recedes from the point $((x_1^+), (x_2^+)_1)$ into the region situated above the line Γ_1'. Since $s_2 > s_1$, then, from what has just been said, $(x_1^+)_2 > (x_1^+)_1$ and the half-path L_2 will also reach the region lying above the line Γ_1', and will be on the right of L_1, since the half-paths L_1 and L_2 cannot intersect each other. Therefore the half-path L_2 will also reach Γ_2 at a point $s_2' > s_1'$.

We can now state that since for $\sigma \geqslant -1$ all paths of slow motions in the regions (II) and (III) reach the line Γ_2, then all points with $s > 0$ have consecutive points s', and, in virtue of the statement (2), $s' > s_0' = \Pi(0)$. If $\sigma < -1$, then a consecutive point will exist only for points $s > s_0$. s_0 is the coordinate of that point $s' = 0$ on Γ_1 which is transformed into the point $s' = 0$ on Γ_2 (for $s < s_0$ the corresponding half-paths L do not reach

on that line,

$$-k < \frac{dx_2^+}{dx_1^+} < 0.$$

To investigate the paths of "slow" motions in (II) consider their intersections with the family of parallel straight lines

$$x_2 + \left(k - \frac{1}{k}\right) x_1 = a = \text{const}, \tag{A}$$

one of which ($a = -(k^2+k-1)/k$) is the line Γ_1' within the region (II). Since, by (10.79b)

$$\frac{d}{dt}\left[x_2 + \left(k - \frac{1}{k}\right) x_1\right] = \sigma - x_2 - \frac{1}{k}(\sigma - x_1),$$

then the paths of "slow" motions in (II) intersect the straight lines (A) from below when below the line

$$x_2 - \sigma = \frac{1}{k}(x_1 - \sigma), \tag{B}$$

and from above when above this line. For $\sigma \geqslant -1$ the straight line (B) lies above the region (II); therefore all paths in (II) (in particular, all half-paths L) intersect the lines (A) from below approaching the half straight line Γ_2. At the points of the line Γ_1', according to (10.80a)

$$\sigma - x_2 - \frac{1}{k}(\sigma - x_1) = \sigma - x_2^+ - \frac{1}{k}(\sigma - x_1^+) = k^2 s + \frac{(1+\sigma)(k-1)}{k}$$

and so for $\sigma < -1$, the region above the line Γ_1' is only reached by those half-paths L in the region (II) that correspond to $s > -(1+\sigma)(k-1)/k^3$; since for the points (x_1^+, x_2^+) in (II) $0 < s < 1/k$, then such half-paths exist for $-(k^2+k-1)/(k-1) < \sigma < -1$ only. To prove that the half-paths L that correspond to $s < -(1+\sigma)(k-1)/k^3$ reach the region below the line Γ_1' and, not intersecting the line Γ_1', do not reach the line Γ_2, we assume that, for $\sigma < -1$, a certain half-path L, leaving the point (x_1^{-+}, x_2^{-+}) and corresponding to $s = \bar{s} < -(1+\sigma)(k-1)/k^3$, intersects the line Γ_1' at least once more. At this point of intersectioin (x_1^+, x_2^+) the half-path L would intersect the line Γ_1' from below which is imposs ble since, $x_1 < 0$ in the region (II) for $\sigma < -1$, and $x_1^{-+} < x_1^{-+}$, i.e. the point $(x^{\prime+}, x_2^{-+})$ would correspond to a value $\bar{\bar{s}} < \bar{s} < -(1+\sigma)(k-1)/k^3$ whereas in fact x^+ is a monotonically increasing function of s.

the half line Γ_2 but approach the singular point (σ, σ) asymptotically (see Fig. 586)).

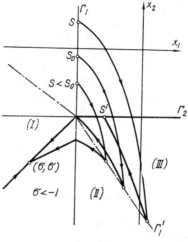

Fig. 586

(3) The correspondence function depends on the parameters of the system and, in particular, on the reduced grid bias σ. Since in (II)

$$\frac{dx_2}{dx_1} = -k + \frac{\sigma - x_2}{\sigma - x_1}$$

and in (III)

$$\frac{dx_2}{dx_1} = -k + \frac{(1+\alpha)(\sigma - x_2)}{\sigma - (1+\beta)x_1},$$

then, differentiating with respect to σ, we obtain in (II)

$$\frac{\partial}{\partial \sigma}\left[\frac{dx_2}{dx_1}\right] = -\frac{x_1 - x_2}{(\sigma - x_1)^2} < 0,$$

and in (III)

$$\frac{\partial}{\partial \sigma}\left[\frac{dx_2}{dx_1}\right] = -(1+\alpha)\frac{(1+\beta)x_1 - x_2}{[\sigma - (1+\beta)x_1]^2} < 0.$$

In both regions, as σ increases, there is a clockwise rotation of the vector field of phase velocities of the paths of slow motions. In (III) $\partial(dx_2/dx_1)/\partial\sigma \approx 0$, as follows from (10.78), and so we will neglect the rotation of the tangents to the paths in this region. Moreover, the coordinates x_1^+ and x_2^+ of the end points of the jumps do not depend on σ, and so the points where the positive half-paths L, corresponding to a fixed value of s, meet the line Γ_2 are shifted to the right as τ increases. Therefore

$$\frac{\partial s'}{\partial \sigma} > 0; \tag{10.81}$$

and, evidently, for all values of s that correspond to half-paths L lying entirely in the region (III), $\partial s'/\partial \sigma \approx 0$. Thus *as σ decreases, the graph of the correspondence function $s' = \Pi(s)$ on Lamerey's diagram (the s, s' plane) either does not vary or else is displaced downwards, if only partially.*

To evaluate the correspondence function $s' = \Pi(s)$ since the jump equations (10.80) and also the differential equations of "slow" motions are piece-wise linear, we divide the interval of variation of s: $0 < s < +\infty$ for $\sigma \geqslant -1$, and $s_0 < s < +\infty$ for $\sigma < -1$ into sections in each of which the equations are linear.

(a) *The point transformation Π for $s \geqslant 1 + (k-1)/(1+\alpha)$.* For sufficiently large values of s the representative point, after jumping to the point (x_1^+, x_2^+), moves with a "semi-rapid" motion in region (III)

$$x_2 + kx_1 = a = \text{const} \geqslant -1$$

and finally reaches the line Γ_2 where $x_1' = (1+a)/k \geqslant 0$. Since

$$a = x_2^+ + kx_1^+ = (1+\alpha)(s-1) - k$$

according to (10.80), this case occurs for

$$s \geqslant 1 + \frac{k-1}{1+\alpha} = \frac{k+\alpha}{1+\alpha}$$

only, and the consecutive point is at $s' = 1 + x_1' = 1 + (1+a)/k$ or

$$s' = \frac{1+\alpha}{k}s - \frac{\alpha}{k} \geqslant 1. \tag{10.82}$$

Note that the point $s = 1 + (k-1)/(1+\alpha)$ has the consecutive point $s' = 1$ and that for $s > 1 + (k-1)/(1+\alpha)$ $ds'/ds = (1+\alpha)/k < 1$. Therefore $s' < s$ and *the point transformation Π cannot have a fixed point with a co-ordinate $s^* \geqslant 1 + (k-1)/(1+\alpha)$.*

(b) *The point transformation Π for $1/k \leqslant s < 1 + (k-1)/(1+\alpha)$ and $\sigma \geqslant -1$.* For $1/k < s < 1 + (k-1)/(1+\alpha)$, the end point of a jump (x_1^+, x_2^+) lies in (*III*), after which the representative point moves along a path of semi-rapid motion

$$x_2 + kx_1 = a = x_2^+ + kx_1' < -1$$

and either reaches the vicinity of a phase path of slow motion

$$x_1 \equiv x_0 = \frac{\sigma}{1+\beta} \approx 0$$

for $\sigma \geqslant 0$, meeting Γ_2 where $x_1' \approx x_0 \approx 0$, or else for $\sigma < 0$, reaches the boundary of (*III*) at the point (0, *a*) and continues into (*II*).

Thus, for $\sigma \geqslant 0$, the consecutive point is at

$$s' = 1 \dagger \tag{10.83}$$

and $s^* = 1$ will be a stable fixed point of the transformation Π.

Now consider the case $-1 \leqslant \sigma < 0$. By integrating equations (10.79b), it is easy to obtain the equations of the path that lies in (*II*) and starts at $t = 0$ from the point (0, *a*)

$$\left.\begin{array}{l} x_1 = \sigma(1-e^{-t}), \\ x_2 = \sigma - (\sigma - a + k\sigma t)e^{-t}. \end{array}\right\} \tag{10.84}$$

The representative point, moving along this path, will reach Γ_2 at $t = \tau > 0$, where

$$s' - 1 = \sigma(1 - e^{-\tau}),$$
$$1 = \sigma - (\sigma - a + k\sigma\tau)e^{-\tau}.$$

Since

$$a = \begin{cases} (1+\alpha)(s-1) - k & \text{for } s \geqslant 1, \\ s - 1 - k & \text{for } \frac{1}{k} \leqslant s \leqslant 1, \end{cases}$$

it follows that the correspondence function of the transformation Π is

for $1 \leqslant s < 1 + \dfrac{k-1}{1+\alpha}$

$$\left.\begin{array}{l} s = 1 + \dfrac{k + \sigma + k\sigma\tau - (1+\sigma)e^\tau}{1+\alpha}, \\ s' = 1 + \sigma(1 - e^{-\tau}); \end{array}\right\} \tag{10.85a}$$

† More precisely, $s' = 1 + g(s)$ and $ds'/ds = g'(s)$, where $g(s)$ and $g'(s)$ are small quantities of the order of $(1+\alpha)/(1+\beta)$ and $\sigma/(1+\beta)$.

SYMMETRICAL MULTIVIBRATOR (WITH GRID CURRENTS)

and for $\frac{1}{k} \leq s \leq 1$

$$\left.\begin{array}{l} s = 1+k+\sigma+k\sigma\tau-(1+\sigma)e^{\tau}, \\ s' = 1+\sigma(1-e^{-\tau}). \end{array}\right\} \quad (10.85b)$$

To the point $s = 1+(k-1)/(1+\alpha)$ there corresponds $\tau=0$ and $s'=1$, to the point $s=1$ the value $\tau=\tau_1$, and to the point $s=1/k$ the value $\tau=\tau_2$, where τ_1 and τ_2 are determined by the equations

$$\left.\begin{array}{l} (1+\sigma)e^{\tau_1} = k+\sigma+k\sigma\tau_1, \\ (1+\sigma)e^{\tau_2} = 1-\dfrac{1}{k}+k+\sigma+k\sigma\tau_2 \end{array}\right\} \quad (10.86)$$

A graphical solution of these equations is shown in Fig. 587 and it is evident that $\tau_2 > \tau_1$.

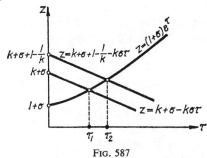

Fig. 587

Since

$$\frac{ds}{d\tau} = \begin{cases} \dfrac{k\sigma-(1+\sigma)e^{\tau}}{1+\alpha} & \text{for } 1 < s < 1+\dfrac{k-1}{1+\alpha}, \\ k\sigma-(1+\sigma)e^{\tau} & \text{for } \dfrac{1}{k} < s < 1, \end{cases}$$

$$\frac{ds'}{d\tau} = \sigma e^{-\tau} < 0, \quad \text{since} \quad \sigma < 0,$$

$$\frac{ds'}{ds} = \begin{cases} (1+\alpha)\dfrac{-\sigma e^{-\tau}}{-k\sigma+(1+\sigma)e^{\tau}} & \text{for } 1 < s < 1+\dfrac{k-1}{1+\alpha}, \\ \dfrac{-\sigma e^{-\tau}}{-k\sigma+(1+\sigma)e^{\tau}} & \text{for } \dfrac{1}{k} < s < 1, \end{cases}$$

then, for $-1 \leq \sigma < 0$, $ds/d\tau < 0$ and as τ varies from 0 to τ_2, s decreases monotonically from $1+(k-1)/(1+\alpha)$ to $1/k$. In addition, from the first of the conditions (10.78)

$$0 < \frac{ds'}{ds} < \frac{1+\alpha}{k} < 1 \quad \text{for} \quad 1 < s < 1 + \frac{k-1}{1+\alpha}$$

$$0 < \frac{ds'}{ds} < \frac{1}{k} < 1 \quad \text{for} \quad \frac{1}{k} < s < 1.$$

Thus, *for* $-1 \leqslant \sigma < 0$, *the point transformation Π cannot have fixed points s^* on the interval* $1 < s < 1 + (k-1)/(1+\alpha)$, since on this interval $s' < 1$, *but can have a single stable fixed point on the interval* $1/k \leqslant s < 1$. The uniqueness and stability of the fixed point follows from the inequality $0 < ds'/ds < 1$, proved above for $1/k < s < 1$. The value $\tau = \tau^*$ for this fixed point is given by (10.85b) as

$$k + k\sigma\tau^* + \sigma e^{-\tau^*} - (1+\sigma)e^{\tau^*} = 0 \tag{10.87}$$

τ^* is also the half-period of the discontinuous oscillations if the durations of rapid and semi-rapid motions are neglected.

The condition for the existence of a fixed point s^* on the interval $1/k \leqslant s < 1$ is, clearly, $(s')_{\tau=\tau_0} \geqslant 1/k$ or, using (10.81)

$$\sigma \geqslant \sigma_1 = \sigma_1(k), \tag{10.88}$$

where σ_1 is a branch value of the parameter σ and is determined by

$$(s')_{\tau=\tau_2} = 1 + \sigma(1 - e^{-\tau_2}) = \frac{1}{k}$$

together with the second of the equations (10.86) that expresses τ_2 as a function of σ and k. Since $\tau_2 > 0$, then $\sigma_1 < 0$; for $k > 1$, and $\sigma_1 = -1$ for $k = k_1 = 2 \cdot 219, \ldots$ A graph of the function $\sigma_1 = \sigma_1(k)$, is shown in Fig. 588. Since for $k > k_1 \approx 2 \cdot 2, \ldots$, $\sigma_1 < -1$, a fixed point s^* exists on the interval $1/k \leqslant s < 1$ for $-1 < \sigma < 0$, if $k > k_1$.

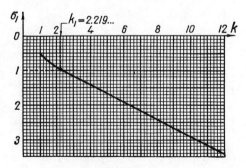

Fig. 588

(c) *The point transformation Π for $0<s<1/k$ and $\sigma \geqslant -1$*. In this case the end point (x_1^+, x_2^+) of a jump starting from a point s on Γ_1 lies, according to (10.80a), inside the region (*II*). Integrating the differential equations (10.79b), it is easy to obtain the following equation for the path L that starts from (x_1^+, x_2^+) at $t=0$ and enters (*II*):

$$x_1 = \sigma + (x_1^+ - \sigma)e^{-t}, \quad x_2 = \sigma + [k(x_1^+ - \sigma)t + x_2^+ - \sigma]e^{-t}. \quad (10.89)$$

For $\sigma \geqslant 1/(k-1) > 0$ this path reaches the boundary of the region (*II*) and then enters region (*III*). Otherwise, L remains in (*II*) and meets Γ_2 where $x_1 = x_1'$ and lies between x_1^+ and τ, and the coordinate $s^t = x_1' - 1$ of the consecutive point will be determined by

$$\left. \begin{array}{l} s'-1 = \sigma + (ks-1-\sigma)e^{-\tau}, \\ -1 = \sigma + [k(ks-1-\sigma)\tau - (k^2-1)s - 1 - \sigma]e^{-\tau} \end{array} \right\} \quad (10.89a)$$

which follow from (10.80a).

Therefore, when $\sigma = -1$, the transit time for all these L paths is $\tau = 1 - 1/k^2$; and the correspondence function (for $0<s<1/k$) will be linear

$$s' = ske^{-\left(1-\frac{1}{k^2}\right)}, \quad (10.90)$$

Now $k_1 = 2\cdot 2, \ldots$, is the single root of the equation $k \exp(-1+k^{-2}) = 1$, so for $k > k_1$ the segment of the correspondence function graph is above $s' = s$, and for $k < k_1$ it is below it[†].

For $\sigma > -1$ the graph of the correspondence function $s' = \Pi(s)$ for $0 < s \leqslant 1/k$ must be everywhere above the straight line (10.90). Therefore, as s varies from 0 to $1/k$, s' increases monotonically (since $ds'/ds > 0$) from a certain value $s_0' = \Pi(0) > 0$ to a value $\Pi(1/k) > \exp(-1+k^{-2})$. From the equations (10.89a) we obtain the correspondence function

$$\left. \begin{array}{l} s = (1+\sigma)\dfrac{1+k\tau-e^\tau}{k^2\tau-(k^2-1)}, \\ s' = (1+\sigma)\dfrac{k^2\tau-(k^2+k-1)(1-e^{-\tau})}{k^2\tau-(k^2-1)}, \end{array} \right\} \quad (10.91)$$

on condition, of course, that $s' \leqslant 1$.

If, however, for some values of s in the interval $0<s<1/k$ the second relation (10.91) gives values $s'>1$, this clearly indicates that these paths L

[†] Hence it follows once more, by virtue of (10.81), that, for $-1<\sigma<0$ and $k \geqslant k_1$, $(s')_{s=1/k} > 1/k$, and the transformation Π has a fixed point $1/k < s^* < 1$.

do not reach the segment of the half straight line Γ_2 lying within (II) but enter the region (III). Therefore, for these values of s, the correspondence function is not expressed by (10.91). This is easily shown to happen when $\sigma \geqslant 1/(k-1)$ if $0 < s \leqslant 1/k$. Therefore the paths L corresponding to such values of s, starting from points (x_1^+, x_2^+) in (II), enter (III) on or near the path $x_1 \equiv x_0 = \sigma/(1+\beta) \approx 0$ meeting Γ_2 where $x_1 = x_1' \approx x_0 \approx 0$. Hence,

$$s' = 1. \tag{10.91a}$$

From the correspondence function (10.91), it is evident that the value $\tau = \tau_2$, determined by the second of the equations (10.86), corresponds to the point $s = 1/k$, and the value $\tau = \tau_3 > 0$ uniquely determined (for $k > 1$) by the equation

$$e^{\tau_3} = 1 + k\tau_3$$

corresponds to the point $s = 0$. It is easily shown that values of τ in the interval $\tau_3 > \tau > \tau_2$ correspond to values of s in the interval $0 < s < 1/k$.

To prove the last assertion consider the auxiliary function

$$\Psi(\tau) = \frac{1 + k\tau - e^\tau}{k^2\tau - (k_2 - 1)},$$

having (for $\sigma > -1$) the same sign as s. The denominator $k^2\tau - (k^2 - 1)$ is negative for $\tau < \tau_4$ and positive for $\tau > \tau_4$, where $\tau_4 = 1 - 1/k^2$, and zero at $\tau = \tau_4$. The numerator $1 + k\tau - e^\tau$, is zero for $\tau = 0$, is positive for $0 < \tau < \tau_3$ (τ_3 has been introduced above),

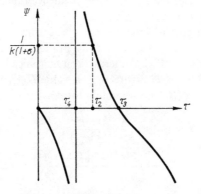

Fig. 589

and is again zero for $\tau = \tau_3$. Also since $0 < \tau_4 < 1$, $k = (1-\tau_4)^{-\frac{1}{2}}$; and for $0 < \tau < 1$, $\tau(1-\tau)^{-\frac{1}{2}} > e^\tau - 1$, then $\tau_4 < \tau_3$. These results suffice to construct the graph of Ψ shown in Fig. 589. It is evident that $\tau_4 < \tau_2 < \tau_3$ since τ_2 satisfies the equation $\Psi(\tau_2) = 1/k(1+\sigma) > 0$.

Thus values of s in the interval $0 < s < 1/k$ correspond to values of the parameter τ in the interval $\tau_3 > \tau > \tau_2$. But $ds/d\tau$ and hence $ds'/d\tau$ are negative in this interval ($ds'/ds > 0$), so that as τ increases from τ_2 to τ_3, s decreases monotonically from $1/k$ to 0, and s' from $(s')_{\tau=\tau_2} > \exp(-1 + k^{-2})$ to $s'_0 = (s')_{\tau=\tau_3} > 0$.

If the point transformation Π has a fixed point s^* in the interval $0 < s < 1/k$, then the parameter $\tau = \tau^*$ ($\tau_2 < \tau^* < \tau_3$) corresponding to it is determined by the condition $s = s' < 1/k < 1$ or, according to (10.91), by

$$e^\tau + k(k-1)\tau - (k^2 + k - 1)(1 - e^{-\tau}) - 1 = 0. \tag{10.92}$$

It can be shown that this equation has a single positive root τ^*, for example, by considering the auxiliary function

$$\Phi(\tau) = e^\tau + k(k-1)\tau - (k^2 + k - 1)(1 - e^{-\tau}),$$

where

$$\Phi'(\tau) = e^\tau + k(k-1) - (k^2 + k - 1)e^{-\tau},$$
$$\Phi''(\tau) = e^\tau + (k^2 + k - 1)e^{-\tau} > 0,$$
$$\Phi(0) = 0, \quad \Phi'(0) = -2(k-1) < 0,$$
$$\Phi(+\infty) = +\infty, \quad \Phi'(+\infty) = +\infty.$$

and whose graph is shown in Fig. 590.

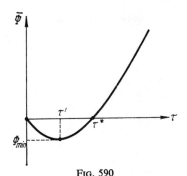

Fig. 590

Since for $k > k_1 = 2 \cdot 2, \ldots$, the straight line (10.90) lies above $s' = s$, the graph of the correspondence function $s' = \Pi(s)$ for $k > k_1$ and $\sigma > -1$ will also lie above this bisector, and for all values of s in the interval $0 < s < 1/k$ the inequality $s' = \Pi(s) > s$ will be valid. Therefore, in the interval $0 < s < 1/k$, there are no fixed points of the transformation Π.

For $\sigma > -1$, but $k < k_1$ (when $\sigma_1(k) > -1$) two cases are possible.

(1) If $\sigma > \sigma_1(k)$ then, as has been shown above (see (10.88)), $(s')_{\tau=\tau_2} > 1/k = (s)_{\tau=\tau_2}$. Therefore the difference $s'-s$, being a continuous function of s' reduces to zero over the interval $0 < s < 1/k$ either never or an even number of times. The latter is impossible, as we have just proved. Therefore, in this case also, the transformation Π has no fixed points on the interval $0 < s < 1/k$.

(2) For $-1 < \sigma < \sigma_1(k)$, $(s')_{\tau=\tau_2} < 1/k$, i.e. the difference $s'-s$ is negative for $s = 1/k$ when $\tau = \tau_2$; for $s = 0$ when $\tau = \tau_3$ this difference is positive as before. Therefore in the interval $0 < s < 1/k$ the difference $s'-s$ reduces to zero and there is a single fixed point s^* of the transformation Π. Because it is the only fixed point it is bound to be *stable*. In fact, if the fixed point s^* were unstable, the inequality $ds'/ds < 1$ would be satisfied for $s = s^*$. Then the difference $s'-s$ would be a positive quantity for $s^* < s < s^* + \varepsilon$, where ε is a small positive number, and, therefore, would reduce to zero at least once more in the interval $s^* + \varepsilon < s < 1/k$, which is impossible.

(d) *The point transformation Π for $\sigma < -1$*. For $\sigma < -1$, there exists in the region (I) a stable state of equilibrium (σ, σ) towards which some paths of slow motions go (see Fig. 585a and Fig. 586). Now, there is a positive s_0 which is a lower bound of all s that have consecutive points (see also page 769). Of course, this lower bound of the interval of existence of the transformation Π must be less than $1 + (k-1)/(1+\alpha)$, since points $s \geq 1 + (k-1)/(1+\alpha)$ have consecutive points s' determined by the function (10.82) for all values of σ and in particular for $\sigma < -1$.

In *deriving* the expressions (10.85a), (10.85b) and (10.91) for the correspondence function of Π, we did not need the condition $\sigma \geq -1$. Therefore these expressions remain valid for $\sigma < -1$ provided, of course, that s is in an interval such that $s > s_0$.

Note, also, that because of (10.81), when $\sigma < -1$ the following inequality is true:

$$s' = \Pi(s) < (s')_{\sigma=-1}, \tag{10.93}$$

for $s_0 < s < 1 + (k-1)/(1+\alpha)$, where

$$(s')_{\sigma=-1} = \begin{cases} ske^{-\left(1-\frac{1}{k^2}\right)} & \text{for } 0 < s \leq \frac{1}{k}, \\ e^{-\left(1-\frac{s}{k}\right)} & \text{for } \frac{1}{k} \leq s \leq 1, \\ e^{-\left(1+\frac{\alpha}{k}-\frac{1+\alpha}{k}s\right)} & \text{for } 1 \leq s \leq 1 + \frac{k-1}{1+\alpha}. \end{cases}$$

is the correspondence function for $\sigma=-1$[†]. It follows from (10.93) that $s'<s$ for $\sigma<-1$, $k<k_1 = 2\cdot 2, \ldots$, and for all values of $s>s_0$. *Therefore in this case the transformation Π has no fixed points*. It can have fixed points for $k>k_1$ only, and, since $s'<1$ for $\sigma<-1$ and $s_0<s<1+(k-1)/(1+\alpha)$ the fixed points s^* can only lie on the interval $s_0<s<1$.

Hence the parameter $\tau=\tau^*$ for these points is determined by the equation (10.87), if $1/k<s^*<1$, and by the equation (10.92) if $s^*<1/k$. The latter equation has been shown to have, at most, one positive root τ^*. Therefore, if on the interval $s_0<s<1/k$ there is a fixed point s^* of the transformation Π, this is the *only one*. Therefore consider the case when $\sigma<-1$ and $k>k_1=2\cdot 2, \ldots$ Since when $s=s_0$ the difference $s'-s = -s_0$ is negative, so the transformation Π has only one fixed point if, for $s=1/k$, the difference

$$s'-s = (s')_{\tau=\tau_2} - \frac{1}{k} > 0$$

i.e. if (see (10.88))

$$\sigma > \sigma_1(k)$$

where $\sigma_1(k)<-1$ for $k>k_1$. At this fixed point $s=s^*$ the difference $s'-s$ varies its sign from negative (for $s<s^*$) to positive (for $s>s^*$) and thus at $s=s^*$ $d(s'-s)/ds>0$ i.e. *the unique fixed point s^*, which exists for $k>k_1$ and $\sigma_1(k)<\sigma<-1$, is unstable on the interval $s_0<s<1/k$.*

If, however, $\sigma<\sigma_1(k)$, then, for $s=1/k$, $s'-s=(s')_{\tau=\tau_2}-1/k<0$, and therefore the difference $s'-s$ either does not reduce to zero on the interval $s_0<s<1/k$, or else reduces to zero an even number of times. Since the latter is impossible, then *the transformation Π has no fixed points on the interval $s_0<s<1/k$ for $\sigma<\sigma_1(k)$*.

Finally consider the correspondence function over the interval $1/k<s<1$. Since $s'>0$ for all values of $s>s_0$, we have, from (10.85b),

$$s' = 1+\sigma(1-e^{-\tau}) > 0, \quad \text{i.e.} \quad \frac{1+\sigma}{\sigma}e^\tau < 1,$$

and

$$\frac{ds}{ds'} = e^\tau\left(k-\frac{1+\sigma}{\sigma}e^\tau\right) > e^\tau(k-1) > e^\tau(k_1-1) > 1$$

[†] Explicit expressions for the correspondence function for $\sigma=-1$ and

$$\frac{1}{k} \leq s \leq 1+\frac{k-1}{1+\alpha}$$

are obtained from (10.85a) and (10.85b): for

$$\sigma=-1, s'=e^{-\tau}$$

while τ is found to be a piece-wise-linear function of s.

(it must be remembered that we are still considering the case $\sigma > -1$ and $k > k_1 = 2\cdot 2, \ldots, > 2$). Thus for all but such that $s > s_0$, $1/k < s < 1$ we have

$$0 < \frac{ds'}{ds} < 1 \quad \text{and} \quad \frac{d}{ds}(s'-s) < 0.$$

Therefore, if there is a fixed point on the interval $1/k < s < 1$ it will be *stable* and the only one on this interval, since the difference $s'-s$ cannot reduce to zero more than once. Since $s'-s < 0$ for $s = 1$, this fixed point only exists if

$$\sigma > \sigma_1(k),$$

when, for $s = 1/k$, the difference $s' - s = (s')_{\tau = \tau_2} - 1/k > 0$.

If, however, $\sigma < \sigma_1(k)$, then, for $s = 1/k$ if $s_0 < 1/k$, or for $s = s_0$ if $1/k < s_0 < 1$, the difference $s'-s < 0$ for all values of s over the interval, since there $d(s'-s)/ds < 0$[†]. Again, therefore, the transformation Π has no fixed points on the interval $1/k < s < 1$.

Thus, for $\sigma < -1$, the transformation Π has no fixed points either for $k < k_1$, or for $k > k_1$ when $\sigma < \sigma_1(k)$, and has two fixed points (one stable on the interval $1/k < s < 1$ and one unstable on the interval $s_0 < s < 1/k$) for $k > k_1$ and $\sigma_1(k) < \sigma < -1$.

4. *Lamerey's diagram*

Soft and hard modes of excitation of discontinuous self-oscillations. It is now possible to construct the graphs of the correspondence functions and to analyse possible modes of operation of the multivibrator. The family of curves of the correspondence function $s' = \Pi(s)$ for various values of σ has the form shown in Fig. 591 for $k > k_1 = 2\cdot 2, \ldots$, and in Fig. 592 for $k < k_1$. We are considering the case $k > 1 + \alpha$ and $(\beta \gg 1 + \alpha, \sigma)$.

For $\sigma > -1$ the transformation Π exists for all positive s and has a unique and stable fixed point s^* $(0 < s^* \leqslant 1)$, to which tend, as is easily verified by constructing Lamerey's ladder giving the sequence of points of intersection of paths with the lines Γ_1 and Γ_2. In the phase space there is a single stable limit cycle approached by all other paths. Thus for $\sigma > -1$ (for $E_g > -u_0$) there is a *soft mode of excitation of discontinuous self-oscillations.*

For $\sigma < -1$ there is a stable equilibrium state and the point transformation Π only exists for $s > s_0$ (s_0 is positive and depends on k and σ). The

---

[†] We omit the analysis of the case $1 < s_0 < 1 + (k-1)/(1+\alpha)$, since then the transformation Π does not exist for $s < 1$.

points $0<s<s_0$ lead to paths which approach the equilibrium state asymptotically without reaching the boundary of the region of "slow" motions. Two cases are possible.

For $k<k_1$, and for $k>k_1$ but with $\sigma<\sigma_1(k)<-1$, the transformation Π has no fixed points, $s'<s$ for all values of s greater than s_0 and, therefore, all sequences of intersection points of paths (on the x_1, x_2 plane) with

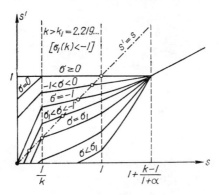

Fig. 591

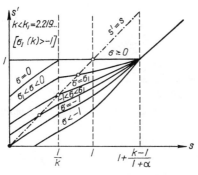

Fig. 592

the lines Γ_1 and Γ_2 are finite, the last point lying on the interval $0<s<s_0$. In this case, after a finite number of discontinuous oscillations, a stable equilibrium state is reached whatever the initial conditions. *The multivibrator cannot generate self-oscillations.*

For $k>k_1=2\cdot 2, \ldots$, and $\sigma_1(k)<\sigma<-1$ the point transformation Π has two fixed points s_1^* and s_2^* ($s_0<s_1^*<1/k<s_2^*<1$), the first of which is unstable and the second stable. In the phase space, there is a stable equilibrium state, and two discontinuous limit cycles, one of which ($s=s_2^*$)

is stable and the other ($s=s_1^*$) is unstable. It is easily seen that all sequences of points of intersection of paths with the lines Γ_1 and Γ_2 from initial points $s > s_1^*$ lead to the stable fixed point s_2^*, while the sequences with initial points $s < s_1^*$ are finite, since the last point $s_N < s_0$ (again, easily verified by constructing Lamerey's ladders).

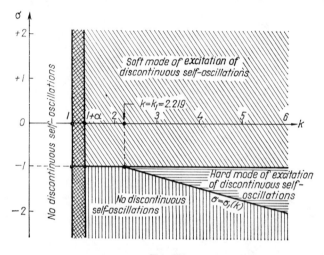

Fig. 593

Thus, in this case, depending on the initial conditions either a stable state of equilibrium or a stable self-oscillating mode of operation is established: so that the multivibrator *has a hard mode of excitation of discontinuous self-oscillations*. For a fixed $k > k_1$, when $\sigma_1 = \sigma_1(k) < -1$ we obtain a "hysteresis-type" dependence of the amplitude of the self-oscillations upon the grid bias E_g (upon parameter σ) which is typical of hard excitation. If σ increases sufficiently slowly from negative values $\sigma < \sigma_1$, then the multivibrator is in an equilibrium state until $\sigma = -1$, at which the equilibrium state becomes unstable. For $\sigma = -1$ the multivibrator generates self-oscillations with a finite amplitude. As the grid bias is increased further, the self-oscillations continue with increasing amplitude. If now the grid bias is decreased so that σ decreases the self oscillations in the multivibrator will continue for $\sigma > \sigma_1$ with decreasing amplitude but still finite at $\sigma \to \sigma_1 + 0$. For $\sigma = \sigma_1$ the self-oscillations stop, since the stable limit cycle merges with the unstable one, and there is an equilibrium state. The parameter τ must, of course, be varied slowly in this experiment. Fig. 593 shows the k, τ stability diagram inside the boundaries $k=1$, $\sigma=-1$ and $\sigma=\sigma_1(k)$ (for $\sigma<-1$). The region $1<k<1+\alpha$ has not been considered, whilst for $k<1$ a stable equilibrium state is always reached.

5. Self-oscillations of the multivibrator for $E_g \geqslant 0$

The case when $E_g \geqslant 0$ is the most interesting from the point of view of practical applications and, as before, we assume the conditions (10.78) are satisfied. For $E_g \geqslant 0$ (for $\sigma \geqslant 0$) the single equilibrium state

$$x_1 = x_2 = \frac{\sigma}{1+\beta} \approx 0$$

is unstable, and all paths approach a single stable discontinuous limit cycle as $t \to +\infty$, corresponding to a unique stable fixed point $s^* \approx 1$ of the point transformation Π. The projections of this limit cycle on the x_1, x_2 and y_1, y_2 planes are shown (qualitatively) in Fig. 594. This limit cycle consists of

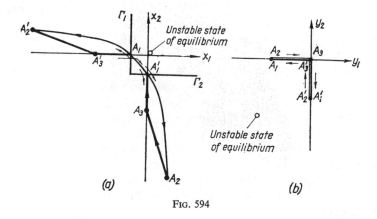

Fig. 594

paths of "rapid" motions A_1A_2 and $A_1'A_2'$ (the projections of these paths on the y_1, y_2 plane are points, since $y_1, y_2 \approx$ const. during a "rapid" motion of the representative point), of sections of paths of "semi-rapid" motions A_2A_3 and $A_2'A_3'$, which are described by the representative point during intervals of time of the order of $(1+\alpha)/(1+\beta) \ll 1$ (in units of dimensionless time), and of sections of paths of "slow" motions A_3A_1' and $A_3'A_1$. The projections of the paths of "slow" and "semi-rapid" motions on the y_1, y_2 plane coincide. They have been somewhat displaced from each other in the drawing for the sake of clarity.

Since the coordinate of the fixed point is $s^* = 1$ for $\sigma \geqslant 0$ (to an accuracy up to terms of the order of $(1+\alpha)/(1+\beta)$ and $\sigma/(1+\beta)$), then the path of "rapid" motion A_1A_2 starts from a point A_1.

$$x_1^{(1)} = -1, \quad x_2^{(1)} = s^* - 1 = 0, \quad y_1^{(1)} = -(k-1), \quad y_2^{(1)} = 0;$$

using (10.75) to evaluate y_1 and y_2. Then the end point A_2 of this path is determined as functions of $x_1^{(1)}$ and $x_2^{(2)}$ by the jump conditions (10.80)

$$x_1^{(2)} = \frac{k-1}{1+\alpha}, \quad x_2^{(2)} = -k\frac{k+\alpha}{1+\alpha}, \quad y_1^{(2)} = -(k-1), \quad y_2^{(2)} = 0.$$

Along the path A_1A_2: $y_1 \equiv y_1^{(1)} = -(k-1)$, $y_2 \equiv y_2^{(1)} = 0$, and so according to (10.77a), this path is determined approximately (μ small but finite) by the equations

$$\left. \begin{array}{l} \mu \dot{x}_1 = k-1-x_1-\alpha\psi(x_1)-k\varphi(x_2), \\ \mu \dot{x}_2 = -x_2-\alpha\psi(x_2)-k\varphi(x_1) \end{array} \right\} \quad (10.94)$$

outside the small neighbourhood of the equilibrium states of these equations, or by the equation of the integral curves

$$\frac{dx_2}{dx_1} = \frac{-x_2-\alpha\psi(x_2)-k\varphi(x_1)}{k-1-x_1-\alpha\psi(x_1)-k\varphi(x_2)}. \quad (10.94a)$$

The paths of the approximate equations (10.94) is shown in Fig. 595. These equations have two states of equilibrium A_1 and A_2: the point $A_1(-1, 0)$—the initial point of a jump—is an unstable multiple singular point of the saddle node type, and the point $A_2((k-1)/(1+\alpha), -k(k+\alpha)/(1+\alpha))$ is a stable node towards which all paths of "rapid" motions go (with $y_1 \equiv -(k-1)$ and $y_2 \equiv 0$). From the point A_1 there is a *separatrix* of this singular point. This separatrix (see Section 3 of this chapter), will be the path of "rapid" motion A_1A_2 that is a part of the discontinuous limit cycle.

Inside the quadrant (IV) (see Fig. 595) $-1 \leqslant x_1 \leqslant 0$, $-1 \leqslant x_2 \leqslant 0$, the equation of the integral curves of "rapid" motions (the equation (10.94a)) can be written

$$\frac{dx_2}{dx_1} = \frac{x_2+k(x_1+1)}{x_1+1+kx_2}.$$

Therefore the separatrix from A_1 will be (within the limits of the region (IV)) the segment of the straight line[†]

$$x_2 = -(x_1+1)$$

[†] The other separatrix that approaches the singular point A_1 is in the region (V): $-1 \leqslant x_1 < 0, x_2 > 0$ and (within this region) is a segment of the straight line $x_2 = \varkappa(x_1+1)$ where

$$\varkappa = \frac{\alpha}{2k} + \sqrt{\left(\frac{\alpha}{2k}\right)^2 + 1}.$$

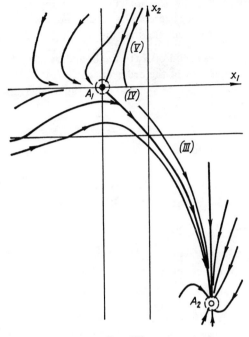

Fig. 595

At the point $(0, -1)$ the separatrix passes into (III) where it is determined by the equation

$$\frac{dx_2}{dx_1} = \frac{-x_2 - k(x_1 + 1)}{k - 1 - (1 + \alpha)x_1}.$$

Integrating and using the initial condition: $x_2 = -1$ at $x_1 = 0$, the equation of the separatrix $A_1 A_2$ (within the limits of the region (III)) is

$$x_2 = -k\frac{k+\alpha}{1+\alpha} - \frac{k}{\alpha}\left(\frac{k-1}{1+\alpha} - x_1\right) +$$

$$+ \left[k\frac{k+\alpha}{1+\alpha} + \frac{k}{\alpha}\frac{k-1}{1+\alpha} - 1\right]\left(1 - \frac{1+\alpha}{k-1}x_1\right)^{\frac{1}{1+\alpha}}.$$

The path of "rapid" motion $A_1 A_2$—the separatrix of the singular point A_1—is shown in Figs. 595 and 594(a).

Also observe that from equations (10.94) the approximate dependence of x_1 and x_2 upon t during a jump-wise motion along A_1A_2 can be deduced. It is easily shown, for example, that the transit time from the boundary of a $0\left(\mu^{\frac{1}{2}}\right)$-neighbourhood of A_1 to the boundary of a $0\left(\mu^{\frac{1}{2}}\right)$-neighbourhood of A_2 is a quantity of the order of $\mu \ln (1/\mu)$.

At A_2 a "rapid" motion becomes a "semi-rapid" one along a path close to the straight line

$$x_2 + kx_1 = \text{const.} = x_2^{(2)} + kx_1^{(2)} = -k;$$

The transit time on this path is of the order of $(1+\alpha)/(1+\beta)$. The point A_3 is

$$x_1^{(3)} = 0, \quad x_2^{(3)} = -k,$$
$$y_1^{(3)} = 0, \quad y_2^{(3)} = 0.$$

Next, there is a path of "slow" motion A_3A_1'

$$x_1 \equiv 0$$

along which $\dot{x}_2 \approx 1$. For this path

$$\dot{x}_2 + x_2 = \sigma$$

(see the equations (10.79c) and, therefore,

$$x_2 = \sigma - (k+\sigma)e^{-t}$$

at $t=0$ the representative point is at A_3). Therefore, after a duration

$$\tau^* = \ln \frac{k+\sigma}{1+\sigma}$$

the representative point reaches Γ_2 at the point A_1', from whence a rapid motion will start again. The second half of the limit cycle $A_1'A_2'A_3'A_1$ is symmetrical to the half $A_1A_2A_3A_1'$ (in the x_1, x_2, y_1, y_2 space, with respect to the plane $x_1 = x_2$, $y_1 = y_2$ and on the x_1, x_2 and y_1, y_2 planes with respect to the bisectors $x_1 = x_2$ and $y_1 = y_2$).

Knowing the limit cycle $A_1A_2A_3A_1'A_2'A_3'A_1$ (Fig. 594) we can easily plot the waveforms of the voltages in a multivibrator during the corresponding self-oscillations. Such waveforms for the voltages u_1, v_1 and $u_{a2} = u_1 + v_1$ are shown in Fig. 596. If we neglect the duration of "rapid" and "semi-rapid" motions the period of the discontinuous oscillations of a multivibrator with grid current when $\sigma \geqslant 0$ ($E_g \geqslant 0$) is

$$T = 2\tau^* = 2 \ln \frac{k+\sigma}{1+\sigma}$$

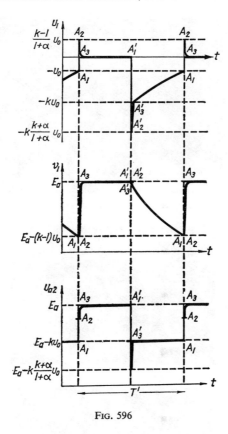

Fig. 596

(in units of dimensionless time) and

$$T' = 2(R_a+R_g)C \ln \frac{\dfrac{SR_a}{1+R_a/R_g}+\dfrac{E_g}{u_0}}{1+\dfrac{E_g}{u_0}}$$

(in ordinary units).

It is a fact that an experimental verification fully confirms the theory developed here. The photograph of a limit cycle on the plane of the voltages u_1 and u_2 (Fig. 597) and the photographs of oscillograms of the voltages u_1, v_1 and u_{a2} (Fig. 598) give support to this statement (to make the "rapid" motions visible in the photograph C_a and C_g were increased artificially).

788 DISCONTINUOUS OSCILLATIONS [X

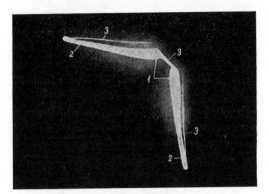

Fig. 597

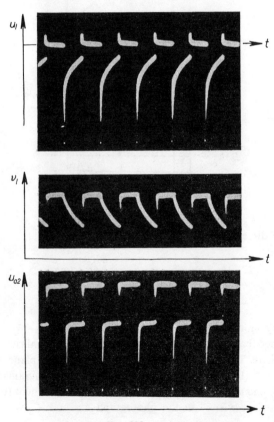

Fig. 598

CHAPTER XII†

COMMENTS ON MORE RECENT WORKS‡ WHICH SUPPLEMENT THE CONTENTS OF THIS TREATISE

THE first edition of this present volume was published in the U.S.S.R. in 1937 by Andronov and Khaikin (Chaikin) but was practically unknown outside Russia until it was freely translated into English by Professor S. Lefschetz and published by the Princeton University Press in 1949. This second edition is now associated with the names of Andronov, Vitt, and Khaikin and was published in Moscow in 1959 and contains much additional material due to Andronov and his collaborators, but it does refer only briefly, at the most, to other work on similar topics published outside the U.S.S.R. and whose origins lie in the appearance of the first edition. The second edition, like the first, is devoted to applications and exposition of the qualitative theory of differential equations originated by Poincaré and the idea of a discontinuous transition in a dynamic process originally due to Mandelshtam. The book does not enter into the quantitative methods of discussing non-linear oscillations due to Krylov and Bogoliubov[31] and later put on a sounder basis by Bogoliubov and Mitropolski[3] although readers will no doubt acquaint themselves with this work which has been brought to the notice of scientists and engineers outside the U.S.S.R. through the efforts of N. Minorsky, and S. Lefschetz and his collaborators.

Minorsky's recent text[44] published in 1961 is an excellent account of Russian, and his own, work in the field of non-linear oscillations. The references show that the theory of oscillations and the stability of oscillations in systems described by non-linear differential or difference differential equations is of world-wide interest to mathematicians, engineers and physicists. Since the publication in English of the first edition of this book, Professor S. Lefschetz has been the most active and eminent mathematician outside the U.S.S.R. contributing to the qualitative theory of differential equations. Due to him there have appeared five publications called *Con-*

† Added by the English Editor.
‡ See additional references.

tributions to the Theory of Non-linear Oscillations[38] which contain contributions to this subject by mathematicians. Another book edited by La Salle and Lefschetz[34] entitled *Non-Linear Differential Equations and Non-Linear Mechanics* gives some idea of the wide range of topics being investigated by mathematicians, particularly in the U.S.A. and U.S.S.R.

Interest in non-linear oscillations is by no means confined to the U.S.A. and U.S.S.R. although most papers do emanate from these countries. Since 1957 it can be estimated from the major sources of abstracts that some 800–900 papers have been published in the more important scientific languages, and the rate of publication is known to be increasing. There are many text books in mechanics and control system theory that use the methods of Andronov and other Russian authors to discuss non-linear oscillations. H. Kauderer is the author of a German text[30] which contains a large section devoted to the nonlinear vibrations of systems with a single degree of freedom with motions damped, undamped, forced and self-excited and makes much use of phase plane methods. W. J. Cunningham[13], J. C. West[53] R. L. Cosgriff[12], Graham and McRuer[22], Thaler and Pestel[49] and other authors have written texts on non-linear control systems and all are indebted to the pioneering work of Andronov. It is interesting to note that although the Russian title of this book is *Theory of Oscillations* it might equally well have been called the *Theory of Oscillators* for there is much discussion, as will have been noted, of just how self-excited oscillations can be maintained and many examples are drawn from electrical engineering wherein much design effort is expended on producing sustained oscillations in active circuits. Of course, the theory can and is used to discuss the decay or non-excitation of oscillations and is quite general in its application to autonomous systems and their stability. The authors deliberately restricted themselves in this work to autonomous systems with one degree of freedom, perhaps because their colleagues, such as Krylov, were engaged in discussing forced oscillations, but possibly also because they felt this work to be of direct and practical use to designers of oscillators. The introduction of Mandelshtam's discontinuous theory enabled them to discuss, quite rigorously, the action of relaxation oscillators such as multivibrators. There is no doubt that since the appearance of this book in the U.S.S.R. there has not been a substantial addition to their work which is of direct use to an engineer engaged in designing oscillators. Much work has been done on specific systems and much mathematical work of interest, at the moment, only to mathematicians has been stimulated, but practical design methods based on recent contributions remain few.

The authors were interested in using the qualitative theory of differential equations as outlined by Poincaré, and this and other Russian work in this field has undoubtedly caused the present activity in this area outside the U.S.S.R. Mathematicians are interested in the existence, uniqueness, and stability of periodic (and other) solutions of differential equations and there is a constant stream of published papers some of which are extremely valuable to the engineer and physicist and many of which are not. An interesting and valuable book to the mathematically inclined reader of this book is a translation by Lefschetz of a Russian work[46] by Nemytskii and Stepanov on the qualitative theory of differential equations. Lefschetz has also written an excellent, but not elementary, text[39] on the same topic. From a different standpoint R. Bellman has written a book[2] concerned with the stability of solutions of differential equations. Earlier still Coddington and Levinson[11] wrote a very readable account of the theory of differential equations which will be of interest to many readers. The stability of physical systems containing energy sources, especially control systems, is of immense practical importance and indeed it is this importance that results in funds in most countries being allocated, in one way or another, to the support of research workers. Stability in this book is discussed mainly in terms of the local stability near an equilibrium (critical) point by means of Liapunov's First Method and in terms of orbital stability of limit cycles. There are powerful methods outside the scope of this book and one of them, Liapunov's Second Method, is now a favourite field of research, particularly in the U.S.S.R. This method has the great advantage of dealing with stability in the large (or over a region of the phase-plane) and is applicable to systems of greater than the second order. Both Andronov and Vitt knew of this work because they published a paper in 1933[1] on this subject but chose not to use it. The unearthing of Liapunov's Second Method for use in non-linear control systems and vibrating systems is usually credited to Chetaev and his students.

Apparently independently of the work of Andronov and his colleagues the same or similar problems were considered by Krylov and Bogoliubov whose account can be read in English in a free translation of their book[31] by S. Lefschetz. They were particularly interested in applied problems and used a method which is also associated with Van der Pol. They chose as a possible solution to the second order non-linear equation a sinusoidal function whose amplitude a and phase Φ are slowly varying time functions. They then obtained expressions for da/dt and $d\Phi/dt$ as power series in μ, where μ is some parameter in the differential equation. The coefficients of the power series are functions of a and Φ and are assumed periodic. This power series may or may not converge but assuming it does then by successive operations first and higher approximations to a and Φ can be found. If convergence is doubtful or if the series is non-convergent then another technique known as asymptotic integration can be used to give a solution of known accuracy. A book[3] by Bogoliubov and Mitropolski gives a rigorous account of the method and is full of worked examples. The work of Krylov, Bogoliubov, Mitropolski and others is complementary to that of Andronov, Vitt and Khaikin and the reader of this book would find it interesting and illuminating. There is a good account in Minorsky's latest text[43, 44]. The methods can be used for autonomous and non-autonomous systems, the latter including forced oscillations. Hayashi[26, 27] has published some detailed work on examples drawn from practice and has compared calculation with experiment.

The work of Andronov, Vitt and Khaikin was directed to the introduction of the methods of Poincaré and Mandelshtam to engineers and physicists and any extensions made to the theory was for this purpose. The mathematical background has received extensive attention during the last fifteen years or so and there have been many practical examples of the use of their methods in vibration analysis and in the theory of control systems. As we shall see, however, there does not seem to have been a new breakthrough providing the practical design engineer with more powerful tools than those demonstrated in this book. We are here discussing the phase-plane (or phase-cylinder) concept. It is unfortunate that systems with orders greater than two usually require a phase-space of three or more dimensions to contain a trajectory. This is difficult to visualise and the methods applicable to a plane are of little use. Nevertheless a knowledge of the methods of this book would seem to be an indispensable pre-requisite to a study of other methods of analysis, such as Liapunov's Second Method.

An extension to the work of Poincaré and hence of this book, was made by I. G. Malkin[40] who published his book in 1956 in Moscow. Malkin considers both autonomous and non-autonomous systems described by an equation

$$\ddot{x} + \omega^2 x = \mu f(t, \dot{x}x)$$

as Andronov has done. The *generating* solution valid as $\mu \to \infty$ plays an important role and he establishes criteria for the existence of solutions in terms of certain roots of the characteristic equation. Malkin gives a thorough treatment of almost periodic solutions, periodic solutions, and stability. He extends his work beyond second order systems and this book is an important one. It contains many examples including the use of special transformations and functions to simplify analysis.

A novel approach to obviate the difficulties in finding the characteristic exponents of a non-linear system, as required in Poincaré's theory for the *existence* of a periodic solution, has been made by Minorsky[43, 44]. The idea is to replace the original, perhaps non-autonomous, differential equation by another autonomous one, such that the existence and stability of its singular point is the criterion of the existence and stability of a periodic solution of the original equation. This method Minorsky calls the *stroboscopic* method because he considers a curve which would be obtained if photographs of the system trajectory in phase-space were taken at fixed intervals of time, the interval being the period of a limit cycle adjacent to the trajectory.

S. Lefschetz[38] has considered the phase portrait of a very general second order non-linear system, whose linearised version has characteristic roots which are both zero. He studies the periodic nature of any solution, and extends previous work. Y. H. Ku[32, 33] in his book uses the acceleration plane rather than the phase plane, which is interesting but does not seem to have great advantages. He also, by numerical methods, discusses third and higher order non-linear systems. Ku's book has many examples drawn from non-linear control systems, and it has an excellent bibliography.

An important paper on applications of the phase-plane and phase-space methods and the point transformations of Andronov is by Gorskaya, Krutava and Rutkovskii[21] describing work carried out during 1954–8. It is concerned with servos in auto-pilots which are fully described and are non-linear. Many problems of these servos are analysed using the phase-plane and there are many phase portraits. The point transformation is used and stability and limit cycles investigated. Certain relay servos which are of the third order are discussed in phase-space. This paper is an interesting account of the practical use made of Andronov's methods in the U.S.S.R.

Another paper of the same type as the one quoted above and well demonstrating the use of qualitative methods is that by Bogusz and Kazimierz[4]. They investigate the equation

$$\ddot{x} + \omega^2 x + a\dot{x}^2 = 0$$

by both qualitative and quantitative methods. Using the velocity hodograph they obtain families of trajectories on the phase-plane and also obtain solutions in closed finite form. The qualitative method is interesting and new.

Ostrovskii[47], Thaler and Pestel[49], and Han and Thaler[25], have made attempts to utilise phase-plane methods in the *design* of control systems, and shew that it is relatively easy to decide how to vary the damping *discontinuously*. In this way a very satisfactory response to step inputs can be obtained and the theory is applied to systems of second and third order.

An unusual book on non-linear differential equations has been written by Struble[48] and it gives a stimulating account of qualitative theory which can be read with profit by engineers and others who have worked their way through this book. Those interested in control systems will find much use of phase-plane, phase-space, and the concepts of trajectories in the works of Flugge-Lotz[16, 17, 18, 19], Tsypkin[50], Hamel[24, 20], and many others who are interested in the use of discontinuous elements, such as relays and contactors, in control systems. In these systems it is necessary to ensure that, whatever the input signal, no stable limit cycle will be approached. Concepts based on the work of Andronov have proved most valuable, and the phase-plane or a derivative of it is constantly used to discuss situations geometrically. Currently there is much use made of Liapunov's Second Method of assessing stability and we refer readers to modern books on non-linear control systems and to Hahn[23], Malkin[41] and La Salle[37], and to the numerous articles in the journals devoted to control systems in English, German and Russian.

The quantitative theory of non-linear differential equations has made much progress since the original work by Krylov and Bogoliubov[31] was written. Bogoliubov and Mitropolski[3], the many works written or edited by Lefschetz, papers by La Salle[see 38], Möser[45, 38], Friedrichs[14, 15], Cartwright and Littlewood[5, 6], Cartwright[7], and Cesari[38, 8, 9] and many others have advanced the subject. There are to-day numerous alternative approaches to the solution of non-linear differential equations but the engineer or physicist finds it difficult to choose a best approach, although the splendid book by Minorsky[44] does help. The quantitative studies still await some genius to do for them what Poincaré did for the qualitative approach.

APPENDIX

BASIC THEOREMS OF THE THEORY OF DIFFERENTIAL EQUATIONS

HERE we present some standard propositions in the theory of differential equations that have been used in the text. Proofs and further extended discussions can be found in treatises on ordinary differential equations such as those by L. Bieberbach, E. L. Ince, E. Kamke, S. Lefschetz, W. Hurewicz, G. Sansone, Coddington and Levinson, I. G. Petrovskii, etc.

Let a system of differential equations be

$$\left.\begin{array}{l}\dfrac{dx_1}{dt} = P_1(x_1, x_2, \ldots, x_n, t), \\ \cdots\cdots\cdots\cdots\cdots\cdots \\ \dfrac{dx_n}{dt} = P_n(x_1, x_2, \ldots, x_n, t)\end{array}\right\} \quad (A.1)$$

(n is an arbitrary integer), where the functions $P_i(x_1, x_2, \ldots, x_n, t)$ defined in a certain open region R, are continuous in this region and have continuous partial derivatives with respect to $x_1, x_2, \ldots, x_n, t$.

THEOREM I (*on the existence and uniqueness of a solution*).

At any point $M_0(t_0, x_1^0, \ldots, x_n^0)$ of R, there exists an interval of t containing $t_0 (t_1 < t < t_2)$ and only one system of functions

$$x_k = \varphi_k(t) \quad (k = 1, 2, \ldots, n),$$

defined in this interval, for which the following conditions are satisfied:

(a) $\varphi_k(t_0) = x_k^0$;

(b) for all values of $t(t_1 < t < t_2)$ the point $M[t, \varphi_1(t), \ldots, \varphi_n(t)]$ belongs to the region R;

(c) $\dot\varphi_k(t) = P_k[t, \varphi_1(t)\varphi_n(t)]$ for all $t_1 < t < t_2$, i.e. our system of functions satisfies the system of differential equations (A.1);

(d) whatever closed region $\tilde R_1$, entirely contained in the region R, we may choose, there are values t' and t'' ($t_1 < t' < t_2$ and $t_1 < t'' < t_2$) such that the points $M_1[t', \varphi_1(t'), \ldots, \varphi_n(t')]$ and $M_2[t'', \varphi_1(t''), \ldots, \varphi_n(t'')]$ lie outside $\tilde R_1$.

It can be shown that the interval $t_1 < t < t_2$ in Theorem I, is, by the condition (d), "the maximum possible interval for definition of the solution" in this sense, that there is no interval (t_1^*, t_2^*) containing the interval (t_1, t_2) on which functions $x_k = \varphi_k^*(t)$ may be defined such as satisfy the conditions (a), (b) and (c) and therefore coincide on the interval (t_1, t_2) with the functions $x_k = \varphi_k(t)$. Often this condition (d) is expressed thus: "the solution of the system can be continued up to the boundary of the region R".

In this book, by a solution of a system of the form (A.1) we always mean the solution defined on the maximum possible interval of t. Usually a solution is defined for all values of t, in the interval $-\infty < t < +\infty$.

In the $t, x_1, x_2, \ldots, x_n$ space the functions $x_k = \varphi_k(t)$ determine an *integral curve*. By Theorem I, through each point $M(t_0, x_1^0, \ldots, x_n^0)$ of the region R there passes one and only one integral curve.

To emphasize, the fact that the solution depends on the initial values $t_0, x_1^0, \ldots, x_n^0$, this is also written as

$$x_k = \varphi_k(t, t_0, x_1^0, \ldots, x_n^0). \tag{A.2}$$

We have, obviously,

$$\varphi_k(t_0, t_0, x_1^0, \ldots, x_n^0) \equiv x_k^0.$$

If $t_0, x_1^0, \ldots, x_n^0$ are considered as arbitrary parameters (but such, of course, that $M(t_0, x_1^0, \ldots, x_n^0)$ belongs to R), the system of functions (A.2) is called *the general solution*. If $t_0, x_1^0, \ldots, x_n^0$ are fixed, the system (A.2) is a particular solution or simply a solution. The following theorem is applied to it.

THEOREM II *(on the continuity of the dependence upon the initial conditions)*.

Let

$$x_k = \varphi_k(t, t^*, x_1^*, x_2^*, \ldots, x_n^*)$$

be a solution of (A.1), defined for t in $t_1 < t < t_2$ and let τ_1 and τ_2 be arbitrary numbers belonging to this interval, $\tau_1 < \tau_2$. Then for an arbitrary positive ε, there is a positive number $\delta = \delta(\varepsilon, \tau_1, \tau_2)$ such that for all values of t_0, $x_1^0, \ldots, x_n^0$ for which

$$|t_0 - t^*| < \delta, \quad |x_i^* - x_i^0| < \delta \quad (i = 1, 2, \ldots, n),$$

the solution

$$x_k = \varphi_k(t, t_0, x_1^0, \ldots, x_n^0)$$

is defined for all values $\tau_1 \leq t \leq \tau_2$ and satisfies the inequalities

$$|\varphi_k(t, t_0, x_1^0, \ldots, x_n^0) - \varphi_k(t, t^*, x_1^*, \ldots, x_n^*)| < \varepsilon.$$

THEOREM III.

If the functions $P_i(t, x_1, \ldots, x_n)$ of (A.1) have continuous partial derivatives with respect to the variables $x_1, x_2, \ldots, x_n$, the functions

$$x_k = \varphi_k(t, t_0, x_1^0, x_2^0, \ldots, x_n^0) \qquad (k = 1, 2, \ldots, n)$$

have *continuous partial derivatives* with respect to the variables $x_1^0, x_2^0, \ldots, x_n^0$ [†]. These partial derivatives, together with the functions φ_k themselves, satisfy a system of differential equations

$$\frac{d\varphi_k}{dt} = P_k(t, \varphi_1, \ldots, \varphi_k),$$

$$\frac{d}{dt}\frac{\partial \varphi_k}{\partial x_i^0} = \frac{\partial P_k}{\partial \varphi_1}\frac{\partial \varphi_1}{\partial x_i^0} + \ldots + \frac{\partial P_k}{\partial \varphi_n}\frac{\partial \varphi_n}{\partial x_i^0},$$

$$i = 1, 2, \ldots, n; \qquad k = 1, 2, \ldots, n.$$

In the case where the functions $P_i(t, x, \ldots x_n)$ are analytic functions of their variables, the following theorem is valid.

THEOREM IV.

If the functions $P_k(t, x_1, x_2, \ldots, x_n)$ are analytic functions of the variables $x_1, x_2, \ldots, x_n$, the functions

$$x_k = \varphi_k(t, t_0, x_1^0, \ldots, x_n^0)$$

are *analytic functions of their arguments* in a neighbourhood of every system of values for which the functions are defined.

Theorems I–IV are employed, in particular, in the investigation of the sequence function. From the method of construction of the sequence function, it is easily seen that when the right-hand sides of the system equations are analytic functions, the sequence function, by Theorem IV, is also an analytic function. When the right-hand sides have continuous derivatives with respect to x and y, it follows from Theorems I, II and III that the sequence function is continuous and has a continuous derivative (see Section 7, Chapter V).

[†] When the right-hand sides of the system (A.1) have continuous partial derivatives with respect to the variables $x_1, x_2, \ldots, x_n$ of orders up to $k \geqslant 1$, the solution of this system has continuous partial derivatives with respect to x_i^0 of the same order k. However, the case where $k > 1$ is not made use of in this book.

Consider now the system (A.1)

$$\frac{dx_k}{dt} = P_k(t, x_1, \ldots, x_n) \qquad (k = 1, 2, \ldots, n)$$

and the "varied" or "perturbed" system

$$\frac{dx_k}{dt} = P_k(t, x_1, \ldots, x_n) + p_k(t, x_1, \ldots, x_n) \qquad (k = 1, 2, \ldots, n)$$
(A.3)

where $p_k(t, x_1, \ldots, x_n)$ are functions defined in the same region R as P_k, are continuous in this region and have continuous partial derivatives with respect to $x_1, x_2, \ldots, x_n$.

Furthermore, let the functions P_k depend on a certain parameter μ, so that

$$\frac{dx_k}{dt} = P_k(t, x_1, \ldots, x_n, \mu) \qquad (k = 1, 2, \ldots, n). \qquad (A.4)$$

At a particular value $\mu = \mu_0$, we have the system

$$\frac{dx_k}{dt} = P_k(t, x_1, \ldots, x_n, \mu_0) \qquad (k = 1, 2, \ldots, n), \qquad (A.5)$$

We can consider the system (A.4) for $\mu \neq \mu_0$ to be a perturbed version of (A.5) and so consider the system (A.4) in the form

$$\frac{dx_k}{dt} = P_k(t, x_1, \ldots, x_n, \mu_0) + p_k(t, x_1, \ldots, x_n),$$

where

$$p_k(t, x_1, \ldots, x_n) = P_k(t, x_1, \ldots, x_n, \mu) - P_k(t, x_1, \ldots, x_n, \mu_0).$$

THEOREM V *(on the continuity of the dependence of the solution on a variation of the right-hand side and of the initial value).*

Let

$$x_k = \varphi_k(t, t_0, x_1^0, \ldots, x_n^0) \qquad (k = 1, 2, \ldots, n)$$

be a solution of (A.1), defined for all values of t in the interval (t_1, t_2) and with $t_1 < t_0 < t_2$. Let τ_1 and τ_2 satisfy the inequality $t_1 < \tau_1 < t_0 < \tau_2 < t_2$. Then for any $\varepsilon > 0$ there is a $\delta > 0$ such that when $|p_k(t, x_1, \ldots, x_n)| < \delta$ ($k = 1, 2, \ldots, n$) in R, and $|x_i^0 - x_i^*| < \delta$ ($i = 1, 2, \ldots, n$) the solution of the system (A.3), corresponding to initial values $t_0, x_1^*, \ldots, x_n^*$, is

$$x_k = \varphi_k^*(t, t_0, x_1^*, \ldots, x_n^*)$$

defined for all t, $\tau_1 \leq t \leq \tau_2$, and satisfies the following inequalities in the interval of t:

$$|\varphi_k^*(t, t_0, x_1^*, \ldots, x_n^*) - \varphi_k(t, t_0, x_1^0, \ldots, x_n^0)| < \varepsilon \quad (k = 1, 2, \ldots, n).$$

COROLLARY. If the right-hand sides of the system (A.4) are continuous functions of μ, the functions $\varphi_k(t, t_0, x_1^0, \ldots, x_n^0, \mu) = x_k$, are also continuous functions of μ.

Suppose that the functions $P_k(t, x_1, x_2, \ldots, x_n)$ and $P_k(t, x_1, \ldots, x_n) + p_k(t, x_1, \ldots, x_n)$ have continuous partial derivatives with respect to the variables $x_1, x_2, \ldots, x_n$. Then, from Theorem III, the functions $\varphi_k(t, t_0, x_1^0, \ldots, x_n^0)$ and $\varphi_k^*(t, t_0, x_1^0, \ldots, x_n^0)$ in the solutions of (A.1) and (A.3) have partial derivatives with respect to $x_1^0, x_2^0, \ldots, x_n^0$:

$$\frac{\partial \varphi(t, t_0, x_1^0, \ldots, x_n^0)}{\partial x_i^0} \quad \text{and} \quad \frac{\partial \varphi^*(t, t_0, x_1^0, \ldots, x_n^0)}{\partial x_i^0}.$$

Let a solution of the system (A.1) be defined in the interval $t_1 < t < t_2$, and let τ_1 and τ_2 be certain numbers that satisfy the inequalities $t_1 < \tau_1 < \tau_2 < t_2$. Then we have the following Theorem.

THEOREM VI.

For any $\varepsilon > 0$ there is a $\delta > 0$ such that, if

$$|p_k(t, x_1, \ldots, x_n)| < \delta, \quad \left|\frac{\partial p_k(t, x_1, \ldots, x_n)}{\partial x_i}\right| < \delta, \quad |x_i^0 - x_i^*| < \delta$$

$$(k = 1, 2, \ldots, n; \; i = 1, 2, \ldots, n),$$

in the region R, then the solution of the system (A.3)

$$x = \varphi^*(t, t_0, x_1^*, \ldots, x_n^*)$$

is defined for all t in the interval $\tau_1 \leq t \leq \tau_2$, where the following inequalities are satisfied:

$$\left|\frac{\partial \varphi(t, t_0, x_1^0, \ldots, x_n^0)}{\partial x_i^0} - \frac{\partial \varphi^*(t, t_0, x_1^*, \ldots, n_n^*)}{\partial x_i^*}\right| < \varepsilon.$$

If $P_k(t, x_1, \ldots, x_n, \mu)$ and its derivatives $\partial P_k(t, x_1, \ldots, x_n, \mu)/\partial x_i$ are continuous functions of μ, and $x_k = \varphi_k(t, x_1^0, \ldots, x_n^0, \mu)$ is a solution of (A.4) then the derivatives

$$\frac{\partial \varphi_k(t, x_1^0, \ldots, x_n^0, \mu)}{\partial x_k^0}$$

are also continuous functions of μ.

APPENDIX

Let us consider one more case, when in the system (A.4) the right-hand sides are analytic functions of all their arguments. The following theorem is valid for this system.

THEOREM VII.

If the functions $P_k(t, x_1, \ldots, x_n, \mu)$ are analytic functions of their arguments, then the functions

$$x_k = \varphi_k(t, t_0, x_1^0, \ldots, x_n^0, \mu)$$

are also analytic functions of all their arguments in a neighbourhood of every system of values $t, t_0, x_1^0, \ldots, x_n^0$ for which they are defined.

COROLLARY. Let the particular solution

$$x_k = \varphi_k(t, t_0, x_1^*, \ldots, x_n^*, \mu^*)$$

be defined for all t in the interval $t_1 < t < t_2$ and let τ_1 and τ_2 be such that $t_1 < \tau_1 < \tau_2 < t_2$. Then the functions

$$x_k = \varphi_k(t, t_0, x_1^0, \ldots, x_n^0, \mu)$$

can be expanded in power series with respect to $(x_i^0 - x_i^*)$ $(i = 1, 2, \ldots, n)$, converging for all t and μ that satisfy

$$\tau_1 \leqslant t \leqslant \tau_2, \quad |\mu - \mu^*| < \delta \qquad (A.6)$$

and for all

$$|x_i^0 - x_i^*| < h_0,$$

where h_0 is a certain constant independent of the t and μ that satisfy the inequalities (A.6). The coefficients of these series are analytic functions of μ in the interval

$$|\mu - \mu^*| < \delta.$$

REFERENCES

(a)

1. AIZERMAN M. A., *Vvedeniye v dinamiku avtomaticheskogo regulirovaniya dvigatelei (Introduction to the Theory of Automatic Control of Motors)*. Mashgiz, (1950).
2. AIZERMAN M. A., Physical Basis of the Application of Small-Parameter Methods to the Solution of Non-linear Problems of the Theory of Automatic Regulation; *Avtomat. i Telemekh.* **14**, 597 (1953).
3. ANDRONOV A. A., Poincaré's Limit Cycles and the Theory of Self-Oscillations; in *Sobraniye trudov A. A. Andronova (Collected Works by A. A. Andronov)*, p. 41, Izd. A. N. SSSR, (1956).
4. ANDRONOV A. A., Mathematical Problems of the Theory of Self-Oscillations; in *Sobraniye trudov A. A. Andronova*, p. 85, Izd. Akad. Nauk SSSR (1956).
5. ANDRONOV A. A., Poincaré's Limit Cycles and the Theory of Oscillations, in *Sobraniye trudov A. A. Andronova*, p. 32, Izd. Akad. Nauk SSSR (1956).
6. ANDRONOV A. A. and VITT A. A., Discontinuous Periodic Solutions and the Theory of Abraham and Bloch's Multivibrator; *Dokl. Akad. Nauk SSSR* **8**, 189, (1930); also in *Sobraniye trudov A. A. Andronova*, p. 65, Izd. Akad. Nauk SSSR (1956).
7. ANDRONOV A. A. and VITT A. A., Contribution to Van der Pol's Theory of Entrainment; *Archiv für Elektrotechnik* **24**, 99 (1930); also in *Sobraniye trudov A. A. Andronova*, p. 51, Izd. Akad. Nauk SSSR (1956).
8. ANDRONOV A. A. and VITT A. A., On Stability in the Sense of Lyapunov, *Zh. eksp. teor. fiz.* **3**, 373 (1933); also in *Sobraniye trudov A. A. Andronova*, p. 140, Izd. Akad. Nauk SSSR (1956).
9. ANDRONOV A. A. and VOZNESENSKII I. N., On the Works of D. C. Maxwell, I. A. Vyshnegradskii and A. Stodola in the Field of the Theory of Regulation of Machines; in Collection: *D. K. Maksvell, I. A. Vyshnegradskii, A. Stodola. Teoriya avtomaticheskogo regulirovaniya (klassiki nauki) (D. C. Maxwell, I. A. Vyshnegradskii, A. Stodola. Theory of Automatic Regulation, Classics of Science)*, Izd. Akad. Nauk SSSR (1949); also in *Sobraniye trudov A. A. Andronova*, p. 490, Izd. Akad. Nauk SSSR (1956).
10. ANDRONOV A. A. and LEONTOVICH YE. A., Contribution to the Theory of Variations of the Qualitative Structure of the Mapping of the Plane by Paths; *Dokl. Akad. Nauk SSSR* **21**, 427 (1938); also in *Sobraniye trudov A. A. Andronova*, p. 217, Izd. Akad. Nauk SSSR (1956).
11. ANDRONOV A. A. and LEONTOVICH YE. A., Certain Cases of the Dependence of Limit Cycles upon a Parameter; Uchenye zapiski Gor'kovskogo Gosudarstvennogo Universiteta, p. 3 (1939); also in *Sobraniye trudov A. A. Andronova*, p. 188, Izd. Akad. Nauk SSSR (1956).
12. ANDRONOV A. A. and LEONTOVICH YE. A. Generation of Limit Cycles from a Non-coarse Focus or Centrepoint and from a Non-coarse Limit Cycle; *Dokl. Akad. Nauk SSSR* **99**, 885 (1954).
13. ANDRONOV A. A. and LEONTOVICH YE. A. Generation of Limit Cycles from a Non-coarse Focus or Centre-point and from a Non-coarse Limit Cycle; *Mat. Sbornik* **40**, 179 (1956).

14. ANDRONOV A. A. and LYUBINA A. G., Application of Poincaré's Theory on "Bifurcation Points" and on the "Change of Stability" to Simple Self-Oscillating Systems; *Zh. eksp. teor. fiz.* **5**, 3–4 (1935); also in *Sobraniye trudov A. A. Andronova*, p. 125, Izd. Akad. Nauk SSSR (1956).
15. ANDRONOV A. A., MANDEL'SHTAM L. I. and PAPALEKSI N. D., *Novye issledovaniya v oblasti nelineinykh kolebanii (New Investigations in the Field of Non-linear Oscillations)*, Radioizdat (1936); also in *Physics of the USSR* **11**, 2–3, 1 (1935).
16. ANDRONOV A. A. and NEIMARK YU. I., On the Motions of an Ideal Model of Clocks Having Two Degrees of Freedom, Model of Pre-Galileian Clocks; *Dokl. Akad. Nauk SSSR* **51**, 17 (1946); also in *Sobraniye trudov A. A. Andronova*, p. 313, Izd. Akad. Nauk SSSR (1956).
17. ANDRONOV A. A. and PONTRYAGIN L. S., Coarse Systems; *Dokl. Akad. Nauk SSSR* **14**, 247 (1937); also in *Sobraniye trudov A. A. Andronova*, p. 181, Izd. Akad. Nauk SSSR (1956).
18. BARKHAUSEN G. G., *Cathode Valves*, vol. II (Russian translation) (1928).
19. BAUTIN N. N., Contribution to the Theory of Synchronization; *Zh. tekh. fiz.* **9**, 510 (1939).
20. BAUTIN N. N., On a Case of Non-harmonic Oscillations; *Uchenye zapiski Gor'kovskogo Universiteta* **12**, 231 (1939).
21. BAUTIN N. N., On a Differential Equation Having a Limit Cycle; *Zh. tekh. fiz.* **9**, 601 (1939).
22. BAUTIN N. N., On the Motion of an Ideal Model of Clock having Two Degrees of Freedom. Model of Galileo-Huygens Clock; *Dokl. Akad. Nauk SSSR* **61**, 17 (1948).
23. BAUTIN N. N., On Mandel'shtam Problem in the Theory of Clocks, *Dokl. Akad. Nauk SSSR* **65**, 279 (1949).
24. BAUTIN N. N., Dynamic Model of Chronometer Motion; *Inzh. Sb. Akad. Nauk SSSR* **12**, 3 (1952).
25. BAUTIN N. N., Dynamic Model of Clock Motion without Natural Period; *Inzh. sb. Akad. Nauk SSSR* **16**, 3 (1953).
26. BAUTIN N. N., On the Periodical Solutions of a System of Differential Equations; *Prikl. Mat. i Mekh.* **18**, 128 (1954).
27. BAUTIN N. N., Dynamic Theory of Clock Motions without Constructional Stop of the Escape Wheel; *Inzh. sb. Akad. Nauk SSSR* **21**, 3 (1955).
28. BAUTIN N. N., Dynamic Models of Free Clock Motions; in *Sbornik pamyati A. A. Andronova (Collected Articles in Commemoration of A. A. Andronov)*, p. 109, Izd. Akad. Nauk SSSR (1955).
29. BEZMENOV A. YE., Barkhausen and Mueller Methods from the Viewpoint of the Rigorous Theory of Self-Oscillations, *Zh. tekh. fiz.* **6**, 467 (1936).
30. BEZMENOV A. YE., Theory of Rukop's Stopping Diagram; *Elektrosvyaz'*, No. 4, p. 13 (1938).
31. BELYUSTINA L. N., Determination of the Qualitative Structure of the Mapping by Paths of the Phase Plane of a Coarse System (to be published).
32. BENDRIKOV G. A. and GORELIK G. S., Application of Braunovskii Valve to the Investigation of the Motion of the Representative Point on the Plane of Van der Pol's Variables; *Zh. tekh. fiz.* **5**, 620 (1935).
33. BESSONOV L. A., *Elektricheskiye tsepi so stal'yu (Electrical Circuit with Steel)* Gosenergoizdat (1948).
34. BIRKHOFF G. D., *Dynamical System* (Russian translation), Moscow—Leningrad (1941).
35. BOGOLYUBOV N. N., *O nekotorykh statisticheskikh metodakh v matematicheskoi fizike (On Certain Statistical Methods in Mathematical Physics)*, Izd. Akad. Nauk SSSR (1945).
36. BOGOLYUBOV N. N. and MITROPOL'SKII YU. A., *Asimptoticheskiye metody v teorii*

nelineinykh kolebanii (Asymptotic Methods in the Theory of Non-linear Oscillations), Gostekhizdat (1955).
37. BREMZEN A. S. and FAINBERG I. S., Analysis of the Operation of two Coupled Relaxation Oscillators; *Zh. tekh. fiz.* **11**, 959 (1941).
38. BULGAKOV B. V., On the Application of Poincaré's Method to Free Pseudolinear Oscillating Systems; *Prikl. Mat. i Mekh.* **6**, 263 (1942).
39. BULGAKOV B. V., On the Application of Van der Pol's Method to the Pseudolinear Oscillations of a System with many Degrees of Freedom; *Prikl. Mat. i Mekh.* **6**, 395 (1942).
40. BULGAKOV B. V., Self-oscillations of Regulating Systems; *Prikl. Mat. i Mekh.* **7**, 97 (1943).
41. BULGAKOV B. V., *Kolebaniya (Oscillations)*, Gostekhizdat (1954).
42. VASIL'EVA A. B., On Differential Equations Containing Small Parameters; *Mat. Sbornik* **31** (73), 587 (1952).
43. VITKEVICH V. V., "Hard" Self-excitation Mode of a Relaxation Generator (Multivibrator); *Zh. tekh. fiz.* **16**, No. 3, 309 (1946).
44. VLASOV N. P., Self-oscillating System with Single-phase Asynchronous Motor; *Zh. tekh. fiz.* **5**, 641 (1935).
45. GAUZE G. F. and VITT A. A., On Periodic Oscillations of Population Numbers; *Izv. Akad. Nauk*, series 7, 1551 (1934).
46. GOL'DFARB L. S., On Certain Non-linearities in Regulating System; *Avtomat. i Telemekh.* **8**, 347 (1947).
47. GOL'DFARB L. S., Method of Investigation of Non-linear Systems Based on the Principle of the Harmonic Balance; in: *Sbornik Teoriya avtomaticheskogo regulirovaniya (Collected Articles "Theory of Automatic Regulation")*, Mashgiz (1951).
48. GORELIK G., Kuzovkin V. and Sekerskaya V., Investigation of Squegging Oscillations; *Tekhnika radio is slabykh tokov* **11**, 629 (1932).
49. GRADSHTEIN I. S., Non-linear Differential Equations with Small Coefficients for Certain Derivatives; *Dokl. Akad. Nauk SSSR* **66**, 789 (1949).
50. GRADSHTEIN I. S., Differential Equations in which Various Degrees of a Small Parameter occur as Coefficients; *Dokl. Akad. Nauk SSSR* **82**, 5 (1952).
51. GRANOVSKII V. L., *Elektricheskii tok v gaze (Electrical Current in a Gas)* vol. I, Gostekhizdat (1952).
52. DORODNITSYN A. A., Asymptotic Solution of Van der Pol's Equations; *Prikl. Mat. i Mekh.* **11**, 313 (1947).
53. YEVTYANOV S. I., Theory of a Self-Oscillator with Grid Leak; *Elektrosvyaz'*, No. 9 66 (1940).
54. ZHEVAKIN S. A., Contribution to the Theory of Sidereal Variability; *Dokl. Akad, Nauk SSSR* **99**, 217 (1954).
55. ZHEVAKIN S. A., On the Phase Shifts between Oscillations of Brightness and Oscillations of Radial Velocity of Variable Stars, *Dokl. Nauk SSSR* **99**, 353 (1954).
56. ZHEVAKIN S. A., On Self-Oscillations of Variable Stars of "Large Recurrency"; in *Sbornik pamiati A. A. Andronova*, p. 629, Izd. Akad. Nauk SSSR (1955).
57. ZHELEZTSOV N. A., Self-modulation of the Self-Oscillations of a Valve Generator with Automatic Bias in the Grid Circuit; *Zh. tekh. fiz.* **13**, 495 (1948).
58. ZHELEZTSOV N. A., Contribution to the Theory of the Symmetrical Multivibrator; *Zh. tekh. fiz.* **20**, 778 (1950).
59. ZHELEZTSOV N. A., Contribution to the Theory of the Valve Generator with a Two-mesh RC-Circuit; *Trudy Gor'kovskogo fiziko-tekhnicheskogo instituta i radiofizicheskogo fakulteta Gor'kovskogo Gosudarstvennogo Universiteta* **35**, 220 (1957).
60. ZHELEZTSOV N. A., Contribution to the Theory of Discontinuous Oscillations in Systems of the Second Order; *Radiofizika* **1**. No. 1 (1958).

REFERENCES

61. ZHELEZTSOV N. A. and RODYGIN L. V., Contribution to the Theory of the Symmetrical Multivibrator; *Dokl. Nauk SSSR* **81**, 391 (1951).
62. ZHELEZTSOV N. A. and FEIGIN M. I., On the Modes of Operation of the Symmetrical Multivibrator; *Radiotekh. i elektron.* **2**, 751 (1957).
63. ZHUKOVSKII N. YE., On the Motion of a Pendulum with Friction at the Point of Suspension; in *Collected Works*, vol. I, p. 290, Gostekhizdat (1948).
64. ZHUKOVSKII N. YE., On Soaring of Birds; *Trudy otdel. fiz. nauk Obshch. lyubit. yestestvz.* **4**, No. 2, 29 (1891); also in *Collected Works*, vol. 4, p. 5, Gostekhizdat (1949).
65. ITSKHOKI YA. S., *Impul'snaya tekhnika (Pulse Techniques)*, Sovradio (1949).
66. KAZAKEVICH V. V., On the Approximate Solution of Van der Pol's Equation; *Dokl. Akad. Nauk SSSR* **49**, 424 (1945).
67. KAZAKEVICH V. V., Multiple Systems and Simple Dynamic Models of Clocks; *Dokl. Akad. Nauk SSSR* **74**, 665 (1950).
68. KAIDANOVSKII N. L., Nature of Mechanical Self-Oscillations Arising in the Presence of Dry Friction; *Zh. tekh. fiz.* **19**. No. 9 (1949).
69. KAIDANOVSKII N. L. and KHAIKIN S. E., Mechanical Relaxation Oscillations; *Zh. tekh. fiz.* **3**, No. 1. (1933).
70. KAMKE E., Handbook of Ordinary Differential Equations (Russian translation), Foreign Literature Publishing House, Moscow (1950).
71. KAPCHINSKII I. M., *Metody teorii kolebanii v radiotekhnike (Methods of the Theory of Oscillations in Radio Engineering)*, Gosenergoizdat (1954).
72. KARMAN T. and BIOT M., *Mathematical Methods in Engineering* (Russian translation), Gostekhizdat (1948).
73. KOBZAREV YU. B., Frequency Stability of a Self-oscillating System; in *Sb. "Pervaya Vsesoyuznaya konferentsiya po kolebaniyam" "Collected Articles" First All-union Conference on Oscillations)* vol. 1, p. 5, Moscow (1933).
74. KOBZAREV YU. B., On a Quasi-linear Method of Analysis of the Phenomena in a Valve Generator (of Quasi-sinusoidal Oscillations): *Zh. tekh. fiz.* **5**, 216 (1935).
75. KOLMOGOROV A. N., Analytical Methods of the Theory of Probabilities; *Uspekhi mat. nauk*, No. 5, 5 (1938).
76. KRYLOV A. N., *Lektsii o priblizhennykh vychisleniyakh (Lectures on Approximate Calculations)*, Izd. Akad. Nauk SSSR (1933).
77. KRYLOV A. N., On the Use of the Method of Successive Approximations for Finding the Solution of Certain Differential Equations of Oscillatory Motion; *Izv. Akad. Nauk SSSR*, p. 1, (1933).
78. KRYLOV N. M. and BOGOLYUBOV N. N., *Novye metody nelineinoi mekhaniki v ikh primenenii k izucheniyu raboty elektronnykh generatorov (New Methods of Nonlinear Mechanics in their Application to the Investigation of the Operation of Elecktronic Generators)*, pt. I, United Scientific and Technical Press (1934).
79. KRYLOV N. M. and BOGOLYUBOV N. N., *Vvedeniye v nelineinuyu mekhaniku (Introduction to Non-linear Mechanics)*, Izd. Akad. Nauk SSSR (1937).
80. LEONTOVICH YE. A. and MAYER A. G., On the Paths that Determine the Qualitative Structure of the Mapping of a Sphere by Paths; *Dokl. Akad. Nauk SSSR* **14**, 251 (1937).
81. LEONTOVICH YE. A. and MAYER A. G., General Qualitative Theory; an Appendix to Chapters 5 and 6 in the Russian translation of Poincaré's book *On Curves Defined by Differential Equations*, Gostekhizdat (1947).
82. LEONTOVICH YE. A. and MAYER A. G., On a Circuit Determining the Topological Structure of the Mapping by Paths; *Dokl. Akad. Nauk SSSR* **103**, 557 (1955).
83. LEONTOVICH M. A., *Statisticheskaya fizika (Statistical Physics)*, Gostekhizdat (1944).

REFERENCES

84. LIAPUNOV A. M., *Obshchaya zadacha ob ustoichivosti dvizheniya (The General Problem of the Stability of Motion)*, Khar'kov (1892); also Gostekhizdat (1950).
85. MAIYER A. G., Investigation of Rayleigh and Van der Pol's Equations; *Izv. Gor'kovskogo Gosudarstvennogo Universiteta*, No. 2 (1936).
86. MAIYER A. G., Contribution to the Theory of Coupled Oscillations of two Self-exciting Generators; *Uchenye zapiski Gor'kovskogo Gosudarstvennogo Universiteta*, No. 2, p. 3 (1935).
87. MALKIN I. G., *Metody Lyapunova i Puankare v teorii nelineinykh kolebanii (Lyapunov's and Poincaré's Methods in the Theory of Non-linear Oscillations)*, Gostekhizdat (1949).
88. MANDEL'SHTAM L. I., Problems of Electrical Oscillating Systems and Radio Engineering; in: Sb. *"Pervaya Vsesoyuznaya Konferentsiya po kolebaniyam" (Collected Articles "First All-union Conference on Oscillations)*, Vol. 1, p. 5, GTTI, (1933).
89. MANDEL'SHTAM L. I., Papaleksi N. D., Andronov A. A., Vitt A. A., Gorelik G. S. and Khaikin S. E., *Novye issledovaniya nelineinykh kolebanii (New Investigations of Non-linear Oscillations)*, Radioizdat (1936).
90. MANDEL'SHTAM L. I. and PAPALEKSI N. D., On the Justification of a Method of Approximate Solution of Differential Equations; *Zh. eksp. teor. fiz.* **4**, 117 (1934).
91. MEYEROVICH L. A. and ZELICHENKO L. G., *Impul'snaya tekhnike (Pulsa Technique)*, Sovradio, Moscow (1953).
92. MIGULIN V. V. and LEVITAS D. M., On the Operation of a Blocking Oscillator; *Zh. tekh. fiz.* **17**, No. 10 (1947).
93. MISHCHENKO YE. F., Asymptotic Evaluation of Periodic Solutions of Systems of Differential Equations Containing Small Parameters for the Derivatives; *Izv. Akad. Nauk SSSR* (ser. matem.) **21**, 627 (1957).
94. MISHCHENDO YE. F. and PONTRYAGIN L. S., Periodic Solutions of Systems of Close-to-Discontinuous Differential Equations; *Dokl. Akad. Nauk SSSR* **102**, 889 (1955).
95. NEIMARK YU. I., *Ustoichivost' linearizovannykh sistem (Stability of Linearized Systems)*, Leningrad (1949).
96. NEMYTSKII V. V. and STEPANOV V. V., *Kachestvennaya teoriya differentsial'nykh uravnenii (Qualitative Theory of Differential Equations)*, Gostekhizdat (1949).
97. NETUSHIL A. V., Contribution to the Theory of "Jumps" in Non-linear Systems; *Zh. tekh. fiz.* **15**, 873 (1945).
98. NIKOL'SKII G. N., Contribution to the Problem of Stability of a Ship on a Given Course; *Trudy tsentral'noi laboratorii provodnoi svyazi*, No. 1 (1934).
99. *Osnovy avtomaticheskogo regulirovaniya (Fundamentals of Automatic Regulation)*, Collected Articles edited by V. V. Solodovnikov, Mashgiz (1954).
100. PAPALEKSI N. D., On Certain Recent Problems in the Theory of Oscillations; *Uspekhi fizicheskikh nauk* **11**, 185 (1931).
101. PAPALEKSI N. D., ANDRONOV A. A., GORELIK G. S. and RYTOV S. M., Certain Investigations in the Field of Non-linear Oscillations Carried out in the USSR Starting from 1935, *Uspekhi fizicheskikh nauk* **33**, 335 (1947).
102. PETROV V. V. and ULANOV G. M., Theory of Two Simple Relay-Type Automatic Control Systems; *Avtomat. i Telemakh.* **11**, 289 (1950).
103. PETROVSKII I. G., *Lektsii po teorii obyknovennykh differentsial'nykh uravnenii (Lectures on the Theory of Ordinary Differential Equations)*, Gostekhizdat (1952); also in English translation. Interscience Publishers (1954).
104. B. VAN DER POL, *Non-linear Theory of Electrical Oscillations* (Russian translation), Svyaz'tekhizdat (1935).
105. PONTRYAGIN L. S., Asymptotic Behaviour of the Solutions of Systems of Differential Equations with a Small Parameter in the Higher-Order Derivatives; *Izv. Akad. Nauk SSSR* (ser. matem.) **21**, 605 (1957).

106. PONTRYAGIN L. S., ANDRONOV A. A. and VITT A. A., On the Statistical Investigation of Dynamic Systems; *Zh. eksp. teor. fiz.* **3**, 165 (1933); also in *Sobraniye trudov A. A. Andronova*, p. 142. Izd. Akad. Nauk SSSR, (1956).
107. PONTRYAGIN L. S., On Close-to-Hamiltonian Dynamic Systems; *Zh. eksp. teor. fiz.* **4**, 883 (1934).
108. POINCARÉ A., *On Curves Defined by Differential Equations* (Russian translation), Gostekhizdat (1947).
109. RZHEVKIN S. N. and VVEDENSKII B. A., Squegging Triode Generator, Its Theory and Applications; *Telegr. i telef. bez provodov*, No. 11, 67 (1921).
110. RUNGE K., *Graphical Methods of Mathematical Calculations* (Russian translation), GTTI (1932).
111. RYTOV S. M., Developments of the Theory of Non-Linear Oscillations in the USSR; *Radiotekh. i elektronika* **2**, 1435 (1957).
112. SKIBARKO A. P. and STRELKOV S. P., Qualitative Investigations of Processes in a Generator with Complex Circuit, *Zh. tekh. fiz.* **4**, 158 (1938).
113. STEPANOV V. V., *Kurs differentsial'nykh uravnenii (Course on Differential Equations)*, Gostekhizdat (1953).
114. STOKER D., *Non-linear Oscillations in Mechanical and Electrical Systems* (Russian translation), Foreign Literature Publishing House, Moscow (1952).
115. STRELKOV S. P., *Vvedeniye v teoriyu kolebanii (Introduction to the Theory of Oscillations)*, Gostekhizdat (1950).
116. STRELKOV S. P., Froude's Pendulum *Zh. tekh. fiz.* **3**, 563 (1933).
117. STRUTT D. V. (Rayleigh) *Theory of Sound* (Russian translation), Vol. 1, p. 235, Gostekhizdat (1955).
118. TEODORCHIK K. F., *Avtokolebatel'nye sistemy (Self-Oscillating Systems)*, Gostekhizdat (1952).
119. TIKHONOV A. M., Systems of Differential Equations Containing Small Parameters for the Derivatives; *Mat. Sbornik* **31** (73), 575 (1952).
120. TOLLE M., *Regulirovaniye silovykh mashin (Regulation of Power Engines)*, Gosenergoizdat (1951).
121. FEL'DBAUM A. A., *Vvedeniye v teoriyu nelineinykh tsepei (Introduction to the Theory of Non-linear Circuits)*, Gosenergoizdat (1948).
122. FEL'DBAUM A. A., Simplest Relay-type Automatic Regulations Systems *Avtomat. i Telemekh.* **10**, 249 (1949).
123. FRANK-KAMENETSKII D. A., Mechanism of Two-Stage Ignition; *Zh. fiz. khim.* **14**, 30 (1940).
124. FRANK-KAMENETSKII D. A., The Problem of Self-Oscillations in the Theory of Variable Stars, in *Sbornik pamiati A. A. Andronova*, p. 691, Izd. Akad. Nauk SSSR (1955).
125. CHAIKIN S. E., Continuous and Discontinuous Oscillations; *Zh. prikl. fiz.* **7**, No. 6, 21 (1930).
126. CHAIKIN S. E., Self-Oscillating Systems, in *Sbornik, Pervaya Vsesoyuznaya konferentsiya po kolebaniyam*, Vol. 1, p. 72, GTTI (1933).
127. CHAIKIN S. E., On the Influence of Small Parameters on the Nature of the Stationary States of a Dynamic System, *Zh. tekh. fiz.* **5**, 1389 (1935).
128. SHISHELOV L. L., *Mekhanika chasovogo mekhanizma (Mechanics of Clock-Work)*, parts I, II, III, Leningrad (1933—1937).
129. EL'SGOL'TS L. Z., *Obyknovennye differentsial'nye uravneniya (Ordinary Differential Equations)*, Gostekhizdat (1954).
130. YUZVINSKII V., On the Self-Oscillations of a System, analogous to the Reed of an Accordion and of a Clarinet; *Zh. tekh. fiz.* **4**, 1295 (1934).
131. ABRAHAM N. and BLOCH E., Mesures en valeur absolue des périods des oscillations électriques de haute fréquence; *Annales de Physique*, Sér. 9, **12**, 237 (1919).

REFERENCES

132. Airy C., On the Regulator of the Clock-work for effecting Uniform Movement of Equatoreales; *Mem. Roy. Astr. Soc.*, London **11**, 249 (1840), **20**, 115 (1850–51).
133. Andrade J., *L'Horlogérie et Chronométrie*, Paris (1925).
134. Appleton E. and Greaves W., On the Solution of the Representative Differential Equation of the Triodes Oscillator; *Phil. Mag. Ser.* **6**, 45 (1923).
135. Appleton E. and Van der Pol B., On a Type of Oscillation-hysteresis in a Simple Triode Generator; *Phil. Mag. Ser.* **6**, **43**, 177 (1922).
136. Barkhausen H., Die Vakuumröhre und ihre technische Anwendung; *Jahrbuch der drahtlosen Telegraphie* **14**, 27 (1919), **16**, 82 (1920).
137. Bendixson I., Sur les courbes définies par des équations différentielles; *Acta Math.* **24**, 1 (1901).
138. Bieberbach L., *Theorie der Differentialgleichungen*, Berlin, J. Springer (1930).
139. Birkhoff G., Quelques théorèmes sur les mouvements der systèmes dynamiques: *Bull. Soc. Math. de France*, 40 (1912).
140. Bowschwerow V. (Bovshverov V.), Experimentelle Untersuchung des Phasenraumes autoschwingender Systeme; *Tech. Phys. of the USSR* **11**, 43 (1935).
141. Chaikin S. (Khaikin S.) and Lochakov L., Oscillations "discontinues" dans un circuit à capacité et selfinduction; *Tech. Phys. of the USSR* **11**, 43 (1935).
142. Cholodenks L., Zur Theorie der Frühaufschen Kipp-Schaltung; *Tech. Phys. of the USSR* **2**, 552 (1935).
143. Le Corbeiller P., *Les systèmes auto-entretenues et les oscillations de relaxation*, Paris, Hermann (1931).
144. Le Corbeiller P., Le mécanisme de la production des oscillations; *Annales des Postes, Télégraphes et Téléphones* **21**, 697 (1932).
145. Debaggis L., *Contributions to the Theory of Non-linear Oscillations*, Princeton 1952; also in Russian in *Uspekhi mat. nauk* **10**, No. 4, 66 (1955).
146. Decaux and Corbeiller P., Sur un système électrique auto-entreténue utilisant un tube à néon; *C.R. Acad. Sc.* (Paris) **193**, 723 (1931).
147. Dulac H., Sur les cycles limités; *Bull. Soc. Math. de France* **51** (1923).
148. Dulac H., Recherche des cycles limités; *C.R. Acad. Sc.* (Paris) **204**, 23 (1937).
149. Fatou P., Sur le mouvement d'un système soumis à des forces à courte période; *Bull. Soc. Math. de France* **56**, 98 (1928).
150. Friedlander E., Steurungsvorgänge durch "Feldzerfall" und Kippschwingungen in Elektronenröhren; *Z. f. techn. Phys.* **7**, 481 (1926).
151. Friedlander E., Ueber Stabilitätsbedingungen und ihre Abhängigkeit von Steuerorganen und Energie-Speichern; *Phys. Zts.* **27**, 361 (1926).
152. Friedlander E., Ueber Kippschwingungen insbesondere bei Elektronenröhren; *Arch. f. Elektrotech.* **17**, 1, 103 (1927).
153. Friedlander E., Einige Berichtigungen und Ergänzungen zum Problem der Kippschwingungen; *Arch. f. Elektrotech.* **20**, 158 (1928).
154. Frommer M., Singulare Punkte; *Math. Ann.* **99** (1928).
155. Frühauf G., Eine neue Schaltung zur Erzeugung von Schwingungen mit linearem Spannungsverlauf; *Arch. f. Elektrotech.* **21**, 471 (1929).
156. Haag J., Sur des oscillations auto-entretenues; *C.R. Acad. Sc.* (Paris) **199**, 906 (1934).
157. Haag J., Sur l'étude asymptotique des oscillations de relaxation; *C.R. Acad. Sc.* (Paris) **292**, 102 (1936).
158. Haag J., Étude asymptotique des oscillations de relaxation; *Ann. Sci. École Norm. Sup.* **60** (1943).
159. Haag J., Examples concrèts d'étude asymptotique d'oscillation de relaxation; *Ann. Sci. École Norm. Sup.* **61** (1944).
160. Den Hartog, *Mechanical Vibrations*, McGraw-Hill Co. (1947).

REFERENCES

161. HEEGNER K., Ueber Schwingungserzeugung mittels eines Elektronenröhrensystems, welche Selbstinduktion nicht erhalten; *Zts. f. Hochfrequenztechnik* **29**, 151 (1927).
162. HEEGNER K. and WATANABE I., Ueber Schwingungserzeugung mittels eines Elektronenröhrensystems, bei welchem die Kapazität von untergeordneter Bedeutung ist; *Zts. f. Hochfrequenztechnik* **34**, 49 (1929).
163. HOPF E., Zwei Sätze ueber den wahrscheinlichen Verlauf der Bewegung dynamischer Systeme; *Math. Ann.* **103**, No. 4–5.
164. HULL, Das Dynatron, eine Vakuumröhre mit der Eigenschaft des negativen elektrischen Widerstandes; *Jahrbuch der drahtlosen Telegraphie* **14**, 47, 157 (1919).
165. IOBST G., Drei Beiträge ueber Schwingungserzeugung; *Telefunken Zeitung*, No. 47, 11 (1927).
166. IKONNIKOV E., On the Dynamics of symmetrical Flight of an Aeroplane; *Tech. Phys. of the USSR* **4**, No. 6, 1 (1937).
167. KAMKE E., *Differentialgleichungen reeller Funktionen*, Leipzig, Akad. Verlagsgesellschaft (1930).
168. KOENIGS, Recherches sur les substitutions uniforms; *Bulletin des Sciences Mathématiques*, 1883.
169. KOENIGS, Recherches sur les équations functionelles; *Annales de l'École Normale*, 1884.
170. KRUGER, Mechanische Schwingungssysteme mit Stossanregung; *Annalen der Physik* **70**, 291 (1923).
171. LANCHESTER F., *Aerodonetics*, London (1908).
172. LEAUTE H., Sur les oscillations à longues périodes dans les machines activiés par des moteurs hydrauliques et sur les moyens de prévenir ces oscillations; *Jour. de l'École Polytechnique* **55**, 1 (1885).
173. LEVINSON N., Perturbations of Discontinuous Solutions of Non-linear Differential Equations; *Acta Math.* **82**, 71 (1951).
174. LIENARD A., Étude des oscillations entretenues; *Revue Générale d'Électricité* **23**, 901, 946 (1928).
175. LOTKA, *Elements of Physical Biology*, Baltimore (1925).
176. MAYER A., On the Theory of Coupled Vibrations of two Self-excited Generators; *Tech. Phys. of the USSR* **2**, No. 5, 1 (1935).
177. MAYER A., A Contribution to the Theory of Forced Oscillations in a Generator with two Degrees of Freedom; *Tech. Phys. of the USSR* **3**, No. 12, 1 (1936).
178. MOLLER H., Quantitative Behandlung der Schwingungen im Röhrengenerator mit Hilfe der Schwingkennlinien; *Jahrbuch der drahtlosen Telegraphie* **14**, 326 (1919).
179. OLLENDORF F. and PETERS W., Schwingungsstabilität Parallelarbeit in der Synchronmaschine; *Wissenschaftliche Veröffentlichungen aus dem Siemens-Konzern* **5** (1926).
180. PAPALEXI N., *Theorie des Elektronenröhrengenerators*, Odessa (1922).
181. POINCARÉ H., *Oeuvries*, vol. 1, Paris, Gauthier-Villars (1928).
182. POINCARÉ H., Sur l'équilibre d'une masse fluide animée d'un mouvement de rotation; *Acta Math.* **7** (1885).
183. POINCARÉ H., *Figures d'equilibre d'une masse fluide*, Paris (1903).
184. POINCARÉ H., Sur le problème de trois corps et les équations de la dynamique; *Acta Math.* **13** (1890).
185. POINCARÉ H., *Les méthodes nouvelles de la mécanique céleste*, Paris, Gauthier-Villars (1892–1899).
186. VAN DER POL B., A Theory of the Amplitude of Free and Forced Triode Vibrations; *Radio Review* **1**, 701 (1920).
187. VAN DER POL B., An Oscillation—Hysteresis in a Triode–generator; *Phil. Mag.* **43**, 177 (1922).
188. VAN DER POL B., On Relaxation Oscillations; *Phil. Mag.* (7) **2**, 978 (1926).

189. VAN DER POL B., Ueber Relaxationschwingungen; *Zts. f. Hochfrequenztechnik* **28**, 178 (1926); **29**, 114 (1927).
190. VAN DER POL B., Forced Oscillations in a Circuit with Non-linear Resistance; *Phil. Mag.* (7) **3**, 65 (1927).
191. VAN DER POL B., Oscillations sinusoidales et de relaxation; *L'onde électrique*, 1930.
192. VAN DER POL B., and VAN DER MARK M., Le battement du coeur considéré comme oscillation et un modèle électrique du coeur; *L'onde électrique* **7**, 365 (1928).
193. ROBB A., On a Graphical Solution of a Class of Differential Equations occurring in Wireless Telegraphy; *Phil. Mag.* (6) **43**, 700 (1922).
194. RUKOP H., Reisdiagramme von Senderröhren; *Telefunken Zeitung* **6**, Juni 1923, 27; September 1923, 20.
195. RUKOP H., Reisdiagramme von Senderröhren; *Zts. f. technische Physik* **5**, 260, 299, 387, 441, 569 (1924).
196. LA SALLE J., Relaxation Oscillations; *Quart. of Appl. Math.* **7**, 1 (1949).
197. SCHUNK H., ZENNECK I., Ueber Schwingungskreise mit Eisenkernspulen; *Jahrbuch der drahtlosen Telegraphie* **19**, 170 (1922).
198. TRICOMI F., Integrazione di una equazione differenziale presentatasi in elettrotecnica; *Annali della R. Scuola Normale di Pisa*, Ser. **11**, **2**, 1 (1933).
199. VOLTERRA V., *Leçons sur la théorie mathématique de la butte pour la vie*, Paris, Gauthier-Villars (1931).
200. WAGNER K., *Der Lichtbogen als Wechselstromerzeuger*, Leipzig, Verlag S. Hirzel (1910).
201. WEBB H. and BECKER G., Theory of Multivibrator; *Journ. of Appl. Phys.* **15**, 825 (1944).

(b) Additional References for Chapter 11

1. ANDRONOV A. A. and WITT A. — see main list.
2. BELLMAN R., *The Stability Theory of Differential Equations*, McGraw-Hill, New York (1954).
3. BOGOLIUBOV N. N. and MITROPOLSKI Y. A., *Asymptotic Methods in Non-Linear Oscillations*, 2nd edition, Moscow, (1958). English edition published by Gordon & Breach, New York (1962).
4. BOGUSZ W. and KASIMIERZ S., The Non-linear System $\ddot{x} + \omega^2 x + a\dot{x}^2 = 0$ investigated in two ways, *Rozprawy Inz.* **8**, No. 2, 189. (1960).
5. CARTWRIGHT M. L. and LITTLEWOOD J. E., *J. Lond. Math. Soc.* (1945).
6. CARTWRIGHT M. L. and LITTLEWOOD J. E., *Ann. of Math.* (1947).
7. CARTWRIGHT M. L., Forced Oscillations in nearly Sinusoidal Systems, *J. I. E. E.*, **95** (3), 88—96 (1948).
8. CESARI L., *Asymptotic Behaviour and Stability Problems*, Springer, Berlin (1959) (in German).
9. CESARI L. and HALE J. K., A New Sufficient Condition for Periodic Solutions of Non-Linear Differential Equations, *Proc. Amer. Math. Soc.* **8**, 757 (1957).
10. CHETAYEV N., *Stability of Motion*, Moscow (1946).
11. CODDINGTON E. A. and LEVINSON N., *Theory of Ordinary Differential Equations*, Van Nostrand, New York (1955).
12. COSGRIFF R. L., *Non-Linear Control Systems*, McGraw-Hill, New York (1958).
13. CUNNINGHAM W. J., *Introduction to Non Linear Analysis*, McGraw-Hill, New York (1958).
14. FRIEDRICHS K. O., *Advanced Theory of Differential Equations*, New York University (1949).
15. FRIEDRICHS K. O., *Fundamentals of Poincaré's Theory*, Proc. Symp. Non-Linear Circuit Analysis, New York, (1953).

16. FLÜGGE-LOTZ I., *Discontinuous Automatic Control*, Princeton, U.P., U.S.A.
17. FLÜGGE-LOTZ I. and TAYLOR C. F. and LINDBERG H. E., Investigation of a Non-Linear Control System, U.S., N.A.C.A., Report 1391 (1958).
18. FLÜGGE-LOTZ I. and YIN M., Optimum Response of Second Order Systems with Contactor Control, *Trans. A. S. M. E.*, **83** D. *(J. Basic Eng.)* No. 1, 59 (March, 1961).
19. FLÜGGE-LOTZ I. and LINDBERG H. E., Studies of Second and Third Order Contactor Control Systems, U.S. N.A.C.A., T.N.D. 107 (Oct. 1959).
20. GILLE J. C., PELEGRIN M. J., and DECAULNE P., *Feedback Control Systems*, McGraw-Hill, New York (1959).
21. GORSKAYA N. S., KRUTAVA I. M. and RUTKOVSKII V. Y., *Dynamics of Non-Linear Servomechanisms*, Moscow (1959) (in Russian).
22. GRAHAM D. and McRUER D., *Analysis of Non-Linear Control Systems*, Wiley, New York (1961).
23. HAHN W., *Theory and Application of Liapunov's Direct Method*, Springer, Berlin (1959); English edition, Prentice-Hall, U.S.A. (1963).
24. HAMEL B. — for account of his work, *see* reference 20, pp. 440 *et seq.*
25. HAN K. W. and THALER G. J., Phase Space Analysis, *Trans. A. I. E. E. (App. & Ind.)*, 196 (Sept. 1961).
26. HAYASHI C., *Forced Oscillations in Non-Linear Systems*, Nippon, Osaka, Japan (1953).
27. HAYASHI C., Stability Investigations of the Non-Linear Periodic Oscillation, *J. App. Phys.* **24**; 344-8 (1953).
28. KALMAN R., Phase-plane Analysis of Automatic Control Systems with Non-Linear Gain Elements, *Trans. A. I. E. E.* **73** (II), 383-90 (1954).
29. KALMAN R. and BERTRAM J. E., Control System Analysis and Design via Liapunov's Second Method, parts I and II, *Trans. A.S.M.E., J. Basic Engineering*, (June, 1960).
30. KAUDERER H., *Nichtlineare Mechanik*, Springer, Berlin (1958).
31. KRYLOV N. M. and BOGOLIUBOV N., *Introduction to Non-Linear Mechanics*, Moscow (1937); English Edition by S. Lepschetz, Princeton U.P., U.S.A. (1947).
32. KU Y. H., *Analysis and Control of Non-Linear Systems*, Ronald Press, New York (1958).
33. KU Y. H., Theory of Non-Linear Control, *J. Franklin Inst.* **271**. No. 2, 108 (Feb. 1961).
34. LA SALLE J. P. and LEFSCHETZ S., *Non-Linear Differential Equations and Non-Linear Mechanics*, Academic Press, New York (1963).
35. LA SALLE J. P. and LEFSCHETZ S., *Recent Soviet Contributions to Mathematics*, Macmillan, New York (1962).
36. LA SALLE J. P. and LEFSCHETZ S., Recent Soviet Contributions to Ordinary Differential Equations and Non-Linear Mechanics, U.S. Air Force Report A.F.O.S.R. TN 59–308. (1959).
37. LA SALLE J. P. and LEFSCHETZ S., *Stability by Liapunov's Direct Method*, Academic Press, New York (1961).
38. LEFSCHETZ S. (Editor) *Contributions to the Theory of Non-Linear Oscillations*, Vols. I to V, Princeton University Press, U.S.A. (1950-9).
39. LEFSCHETZ S., *Differential Equations: Geometric Theory*, Interscience, New York (1957).
40. MALKIN I. G., *Some Problems in the Theory of Non-Linear Oscillations*, Moscow, (1956).
41. MALKIN I. G., *Theory of Stability of Motion*, Moscow (1952); German edition by R. Oldenbourg, Munich (1959); English edition by U.S. Atomic Energy Commission, AEC tr — 3352 (1958).
42. MANDELSTAM L. and PAPALEXI N., *Zeit. für Physik* **73** (1932).

REFERENCES

43. MINORSKY N., *Introduction to Non-Linear Mechanics*, Edwards, Ann Arbor, U.S.A. (1947).
44. MINORSKY N., *Non Linear Oscillations*, Van Nostrand, New York (1962).
45. MÖSER J., *On Non-Linear Electric Circuits*, I.B.M. Research Report No. RC. 458 (1961).
46. NEMYSTSKII V. V. and STEPANOV V. V., *Qualitative Theory of Differential Equations*, 2nd edition, Moscow (1949) English edition by S. Lefschetz, Princeton, U.P. U.S.A. (1960).
47. OSTROVSKII M., Increasing the Response Speed of Automatic Control Systems by means of Non-Linear Devices, *Automat. i. Tel. Moscow*, **19**, 3 (0000).
48. STRUBLE R. A. *Non Linear Differential Equations*, McGraw-Hill, New York (1962).
49. THALER G. J. and PESTEL M. P., *Analysis and Design of Non-Linear Feedback Control Systems*, McGraw-Hill, New York (1962).
50. TSYPKIN J. Z., *Theory of Relay Control Systems*, Moscow (1956). Published in German by Oldenbourg and Technik (1958).
51. VOGEL T., *Ann. des Telecomm.*, **6**, (1961).
52. VOGEL T., see *Colloquium Int. Porquerolles* (1951) and an account by N. Minorsky in reference 44.
53. WEST J. C., *Analytical Techniques for Non-Linear Control Systems*, English Universities Press, London. (1960).

INDEX

Amplitude 3
Arc, electrical 219, 281, 652
Automatic Pilot
  two-position 501
  two-position with parallel feedback, 502, 512
  two-position with spatial delay 519, 517
  two-position with time delay 517, 528
  two-position with velocity correction 502

Balance diagram, energy– 80
Bendixson's criterion 305
  for a cylinder 422
Bifurcation *see* Branch
Blocking oscillator 730
Blocking oscillator
  capacitive restoration of 743
  inductive restoration of 743
Branch or bifurcation diagram 100, 221, 640
Branch or bifurcation value of a parameter 99, 221, 406, 408
Branching or bifurcation in a self-oscillating system 406, 408, 635, 693

Capacitances, parasitic 249
Cauchy's theorem 795, 796 etc.
Cell
  simply connected and doubly connected 372
  topological structure of its mapping by paths 370, 398
Centre 7, 82, 265
Circuit
  "universal" 268, 724
  with neon tube 239, 696
Clocks 168, 182
Coarse system 352, 374
Coulomb law of friction, 152
Curve
  contact 335
  integral 7, 30, 254
Cycle
  coarse limit 385

  limit 160, 162, 287
  orbitally stable limit 289
  semi-stable multiple limit 386
  stability in the sense of Liapunov of a limit 289
  stable – unstable limit 289, 409, 412
  without contact 317, 332
Cycles, stability condition of limit 289, 296
Cylinder
  development of phase 95, 96, 147
  phase 95, 147, 419, 561

Decrement, logarithmic 18
Degrees of freedom, number of xix
Dissipation conditions, energy– 147, 148
Dulac's criterion 305, 333
  for the cylinder 422, 440

Energy
  integral 110, 118
  of system 75, 79
Engine, steam– 559
Escapement of clock 169, 183

Factor, integrating 131
Focus
  multiple 382
  stable 25, 26, 265
  unstable 58, 265
Force, generalized 146
Frequency, angular 3
Friction
  "dry" Coulomb xxiii, xxiv, 174
  "fluid" viscous linear xxiii, 147
  "negative" 50
  "square-law" 149, 151, 555
Froude–Zhukovskii's pendulum 53
Frühhauf's circuit 702
Function
  correspondence 444
  sequence 161, 291, 293 et seq., 444

Generator
  dynatron 68, 267, 699

relaxation 239
  with a characteristic without saturation, valve 446, 628
  with a discontinuous ∫ characteristic, valve 157, 172, 627
  with biassed discontinuous characteristic, valve 468, 626
  with grid currents, valve 632
  with inductive feedback, valve 157, 514, 601, 623, 627
  with symmetrical valve characteristic, valve 461
  with two-mesh RC circuit, valve 343, 480, 611
Glider flight 436

Half-path
  orbitally stable (non-singular) 364
  orbitally unstable (singular) 364
  positive-negative 363
Hamilton's equation 119
Hamilton's function 119

Increment, logarithmic 59
Invariant, integral 133
Isocline 8, 23
Isoclines, method of 23, 341

Jump conditions 41, 43, 249, 674, 705
Jumps
  conservative 48
  non-conservative 49

Koenigs's theorem 294

Lagrange–Maxwell's equations 118
Lagrange's
  equations 118, 146
  function 118
  theorem 88
Lamerey's diagram 161, 178, 195, 293
Lejeune–Dirichlet's theorem 88
Liapunov's theorem 272, 274
  converse of 88
Limit cycle see Cycle
Line, phase 37, 213, 236
Line (of a dynamic system), singular 352

Model
  dynamical xviii
  mathematical xv
Motion
  escaping (or run-away) 89
  escaping limitation 93
  limitation 67, 93
  periodic 5, 79
Motion
  "rapid" 660
  slip – 501, 512 et seq, 527, 533
  "slow" 661
  stationary xxvii
Motor, single-phase asynchronous 229
Multivibrator
  symmetrical 750
  with grid currents 758
  with inductance in the anode circuit 712
  with one R.C. circuit 246, 655, 680

Node, 34
  stable 34, 261
  unstable 60, 261

Oscillations
  aperiodically damped 27
  damped 16
  discontinuous 249, 491, 645
  discontinuous mechanical 690
  periodic 5, 15, 82, 234
  relaxation 239
Oscillator
  harmonic 1
  with Coulomb dry friction 151
  with linear friction 15, 169
  with square-law friction 150, 555

Parameters parasitic 248, 659
Path
  entire (complete) phase 4, 353, 354
  limit phase 355
  orbitally stable (non-singular) phase 366, 368
  orbitally unstable (singular) phase 366, 367
  phase 4, 34, 255
  self-limiting phase 355
Pendulum (large deviations) 96
Period, conditional 17
Pfaff's equation 133, 138

INDEX

Phase of oscillations, initial 3
Phugoid 439
Plane, phase 5
Poincaré–Bendixson Theorem 361
Poincaré's
  indices 300
  method 613
  sphere 325
Point
  of half-path, Limit 353, 355
  of path, Limit 354
  (of point transformation), fixed 161, 292
  representative 4
  simple (multiple) singular 280
  singular 7
  stability of fixed 294 et seq.

Regulator
  frictional 230
  two-position temperature 235
Relay, valve 224, 306, 536
Runge's method 350

Saddle-point 67, 70, 84
Segment without contact (transversal) 291, 357
Self-oscillations 162, 199, 291
  almost sinusoidal 200, 583
  discontinuous 249, 250, 645
  hard mode of excitation of 166, 552, 608, 640, 780
  soft mode of excitation of 166, 604, 607, 638, 780
Self-resonance 166
Separatrix 90, 369
Set
  of limit points 355
  simply connected closed 355
Space, functional phase 529
Stability
  absolute 27, 547
  of clock motion 199
  of fixed point 293
  of periodic motion (in the sense of Liapunov) 126, 289
  of state of equilibrium (in the sense of Liapunov) 11, 214, 279
  orbital 128, 288
  structural 352, 375
State
  of equilibrium 11, 279
  of equilibrium, coarse 279, 377
  of equilibrium, stability of 11, 214, 279
Surface
  cylindrical phase 95, 419, 561
  with many sheets, phase 185, 519, 540, 551, 630
System
  autonomous xxviii
  coarse xxix, 374, 375, 376
  conservative 75, 125, 128
  dissipative 146
  dynamic xviii
  linear 1
  piece-wise conservative 555
  piece-wise linear 443
  relay 536
  self-oscillating 162, 199, 287, 288
  varied 375

Topological invariant properties 363
Transformation
  parametrical representation of point 444
  point 161, 291
Transversal *see* Segment without contact
Tube, neon 239

Van der Pol's
  equation 342
  method 585
  variables 586
Velocity, phase 4, 9

Zhukovskii's problem 436

A CATALOG OF SELECTED
DOVER BOOKS
IN SCIENCE AND MATHEMATICS

A CATALOG OF SELECTED
DOVER BOOKS
IN SCIENCE AND MATHEMATICS

QUALITATIVE THEORY OF DIFFERENTIAL EQUATIONS, V.V. Nemytskii and V.V. Stepanov. Classic graduate-level text by two prominent Soviet mathematicians covers classical differential equations as well as topological dynamics and ergodic theory. Bibliographies. 523pp. 5⅜ x 8½. 65954-2 Pa. $14.95

MATRICES AND LINEAR ALGEBRA, Hans Schneider and George Phillip Barker. Basic textbook covers theory of matrices and its applications to systems of linear equations and related topics such as determinants, eigenvalues and differential equations. Numerous exercises. 432pp. 5⅜ x 8½. 66014-1 Pa. $10.95

QUANTUM THEORY, David Bohm. This advanced undergraduate-level text presents the quantum theory in terms of qualitative and imaginative concepts, followed by specific applications worked out in mathematical detail. Preface. Index. 655pp. 5⅜ x 8½. 65969-0 Pa. $14.95

ATOMIC PHYSICS (8th edition), Max Born. Nobel laureate's lucid treatment of kinetic theory of gases, elementary particles, nuclear atom, wave-corpuscles, atomic structure and spectral lines, much more. Over 40 appendices, bibliography. 495pp. 5⅜ x 8½. 65984-4 Pa. $13.95

ELECTRONIC STRUCTURE AND THE PROPERTIES OF SOLIDS: The Physics of the Chemical Bond, Walter A. Harrison. Innovative text offers basic understanding of the electronic structure of covalent and ionic solids, simple metals, transition metals and their compounds. Problems. 1980 edition. 582pp. 6⅛ x 9¼. 66021-4 Pa. $16.95

BOUNDARY VALUE PROBLEMS OF HEAT CONDUCTION, M. Necati Özisik. Systematic, comprehensive treatment of modern mathematical methods of solving problems in heat conduction and diffusion. Numerous examples and problems. Selected references. Appendices. 505pp. 5⅜ x 8½. 65990-9 Pa. $12.95

A SHORT HISTORY OF CHEMISTRY (3rd edition), J.R. Partington. Classic exposition explores origins of chemistry, alchemy, early medical chemistry, nature of atmosphere, theory of valency, laws and structure of atomic theory, much more. 428pp. 5⅜ x 8½. (Available in U.S. only) 65977-1 Pa. $11.95

A HISTORY OF ASTRONOMY, A. Pannekoek. Well-balanced, carefully reasoned study covers such topics as Ptolemaic theory, work of Copernicus, Kepler, Newton, Eddington's work on stars, much more. Illustrated. References. 521pp. 5⅜ x 8½. 65994-1 Pa. $12.95

PRINCIPLES OF METEOROLOGICAL ANALYSIS, Walter J. Saucier. Highly respected, abundantly illustrated classic reviews atmospheric variables, hydrostatics, static stability, various analyses (scalar, cross-section, isobaric, isentropic, more). For intermediate meteorology students. 454pp. 6⅛ x 9¼. 65979-8 Pa. $14.95

CATALOG OF DOVER BOOKS

RELATIVITY, THERMODYNAMICS AND COSMOLOGY, Richard C. Tolman. Landmark study extends thermodynamics to special, general relativity; also applications of relativistic mechanics, thermodynamics to cosmological models. 501pp. 5⅜ x 8½. 65383-8 Pa. $13.95

APPLIED ANALYSIS, Cornelius Lanczos. Classic work on analysis and design of finite processes for approximating solution of analytical problems. Algebraic equations, matrices, harmonic analysis, quadrature methods, much more. 559pp. 5⅜ x 8½. 65656-X Pa. $13.95

INTRODUCTION TO ANALYSIS, Maxwell Rosenlicht. Unusually clear, accessible coverage of set theory, real number system, metric spaces, continuous functions, Riemann integration, multiple integrals, more. Wide range of problems. Undergraduate level. Bibliography. 254pp. 5⅜ x 8½. 65038-3 Pa. $8.95

INTRODUCTION TO QUANTUM MECHANICS With Applications to Chemistry, Linus Pauling & E. Bright Wilson, Jr. Classic undergraduate text by Nobel Prize winner applies quantum mechanics to chemical and physical problems. Numerous tables and figures enhance the text. Chapter bibliographies. Appendices. Index. 468pp. 5⅜ x 8½. 64871-0 Pa. $12.95

ASYMPTOTIC EXPANSIONS OF INTEGRALS, Norman Bleistein & Richard A. Handelsman. Best introduction to important field with applications in a variety of scientific disciplines. New preface. Problems. Diagrams. Tables. Bibliography. Index. 448pp. 5⅜ x 8½. 65082-0 Pa. $12.95

MATHEMATICS APPLIED TO CONTINUUM MECHANICS, Lee A. Segel. Analyzes models of fluid flow and solid deformation. For upper-level math, science and engineering students. 608pp. 5⅜ x 8½. 65369-2 Pa. $14.95

ELEMENTS OF REAL ANALYSIS, David A. Sprecher. Classic text covers fundamental concepts, real number system, point sets, functions of a real variable, Fourier series, much more. Over 500 exercises. 352pp. 5⅜ x 8½. 65385-4 Pa. $11.95

PHYSICAL PRINCIPLES OF THE QUANTUM THEORY, Werner Heisenberg. Nobel Laureate discusses quantum theory, uncertainty, wave mechanics, work of Dirac, Schroedinger, Compton, Wilson, Einstein, etc. 184pp. 5⅜ x 8½. 60113-7 Pa. $6.95

INTRODUCTORY REAL ANALYSIS, A.N. Kolmogorov, S.V. Fomin. Translated by Richard A. Silverman. Self-contained, evenly paced introduction to real and functional analysis. Some 350 problems. 403pp. 5⅜ x 8½. 61226-0 Pa. $10.95

PROBLEMS AND SOLUTIONS IN QUANTUM CHEMISTRY AND PHYSICS, Charles S. Johnson, Jr. and Lee G. Pedersen. Unusually varied problems, detailed solutions in coverage of quantum mechanics, wave mechanics, angular momentum, molecular spectroscopy, scattering theory, more. 280 problems plus 139 supplementary exercises. 430pp. 6½ x 9¼. 65236-X Pa. $13.95

CATALOG OF DOVER BOOKS

ASYMPTOTIC METHODS IN ANALYSIS, N.G. de Bruijn. An inexpensive, comprehensive guide to asymptotic methods—the pioneering work that teaches by explaining worked examples in detail. Index. 224pp. 5⅜ x 8½. 64221-6 Pa. $7.95

OPTICAL RESONANCE AND TWO-LEVEL ATOMS, L. Allen and J. H. Eberly. Clear, comprehensive introduction to basic principles behind all quantum optical resonance phenomena. 53 illustrations. Preface. Index. 256pp. 5⅜ x 8½.
65533-4 Pa. $8.95

COMPLEX VARIABLES, Francis J. Flanigan. Unusual approach, delaying complex algebra till harmonic functions have been analyzed from real variable viewpoint. Includes problems with answers. 364pp. 5⅜ x 8½. 61388-7 Pa. $9.95

ATOMIC SPECTRA AND ATOMIC STRUCTURE, Gerhard Herzberg. One of best introductions; especially for specialist in other fields. Treatment is physical rather than mathematical. 80 illustrations. 257pp. 5⅜ x 8½. 60115-3 Pa. $7.95

APPLIED COMPLEX VARIABLES, John W. Dettman. Step-by-step coverage of fundamentals of analytic function theory—plus lucid exposition of five important applications: Potential Theory; Ordinary Differential Equations; Fourier Transforms; Laplace Transforms; Asymptotic Expansions. 66 figures. Exercises at chapter ends. 512pp. 5⅜ x 8½. 64670-X Pa. $12.95

ULTRASONIC ABSORPTION: An Introduction to the Theory of Sound Absorption and Dispersion in Gases, Liquids and Solids, A.B. Bhatia. Standard reference in the field provides a clear, systematically organized introductory review of fundamental concepts for advanced graduate students, research workers. Numerous diagrams. Bibliography. 440pp. 5⅜ x 8½. 64917-2 Pa. $11.95

UNBOUNDED LINEAR OPERATORS: Theory and Applications, Seymour Goldberg. Classic presents systematic treatment of the theory of unbounded linear operators in normed linear spaces with applications to differential equations. Bibliography. 199pp. 5⅜ x 8½. 64830-3 Pa. $7.95

LIGHT SCATTERING BY SMALL PARTICLES, H.C. van de Hulst. Comprehensive treatment including full range of useful approximation methods for researchers in chemistry, meteorology and astronomy. 44 illustrations. 470pp. 5⅜ x 8½.
64228-3 Pa. $12.95

CONFORMAL MAPPING ON RIEMANN SURFACES, Harvey Cohn. Lucid, insightful book presents ideal coverage of subject. 334 exercises make book perfect for self-study. 55 figures. 352pp. 5⅜ x 8¼. 64025-6 Pa. $11.95

OPTICKS, Sir Isaac Newton. Newton's own experiments with spectroscopy, colors, lenses, reflection, refraction, etc., in language the layman can follow. Foreword by Albert Einstein. 532pp. 5⅜ x 8½. 60205-2 Pa. $12.95

GENERALIZED INTEGRAL TRANSFORMATIONS, A.H. Zemanian. Graduate-level study of recent generalizations of the Laplace, Mellin, Hankel, K. Weierstrass, convolution and other simple transformations. Bibliography. 320pp. 5⅜ x 8½.
65375-7 Pa. $8.95

CATALOG OF DOVER BOOKS

THE ELECTROMAGNETIC FIELD, Albert Shadowitz. Comprehensive undergraduate text covers basics of electric and magnetic fields, builds up to electromagnetic theory. Also related topics, including relativity. Over 900 problems. 768pp. 5⅜ x 8¼. 65660-8 Pa. $18.95

FOURIER SERIES, Georgi P. Tolstov. Translated by Richard A. Silverman. A valuable addition to the literature on the subject, moving clearly from subject to subject and theorem to theorem. 107 problems, answers. 336pp. 5⅜ x 8½. 63317-9 Pa. $9.95

THEORY OF ELECTROMAGNETIC WAVE PROPAGATION, Charles Herach Papas. Graduate-level study discusses the Maxwell field equations, radiation from wire antennas, the Doppler effect and more. xiii + 244pp. 5⅜ x 8½. 65678-0 Pa. $6.95

DISTRIBUTION THEORY AND TRANSFORM ANALYSIS: An Introduction to Generalized Functions, with Applications, A.H. Zemanian. Provides basics of distribution theory, describes generalized Fourier and Laplace transformations. Numerous problems. 384pp. 5⅜ x 8½. 65479-6 Pa. $11.95

THE PHYSICS OF WAVES, William C. Elmore and Mark A. Heald. Unique overview of classical wave theory. Acoustics, optics, electromagnetic radiation, more. Ideal as classroom text or for self-study. Problems. 477pp. 5⅜ x 8½. 64926-1 Pa. $13.95

CALCULUS OF VARIATIONS WITH APPLICATIONS, George M. Ewing. Applications-oriented introduction to variational theory develops insight and promotes understanding of specialized books, research papers. Suitable for advanced undergraduate/graduate students as primary, supplementary text. 352pp. 5⅜ x 8½. 64856-7 Pa. $9.95

A TREATISE ON ELECTRICITY AND MAGNETISM, James Clerk Maxwell. Important foundation work of modern physics. Brings to final form Maxwell's theory of electromagnetism and rigorously derives his general equations of field theory. 1,084pp. 5⅜ x 8½. 60636-8, 60637-6 Pa., Two-vol. set $25.90

AN INTRODUCTION TO THE CALCULUS OF VARIATIONS, Charles Fox. Graduate-level text covers variations of an integral, isoperimetrical problems, least action, special relativity, approximations, more. References. 279pp. 5⅜ x 8½. 65499-0 Pa. $8.95

HYDRODYNAMIC AND HYDROMAGNETIC STABILITY, S. Chandrasekhar. Lucid examination of the Rayleigh-Benard problem; clear coverage of the theory of instabilities causing convection. 704pp. 5⅜ x 8¼. 64071-X Pa. $14.95

CALCULUS OF VARIATIONS, Robert Weinstock. Basic introduction covering isoperimetric problems, theory of elasticity, quantum mechanics, electrostatics, etc. Exercises throughout. 326pp. 5⅜ x 8½. 63069-2 Pa. $9.95

DYNAMICS OF FLUIDS IN POROUS MEDIA, Jacob Bear. For advanced students of ground water hydrology, soil mechanics and physics, drainage and irrigation engineering and more. 335 illustrations. Exercises, with answers. 784pp. 6⅛ x 9¼. 65675-6 Pa. $19.95

CATALOG OF DOVER BOOKS

NUMERICAL METHODS FOR SCIENTISTS AND ENGINEERS, Richard Hamming. Classic text stresses frequency approach in coverage of algorithms, polynomial approximation, Fourier approximation, exponential approximation, other topics. Revised and enlarged 2nd edition. 721pp. 5⅜ x 8½. 65241-6 Pa. $15.95

THEORETICAL SOLID STATE PHYSICS, Vol. 1: Perfect Lattices in Equilibrium; Vol. II: Non-Equilibrium and Disorder, William Jones and Norman H. March. Monumental reference work covers fundamental theory of equilibrium properties of perfect crystalline solids, non-equilibrium properties, defects and disordered systems. Appendices. Problems. Preface. Diagrams. Index. Bibliography. Total of 1,301pp. 5⅜ x 8½. Two volumes. Vol. I: 65015-4 Pa. $16.95
Vol. II: 65016-2 Pa. $16.95

OPTIMIZATION THEORY WITH APPLICATIONS, Donald A. Pierre. Broad spectrum approach to important topic. Classical theory of minima and maxima, calculus of variations, simplex technique and linear programming, more. Many problems, examples. 640pp. 5⅜ x 8½. 65205-X Pa. $16.95

THE CONTINUUM: A Critical Examination of the Foundation of Analysis, Hermann Weyl. Classic of 20th-century foundational research deals with the conceptual problem posed by the continuum. 156pp. 5⅜ x 8½. 67982-9 Pa. $6.95

ESSAYS ON THE THEORY OF NUMBERS, Richard Dedekind. Two classic essays by great German mathematician: on the theory of irrational numbers; and on transfinite numbers and properties of natural numbers. 115pp. 5⅜ x 8½.
21010-3 Pa. $5.95

THE FUNCTIONS OF MATHEMATICAL PHYSICS, Harry Hochstadt. Comprehensive treatment of orthogonal polynomials, hypergeometric functions, Hill's equation, much more. Bibliography. Index. 322pp. 5⅜ x 8½. 65214-9 Pa. $9.95

NUMBER THEORY AND ITS HISTORY, Oystein Ore. Unusually clear, accessible introduction covers counting, properties of numbers, prime numbers, much more. Bibliography. 380pp. 5⅜ x 8½. 65620-9 Pa. $10.95

THE VARIATIONAL PRINCIPLES OF MECHANICS, Cornelius Lanczos. Graduate level coverage of calculus of variations, equations of motion, relativistic mechanics, more. First inexpensive paperbound edition of classic treatise. Index. Bibliography. 418pp. 5⅜ x 8½. 65067-7 Pa. $12.95

MATHEMATICAL TABLES AND FORMULAS, Robert D. Carmichael and Edwin R. Smith. Logarithms, sines, tangents, trig functions, powers, roots, reciprocals, exponential and hyperbolic functions, formulas and theorems. 269pp. 5⅜ x 8½.
60111-0 Pa. $6.95

THEORETICAL PHYSICS, Georg Joos, with Ira M. Freeman. Classic overview covers essential math, mechanics, electromagnetic theory, thermodynamics, quantum mechanics, nuclear physics, other topics. First paperback edition. xxiii + 885pp. 5⅜ x 8½. 65227-0 Pa. $21.95

CATALOG OF DOVER BOOKS

ORDINARY DIFFERENTIAL EQUATIONS, Morris Tenenbaum and Harry Pollard. Exhaustive survey of ordinary differential equations for undergraduates in mathematics, engineering, science. Thorough analysis of theorems. Diagrams. Bibliography. Index. 818pp. 5⅜ x 8½. 64940-7 Pa. $18.95

STATISTICAL MECHANICS: Principles and Applications, Terrell L. Hill. Standard text covers fundamentals of statistical mechanics, applications to fluctuation theory, imperfect gases, distribution functions, more. 448pp. 5⅜ x 8½. 65390-0 Pa. $11.95

ORDINARY DIFFERENTIAL EQUATIONS AND STABILITY THEORY: An Introduction, David A. Sánchez. Brief, modern treatment. Linear equation, stability theory for autonomous and nonautonomous systems, etc. 164pp. 5⅜ x 8½. 63828-6 Pa. $6.95

THIRTY YEARS THAT SHOOK PHYSICS: The Story of Quantum Theory, George Gamow. Lucid, accessible introduction to influential theory of energy and matter. Careful explanations of Dirac's anti-particles, Bohr's model of the atom, much more. 12 plates. Numerous drawings. 240pp. 5⅜ x 8½. 24895-X Pa. $7.95

THEORY OF MATRICES, Sam Perlis. Outstanding text covering rank, nonsingularity and inverses in connection with the development of canonical matrices under the relation of equivalence, and without the intervention of determinants. Includes exercises. 237pp. 5⅜ x 8½. 66810-X Pa. $8.95

GREAT EXPERIMENTS IN PHYSICS: Firsthand Accounts from Galileo to Einstein, edited by Morris H. Shamos. 25 crucial discoveries: Newton's laws of motion, Chadwick's study of the neutron, Hertz on electromagnetic waves, more. Original accounts clearly annotated. 370pp. 5⅜ x 8½. 25346-5 Pa. $10.95

INTRODUCTION TO PARTIAL DIFFERENTIAL EQUATIONS WITH APPLICATIONS, E.C. Zachmanoglou and Dale W. Thoe. Essentials of partial differential equations applied to common problems in engineering and the physical sciences. Problems and answers. 416pp. 5⅜ x 8½. 65251-3 Pa. $11.95

BURNHAM'S CELESTIAL HANDBOOK, Robert Burnham, Jr. Thorough guide to the stars beyond our solar system. Exhaustive treatment. Alphabetical by constellation: Andromeda to Cetus in Vol. 1; Chamaeleon to Orion in Vol. 2; and Pavo to Vulpecula in Vol. 3. Hundreds of illustrations. Index in Vol. 3. 2,000pp. 6¼ x 9¼. 23567-X, 23568-8, 23673-0 Pa., Three-vol. set $44.85

CHEMICAL MAGIC, Leonard A. Ford. Second Edition, Revised by E. Winston Grundmeier. Over 100 unusual stunts demonstrating cold fire, dust explosions, much more. Text explains scientific principles and stresses safety precautions. 128pp. 5⅜ x 8½. 67628-5 Pa. $5.95

AMATEUR ASTRONOMER'S HANDBOOK, J.B. Sidgwick. Timeless, comprehensive coverage of telescopes, mirrors, lenses, mountings, telescope drives, micrometers, spectroscopes, more. 189 illustrations. 576pp. 5⅜ x 8¼. (Available in U.S. only) 24034-7 Pa. $11.95

CATALOG OF DOVER BOOKS

SPECIAL FUNCTIONS, N.N. Lebedev. Translated by Richard Silverman. Famous Russian work treating more important special functions, with applications to specific problems of physics and engineering. 38 figures. 308pp. 5⅜ x 8½. 60624-4 Pa. $9.95

OBSERVATIONAL ASTRONOMY FOR AMATEURS, J.B. Sidgwick. Mine of useful data for observation of sun, moon, planets, asteroids, aurorae, meteors, comets, variables, binaries, etc. 39 illustrations. 384pp. 5⅜ x 8¼. (Available in U.S. only) 24033-9 Pa. $8.95

INTEGRAL EQUATIONS, F.G. Tricomi. Authoritative, well-written treatment of extremely useful mathematical tool with wide applications. Volterra Equations, Fredholm Equations, much more. Advanced undergraduate to graduate level. Exercises. Bibliography. 238pp. 5⅜ x 8½. 64828-1 Pa. $8.95

POPULAR LECTURES ON MATHEMATICAL LOGIC, Hao Wang. Noted logician's lucid treatment of historical developments, set theory, model theory, recursion theory and constructivism, proof theory, more. 3 appendixes. Bibliography. 1981 edition. ix + 283pp. 5⅜ x 8½. 67632-3 Pa. $8.95

MODERN NONLINEAR EQUATIONS, Thomas L. Saaty. Emphasizes practical solution of problems; covers seven types of equations. ". . . a welcome contribution to the existing literature...."–*Math Reviews*. 490pp. 5⅜ x 8½. 64232-1 Pa. $13.95

FUNDAMENTALS OF ASTRODYNAMICS, Roger Bate et al. Modern approach developed by U.S. Air Force Academy. Designed as a first course. Problems, exercises. Numerous illustrations. 455pp. 5⅜ x 8½. 60061-0 Pa. $10.95

INTRODUCTION TO LINEAR ALGEBRA AND DIFFERENTIAL EQUATIONS, John W. Dettman. Excellent text covers complex numbers, determinants, orthonormal bases, Laplace transforms, much more. Exercises with solutions. Undergraduate level. 416pp. 5⅜ x 8½. 65191-6 Pa. $11.95

INCOMPRESSIBLE AERODYNAMICS, edited by Bryan Thwaites. Covers theoretical and experimental treatment of the uniform flow of air and viscous fluids past two-dimensional aerofoils and three-dimensional wings; many other topics. 654pp. 5⅜ x 8½. 65465-6 Pa. $16.95

INTRODUCTION TO DIFFERENCE EQUATIONS, Samuel Goldberg. Exceptionally clear exposition of important discipline with applications to sociology, psychology, economics. Many illustrative examples; over 250 problems. 260pp. 5⅜ x 8½. 65084-7 Pa. $8.95

LAMINAR BOUNDARY LAYERS, edited by L. Rosenhead. Engineering classic covers steady boundary layers in two- and three- dimensional flow, unsteady boundary layers, stability, observational techniques, much more. 708pp. 5⅜ x 8½. 65646-2 Pa. $18 95

LECTURES ON CLASSICAL DIFFERENTIAL GEOMETRY, Second Edition, Dirk J. Struik. Excellent brief introduction covers curves, theory of surfaces, fundamental equations, geometry on a surface, conformal mapping, other topics. Problems. 240pp. 5⅜ x 8½. 65609-8 Pa. $8.95

CATALOG OF DOVER BOOKS

ROTARY-WING AERODYNAMICS, W.Z. Stepniewski. Clear, concise text covers aerodynamic phenomena of the rotor and offers guidelines for helicopter performance evaluation. Originally prepared for NASA. 537 figures. 640pp. 6⅛ x 9¼.
64647-5 Pa. $16.95

DIFFERENTIAL GEOMETRY, Heinrich W. Guggenheimer. Local differential geometry as an application of advanced calculus and linear algebra. Curvature, transformation groups, surfaces, more. Exercises. 62 figures. 378pp. 5⅜ x 8½.
63433-7 Pa. $9.95

INTRODUCTION TO SPACE DYNAMICS, William Tyrrell Thomson. Comprehensive, classic introduction to space-flight engineering for advanced undergraduate and graduate students. Includes vector algebra, kinematics, transformation of coordinates. Bibliography. Index. 352pp. 5⅜ x 8½. 65113-4 Pa. $9.95

A SURVEY OF MINIMAL SURFACES, Robert Osserman. Up-to-date, in-depth discussion of the field for advanced students. Corrected and enlarged edition covers new developments. Includes numerous problems. 192pp. 5⅜ x 8½. 64998-9 Pa. $8.95

ANALYTICAL MECHANICS OF GEARS, Earle Buckingham. Indispensable reference for modern gear manufacture covers conjugate gear-tooth action, gear-tooth profiles of various gears, many other topics. 263 figures. 102 tables. 546pp. 5⅜ x 8½.
65712-4 Pa. $14.95

SET THEORY AND LOGIC, Robert R. Stoll. Lucid introduction to unified theory of mathematical concepts. Set theory and logic seen as tools for conceptual understanding of real number system. 496pp. 5⅜ x 8¼. 63829-4 Pa. $12.95

A HISTORY OF MECHANICS, René Dugas. Monumental study of mechanical principles from antiquity to quantum mechanics. Contributions of ancient Greeks, Galileo, Leonardo, Kepler, Lagrange, many others. 671pp. 5⅜ x 8½.
65632-2 Pa. $14.95

FAMOUS PROBLEMS OF GEOMETRY AND HOW TO SOLVE THEM, Benjamin Bold. Squaring the circle, trisecting the angle, duplicating the cube: learn their history, why they are impossible to solve, then solve them yourself. 128pp. 5⅜ x 8½. 24297-8 Pa. $4.95

MECHANICAL VIBRATIONS, J.P. Den Hartog. Classic textbook offers lucid explanations and illustrative models, applying theories of vibrations to a variety of practical industrial engineering problems. Numerous figures. 233 problems, solutions. Appendix. Index. Preface. 436pp. 5⅜ x 8½. 64785-4 Pa. $11.95

CURVATURE AND HOMOLOGY, Samuel I. Goldberg. Thorough treatment of specialized branch of differential geometry. Covers Riemannian manifolds, topology of differentiable manifolds, compact Lie groups, other topics. Exercises. 315pp. 5⅜ x 8½. 64314-X Pa. $9.95

HISTORY OF STRENGTH OF MATERIALS, Stephen P. Timoshenko. Excellent historical survey of the strength of materials with many references to the theories of elasticity and structure. 245 figures. 452pp. 5⅜ x 8½. 61187-6 Pa. $12.95

CATALOG OF DOVER BOOKS

A CONCISE HISTORY OF MATHEMATICS, Dirk J. Struik. The best brief history of mathematics. Stresses origins and covers every major figure from ancient Near East to 19th century. 41 illustrations. 195pp. 5⅜ x 8½. 60255-9 Pa. $8.95

A SHORT ACCOUNT OF THE HISTORY OF MATHEMATICS, W.W. Rouse Ball. One of clearest, most authoritative surveys from the Egyptians and Phoenicians through 19th-century figures such as Grassman, Galois, Riemann. Fourth edition. 522pp. 5⅜ x 8½. 20630-0 Pa. $11.95

HISTORY OF MATHEMATICS, David E. Smith. Nontechnical survey from ancient Greece and Orient to late 19th century; evolution of arithmetic, geometry, trigonometry, calculating devices, algebra, the calculus. 362 illustrations. 1,355pp. 5⅜ x 8½. 20429-4, 20430-8 Pa., Two-vol. set $26.90

THE GEOMETRY OF RENÉ DESCARTES, René Descartes. The great work founded analytical geometry. Original French text, Descartes' own diagrams, together with definitive Smith-Latham translation. 244pp. 5⅜ x 8½. 60068-8 Pa. $8.95

THE ORIGINS OF THE INFINITESIMAL CALCULUS, Margaret E. Baron. Only fully detailed and documented account of crucial discipline: origins; development by Galileo, Kepler, Cavalieri; contributions of Newton, Leibniz, more. 304pp. 5⅜ x 8½. (Available in U.S. and Canada only) 65371-4 Pa. $9.95

THE HISTORY OF THE CALCULUS AND ITS CONCEPTUAL DEVELOPMENT, Carl B. Boyer. Origins in antiquity, medieval contributions, work of Newton, Leibniz, rigorous formulation. Treatment is verbal. 346pp. 5⅜ x 8½. 60509-4 Pa. $9.95

THE THIRTEEN BOOKS OF EUCLID'S ELEMENTS, translated with introduction and commentary by Sir Thomas L. Heath. Definitive edition. Textual and linguistic notes, mathematical analysis. 2,500 years of critical commentary. Not abridged. 1,414pp. 5⅜ x 8½. 60088-2, 60089-0, 60090-4 Pa., Three-vol. set $32.85

GAMES AND DECISIONS: Introduction and Critical Survey, R. Duncan Luce and Howard Raiffa. Superb nontechnical introduction to game theory, primarily applied to social sciences. Utility theory, zero-sum games, n-person games, decision-making, much more. Bibliography. 509pp. 5⅜ x 8½. 65943-7 Pa. $13.95

THE HISTORICAL ROOTS OF ELEMENTARY MATHEMATICS, Lucas N.H. Bunt, Phillip S. Jones, and Jack D. Bedient. Fundamental underpinnings of modern arithmetic, algebra, geometry and number systems derived from ancient civilizations. 320pp. 5⅜ x 8½. 25563-8 Pa. $8.95

CALCULUS REFRESHER FOR TECHNICAL PEOPLE, A. Albert Klaf. Covers important aspects of integral and differential calculus via 756 questions. 566 problems, most answered. 431pp. 5⅜ x 8½. 20370-0 Pa. $8.95

CATALOG OF DOVER BOOKS

CHALLENGING MATHEMATICAL PROBLEMS WITH ELEMENTARY SOLUTIONS, A.M. Yaglom and I.M. Yaglom. Over 170 challenging problems on probability theory, combinatorial analysis, points and lines, topology, convex polygons, many other topics. Solutions. Total of 445pp. 5⅜ x 8½. Two-vol. set.
Vol. I: 65536-9 Pa. $7.95
Vol. II: 65537-7 Pa. $7.95

FIFTY CHALLENGING PROBLEMS IN PROBABILITY WITH SOLUTIONS, Frederick Mosteller. Remarkable puzzlers, graded in difficulty, illustrate elementary and advanced aspects of probability. Detailed solutions. 88pp. 5⅜ x 8½.
65355-2 Pa. $4.95

EXPERIMENTS IN TOPOLOGY, Stephen Barr. Classic, lively explanation of one of the byways of mathematics. Klein bottles, Moebius strips, projective planes, map coloring, problem of the Koenigsberg bridges, much more, described with clarity and wit. 43 figures. 210pp. 5⅜ x 8½. 25933-1 Pa. $6.95

RELATIVITY IN ILLUSTRATIONS, Jacob T. Schwartz. Clear nontechnical treatment makes relativity more accessible than ever before. Over 60 drawings illustrate concepts more clearly than text alone. Only high school geometry needed. Bibliography. 128pp. 6⅛ x 9¼. 25965-X Pa. $7.95

AN INTRODUCTION TO ORDINARY DIFFERENTIAL EQUATIONS, Earl A. Coddington. A thorough and systematic first course in elementary differential equations for undergraduates in mathematics and science, with many exercises and problems (with answers). Index. 304pp. 5⅜ x 8½. 65942-9 Pa. $8.95

FOURIER SERIES AND ORTHOGONAL FUNCTIONS, Harry F. Davis. An incisive text combining theory and practical example to introduce Fourier series, orthogonal functions and applications of the Fourier method to boundary-value problems. 570 exercises. Answers and notes. 416pp. 5⅜ x 8½. 65973-9 Pa. $11.95

AN INTRODUCTION TO ALGEBRAIC STRUCTURES, Joseph Landin. Superb self-contained text covers "abstract algebra": sets and numbers, theory of groups, theory of rings, much more. Numerous well-chosen examples, exercises. 247pp. 5⅜ x 8½.
65940-2 Pa. $8.95

STARS AND RELATIVITY, Ya. B. Zel'dovich and I. D. Novikov. Vol. 1 of *Relativistic Astrophysics* by famed Russian scientists. General relativity, properties of matter under astrophysical conditions, stars and stellar systems. Deep physical insights, clear presentation. 1971 edition. References. 544pp. 5⅜ x 8½.
69424-0 Pa. $14.95

Prices subject to change without notice.
Available at your book dealer or write for free Mathematics and Science Catalog to Dept. GI, Dover Publications, Inc., 31 East 2nd St., Mineola, N.Y. 11501. Dover publishes more than 250 books each year on science, elementary and advanced mathematics, biology, music, art, literature, history, social sciences and other areas.